Contents

Foreword

The manufacturing arena is a vast area consisting of a multitude of fabrication processes utilized to produce the products that we use on a daily basis. Starting with the raw materials, manufacturing processes will transform the material into the component parts, which get assembled into products from components that are extermely large to components that are extremely small. Large ships, truck, cars, firearms, mechanical pencils, pens, and all other items we both see and use are all the result of a manufacturing facility. Even the clothing that we wear has been produced from various processes. Being able to produce these products in an efficient and economical manner is the function of manufacturing facilities allowing them to be purchased at a reasonable price, which is affordable by the customers. This all begins with the products being designed with the fabrication processes in mind, allowing the fabrication processes to be as efficient as possible, producing the products as rapidly as possible and at the highest quality levels.

Manufacturing dates back to the beginning of civilization as we know it. From early man developing their stone tools and arrowheads, fabrication has been part of our lives. Looking at some of the early advanced civilizations such as the Egyptians and Romans, they built some of the most amazing structures the world has ever seen. Along with those, the statues and jewelry that they produced required manufacturing. Processes such as casting were utilized to form these unique shapes from the molten gold and silver they used to create those amazing configurations. Although these items were not mass-produced, the processes utilized for creating them are still used today. As time progressed, so did the manufacturing processes. New materials have been developed, allowing for a wider variety for products and the creation of new processes to shape the materials. At first, the materials and processes were crude and difficult to produce quality components, but refining the materials and processes have allowed for high-quality parts being produced from new and exciting materials. A good example of this is the introduction of plastics. At first, products produced from plastics were considered "cheap items" because of the quality of the material and the crude shaping processes. Today, plastics are utilized in almost every product we use daily.

This dynamic arena of manufacturing is constantly changing and evolving. With new technology and materials being developed constantly, the fabrication world is changing to meet the new requirements for products. From the production of products consisting of mainly metals to the production of microchips, manufacturing has improved to the high level of technology we see today. As new and more intricate requirements are developed for products, manufacturing will respond to the demands with new processes to provide them. This exciting and evolving area will change in ways that cannot even be imagined today. From the conception years ago, to what we see today, to whatever the future will bring, manufacturing will remain the mainstay of production of products of every type.

About the Author

John Davis is an Associate Teaching Professor in the Mechanical and Industrial Engineering Department at Montana State University's College of Engineering. Prior to becoming a professor, he worked for 26 years for a fortune 150 company in both the manufacturing and product development arenas, where he was involved in both the technical and business functions resulting in a full view of the processes from all angles. Along with his professorship, John operates a consulting business where he aids local small businesses to improve their operations and throughput for maximizing overall profits.

He has earned a Bachelor of Science degree in Mechanical Engineering, a Master of Business Administration degree, and a Doctorate of Science degree in Management Systems Engineering. John is very active in the community supporting several not-for-profit organizations and is a life member of the Masonic Fraternity. He makes his home with his wife Joanne and his son Chandler in Bozeman Montana, where they all enjoy the outdoors and beauty that Big Sky Country has to offer.

CHAPTER

1

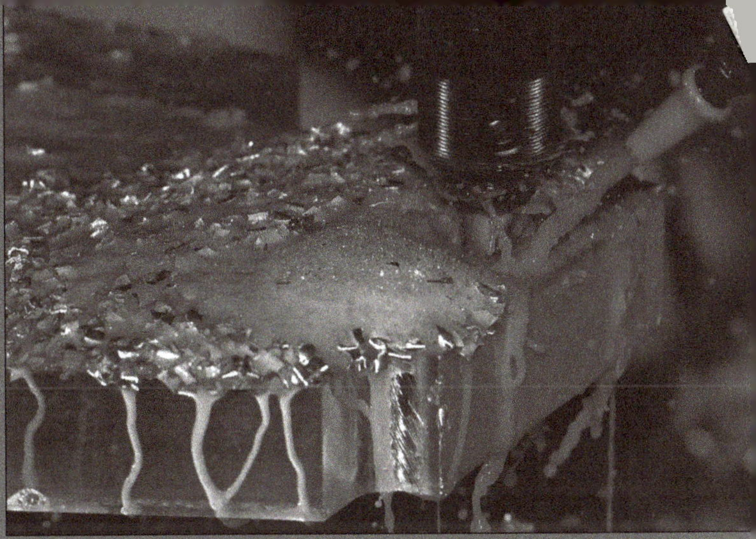

Introduction to Manufacturing

1.1 INTRODUCTION

Manufacturing is a transformation process where materials are transformed into the products that we come in contact with on a daily basis. From the vehicles we drive, to the computers we use, to the simple pens and pencils we write with, manufacturing brings all these items to us. If one was to look at the complexity of even the simplest of products we use and their low cost, it amazes most people as to how they can be fabricated and sold at such a reasonable cost to us. This is all due to the structure of the factory and the efficiency which it operates with. Every factory will be specialized in what they are producing and does not vary much from their expertise. For example, a manufacturing plant that is producing computers will not be able to start producing ball point pens. Even though a ball point pen involves far less the technology than a computer does, the factory is not set up to produce pens. This is the specialization of a manufacturing factory, do what you are designed to do, and do it better than your competitors.

This chapter brings the manufacturing requirements and structures into view. Various industries are structured to service different requirements and are set up differently. Besides these, the support systems and indicators will be briefly touched upon along with how they interface with the production processing.

1.2 MANUFACTURING SYSTEM

Manufacturing is a system consisting of series of various inputs, processing operations, and outputs. The fabrication process takes countless forms depending on what is being manufactured in a given factory. Regardless of the products being produced, they all follow the same system for producing the products. Figure 1.1 shows

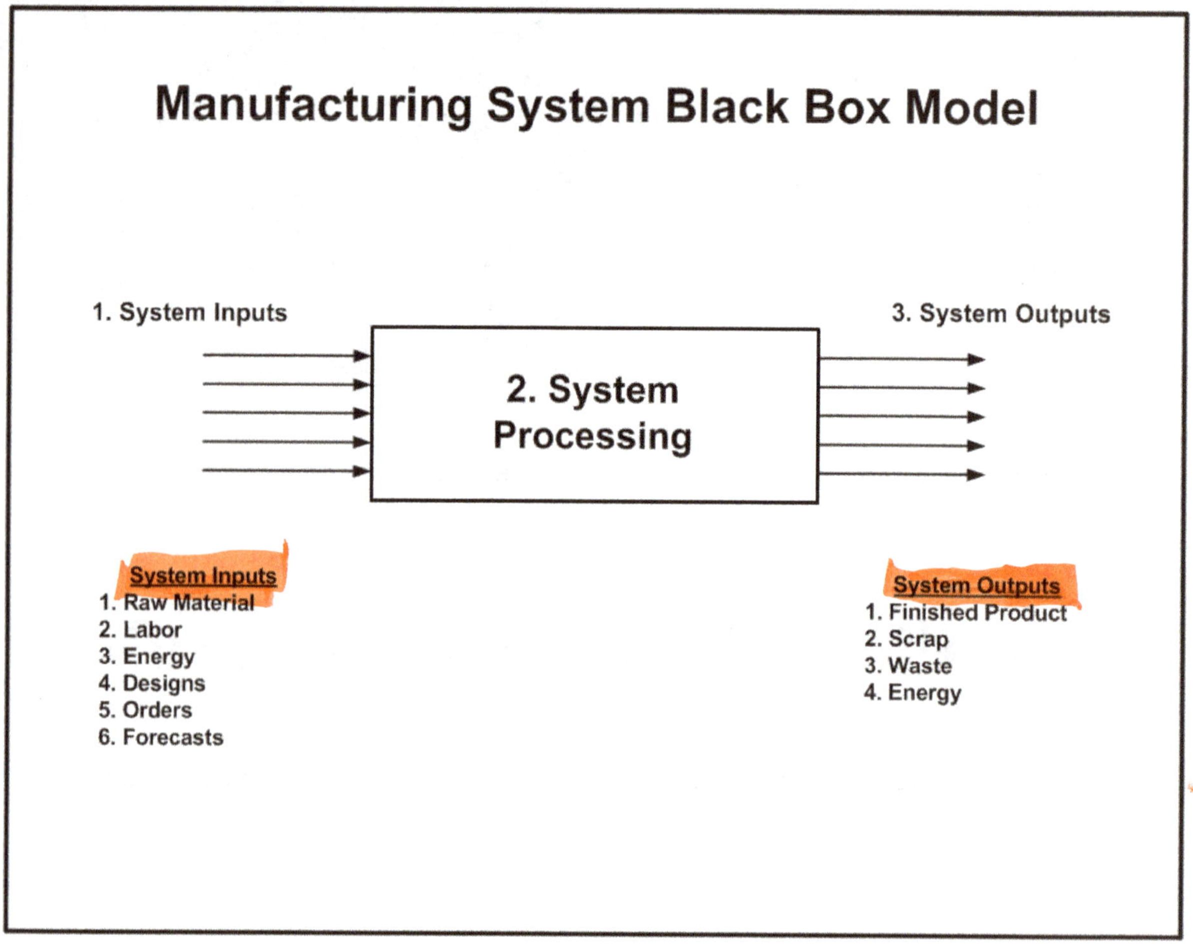

Figure 1.1 Manufacturing Black Box Model

a black box model of any manufacturing system. This shows the various inputs into the system and the outputs from it. The box in the middle is where all the processing is performed to transform the inputs into the outputs. As demand increases for more outputs, the inputs into the system will correspondingly increase, and likewise, as demand decreases, so will the required inputs.

To expand on the processing side of the system, there are several factors that need to be synchronized to keep the operation running smoothly. Figure 1.2 shows the areas that need to be balanced during processing. These are explained as follows.

Work-in-Process (WIP): Work in process relates to the materials that are being processed at any given time. It is no longer a raw material, nor is it finished product. This is the processing that is being performed at various stages leading to the final assembly where it becomes the finished product after assembly. Once the product is completed, it becomes finished product inventory and will have a higher carrying cost to the company.

Labor Force: The amount of labor required can be a very tricky area to balance. If the customer demand drops, then less labor is required in the operation. But, if the company reduces the labor force, two things can happen, which can cause issues. First, if the demand ramps back up, there will not be enough labor to perform the work. Second, if labor is reduced, experienced workers will be released and you may be able to replace a person, but not the experience that was released.

Process Flow: How the process from the beginning to the end flows through the factory will determine how efficient the operation is running. If the sequences of operations are not in a steady flow through the

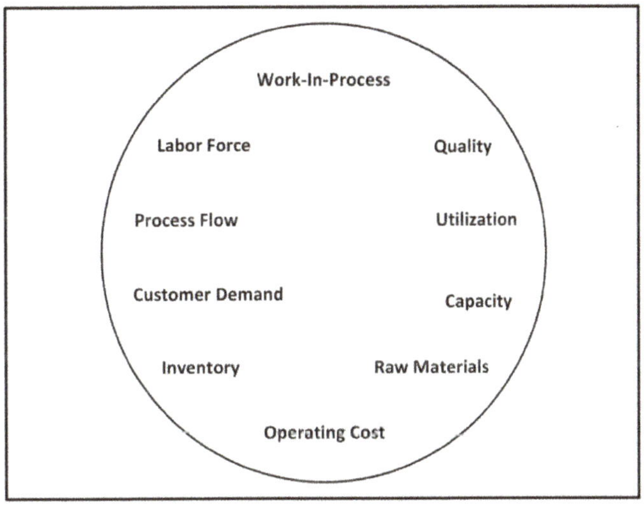

Figure 1.2 Manufacturing Balance and Control Areas

operation, then time is wasted in moving the parts from one point to the other. By having a smooth process flow from one operation to the other throughout the factory, the work in process will be able to move smoothly from one phase to another.

Customer Demand: The customer demand will determine all of the factors listed here. As the demand for products fluctuates, so will the work in process, labor force, inventory, raw materials, and operating cost. This demand can come in two ways. First, some companies operate against a forecast, which predicts the amount of products required and when. Second is an order-based company. In an order-based company, no products are started until an order is received, which can be much more difficult to predict.

Inventory: When a product is completed, it becomes inventory. Also, some commercially purchased materials that are going to be assembled into the products are inventory as well. Basically, inventory is money that is tied up until the products get sold. If the inventory levels are too high, the company has most of its capital tied up there and can cause a burden on the financial line.

Quality: Every customer wants a product that works properly with no problems. The quality needs to be controlled from the very first stage of the process through the product getting into the customer's hands. By implementing a structured quality program, every step of the process will have checks and balances to ensure that the final result will be a quality product. This will be covered in more detail later.

Utilization: Equipment is required in every manufacturing operation regardless of what they are producing. How much the equipment is being utilized is a measure of the efficiency of the operation. For example, if a given machine is being used to produce several different parts where the majority of the downtime is when it is being broken down from one part and being set up for another, it is being highly utilized. However, if a given machine is set up to produce one part only, if there is no demand for that part, the machine sits idle and the utilization is very low.

Capacity: How much work can be flowed through a manufacturing factory determines what the capacity is. If, say, a factory is working 5 days a week, 8 h per day, they are operating at 24% of their total capacity. However, if the factory is working 7 days a week at 24 h per day, they are operating at 100% of their capacity. Once the capacity reaches 100%, no more work can be scheduled in the operation.

Raw Materials: Regardless of what is being fabricated, raw materials are required. How these materials are procured depends on the type of products being fabricated. If a company is working against a marketing forecast, the materials are ordered and stored at a minimum and utilized as required. If the company is working against orders, the raw materials will be procured on an order-by-order basis as needed. Carrying too much raw materials in inventory ties up the capital for the operation, which could be utilized elsewhere in the business.

Operating Cost: Every operation has a cost that it operates within. The cost encompasses the total amount it costs to run the company. From direct labor costs, indirect labor cost, energy costs, and so on. they all get totaled up, which gives the operation a cost to produce the products. This total cost gets divided by the amount of hours the operation works per year, giving them an hourly operating cost. The more hours that are worked, the lower the operating cost the factory will realize.

1.3 MANUFACTURING TRANSITION

The entire purpose of manufacturing is to transition materials into products. This transition is viewed in two segments, which need to be synchronized together for the operation to be successful. The two areas of transition are *Technological Transition* and *Economic Transition*. These are discussed as follows in detail, including how they must be performed together for a successful operation.

1.3.1 Technological Transformation

Technological transformation is the application of various technologies to transform the raw materials into different shapes. This may include very low technology like manual operations or very high technology such as fully automated operations. The purpose is to create the parts required for the products being marketed. The altering of geometry, physical properties, and appearance is the function of manufacturing in general. The industry and the products being produced will determine what technology would be required to accomplish the final results desired.

1.3.2 Economic Transformation

Economic transition is a simple concept. Purchase the inputs for the system; add additional cost for the technological transition; then sell the products for a greater value. The issue here is to manage the operation in order to accomplish this transition. If enough profit is not being realized, the company will run the risk of failure. There needs to be very tight controls on the processing and overall cost of the operation. Companies are all about making a profit, and this transition needs to minimize the cost and maximize the revenue. This balance within the operation can be seen in Figure 1.2.

1.4 MANUFACTURING INDUSTRIES

Manufacturing industries are constructed to serve their customer base in the most effective and efficient manner possible. The structure of these operations is mainly dependent on the volume of products they will produce and the overall size of the products. Each is set up to handle the volume and requires a different workforce to produce the end products. They are divided into three categories: *Low Production*, *Medium Production*, and *High Production*. These are explained as follows.

1.4.1 Low Production Industries

A low production industry handles products from 1 to 100 units per year. These companies work against orders and normally carry no inventory of products or raw materials until an order is received. They are set up like a job shop where the product stays stationary and everything required is fabricated in other locations and fed into the assembly area. Because many of these products are custom-built to the customer's specifications, the workforce is very technically trained. Many of them are Machinists capable of fabricating the required custom parts and engineers to calculate the performance requirements. Figure 1.3 shows how a low production industry is structured.

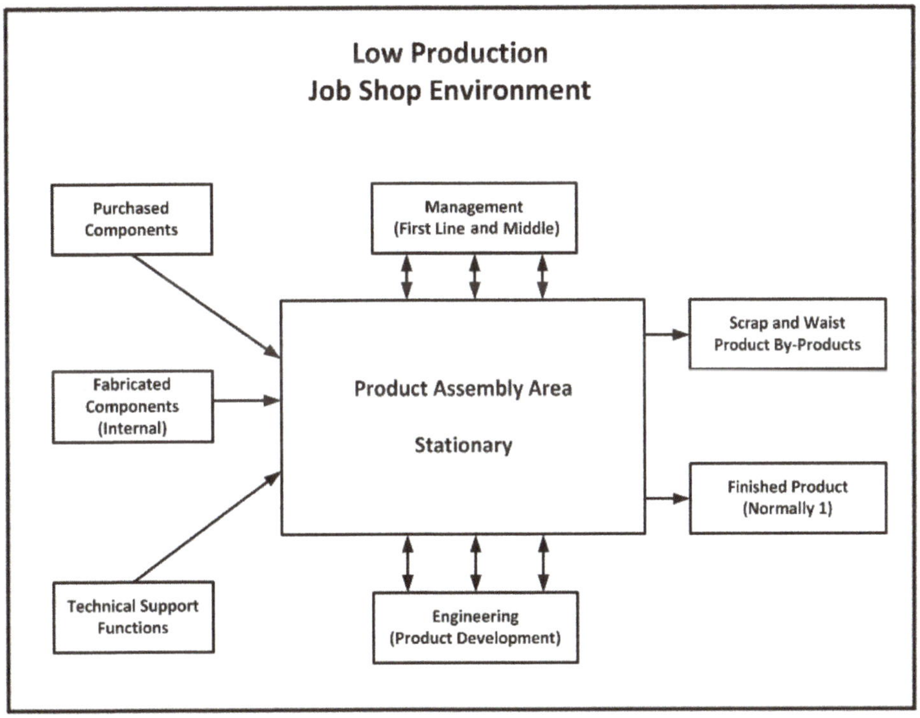

Figure 1.3 Low Production Industry Structure

1.4.2 Medium Production Industries

In a medium production industry, they normally handle from 100 to 10,000 units per year. Most of them produce several different products with some commonalty across them. These operations normally work against an annual forecast for their products and a balance of raw materials is maintained in inventory along with purchased components. This inventory will fluctuate based on the forecast requirements. An industry for medium production is structured much differently than a low production industry. In this setup, the component parts are fabricated away from the assembly area in batch processing. This means that one operation performs some work on the part, and then they move to the next, and so on. From there, they get moved to the assembly area at the various locations they are required at. The product is moved down an assembly line manually and at each station, the assembly builds to the final product. This assembly process needs to be very well timed so that each station takes about the same amount of time to avoid a bottleneck in the process. The labor force in this industry can be a combination of skilled and unskilled. As the equipment is set up to perform the required operation by a skilled person, an unskilled operator can run the equipment with no issues. Figure 1.4 shows the structure of a medium production industry.

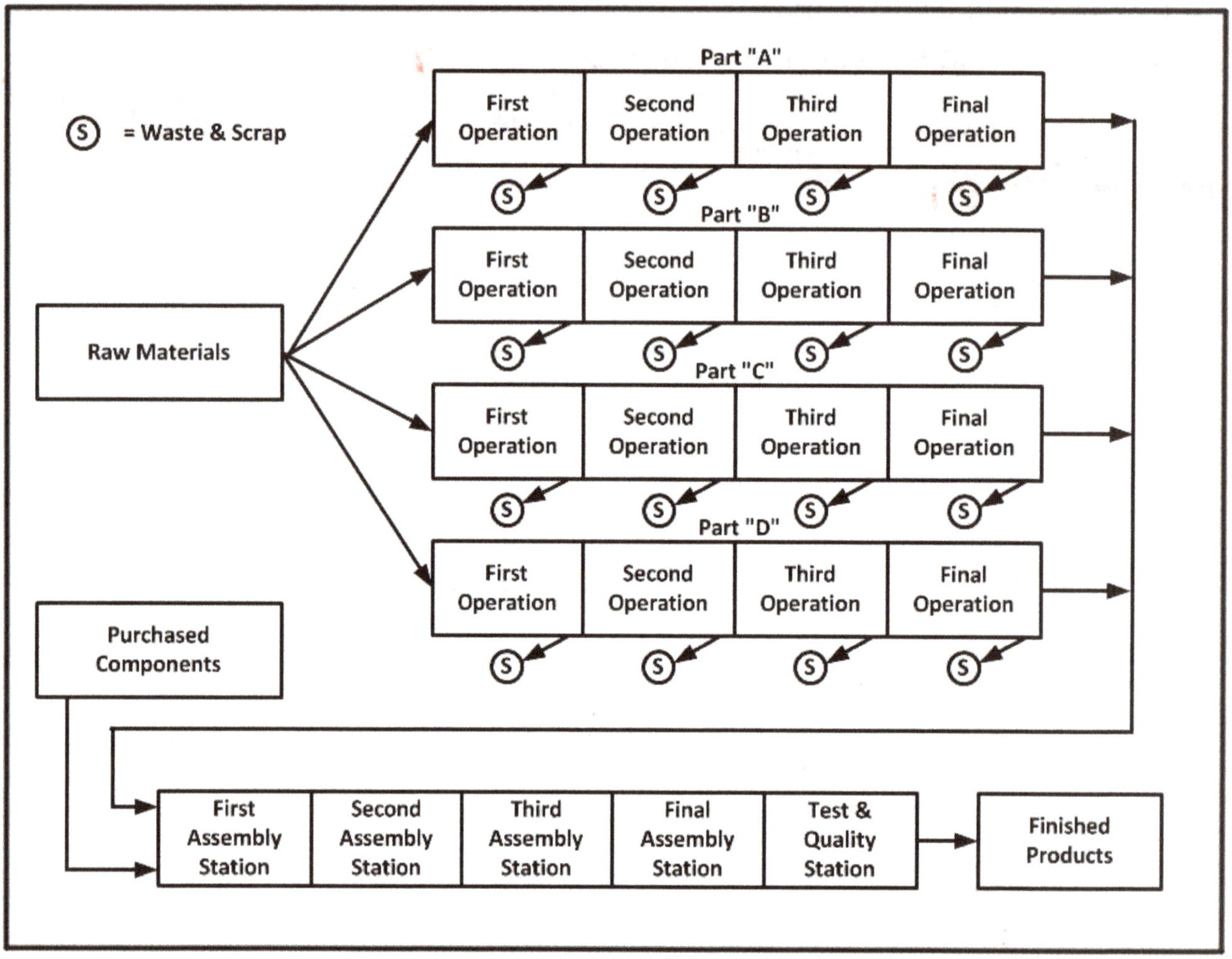

Figure 1.4 Medium Production Industry Structure

1.4.3 High Production Industries

High production industries handle huge volumes of units per year, from 10,000 and up. Industries of this nature are very highly automated with vast amounts of money invested in the equipment. Mainly, a high volume industry will be producing one family of products having very little variation in them. The component parts are produced offline or procured from suppliers and get delivered to the assembly area daily or several times per day. The product is moved down an assembly line automatically and the assembly is performed on the moving product. All the labor that needs to be performed is divided down to the smallest amount where each operator will perform only one or two tasks. With this limited amount of work being performed, the direct labor force does not have to possess any specialized skill, only what their job function required. For example, if an automaker has the vehicles moving down the assembly line, one operator may be placing the tires on one side of the car, a second operator would be responsible for assembling the lug nuts and tightening them. This process continues from the very first operation to the last where the vehicle is completed. Figure 1.5 shows the structure of a high production industry.

This is how a high volume industry can produce vast amounts of products in a very short amount of time. If we were to look at the rate at which vehicles are coming off the assembly line in Detroit, the rate is one vehicle every 58 s. This highly automated system needs to run in perfect timing in order for this to be accomplished, and if there is any interruption in the system, everything will come to a halt.

To bring things into perspective between the three levels of production industries, let us look at producing a car. In a high production environment, a new vehicle costs somewhere between $20,000 and $30,000. However, if we looked at a low production industry such as Lamborghini, their top model is custom-built, where

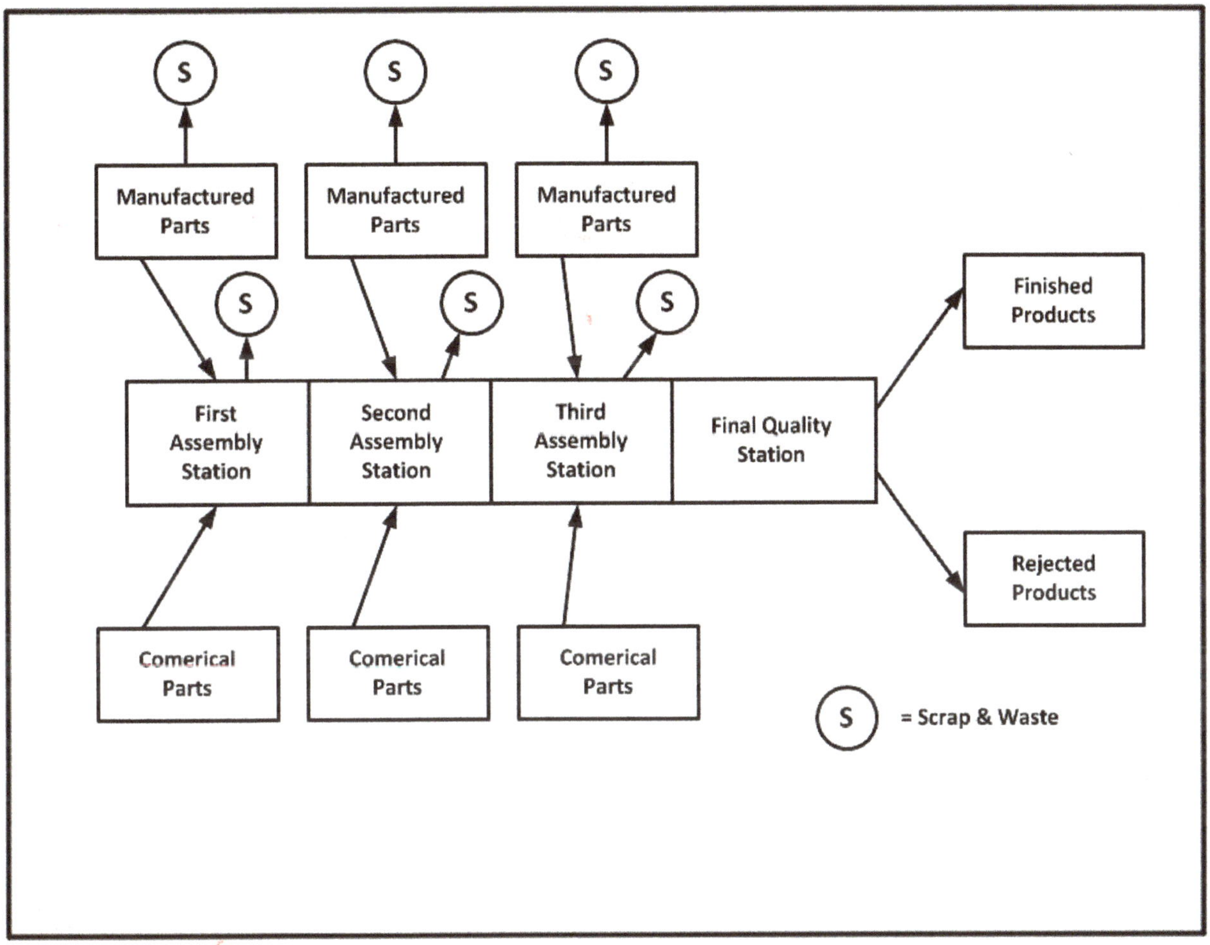

Figure 1.5 High Production Industry Structure

about 20–30 vehicles are produced annually with a cost of between $500,000 and $750,000 per vehicle. The function of both vehicles is the same, but how they are being produced and the volume of the vehicles determines their cost.

1.5 PRODUCT VARIETY

All manufacturing operations are geared to producing a product that the factory is set up to make. If a company is producing several different products, they may be divided into two categories, which are *Soft Variety* and *Hard Variety*. Even though the products may look and perform differently, there may be a lot of similarity between them from a manufacturing perspective. Sometimes, it is not easy to tell the difference between the two; however, there is a vast difference between a soft and a hard variety in products. This is explained as follows.

1.5.1 Soft Variety

Soft variety products may appear totally different, but have a lot of similarity from a manufacturing standpoint. What determines a soft variety product is the amount of common parts that are used across the various models that the operation is producing. By having several common parts and technology in the products being produced, the operation can standardize the components and mass-produce them, which will give them an economy of scale in manufacturing.

If we were to look at the automotive industry, we notice that each auto manufacturer produces a variety of vehicles that they sell. Each vehicle looks totally different to the customer, but at a closer examination, there is a vast amount of similarity. With the exception of the body style and interior of the vehicles, for the most part, the drive train is the same. They share the same steering system, engines, transmissions, and rear drive axles. This type of product difference classifies it as a soft product variety.

1.5.2 Hard Variety

Hard variety product has little to no commonality across the products, even if the technology is similar. Looking at the automotive industry again, if the operation is producing cars and then decides to start producing large trucks, with the exception of the instrumentation panel on the dashboard, the entire drive train is totally different in a large truck as opposed to a passenger car, even though the technologies of both are similar. To make a transition of this nature would require the operation to completely set up a different manufacturing system to produce and assemble the trucks. This type of product difference is a hard product variety even though it is within the same industry and the technologies are similar.

1.6 MANUFACTURING CAPABILITY

Every manufacturing operation will have certain capabilities depending on the products they are producing. This capability has to do with three major factors as to what they can and cannot produce. Before any manufacturing operation can venture into any additional fabrication of products, they need to evaluate their capability closely. The three areas requiring evaluation are explained as follows. For example, if a given manufacturing operation is producing large construction equipment such as bull dozers, and they want to start to produce small garden tractors, the layout of the factors is scaled much too large to handle the smaller product.

1.6.1 Technological Processing

The technology that an operation produces is critical to what type of work they can do. Technology can be segmented into several categories such as, mechanical, electrical, and overall technology. The manufacturing of computers requires a much different technology than do, say, cell phones. Even though both are electronic devises, the technology that allows them to function is totally different. Every manufacturing operation is geared toward a given technology and needs to stay within their expertise in order to be successful in the market place.

1.6.2 Product Size and Weight

The size and weight of the products that an operation produces will determine what the factory is capable of producing. When an operation is set up for products of a given size and weight, any new work needs to fall into a similar category in order for production to be smooth. The size and weight of the new product would not fit the way the factory is scaled to accept a much small and lighter product. The same would be in the reverse situation. Every manufacturing operation is scaled to the products they produce and a scale-up or -down will not be a functional fit for the factory.

1.6.3 Production Capacity

The capacity of an operation determines how much work can be performed within it. Every operation has a limit on how much work can be flowed through it, and when it reaches its maximum, it cannot accept any more. For example, if the operation is working either a 3 h shift per day or 7 days per week, their manufacturing capacity is at 100%. If any additional work needs to be performed, the operation has two options. Option 1 is to send the new work out to a secondary facility to produce the required parts or products. Option 2 would be to off load some of the work they are currently producing and then accept the new work to be done.

1.7 MANUFACTURING SUPPORT SYSTEMS

Every manufacturing system requires a vast amount of support to ensure that everything flows smoothly. All of these support functions operate separately, but need to interface with each other in order to keep the system flowing in an effective and efficient manner. A brief description of the main support functions is provided as follows.

1.7.1 Process Engineering

Manufacturing engineering is possibly the most important support function in any manufacturing operation. They determine how the parts are going to be fabricated and assembled in the most efficient and effective manner. Aside from the fabrication of the parts, the manufacturing engineering function is also responsible for the quality of both the fabricated parts and the functions of the finished products. This function is covered in more detail in Chapter 2.

1.7.2 Production Planning

Determining how many parts and when they need to start being produced so the operation runs smoothly and there are no interruptions in the assembly process is the function of production planning. This requires knowing exactly how much work is already in process, what is the total lead time to fabricate them, and the demand for the products. Besides managing the work that is being produced, they also have to account for all the suppliers that produce parts and ship them to the operation for assembly. This exact balance is critical to ensuring that the operation will run in a seamless manner and the required amount of products is available when required for the market place.

1.8 SUMMARY

A manufacturing operation is a very-well-timed machine. As the inputs are flowed into it and the processing begins, it is a seamless flow from the beginning to the end where the products are completed. This well-oiled machine requires a vast amount of planning for it to run with no interruptions. Even when a new product is to be introduced, the balance of the existing products needs to be closely evaluated to ensure a steady, even flow of products. Understanding how complex the manufacturing system is ensures that the end results realized are produced in the most effective, efficient, and economical manner possible.

1.9 QUESTIONS

1. Within the Manufacturing System, why is important to balance the inputs and the outputs?

2. What is Production Capacity, and what options would a manufacturing operator have if they were at 100% and more work was required to produce additional products?

3. What is meant by work in process (WIP)?

4. What is utilization in a manufacturing operation, and why is it important to measure?

5. What is the difference between a Technological Transformation and an Economic Transformation of a product?

6. Manufacturing Industries can be Low, Medium, or High production faculties. Explain the difference between the three and what the advantages and limitations of each are.

7. Every operation has a given Manufacturing Capability. What are the three areas that determine this capability? Explain each.

CHAPTER

2

Process Engineering and Technical Support

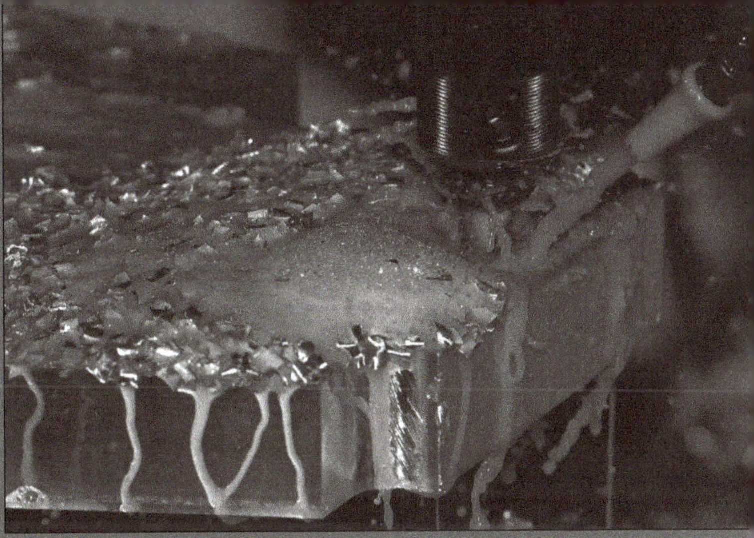

2.1 INTRODUCTION

Every manufacturing factory requires a vast amount of technical support to ensure that the products are being produced to the customers' requirements and delivered in a timely manner. This support is divided into several areas where each is responsible for specific aspects of the operation, but must all work in unison to ensure that everything is operating as effectively, efficiently, and economically as possible. From the beginning of a new part or product being introduced, to existing parts and products being fabricated, everything needs to be accounted for and synchronized perfectly. Any disruption to the system flow will result in either a subquality part or product or failure to meet the required delivery date.

The critical areas that support a manufacturing operation on a daily basis are *Process Engineering*, *Production Planning*, *Quality Control*, and *Design Engineering*. These support functions are covered in this chapter outlining their individual functions and how they all interface to work together within the operation.

2.2 PROCESS ENGINEERING

The process engineering function within a manufacturing operation is a key area taking in a wide variety of functions. They are responsible for methodization of parts and assemblies within the operation and resolving the day-to-day issues that occur. The many hats that they wear are *Process Planning*, *Tooling Requirements*, *Cost Estimating*, and *Factory Flow and Efficiency*. Each of these functions is explained as follows.

2.2.1 Process Planning

Preparing process plans is a major function performed by the process engineers. This involves deciding how the part will be fabricated, what sequence of operations will be performed, which machines would be utilized for each operation, and what tooling is required. These process plans are extremely detailed, providing a step-by-step sequence for each operation. Along with preparing process plans for fabricating the component parts, process plans for the assembly of the products are also generated. This total sequence of events will start from the raw materials utilized and end with the product being completely assembled and tested.

All processes in manufacturing follow a basic flow from the beginning to the end. This sequence contains four different levels of processing: *Initial Operation*, *Secondary Operation(s)*, *Physical Enhancement Operation(s)*, and *Finishing Operation(s)*. This basic flow is an outline, and not every part being fabricated will require to be processed through all of them depending on the unit itself. Figure 2.1 shows the basic flow for parts being produced, and the operations are explained as follows.

2.2.1.1 Initial Operation

The initial operation is the first processing performed on the workpiece. If it were a part requiring machining, it would be the first cut made on the part to establish a datum surface for locating it for other operations to follow. For a component being molded out of plastic, this would be where the part is molded in the machine. The initial operation would always be the first step in the process plan.

2.2.1.2 Secondary Operation(s)

Almost every part being fabricated will require some secondary operations after the initial step. For parts being machined to accomplish the final results, there could be several different cutting operations performed on several different machines depending on the configuration being accomplished. Even for a part being molded in plastic, a secondary operation may consist of removing the filling sprue where the plastic filled the cavity that

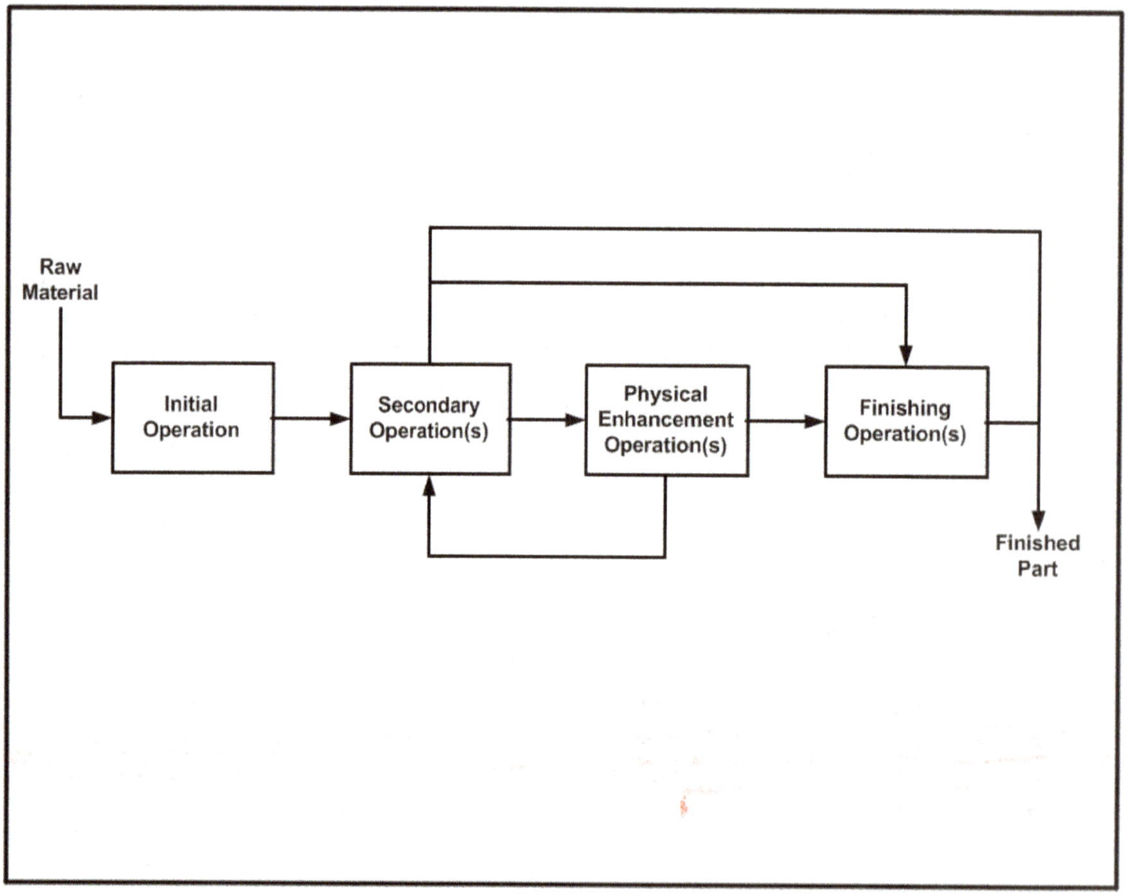

Figure 2.1 Basic Process Flow of Fabricated Parts

created the part. Depending on the processes required to create the part, secondary operations can be from one to several depending on the requirements to complete the component's geometric requirements.

2.2.1.3 Physical Enhancement Operation(s)

Physical enhancement operations may or may not be required on the parts being produced. For metal parts, this could include any heat-treating processes required, such as hardening of the part, tempering, and annealing. From this step, the parts may or may not require more secondary operations before they are completed. So, from any physical enhancement operations, if required, the parts could move on to the finishing operations or back to more secondary operations.

2.2.1.4 Finishing Operation(s)

The final step in the basic process flow is the finishing operation(s). Almost all metal parts would require some type of finishing on them, either for corrosion protection or for decorative purposes. These processes could include plating, painting, anodizing, and powder coating. Many of these processes required several steps or operations to accomplish the final result, and after finishing, they would not go back for additional operations. However, plastic parts for the most part require no finishing and would bypass this step in the process flow.

Looking at what is required for generating a detailed process plan, the process is to lay out a very detailed sequence of operations from the raw material to the finished part. This begins with receiving the part drawing of what needs to be processed. Figure 2.2 shows a typical part print received in manufacturing for a unit that is required. From there, the process engineer would produce a process plan sketch where the dimensions are changed to letter values. This is done to reduce the confusion regarding the entire numerical dimension displayed on the part drawing, and the process plan will reference the letter designations on the sketch to the numerical dimensions on the part drawing in the process plan. Figure 2.3 shows an example of how the part print dimensions are changed to letter designations. From here, the process plan is generated with the operation number, the department that the operation will be performed in along with the machines that would be utilized for each step. Along with this, any and all tooling required for that particular operation is listed. Figure 2.4 shows an example of the first three operations required for producing the part specified on the part drawing.

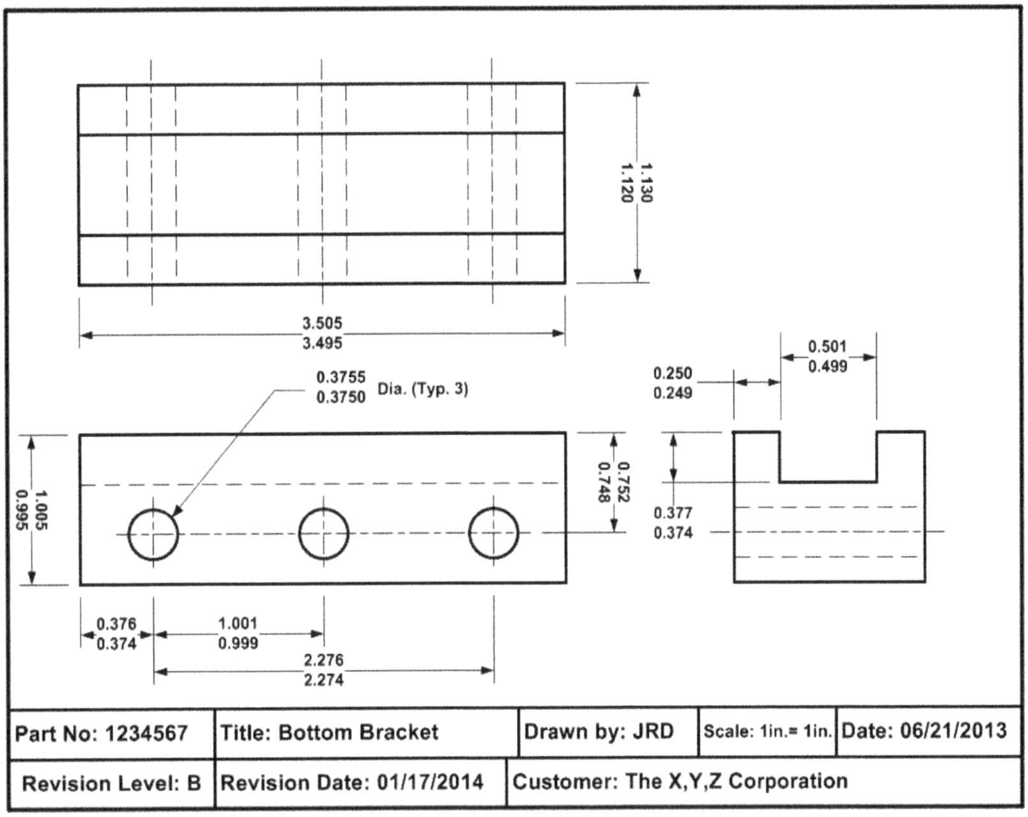

Figure 2.2 Part Print of Unit to Be Fabricated

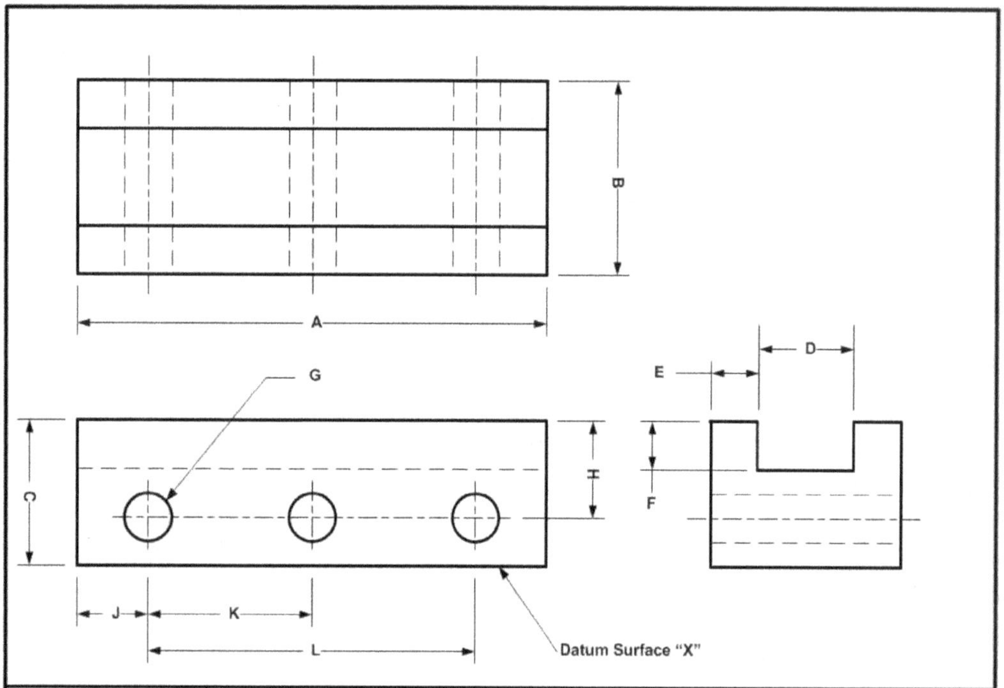

Part No: 1234567	Title: Bottom Bracket	By: CKD	Date: 07/08/2013
Revision Level: B	Revision Date: 2/03/2014	Customer: The X,Y,Z Corporation	

Figure 2.3 Process Plan Sketch of the Part to Be Fabricated

<table>
<tr><td colspan="2">Part No: 1234567</td><td colspan="2">Title: Bottom Bracket</td><td colspan="2">Customer: The X,Y,Z Corporation</td></tr>
<tr><td>By: CKD</td><td>Date: 02/08/2013</td><td>Revision: B</td><td>Revision Date: 07/08/2014</td><td>Total Sheets: 7</td><td>Sheet 1 of 7</td></tr>
<tr><td>Oper</td><td>Dept</td><td>Mach</td><td colspan="2">Description</td><td>Tooling</td></tr>
<tr><td>100</td><td>4500</td><td>45-032</td><td colspan="2">Grind Surface X to Establish Datum "X".
Hold Surface finish to Ra 32-64</td><td>Std. Holding Fixture
Wheel #45-173</td></tr>
<tr><td>200</td><td>1200</td><td>12-027</td><td colspan="2">Rough Mill Dimension A to 3.525 / 3.515.
Rough Mill Dimension C to 1.025 / 1.015
Finish Mill Dimension A to 3.505 / 3.495.
Finish Mill Dimension C to 1.005 / 0.995.</td><td>Holding Fixture 1234567-A
Roughing Cutter #12-184
Finishing Cutter #12-069</td></tr>
<tr><td>300</td><td>1200</td><td>12-018</td><td colspan="2">Rough Mill Dimension B to 1.150 / 1.140.
Finish Mill Dimension B to 1.130 / 1.120.
Rough Mill Dimension E to 0.270 / 0.255.
Rough Mill Dimension F to 0.390 / 0.385.
Rough Mill Dimension D to 0.525 / 0.515.
Finish Mill Dimension E to 0.250 / 0.249.
Finish Mill Dimension F to 0.377 / 0.374
Finish Mill Dimension D to 0.501 / 0.499.</td><td>Holding Fixture 1234567-B
Roughing Cutter #12-184
Finishing Cutter #12-069
Roughing Cutter # 12-163
Finishing Cutter #12-031</td></tr>
</table>

Figure 2.4 Detailed Process Plan for Fabricating the Unit

Process plans are the road map to producing the parts in a manufacturing operation. They are constantly updated based on the changes to the part drawing to the new revision level, along with any manufacturing improvements in the processes as better methods are found. From this, production planning can generate the orders for the parts and the fabrication process will begin. Accurate process planning is the overall key to successful fabrication in any manufacturing operation.

2.2.2 Tooling Requirements

Requirements for tooling fall into two categories. The first is specifying which standard cutting tools would be used for each operation in the process plan. This involves researching new cutting tools available and testing them before releasing them to be used in production. New cutting tools and materials are being offered all the time, and detailed testing for tool life against price is required to ensure that they will perform better and more economically than the current tools.

The second category is a specially designed tooling required for the parts being produced at the given operations. This type of tooling is custom-designed and dedicated to a given part for a given operation. Although the process engineer may not be the individual performing the actual design of the tool, they will specify what functions the tool needs to perform for that operation. Tooling of this nature could be injection molds for producing plastic parts, punch and die sets for stamping out sheet metal parts, or holding and drilling fixtures to hold and locate the parts during milling and drilling operations. As the process plan continues through the required operation, inspection steps would be included to check for critical areas on the part throughout the entire process.

2.2.3 Factory Flow and Efficiency

Along with process planning, the process engineering is responsible of how efficient and effectively the parts will flow through the operation. Traditionally, this is an industrial engineering function, but the process engineer will get heavily involved in the implementation. This may be viewed on a micro- or a macrolevel. Both micro- and macrolevel are important for increasing the efficiency of the operation, which is normally performed by evaluating the microlevel first and then moving onto the macrolevel.

For microlevel of process improvement, it would look at just the individual operation being performed to reduce part handling from loading the machine, actual machining time, and the unloading of the machine as the parts are being worked on. By reducing the time required to perform that particular operation, the workpieces can move through that given operation and on to the next in a timely manner.

A macrolevel of process improvement would look at the entire sequence of operations to improve the flow and throughputs. The ultimate goal here is to have the operations required for the parts located as closely as possible to reduce part handling time between them. After each operation has been evaluated for the most efficient manner to perform that given operation, the overall flow is then evaluated. For example, two parts ("A" and "B") are being produced in a given manufacturing facility; the flow of each is evaluated for how they move from one operation to the next. Figure 2.5 shows how part "A" flowed originally and through macroimprovements, the equipment required for the operations were rearranged to reduce the traveling between machines. Likewise, Figure 2.6 shows the similar improvement for part "B." The final result is shown in Figure 2.7 where both parts are combined in a common area, reducing the overall flow travel and opening up floor space, which can be utilized for additional manufacturing.

Working on factory flow and efficiency is not a one-time event, but rather a continuous process, which is not attic but dynamic in nature. The continuous improvement is a daily task for process engineering. The more efficiently and effectively the product can flow through the operation, the more economically the parts can be produced. This results in a higher profit margin for the operation, which is what the final goal of any business is.

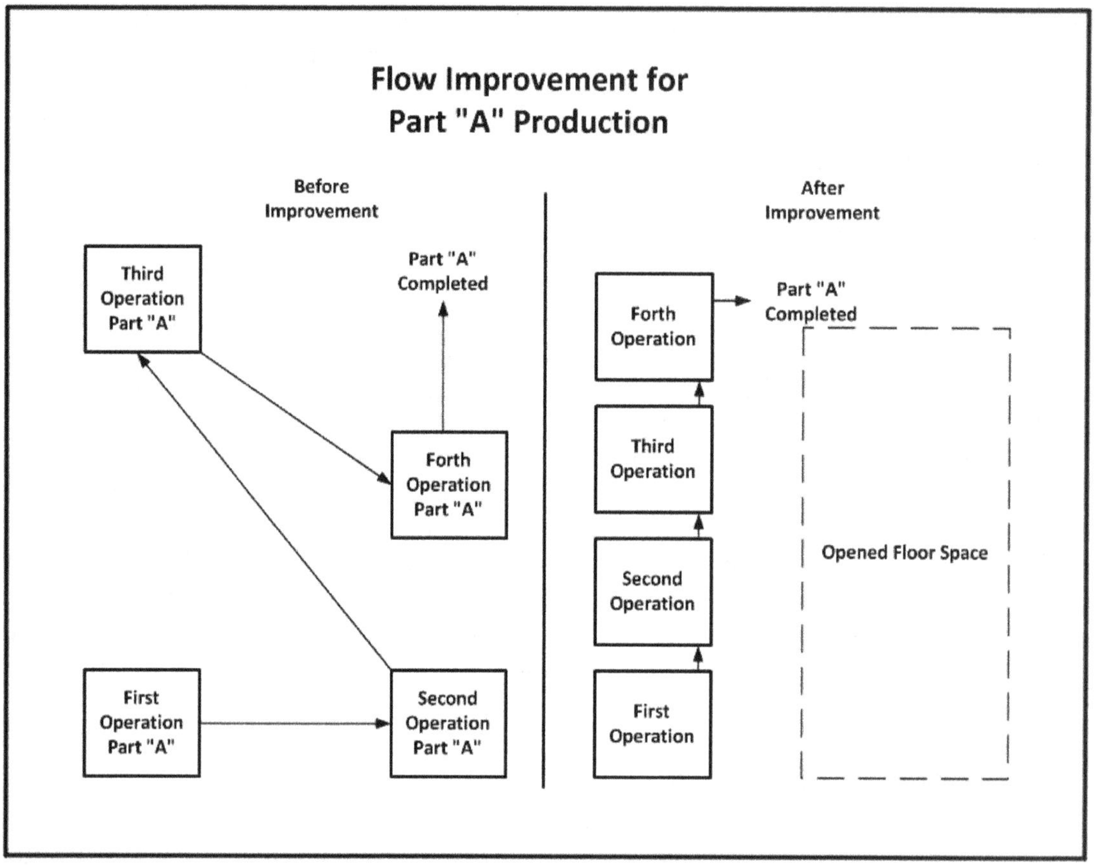

Figure 2.5 Flow Improvement for Part "A"

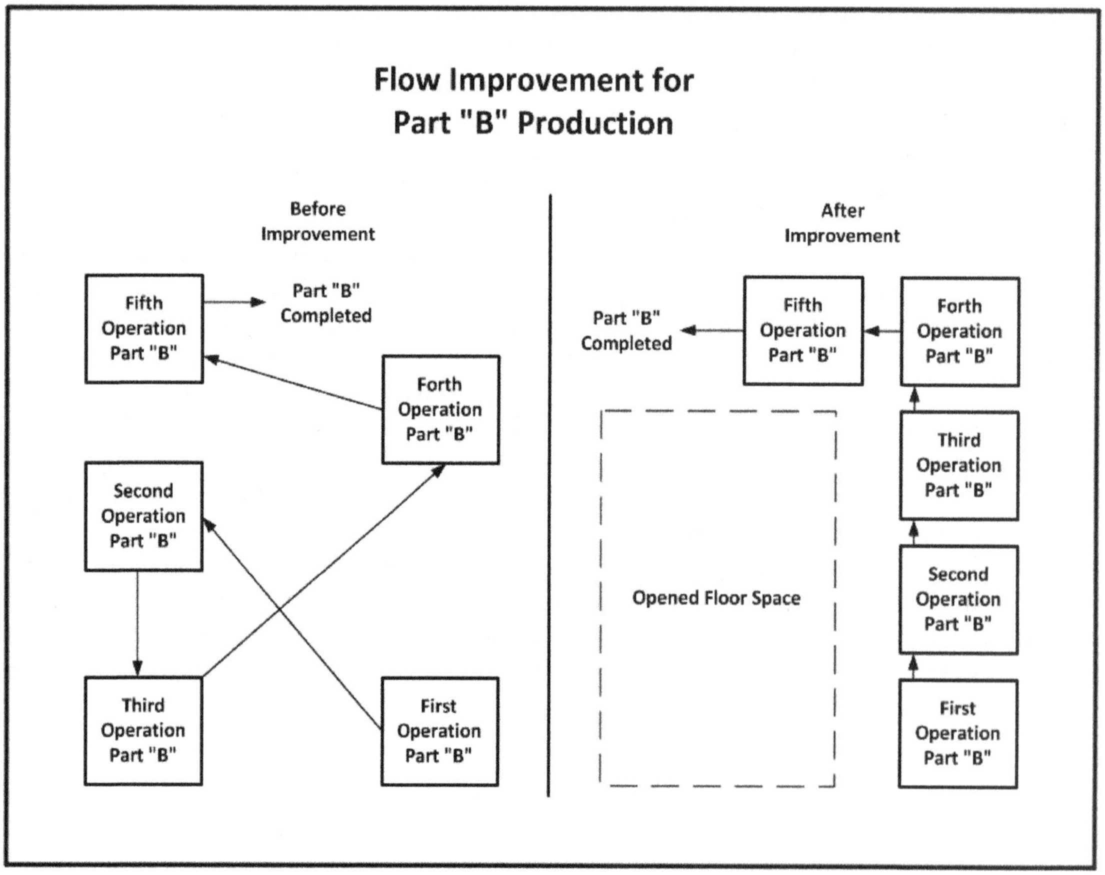

Figure 2.6 Flow Improvement for Part "B"

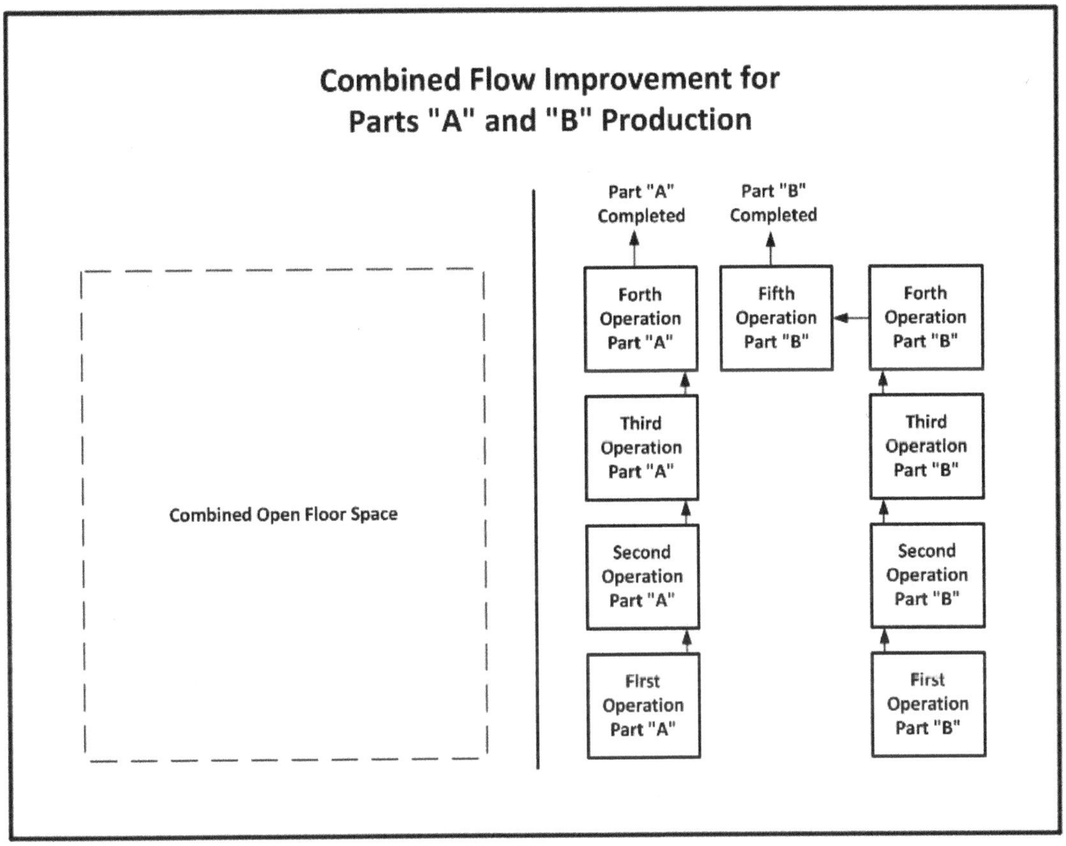

Figure 2.7 Combined Flow Improvement for Parts "A" and "B"

2.2.4 Cost Estimating

Knowing exactly what the cost to produce the products would be is critical to any operation. By estimating the cost for each operation from the very first through the final assembly, the total cost can be rolled up into a final product cost. This will determine what the final product can be sold for after the operation applies the markup to the customer. So, the more efficiently the product can flow through the factory, the lower the manufacturing cost would be and a higher profit margin can be realized by the business. This is covered in detail in Chapter 18.

2.3 PRODUCTION PLANNING

Knowing when the work needs to be scheduled into the operations is the function of production planning. This group is responsible for ensuring that every component part that needs to be produced, along with any commercially purchased components, will be available when required for assembly of the product. This requires knowing what the lead time for producing the parts is along with the delivery times for commercial components. For the fabricated components, the ordering of the raw materials required is also a function of production planning. Basically, they are responsible for ensuring that everything required for the finished product in the quantity required is ordered and scheduled in time for the end result to happen. When an order is released to fabricate parts, each part would be accompanied by a route card. This follows the parts through each operation and is signed off when the operation is completed. A route card is a very abridged version of a process plan only stating what process is to be performed at each operation. It also accounts for the amount of parts entering the operation and exiting. An example of a route card is shown in Figure 2.8.

Route Card						
Part Number: 1234567	Revision: B	Order Number: 28		Quantity: 1500	Date: 02/23/2014	
Operation	Description	Department	Start Quantity	Start Date	End Quantity	End Date
100	Grind	4500	1600	02/24/2014	1595	02/26/2014
200	Mill	1200	1595	02/27/2014	1585	03/02/2014
300	Mill	1200	1585	03/04/2014	1565	03/06/2014
400	Drill	2700	1565	03/06/2014	1560	03/08/2014
500	Inspect	0000	1560	03/10/2014	1557	03/12/2014
600	Polish	1800	1557	03/15/2014	1548	03/17/2014
700	Plate	3500	1548	03/18/2014	1537	03/19/2014
800	Inspect	0000	1537	03/20/2014	1531	03/21/2014
900	Package	3500	1531	03/23/2014	1531	03/24/2014

Figure 2.8 Route Card that Accompanies the Part through the Fabrication Process

Depending on the business, this would be performed either against a forecast or on a build-to-order basis. In low production operations, they typically run against receiving orders and per order, production planning would order and schedule everything on receiving customer orders. For medium and high production operations, they would be running against a forecast. These forecasts would be viewed either monthly, or, in high production operation, weekly or even daily. The difference in forecasts and build-to-order scheduling is explained as follows.

2.3.1 Build to Order

In operations that are building the products based on customer orders, very little, if any, inventory is held. As an order is received, the production planning organization will begin the process by ordering all the raw materials and commercial components required to build the number of units stated on the order. Upon receiving the order, they will notify the customer as to the delivery date based on the lead times for receiving materials and fabrication. If additional orders get received from other customers, the process begins over for that particular order. The fabrication process starts and each part for a particular order is monitored through the process until the customer's order is completed and shipped. This process requires strict control of the work being produced, and for the most part, in a low production operation, which mainly operates on a build-to-order system, the parts are being fabricated manually and is time-consuming. Figure 2.9 shows the typical flow of a build-to-order system from receiving a customer order through completion of the product.

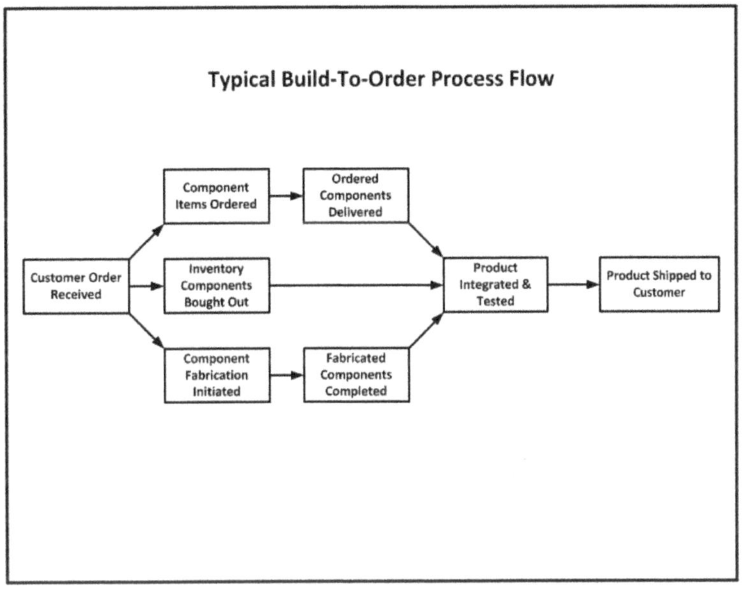

Figure 2.9 Typical Build-to-Order Process Ordering Flow

2.3.2 Build to Forecast

For operations in the medium-to-high volume category, they mainly operate against a marketing forecast, which can be either monthly requirement or broken down to weekly or daily requirements. This system is more dynamic in nature where the same parts are fabricated, but the volumes of the orders vary based on the forecast requirements. From the information on the process plans, production planning knows when to start the orders for the various fabricated parts and when to order commercially purchased parts based on their delivery lead times. If we look at Figure 2.10, which shows an annual forecast broken down monthly, it shows what the overall requirements are for monthly quantity along with how many parts are either fabricated or purchased commercially. Also, there is additional information given for what the longest lead time is for a component. This is critical to know so that they can be ordered well in advance. The historical quality levels are stated, so the order of the parts can be increased to compensate for the defective parts and the parts consumed in the setup of various operations.

Monthly Forecast for Product XYZ

Month	Quantity	Fabricated Parts	Commercial Parts	Longest Lead Time	Quality Level	Order Quantity
January	143	536	1285	65 Days	5.3%	150
February	128	536	1285	55 Days	5.1%	134
March	163	536	1285	55 Days	4.9%	171
April	185	536	1285	62 Days	5.3%	194
May	191	536	1285	68 Days	4.6%	201
June	215	536	1285	51 Days	5.0%	226
July	228	536	1285	54 Days	4.9%	239
August	253	536	1285	72 Days	5.2%	266
September	205	536	1285	57 Days	4.7%	215
October	165	536	1285	60 Days	4.9%	173
November	138	536	1285	58 Days	5.2%	145
December	124	536	1285	63 Days	5.5%	130

Figure 2.10 Typical Monthly Forecast Ordering Flow

2.4 QUALITY CONTROL

Ensuring the quality of the products being produced is critical for customer satisfaction. No one wants to purchase a product that has defects or does not function properly. The quality of a product begins with the very first step in producing it, not when it is completed. Every operation and step along the way will have checks and balances, checking not only the parts being produced but also the process. By ensuring that the process is capable of producing the components, the end result will be satisfactory. Along with ensuring the quality of the fabricated part, the commercially purchased parts need to be guaranteed that they meet the requirements as well.

The two main quality functions that are conducted during fabrication and assembly are *Inspection* and *Quality Control.* Both are put in place to ensure the quality of the fabricated parts and product functionality, but are different in how they accomplish their tasks. By working together, the quality can be ensured during the entire process. These quality functions are covered in detail in Chapter 28, and a description of each is provided as follows.

2.4.1 Inspection

The purpose of inspection is to verify that the fabricated parts are to the dimensions specified on the part drawing. As the parts are being produced, they would be measured to check that the sizes of the parts are at the dimensions and within the tolerances required. This is also used to measure the commercially purchased components as well in a function called incoming inspection. Depending on the criticality of the particular dimension being inspected, the operation may choose to inspect every part, which would be 100% inspection, or a sample from the entire lot. The decision to inspect 100% or a sample is determined on the function of the part and how a defect would affect the overall function of the product or if a defect in that area could cause harm or injury to the operator. Figure 2.11 shows a typical inspection card that would be used to inspect the dimensions

Inspection Record

Part Number: 1234567	Revision: B	Order Number: 28	Quantity: 1500	Date: 02/13/2014

Inspection Description	Frequency	Starting Quantity	Ending Quantity	Date
Inspect (3) Holes G to 0.3775 / 0.3750 Dia.	100% Sample	1560	1557	03/10/2014
Inspect Dimension H to 0.752 / 0.748	20% Sample	1557	1557	03/11/2014
Inspect Dimensions J, K & L to 0.376 / 0.374; 1.001 / 0.999: 2.276 / 2.375	25% Sample	1557	1557	03/11/2014
Inspect Dimensions D, E & F to 0.501 / 0.499; 0.250 / 0.249; 0.377 / 0.374	50% Sample	1557	1557	03/12/2014
Inspect Dimensions A, B & C to 3.505 / 3.495; 1.130 / 1.120; 1.005 / 0.995	10% Sample	1557	1557	03/12/2014

Figure 2.11 Inspection Form for Part Number 1234567

of part 1234567, which was being produced. You will notice that the three hole diameters are being inspected 100%, while the other dimensions have a different sample size. What would occur here is, for example, the "H" dimension of 0.752/0.748 would require a sample check of only 20%. This means that of the 1,557 parts in the entire lot, only 311 parts would be randomly selected and checked. This is called lot sampling, and based on the readings obtained; statistical analysis would be performed to determine what the entire lot would contain. How this is performed is covered in Chapter 28.

2.4.2 Process Control

By controlling the process that is fabricating the part is critical for avoiding bad parts. Process control performs that function by collecting data while the parts are being produced and analyzing the results. As the parts are being fabricated, a determined sample lot is gathered during the process at set time intervals and inspected. This is performed on an ongoing basis, and as the data is collected and analyzed, a picture begins to appear of exactly what the process is doing. This real-time view of the process can be used to predict when the dimension being checked may exceed the tolerance specified. Figure 2.12 shows the data that was collected for part number 1234567 for dimension "B" at 1.025/1.015. A sample of five parts is inspected every half hour and analyzed. From this data, it begins to show what is happening in the process as the averages of each data set are increasing slightly. The changes in the averages for the sets are listed as follows along with the ranges. The range of each set is the difference between the highest and lowest values in the set, which shows how tight the process is holding the parts being fabricated. To view this data in a pictorial manner, Figure 2.13 shows a control chart where the averages of each set are plotted with the upper tolerance and lower tolerance limits shown. Figure 2.14 shows a range chart of the same data displaying the difference in the ranges of each sample group at the inspection times. This information provides a real-time view of the process and is covered in detail in Chapter 28.

SPC Inspection Form

Part Number: 1234567	Revision: B	Order Number: 28	Operation: 300	Machine 12-018	Date: 02/13/2014
Dimension B: 1.025 / 1.015		Lot Inspection Size: 5		Inspection Time Frame: 0.5 Hours	

Time	8:00	8:30	9:00	9:30	10:00	10:30	11:00	11:30
1	1.0174	1.0178	1.0182	1.0186	1.0189	1.0195	1.0199	1.0205
2	1.0172	1.0176	1.0183	1.0185	1.0192	1.0197	1.0201	1.0207
3	1.0175	1.0177	1.0179	1.0188	1.0191	1.0195	1.0203	1.0206
4	1.0174	1.0179	1.0184	1.0186	1.0194	1.0198	1.0201	1.0208
5	1.0174	1.0178	1.0182	1.0188	1.0193	1.0196	1.0202	1.0206
Average	1.01736	1.01776	1.01820	1.01866	1.01918	1.01962	1.02012	1.02082
Standard Deviation	0.000102	0.000167	0.000167	0.000120	0.000172	0.000117	0.000133	0.000102
Maximum	1.0175	1.0179	1.0184	1.0188	1.1094	1.0198	1.0208	1.0205
Minimum	1.0172	1.0176	1.0179	1.0185	1.10989	1.0195	1.0199	1.0201
Lot Range	0.0003	0.0003	0.0005	0.0003	0.0005	0.0003	0.0004	0.0003
Change		+0.00040	+0.00044	+0.00046	+0.00052	+0.00044	+0.00050	+0.00052

Figure 2.12 Process Control Data for Part Number 1234567

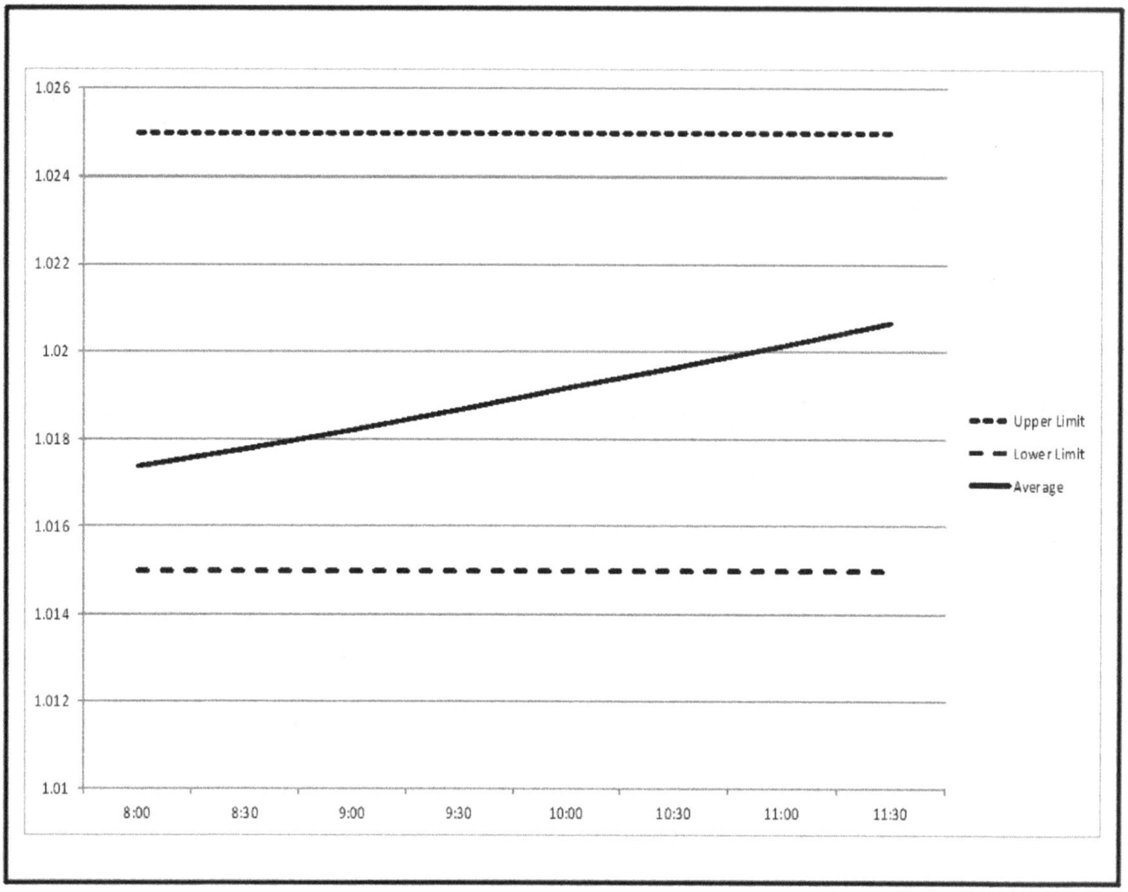

Figure 2.13 Process Control Chart for Part Number 1234567

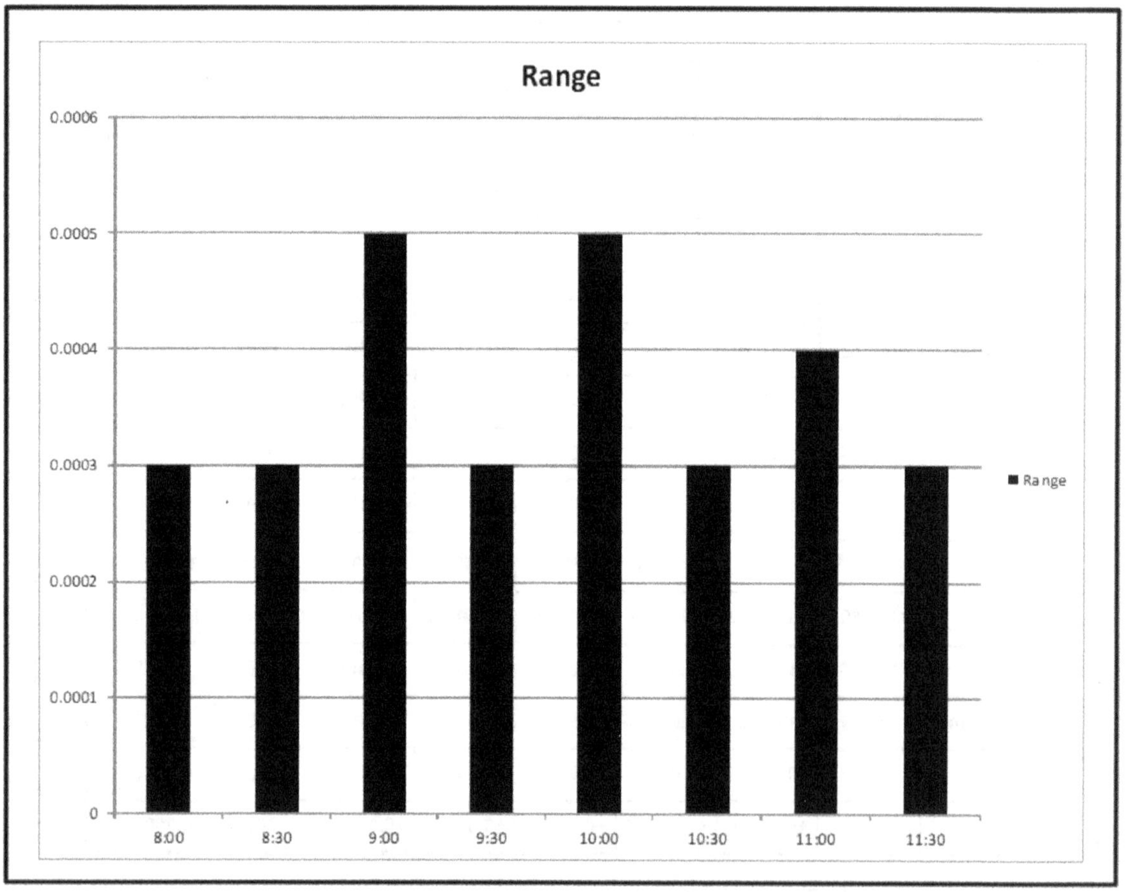

Figure 2.14 Range Chart for Part Number 1234567

2.5 DESIGN ENGINEERING

The organization responsible for designing the parts and products plays a big role in successful manufacturing. In the past, design engineering did not interface with manufacturing, but rather designed the product in a silo and passed it over to manufacturing. This lead to a lot of parts that could not be fabricated or would have a high cost associated with them. This resulted in a lot of design changes, which would slow the entire process down. Figure 2.15 shows how this silo design environment operated and the corresponding constant passing requests and design changes back and forth.

In today's environment, the wall between design engineering and manufacturing has been removed and as the product is being designed, manufacturing is providing input before the drawings are finalized. With this open line of communication, the amount of requests for changes from manufacturing is greatly reduced, resulting in a much quicker release of a stable design that is able to be manufactured. Figure 2.16 shows this open line of communication process between design engineering and manufacturing.

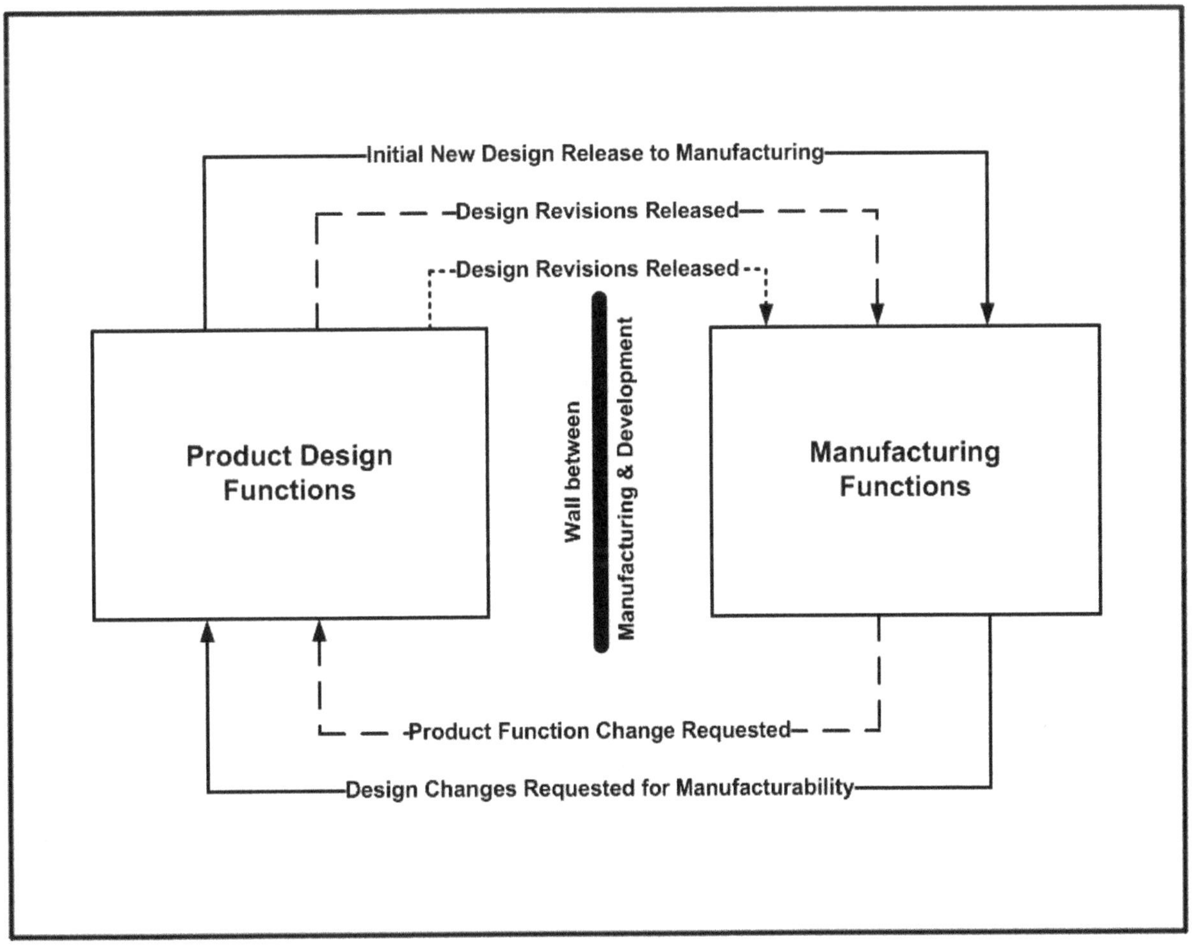

Figure 2.15 Product Development in a Silo Environment

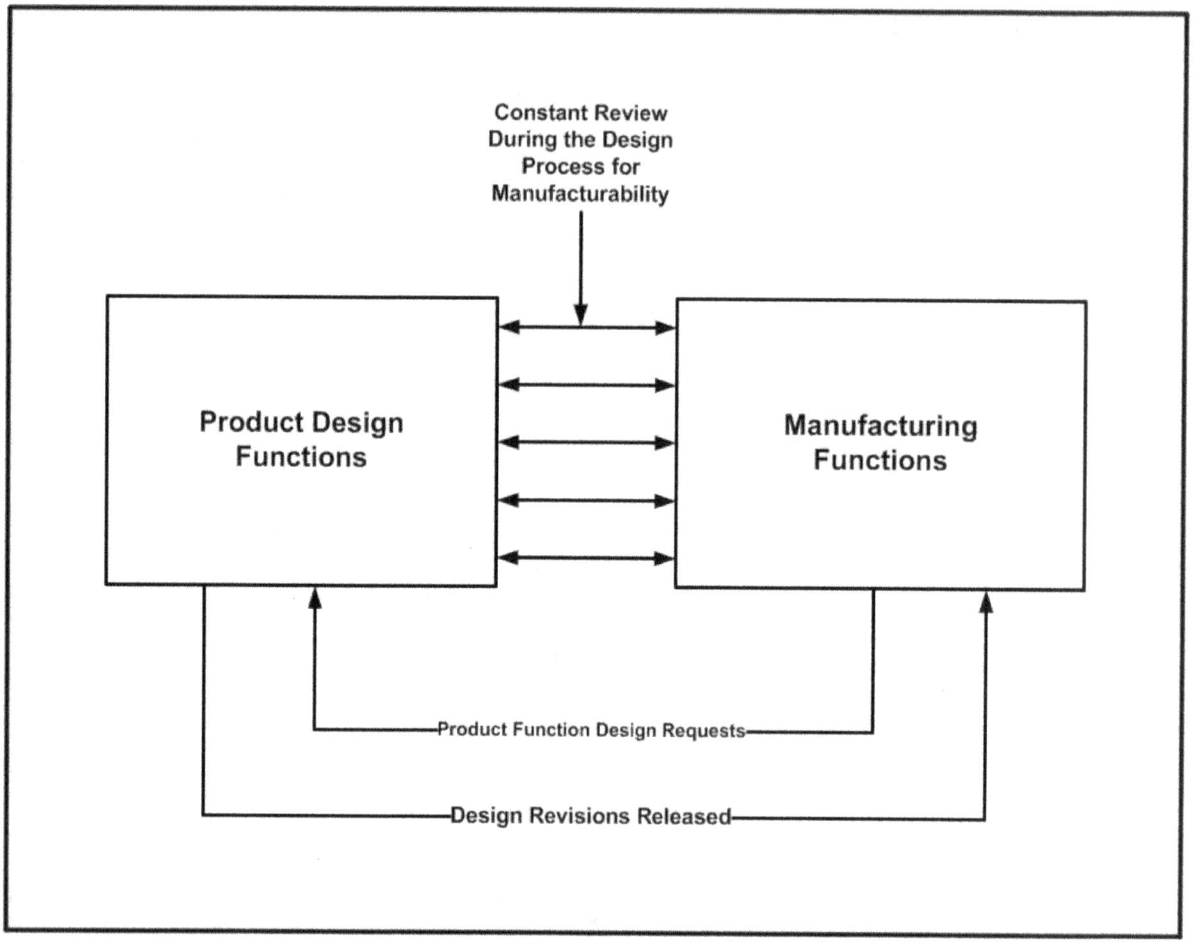

Figure 2.16 Product Development in a Concurrent Environment

2.6 CONCURRENT ENGINEERING

As described earlier, with the interface between design engineering and manufacturing is known as concurrent engineering. In a fully concurrent engineering environment, there is more than just communication between design engineering and manufacturing, but rather all disciplines that would be in the product from every aspect. This is normally done by having a leader of the project and a member of each discipline involved from the very onset of the project. Each of the representatives on the team would have a responsible group reporting to them working on the project. Figure 2.17 shows the structure of a concurrent development project with the disciplines involved. The function of each discipline is outlined as follows.

Product Development: This group is responsible for the actual designing of the new product. This may consist of various engineering disciplines such as mechanical engineering, electrical engineering, software engineering, and systems engineering.

Manufacturing: Manufacturing is responsible for ensuring that the design is able to be fabricated in the most efficient and effective manner. They would develop the process plans and determine the most economical way to produce the product based on the volume requirements.

Marketing: Marketing is actually the beginning of the new product. From the information collected from customers, they develop an outline for the new product based on customer requirements. This starts the project moving forward, and as it does, marketing provides the volume requirements for production.

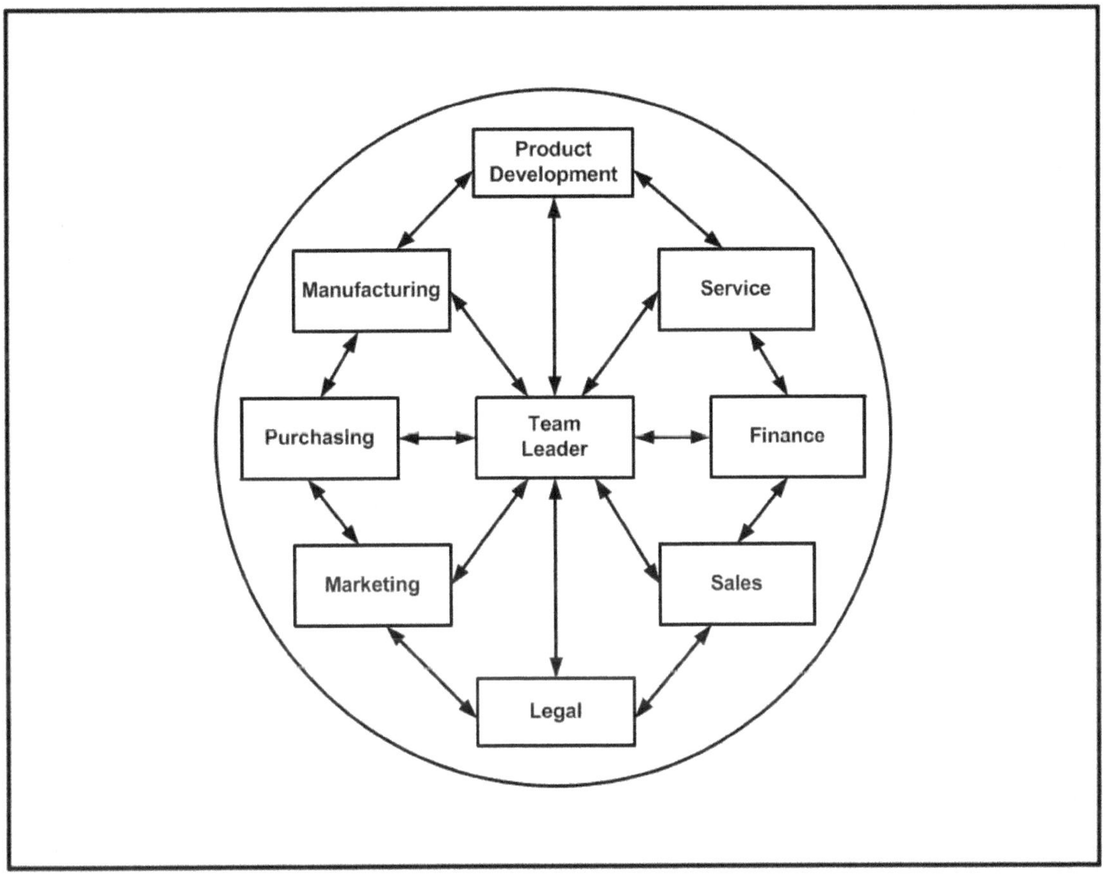

Figure 2.17 Typical Concurrent Product Development Structure

Purchasing: Once the design begins, purchasing would be responsible for securing vendors required for all commercial components in the design. Based on the quantity per year, they would negotiate both delivery and pricing. They are also responsible for ensuring that the vendors being used are capable of delivering the quantity required on time and to the prescribed quality levels.

Service: Service is involved to ensure that whatever service is required on the product, the design allows for this to be performed as easily as possible. They would also be responsible for providing a list of service parts that can be inventoried and bought out as required.

Sales: Sales is responsible for actually getting the product into the customer's hands. They would be setting up the sales and distribution programs and any promotional programs needed based on the customer base being served and the volume of products projected by marketing.

Finance: During the project, finance would be responsible for the budgets and ensuring that any overrun on the budgets can be justified. They would monitor all development costs associated with the project along with sales promotional and advertising costs.

Legal: The legal functions on the project are to check if any existing patents are held by other companies for similar technology being used. Along with this, they would be applying for patents on new technology being developed as the project progresses.

2.6.1 Design for Manufacturability

Part of concurrent engineering is designing the product for manufacturability. These "Design for" techniques get applied during the design phase of the development and will save both revisions to the design and increase manufacturing time and quality. Each of these rules evaluates the parts and assemblies while it is still in the design stages by members of the concurrent engineering team during the process for their input. This ensures that the final design would be able to incorporate the requests prior to the final release, thus eliminating change requests later. The three major "Design for" techniques utilized are *Design for Fabrication, Design for Assembly,* and *Design for Service,* which are explained as follows.

Design for Fabrication: When components are to be manufactured, understanding the overall process required is critical during designing to ensure that the part can be fabricated in the most efficient and effective manner. By evaluating how the part would be located for each operation and identifying areas where the processes are not capable of producing the required results will increase the chances of producing an acceptable part. Areas such as overall configuration, dimensions and tolerances, and surface finishes are looked at for their manufacturability. This requires not just looking at the individual component, but rather evaluation of the assembly and what the function of the part is in the product. Upon this evaluation, the changes may be to design a complicated component in two or three different parts and assemble them to the configuration required. Or, an assembly may be changed to one larger part if the processes being utilized are capable of producing it.

Design for Assembly: Evaluating the overall assembly and subassemblies in the design results in a product that can be assembled with ease and defect-free. Some of the considerations here would be evaluating the overall ease of assembly for the operator where very little special tooling and fixturing would be required and having the assembly designed so it is not possible to assemble components in the wrong configuration. When designing the components, making them symmetrical even if the configuration on the opposite side is not functional avoids an improper assembly. In the same manner, designing the parts symmetrically reduces the number of orientations required when assembling, which reduces the probability for errors. If threaded fasteners are being used, reducing the number and type of fasteners will also eliminate the potential of assembling the wrong fastener in the wrong location.

Design for Service: Service is an area that very often gets overlooked during the design stage of a product. In most products, service will be required to replace worn or broken components. By allowing good access to those areas, a service call can be made much easier and take far less time to perform. Also, knowing which parts in the design will be required for service, the design can accommodate the replacement of these items even by the customer, thus reducing the amount of service calls required.

2.7 SUMMARY

Manufacturing is a very complex system where far more than just the actual fabrication and assembly of the products is performed. By having all of the technical support functions working in unison, the operation will flow smoothly, resulting in one that is efficient, effective, and economical. As changes are made to the inputs of the system, the processing sector will adjust accordingly to ensure that the manufacturing runs smoothly. By having the required support groups all performing their individual functions, and interfacing constantly with each other, the final products will be completed as economically as possible.

2.8 QUESTIONS

1. When developing a Process Plan for producing components, what are the four steps to follow from the raw material to the finished component?

2. What is a Physical Enhancement operation, and is it always required when fabricating component parts?

3. What is the difference between a Process Plan and a Route Card, and why are both required when fabricating component parts?

4. Why is a constant evaluation of Factory Flow important to the operation?

5. For an operation performing Build-to-Order manufacturing, how is this different from an operation performing Build to Forecast?

6. Why is Process Control important in a manufacturing operation, and what does it do?

7. How has Concurrent Engineering helped in both the product design and manufacturing arenas?

8. With Concurrent Engineering, why do they need representation from areas not directly part of either design engineering or manufacturing engineering? Explain what these other functions provide to the project.

9. Design for Manufacturing ensures that the product being designed can also be fabricated as easily as possible. Why is Design for Service important if that has nothing to do with fabrication?

CHAPTER

3

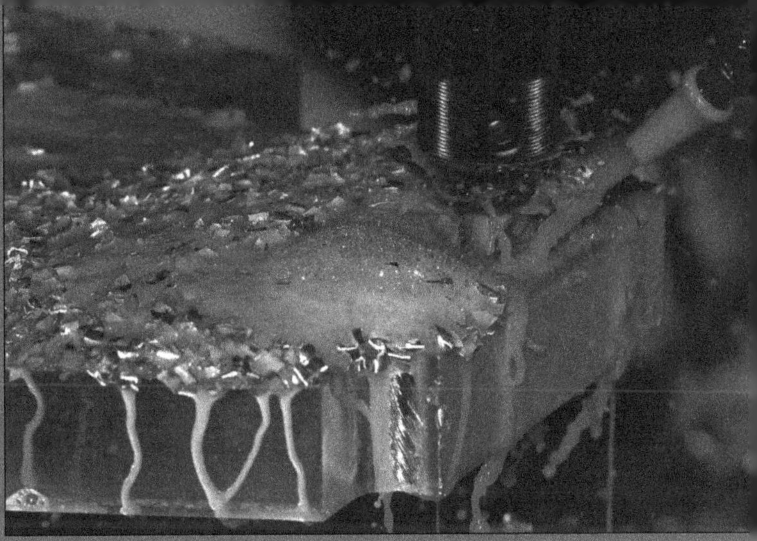

Dimensions, Tolerances, and Surfaces

3.1 INTRODUCTION

How a component is going to be fabricated is very much dependent on the configuration and the tolerances, along with what surface finish requirements are desired. Of the variety of manufacturing processes, each has certain capabilities and drawbacks. Depending on what is required to make the part based on the dimensions, tolerances, and surface finish requirements, the optimum process can be selected. Knowing how dimensions and tolerances are applied to a part drawing, along with any surface requirements, is critical in selecting the proper process for fabrication. This chapter provides an explanation of these important design concepts and how they both should and should not be applied.

3.2 DIMENSIONS AND TOLERANCES

Understanding what a dimension and a tolerance are and how they are applied is important for fabricating any component part. By evaluating the dimensions for the configuration of the part along with the tolerances of each dimension, the optimum process can be selected for fabrication. Dimensions and tolerances are defined as follows.

Dimension: A dimension is the exact length that is specified on the part drawing. This is a numerical value that is either a linear length or an angular value. This represents the starting point for processing the parts in fabrication. Every length, width, height, hole diameter, and angular configuration on the part drawing will have a dimension and a relationship to some location on the part drawing.

Tolerance: A tolerance is how much the fabricated part can deviate from the specified dimension. What this represents is the maximum and minimum that the measurement can be from the dimension. Nothing can be manufactured exactly to a dimension, so the tolerance provides the window for the dimension for any manufacturing process.

Dimensions and tolerances can be applied to a part drawing in three accepted manners. Figure 3.1 shows the three methods of application on a part drawing. These are covered in more detail as follows.

Bilateral Dimension: This type of dimensioning shows the numerical value and a range both up and down from it. In Figure 3.1, with the dimension and tolerance shown, the size can range from as large as 3.755 to as small as 3.748.

Unilateral Dimension: This type of dimensioning shows a maximum or minimum numerical value and the range it can vary in one direction. Figure 3.1 shows that the maximum value is 3.755 and the minimum value is 3.748.

Limit Dimension: Limit dimensions do not show a tolerance from a given value, but rather specify the high and low range in values. Figure 3.1 shows this range to be from 3.755 to 3.748.

Looking at all three of the accepted methods of applying dimensions and tolerances in Figure 3.1, each shows the exact same range that the length can be.

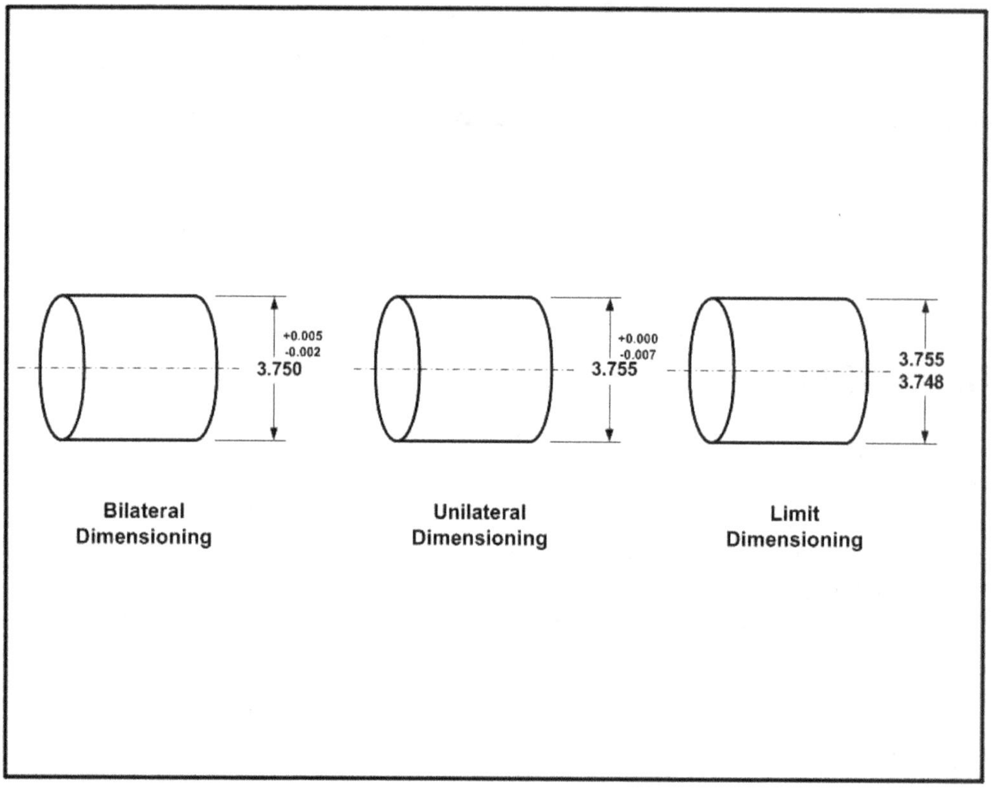

Figure 3.1 Dimension and Tolerance Types

3.2.1 Correct Dimensioning

Dimensions need to be applied properly on a part drawing in order for the component to be able to be manufactured. When a part is dimensioned, there is always an overall length specified and other dimensions and tolerances are encapsulated within that. However, there always needs to be an area that is left open so the overall dimension and tolerance can be attained. Looking at Figures 3.2 and 3.3, you can see that the overall length is specified at 5.125 with a ±0.001 tolerance. With two of the three sections having dimensions and tolerances, the open section can be adjusted to bring the overall length to the size and tolerance specified. However, Figure 3.4

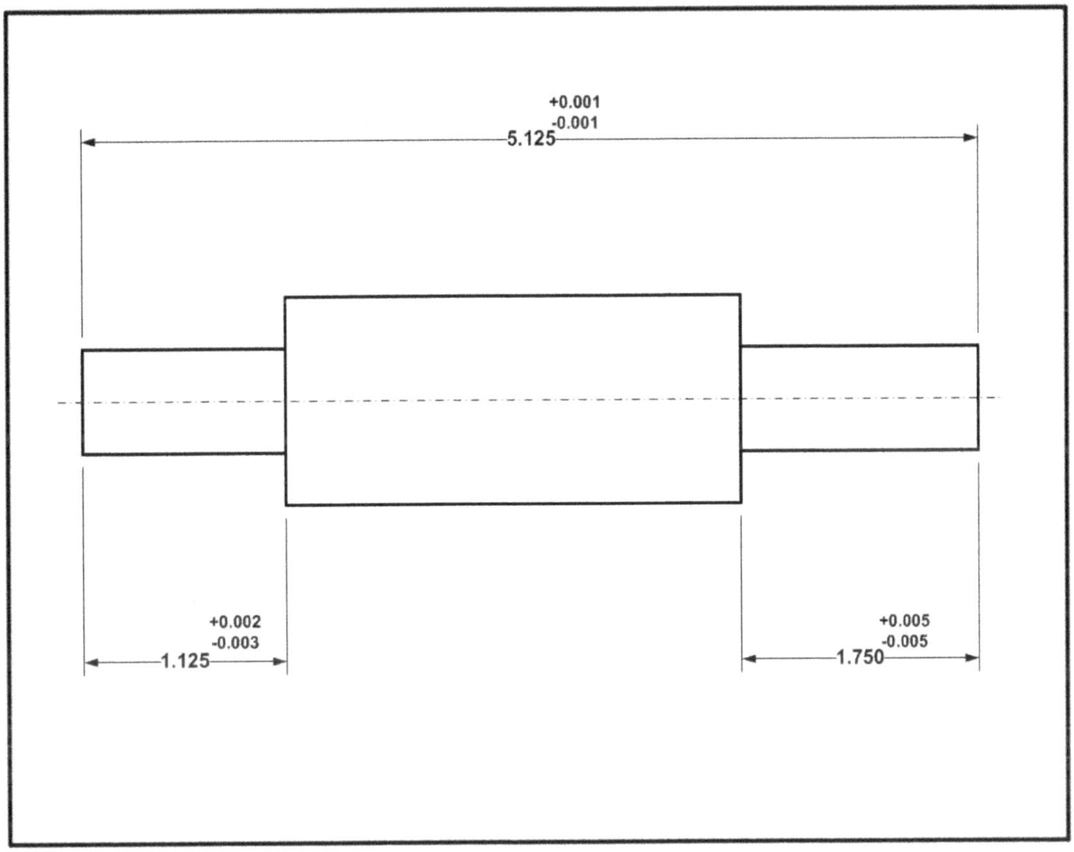

Figure 3.2 First Example of Correct Dimension and Tolerance Application

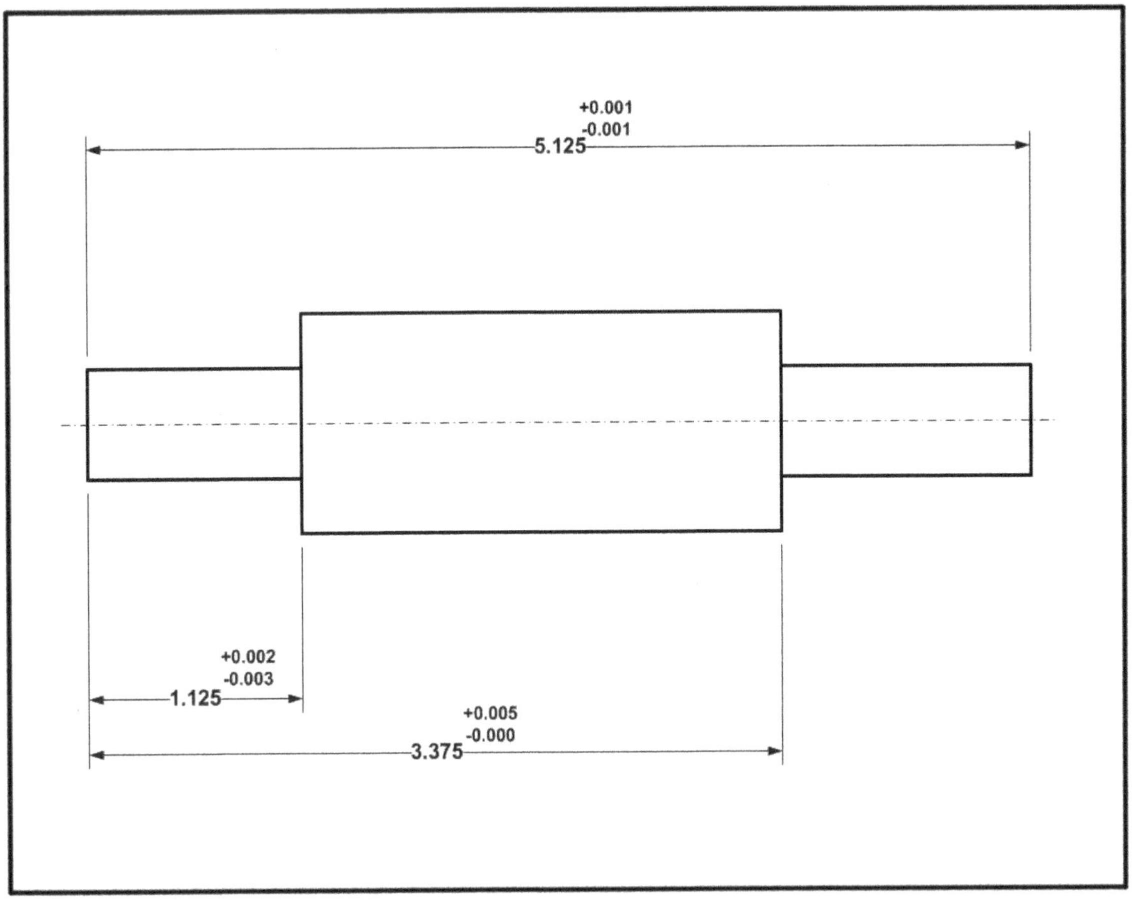

Figure 3.3 Second Example of Correct Dimension and Tolerance Application

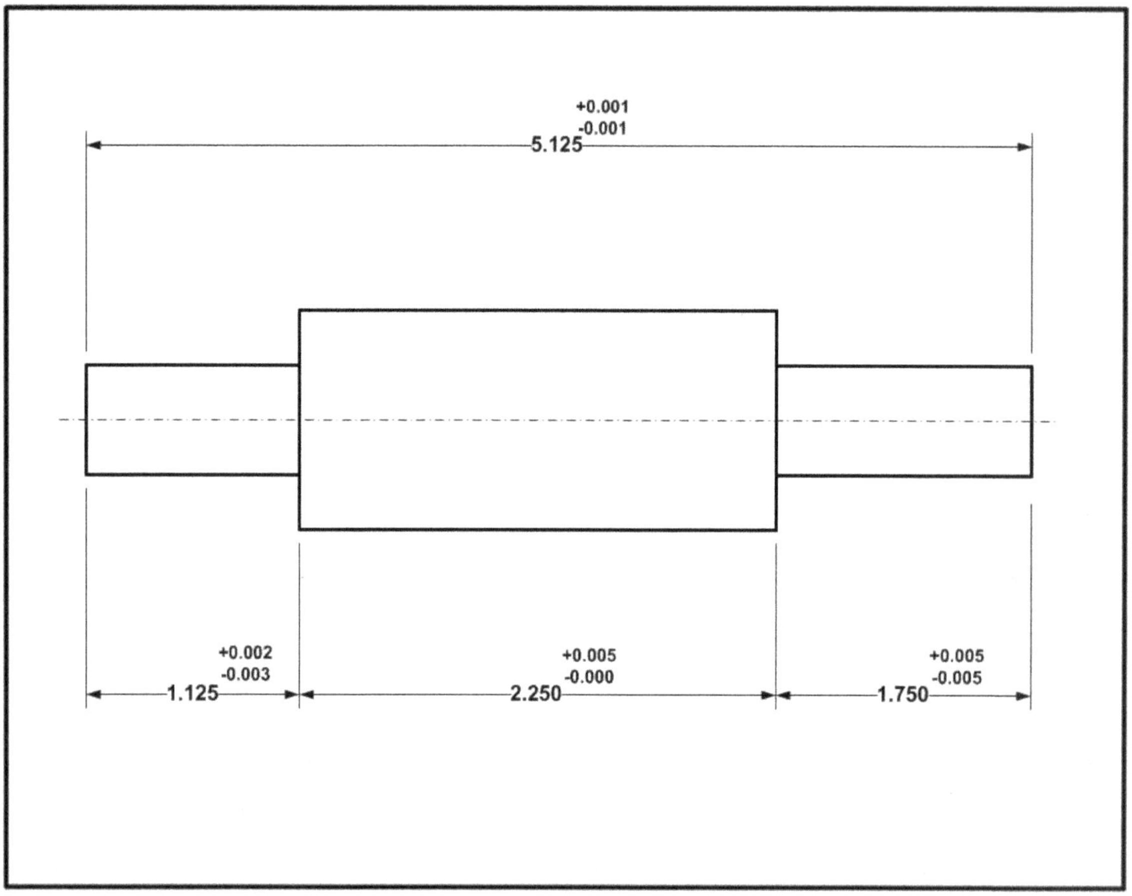

Figure 3.4 Example of Incorrect Dimension and Tolerance Application

shows an example of improper dimensioning. If all three of the dimensions and tolerances that are specified for the three sections of the overall length, 5.125 ± 0.001 could never be held. If all three of the dimensions were at their high limit, the overall length would be 5.137, and if they were at the low limit, it would be 5.117. This is known as double dimensioning on a part drawing.

3.3 SURFACES

Every manufacturing process will affect the surface of a fabricated part. As the process is being performed, the surface structure is changed, leaving an altered state, along with a degree of roughness on it. This section explains *Surface Structure*, *Surface Roughness*, *Surface Waviness*, and *Surface Lay* along with how they are measured and controlled during processing.

3.3.1 Surface Structure

Regardless of the fabrication process utilized, the surface structure of a part will be altered during processing. As the workpiece is being fabricated, the structure of the surface undergoes changes. Figure 3.5 shows a sectional view of a part, exposing the substrate, altered layer, and the surface texture. The substrate is the unaltered material, which is not affected by the fabrication process. This maintains the original microstructure of the material. The altered layer is where the material has been affected by the processing and will have a different microstructure than the original material. Every fabrication process will produce an altered layer on the part, and how aggressive the fabrication process is will determine how deeply the altered layer of the part will be and the microstructure of it as well.

Looking at how various processes will affect the altered layer of a part, it occurs in various ways. For example, if a part is being machined, the friction and heat generated from the cutter will cause the layer to be altered

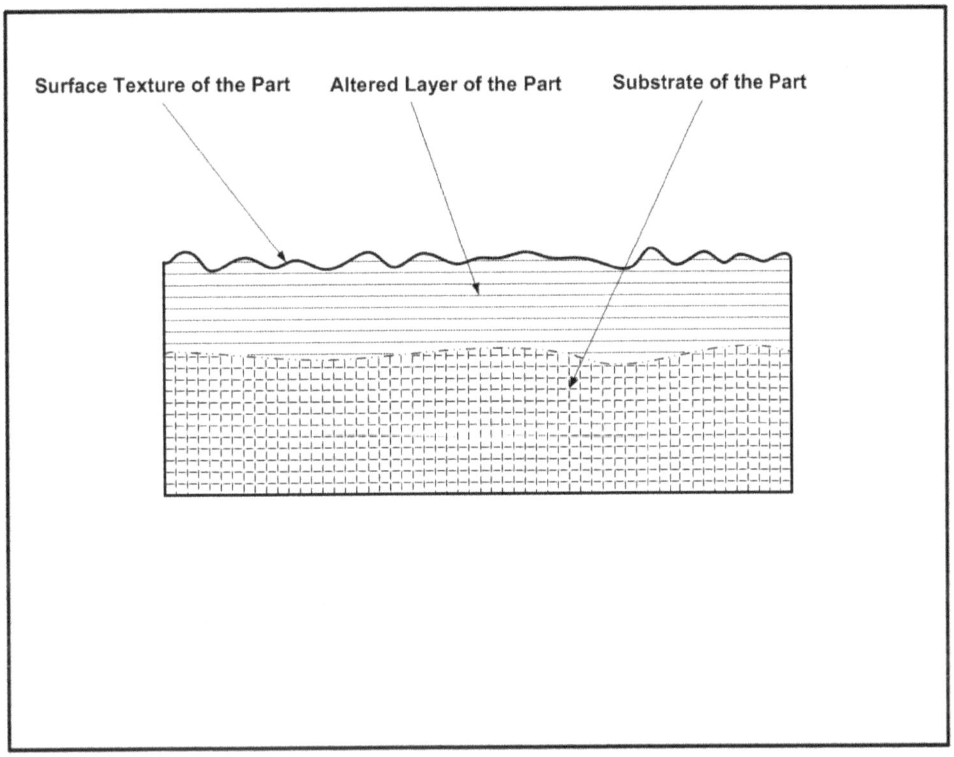

Figure 3.5 Surface Texture and Substrate Structure

to a certain depth depending on how aggressive the cutting is and how much coolant is being applied at that time. In the same manner, if a part is being produced in a casting operation, the molten metal poured into the mold will start to solidify first, cooling faster than the core of the part will. This produces a different structure on the surface than in the center of the part and also produces surface stresses.

3.3.2 Surface Texture

When performing any fabrication process to a part, the texture of the surface which it is being performed on will be altered as well. Depending on the process being utilized, the texture can be either greatly improved or worsened on the surface of the part. For some parts, the surface texture is not critical to the functionality, but on other parts, it is. The surface texture is divided into three categories, which are *Surface Roughness*, *Surface Waviness*, and *Surface Lay*. Each of these is explained as follows.

3.3.2.1 Surface Roughness

The roughness of a surface is what you can feel when you rub your finger across it. This denotes the ridges that are left on the surface of the part from the fabrication process while manufacturing the component part. Roughness is a measure of very finely spaced ridges on the part, which are to a given depth. Figure 3.6 shows the small deviations from the surface, which constitute the roughness. These are measured in depth in micro inches in what is called an Ra scale. This measures the depth of the roughness of the part to determine the amount of roughness.

3.3.2.2 Surface Waviness

Surface waviness is different from surface roughness. While roughness is a measure of small ridges left behind from the fabrication process, waviness is a measure of deviations of a larger and longer nature. The waviness does not affect the roughness of a part, but rather the overall flatness of it. However, the waviness in a part is caused by either the fabrication process or the setup of the part being manufactured. Figure 3.6 shows the waviness spacing and the depth from the base line of a part.

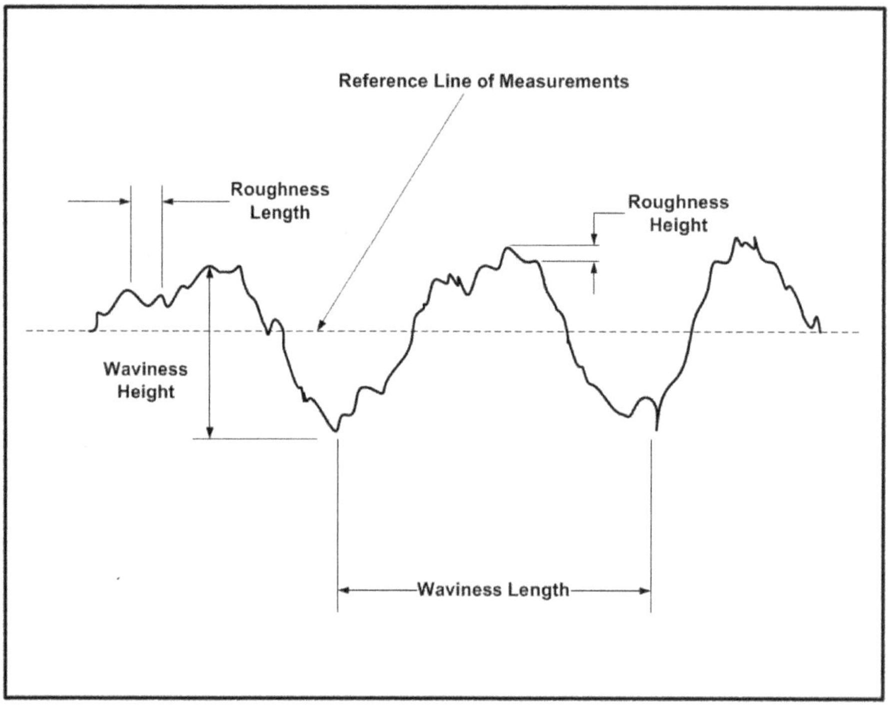

Figure 3.6 Surface Roughness and Waviness

3.3.2.3 Surface Lay

The lay of a surface is very simple; it is the direction that the roughness is in the fabricated part. In some cases, the direction is not critical to the function, but in other cases, it is. If the part were to be used as a friction means in the function of the design, then a given roughness would be called out with a direction required. Likewise, if the part were to rub against another part in the design and friction needed to be minimized, the lay would be called out in the opposite direction. Figure 3.7 shows how the direction of the roughness, or the lay, would be on a fabricated part.

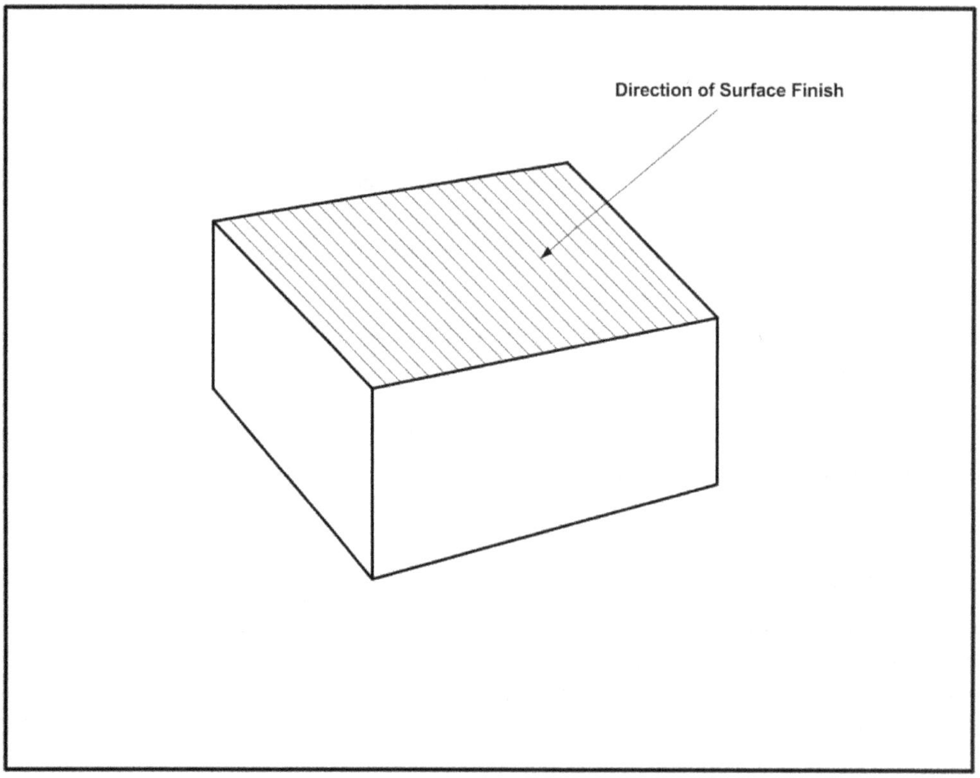

Figure 3.7 Surface Lay Direction

3.4 SURFACE DIMENSIONING

Surface requirements are called out on a part drawing to specify the roughness, waviness, and direction of the lay along with other pertinent information. The symbol used to specify the surface requirements is seen in Figure 3.8. This shows everything required to produce the surface finish needed on the part. Various symbols are used to indicate the direction of the lay on the part as shown in Figure 3.9. The symbol is placed on the

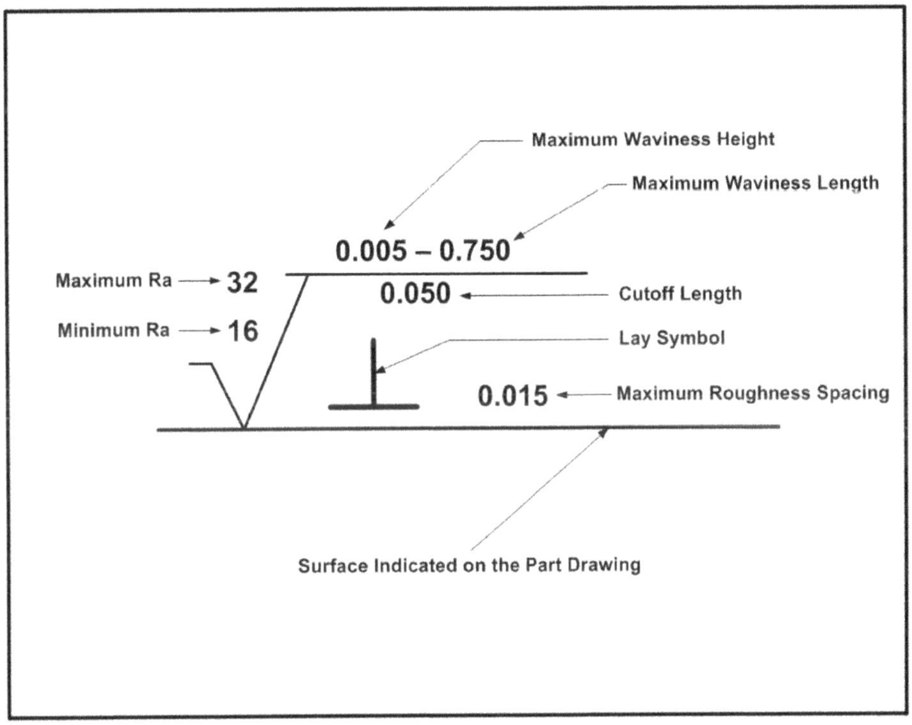

Figure 3.8 Surface Requirement Drawing Symbol

Symbol	Description
═	This symbol indicates that the lay of the surface is parallel to the surface indicated on the part drawing.
⊥	This symbol indicates that the lay of the surface is perpendicular to the surface indicated on the part drawing.
X	This symbol indicates that the lay of the surface is angular to the surface indicated on the part drawing.
C	This symbol indicates that the lay is circular to the center of the surface indicated on the part drawing.
R	This symbol indicates that the lay of the surface is radial to the center of the surface indicated on the part drawing.
P	The symbol indicates that the lay of the surface is nondirectional to the surface indicated on the part drawing.

Figure 3.9 Surface Lay Symbols

part drawing at the surface where the requirements need to be held. Figure 3.10 shows an example of how this surface requirement symbol is indicated on a part drawing for fabrication. Some examples of surface finishes on raw material and fabricated parts utilizing various manufacturing processes are shown in Figure 3.11.

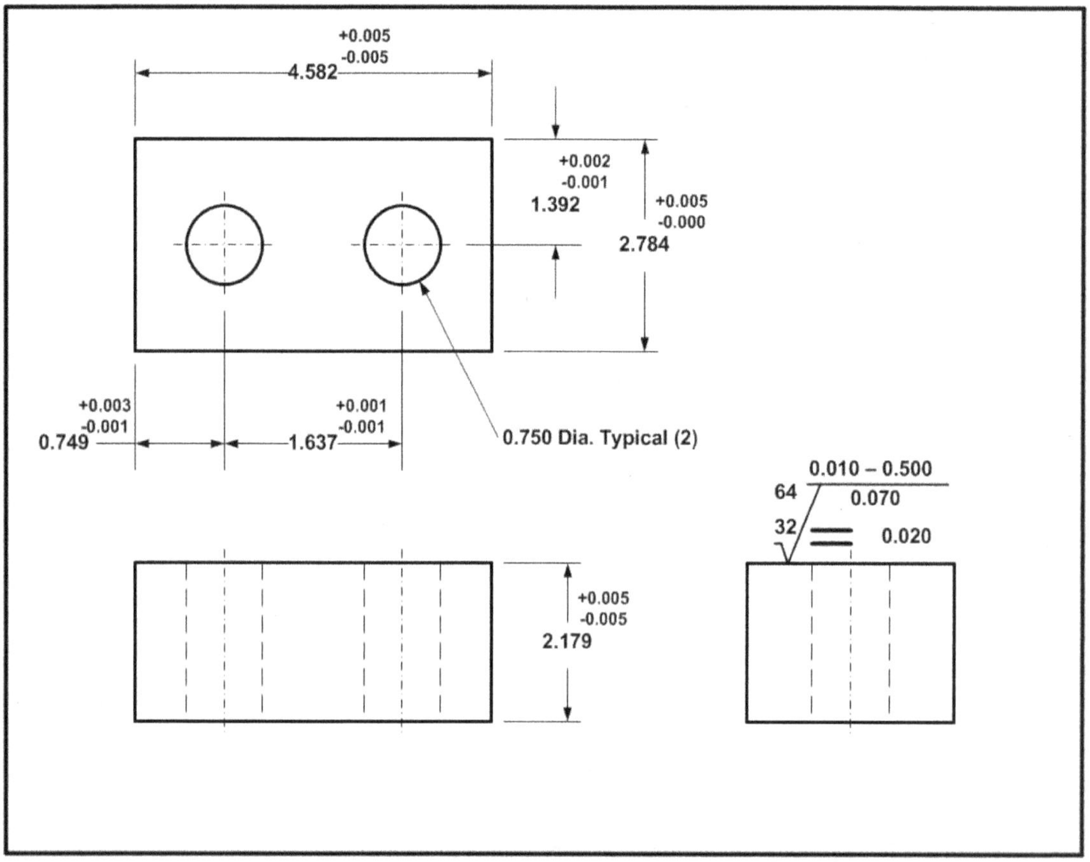

Figure 3.10 Surface Requirement Symbol Properly Displayed on the Part Drawing

Examples of Surface Finishes	
Material Measured	**Ra Reading**
Aluminum Sand Casting	170.0 – 141.5
Aluminum Die Casting	14.8 – 12.3
Roughing Milled Aluminum Surface	63.3 – 59.4
Finished Milled Aluminum Surface	14.1 – 12.4
Ground Steel Surface Perpendicular to the Lay	61.8 – 58.5
Ground Steel Surface Parallel to the Lay	22.6 – 19.5
Roughing Turned Surface	267.2 – 245.6
Finished Turned Surface	20.2 – 17.8
Cold Drawn Aluminum Stock	52.3 – 51.6
Extruded Plastic Stock	8.9 – 8.1
Window Glass	1.3 – 0.9
Super Finished Gage Blocks	0.7 – 0.6

Figure 3.11 Examples of Surface Roughness for Various Materials and Processes

3.5 TOLERANCE AND SURFACE IMPACT ON PART COST

Both tolerances and surface finish requirements have a direct impact on the overall cost of a part being produced. As the tolerance requirements get tighter and the surface finish requirements get smoother, the cost to fabricate the part will go up dramatically. The tighter the tolerances and the smoother the surface finishes, the longer it will take to manufacture the part. Figures 3.12 and 3.13 show the part cost vs. the tolerances and surface finish requirement. As a general rule, the tolerances should be kept as liberal as possible along with the surface finish requirements, unless it is critical to the function of the completed product.

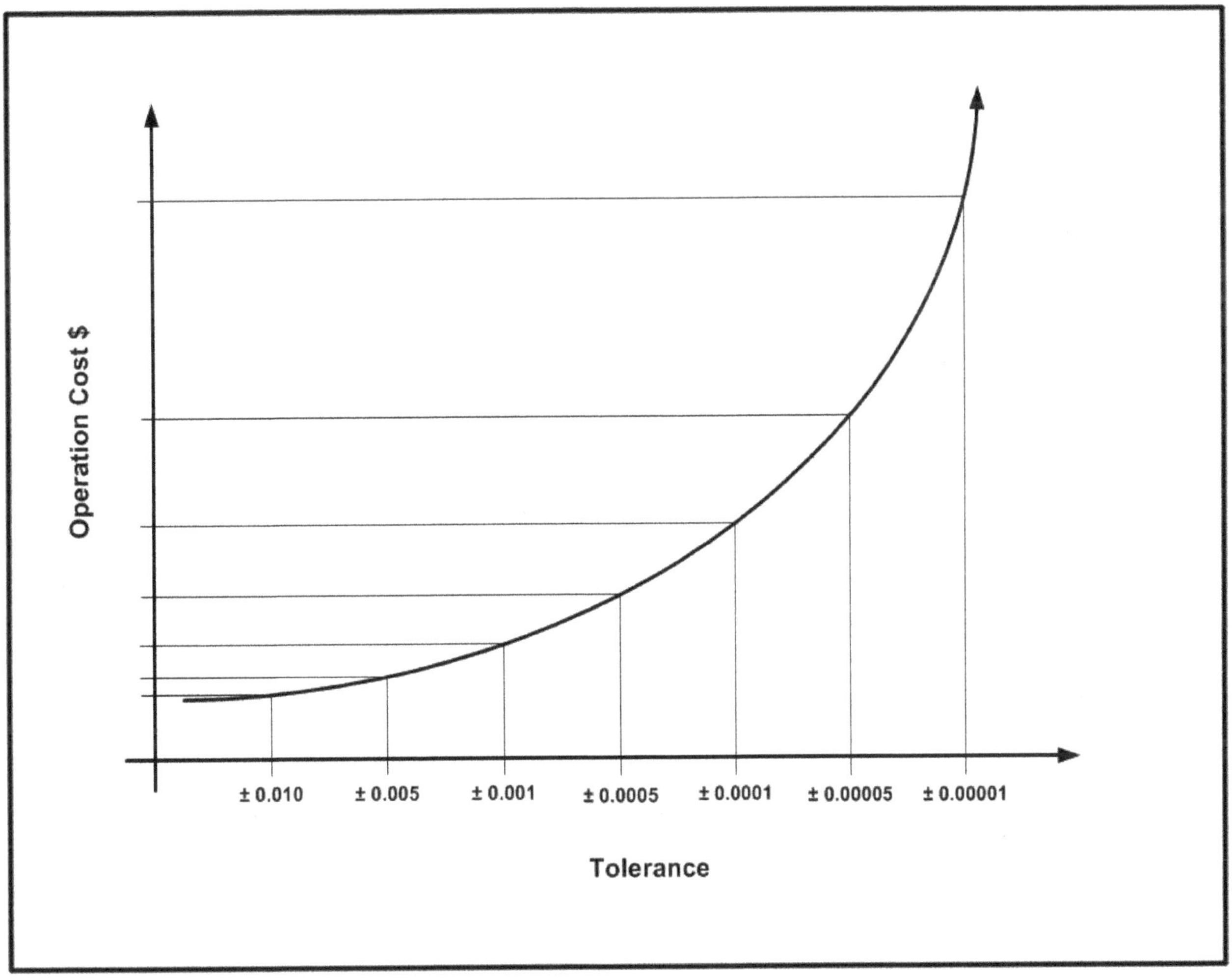

Figure 3.12 Tolerance Impact on Operation Cost

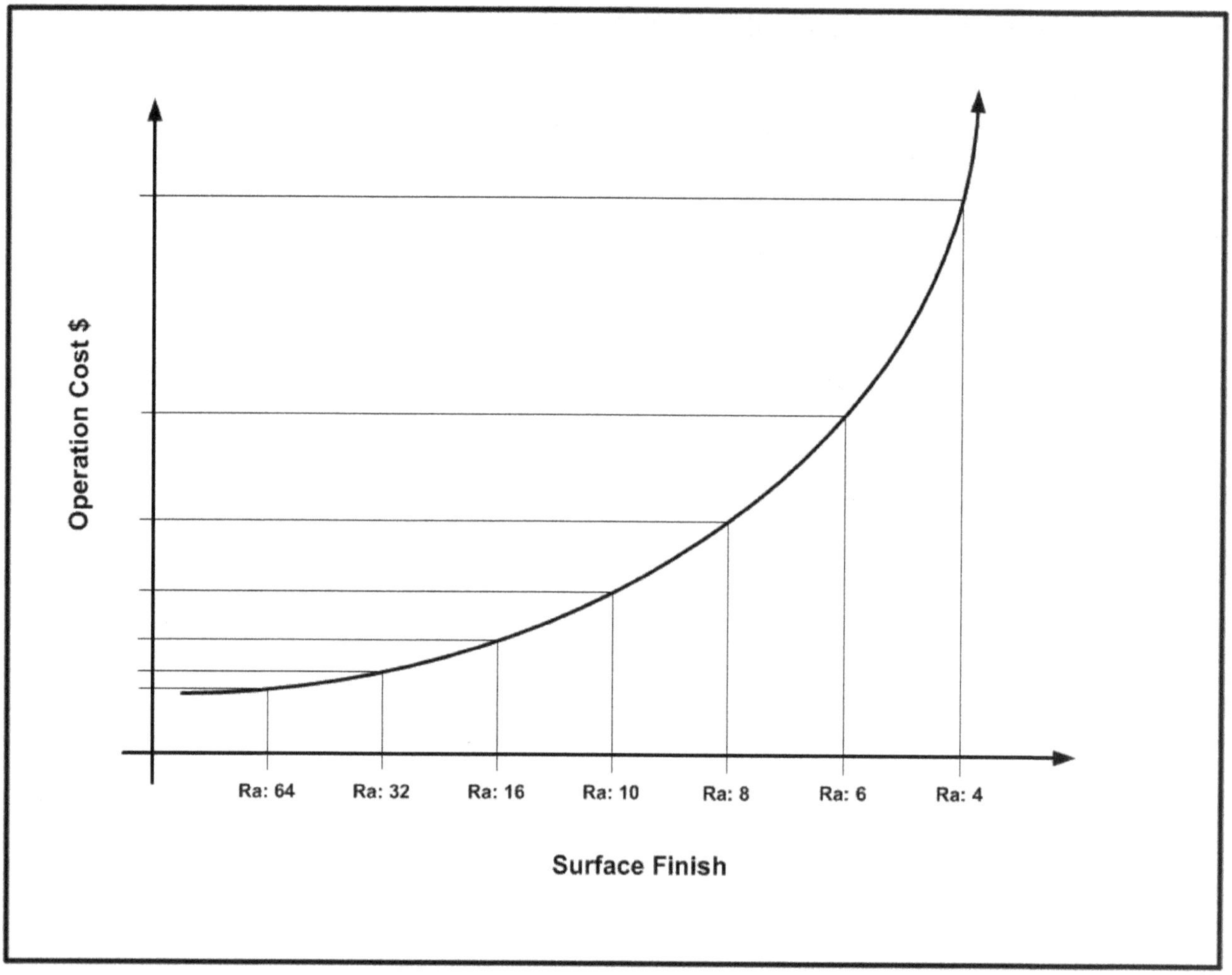

Figure 3.13 Surface Finish Impact on Cost

3.6 SUMMARY

Understanding the importance of correct dimensioning and tolerance applications along with the surface finishing requirements is critical to effective manufacturing of components. Knowing when close tolerances and smooth surface finishes should be applied will greatly improve the manufacturing efficiency. By keeping the noncritical dimensions with tolerances as liberally as possible, and only specifying surface requirements where absolutely necessary, the fabrication process will be as efficient and economical as possible, thus resulting in a lower cost in manufacturing.

3.7 QUESTIONS

1. Define the difference between a dimension and a tolerance, and explain why both are required on a part drawing.

2. Which type of dimensioning has a tolerance that only allows a variance in one direction from the dimension?

3. Why is double dimensioning incorrect on a part drawing, and what problems can it cause in manufacturing?

4. Surfaces are called out on a part drawing as well as dimensions and tolerances. What is the difference between Surface Roughness and Surface Waviness?

5. The Lay of a surface indicates the direction the finish needs to be in on the part. What is the symbol for having the surface perpendicular?

6. Why is it important to keep the tolerances and surface finish requirements as liberal as possible, and what impact can this have on the manufacturing of the component parts?

CHAPTER

4

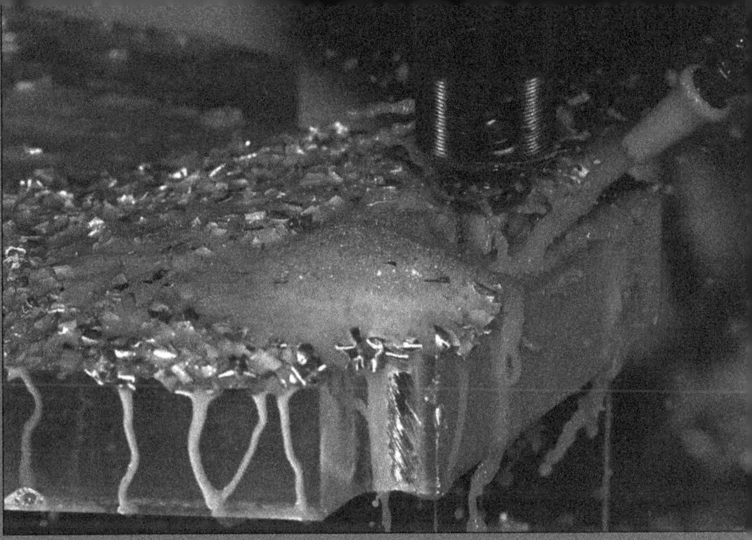

Theory of Metrology

4.1 INTRODUCTION

Metrology is the art and science of measurement. This is a very broad field encompassing a wide range of techniques and processes to accomplish the desired results. It plays a critical role in the manufacturing arena to ensure that whatever is being produced will be to the required dimensions and tolerances. In this chapter, various terms used in metrology are discussed and defined along with processes and techniques utilized. Understanding how and why these techniques are implemented during the fabrication process will provide a better understanding of the processes that will follow.

4.2 METROLOGY

The world of metrology is a vast arena of techniques used for measuring and testing the processes, components be fabricated, and the finished products. *By definition, metrology is the science of measurement.* Metrology is not just limited to the manufacturing area, but also encompasses measurement of a variety of things on a daily basis. Besides what would be measured and tested in manufacturing, metrology also includes, but is not limited to:

> *Length, Weight, Mass, Roughness, Volume, Light, Temperature, Time, Radiation, Molecular Structure, Wind, pH Levels, Air Quality, Voltage, Current, Speed, Velocity, and Acceleration.*

If we think about the time we follow every day, Greenwich England monitors the time that is followed by us and the entire world constantly. They monitor the rotation of the Earth to ensure that the time we follow is always correct. This is done so precisely that when the major underwater earthquake occurred, causing the massive tsunami in the Far East, it was reported that the rotation of the Earth was affected and the time was altered by a few nanoseconds. In the same manner, the meteorologists are constantly measuring wind speeds and ocean currents in order to better predict and forecast the weather around the world.

In manufacturing, metrology is utilized to measure and inspect both components and products. Also, a higher level of metrology is used to inspect and calibrate the instruments that are used to perform these functions. Without this critical measure, there would be no assurance that the parts and products being produced would meet the design requirements.

4.2.1 Accuracy and Precision

Two terms that always get misused are accuracy and precision. Most people think they mean the same thing; however, they are totally different in the world of measurement. Following are the definitions for both.

Accuracy: This is the degree with which the measured value agrees with the true value. If the true value on a part is 1.000 in., how close is the measured value(s) to the 1.000 in. dimension? Several measurements would be taken and compared to the true value that is desired.

Precision: This is a measure of how repeatable the measured readings are. Again, several readings are taken and are compared against each other to examine how repeatable the readings are.

If we were to test two target rifles to see how accurate and precise they are, Figure 4.1 demonstrates a test of this nature. Looking at the results against the definitions stated, you can see that rifle #1 is totally accurate. Looking at the readings from the five shots taken, they are: 0, +5, −5, +5, −5. The average against the true value, which is the center of the target, is zero. From an "accuracy" standpoint, this could not be any more accurate. If we look at rifle #2, it did not even come close to hitting the target at all. But if we look at the tight grouping that it made, it is extremely repeatable, making it very precise.

In the world of metrology and measurement, precision is far more important than is accuracy. Referring to the two targets in Figure 4.1, nothing can be done to improve the precision for target #1. However, with target #2, the precision is very good, and by either adjustments or compensation, the accuracy can be dialed into the center of the target, making it both precise and accurate.

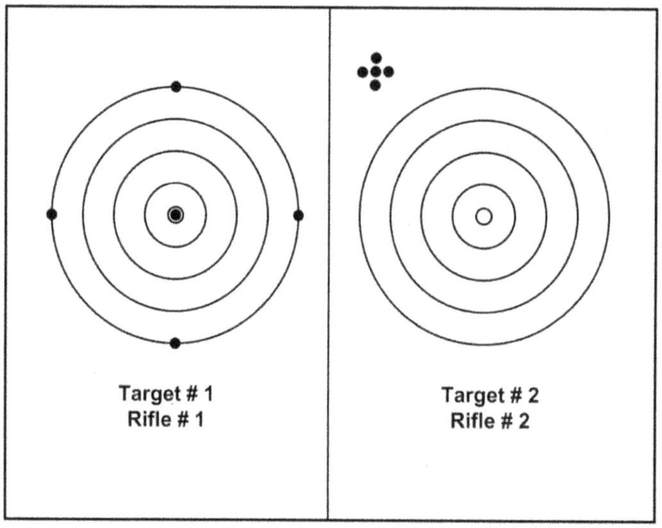

Figure 4.1 Accuracy vs. Precision in Target Shooting

4.3 INSPECTION

During the manufacturing processes, inspection is used as a check and balance for monitoring the components being produced. By definition, *Inspection is an act where an item is compared against a known standard for confirmation.* In simpler terms, a component part (unknown quantity) is compared against the dimensions and tolerances on the part drawing (known standard) to see if everything is within the prescribed limits. As the parts are being processed, critical dimensions are checked to confirm that they meet the print specifications. This is not something that is performed when the part has been completed, although it is performed then as well, but rather during and after the various processing operations.

Looking at Figure 4.2, you see a typical inspection report sheet. The columns are titled and are explained as such:

Step: This is a letter designation for the dimension that will be checked. Traditionally, these will coincide with the letters used on the process plan sketch for fabricating the components. You will notice that two of the letters are purposely left out, which are I and O. This is done so that there will not be any possibility of confusing the letter I with the number 1, or the letter O with the number 0.

Dimension: This is the physical dimension that is on the part drawing. Each dimension on the part drawing will always coincide with a letter designation on the process plan sketch and the inspection report sheet. For Bilateral and Unilateral dimensioning, the dimension is a single number. However, if Limit Dimensioning is used, the middle of the range is calculated and that number is placed in the dimension column.

Tolerance: Tolerance is how much the dimension can deviate. For Bilateral and Unilateral dimensions, these are specified above the dimension. For Limit dimensions, after the middle of the limits is calculated, the tolerance is then placed in the column and will always be equal in both the plus and minus fields.

Reading: Once the inspection has begun, a reading is obtained for each of the steps and entered in this field. This is the actual size of the area being inspected and will then be compared against the corresponding dimensions and tolerances.

Metrology Inspection Report									
Part Number:		Revision Level:	Inspector:			Date:		Sheet of	
Step	Dimension	Tolerance	Reading	Out of Range	Step	Dimension	Tolerance	Reading	Out of Range
A		+ -			P		+ -		
B		+ -			Q		+ -		
C		+ -			R		+ -		
D		+ -			S		+ -		
E		+ -			T		+ -		
F		+ -			U		+ -		
G		+ -			V		+ -		
H		+ -			W		+ -		
J		+ -			X		+ -		
K		+ -			Y		+ -		
L		+ -			Z		+ -		
M		+ -			AA		+ -		
N		+ -			AB		+ -		

Figure 4.2 Metrology Inspection Report Sheet

Out of Range: After the reading has been obtained, it will be compared against the corresponding dimension and tolerance for that step. This is done by taking the dimension and either adding or subtracting the tolerance to it, then seeing if the reading falls within those boundaries.

Example 1: Dimension = 3.7500 Tolerance = +0.003 / −0.001 Reading = 3.7517
3.7500 + 0.003 = 3.7530 3.7517 falls within the limits

Example 2: Dimension = 3.7500 Tolerance = 0.003 / −0.001 Reading = 3.7482
3.7500 − 0.001 = 3.7490 3.7490 − 3.7482 = 0.0008 out of range on the low side

Using the same part print as in the chapter on Process Engineering, Figure 4.3 shows the print and Figure 4.4 shows the process plan sketch with the letter designations. After changing the limit dimensions to a middle reading, the letter designations, dimensions, and tolerances were applied to the inspection report sheet as shown in Figure 4.5, and readings were taken for each step and compared against the dimensions and tolerances. Upon viewing this, you can see that most fell within the limits and were given a "Pass" grade in the out of range column. But four of the dimensions were beyond the limits and were noted as to how much they were over or under the limits set on the part drawing.

This inspection is performed to see if the readings taken on the part correspond to the dimensions on the part drawing. For the readings that fell outside the limits, they will be reviewed to see if a deviation has occurred or if the parts are unusable in the final product.

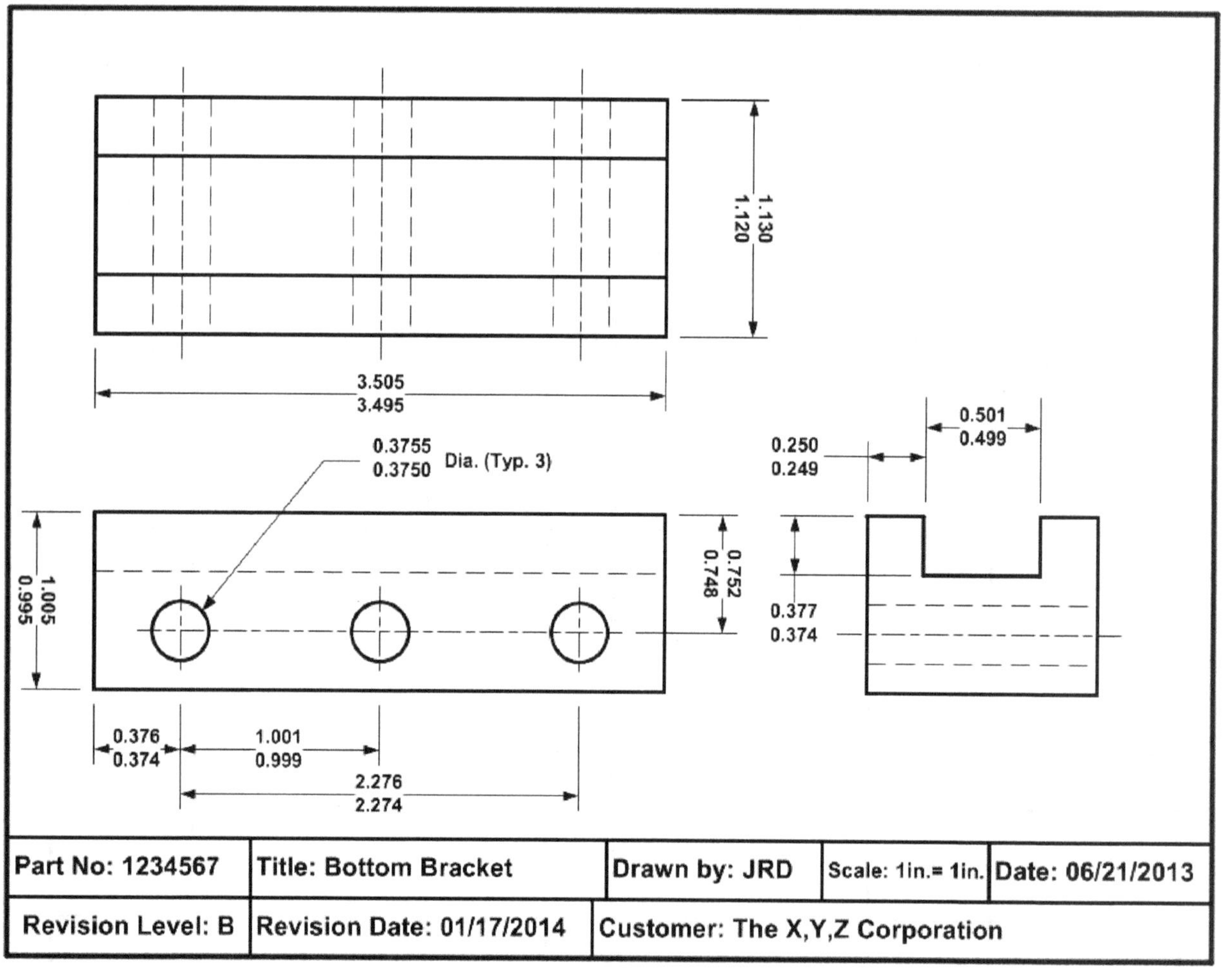

Figure 4.3 Part Drawing for the Component to Be Inspected

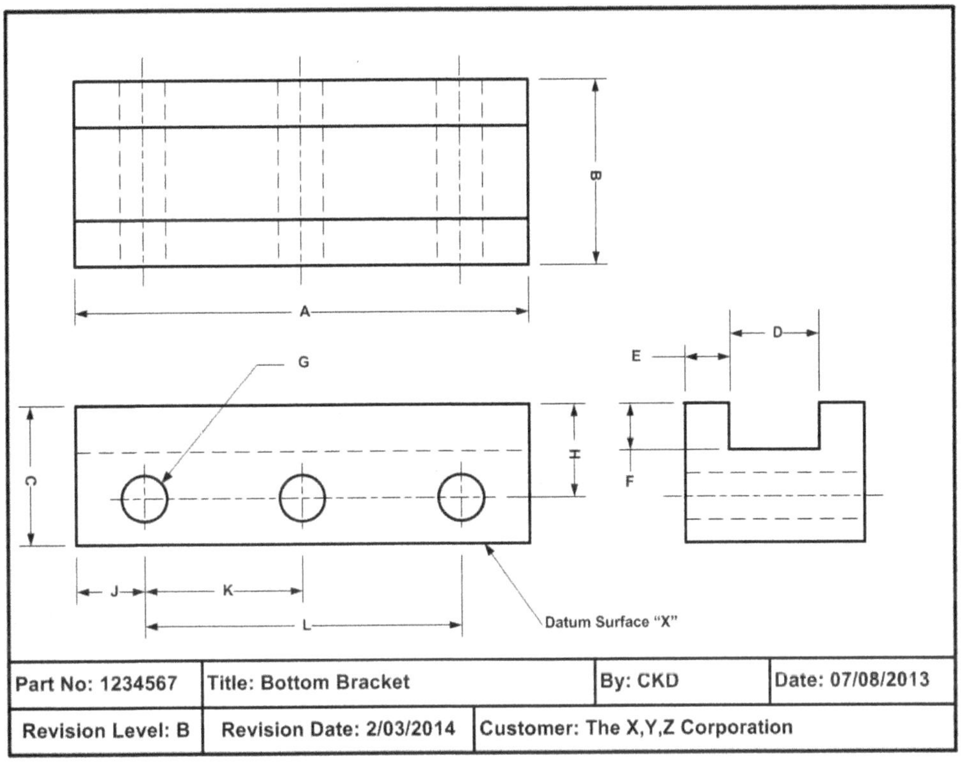

Figure 4.4 Process Plan Sketch of the Component to Be Inspected

Metrology Inspection Report

Part Number: 1234567	Revision Level: B	Inspector: Joanne Davis	Date: 03/14/2014	Sheet 1 of 1

Step	Dimension	Tolerance	Reading	Out of Range	Step	Dimension	Tolerance	Reading	Out of Range
A	3.500	+ 0.005 - 0.005	3.5013	Pass	P		+ -		
B	1.125	+ 0.005 - 0.005	1.1237	Pass	Q		+ -		
C	1.000	+ 0.005 - 0.005	9.9992	Pass	R		+ -		
D	0.500	+ 0.001 - 0.001	0.5027	Pass	S		+ -		
E	0.2495	+ 0.0005 - 0.0005	0.2458	+ 0.0008	T		+ -		
F	0.3755	+ 0.0015 - 0.0015	0.3748	Pass	U		+ -		
G	0.375525	+ 0.00025 - 0.00025	0.37495	- 0.00005	V		+ -		
H	0.750	+ 0.002 - 0.002	0.7534	+ 0.0014	W		+ -		
J	0.375	+ 0.001 - 0.001	0.3758	Pass	X		+ -		
K	1.000	+ 0.001 - 0.001	0.9993	Pass	Y		+ -		
L	2.275	+ 0.001 - 0.001	2.2762	+ 0.0002	Z		+ -		
M		+ -			AA		+ -		
N		+ -			AB		+ -		

Figure 4.5 Completed Inspection Report Sheet for Part Number 1234567

4.4 MEASUREMENT

Measurement is similar to inspection, and it is performed utilizing the same processes. However, there is a major difference between the two. By definition, *Measurement is an act where an unknown quantity is compared against a known standard.* In simpler terms, you have a part (unknown quantity) and you will measure it with a precision measuring device (known standard). All you have is a part with no part print to compare it against and by measuring it, you will know the exact size of the given component. Figure 4.6 shows a typical measurement report sheet, noting that there are no columns for dimensions, tolerances, or out of range because there is nothing to compare the component part to.

The first step in a measurement is to develop a sketch of the part as shown in Figure 4.7. From there, letter designations are applied as shown in Figure 4.8. From learning about correct dimensioning, you will notice that the letter designations violate this rule and it looks like there are double dimensions applied. This is done because without a part print to compare it against, you want to measure everything possible so that a critical area will not be left out. Figure 4.9 shows a completed measurement report sheet for the component part that was checked.

Metrology Measurement Report							
Inspector:		Date:		Requested By:		Sheet 1 of	
Step	Reading	Step	Reading	Step	Reading	Step	Reading
A		P		AC		AR	
B		Q		AD		AS	
C		R		AE		AT	
D		S		AF		AU	
E		T		AG		AV	
F		U		AH		AW	
G		V		AJ		AX	
H		W		AK		AY	
J		X		AL		AZ	
K		Y		AM		BA	
L		Z		AN		BB	
M		AA		AP		BC	
N		AB		AQ		BD	

Figure 4.6 Metrology Measurement Report Sheet

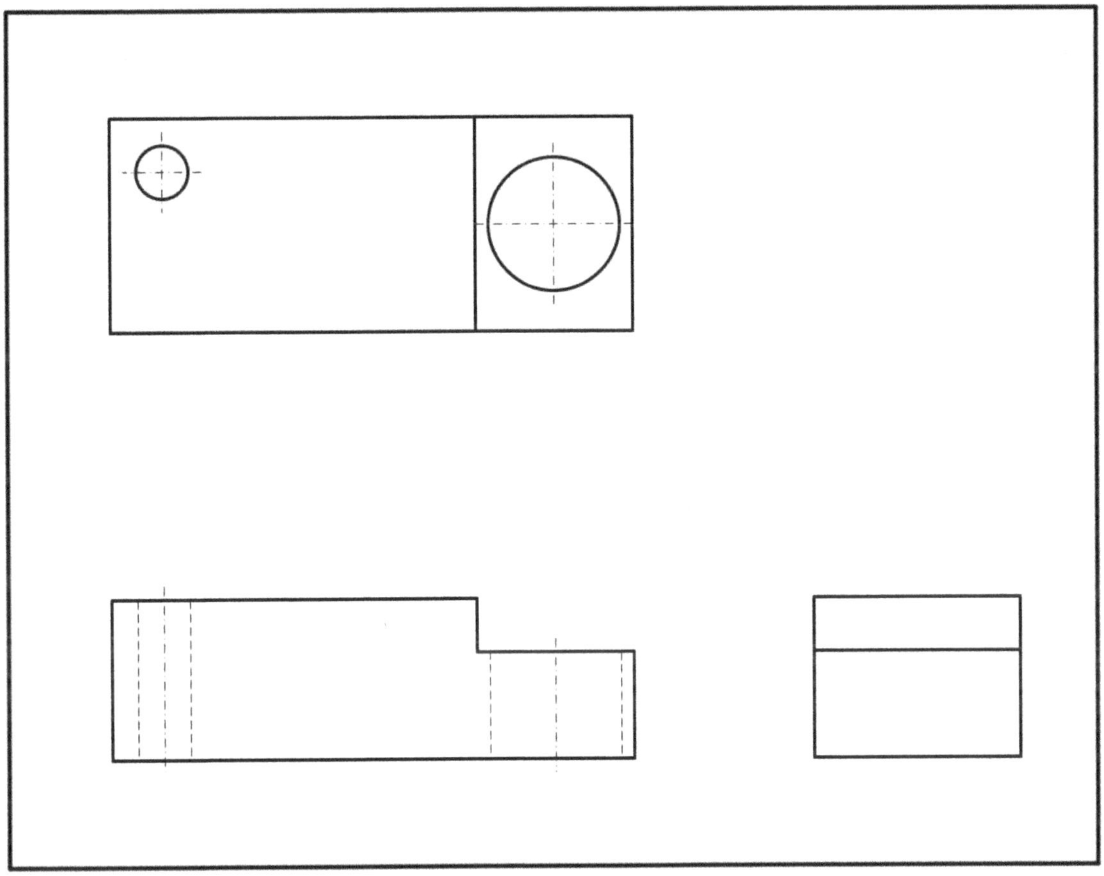

Figure 4.7 Sketch of the Component to Be Measured

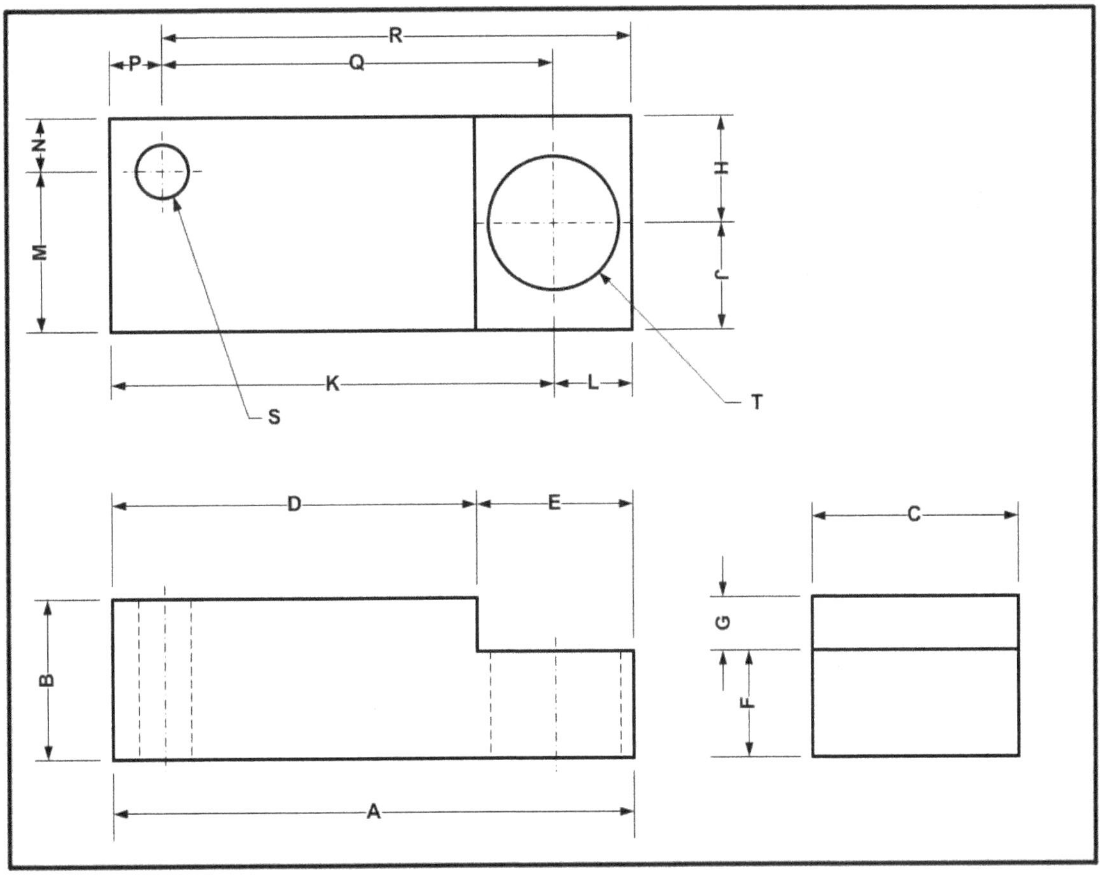

Figure 4.8 Sketch of the Component Part to Be Measured with Step Letters Added

Metrology Measurement Report							
Inspector: Joanne Davis		Date: 03/14/2014		Requested By: Chandler Kane		Sheet 1 of 1	
Step	Reading	Step	Reading	Step	Reading	Step	Reading
A	4.7836	P	0.4779	AC		AR	
B	1.2539	Q	3.1505	AD		AS	
C	1.7562	R	4.3057	AE		AT	
D	2.8461	S	0.4371 Dia.	AF		AU	
E	1.9375	T	1.1273 Dia.	AG		AV	
F	0.7258	U		AH		AW	
G	0.5281	V		AJ		AX	
H	0.8735	W		AK		AY	
J	0.8781	X		AL		AZ	
K	3.6284	Y		AM		BA	
L	1.1552	Z		AN		BB	
M	1.2836	AA		AP		BC	
N	0.4726	AB		AQ		BD	

Figure 4.9 Completed Measurement Report Sheet for Requested Component

4.5 INSPECTION AND MEASUREMENT RULES, GUIDELINES, AND TERMS

When performing both inspections and measurements, there are certain rules and guidelines that need to be adhered to. First, you have to note the tolerances that are applied to the part print for inspection. The measuring device that you will use needs to be precise enough to correctly measure the part with the applied tolerance range. For a measurement, you will want to use the most precision possible to obtain the best readings. Following is a list of rules, guidelines, and terms that apply to any inspection or measurement.

Rule of 10: The rule of 10 implies that the measuring device being utilized needs to be able to read 10 times greater than the specified tolerance on a part. If it cannot, then the readings that you will be obtaining would be a "best guess" of what it actually is.

Example: Dimension = 2.750 Tolerance = ±0.001
Precision of measuring instrument = ±0.0001

Calibration: This is the process where the measuring instrument is checked to ensure that the reading it is giving you is truly the reading it states. Calibration will check the instrument at several readings against a known standard to confirm that it is delivering the proper reading every time. Basically it determines that 0.500 in. does equal 0.500 in. every time it is used. Depending on how precise a measurement is being taken will determine how often it would need to be calibrated. For example, a tape measure that measures to 1/32 of an inch would never need to be calibrated. In the same manner, a micrometer that can measure to ±1/10,000 of an inch may need to be calibrated every 3–6 months to ensure that it is providing a proper reading. However, a high-precision instrument like a laser inspection device capable of measuring to ± one-millionth of an inch may need to be calibrated every time it is going to be used.

Stability: Stability determines how well an instrument holds it calibration. Several factors come into play with stability, such as the overall quality of the measuring device, where it is being used, how it is being used, and the overall age of the device itself. Older measuring devices should be replaced periodically to ensure that they are capable of holding the calibration for a longer period of time.

Drift: Drift is the amount of loss of calibration over time. Every instrument will witness drift over time, and this will determine how often calibration would be required. This also has to do with the overall quality of the measuring device, where it is being used, how it is being used, and the overall age of the device itself.

4.5.1 Inaccuracy Factors

Obtaining a precise reading while performing measurements or inspections is critical in order to collect correct results. If a false reading is taken, component parts that are not to the prescribed standard can be passed through the system, causing issues in either the assembly or performance of the product. Listed in the following are the 10 most common inaccuracy factors that can result in an incorrect reading.

Feel or Pressure: When utilizing a manual measuring device, the amount of pressure exerted while obtaining a reading can result in an incorrect reading. The inspector needs to know what pressure the tool requires, and this is obtained in the feel of the tool as it is placed on the component be checked.

Dirt and Burrs: If the component part has dirt and/or burrs on it while a reading is being taken, the measuring device will be placed over them, causing an incorrect reading. Always ensure that the burrs have been removed and that the component is free of any dirt on the surface before performing a reading.

Alignment: Alignment to the component with the measuring device is critical. If the tool is not placed squarely on the surfaces being checked, the reading will be false. Having a measuring device that is angled to the surfaces being checked will result in an incorrect reading.

Parallax: Parallax is where the measuring device is not being sighted properly by the inspector. If the display on a manual measuring tool is being sighted at a certain angle, the reading being obtained may not be correct, resulting in a false reading.

Heat: Heat causes a material to expand. When a component part that is heated is being measured, the reading taken may be false if the material has expanded. This is an important factor for components requiring extremely close tolerances.

Tool Wear: As measuring devices are used repeatedly, they can wear out. Ensuring that the tool being used to obtain readings is in good functioning condition is important for obtaining correct readings. This is something that is checked and verified during tool calibrations.

Calibration: As mentioned previously, calibration of the measuring devices is critical to ensure that what the tool is reading is truly the size of the component. The precision required for obtaining readings will determine how often the measuring device would require to be calibrated. For readings that require tolerances within a few millionths of an inch, the device is calibrated before every reading.

Damage: A damaged measuring device will never allow for a true reading. Any measuring device that is damaged in any manner needs to be either repaired or discarded.

Environment: The measuring environment needs to be one where no foreign matter is present. Trying to obtain precise reading when dirt, dust, and other contaminations are present will never allow for a correct reading. For extreme-precision readings of extremely close tolerance components, it is sometimes performed in a clean room where the temperature, humidity, and air quality are all closely controlled.

Bias: Bias for a result of a reading simply means that the measuring device is read incorrectly. Also, some readings will require calculations in order to complete them. If these calculations are not correct, the reading will not be correct.

4.6 INSPECTION AND MEASUREMENT TECHNIQUES

Inspections and measurements can be performed in several ways. How critical or important the inspection or measurement is depends on what process and measurement instruments would be utilized. Some inspections are performed while the process is running to observe any deviations occurring, while other inspections would be performed after the parts have been completed. These can be performed either manually or using various automated techniques depending on what needs to be checked and how precise the tolerances are. Both manual and automated techniques are explained as follows along with the various measurement instruments used.

4.6.1 Manual Techniques

Techniques for measuring and inspecting components manually can be used either after the part is fabricated or during the process. Depending on the precision required, measuring instruments can render measurements from 1/64 to 1/10000 of an inch. A variety of measuring instruments are available that can be used either after or during processing, and some of the more common are listed as follows.

Figure 4.10 shows very nonprecise measuring scales. Depending on the application, these may be all that is required or can be used to roughly check the components. The top scale is a 12 in. machinist scale and has graduation of 1/100 of an inch where the bottom tape measure has graduations of 1/16 of an inch.

Figure 4.11 shows calipers. These can be used for measuring outside dimensions, inside dimensions, and depth dimensions. Calipers have a precision of 1/1000 of an inch.

Figure 4.12 shows two different types of micrometers. The top micrometer is for measuring the depth of a component and is precise to 1/1000 of an inch. The bottom micrometer is used for measuring the outside dimensions and depending on the type, it can be precise to either 1/1000 or 1/10000 of an inch.

Figure 4.13 shows a protractor for measuring the angles on a component. It has graduations of 1°.

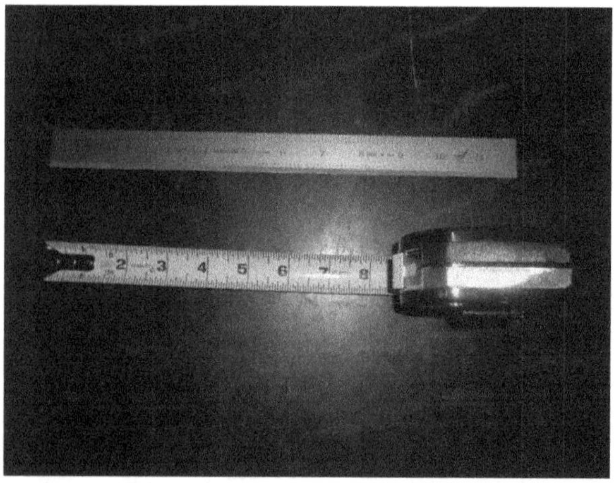

Figure 4.10 Low-Precision Hand Measuring Instruments

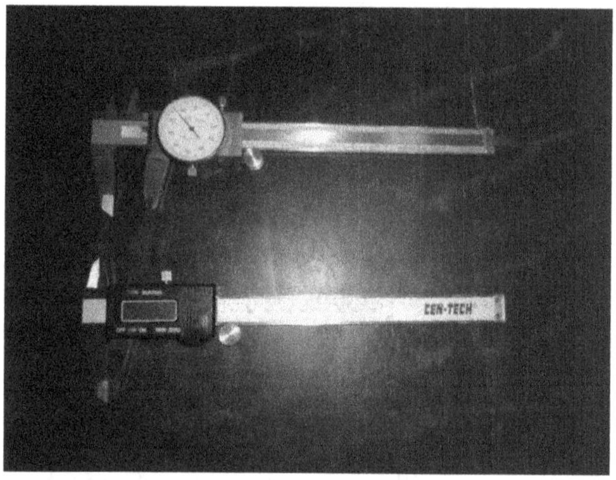

Figure 4.11 Hand Measuring Calipers

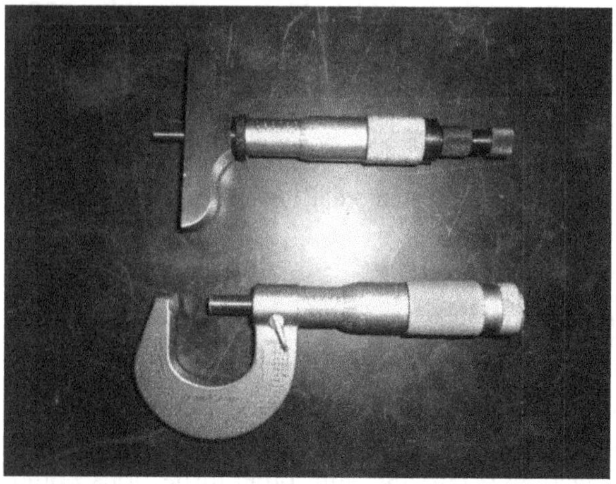

Figure 4.12 Hand Measuring Micrometers

Figure 4.13 Hand Measuring Angle Protractor

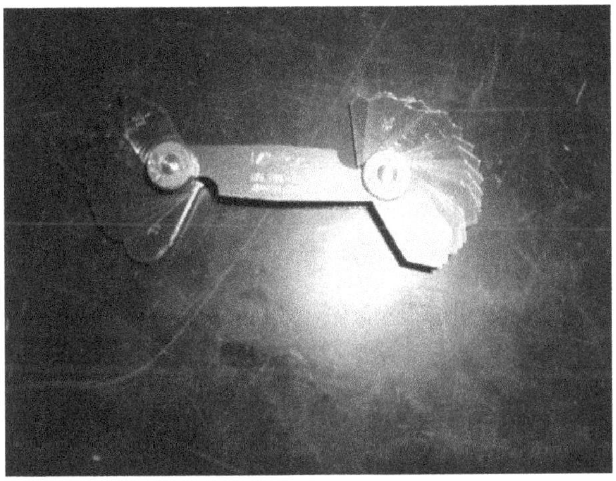

Figure 4.14 Hand Measuring Radius Gage

Figure 4.15 Garnet Surface Plate Inspection/ Measurement Setup

Figure 4.14 shows a radius gage, which has both internal and external radii in steps of 1/64.

Another manual measurement and inspection technique is a setup on a garnet surface plate. This is performed utilizing a variety of instruments that do not function alone, but when combined, build a powerful measuring setup.

Figure 4.15 shows the components in this type of setup, which can measure dimensions within 1/10000 of an inch along with angles within fractions of a second.

4.6.2 Automated Techniques

Some inspection and measurement systems are either fully automated or semiautomated. With a fully automated system, a program is written for the component that is to be inspected, and it will move around the part using a tactile sensor, touching the part in various locations. Figure 4.16 shows an automated coordinate measuring machine. This requires writing a program for what is to be measured with the dimensions and tolerances. The component part is loaded in the machine, and it will automatically index around using a tactile sensor on the arm and will produce a full inspection report for every dimension and on whether they were in tolerance or not.

A semiautomated system is one that needs to be moved manually and the readings are taken from a digital display. What makes it semiautomated is that you do not have to physically measure anything, just read the findings off the displays. Figure 4.17 shows a semiautomated device called an optical comparator. This works by loading the component into it and a magnified shadow is produced. By moving the table against the projection screen, dimensions are read from the digital readout on both the "X" and "Y" axes.

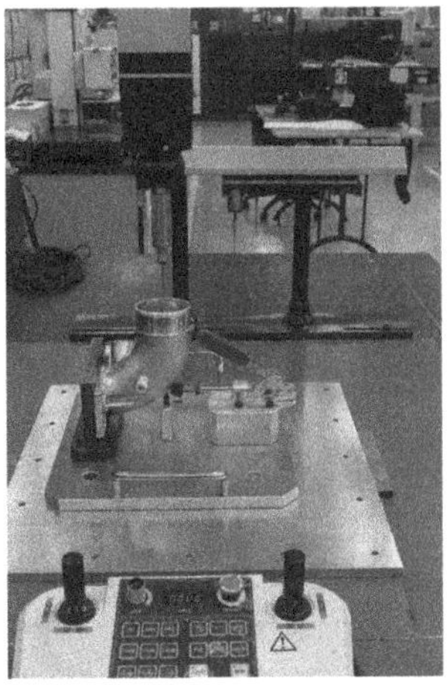

Figure 4.16 Automated Coordinate Measuring Machine (CMM). © sorapol/Shutterstock.com

Figure 4.17 Optical Comparator Measuring Machine

4.7 TESTING

Unlike inspection and measurement, testing looks at the functions of a component part or a product. By definition, *Testing observes a part or product under the functional conditions it will encounter during the operational design parameters.* Every component part in a product performs a given function, and testing validates that it will perform that function without failing. Depending on how critical the function of a given component or a given function in the product is will determine how much, or if any, testing is required. For example, if a component part is a bracket that supports a motor in the assembly, it may not be tested as an individual unit, but rather observed in the testing of the product itself. Likewise, if the same bracket is used in an area of the assembly and must stand up to a heavy load without breaking or deflecting, it would be tested under those conditions.

The amount of testing and the severity of the testing depend on some critical areas in the design. If the failure of the component part in the product, or the failure of the function within the product, can cause danger to the operator, testing in those areas becomes critical. Some of these crucial areas are listed as follows.

Operator Danger: If a function failure, or a component part failure on the product, can cause injury or death to the user during operation, this area needs to be heavily tested. An example of this would be the breaking system in a vehicle. A failure to this part of the product would be very detrimental to the operator and needs to be very thoroughly tested to ensure that the function is guaranteed. However, if a window in the same vehicle fails to roll up or down, this would not cause any danger to the operator and testing would not be required.

Catastrophic Failure: A catastrophic failure may or may not cause injury to the operator, but would result in either a nonfunctioning product or destruction of that product. For example, if a new satellite were launched into space and upon deployment, the functions that it was intended to perform failed, the entire purpose of the project is lost. Likewise, if the same satellite were to explode in space, no one would be injured, but the result would be the same as it was not functioning. There are cases where a catastrophic failure would cause injury or death. If a bridge were to collapse after years of service, those on the bridge at that time would be affected.

Testing in cases were a failure of this nature could either put the entire project in jeopardy or cause injury or death over time is critical. For areas like bridges or structures that the public utilizes daily, initial testing is performed along with periodic testing over time to ensure that the functions are still safe.

Premature Failure: Here, a product is released with areas that can potentially have failures, but over a given period of time. Nothing will last forever, and being able to predict when a failure of the product will occur is important to warrantee the reasons. These areas in the product will be tested to establish the life cycle of that given function over a period of time or over a given number of cycles. It is known upfront in the design that these areas will eventually fail, but through testing, the design can be made to provide the user with the cycles expected.

Testing is performed on component parts and products in two ways. One is through destructive testing where the product is tested with various methods until a failure occurs. This involves applying various loads, cycles to the functions, and performing this until a failure occurs, then evaluating what caused the failure. The other method is nondestructive testing where tests are performed, but the part or product is not destroyed in the process. These two testing methods are explained as follows.

4.7.1 Destructive Testing

Destructive testing implies exactly what it states. The component part or product will be tested until a failure occurs and cannot be used again. Here, a critical area of a part or product is identified and a test plan against the design functions is constructed. Figure 4.18 shows an assembly requiring testing. Here, a part in the function will be shuttling along another part with a loaded force of 275 pounds at a speed of 5.75 in./s. The design requirements are to have a maximum wear at the contact surface of no more than 0.0005 in. and a maximum deflection of no more than 0.0002 in. over 750,000 cycles. A test plan is established against these parameters as shown in Figure 4.19 to perform this destructive test. You will notice that the first test is performed against the exact functional design parameters, and then these are expanded for the other tests that will be performed. This is done to see what would happen if either the load is increased or the speed is increased. These are known as expanded testing boundaries, which are used to establish safety margins in the design.

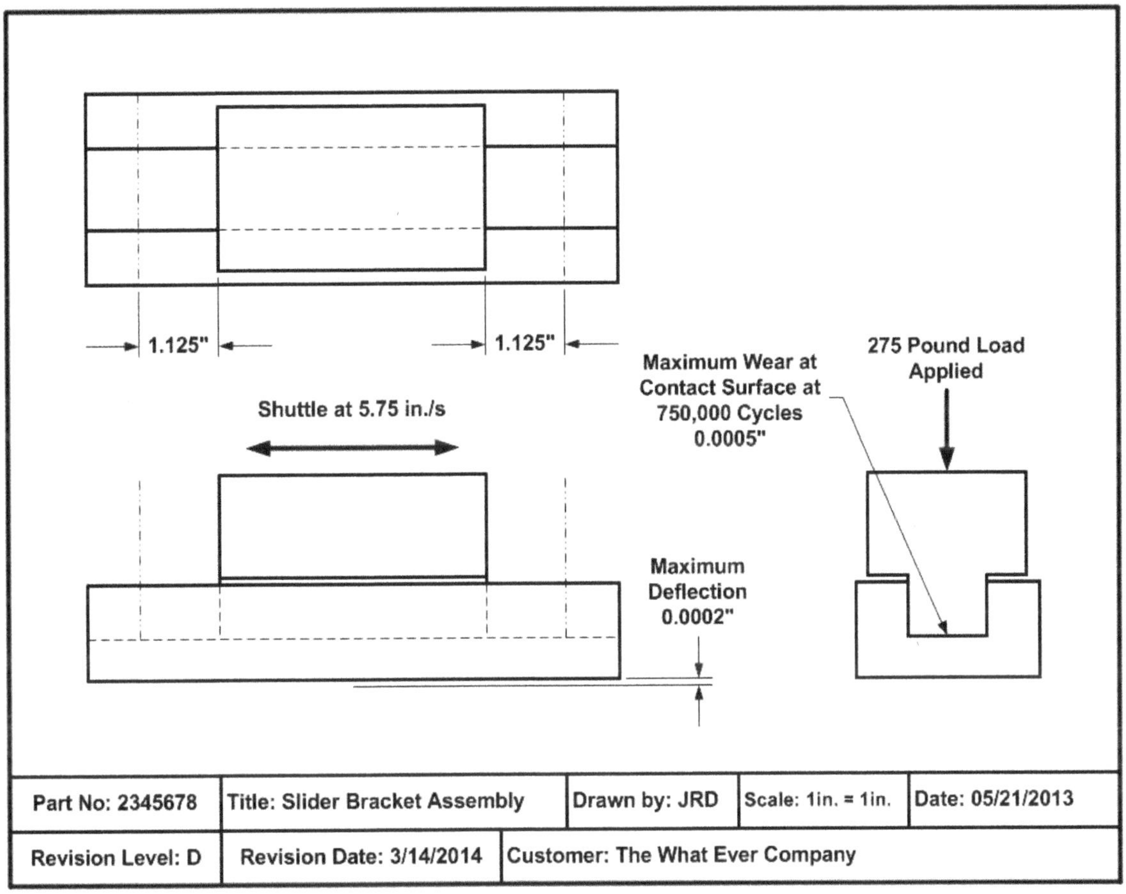

Figure 4.18 Assembly Drawing with Specified Testing Requirements

Test Plan Results						
Part Number: 2345678		Revision Level: D	Inspector: Chandler Kane		Date: 03/14/2014	Sheet 1 of 1
Test No	Shuttle Speed in./s	Applied Load Lbs.	Cycles	Deflection	Surface Wear	Comments
1	5.75	275	750,000			
2	5.75	300	750,000			
3	5.75	325	750,000			
4	5.75	350	750,000			
5	6.00	275	750,000			
6	6.25	275	750,000			
7	6.50	275	750,000			
8	6.75	275	750,000			
9	6.75	300	750,000			
10	6.75	325	750,000			
11	6.75	350	750,000			
12	7.00	350	750,000			
13	7.00	375	750,000			

Figure 4.19 Test Plan Established with Expanded Parameters

Test Plan Results						
Part Number: 2345678		Revision Level: D	Inspector: Chandler Kane		Date: 03/14/2014	Sheet 1 of 1
Test No	Shuttle Speed in./s	Applied Load Lbs.	Cycles	Deflection	Surface Wear	Comments
1	5.75	275	750,000	0.000076"	0.000086"	Wear and Deflection Parameters Passed
2	5.75	300	750,000	0.000089"	0.000104"	Wear and Deflection Parameters Passed
3	5.75	325	750,000	0.000104"	0.000273"	Wear and Deflection Parameters Passed
4	5.75	350	750,000	0.000163"	0.000397"	Wear and Deflection Parameters Passed
5	6.00	275	750,000	0.000075"	0.000093"	Wear and Deflection Parameters Passed
6	6.25	275	750,000	0.000075"	0.000142"	Wear and Deflection Parameters Passed
7	6.50	275	750,000	0.000077"	0.000217"	Wear and Deflection Parameters Passed
8	6.75	275	750,000	0.000076"	0.000415"	Wear and Deflection Parameters Passed
9	6.75	300	750,000	0.000092"	0.000483"	Wear and Deflection Parameters Passed
10	6.75	325	750,000	0.000107"	0.000505"	Wear Parameter Failed and Deflection Parameter Passed
11	6.75	350	750,000	0.000183"	0.000572"	Wear Parameter Failed and Deflection Parameter Passed
12	7.00	350	750,000	0.000193"	0.000625"	Wear Parameter Failed and Deflection Parameter Passed
13	7.00	375	750,000	0.000204"	0.000686"	Wear Parameter Failed and Deflection Parameter Failed

Figure 4.20 Test Plan Conducted with Results

The test is performed, and Figure 4.20 shows the results. From observing the testing results, the original design parameters in test #1 showed no failure. As the load or speed was increased, there was still no failure at a speed of 6.75 in./s and a load of 300 pounds. As the load increased to 325 pounds and the speed remained at 6.75 in./s, the wear on the contact surface exceeded the design parameters, but the deflection remained within. From there, the testing parameters were further expanded, and at a load of 350 pounds and a speed of 6.75 in./s, both the wear and deflection exceeded the function parameters. All of these tests were run for the stated 750,000 cycles, and each test required using a new assembly.

Testing of this nature is performed on all new products prior to releasing them for sale. Also, if a major design change is made in a particular area, a destructive test will be performed in the critical areas to ensure that the changes made did not affect that functional area. If we look at the automotive industry, they perform countless tests on their vehicles. One that we have all seen is the crash test where the vehicle is run down a track and crashed into a wall. After the test is performed, the entire vehicle is evaluated to see how the frame absorbed the impact, how the air bags were deployed, and what effect all of this had on the crash dummies through the thousands of sensors that are imbedded in them.

Destructive testing is extremely expensive due to the fact that a component part or an entire product is destroyed in the process. However, it is extremely important to perform testing, so alterations to the design can be made if it did not meet the specifications and safety factors can be implemented in the design so that the product would always exceed the desired design parameters. But in the process, several parts and products need to be destroyed in order to collect this real-time information.

4.7.2 Nondestructive Testing

In the nondestructive arena, testing is performed but the component part or product is not damaged in any way. Here, various methods are utilized to test if the part or product meets the design specifications. Nondestructive testing traditionally will be performed on extremely critical areas, both before and after a destructive test has been performed. One of the areas that a test of this nature is performed is on welded assemblies. First, an assembly is welded together and a nondestructive test such as an X-ray is performed to ensure that the weld is perfect. From there, the welded assemblies are put through a destructive test to examine how they hold up to the design specifications. Upon reviewing the destructive test results, the welded assemblies are then produced to go into the product, and depending on how critical the function of that given assembly is, they may be tested at a frequency, or every single assembly would be tested to ensure that a failure will not occur.

When the Alaskan pipe line was being installed, every single weld made joining the pipes together was X-rayed, evaluated, and cataloged to document the entire length of the pipe. This assured that every weld on the pipe was up to the standards and that there were no defects internally that could cause a potential failure in the future. All though this was very time-consuming and costly, it proved to be worthy of both time and money well spent for not having future issues.

Looking at what a typical testing a welded joint would require, it would be performed in the following sequence. First, the welded joint would be X-rayed in a nondestructive manner to ensure that the weld had no defects. From there, a destructive test would be performed as described earlier to physically test the design parameters of the joint. Figure 4.21 shows a welded joint and the loads it needs to withstand during service. Upon the completion of the destructive test, future welded joins may undergo a nondestructive X-ray test to confirm that the welds are satisfactory and contain no defects. Figure 4.22 shows an X-ray of a welded joint that is defect-free. Likewise, Figure 4.23 shows an X-ray of a welded joint that contains porosity voids beneath the weld, which make it much weaker than a perfect weld, and it may not withstand the loads and forces required.

Nondestructive testing is costly and time-consuming to perform. This type of testing is performed where any failure could cause a dangerous or catastrophic situation. It always follows or precedes a destructive test, and the frequency of the tests would be dependent on the importance of the product.

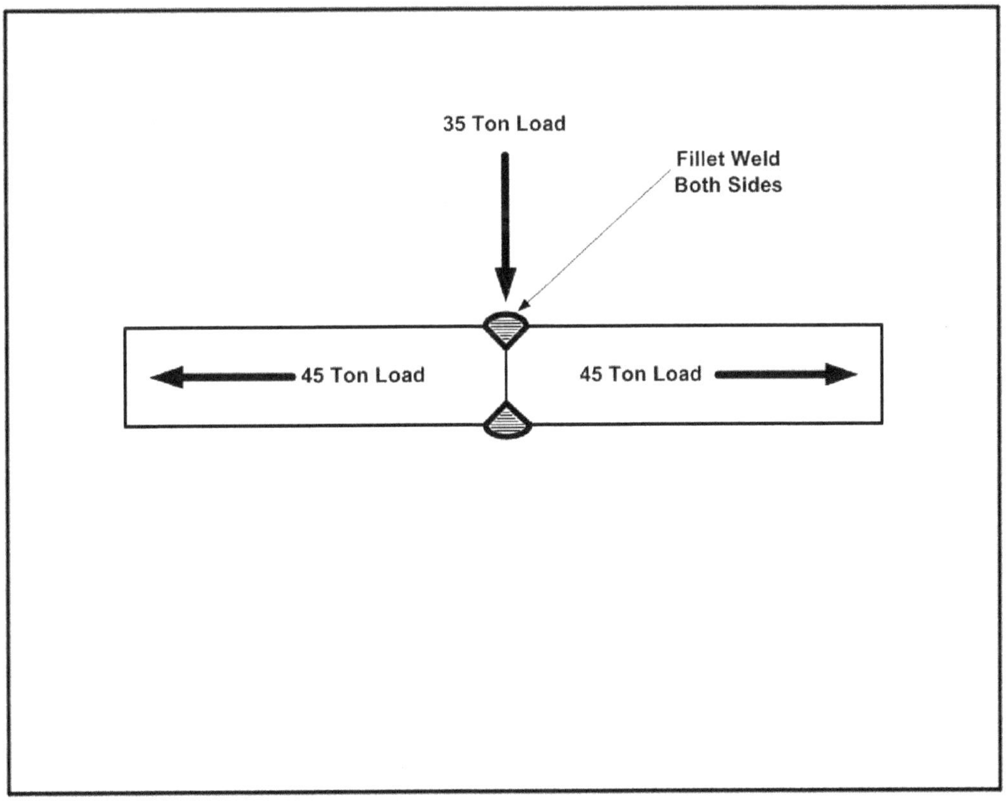

Figure 4.21 Welded Assembly for Both Destructive and Nondestructive Testing

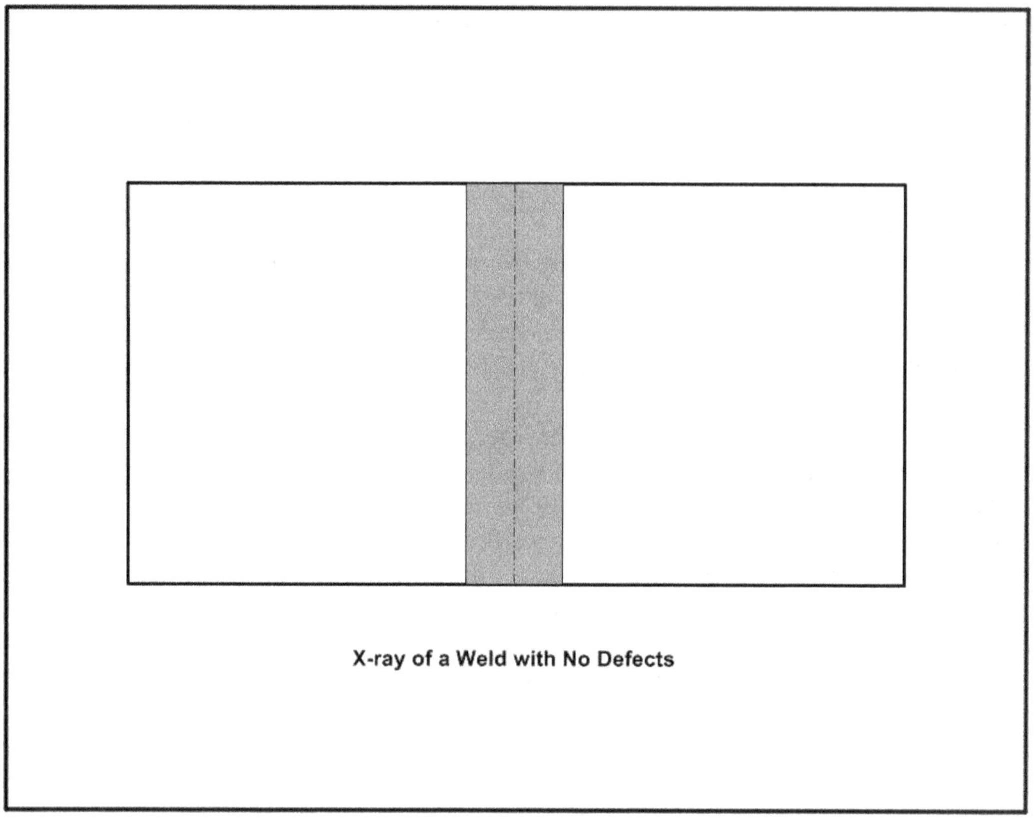

Figure 4.22 Nondestructive X-ray Testing of a Good Welded Joint

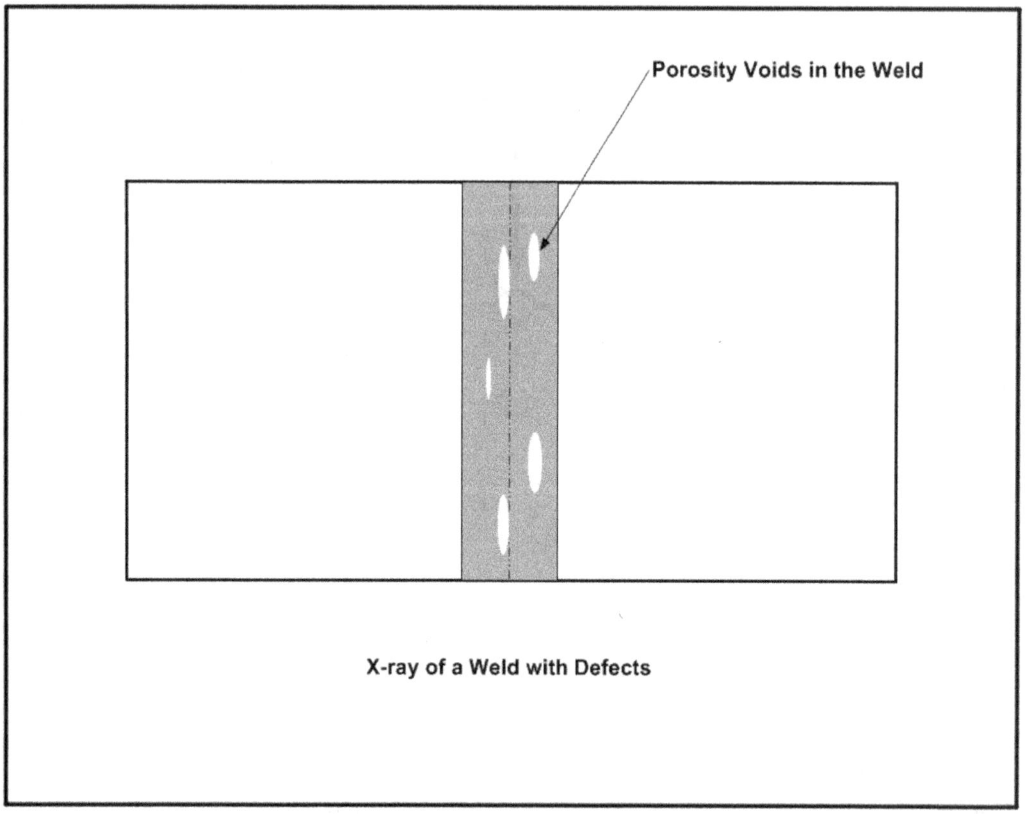

Figure 4.23 Nondestructive X-ray Testing of a Defective Welded Joint

4.7.3 Life Cycle Testing and Designed Obsolescence

As everyone knows, nothing lasts forever. This is not by chance, but rather by design. When a new product is being considered, certain areas of the product are being designed to fail over either a certain period of time or after a given number of cycles. This is a very difficult task to accomplish because if the failure occurs too early, either the customers will be dissatisfied with the product, or, if the product has a warrantee, the company will be spending larger amounts of money to repair the product. The key is to have the products fail at the right time as to not upset the customer or force the company to repair the product while it is under warrantee. To accomplish this correctly, a process called life cycle testing is used to determine the time of failure.

To perform a life cycle test, the product is only run with the normal operating parameters, not the expanded parameters, the way a destructive test was performed. What is being evaluated here is the time or cycles that the product will function before failure occurs. Looking at the automotive industry, they will test everything on the vehicle for its life cycle. This includes areas that are the least important, from door latches and glove compartments, to the break systems and engines. This gives them the basis for establishing the warrantee on the vehicle. It is not by accident that after the warrantee on a vehicle expires, it may need a break job or a clutch replacement. This obsolescence is designed into the product. Besides, if everything lasted forever, there would be no reoccurring sales of new products.

4.8 SUMMARY

Metrology is a very important and critical part of the manufacturing process. Knowing how to develop inspection procedures to ensure that the components being fabricated are to the desired dimensions and tolerances will ensure that the products will assemble and function properly. Along with inspection and measurements, testing is a vital component in manufacturing to isolate critical areas where a failure can cause major issues. Whether it is a destructive test to determine design parameters or a nondestructive test to validate critical components, it will prevent any major failures in the future. By implementing various inspections and testing programs within the manufacturing operation, the final outcome can only be a positive one.

4.9 QUESTIONS

1. By definition, what is Metrology and why is it an important function in the manufacturing arena?

2. Accuracy and Precision are two terms that get misused all the time. What is the difference between Accuracy and Precision?

3. What is the difference between Measurement and Inspection?

4. What is the Rule of 10, and why is it important to adhere to when performing inspections?

5. What is Testing? List some of the reasons why it is performed.

6. Define the difference between Destructive Testing and Nondestructive Testing and where each would be used.

SECTION 1

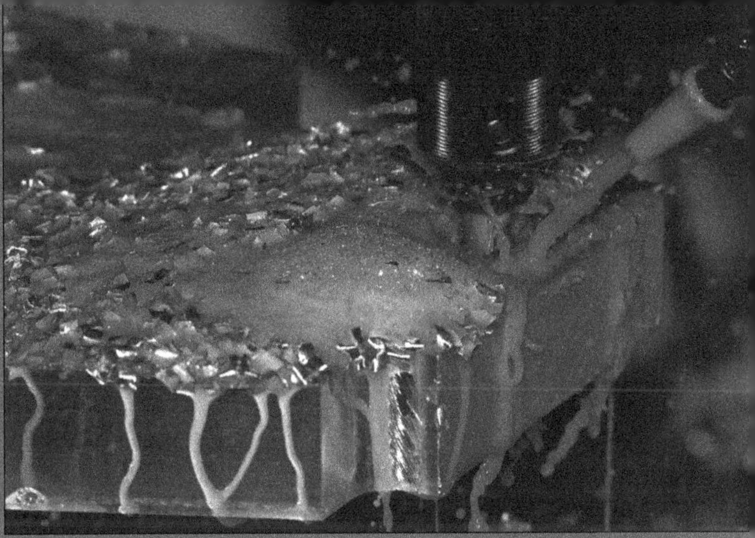

SOLIDIFICATION PROCESSES

Section 1 involves processes dealing with the solidification of the material being used to fabricate the parts. Here, the material being utilized is either completely liquefied or melted down to some consistency in order to either pour it into a given shape or model it to the desired shape. Figure 1 shows the process flow for a typical solidification process from the selection of the material to be utilized, the melting it to a given temperature, shaping the molten material in a selected process, and rendering a finished component.

When dealing with molten materials, there are two important terms that need to be understood, which are Viscosity and Fluidity. With most materials, when they melt or are heated, both the viscosity and fluidity will change. These are defined as follows.

Viscosity: Viscosity refers to a liquid's resistance to flow. For most materials, this is temperature-dependent, and as the material is heated, the flow increases, so the viscosity of the material decreases.

Fluidity: Fluidity refers to a liquid's ability to flow. For most materials, this is temperature-dependent, and as the material is heated, the flow increases, so the fluidity also increases.

A good example of how viscosity and fluidity change with temperature would be cooking oil. If a bottle of cooking oil is placed in the refrigerator, it becomes very thick, almost like a paste. At this stage, the viscosity is extremely high while the fluidity is extremely low. Take the bottle out of the refrigerator and let it set to room temperature, and it changes. The oil will pour out of the bottle, but it is thick. In this state, the viscosity has been greatly reduced and the fluidity is greatly increased from the previous state. Now, heat the oil up in a pan, and it becomes like water where the viscosity is extremely low while the fluidity is extremely high.

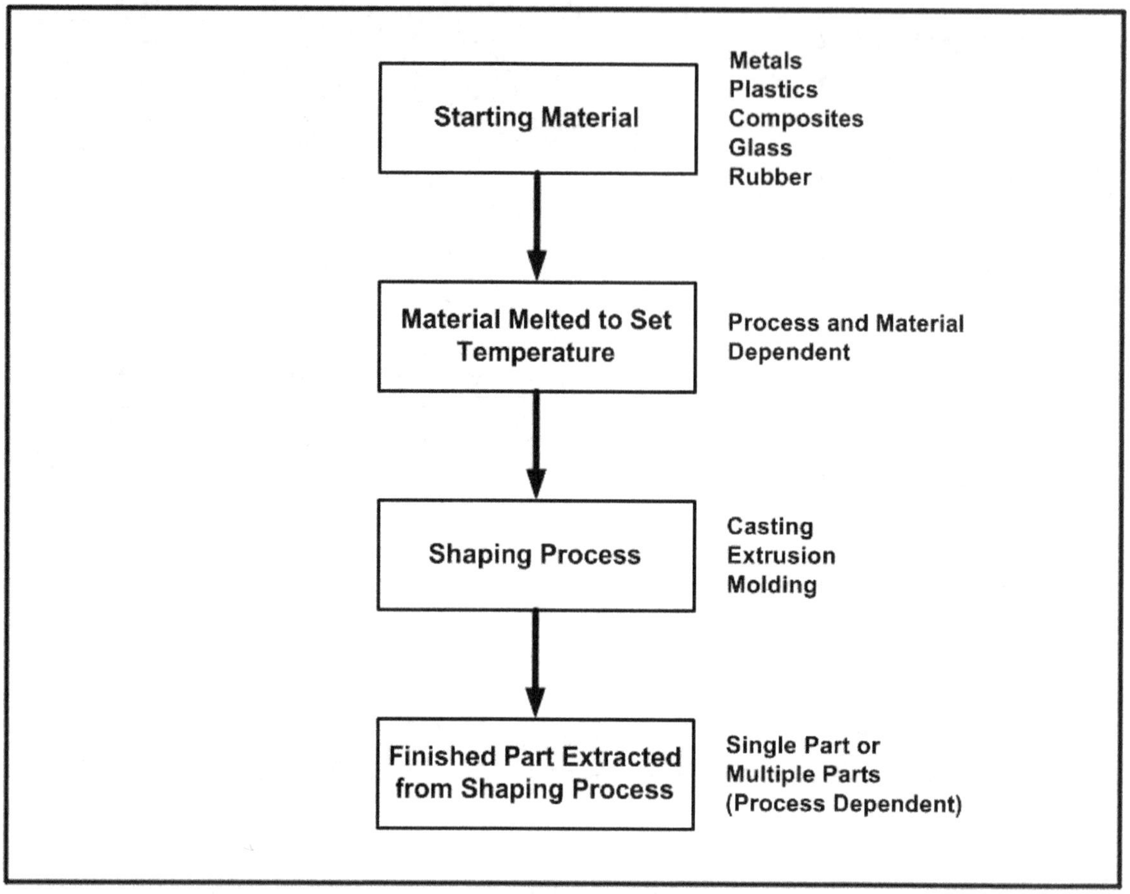

Figure 1 Typical Flow for Solidification Processes

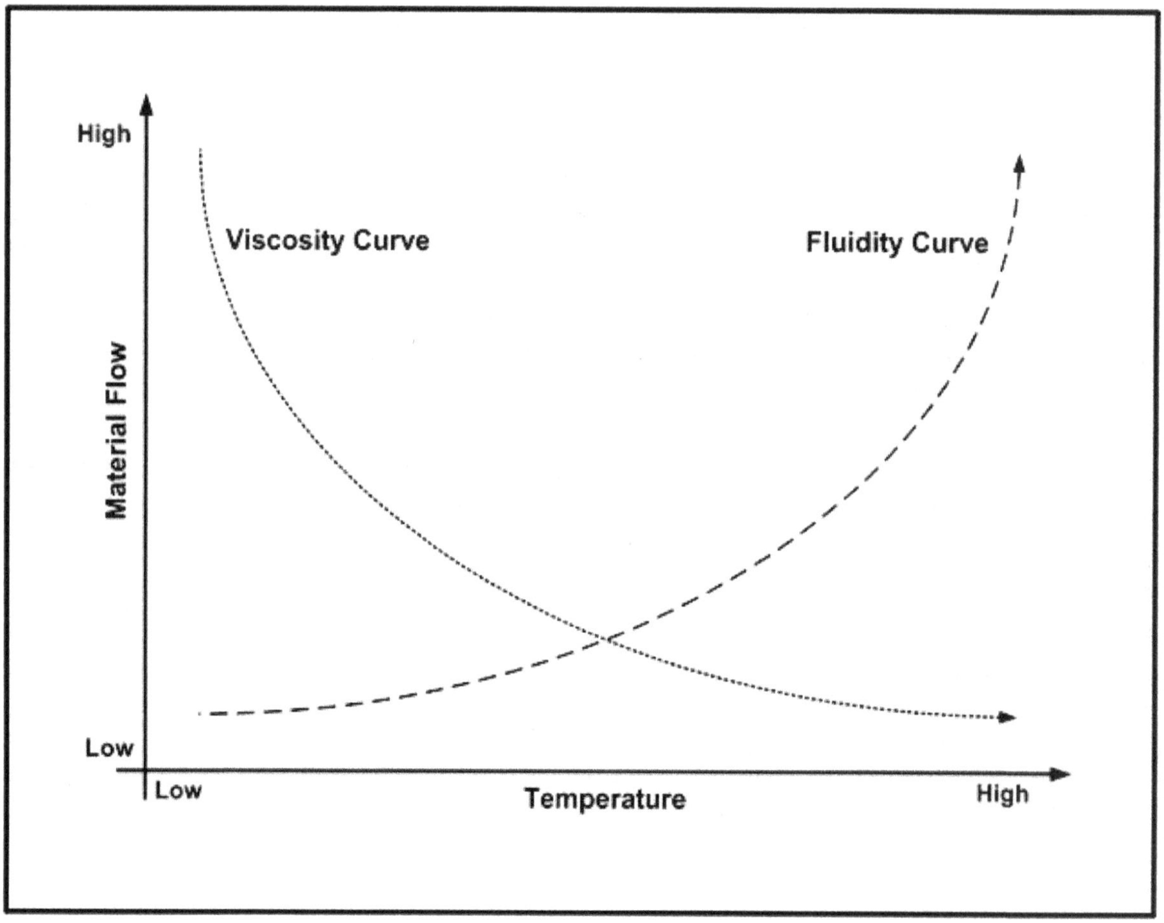

Figure 2 Material Flow of Fluidity and Viscosity against Temperature

Not all materials will go through such an even state where viscosity and fluidity vary so greatly, but when they go from a solid to a liquid state, there will be changes to both. Figure 2 shows a very generic chart of how the viscosity and fluidity of materials will be affected by the change in temperature. Understanding how materials react when melted is critical to understanding the solidification processes.

In this section, we cover a variety of solidification processes such as Casting, Plastic Shaping, Glass Shaping, and Rubber Forming Processes. Each of these entails different shaping processes and will render part configurations specific to those processes. The details of these processes are covered in detail in the following chapters.

CHAPTER

5

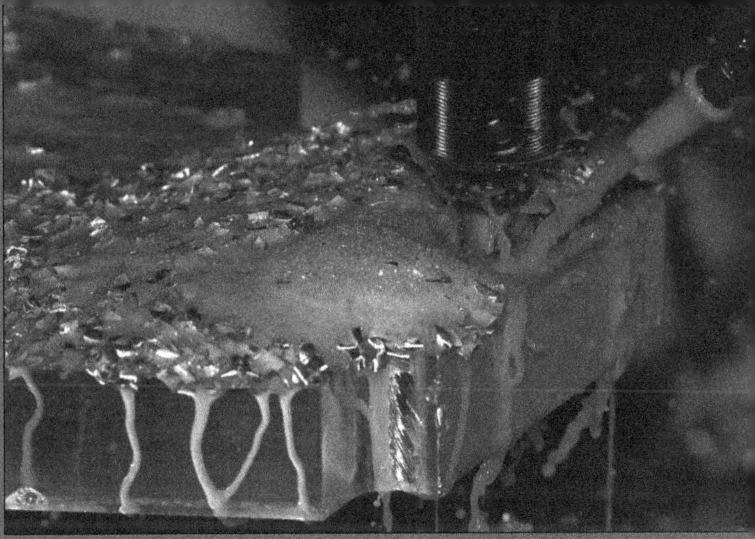

Metal Casting Principles

5.1 INTRODUCTION

The metal casting process is one of the oldest known methods for manufacturing parts. It dates back some 6,000 years to the ancient civilizations such as the Egyptians and Aztecs who cast their beautiful statues and jewelry. They would create molds of the shapes that they desired, mainly out of sand, and pour the molten gold into them to give them the basic shape and configurations desired. From there, they would work the basic shapes, carving and engraving them and ending with polishing the beautiful shapes that are seen today. Later, in the Medieval period, castings were being made from iron and other metals to make heavy boat anchors along with brass and bronze items. As time went on, the casting process has had several technological improvements, but the concept remained the same—pouring molten metal into a cavity to create a shape.

Today, we can see castings being utilized in several of the products that we come into contact with on a daily basis. For example, the engine in your vehicle has a casting made from cast iron for the block.

In this chapter, we discuss the technical aspects of the process—how a mold is designed for various processes and the properties of the material as it goes from a liquid to a solid and then cools to ambient temperatures. A variety of the more common casting processes are covered along with the design considerations that need to be implemented when designing a part that will be produced in any of the casting processes.

5.2 METAL CASTING THEORY

Metal casting is for all intensive principles a simple procedure. A mold of some type is constructed and molten metal is poured into the mold to create a desired shape. However, there are several areas that need to be considered in the process to ensure that the cast part will be of proper size and configuration. These areas include the material being cast, the amount of shrinkage that will occur, and what type of mold and process need to be selected along with the overall design guidelines for casting. With these considerations, a proper process can be selected and the parts can be cast to the configurations and sizes desired.

5.2.1 Mold Filling Requirements

In any casting process, how a mold is filled is critical in order to produce an acceptable part. Regardless of what casting process or mold is being filled, there are four important factors that need to be adhered to, which are *Pouring Temperature, Pouring Rate, Flow Turbulence,* and *Mold Erosion.* These are dependent on both the construction and the overall design of the mold or the processes performed prior to filling it. If these factors are all met, the mold should produce the desired part during the process.

5.2.1.1 Pouring Temperature

The temperature of the molten metal is very important in the filling of a mold. When heating the material, it needs to be *Superheated*, which means that the temperature of the molten metal must be way above the melting temperature. What this does is maintains the molten material as a liquid throughout the entire filling process. If the material is heated only slightly above the melting point, it may start to solidify during filling, which can result in the mold not filling completely or other defects in the part.

For example, if a casting was to be poured with Copper Casting Alloy, which has a melting temperature between 1,941 and 1,981 °F. The temperature that it needs to be heated to must be much higher than that. A good rule of thumb is to heat the material to 20–25% above the melting temperature, which would raise it to between 2,329 and 2,377 °F for 20% and between 2,426 and 2476 °F for 25%. This gives a safety margin while pouring the molten metal into the mold so that it will not start to solidify during filling.

5.2.1.2 Pouring Rate

How quickly the mold is filled is critical to ensure that the material does not start to solidify during filling and to ensure that the mold is completely filled. Based on what casting process is being utilized to produce the part and the construction of the mold itself, two factors are calculated to optimize the pouring rate. These are the *Flow Velocity* and the *Volumetric Flow Rate.* The Flow Velocity determines how fast the material will flow into the mold based on the height that it is entering the mold cavity and the gradational force. The Volumetric Flow Rate determines how much material can flow into the mold under optimum conditions. From both of these factors, the optimum mold filling time can be derived. Both of these are covered in detail later in this chapter.

Melting Temperature of Select Metals

Material	Melting Temp °F ①	Melting Temp °C ①
Magnesium	1,201	649
Aluminum	1,250	677
Zinc Alloy	1,470	800
Gold	1,740	950
Brass Alloy	1,880	1,027
Bronze Alloy	1,970	1,077
Silver	2,140	1,171
Copper Alloy	2,260	1,238
Alloy Cast Iron	2,390	1,310
Low C Steel	2,600	1,427

① Highest Melting Temperatures Listed. Not Exact Material Specific. Source: www.matwab.com

Figure 5.1 Melting Temperatures for Select Casting Metals

5.2.1.3 Flow Turbulence

Turbulence is a factor that is controlled in the overall design of the mold. When a mold is being filled, what is desired is a smooth laminar flow of the molten metal throughout the filling step. By designing the mold filling portion of the mold, the way that the pouring cup and downsprue are positioned will determine how the material will flow. Figure 5.2 shows how a turbulent material flow would occur during filling. By having the molten metal poured straight into the mold cavity, it enters and splashes around. Figure 5.3 shows how this problem can be eliminated by having the molten metal poured into a downsprue, which redirects the flow through a runner into the cavity in a smooth, laminar flow.

5.2.1.4 Mold Erosion

Mold erosion is caused by the molten metal hitting the mold cavity and eroding it away. This can be a factor in any type of molding process and, either instantly or over a period of time, can cause erosion to the cavity producing the part. By designing the mold in such a way that if any erosion is going to occur, it will be in an area not critical to either the mold or the cavity. Figure 5.2 shows how the mold cavity will erode as the material is poured directly onto the surface of the cavity. In Figure 5.3, you can see that erosion is occurring, but in a noncritical area, which does not affect the cavity in any way.

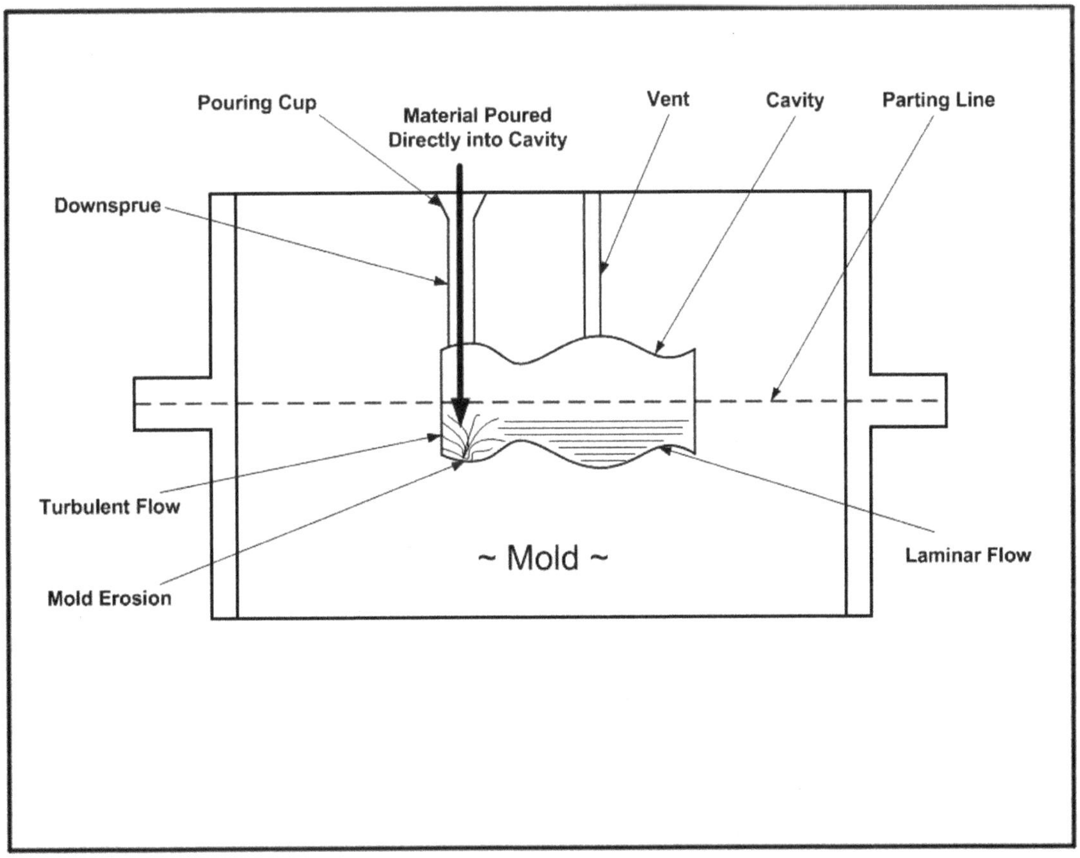

Figure 5.2 Bad Mold Design Filling Issues

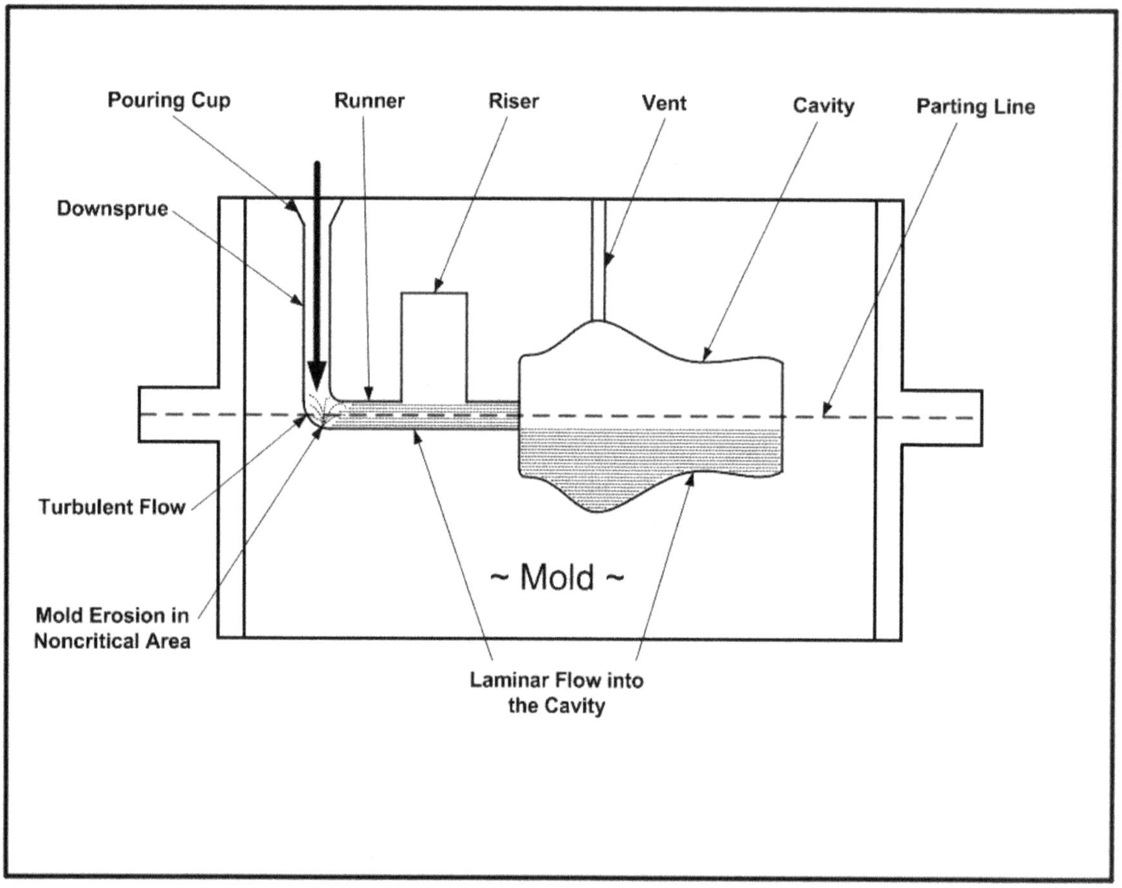

Figure 5.3 Good Mold Design Filling (Issues Resolved)

5.2.2 Mold Types

All molds contain one common element, the mold cavity. This is the area that the molten material will be poured into, creating the shape. Various molds will have different types and configurations of the cavity, but they will all contain one. Depending on the type of molding process being used, the cavity can either be one side in the mold or be split in two where each side of the mold will contain a portion of the cavity. A mold can be either an open mold or a closed mold. The difference between these two types is with an open mold, the cavity is exposed and the molten metal is poured directly into it until filled. With a closed mold, the cavity is contained between two mold halves and the molten material is poured into a chamber that leads to the cavity.

Molds can have a variety of features in their construction and usage. Depending on the casting process being utilized, molds will have a combination consisting of these five features.

5.2.2.1 Permanent Mold

The mold being used to create the casting can be used over and over. It is permanently constructed of hardened steel. This type of mold is used for very high production and is somewhat limited to the configuration of the parts.

5.2.2.2 Expendable Mold

Here, the mold gets destroyed every time it is used and a new mold must be constructed in order to make a second part. The molds can be built out of a variety of materials, but sand and ceramic are the most common. This type of casting is used for both prototyping and production. Even though this may sound like a very inefficient method for casting parts, the processes can be highly automated and some part configurations that are not possible with a permanent mold can be achieved.

5.2.2.3 Permanent Pattern

With expendable molds, a pattern is required to create the cavity for casting the parts. Some patterns are constructed of wood or steel and can be reused over and over to build the expendable molds. The pattern is constructed and the expendable molding material is placed around it, then the pattern is removed, and the mold is ready to be poured with the molten metal.

5.2.2.4 Expendable Pattern

In this condition, the pattern is constructed of a material such as wax or a type of plastic and gets destroyed every time an expendable mold is created. If, say, 100 parts are required, then 100 expendable patterns would be required. Here, an expendable mold is constructed and then the pattern would get melted out of the cavity area before the mold is poured with the molten metal.

5.2.2.5 No Pattern

A 'no pattern' condition is utilized in permanent molds where the configuration of the part is cut into the hard steel of the mold. The geometry of the part is taken from the part prints and then machined into the mold with no pattern being constructed. The processes that utilize a "No Pattern" condition are very highly automated and used for high production of parts in the process.

5.2.2.6 Open Molds

An open mold is one where the cavity is opened and visible to the operator. The molten metal is then poured directly into the mold filling the cavity. Once the molten material reaches the top of the cavity, the mold is filled and allowed to cool. Figure 5.4 shows a typical open mold structure. Some of these molds are permanent where the cavity is machined into a steel block, while others are expendable molds where a pattern is imbedded into sand and the mold is destroyed when retrieving the part.

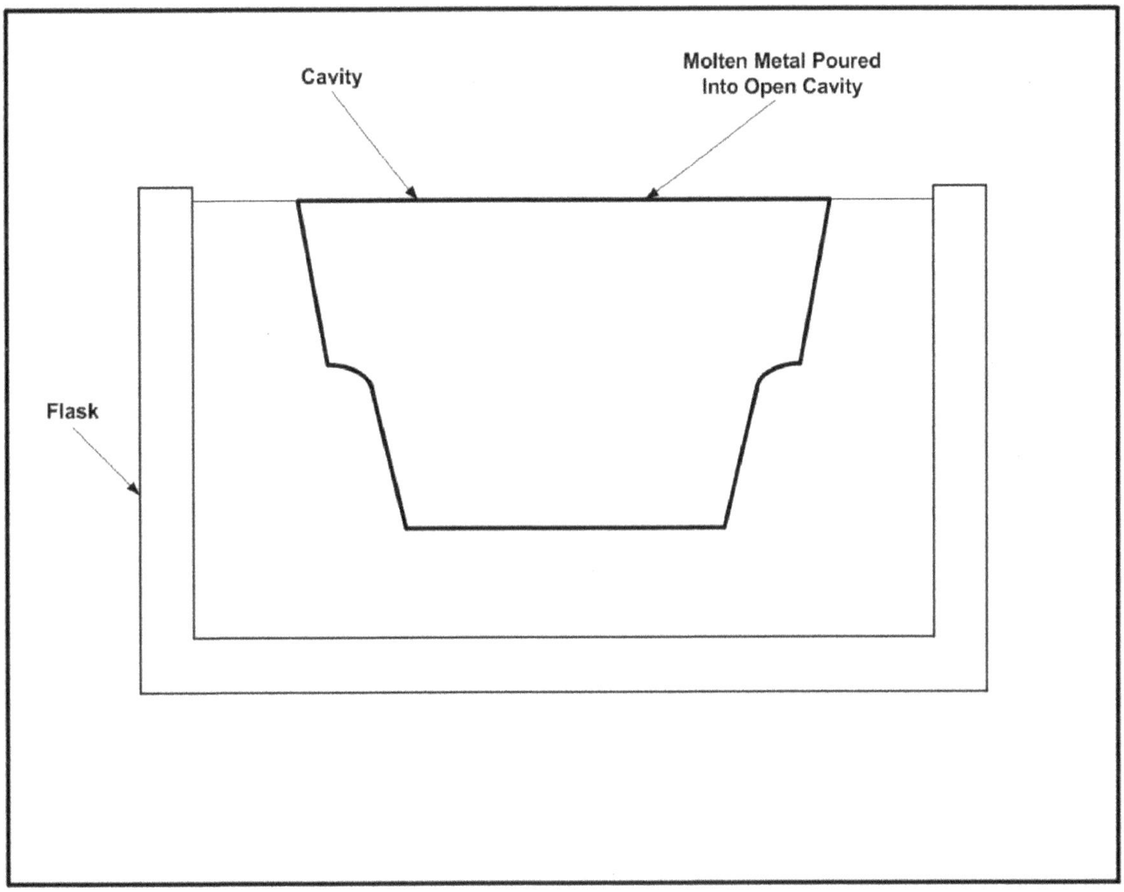

Figure 5.4 Open Mold Design

Open molds are used when creating parts that have a very special configuration where one side of the parts is flat. For example, the open molding process is used when pouring gold bars where the length and sides are shaped, but the bottom of the bars remains flat. The amount of shrinkage is very closely calculated so that the final weight of the bars is very accurate.

5.2.2.7 Closed Molds

A closed mold is one where there are two sections assembled and the cavity creating the part is contained in the center. This type of mold can be either permanent or expendable and may or may not have a required pattern. Figure 5.5 shows a typical closed mold construction, which has several other components to it as opposed to an open mold. These components are as follows:

Pouring Cup: The pouring cup is a funneled out area on top of the downsprue. The molten metal is poured into this area to fill the mold.

Downsprue: The downsprue is what flows the molten material down and into the runner.

Runner: The runner turns the molten metal 90° and flows it into the mold in a smooth laminar flow.

Riser: Some molds will not have a riser in them. The function of the riser is to act as a reserve of material so that when the part cools, additional material can be drawn from the riser. Also, the riser will capture some of the gasses that will be produced during the casting process.

Vent: The vent is necessary to allow the air to escape from the cavity when the material flows into it.

Parting Line: This is where the two sections of the closed mold mate together. After the mold is poured and the material has solidified, the two sections are disassembled and the finished part is retrieved.

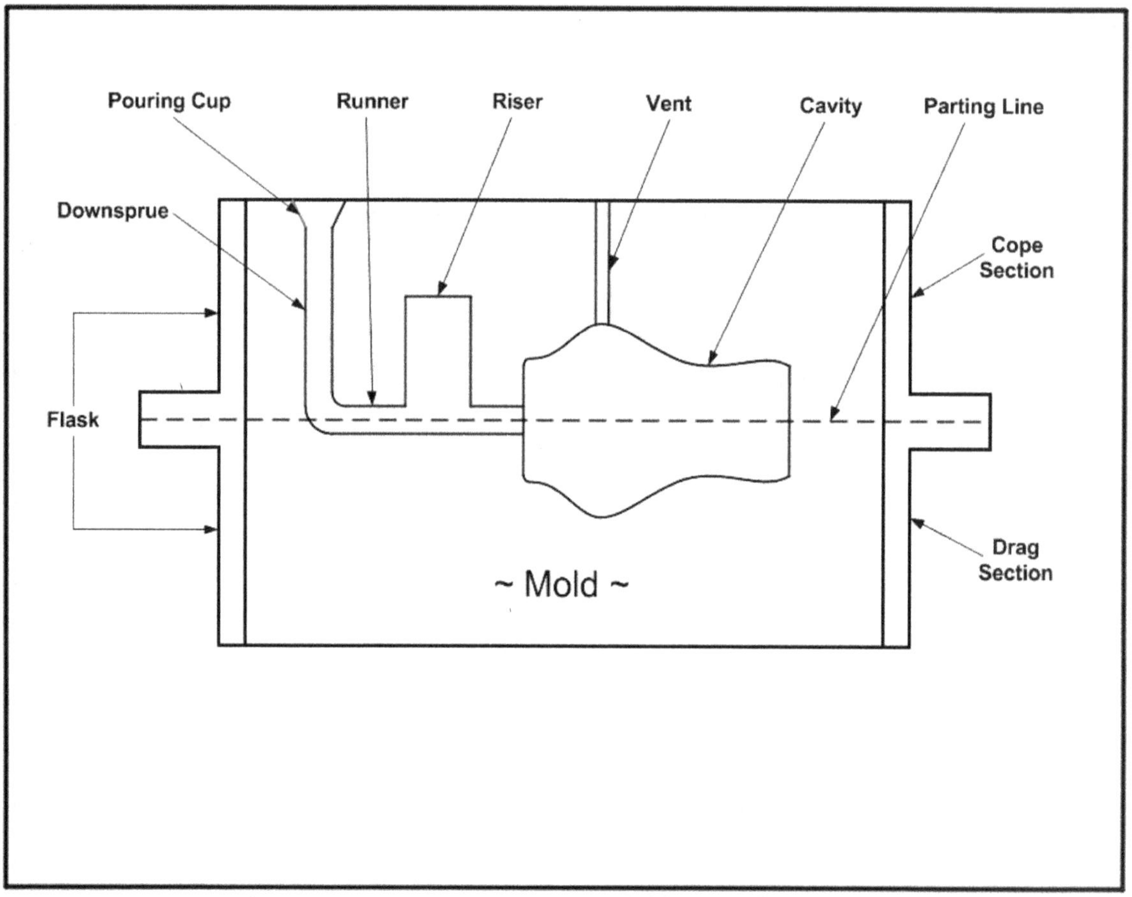

Figure 5.5 Closed Mold Design

5.3 MATERIAL BEHAVIOR DURING CASTING

Understanding how molten metal behaves is critical in calculating what is required for the casting process. From a Superheated melt, through the pouring step, to the solidification step, the material will pass through various changes that need to be considered in order to produce an acceptable part at the end. The factors that need to be considered when calculating a casting pour are: *Shrinkage, Flow Velocity, Volumetric Flow Rate, Mold Filling Time*, and *Total Solidification Time.*

5.3.1 Material Shrinkage

With all metals, there are two different stages of shrinkage that need to be accounted for during the process. These are *Solid Solidification* and *Solid Thermal Contraction*, and both occur at a different time in the process. Figure 5.6 shows the cooling curve from when the material is poured from the furnace into the mold. You will notice that from the pouring temperature, the cooling is at a rapid rate, which is why it needs to be superheated so that solidification will not start during the pouring step. Once it reaches what is called the freeze point, the material will begin to solidify until it becomes solid. This is where the Solid Solidification will occur and the metal will shrink by a given percentage. From that point, the solid metal in the cavity is still extremely hot (slightly below the melting temperature), and as it cools, the metal will begin to contract until it reaches room temperature. This second stage of shrinkage is called Solid Thermal Contraction. Each material has a different percentage factor for solid solidification and solid thermal contraction based on the properties of the material and the freeze point of the material. Figure 5.7 shows these values for some selected materials to be used in the calculations of overall shrinkage.

The shrinkage process happens over time and is displayed in Figures 5.8–5.11. Figure 5.8 shows an open mold poured until it is filled to the top. In Figure 5.9, you see the molten material beginning to solidify from the outside of the part while the center remains a liquid. Solid Solidification begins, and the top of the liquid lowers. Figure 5.10 shows the solidification continuing and further shrinkage is noticed at the top of the part that is now a solid. However, the center of the part has not fully solidified and a shrinkage cavity begins to appear.

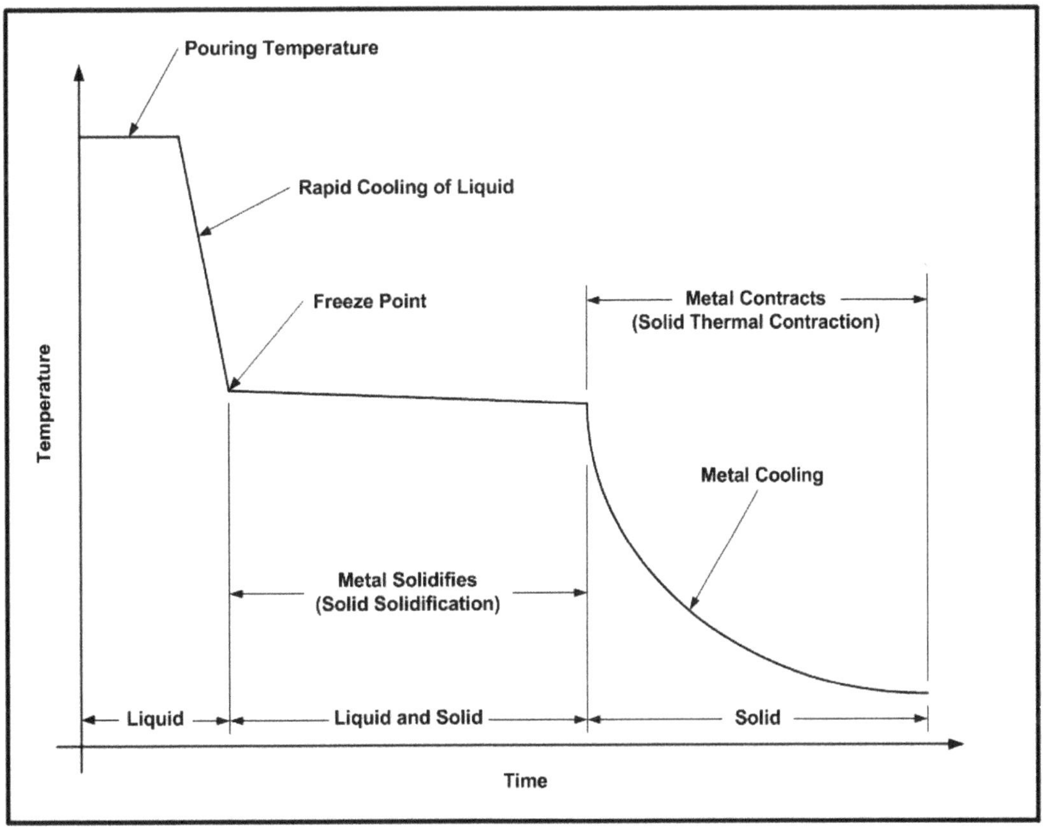

Figure 5.6 Material Cooling Curve Showing Solidification and Contraction

Volumetric Shrinkage and Contraction for Some Metals		
Metal	**Solidification Shrinkage (%)**	**Solid Thermal Contraction (%)**
Aluminum	7.0	5.6
Al alloy (typical)	7.0	5.0
Gray Cast Iron	1.8	3.0
Gray Cast Iron, High C	0.0	3.0
Low C Cast Steel	3.0	7.2
Copper	4.5	7.5
Bronze (Cu-Sn)	5.5	6.0

Figure 5.7 Solid Solidification and Solid Thermal Contraction Values

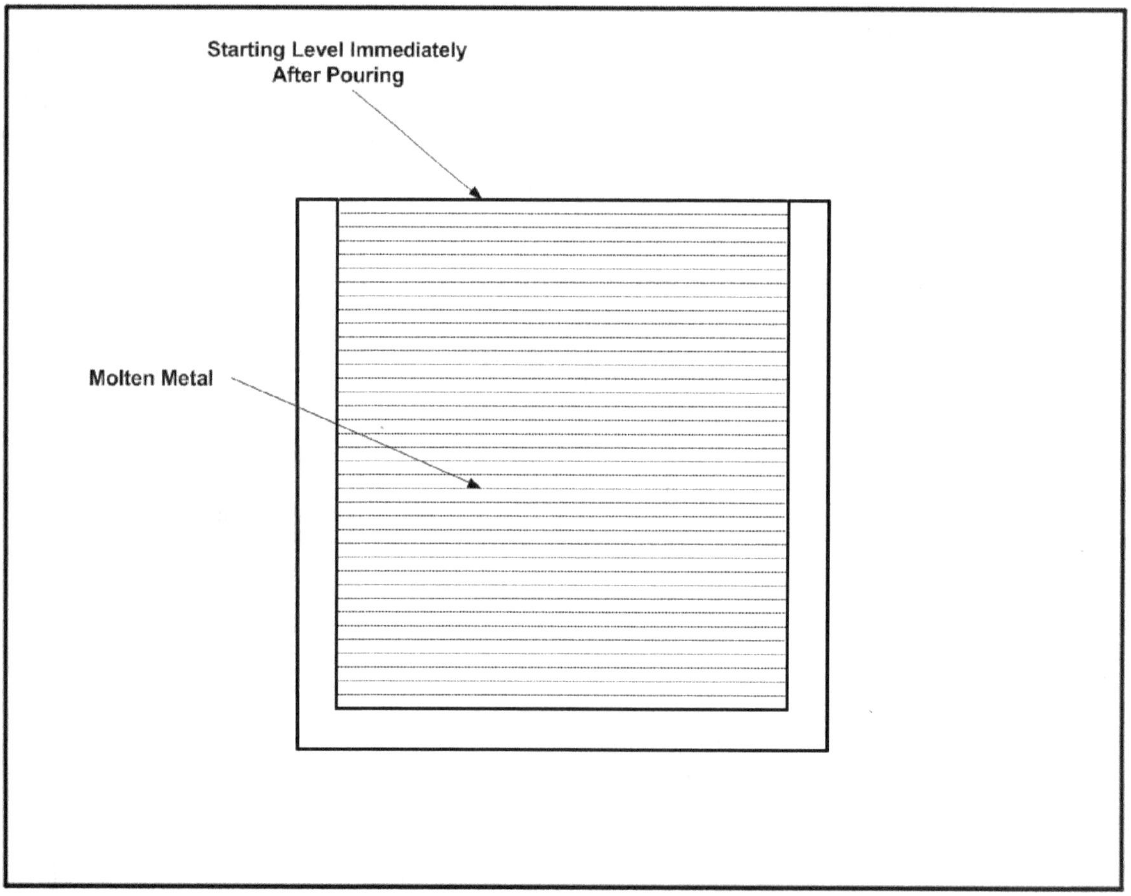

Figure 5.8 First Stage in Material Shrinkage (Initial Pouring)

Figure 5.11 is the final stage of shrinkage where the part is fully solidified, and as it cools to room temperature, it contracts due to solid thermal contraction. This occurs three-dimensionally on the part as the figure notes that it is noticeable on the top and the sides of the part in the mold cavity.

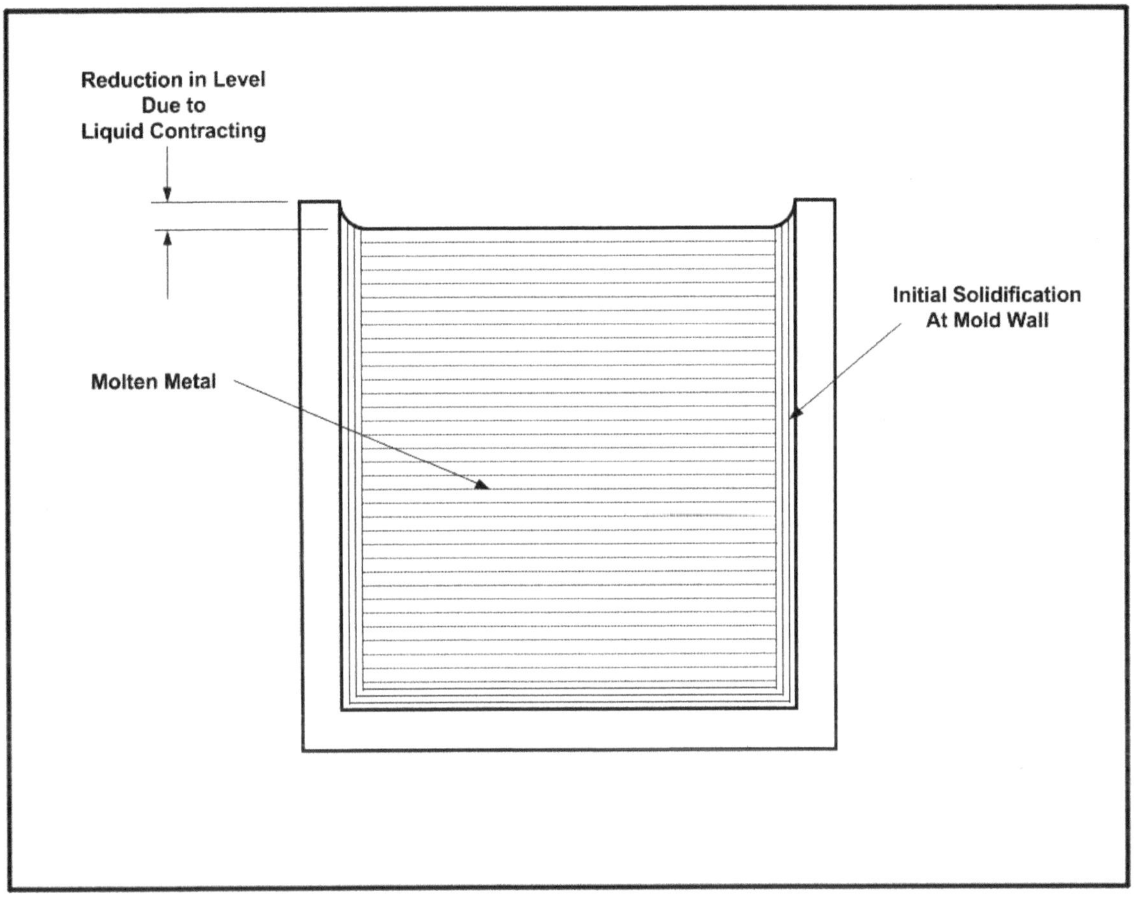

Figure 5.9 Second Stage in Material Shrinkage (Solidification Beginning)

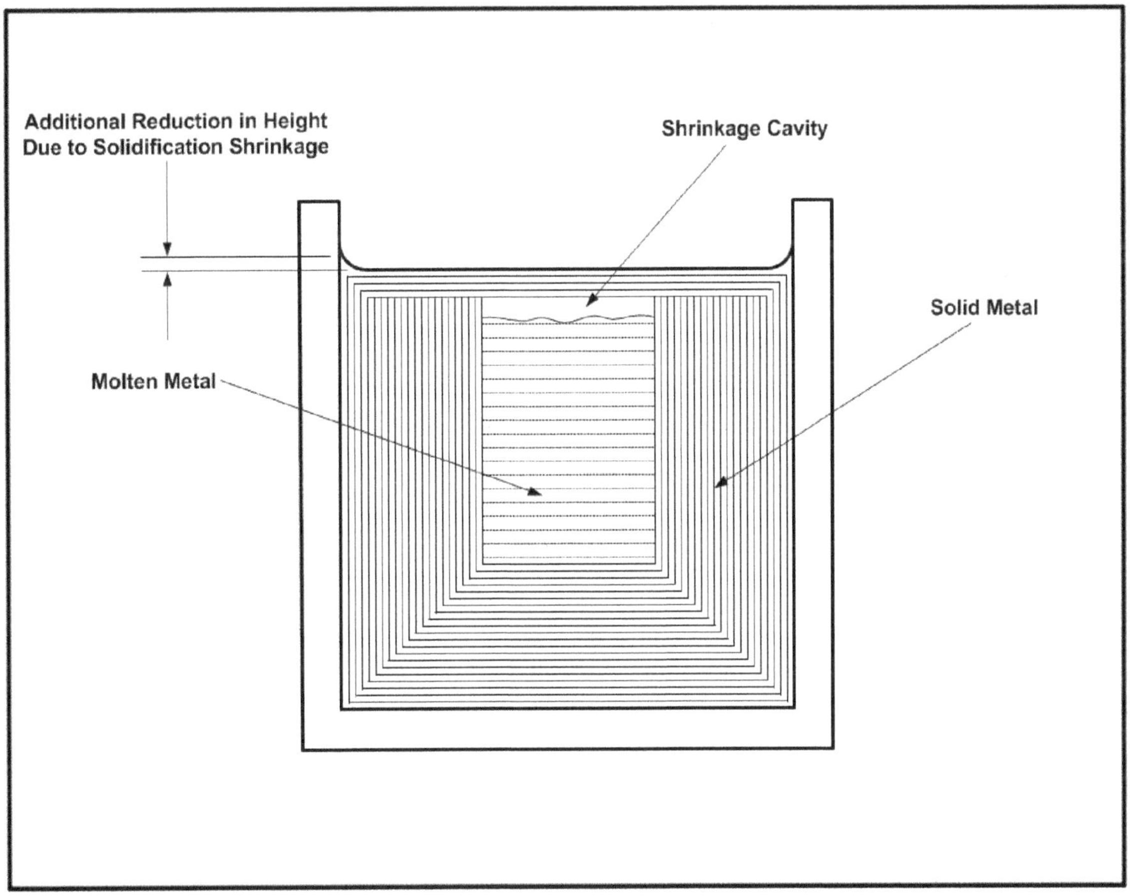

Figure 5.10 Third Stage in Material Shrinkage (Molten Center with Cavity Shrinkage)

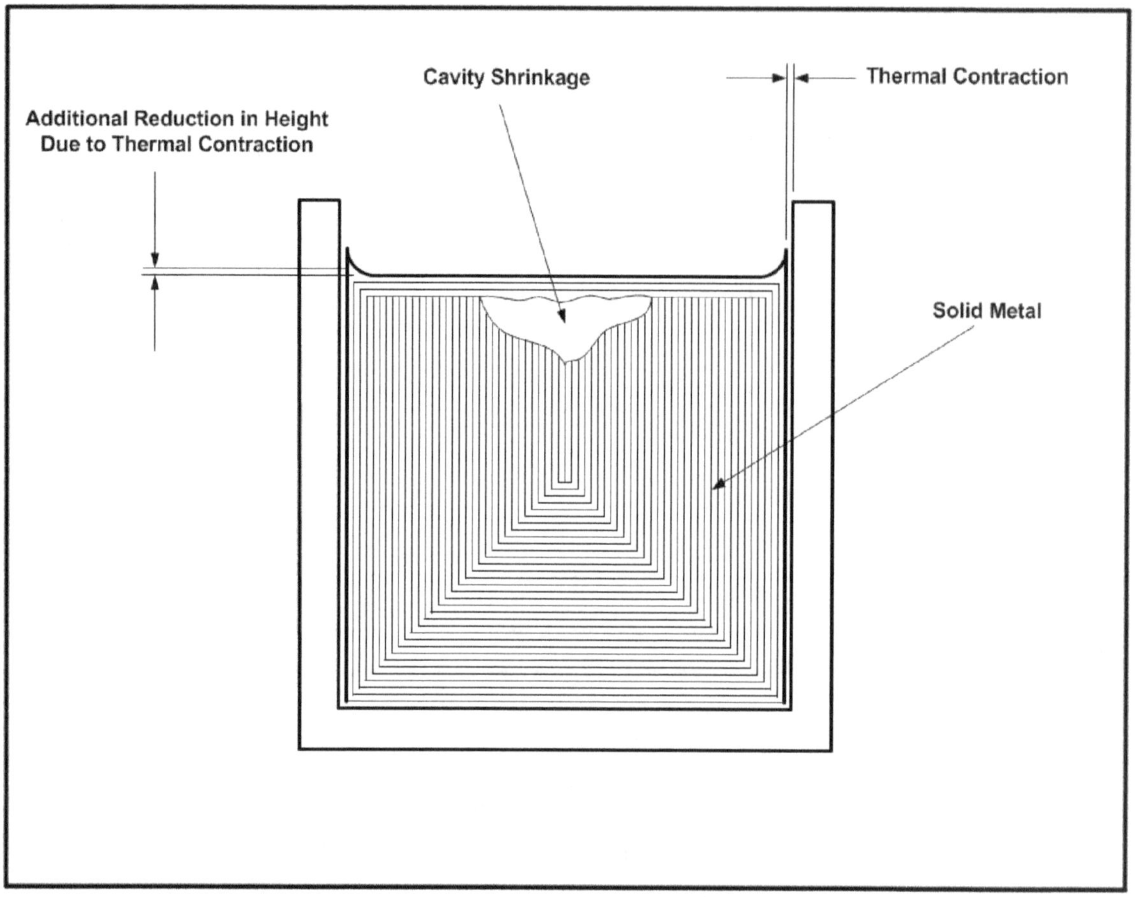

Figure 5.11 Final Stage in Material Shrinkage (Solid Piece with Cavity Shrinkage)

Calculating shrinkage is a three-dimensional value to account for the material shrinking in all three directions. It is calculated by the following equation:

$$S_v = [(1-SS)(1-STC)]^{1/3}$$

S_v = Shrinkage Value
SS = Solid Solidification Value
STC = Solid Thermal Contraction Value

Example: If were we to calculate the shrinkage value for copper, from the table in Figure 5.7, our Solid Solidification is 4.5% and the Solid Thermal Contraction is 7.5%. So, those values would be applied into the equation as follows:

$$S_v = [(1-SS)(1-STC)]^{1/3}$$
$$S_v = [(1-0.045)(1-0.075)]^{1/3}$$
$$S_v = 0.959507$$

This value would be applied to the cavity against the desired part dimensions to determine how much larger the cavity needs to be in order to produce a part to the desired size. Likewise, if the cavity sizes are known, this value is applied to determine what the size of the part would be.

5.3.2 Flow Velocity

The Flow Velocity determines the velocity at which the material will be entering the cavity of the mold. Knowing the velocity of the material entering the mold is one of the steps required in calculating the optimum time required to fill the mold. This is calculated by the following equation:

$$\upsilon = \sqrt{2gh}$$

v = Flow Velocity: in./s; cm/s
g = Gradational Acceleration: 386 in./s^2; 981 cm/s^2
h = Downsprue Height

Example: If the downsprue has a height of 4.500 in. from the mold cavity, it is applied in the equation as follows:

$$\upsilon = \sqrt{2gh}$$

$$v = \sqrt{(2)(386)(4.500)}$$
$$v = 58.94065 \text{ in./s}$$

This informs us of the velocity at which the molten metal will be entering the mold from the distance it needs to travel to reach the cavity or the bottom of the downsprue.

5.3.3 Volumetric Flow Rate

Once the Flow Velocity is calculated, the Volumetric Flow Rate is calculated to determine how much material can enter the mold under optimum conditions. This is determined by the cross-sectional area of the downsprue and the flow velocity. By knowing this, the maximum amount of molten metal that can be filling the mold is determined. The equation for this calculation is as follows:

$$Q = vA$$

Q = Volumetric Flow Rate: in.^{3}/s ; cm^3/s
A = Cross-Sectional Area of the Downsprue: in.^{2}; cm^2
v = Flow Velocity: in./s; cm/s

Example: If the downsprue has a diameter of 0.750 in. and uses the same Flow Velocity as that calculated, it is applied in the equation as follows:

Step 1: Cross-Sectional Area

$$A = \pi r^2$$
$$A = \pi(0.750/2)^2$$
$$A = 0.441786 \text{ in.}^2$$

Step 2: Volumetric Flow Rate

$$Q = vA$$
$$Q = (58.94065)(0.441786)$$
$$Q = 26.03915 \text{ in.}^3\text{/s}$$

This informs us of the volume of molten metal that can enter the cavity of the mold.

5.3.4 Mold Filling Time

Filling the mold as quickly as possible is one of the four important filling requirements covered earlier. The material will begin to cool rapidly once it leaves the furnace, so knowing what the optimum mold fill time is would be an important factor. This is determined by the Volumetric Flow Rate, which informs us of the maximum volume of material that can enter the mold and the Volume of the cavity being filled. The equation for this calculation is as follows:

$$MFT = V/Q$$

MFT = Mold Fill Time: s
V = CAVITY Volume: in.^{3}; cm^3
Q = Volumetric Flow Rate: in.^{3}/s; cm^3/s

Example: A rectangular part is being cast and the CAVITY dimensions are L = 4.250 in., W = 3.750 in., H = 1.625 in. Using the calculated Volumetric Flow Rate, it is applied in the equation as follows:

Step 1: CAVITY Volume

$$V = (L)\,(W)\,(H)$$
$$V = (4.250)\,(3.750)\,(1.625)$$
$$V = 25.89844 \text{ in}^3$$

Step 2: Mold Fill Time

$$MFT = V/Q$$
$$MFT = 25.89844/26.03915$$
$$MFT = 0.994596 \text{ s}$$

From the information on the mold design, we now know the optimum time that would be required to fill the mold.

5.3.5 Solidification Time

After the mold has been filled, there is a time required for the cast part to solidify in the cavity. This is important so the mold will not be opened too soon while the part is still solidifying and so there will be minimal time wasted in waiting for the mold to be opened. Efficiency is a critical part of all manufacturing processes, and the faster the various processes and operations can be performed, the less downtime there will be. This is determined by the Volume of the mold cavity, the Surface Area of the mold cavity, the mold constant, and the exponential factor. The Mold Constant and the exponential factor are determined by the construction of the mold and the temperature of the molten metal being poured into it. The equation for this calculation is as follows:

$$TST = C_m(V/A_s)^n$$

TST = Total Solidification Time: min

C_m = Mold Constant: min/in; min/cm

V = CAVITY Volume: in.^{3}; cm^3

A_s = CAVITY Surface Area: in.^{2}; cm^2

n = Exponential Factor.

Example: Using the dimensions of the cavity stated and the calculated cavity volume, if our mold constant (C_m) is 12.798 min/in.^{2} with an exponential factor (n) of 2.35, the process for the solidification time is as follows:

Step 1: Cavity Surface Area

$$A_s = 2[(L \times W) + (L \times H) + (W \times H)]$$
$$A_s = 2[(4{,}250 \times 3.750) + (4.250 \times 1.625) + (3.750 + 1.625)]$$
$$A_s = 57.8750 \text{ in.}^2$$

Step 2: Solidification Time

$$TST = C_m(V/As)^n$$
$$TST = 12.793(25.89844/57.8750)^{2.35}$$
$$TST = 1.933355 \text{ min}$$

Once the solidification time has been determined, the operator will know how long they need to wait before opening the mold and retrieving the part from the cavity or breaking the mold apart if an expandable mold is being used.

Example Problem

A casting is to be produced according to the print shown in Figure 5.12. The print shows the FINISHED SIZES desired from the casting and the process will have a Mold Constant (Cm) of 23.835 min/in.^{2} with an Exponential Factor (n) of 1.82. The casting will be made of Low C Cast Steel. Determine the following:

- Shrinkage Factor
- Cavity Sizes
- Flow Velocity
- Volumetric Flow Rate
- Mold Fill Time
- Solidification Time

Step 1: Determine the Solidification Shrinkage and Solid Thermal Contraction values from the table in Figure 5.7 for Low C Cast Steel.

$$SS = 3.0\% \qquad STC = 7.2\%$$

Step 2: Calculate the Shrinkage Value.

$$S_v = [(1-SS)(1-STC)]^{1/3}$$
$$S_v = [(1-0.030)(1-0.072)]^{1/3}$$
$$S_v = 0.9655$$

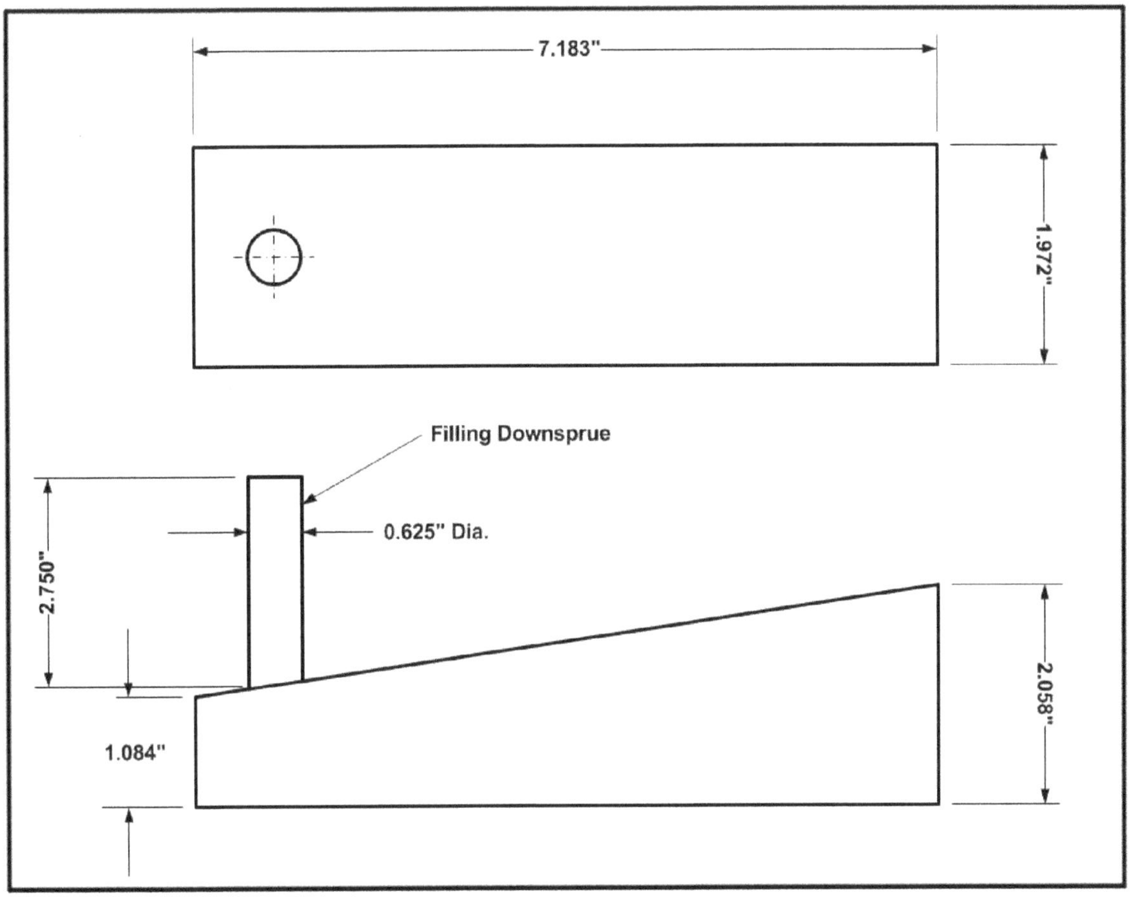

Figure 5.12 Example Problem Information

Step 3: Calculate the CAVITY Dimensions.

$$L = 7.183/0.9655 = 7.4393 \text{ in.}$$
$$H_1 = 1.084/0.9655 = 1.1227 \text{ in.}$$
$$H_2 = 2.058/0.9655 = 2.1314 \text{ in.}$$
$$W = 1.972/0.9655 = 2.0434 \text{ in.}$$

Step 4: Calculate the Flow Velocity.

$$\upsilon = \sqrt{2gh}$$
$$v = \sqrt{(2)(386)(2.750)}$$
$$v = 46.0760 \text{ in./s}$$

Step 5: Calculate the Volumetric Flow Rate.

Step 5a: Calculate the Cross-Sectional Area of the Filling Downsprue.

$$A = \pi r^2$$
$$A = \pi(0.625/2)^2$$
$$A = 0.3068 \text{ in.}^2$$

Step 5b: Calculate the Volumetric Flow Rate.

$$Q = Av$$
$$Q = (0.306796)(46.07602)$$
$$Q = 14.1359 \text{ in.}^3/\text{s}$$

Step 6: Calculate the Mold Fill Time.

Step 6a: Calculate the CAVITY Volume.
See Figures 5.13 and 5.14 for Section Breakdown

$$V_1 = 2(L \times H) + 2(W \times H) + (L \times W)$$
$$V_1 = 2(7.4393 \times 1.1227) + 2(2.0434 \times 1.1227) + (7.4393 \times 2.0434)$$
$$V_1 = 17.0578 \text{ in.}^3$$
$$V_2 = L(H2 - H1)/_2 \times W$$
$$V_2 = 7.4393 \times (2.1314 - 1.1267)/2 \times 2.0424$$
$$V_2 = 7.6634 \text{ in.}^3$$

$$V_t = V_1 + V_2$$
$$V_t = 17.0578 + 7.6634$$
$$V_t = 24.7212 \text{ in.}^3$$

Step 7: Calculate the Mold Fill Time.

$$MFT = Vt/Q$$
$$MFT = 24.7212/14.1359$$
$$MFT = 1.7488 \text{ s}$$

Step 8: Calculate the Solidification Time.

Step 7a: Calculate the CAVITY Surface Area.
See Figures 13 and 14 for Section Breakdown

$$As_1 = [(L \times H_1) + (W \times H_1)]\ 2 + (L \times W)$$
$$As_1 = [(7.4393 \times 1.1227) + (2.0424 \times 1.1227)]\ 2 + (7.4393 \times 2.0424)$$
$$As_1 = 36.4836 \text{ in.}^2$$

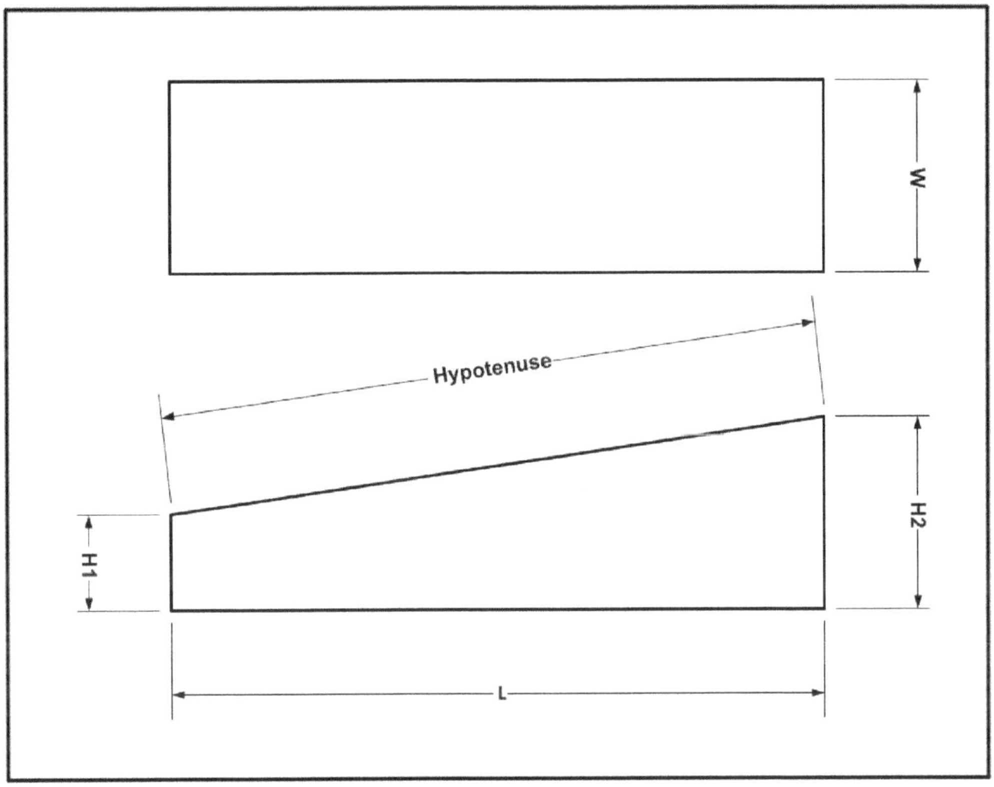

Figure 5.13 Volume and Surface Area Breakdown (1)

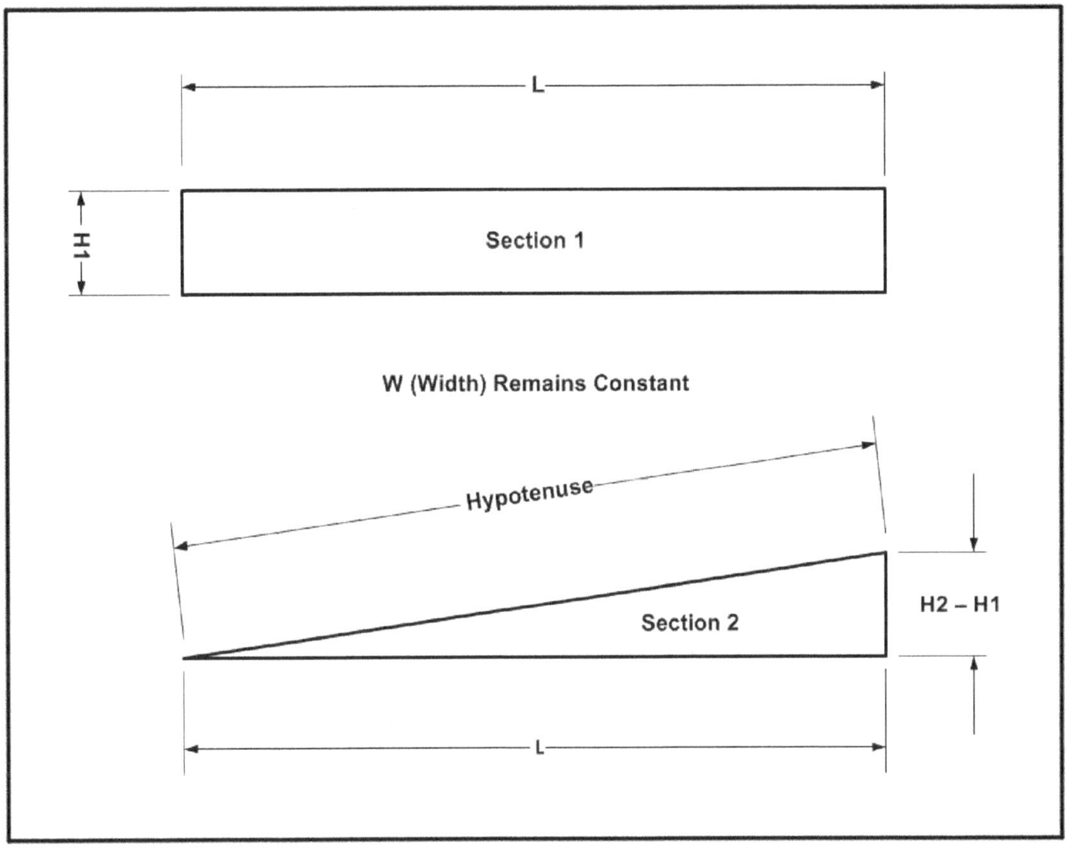

Figure 5.14 Volume and Surface Area Breakdown (2)

$As_2 = [L \text{ x } (H2–H2)/2 \times 2] + [W \times (H_2–H_1)]$
$As_2 = [7.4393 \times (2.1314–1.1227)/2 \times 2] + [2.0424 \times (2.1314 – 1.1227)]$
$As_2 = 9.5647 \text{ in.}^2$

$As_3 = W \times [L^2 + (H_2–H_1)^2]^{0.5}$
$As_3 = 2.0424 \times [7.4393^2 + (2.1314 – 1.1227)^2]^{0.5}$
$As_3 = 15.3328 \text{ in.}^2$

$As_t = As_1 + As_2 + As_3$
$As_t = 36.4836 + 9.5647 + 15.3328$
$As_t = 61.3811 \text{ in.}^2$

Step 8b: Calculate the Solidification Time.
$TST = C_m(V/As_t)^n$
$TST = 23.835(24.7212/61.3811)^{1.82}$
$TST = 4.5538 \text{ min}$

5.4 CASTING DEFECTS

As in any manufacturing process, defects in the parts can and will occur. Even under extremely controlled conditions, things can go wrong or variables in the process can cause defects in the process. Some of the most common defects that can occur during the casting process are listed as follows.

Cold Shot: This defect is mainly caused by the design of the mold itself. As the material enters the mold improperly, it splashes abound in a turbulent manner, causing solid droplets of the molten metal to harden, get entrapped in the cavity, and solidify before the remainder of the material does. A proper filling method, which provides a smooth laminar flow into the cavity, would reduce or eliminate this defect.

Cold Shut: This defect occurs when the filling material flows together in the cavity, but no fusion occurs. Some molds get filled from more than one location, or the material flows around a core in the mold. This is mainly caused from the molten metal not being heated enough or the mold filling time is too slow, and as it flows together, it reaches the freeze point and does not fuse together.

Misruns: This defect occurs when the mold does not fill completely. There are a few reasons that can cause this. If the material is not hot enough, it will solidify during the pouring step, resulting in the cavity not filling all the way. Also, if the pouring rate is too slow, the material can begin to solidify before the cavity is completely filled.

Porosity: Porosity is small air pockets or bubbles that are trapped in the casting. For the most part, porosity in the casting process is unavoidable but can be minimized or isolated in a noncritical portion of the part. By utilizing proper venting and filling of the cavity, this defect can be greatly reduced. Also, with the venting and filling locations, the porosity can be isolated in a noncritical area of the part.

Shrinkage: In every casting process, shrinkage will occur. What is desired is uniform shrinkage throughout the part. Nonuniform shrinkage can be caused by producing a casting where the casting size varies. What can happen is that a depression will form on the top surface of the part, which is commonly called a *Sink Hole*. As the material shrinks, the thicker portions of the casting will take longer and cause this depression to appear. By maintaining all the sections of a casting uniform in thickness, this defect can be reduced.

Hot Tearing: This defect is primarily caused by a combination of the overall design of the part and the design of the mold. What happens is that as the part solidifies and contracts in the mold, it is contained in the cavity, and as it is removed, the sides of the part are scratched and torn. Some of the causes for this are nonuniform wall thickness in the part and too small draft angle on the sides of the part.

5.5 CASTING DESIGN CONSIDERATIONS

If a part is going to be designed, which will be produced in a casting process, there are some guidelines that need to be followed in order to ensure an acceptable part. By following the design considerations listed as follows, the component part that you are designing will have a much higher chance of being produced defect free and the mold being used would be far less expensive as well. Figure 5.15 shows an incorrect design for a part to be cast, while Figure 5.16 demonstrates a correct casting designed part. These rules are listed as follows for designing a casting part.

Geometric Simplicity: Even though very complex castings are produced, it is always a good rule of thumb to keep the geometry as simple as possible. Very complex part designs will result in a very costly mold or mold pattern and can lead to more defects during the process. If a complex part is required, it may be possible to break the design into two or three less complicated parts, which can be assembled together. Complexity will lead to issues in the casting process and result in a more costly part overall.

Uniform Wall Thickness: As we know, shrinkage will always occur in a casting as the material solidifies and contracts. By maintaining all the walls and sections of the casting design uniform, the shrinkage will also be uniform throughout the part. If additional strength is required in certain areas of the casting, a strengthening rib of the same wall thickness of the entire casting can be implemented in that area. See Figure 5.17 for a strengthening rib design.

Draft Angles: A part designed for a casting process, or a pattern that will be used to create an expendable mold, must be able to be removed without damaging the mold or the part itself. Draft angles are incorporated to allow for ejection of the part or pattern. The sides of the part or pattern should contain between 2° and 3° angle at a minimum to allow this to occur. If a larger angle can be incorporated, it should. Also, if

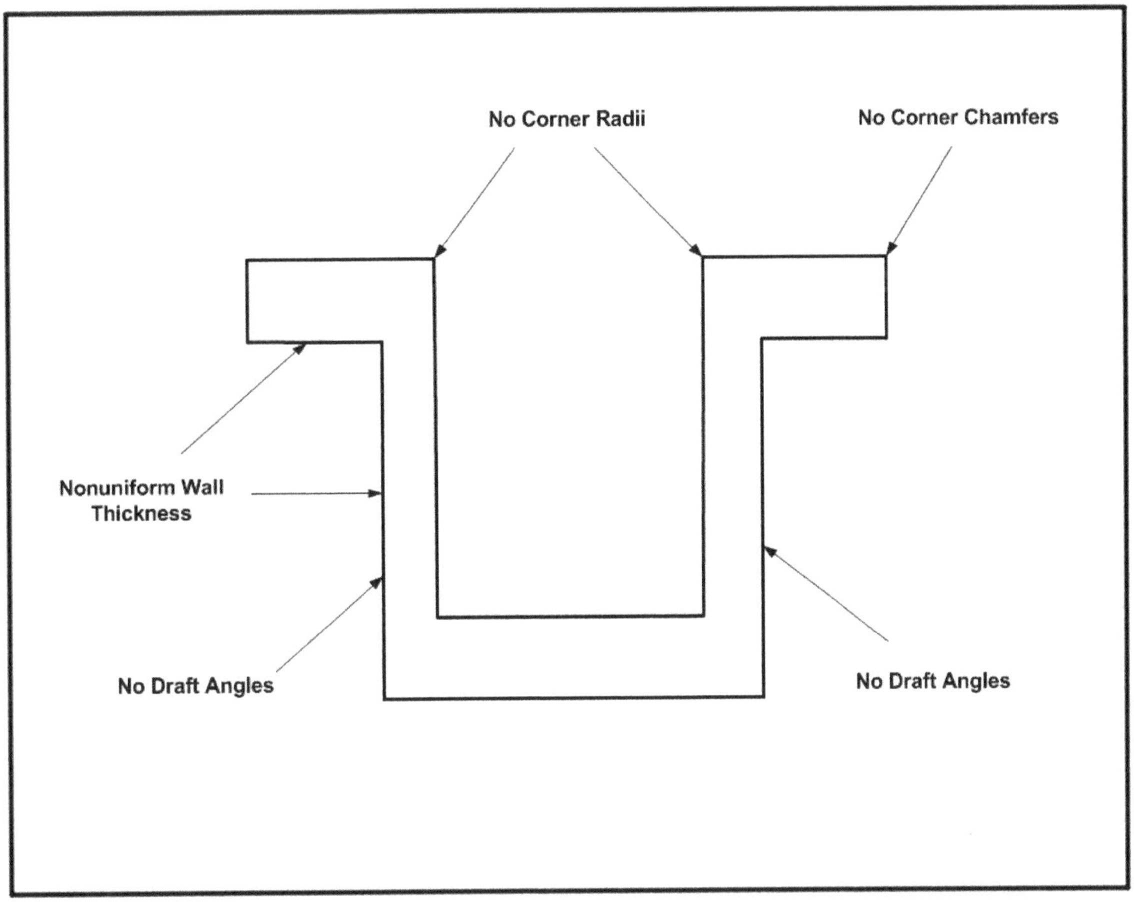

Figure 5.15 Incorrect Casting Part Design

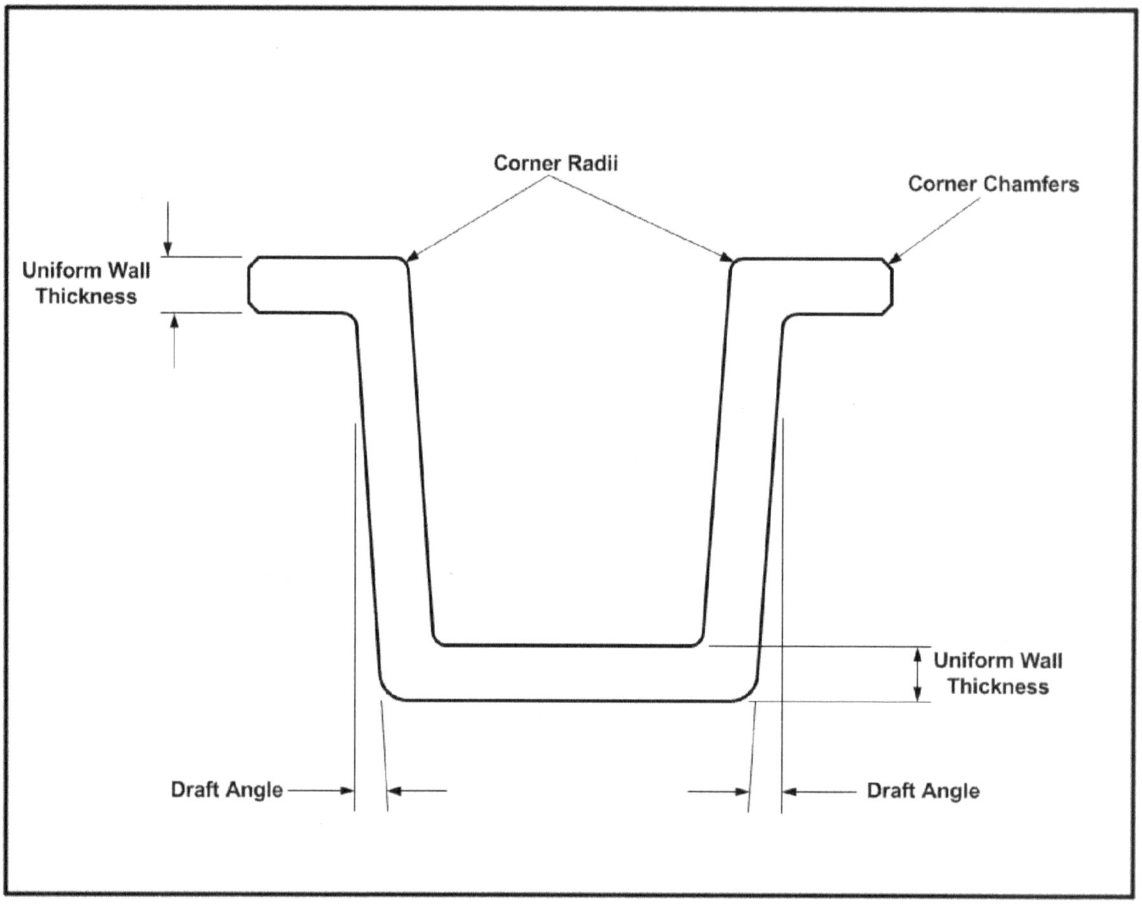

Figure 5.16 Correct Casting Part Design

a straight wall on the casting side is required, it needs to be machined in later. Figure 5.15 shows the draft angle on the sides of the cast part design.

Corners: On a casting design, all the corners need to contain either a chamfer or a radius. Molten metal does not flow into a sharp corner very well, and a sharp corner does not allow for part or pattern ejection well. As with a draft angle, the larger the chamfer or radius on the part, the better it is. If sharp 90° corners are required, they need to be machined in later.

Tolerances and Finishes: Every casting process is capable of producing parts to a given tolerance and surface finish. Knowing what every process is capable of repeatedly is important when designing a part for a particular process, so acceptable parts can be attained. If closer tolerances or surface finishes are required, they must be machined in after the part has been cast.

Machining Allowances: Adding additional material to the casting design to allow for secondary machining later is important if closer tolerances, surface finishes, or configurations are required. Secondary operations are very common on cast parts, and allowing enough material to accommodate this is normal when designing the cast part. The finished functional sizes, configurations, and surface finishes are evaluated, and the casting design incorporates the additional material to be removed in a secondary operation.

Use of Cores: Cores are inserts placed in the mold cavity to allow for either internal structures or large holes in the casting. These are very common in the sand casting process. These cores are inserted into the mold after the cavity has been established and removed from the part after casting. An example of where cores are used is in the sand casting process producing engine blocks. Cores are inserted after the main configuration of the block is made for the piston holes and the water jackets, which are an internal structure. After the part is removed, the cores are removed (which, in the sand casting process, are expendable), allowing for the piston holes to be machined and the internal water jackets to be functional.

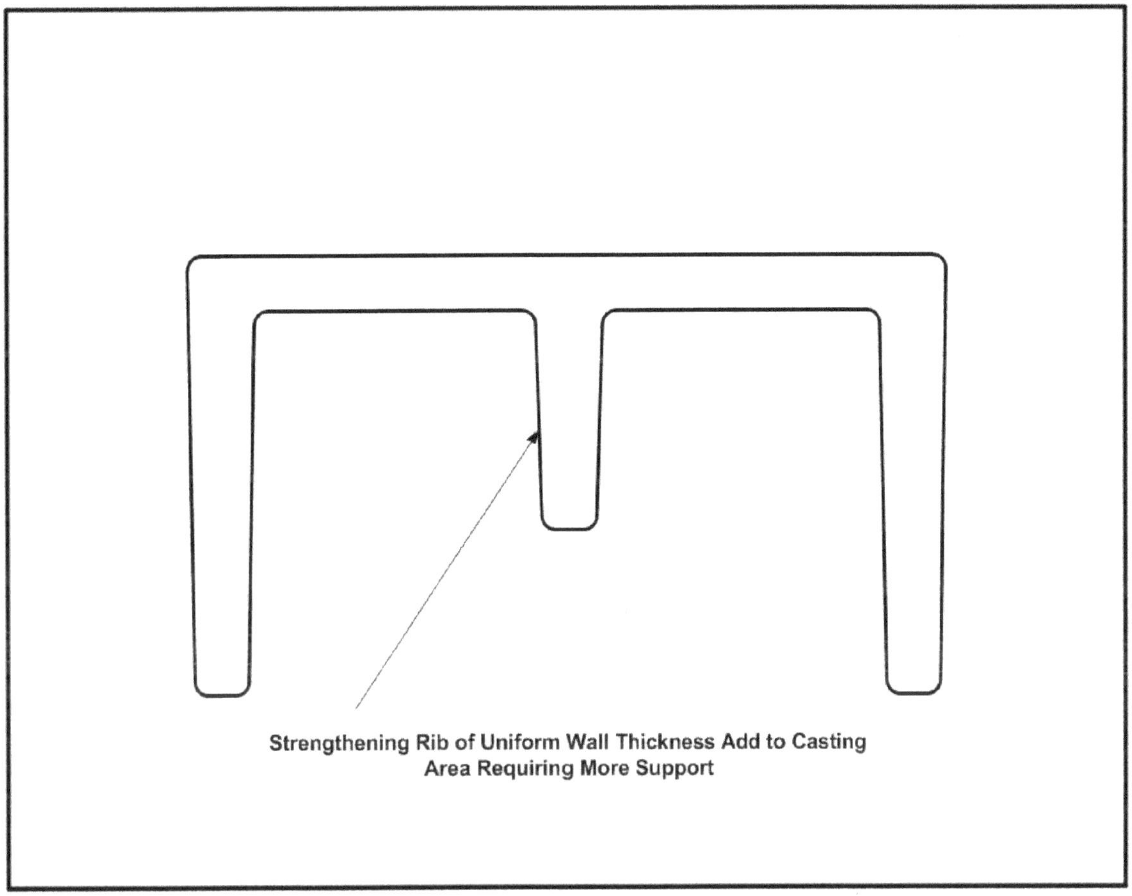

Figure 5.17 Strengthening Rib Design in a Casting Requiring More Strength in a Critical Area

5.6 CASTING PROCESSES

Castings are produced in variety of processes, each having both advantages and disadvantages that need to be understood before selecting a process. Depending on the process, they will have a combination of the following in the mold as mentioned earlier: Permanent Mold, Expendable Mold, Permanent Pattern, Expendable Pattern, or No Pattern. Four of the more common casting processes are discussed in this section.

5.6.1 Sand Casting Process

The sand casting process involves construction of a mold from sand. This sand is a mixture of fine sand and oil. The oil in the sand acts as a binder material, which will hold the sand together when it is packed around a pattern. The process steps are as follows:

Step 1: The mold is constructed consisting of cope (top) and drag (bottom) sections with a pattern plate between them. See Figure 5.18.

Step 2: The sand/oil mixture is packed firmly around the pattern in both the cop and drag sections around the pattern plate. See Figure 5.19.

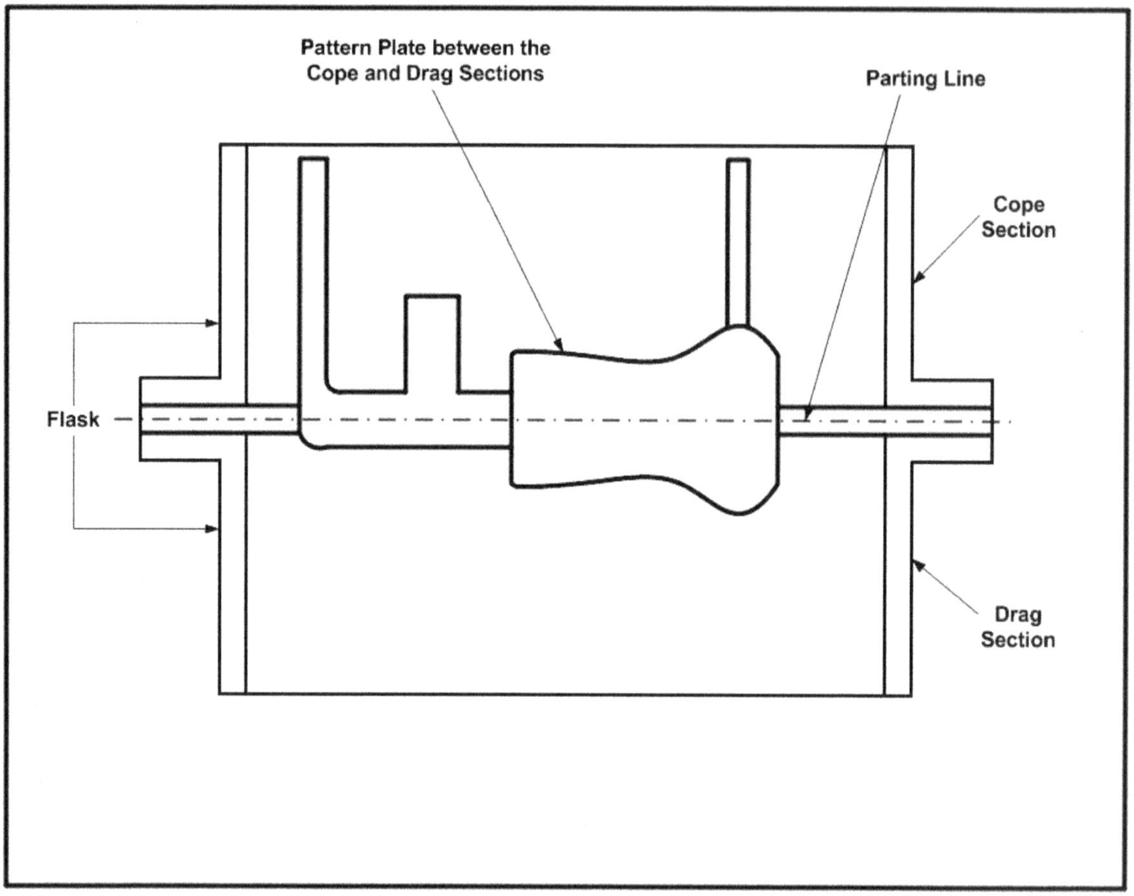

Figure 5.18 Step 1 for a Sand Casting: Pattern Plate Is Assembled between the Cope and Drag Sections

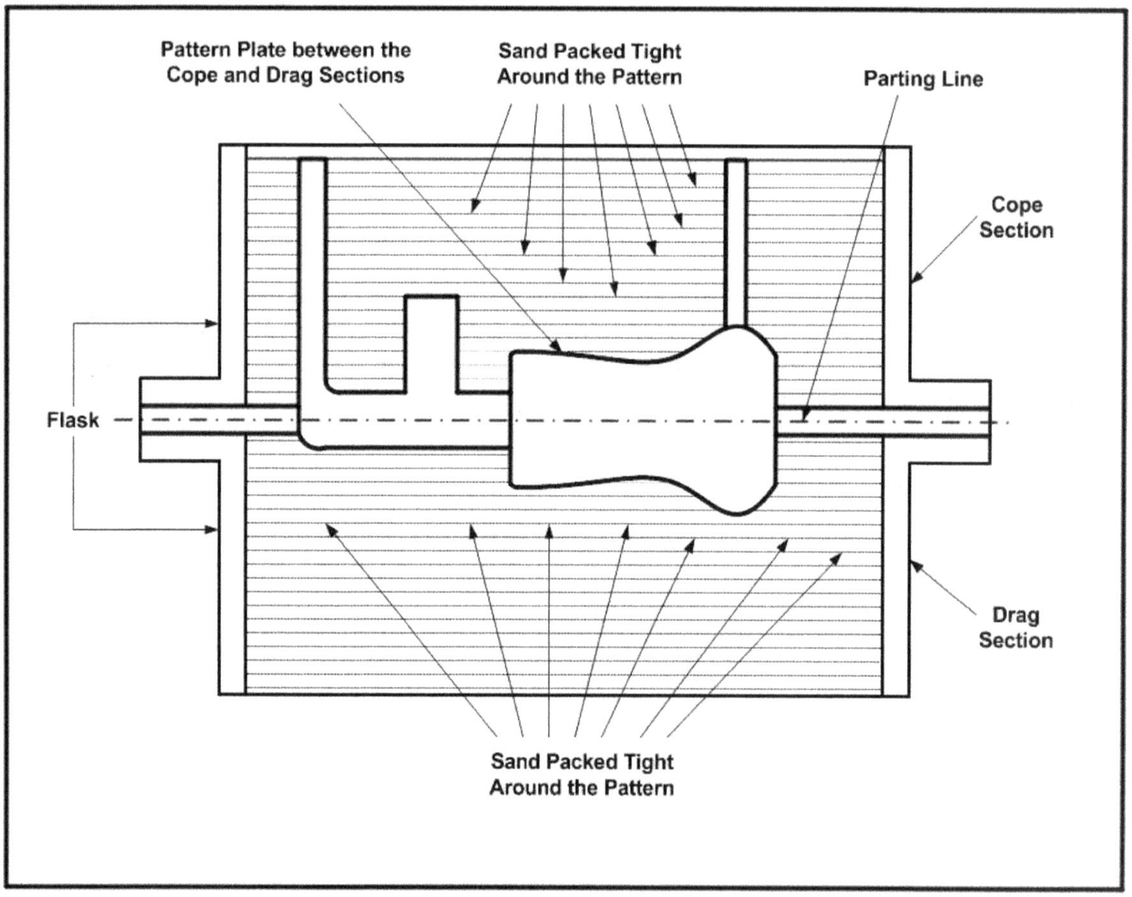

Figure 5.19 Step 2 for a Sand Casting: Sand Is Tightly Packed Around the Pattern Plate

Step 3: The cope and drag sections are disassembled and the pattern plate is removed. A pouring cup is carved into the downsprue and the cope and drag are reassembled. See Figure 5.20.

Step 4: The mold is poured with the molten metal into the cavity until it is filled. When the material comes out of the vent, the mold is filled properly. See Figure 5.21.

Step 5: After allowing the molten material to fully solidify, the cope and drag sections are taken apart and the sand it removed from the part, and the part is cleaned and allowed to cool to room temperature. See Figure 5.22.

The pattern is then reused to construct another sand casting mold, obtaining an expendable mold and a permanent pattern from this process. Cores can be inserted into the mold cavity between steps 4 and 5 if required. The vents, downsprue, and runner are cut off of the part later.

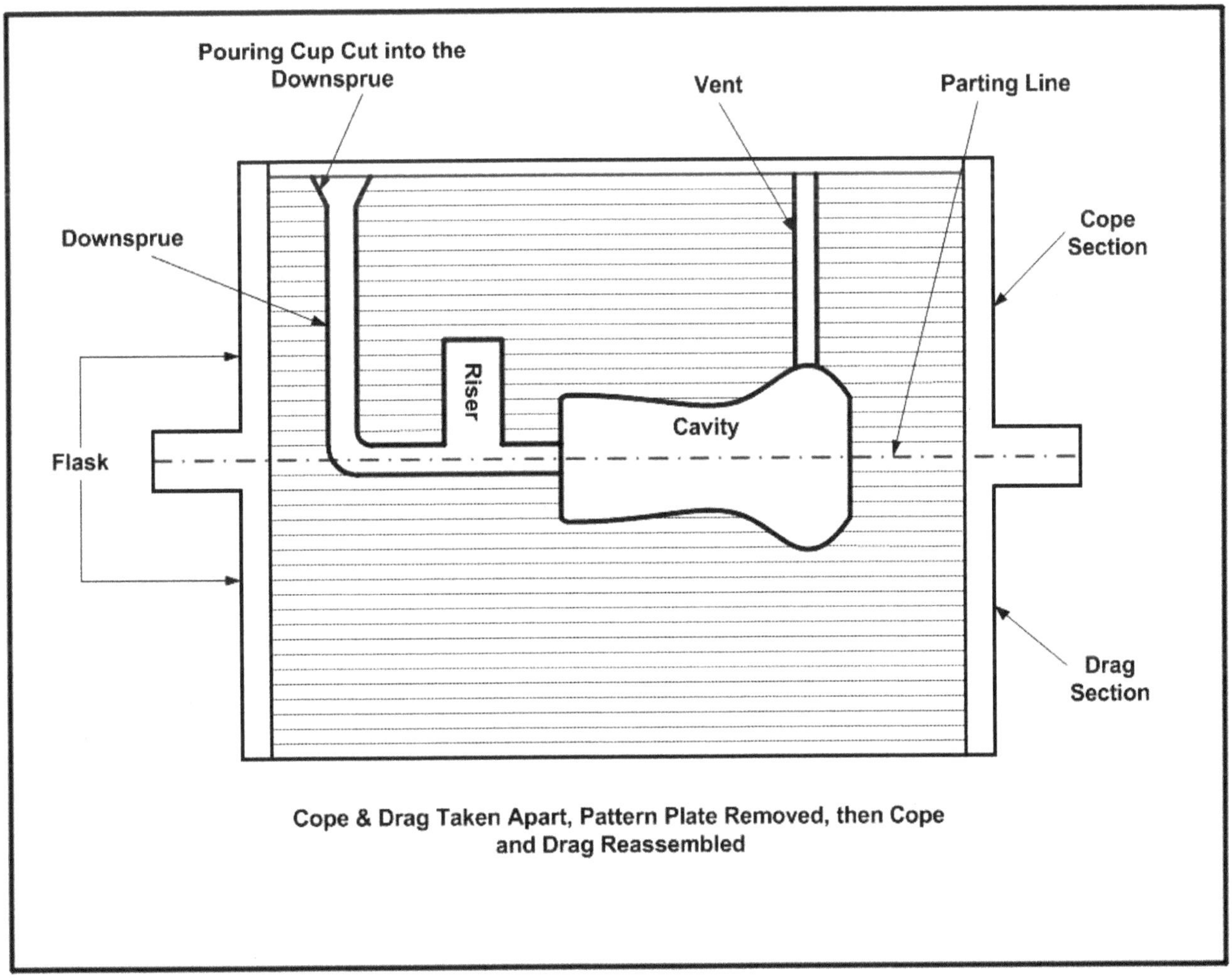

Figure 5.20 Step 3 for a Sand Casting: The Pattern Plate Is Removed and the Cope and Drag Are Reassembled

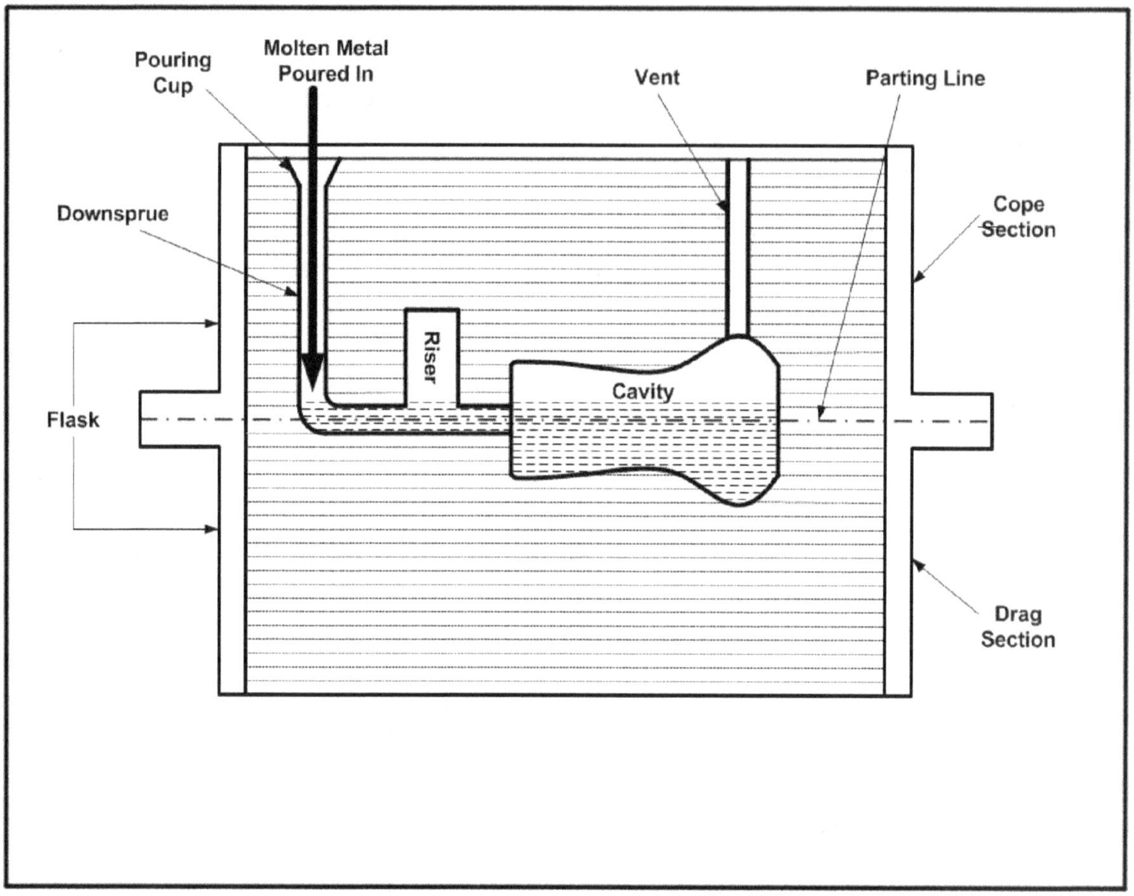

Figure 5.21 Step 4 for a Sand Casting: Molten Metal Is Poured into the Pouring Cup

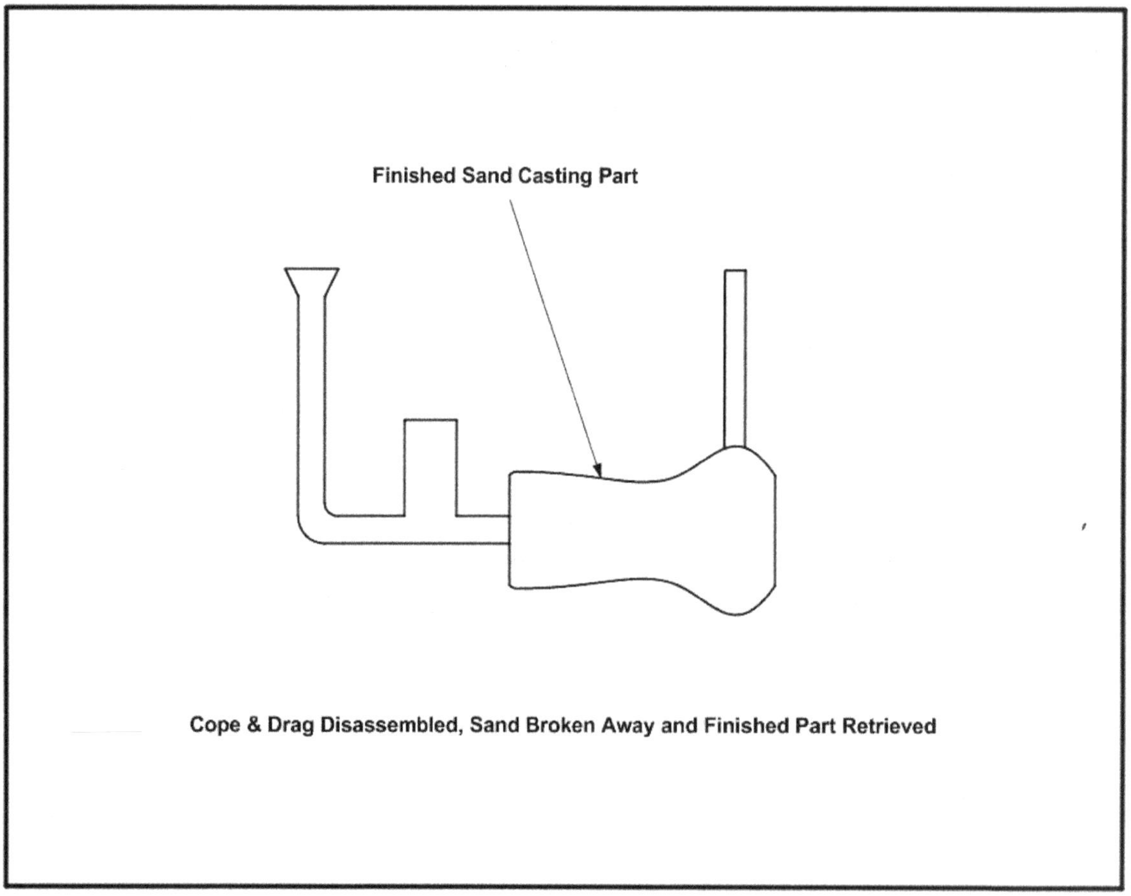

Figure 5.22 Step 5 for a Sand Casting: The Sand Is Broken Away and the Part Is Retrieved

5.6.2 Investment Casting Process

Investment casting is a unique process capable of producing part configurations that other casting processes cannot. It involves a pattern constructed of wax and is used heavily in the casting of jewelry. Although the process is more involved than the sand casting process, it is very widely used throughout the industry today. The process steps for investment casting are as follows:

Step 1: A pattern is constructed of a wax material with the downsprue and filling cup included. See Figure 5.23.

Step 2: The wax pattern is dipped several times into a ceramic material and allowed to dry between drippings. See Figure 5.24.

Step 3: Once the ceramic has dried after the final dipping, it is placed into a kiln and heated to cure the ceramic. During this step, the wax pattern is melted out of the cavity of the mold. See Figure 5.25.

Step 4: Molten metal is poured into the ceramic mold until filled and allowed to solidify. See Figure 5.26.

Step 5: The ceramic is broken away and the finished part is retrieved. See Figure 5.27.

With the pattern being destroyed every time a part is made, we obtain an expendable pattern in this process. Also, the ceramic mold is broken away from the part to retrieve it, and the mold is expendable as well. If, say, 100 parts are required to be fabricated, then 100 patterns need to be made.

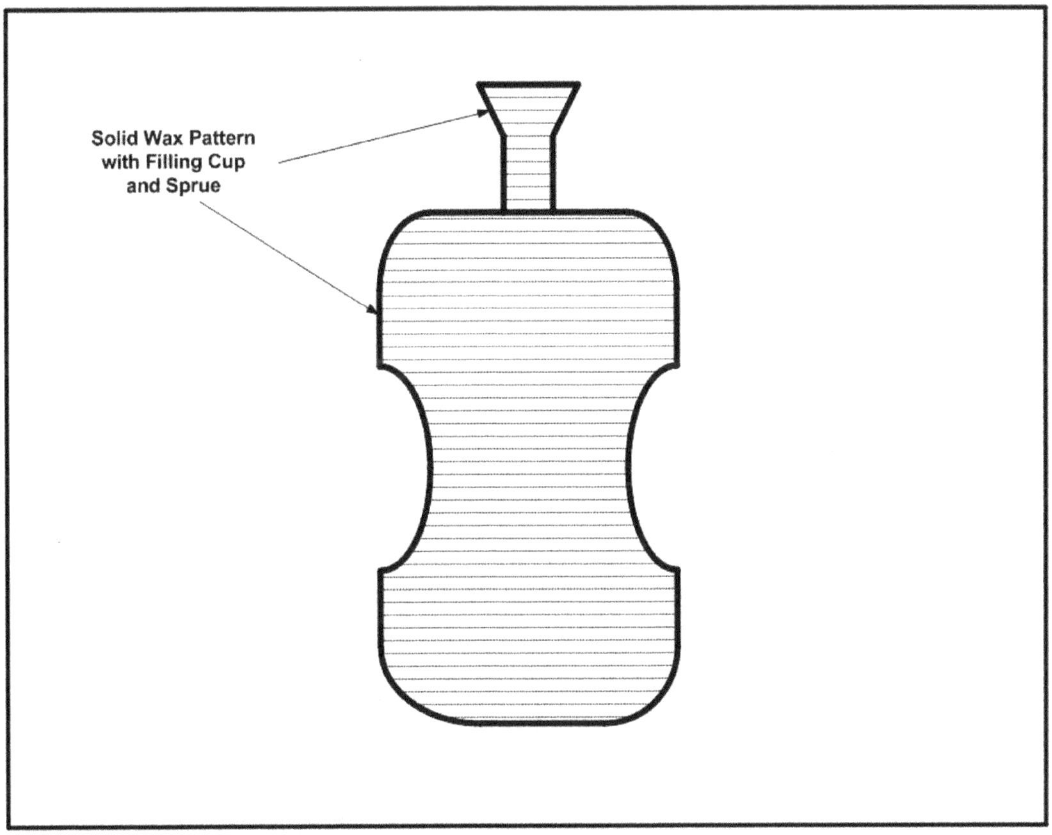

Figure 5.23 First Step in Investment Casting Process (Wax Pattern Constricted)

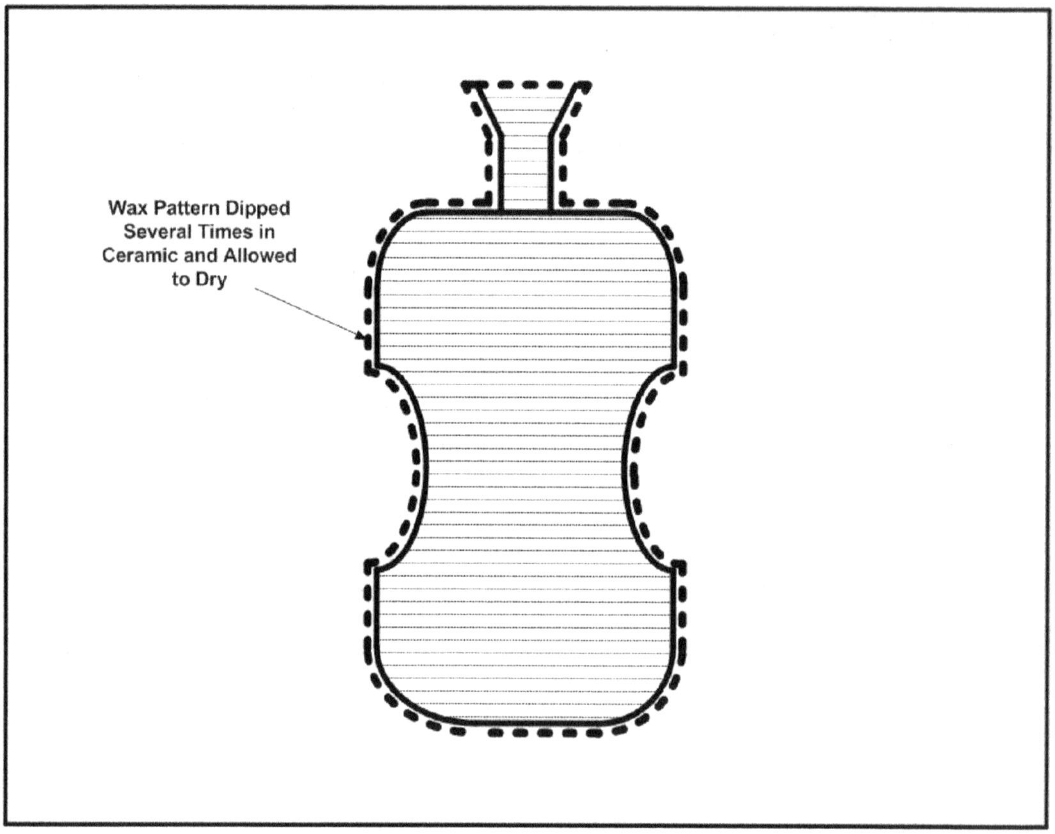

Figure 5.24 Second Step in Investment Casting Process (Wax Pattern Coated with Ceramic)

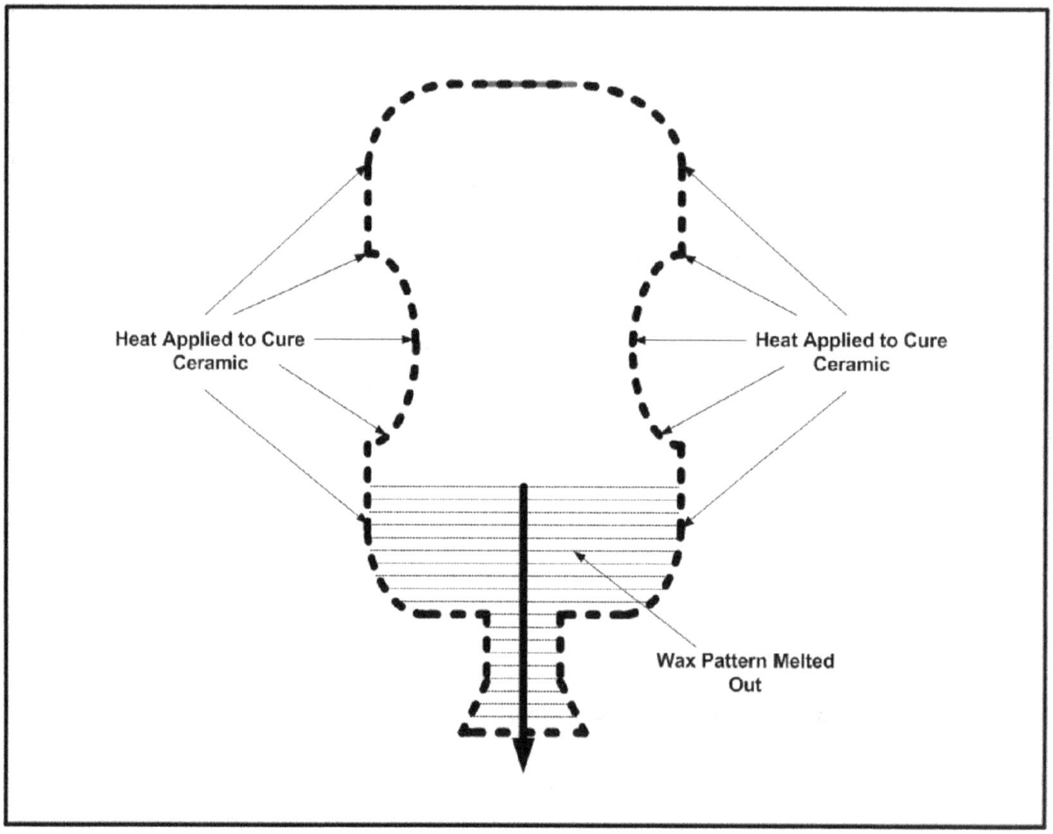

Figure 5.25 Third Step in Investment Casting Process (Ceramic Cured and Wax Pattern Melted Out)

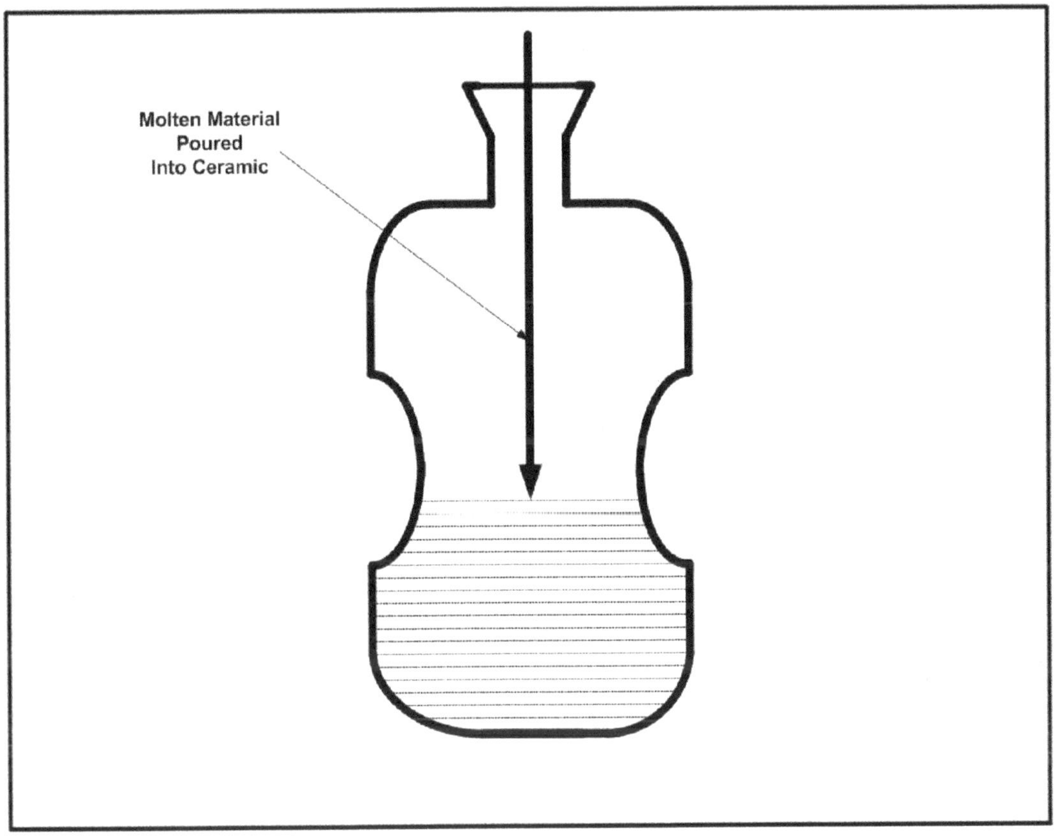

Figure 5.26 Fourth Step in Investment Casting Process (Material Poured into Ceramic Mold)

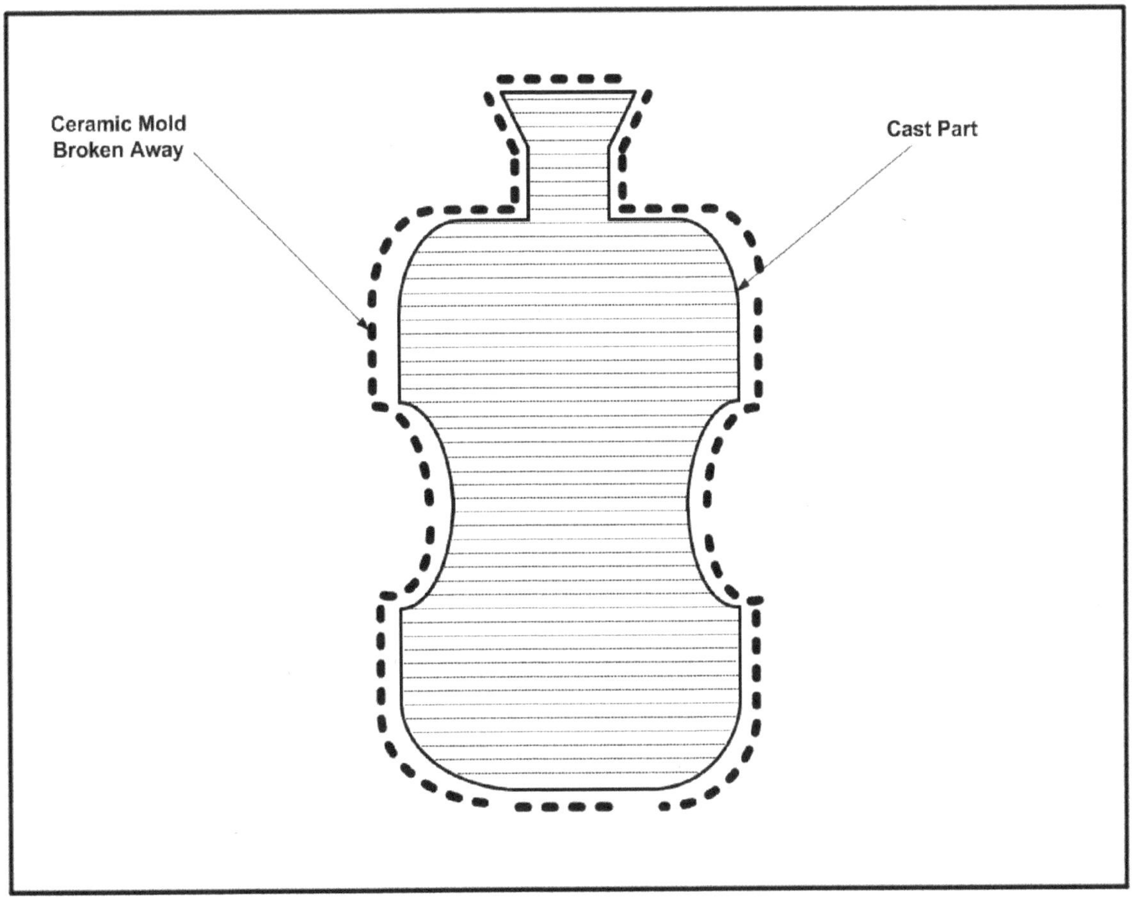

Figure 5.27 Final Step in Investment Casting Process (Ceramic Mold Broken Away to Retrieve the Finished Part)

5.6.3 Die Casting Process

Die casting is a process used for very high production and utilizes mainly zinc or aluminum as the material. This is very highly automated with very expensive tooling. The mold is called a die and has the cavity configuration machined into it and is made of hardened steel. Die casting is not used for thick parts such as an engine block, and the process is very similar to that of producing plastic parts in a mold. The process steps for the die casting process are as follows:

Step 1: The die (mold) is mounted into an injection mold machine. The die closes and the molten material is injected into the mold cavity under high pressure. See Figure 5.28.

Step 2: The molten material solidifies, the movable platen moves back while the ejector pin plate moves in, and the finished part falls free of the cavity. See Figure 5.29.

The process keeps repeating Steps 1 and 2 and parts are produced in a mass production manner. With the die (mold) being constructed of hardened steel, the mold is permanent. With the cavity being machined into the hardened steel, there is no pattern utilized in the construction of the mold.

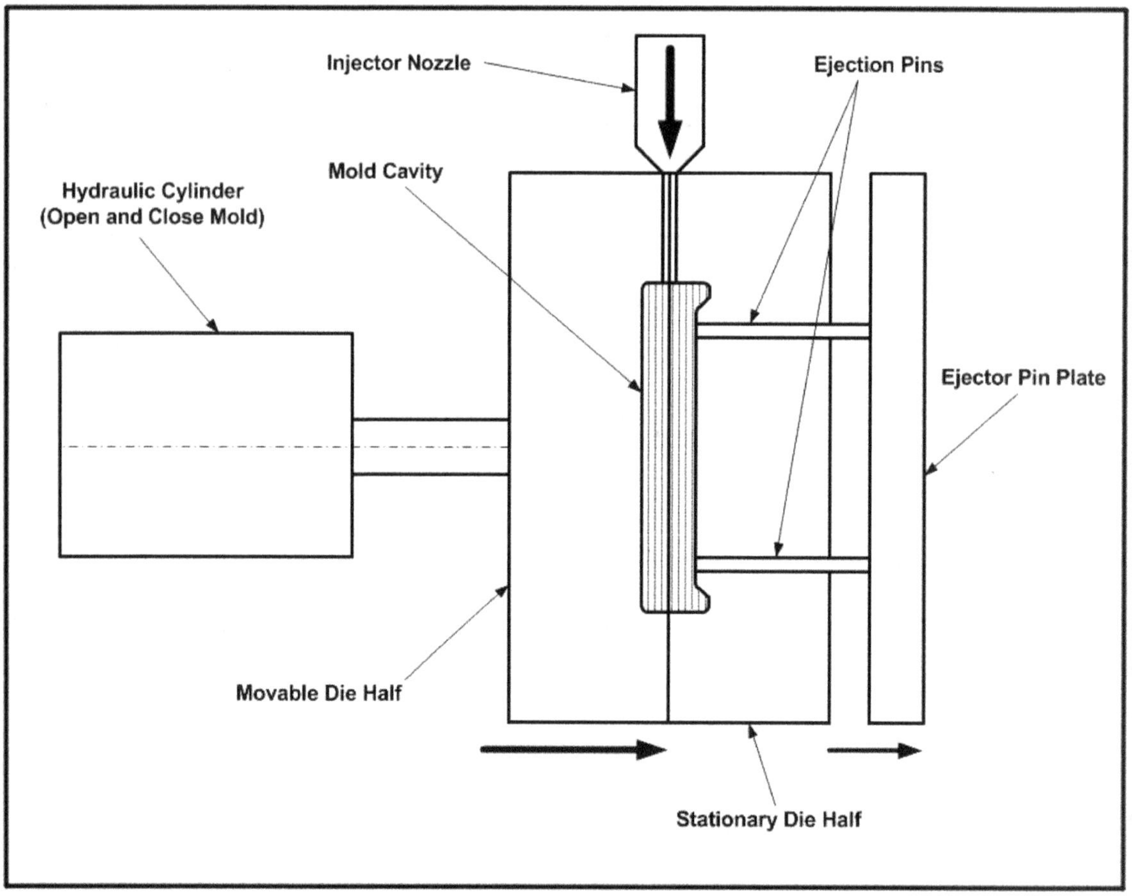

Figure 5.28 Step 1 for Die Casting: The Mold Is Closed and Molten Metal Is Injected under Pressure

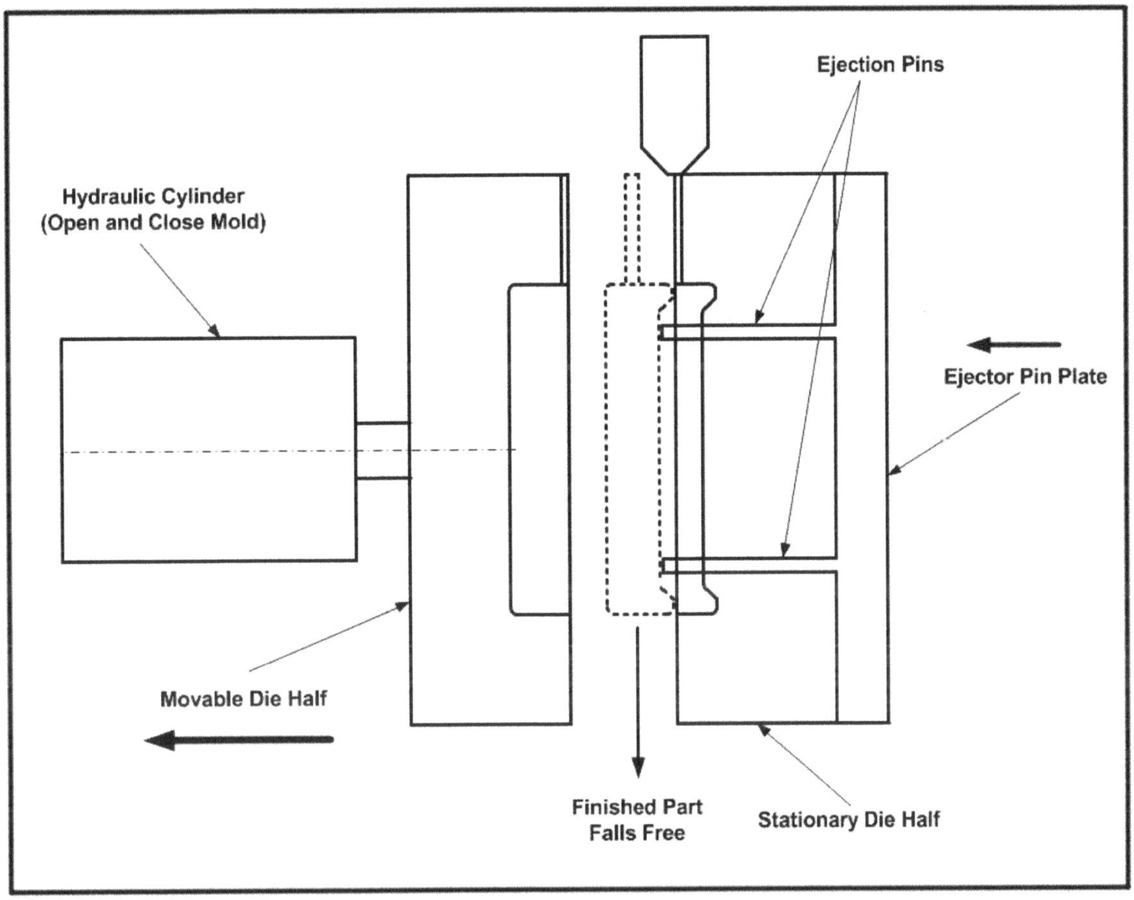

Figure 5.29 Step 2 for a Die Casting: The Mold Opens and the Ejector Pins Push the Part Out of the Cavity

5.6.4 Centrifugal Casting

The centrifugal casting process is designed for casting round, hollow parts. The mold is a two-piece round cavity and is heated to a given temperature. This mold is rotated at a certain revolution as the molten metal is poured into one end. As the mold rotates, the molten metal is cast up against the sides of the mold, creating a round, hollow casting. To obtain the desired wall thickness of the casting, there are three factors that need to be determined. First is the temperature of the mold where the molten metal is poured into. Second is the temperature of the material being poured, and third is the speed at which the mold is being rotated. Unlike the molds used in the other processes mentioned, filling a centrifugal mold is different. With the other processes, the mold is filled to its capacity, whereas with centrifugal casting, the amount of material being poured into the rotating mold needs to be very accurately calculated. If too little material is poured, the wall will not be thick enough, and in the same manner, if too much material is poured, the walls will be too thick. Figure 5.30 shows the concept of a centrifugal casting process.

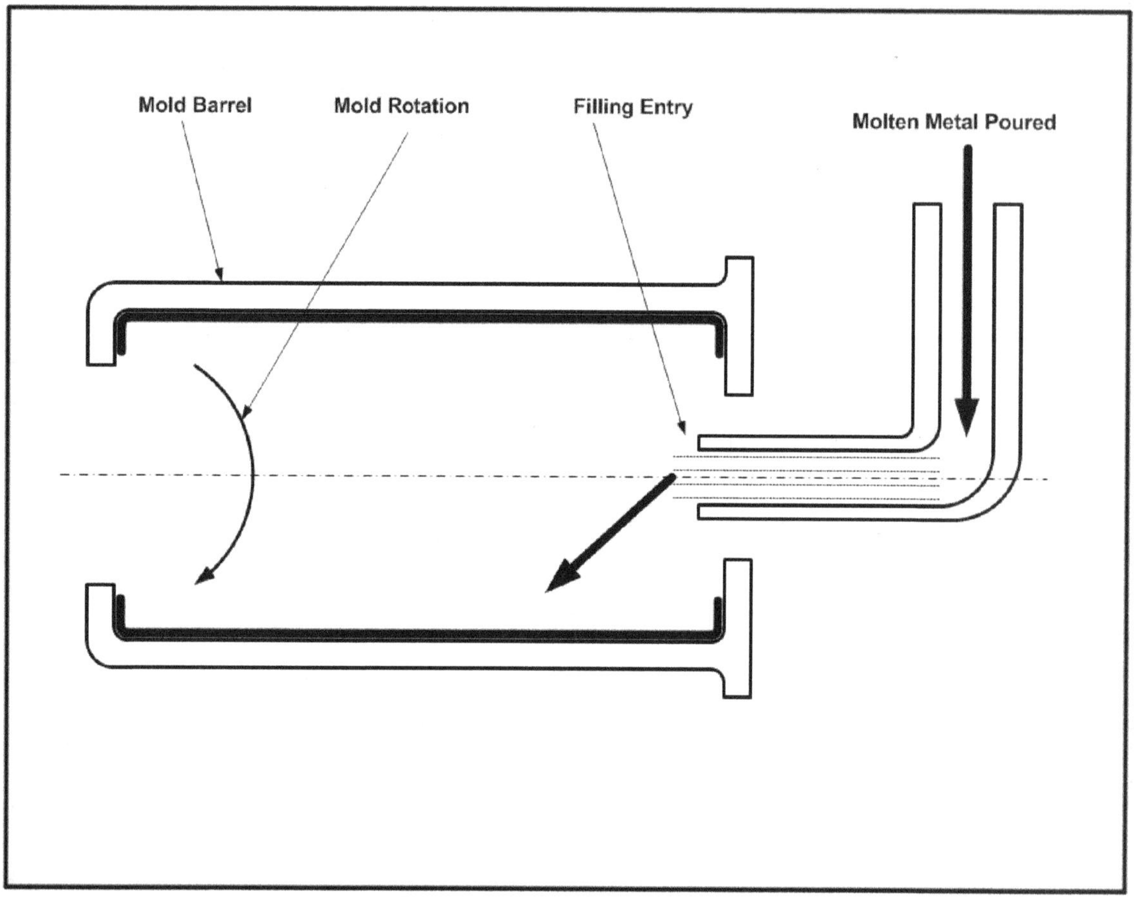

Figure 5.30 Centrifugal Casting Process

5.7 SUMMARY

Metal casting is a very viable manufacturing process, which is utilized extensively in many facets of production worldwide. Being able to produce raw metal parts that are very close to net shapes gives casting an advantage over many other processes. When a complex metal shape is required, machining that from a solid block of metal would be extremely time-consuming and highly wasteful of material. By producing the raw shape from a casting, only 5% to 15% of the material may have to be removed to bring the casting into the net shape desired.

From understanding the principles of the process, we understand how it works and what can be expected in the final result. With a proper casting design from the start, defects can be either eliminated or greatly reduced. By implementing these principles, and understanding what each casting process can, and equally important, cannot do, the results realized will far outweigh alternative manufacturing methods.

5.8 QUESTIONS AND PROBLEMS

Questions

1. Define the difference between Fluidity and Viscosity.

2. What is Mold Erosion, and how can it be avoided in the casting process?

3. Why is it important to have the molten material superheated when pouring a casting mold?

4. What are the two factors that need to be considered in the shrinkage of a casting and the difference between them?

5. When pouring a casting mold, what is the Volumetric Flow Rate and what factors are used to calculate it?

6. Why is it important to have uniform wall thickness throughout a cast part?

7. Why are draft angles used in the casting process?

8. What is porosity in a casting, and how can it be minimized?

9. What is the difference between an expendable mold and a permanent mold? Give an example of each.

10. In the Investment Casting process, how many molds and patterns need to be constructed if 75 parts were to be fabricated and why?

Problems

1. A casting is to be poured from Low C Cast Steel. What is the three-dimensional shrinkage value for this material after solidification and thermal contraction?

2. A rectangular component part is going to be cast from copper material utilizing the sand casting process. If the finished dimensions of the part are 3.274 in. L × 1.946 in. W × 0.835 in. H and a pattern needs to be constructed, what would the pattern dimensions need to be to produce an acceptable part?

3. Using the information from problem #2, what would be the mold filling time if the filling sprue was 0.750 in. in diameter and had a height of 2.750 in.?

4. If the mold utilized in problems #3 and #4 had a mold constant of 13.824 min/in.^{2} and an exponential value of 1.62, what would the total solidification time be?

5. A wedge-shaped component part is going to be cast from Aluminum Alloy material. The cavity sizes are 4.617 in. long, 2.114 in. wide, and 1.738 in. high with a filling sprue that is 0.375 in. in diameter having a height of 3.750 in. with a mold constant of 11.728 min/in.^{2} and an exponential value of 2.17, what would the casting sizes be along with the mold filling time and total solidification time?

6. If problem #5 required a total solidification time of exactly 1 min, what would the mold constant need to be?

7. An investment casting mold is going to be filled with molten Bronze Alloy material, which has been superheated to a temperature of 2,800 °F. Once the material starts to be poured, it cools down at a rate of 35 °F per second. How much time does the pourer have before the material begins to solidify?

8. A rectangular block will be poured from copper material and the desired cast dimensions are 3.750 in. Long × 2.375 in. Wide × 1.625 in. High. The mold has a filling sprue diameter of 0.250 in. and a height of 2.125 in. If the material was superheated to 2,600 °F and cools at a rate of 45 °F per second when being poured, would there be enough time to fill the mold before solidification occurs?

9. A component part is currently being produced from Aluminum. The part is a rectangular block and the cavity dimensions are 4.125 in. L × 3.250 in. W × 1.875 in. H. If all of the finished dimensions of the part have a tolerance of ±0.010 in., could the same mold be utilized if the material were to be changed to Copper?

CHAPTER

6

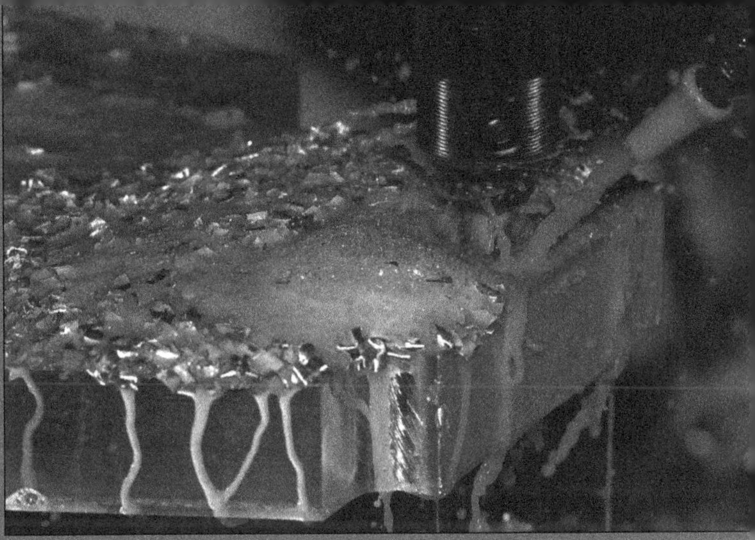

Plastic Shaping Concepts

6.1 INTRODUCTION

Plastic has become a very widely used material in the manufacturing arena. As new materials are developed and shaping processes are perfected, fabricating components out of plastics has become more and more popular. With materials becoming as strong and rigid as some metals and the processes capable of rendering parts requiring no additional operations, the world of plastics now plays a major role in producing parts very economically.

With the selection of shaping process combined with the vast selection of materials, fabricating components from plastics has taken over the manufacturing industry. If we were to look back to the 1960s at vehicles, the entire inside of the car was fabricated from metal. The dashboard, the center console, and just about everything were made of sheet metal. Today, all of those items are fabricated from plastic. One would be hard-pressed to try and stick a magnet anywhere inside of a car now.

This chapter discusses the types of plastic and various shaping processes utilized to fabricate parts. With new materials and process innovations being developed almost daily in this fast-changing arena, the ability to produce new part configurations has given the plastic industry huge advantages over other more costly processes. Some of the more readily used processes and their advantages and disadvantages are covered as follows.

6.2 MATERIAL PROPERTIES

Plastic materials fall into two major categories. There are numerous plastic materials, but each will be in one of the two categories of material type. Both of these material types are plastic, but completely different in how they are formed and shaped. These two categories are *Thermoplastics* and *Thermosets* and are defined as follows.

Thermoplastics: A thermoplastic is a material that can be shaped, remelted, and reshaped over and over. The fabrication facility receives the raw material in pellets, which they would melt down and shape it through various processes to get the desired shapes and configurations. This material, because it can be reused, is extremely recyclable, making it environmentally friendly.

Some materials in this arena can be reactive by exposing it to other elements. An example of that is expandable polystyrene. This material is a thermoplastic, but when the pellets are exposed to high-pressure steam, they expand. The raw material is received as very small, hard pellets. When they are exposed to high-pressure steam, they expand in size. The pellets are placed into an expander and allowed to expand about three-quarters of their expansion capability. From there, they are blown into a mold, and the mold is hit with high-pressure steam causing the pellets to expand to their full size. As they expand in the mold, they bond together causing them to create a shaped part.

Thermosets: This plastic material is created by combining two elements together to create a third. Basically, a select amount of material "A" is mixed with a select amount of material "B" to create material "C"; (A + B = C). The advantage of a material of this type is that it will render properties that a thermoplastic will not. The finished material for the most part will be far more rigid and not as prone to heat affecting it as much because it cannot be remelted. Also, in general, the shrinkage of these materials is less than that of the thermoplastics, allowing them to be fabricated to closer tolerances. From a negative point, these materials are not recyclable due to the fact that the two starting elements cannot be separated once combined.

6.3 SHAPING PROCESSES

Plastic is a very flexible material in terms of shaping. A variety of methods and processes are used to create the various shapes we come in contact with on a daily basis. Working with plastics involves a combination of heat and pressure. Some of the processes for shaping require the material to be melted and pushed either into a mold or through an opening to create a shape. When utilizing plastic in its melted state, the viscosity is very high and will not flow like a liquid. Figure 6.1 shows the Flow Velocity pattern for plastic as it is pushed either into a mold or through an opening. This resistance to flow needs to be accounted for when determining how much pressure is required to complete the process successfully. Some of the more common plastic shaping processes are described as follows.

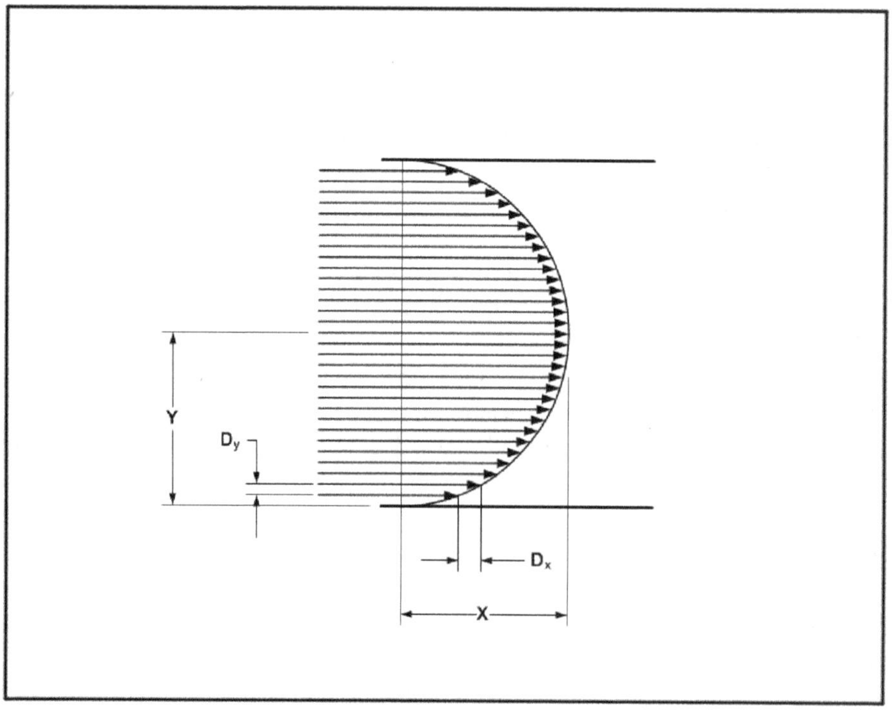

Figure 6.1 Flow Velocity Pattern for Plastic

6.3.1 Extrusion Process

The extrusion process involves pushing the plastic material through an opening to create a given shape. The opening that the plastic material is pushed through is called the die, which has the configuration of the desired shape. Figure 6.2 shows the concept of extruding as the material is forced through the die opening. With the extrusion process, the configuration that is being created is two-dimensional where the X and Y shapes are controlled by the die shape. For all intents and purposes, the length of the part is unlimited. As the part is being extruded from the die, it will be cut off to the desired length. Plastic material does have a memory in the extrusion process and as it exits the die opening, it will expand. This expansion is called viscoelasticity. Figure 6.3 shows how the extruded part will expand as it exits the die. The amount that the material will expand

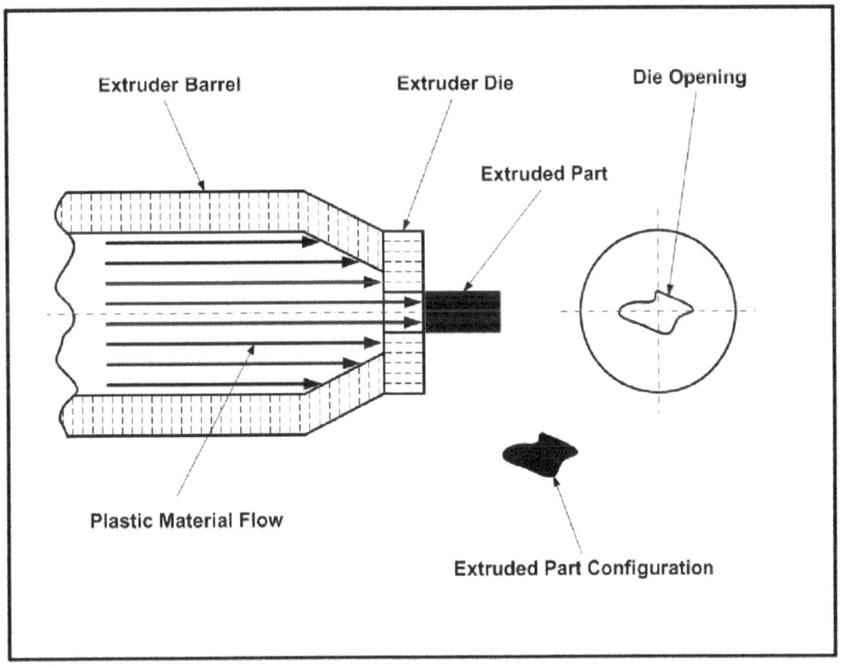

Figure 6.2 Extrusion Process Concept

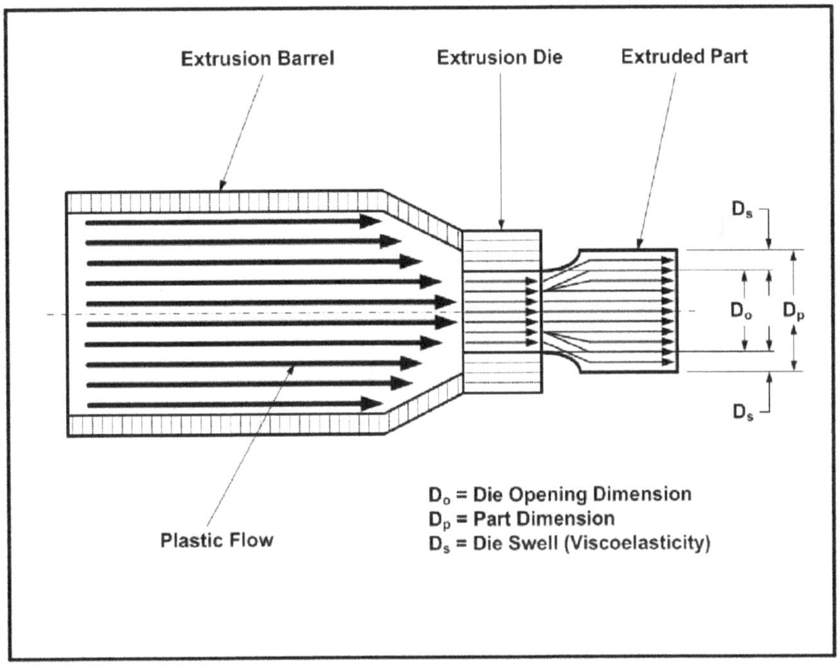

Figure 6.3 Viscoelasticity (Die Swell) of Extruded Material

is dependent on the material being extruded and needs to be calculated very closely in order to obtain the desired configuration.

The unit that pushes the material through the die opening is called the barrel. This is a cylinder with a rotating blade with fins on it very similar to a meat grinding unit. As the blade rotates, the material is forced forward and through the die opening. Figure 6.4 shows the configuration of the extrusion barrel with the factors required to calculate the resulting material flow. The pressure that the material is being forced through the die opening is dependent on two factors, which are *Drag Flow* and *Back Pressure.* Drag Flow is the forward flow of the material as it is pushed through the extruder barrel, and Back Pressure is the resistance pushing back. The combination of these two factors will produce the resulting material flow. Figure 6.5 shows the equations required for calculating the *Flight Angle, Drag Flow, Back Pressure,* and *Resulting Flow Rate.*

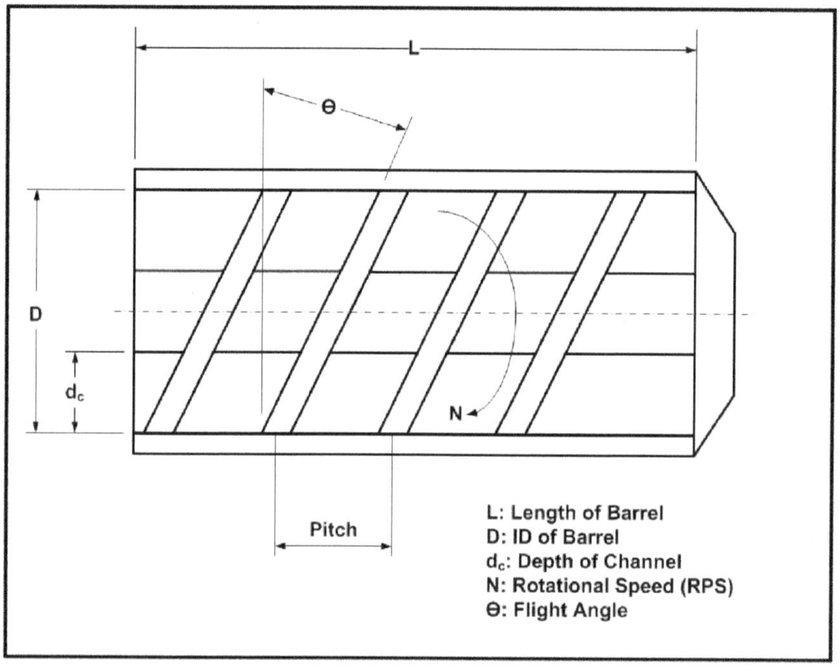

Figure 6.4 Extrusion Barrel Components and Terminology

Flight Angle

$$\tan\theta = \frac{\text{Pitch}}{\pi D}$$

Drag Flow

$$Q_d = 0.5\,\pi^2 D^2 N d_c \sin\theta \cos\theta$$

Back Pressure

$$Q_b = \frac{\rho\,\pi D d_c^3 \sin^2\theta}{12\,\eta L}$$

Resulting Flow Rate

$$Q_x = Q_d - Q_b$$

D = Barrel ID
N = Rotational Speed (RPS)
d_c = Channel Depth
L = Barrel Length
ρ = Head Pressure in the Barrel (PSI)
η = Viscosity (# - s/in.^{2})
θ = Flight Angle

Figure 6.5 Extrusion Equations for Drag Flow, Back Pressure, and Resulting Flow

An example of calculating the resulting flow rate is outlined as follows.

Example Problem

Data: Barrel Diameter (D) = 4.0 in. Length of the Barrel (L) = 30 in.
Rotation Speed (N) = 3.14 RPS Depth of the Channel (d_c) = 0.75 in.
Pitch = 2.671 in. Head Pressure (p) = 500 psi Viscosity (η) = 140×10^{-4} Lbs-Sec/In^2

Step 1: Calculate the Flight Angle θ.

$$\tan \theta = \text{Pitch} / \pi D$$
$$\tan \theta = 2.671 / \pi\, 4.0$$
$$\theta = 12.2°$$

Step 2: Calculate the Drag Flow of the Material.

$$Q_d = 0.5\, \pi^2 D^2 N\, dc \sin \theta \cos \theta$$
$$Q_d = (0.5)\, \pi^2 (4.0^2)\, (3.14)\, (0.75)\, (\sin 12.2)\, (\cos 12.2)$$
$$Q_d = 38.4063 \text{ in.}^3/\text{s}$$

Step 3: Calculate the Back Pressure Flow of the Material.

$$Q_b = p\pi D\, d_c^{\,3} \sin^2 \theta / 12\eta L$$
$$Q_b = (500)\, \pi\, (4)\, (0.75^2) \sin^2 12.2 / (12)\, (140 \times 10^{-4})\, (30)$$
$$Q_b = 23.4885 \text{ in.}^3/\text{s}$$

Step 4: Calculate the Resulting Flow Rate.

$$Q_x = Q_d - Q_b$$
$$Q_x = 38.4063 - 23.4885$$
$$Q_x = 14.9178 \text{ in.}^3/\text{s}$$

6.3.1.1 Extrusion Defects

The extrusion process can relinquish to major defects. These are Melt Fracture and Sharkskin (Bambooing). Each of these is caused by either the melt temperature, or the condition of the extrusion die, or uneven material feeding through the die. Figure 6.6 shows what these defects look like, which are explained as follows.

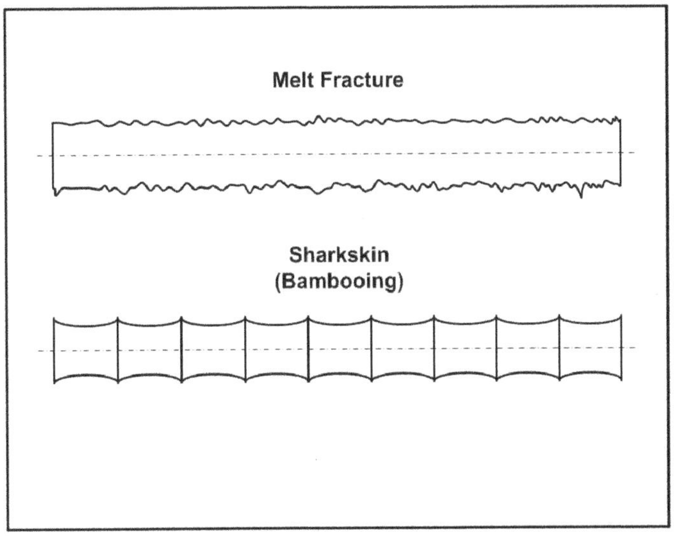

Figure 6.6 Extrusion Defects

Melt Fracture: A melt fracture is caused by either the melt temperature not being high enough or the extrusion die needing to be sharpened. If the material is not heated enough, it will tear as it is pushed through the die leaving a very rough finish on the part being extruded. Also, if the extrusion die needs to be sharpened, it will also cause the material to tear as it exits the die, causing the same aforementioned rough finish on the part.

Sharkskin (Bambooing): Sharkskin is an uneven thickness on the part as it is being extruded through the die opening. This is caused by the barrel feeding the material in an uneven flow, causing it to move faster rather than slowing down. This causes the material from a uneven flow to expand differently as it exits the die and cools down.

6.3.2 Tubing Extrusion Process

Tubing extrusion is performed exactly the same way as general extruding outlined earlier, but the barrel will have a mandrel mounted in front of the die opening. The material is extruded around the mandrel and through the die opening, creating a part with a hole in the center. In this process, the tubing is extruded and then rolled up on a coil, resulting in a long part, which can be cut to a certain size when being used. Figure 6.7 shows the extrusion barrel with the mandrel at the center, creating a hole in the tubing. In this process, the viscoelasticity needs to be compensated for both the outside and inside dimensions.

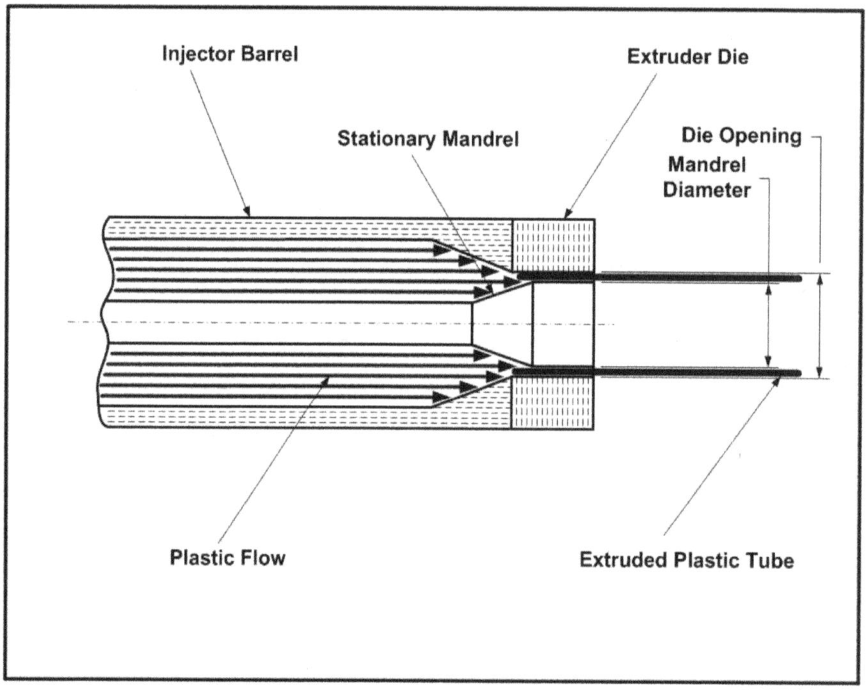

Figure 6.7 Extrusion of Plastic Tube

6.3.3 Injection Molding Process

Injection molding is one of the major processes for producing complex plastic components. As plastic materials have improved and the overall process has been refined, injection molding has become capable of producing parts to better tolerances and more complex configurations. The principle of injection molding is similar to that of the die casting process. Plastic is melted and injected into a mold cavity producing the desired part. Figure 6.8 shows the components of a typical injection molding setup. The mold is closed via the movable platen for the first stage. In Figure 6.9, the plastic material is heated and injected into the mold cavity filling it completely. After a cooling time, the mold opens by moving the movable platen back and the ejector pin frees the molded part as shown in Figure 6.10.

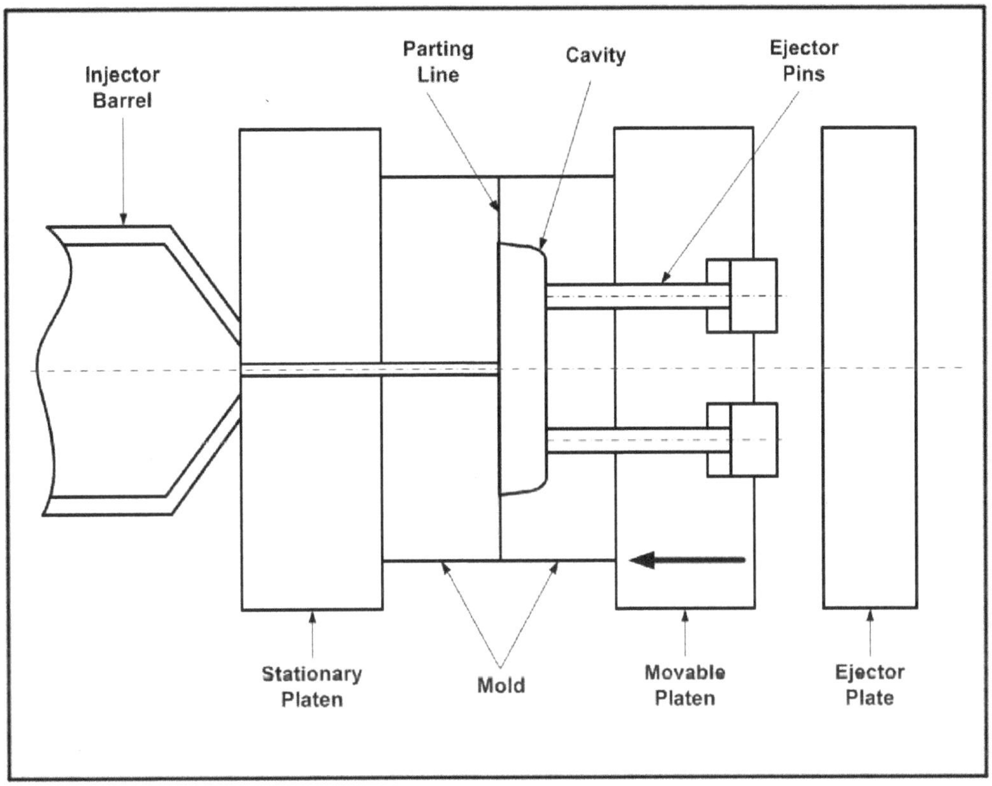

Figure 6.8 Typical Plastic Injection Molding Setup

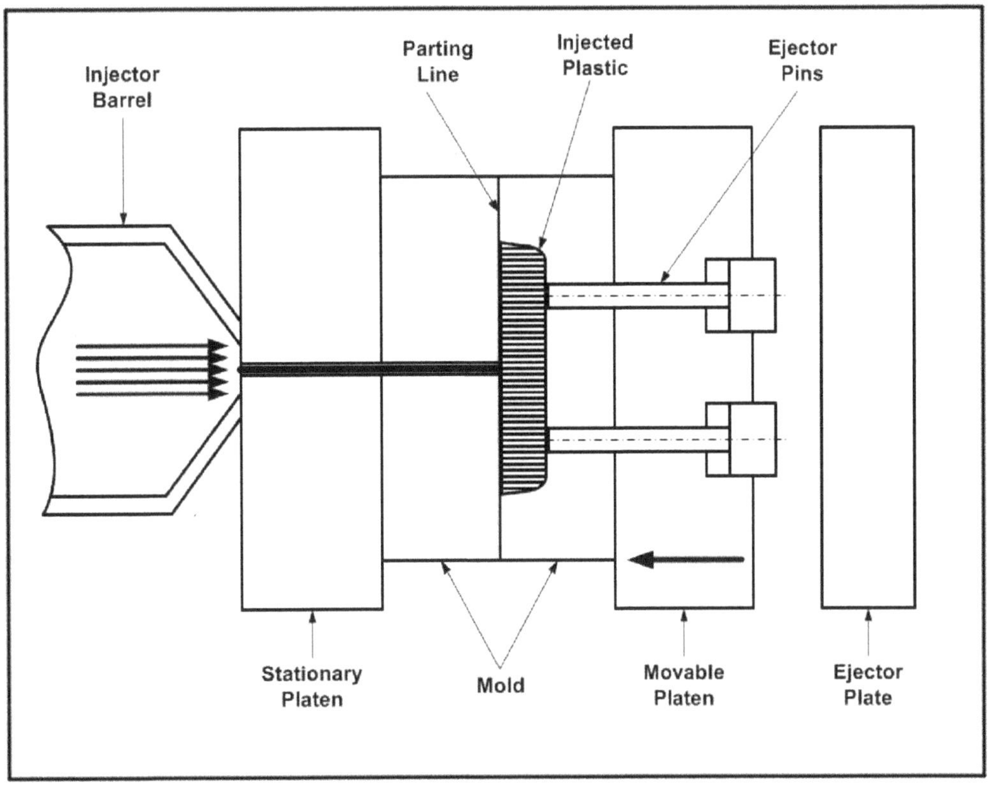

Figure 6.9 Plastic Injected into the Mold Cavity

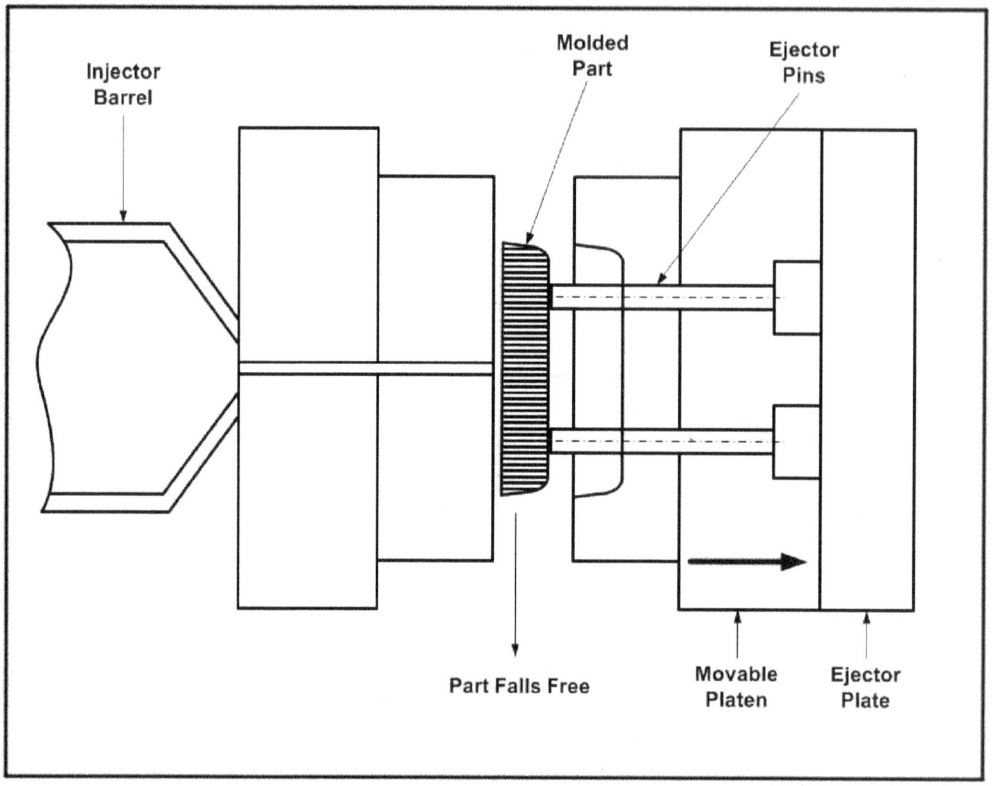

Figure 6.10 Molded Part Ejected from the Cavity

Large components are normally produced in a single cavity mold, whereas smaller components may have a mold with multicavities. These normally either will produce several of the same parts or may produce several different parts of the same material. Figure 6.11 shows the construction of a multicavity mold. Here, eight cavities are contained in a single mold and the plastic material is injected in the center and then carried to the

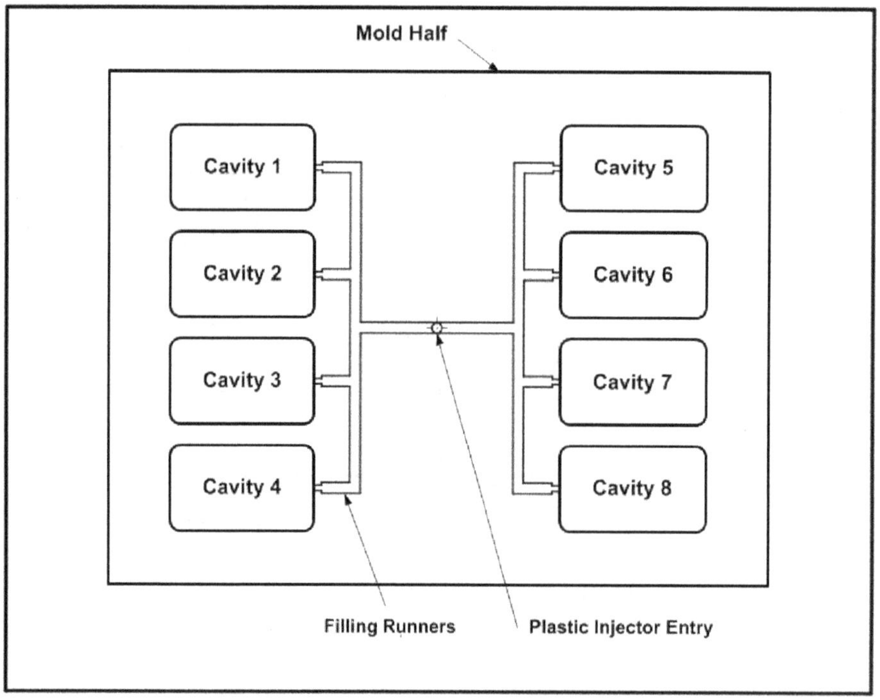

Figure 6.11 Layout of an Eight-Cavity Injection Mold

various cavities via runners to fill them. In a multicavity mold, each cavity would be marked to indicate which part it produced. If an issue arises from a particular cavity, it can be blocked off from the rest of the cavities. This allows the mold to continue running with only one less part being produced.

A feature in injection-molded components that is very helpful in the overall process are date clocks. The purpose of date clocks in a molded part is to be able to identify when a given component was produced and to be able to identify when parts may have had issues with the molding process. These are changeable inserts located in the mold cavity and are adjusted throughout the day as the process is running. Figure 6.12 shows three different date/time clocks that can be located in the cavity. The year/month clock shows the year in the center that the part was produced. The numbers 1 through 12, which the arrow points to, indicate the approximate month that the part was produced. From there, you can see that it was produced in 2013, sometime in the middle of July. The day clock indicated which day in the month that the part was produced. From the figure, it is evident that the part was produced on July 16th. The third time clock shows the time that the part was produced. To read this, 1 is 12:00 a.m., 2 is 6:00 a.m., 3 is 12:00 p.m., and 4 is 6:00 p.m. From there you can read that the part was produced at 10:00 a.m. Totaling this together, the clocks indicate that the component part was produced in 2013, on July 16th at 10:00 a.m.

When working with a thermoset plastic, the concept is the same, but the preparation of the material is different. For thermoplastic material, a large amount of material can be melted in the injection barrel and the process continues to run automatically. However, with a thermoset material, each part that is going to be molded needs to have the exact amount of material missed in the injection barrel. This is due to the fact that a thermoset plastic involves a reaction of mixing two elements together, and a chemical reaction causes it to harden. Each run of the molding machine requires a mixing of the material and then injection into the mold. If 100 parts were required, this mixing stage needs to be repeated 100 times to produce the components.

As with any solidification process, shrinkage will be a factor that ensures that the final dimensions will be attained. Unlike the metal casting process, the only factor that needs to be considered is the shrinkage from thermal contraction. Figure 6.13 shows some shrinkage values for selected thermoplastics and thermosets. This is given in values of in./in. and needs to be applied to every dimension on the part print to ensure that the mold cavity is produced larger, so after shrinkage occurs, the part will be at the desired size. To calculate what the cavity sizes need to be, the following equation is used:

$$D_c = D_p + D_p S + D_p S^2$$

D_c = Cavity Dimension
D_p = Part Dimension (Size Specified on the Part Print)
S = Shrinkage Value in./in. (From Figure 6.13)

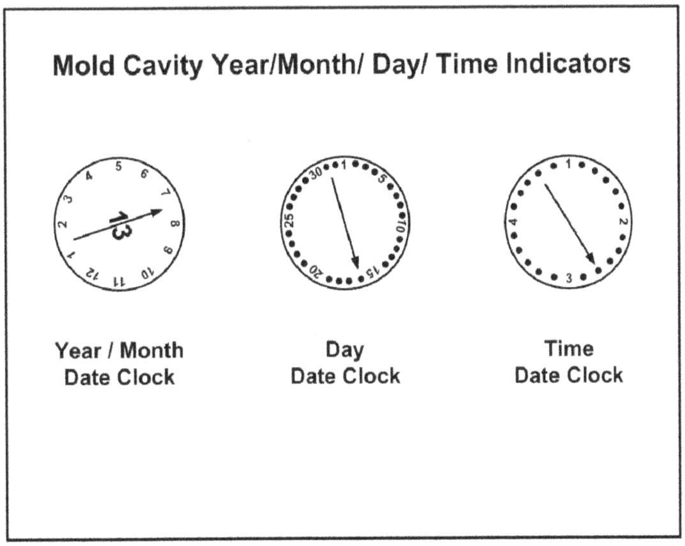

Figure 6.12 Date Clocks in an Injection Mold Cavity

Shrinkage Values for Some Materials			
Thermoplastic Materials		Thermoset Materials	
Material	Shrinkage in./in. ①	Material	Shrinkage in./in. ①
ABS Polymer	0.005	Epoxy	0.001
Acetal	0.021	Melamine	0.007
Nylon 4.6	0.012	Polyester	0.018
Propronate	0.004	Polyrethane	0.005
Polycarbonate	0.009	Vinyl Ester	0.0005
Polystyrene	0.005	Silicone	0.004
Ionomer	0.012	Bakelite	0.003
Polyurethane	0.006	Phenolic	0.007
PVC	0.023		
Polyarylsulfone	0.005		
Polyethylene	0.020		

① Average Shrinkage Values; Not Exactly Material Specific

Figure 6.13 Linear Shrinkage Values for Selected Plastic Materials

Following is an example problem of how this is applied to determine the cavity sizes for a given material.

Example Problem

Figure 6.14 shows a part that will be injection-molded out of a Nylon 4.6 thermoplastic. Looking at the linear shrinkage values in Figure 6.13, it shows that the S value for that material is 0.012 in./in. Utilizing the shrinkage equation of $D_c = D_p + D_pS + D_pS^2$, each dimension on the part print will be applied to render the dimensions for the mold cavity.

$$D_c = 2.483 + (2.483)\ (0.012) + (2.483)\ (0.012^2) = 2.5132 \text{ in.}$$
$$D_c = 1.174 + (1.174)\ (0.012) + (1.174)\ (0.012^2) = 1.1885 \text{ in.}$$
$$D_c = 1.079 + (1.079)\ (0.012) + (1.079)\ (0.012^2) = 1.0923 \text{ in.}$$
$$D_c = 0.693 + (0.693)\ (0.012) + (0.693)\ (0.012^2) = 0.7016 \text{ in.}$$
$$D_c = 0.626 + (0.626)\ (0.012) + (0.626)\ (0.012^2) = 0.6337 \text{ in.}$$
$$D_c = 0.135 + (0.135)\ (0.012) + (0.135)\ (0.012^2) = 0.1367 \text{ in.}$$
$$D_c = 0.375 + (0.375)\ (0.012) + (0.375)\ (0.012^2) = 0.3796 \text{ in.}$$

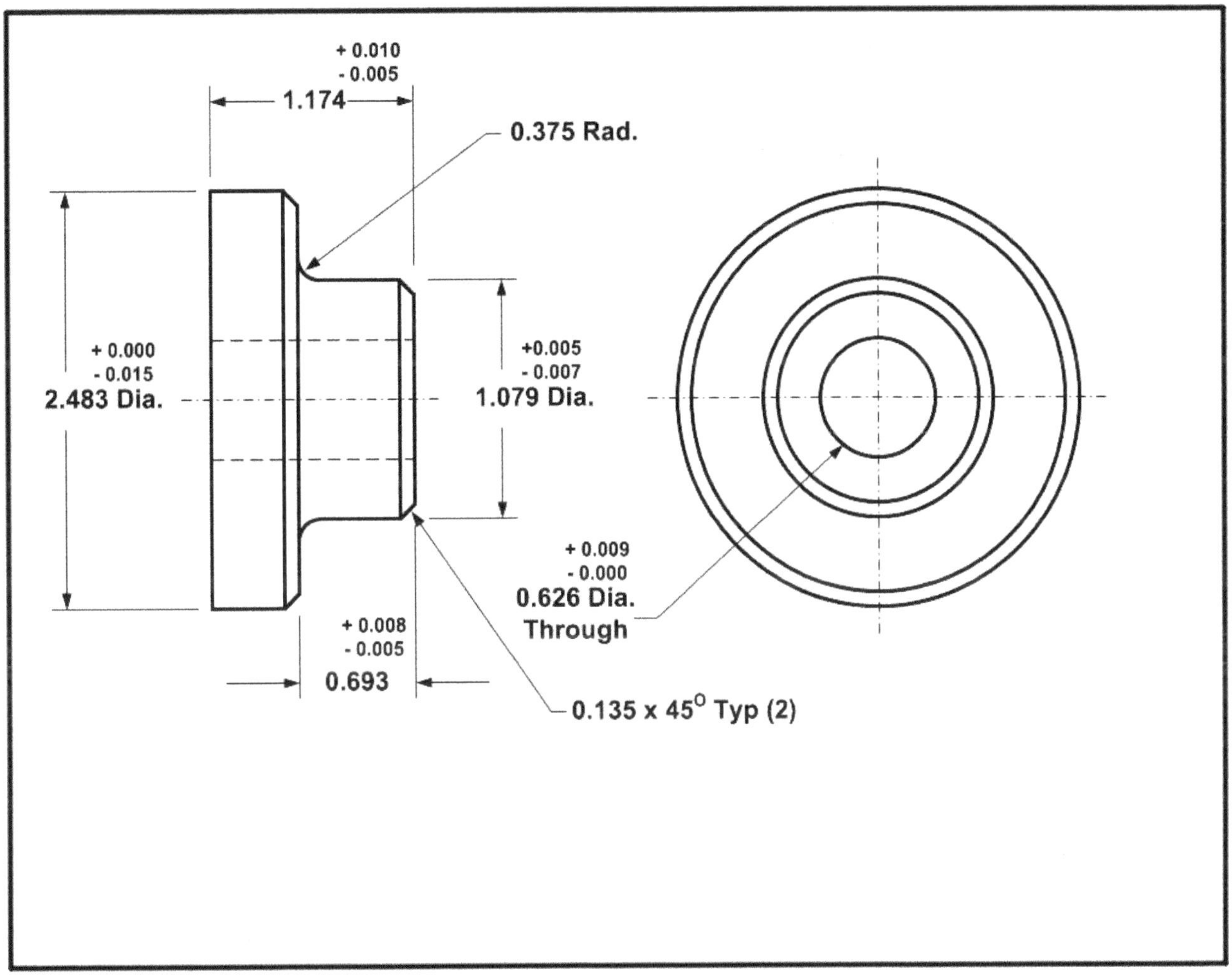

Figure 6.14 Part Print for Example Injection Molding Problem

The danger here is that if the material that is slated to be molded changes, the cavity sizes will not produce a part to the desired sizes. The aforementioned example problem was calculated for Nylon 4.6, which has an "S" value of 0.012 in./in. If the material were to change to ABS Polymer, the "S" value would be 0.005 in./in. and all of the cavity sizes would be too large to produce an acceptable part. Injection molds are very expensive, with some costing over a million dollars. By changing the molding material, the mold would need to be reworked if possible. If it cannot be altered, a new mold would need to be constructed.

6.3.3.1 Injection Molding Defects

Injection molding can produce some defects if the process is not running properly. These can be caused by either the component part not designed properly for molding, or the mold needing to be repaired, or the molding process itself. Following are some of the more common defects in injection molding and their probable cause:

Sink Marks and Voids: Sink marks and voids are depressions in the surface of the component part. These may be visible on the top or bottom of the part or on both opposing sides. The main cause of this defect is having nonuniform wall thickness throughout the part, causing more material shrinkage in various portions of the component part. Figure 6.15 shows sink marks on the surfaces of the component part.

Flashing: Flashing occurs when the plastic material bleeds out between the mold halves causing a thin layer of plastic to create a paper-thin flashing of plastic alone where the parting line of the mold is. This defect can be caused by three different reasons. One of the two process-related reasons is that the material is melted at too high of a temperature, decreasing the viscosity and allowing it to escape from where the mold is closed. The second process-related cause is that if the injection pressure is too high, it causes the same aforementioned phenomenon. Another cause is related to the mold itself. If the mold is not closing properly, this defect will be present even if the melt temperature and injection pressure are set correctly. For the third reason, the mold my need to be reworked to allow it to close properly in the machine. Figure 6.15 shows flashing along the parting line of the mold.

Short Shots: A short shot occurs when the mold cavity does not fill completely, creating an incomplete part. Some reasons for this defect could be that the injection pressure is not high enough, or the melt temperature is too low. Also, if the injector barrel is not filled, the mold cavity would not be completely filled.

Weld Lines (Knit Lines): Weld lines or knit lines are caused when the material entering a mold flows around a core and does not fuse together resulting in a weak portion in the component part. This can be caused by either the melt temperature of the plastic not being high enough, or the injection pressure being too low, or a combination of both. Another cause can be the result of the mold construction itself. If the plastic material is entering the wrong location, even if the pressure and temperature are correct, this defect can still occur. Figure 6.16 shows how the material flows around a core in the mold and creates the weld line with no fusion.

Tare Lines: Tare lines are scrapes on the side of the part from being ejected from the mold. This occurs when the component part gets stuck in the mold, and when it is ejected, the sides of the part become scraped. The main cause of this is improper design by not incorporating enough of a draft angle on the sides of the part.

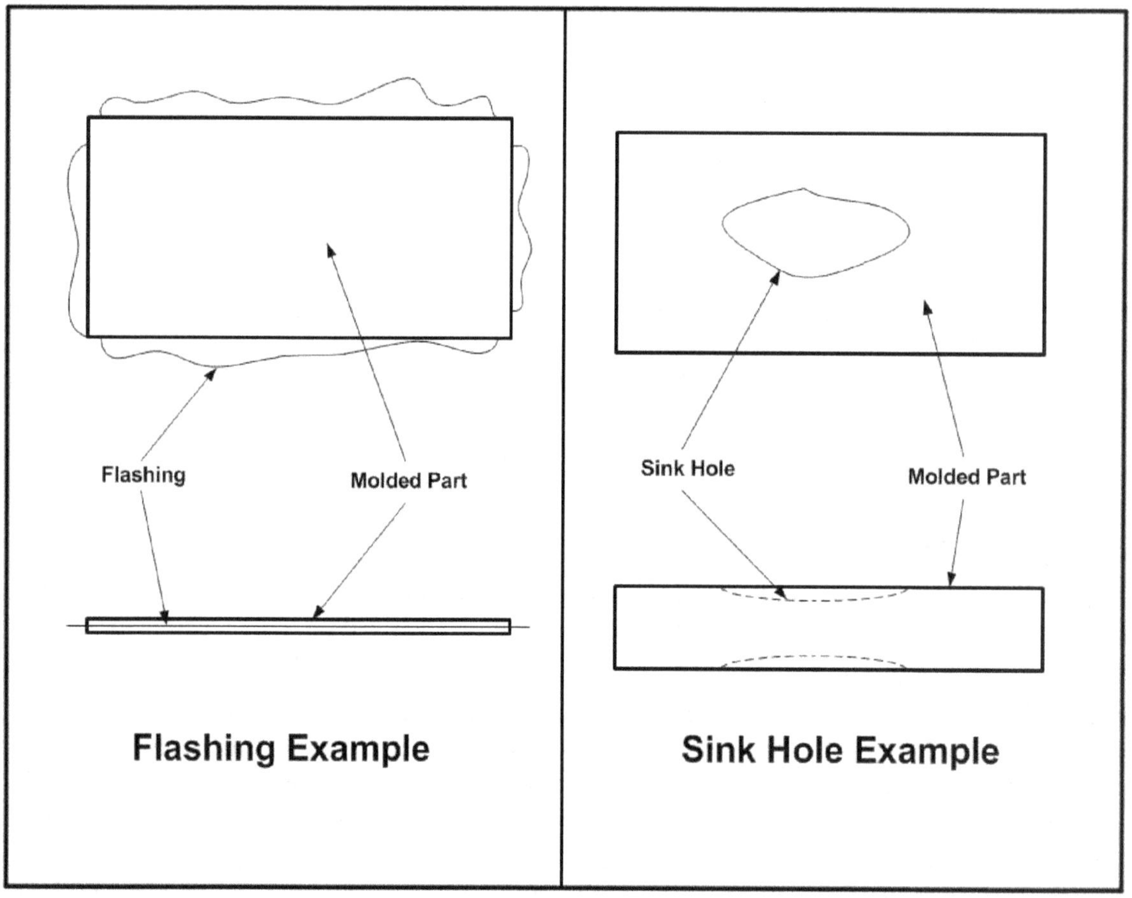

Figure 6.15 Injection Molding Flashing and Sink Marks

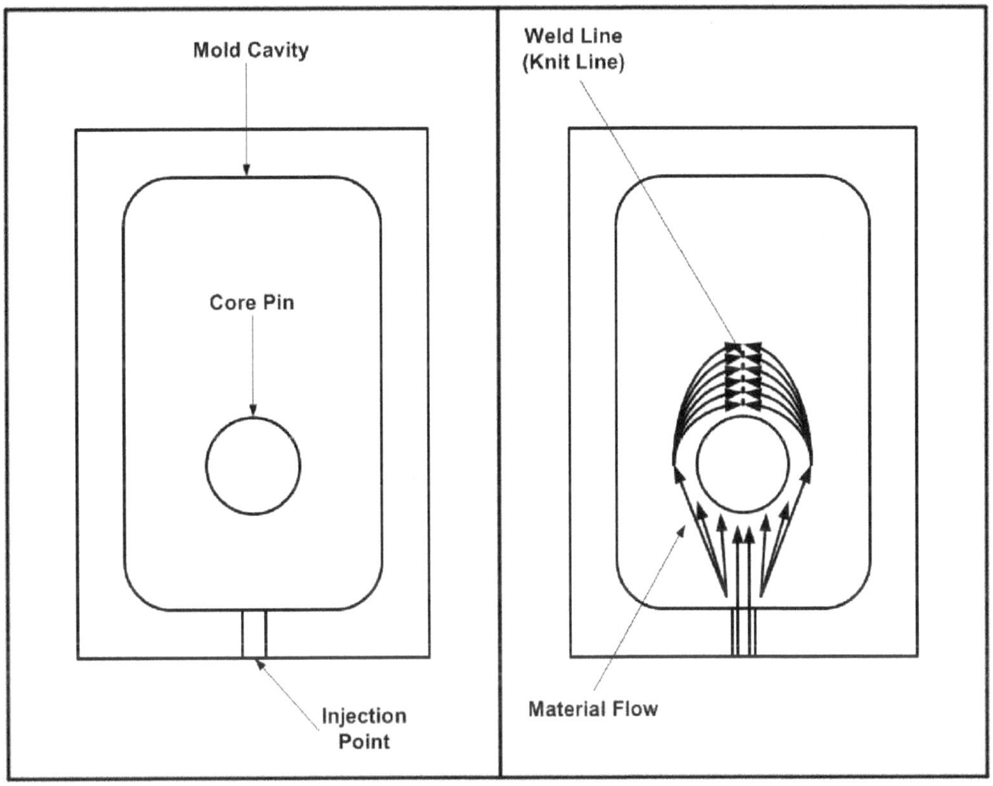

Figure 6.16 Weld Lines (Knit Lines) Caused by Material Flow

6.3.3.2 Injection Molding Design Considerations

As with any manufacturing process, there are design considerations and guidelines that need to be followed. Every process is different, and by designing the component part with these considerations in mind, the end results will be favorable. Following are the major considerations and guidelines for injection molding of plastic materials:

Strength and Stiffness: All plastic materials will render a given strength and stiffness. An evaluation of where the component part will be used is required in order to select the proper material for the application. Strength is always an important factor, but stiffness may be desired or not. If the component part will be subject to some flexing in the assembly, then stiffness may be a determinant and cause a failure over time.

Impact Resistance: Another material selection design parameter is the impact resistance. Some materials resist impact very well while others deform or shatter. Knowing the function of the component part in the assembly is critical for a proper material selection.

Service Temperatures: All plastic materials have a much lower melting temperature than do metals. If the component part is going to be used in environments where the temperature is going to be very high or very low, a failure can occur. With an elevated temperature, the component part will become weak as the material starts to soften up and can fail. In the same manner, if the temperature is very low, the component can become brittle and fail as well. Understanding the environment where a plastic component will operate in will determine the material selection.

Economic Production Quantities: Production of plastic components requires a substantial investment. In order to capitalize on this, the quantity of components must be high enough to offset the cost to produce the parts. Injection molds can be extremely costly, and some cost over a million dollars. Before considering a component part to be molded, the annual quantity needs to be substantial.

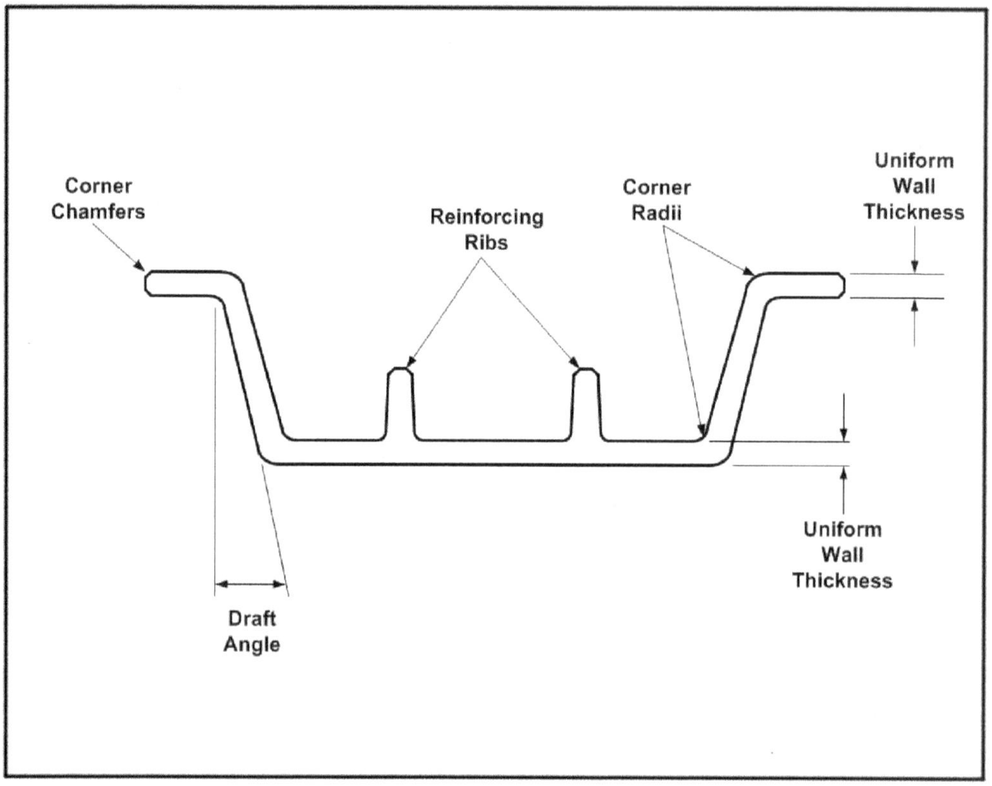

Figure 6.17 Correct Plastic Part Design

Uniform Wall Thickness: In plastic components, maintaining a uniform wall thickness is important to control the shrinkage and keeping it uniform throughout the part. Plastic parts where the wall thickness varies a lot can result in nonuniform shrinkage and cause sink marks and voids. Figure 6.17 shows how the wall thickness should be maintained throughout the part to render an acceptable component in molding.

Reinforcing Ribs: If, in a component part, additional strength is required in specific areas, rather than making the material thicker, reinforcing ribs are designed into those locations. These ribs will also be of the same uniform thickness in most cases to ensure uniform shrinkage. Figure 6.17 shows how reinforcing ribs can be implemented to increase the strength of the plastic component in specific areas.

Corner Radii and Chamfers: Sharp corners cannot be molded into a plastic part. Any of the 90° corners will cause the part to hang up in the mold when being ejected, so all the corners need to contain either a chamfer or a radius. Figure 6.17 shows the implementation of corner radii and chamfers throughout the component part for injection molding.

Draft Angles: No injection-molded part can have straight sides. A draft angle of between 2° and 3° needs to be designed into the part to allow for ejection from the mold. If there is not enough angle on the sides of the part, tare lines till occur on the sides. If the component part requires straight sides, a secondary operation would be required to machine the sides straight. Figure 6.17 shows the draft angles on the part throughout, even on the reinforcing ribs.

Holes: Holes are very possible in injection molding. These require a core pin, which the plastic material flows around, and it is retracted when the mold opens. Proper filling is required to avoid any weld lines on the component part, which needs to be studied to ensure proper filling. Also, the implementation of core pins for molding holes will greatly increase the overall cost of the mold.

Tolerances: Every manufacturing process has limitations on what tolerances it can and cannot hold. Understanding what the plastic processes can attain is critical to ensure that the tolerances required can be maintained. If closer tolerances are required, secondary machining operations can be employed.

Future Disposal and Recycling: Recycling is a very critical element in this day and age. Plastic materials will not decompose and will remain in their configuration forever in a landfill. Pretty much all of the thermoplastic materials are able to be recycled and can be reused over and over. Figure 6.18 shows the seven recycling symbols that are molded into the component parts, indicating what family of plastics they are so that they can be sorted and reused again.

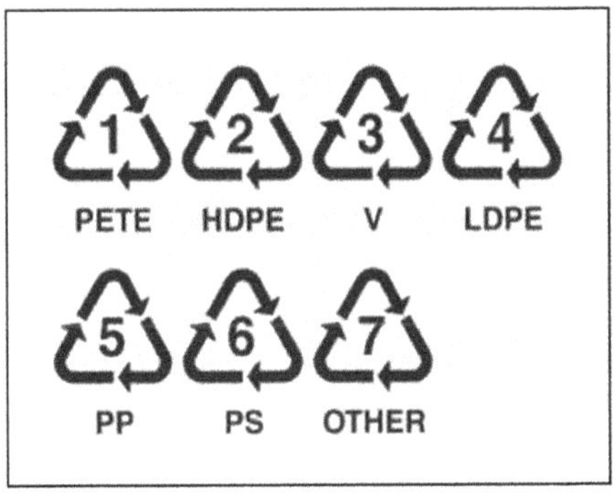

Figure 6.18 Recycling Symbols on Plastic Parts

6.3.4 Blow Molding Process

The blow molding process is utilized for producing components like plastic bottles. This process is a combination of tube extrusion and injection molding with air pressure added in. Figure 6.19 shows the mold receiving the molding material. A tube of plastic is extruded into the open mold to a given length. Once the material has been extruded, the mold closes shut, cutting off the top of the extruded tube, sealing it, and the air injector moves up as shown in Figure 6.20. The next phase is to inject air into the center of the tube to push the plastic material against the sides of the closed mold as shown in Figure 6.21. After the cooling has occurred, the mold opens and the air injector pulls down, allowing the completed component to fall free as shown in Figure 6.22. Blow molding is the primary process utilized for producing plastic bottles of every type. This has replaced a vast amount of glass containers and is much faster and less expensive.

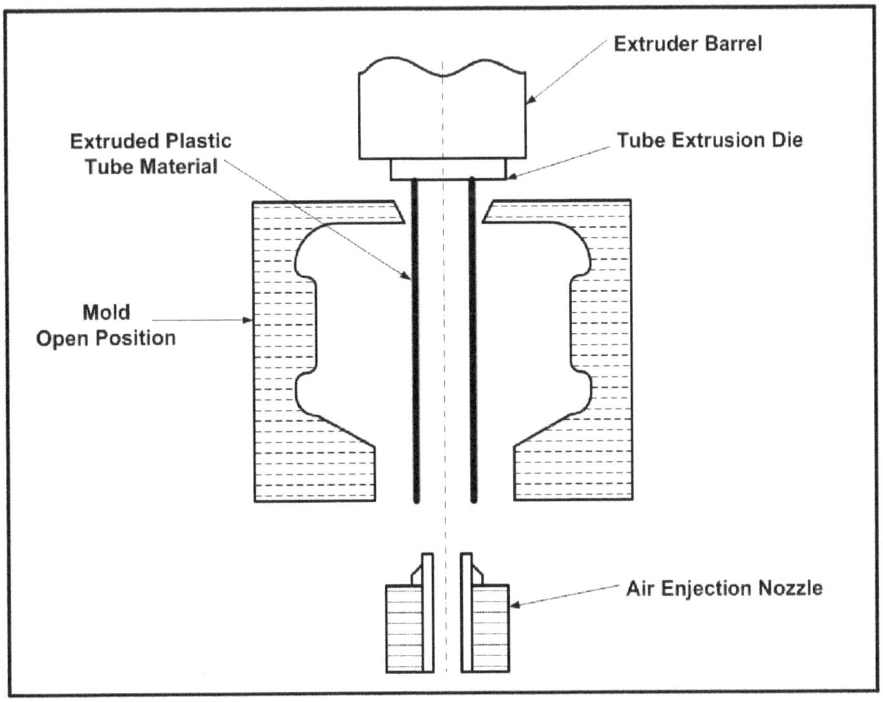

Figure 6.19 Blow Molding Step 1; Mold Opens, and Extruded Tube Runs into the Mold

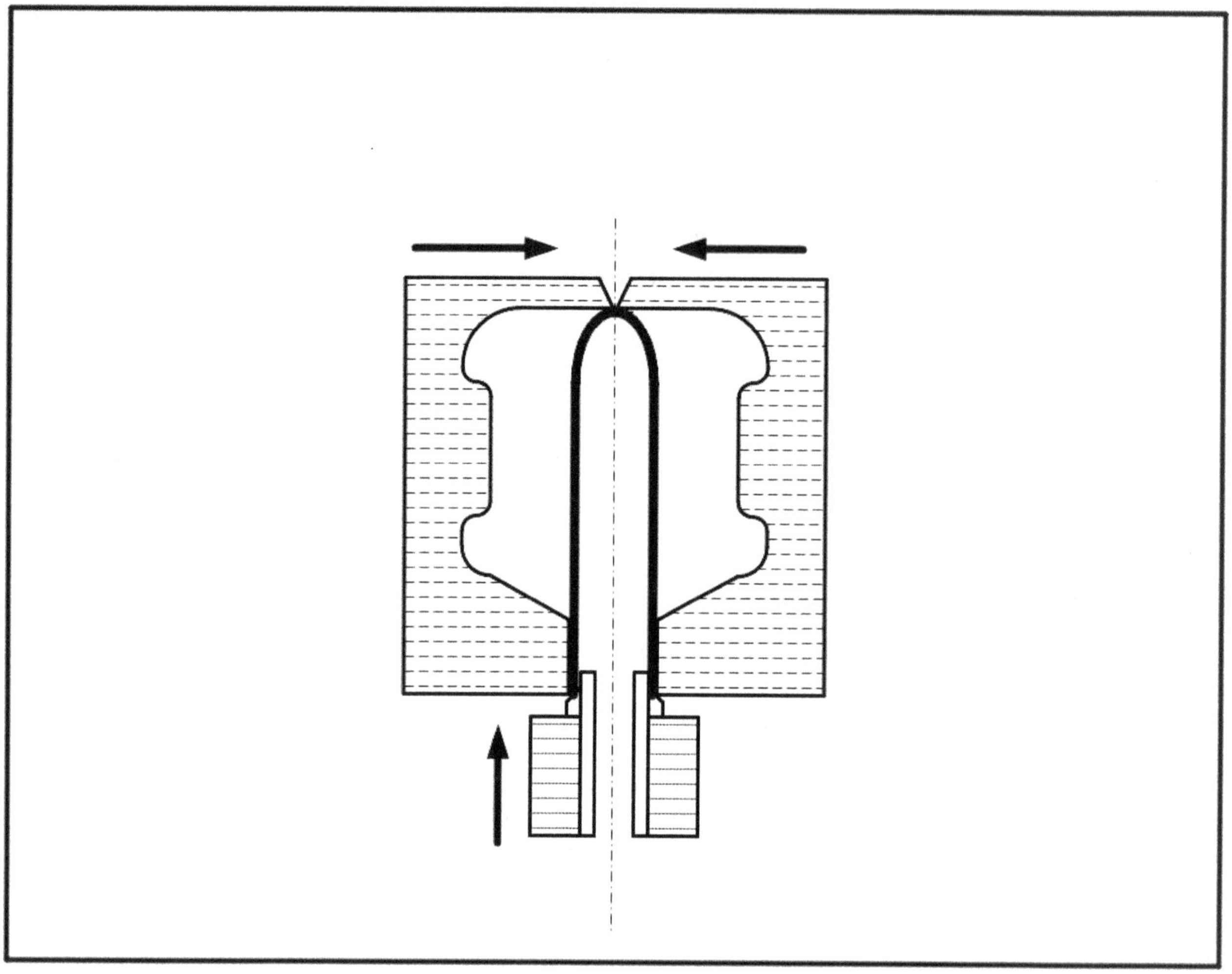

Figure 6.20 Blow Molding Step 2; Mold Closes, Cuts Tube and Seals, Air Injector Seals Bottom

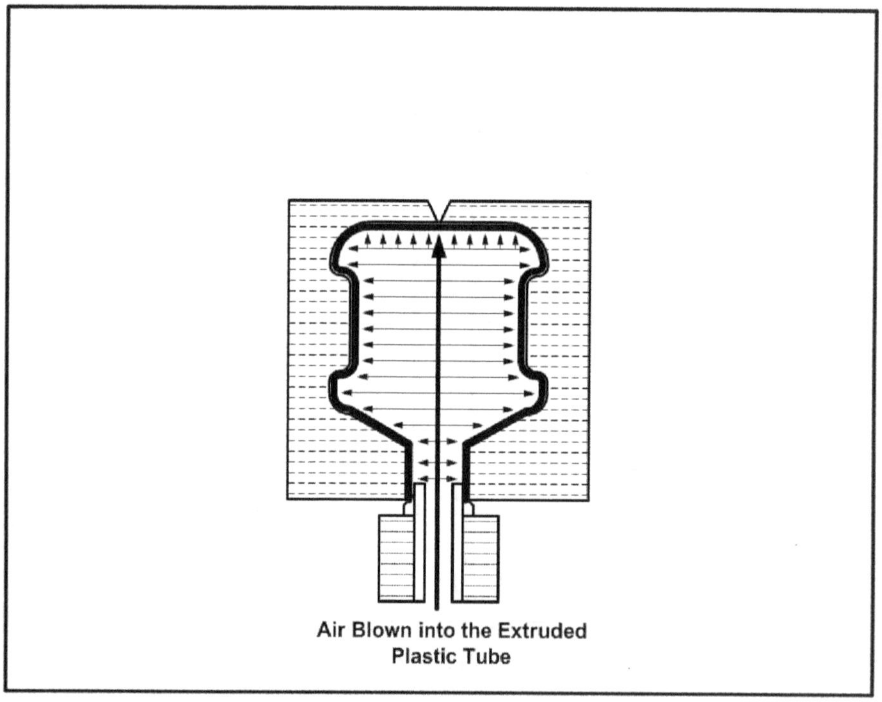

Figure 6.21 Blow Molding Step 3; Air Blown into the Mold from Bottom, Forming the Shape

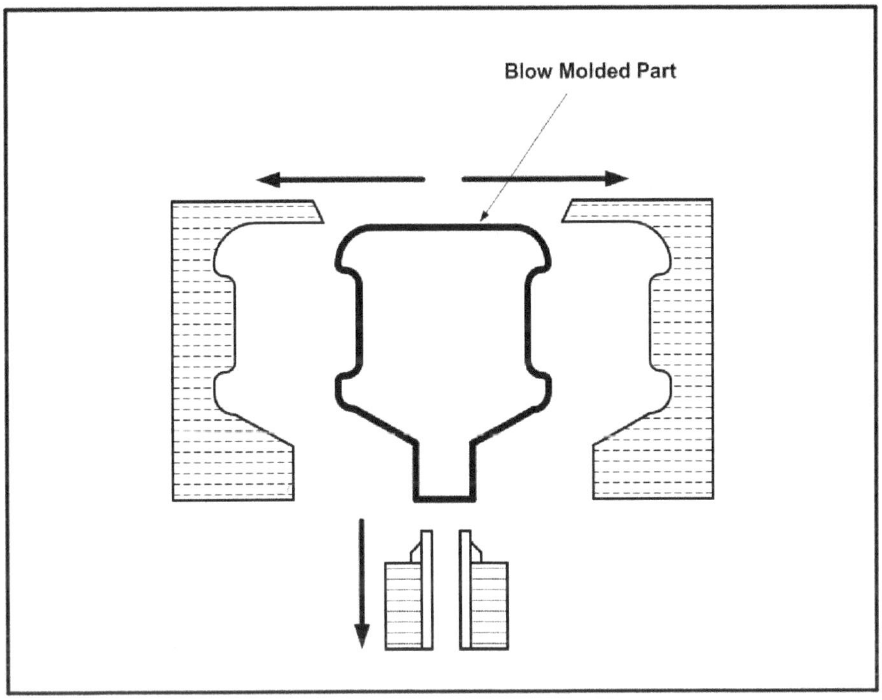

Figure 6.22 Blow Molding Step 4: Injector Retracts, Mold Opens, Part Falls Free

6.3.5 Vacuum Forming Process

Vacuum forming is a process that utilizes thin sheets of plastic. The main products from vacuum forming are packaging containers used to hold parts or products in place. Figure 6.23 shows the configuration of a vacuum molding setup. The top movable platen has an open cavity where the bottom stationary platen obtains its desired shape. There are airtight seals around the mold halves to ensure an airtight seal when the mold is closed. A heated thin sheet of plastic is placed between the platens to be shaped. The mold is closed as shown in Figure 6.24, and a vacuum is applied through the bottom platen through very small holes in the die section.

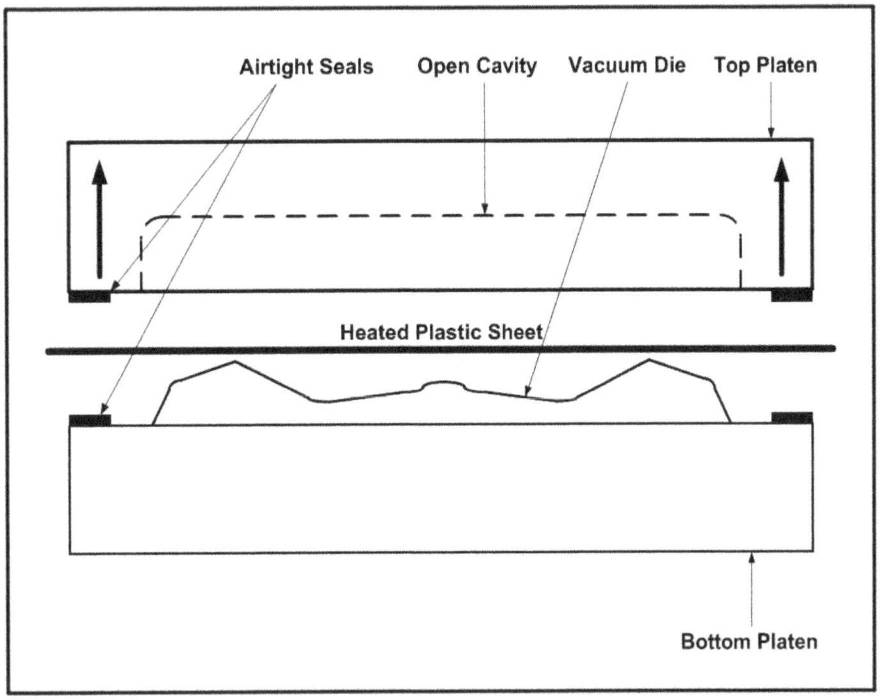

Figure 6.23 Vacuum Forming Step 1; Heated Plastic Sheet Placed in the Mold

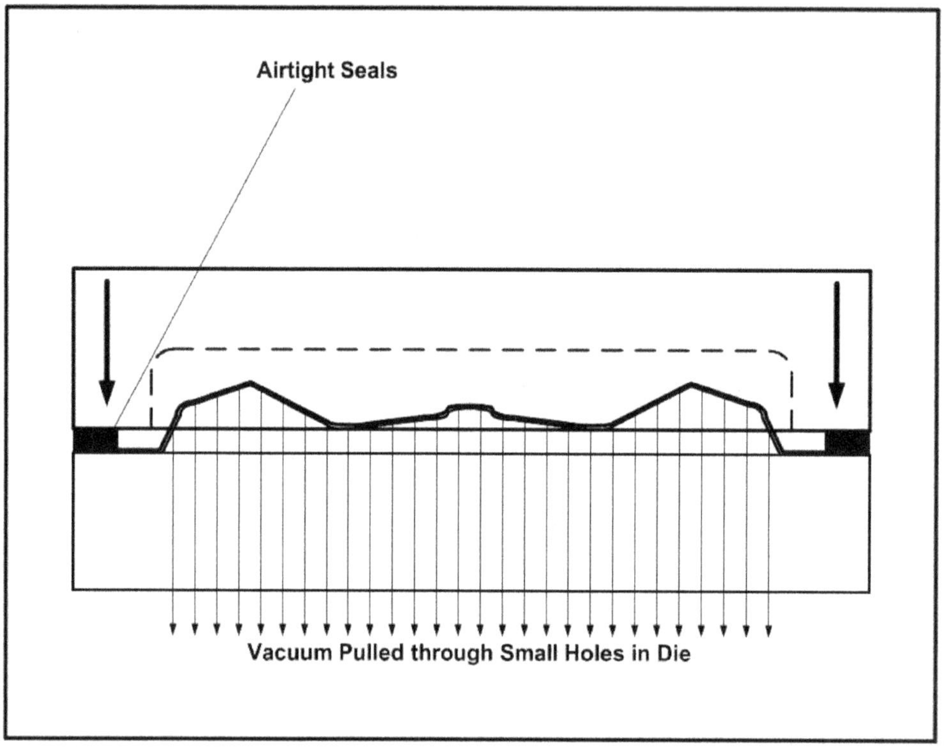

Figure 6.24 Vacuum Forming Step 2; Mold Closes and a Vacuum Pulls the Plastic onto the Mold

The heated plastic is pulled down against the die pattern, causing it to obtain the desired shape. Each run of this process requires the plastic material to be handloaded and time for it to obtain the required heat for shaping.

6.3.6 Hot Pressing Process

Hot pressing utilizes a mold where there is a top and a bottom form. This process can be performed with thicker sheets of plastic than can vacuum forming, rendering it capable of producing much heavier components. The plastic sheet is heated and placed between the top and bottom platen as shown in Figure 6.25. The

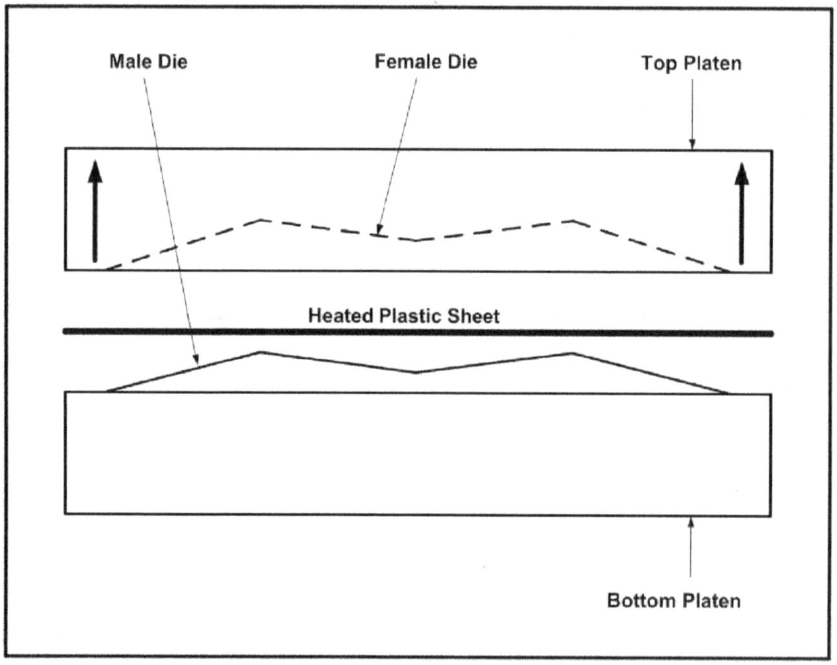

Figure 6.25 Heat Pressing Step 1; Heated Plastic Sheet Loaded into the Mold

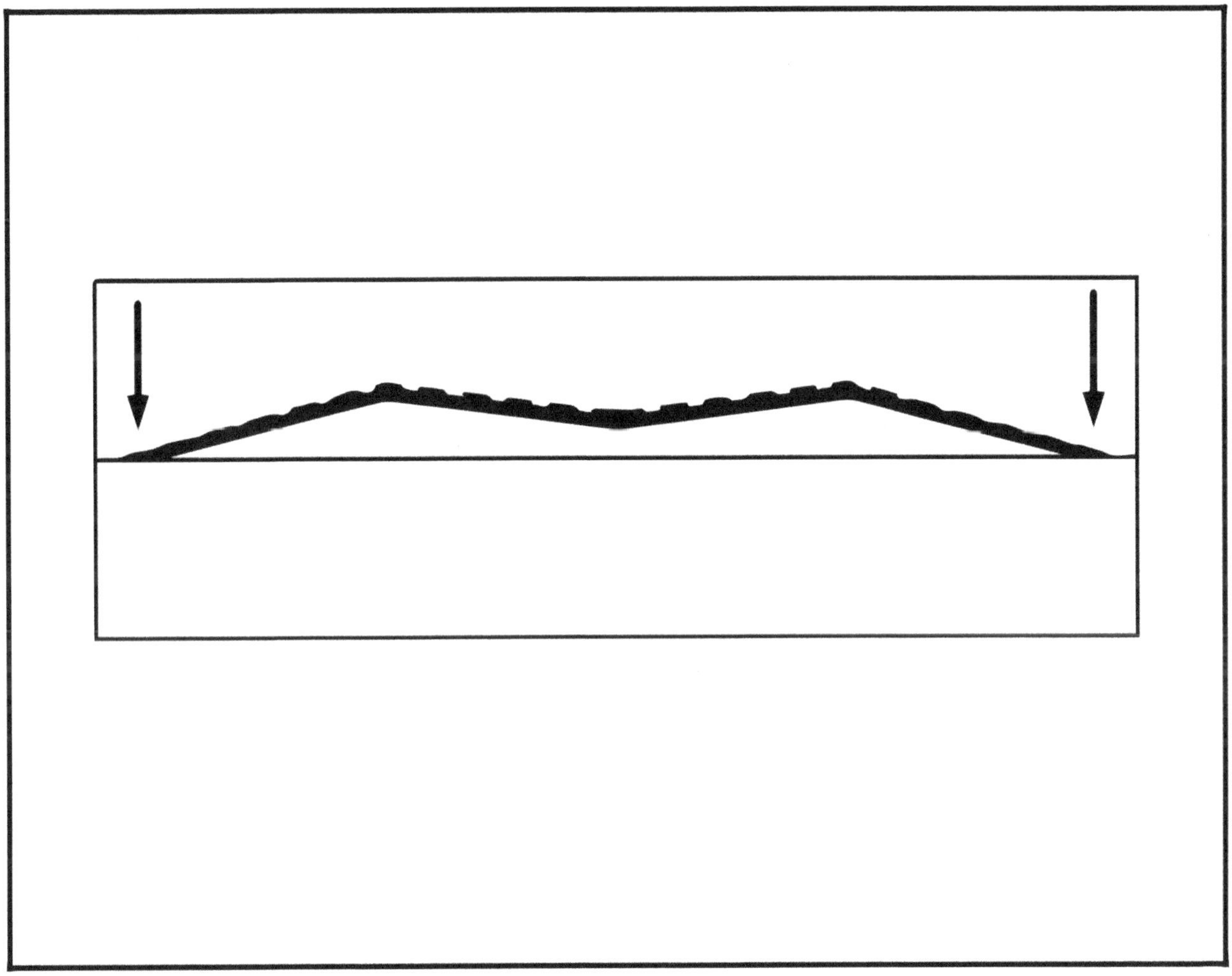

Figure 6.26 Heat Pressing Step 2; Mold Closes and Presses the Plastic between the Male and Female Dies

upper platen moves down forcing the sheet between the upper and lower die shapes shown in Figure 6.26. After a cooling time, the upper platen moves up and the component is removed. This is repeated the same way as vacuum forming, loading a new sheet of plastic in the mold each time.

6.3.7 Sheet Rolling

Rolling of plastics is how plastic sheets of all sizes are produced. This process is simple, but several factors need to be considered. The heated plastic material is sent through a series of rollers, reducing it in thickness each time as it is reheated between rolls. Figure 6.27 shows the rolling process reducing the thickness of the plastic. As it is rolled out, the material becomes thinner and thinner until the desired thickness is obtained. For very thin plastic sheet, it can be rolled up in a coil if desired, allowing it to be unrolled and cut to the desired length at the time of use.

The areas that need to be considered in the rolling process are critical to a successful process. First is the viscoelasticity of the material. The material in rolling will react similar to that in the extrusion process, and as it is fed through the rollers, it will expand as it exits. This needs to be compensated for in order to obtain the proper thickness. The other consideration is the volume of material that the rolling process starts with. In rolling, the volume of the starting material will not change ($V_1 = V_2 = V_3 = \ldots V_n$). Also, as the material becomes thinner, it will exit the rollers faster. With the material volume not changing, the velocity of it exiting the rollers has to increase ($v_o, t_o, w_o = v_f, t_f, w_f$). At first, the exiting velocity will not be that noticeable, but as the thickness reduces further and further, it will increase dramatically. These considerations need to be accounted for in order for the rolling process to be successful.

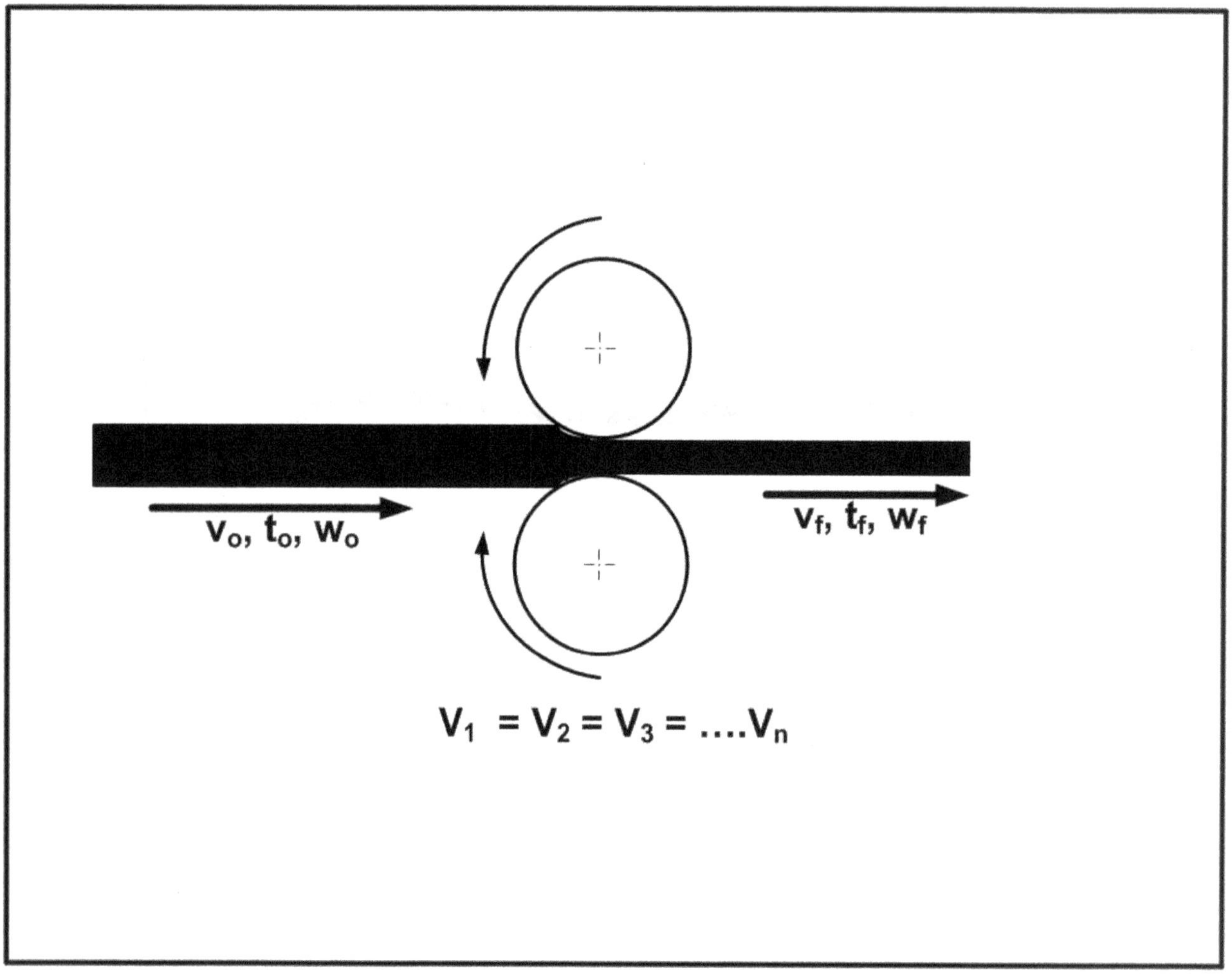

Figure 6.27 Plastic Rolling Concept

6.4 SUMMARY

Plastics have, without a doubt, become a major player in the component fabrication sector. As new materials are invented and process innovations improve the processes, more and more components are being fabricated from plastics. Several components previously made from castings, powder metal, and glass have been converted to plastics because of the economics of the processes. This will only increase as innovations in materials and processes increase.

6.5 QUESTIONS AND PROBLEMS

Questions

1. What is the difference between a Thermoplastic material and a Thermoset material?

2. In the Extrusion process, what are Drag Flow and Back Pressure?

3. The two defects that can occur in the extrusion process are Melt Fractures and Sharkskin. What is the difference between the two, and how are they caused?

4. In the Injection Molding process, what are Date Clocks and what information do they provide and why is this important to know?

5. For injection-molded parts, common defects are Sink Marks and Voids. How these caused, and what are some ways to prevent them from occurring?

6. With all plastic components, service temperatures are a concern. Why is this important to know?

7. Another area that needs to be evaluated for plastic components is Economical Production Quantities. Why is that a concern?

8. Recycling of plastic components is important. Can all plastic materials be recycled? If not, why?

9. When performing sheet rolling of plastic material, what is a critical factor to ensure that the finished product will be of the desired size and why?

Problems

1. An extrusion is being produced from a plastic material. If the information is as follows: Barrel Diameter = 4.250 in.; Channel Depth = 1.750 in.; Rotational Speed = 2.25 RPS; Flight Angle = 14.75°; Barrel Length = 38 in.; Material Viscosity = 184×10^{-4} #-s/in.^{2}; and the Head Pressure = 150 PSI, what would the Resulting Flow Rate be?

2. For problem #1, if the material cooled down a little and the Head Pressure changed from 150 psi to 160 psi, would the process still produce an extrusion?

3. An extrusion is being produced in a process with the following information: Barrel Diameter = 5.250 in.; Channel Depth = 2.125 in.; Pitch = 3.492 in.; Barrel Length = 5 ft 7 in.; Material Viscosity = 213×10^{-4} #-s/in.^{2}; and the Head Pressure = 200 psi. What would the rotational speed need to be if a Resulting Flow Rate of 58 in^3/s was required?

4. With the information in problem #3, if a Head Pressure of 200 psi was a result of the die opening being 1.65 in.^{2}, what would the new Head pressure be if the die opening was reduced to 1.35 in.^{2} and would it still have a positive Resulting Flow Rate?

5. An extrusion process is set up with the following information: Barrel Diameter = 7.125 in.; Channel Depth = 3.250 in.; Barrel Length = 8 ft 8 in.; Material Viscosity = 387×10^{-4} #-s/in.^{2}; and the Head Pressure = 450 psi. If the Pitch were to change from 2.781 in. to 3.428 in., how much would the Resulting Flow Rate increase?

6. A rectangular component part is going to be injection-molded in a single-cavity mold. The desired part size is 5.750 in. L × 3.250 in. W × 0.625 in. H. If the material being used is

   Ionomer Thermoplastic, what size would the cavity need to be?

7. A four-cavity mold is going to be utilized to injection-mold parts that are spherical in shape. The diameter of the spheres is 1.625 in. and it is to be molded from Polyester Thermoset material. What cavity sizes need to be in the mold, and how much material needs to be prepared for each run of the mold, neglecting the filler runners?

8. A 10-cavity mold will be producing component parts of a cylinder shape. The diameter of the cylinder is 2.125 in. with a height of 3.875 in. With Nylon 4.6 Thermoplastic material being used, what cavity sizes are required and what volume of material will be consumed with each run of the mold, neglecting the filler runners?

9. If the same mold in problem #8 is going to be used with Polycarbonate Thermoplastic material, and the tolerance on the height and diameter is 0.005, would it still produce an acceptable part?

10. A new component part is being considered to be injection-molded. It is a rectangular-shaped part that is 4.125 in. L × 3.625 in. W × 1.875 in. H. The two materials that would function are ABS Polymer Thermoplastic and PVC Thermoplastic. Cost is critical, so an evaluation of material volume for producing 250,000 parts needs to be performed. With ABS plastic costing \$0.206/in.^{3} and PVC costing \$0.193/in.^{3}, what would the cost difference be between the two materials and which would be the most economical to use?

CHAPTER

7

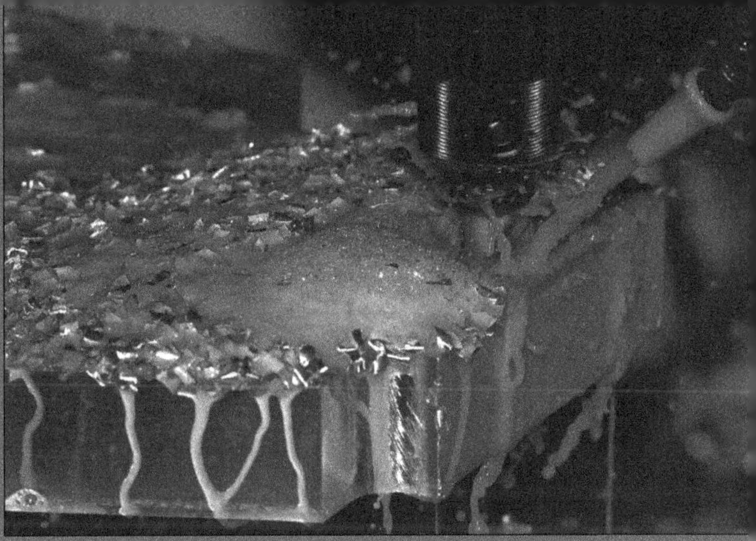

Rubber Shaping Principles and Composites

7.1 INTRODUCTION

Rubber is a material that has been used for hundreds of years since its discovery. Originally this was a natural substance that was taken from nature and refined into a product possessing unique qualities. Over time, synthetic rubber, not derived from natural sources has become more popular for manufacturing products owing to its stronger and more versatile features. By adding additives to the synthetic rubber and running it through various chemical processes, the material can be utilized in several applications that the natural rubber cannot.

Composite materials are a combination of two or more materials resulting in a "super material" that possesses qualities far superior to any of the material alone. This new and developing arena of material science is changing on a daily basis as new composite materials are being developed. Rubber processing is included in the composite material sector because of the additives being mixed with it and the processing is very similar to composite processing. This chapter covers the basics of rubber processing and composite processing along with the more common manufacturing processes utilized to shape the products.

7.2 TYPES OF RUBBER

Rubber types are broken down into two categories which are natural rubber and synthetic rubber. Both have almost identical molecular structures, but are very different in how they are produced and their applications for component parts. A breakdown of both natural and synthetic rubber is outlined below.

7.2.1 Natural Rubber

Natural rubber is derived from elements procured from nature. Rubber trees are tapped and the sap which is liquid latex is extracted for the base of the rubber. This sap is then diluted 50% with water and various acids are added to allow the latex sap to coagulate into chunks of latex in the tank. The two main acids that are used to coagulate the latex in the tank are formic acid or acetic acid, and this process requires up to 12 h for complete coagulation to take place. Once the latex has completely coagulated into chunks in the tank, it is removed and allowed to dry. After some drying time, the latex is rolled out into slabs to remove additional moisture and to produce thinner slabs. The final step is drying it to remove all the moisture in the slabs in a smoke house. Slabs are hung on racks and wood is burned allowing the slabs to dry with wood soaked in creosote. After smoking the latex, the finished material is ready to be utilized in production as a raw material for producing rubber components. Figure 7.1 shows a graphic depiction of the process for producing natural rubber material.

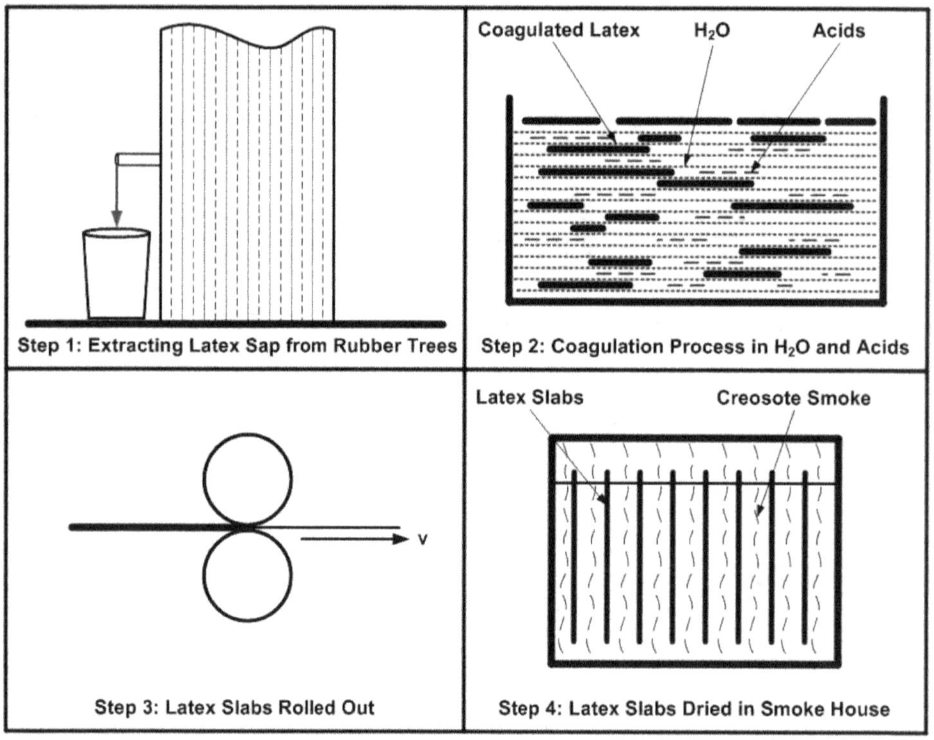

Figure 7.1 Natural Rubber Production Process

7.2.2 Synthetic Rubber

Synthetic rubber is not a product derived from nature, but is a petroleum product where a mixture of petroleum and other chemicals form a rubber material. This is the most common and widely used rubber today and is produced in a multitude of compositions. The process for producing synthetic rubber is very similar to the process used to produce thermoplastics. However, while thermoplastics are produced in small pellets, synthetic rubber is produced in large, bail-size blocks.

7.3 PRODUCTION PROCESS

Working with rubber materials, either natural or synthetic requires four steps in order to produce component parts properly. These four steps involve compounding, mixing, shaping, and vulcanizing. Each of these process steps are outlined below along with their purpose in the fabrication processes.

7.3.1 Compounding

This step involved mixing various additives into pure rubber to give it different properties required for the function which will be required to perform in the component parts. These additives will give the rubber different properties for strength, stiffness, and cycle life for the desired application of the parts being fabricated. Some of the more common additives utilized in compounding are listed below and their purpose in the process.

7.3.1.1 Additives

Carbon Black: Carbon black (soot) is an additive used to increase the strength of the rubber. It is the most important filler and is used in just about every rubber compounding process. This will increase the resistance to abrasion of the product and also increase the tensile strength of the rubber material. Because carbon is black in color, this gives the rubber its black color as seen in tires, most belts, and a variety of other rubber products produced.

String Fibers: Another strengthening additive used in rubber is string fibers. These are not long fibers, but short pieces of string that get mixed into the rubber compound. As with carbon black, this increases the strength of the rubber and adds resistance to wear in applications such as tires and drive belts. Full-length string fibers will be used in various molding processes to reduce the elasticity of the rubber for certain applications.

Fiberglass: Fiberglass string pieces get mixed with the rubber for two reasons. The first reason is to increase the strength of the material and to also increase the rigidity. The second reason is to increase the wear resistance of the material for much longer life of the product. With glass not having the flexibility of string fibers, the material will not be as ductile and can crack if caused to flex repeatedly.

Foaming Agents: For certain rubber product requirements, a foaming agent is added to allow the rubber to foam up during shaping. This is utilized for producing foam rubber components where the material is not as dense with air bubbles internal on the parts. Applications required a fare amount of shock absorbing, and the material is foamed up to allow for the entrapped bubbles to compress during usage of the component parts.

Vulcanization Agents: Vulcanizing agents are added to the rubber if vulcanization is required for the component parts. These include zinc oxide, stearic acid, and a small amount of sulfur. Although they are part of the compounding process, they may be added later in the process to prevent premature vulcanization from taking place.

Other Additives: Some other additives that get used in rubber compounding are diamond dust, metal particles, aluminum oxide, etc. These are used for very special applications, and are not as common as those listed above.

7.3.2 Mixing

After the compounding of the various additives, the rubber must be very thoroughly mixed. Unlike other materials, rubber will maintain a very high viscosity level and not become fluid. This mixing process is traditionally performed in two separate stages. The first is for all the desired additives except the vulcanization agents. Heating the rubber to around 300°F and mixing the additives thoroughly into the rubber will ensure an even distribution throughout the material. The second stage of mixing is when the vulcanization agents are mixed in at a lower temperature. This is to ensure that the material will not start to vulcanize prematurely and render the material not capable of being shaped of formed. Figure 7.2 shows a typical roller mixing unit where the additives are being combined with the rubber together.

7.3.3 Shaping

Rubber can be shaped in different manners depending on what configuration and purpose the component parts will require. Many of the processes can be utilized with other materials, but the more common rubber-shaping processes are detailed below.

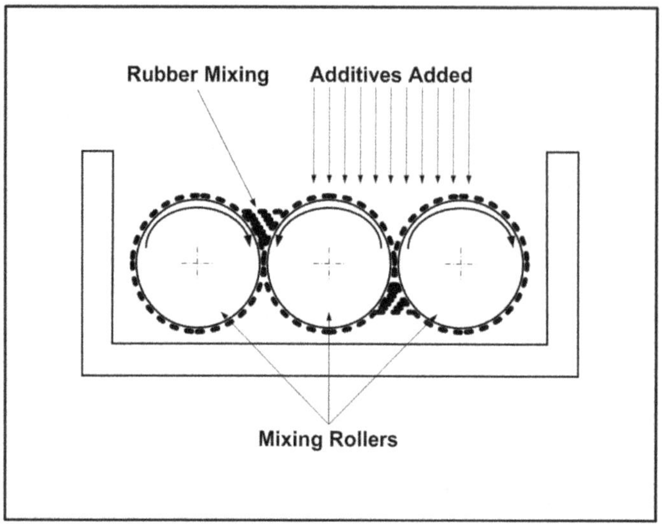

Figure 7.2 Rubber Mixing Process

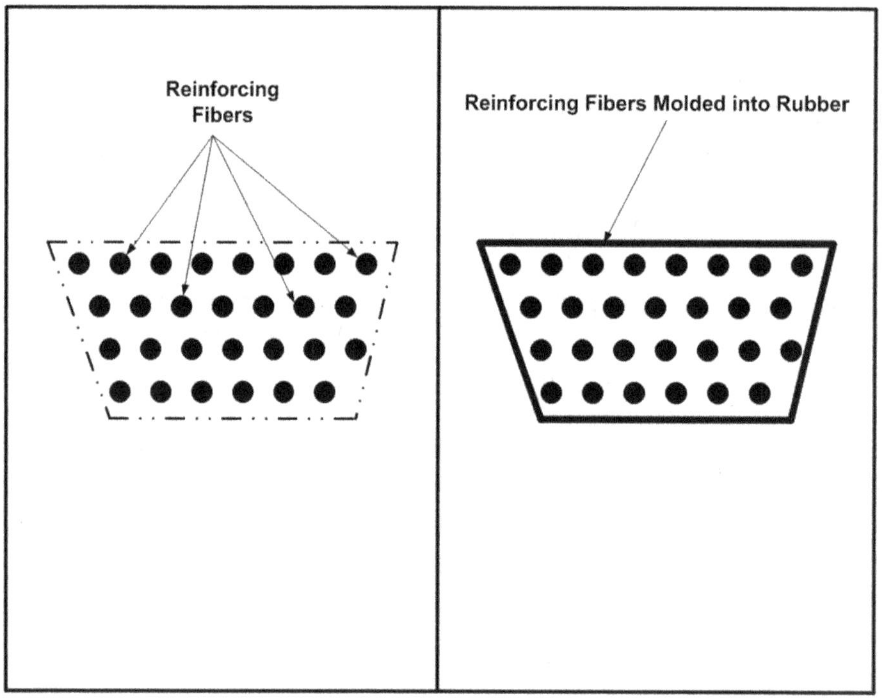

Figure 7.3 V-Belt Construction with Rubber Molded around String Fibers

7.3.3.1 Fiber Molding

Fiber molding is a process where the rubber material is molded around fibers of string or metal. This is very common in the production of drive belts and tires for added strength and reducing the elasticity of the material. The fibers are suspended in the mold and the rubber material is injected so that it completely incases the fibers. Once the rubber is molded, the component part will have very little elasticity and be extremely strong. Figure 7.3 shows the cross section of before and after a fiber molding process has been performed.

7.3.3.2 Dip Casting

Dip coating is a unique process consisting of a pattern and molten rubber. This process is utilized for producing products such as rubber boots and gloves where they are produced with no seam lines at all. A vat containing molten rubber is used along with a pattern as shown in Figure 7.4. The pattern is dipped, sometime several times into the molten rubber vat shown in Figure 7.5. Once the rubber has cooled, the dip cast rubber part is removed from the pattern as shown in Figure 7.6, producing a rubber component with no seams. From there, secondary operations can be performed such as attaching the heal and tread section to rubber boots.

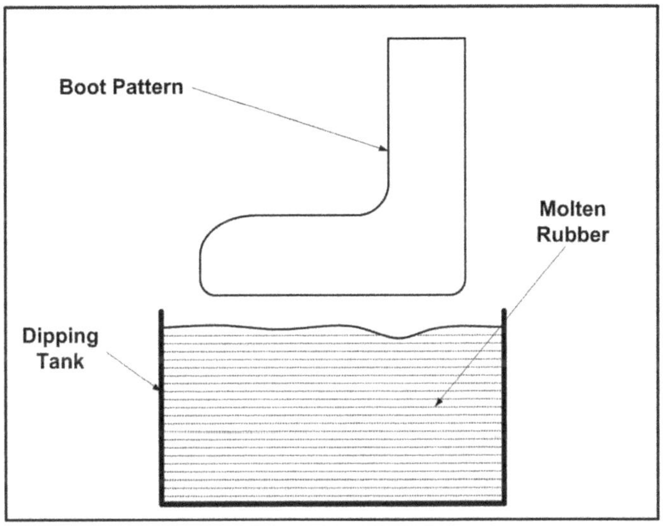

Figure 7.4 Dip Casting Process; Pattern to Be Dipped into Molten Rubber Bath

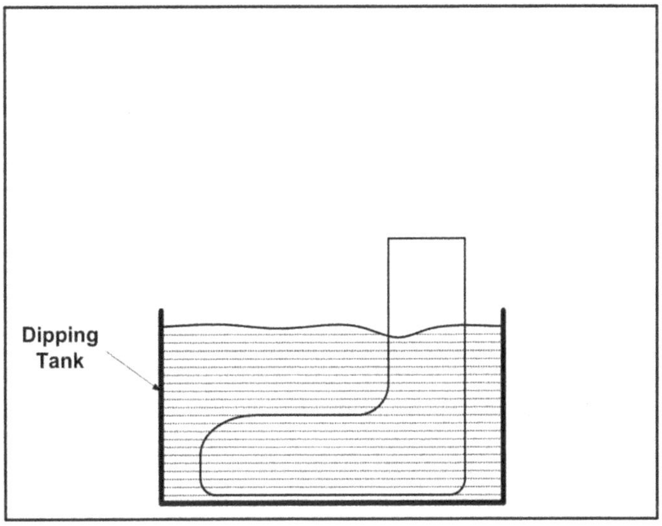

Figure 7.5 Pattern Dipped into Molten Rubber Bath Several Times if Required

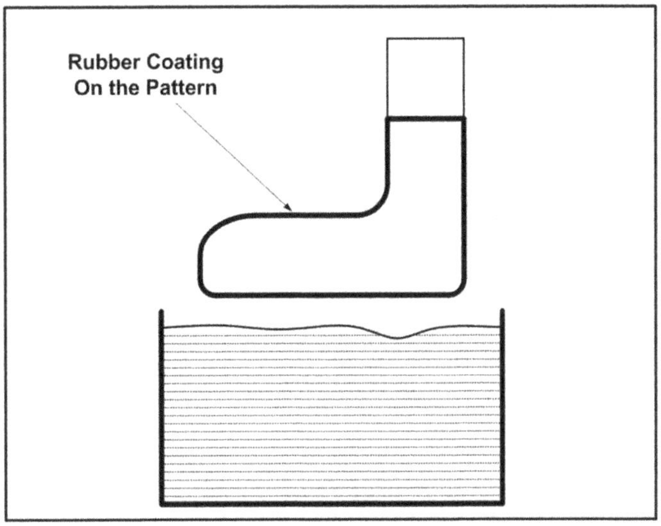

Figure 7.6 Pattern Coated with Rubber to Be Removed

7.3.3.3 Hot Powder Coating

Powder coating is a process that can be performed in two different methods which are electrostatic and hot coating. Hot powder coating is utilized for protection, and is a non-decorative coating. In hot powder coating, the rubber material is ground into very small particles which is placed in a vat at room temperature. The vat will have air injected through the bottom to suspend the rubber particles in air as illustrated in Figure 7.7. A work piece is heated and then run through the room temperature rubber particles suspended in the vat, and they will melt onto the work piece. Figure 7.8 shows how this process is performed. The work piece will cool and the rubber which has melted to it will provide a rubber coating on the piece as in Figure 7.9.

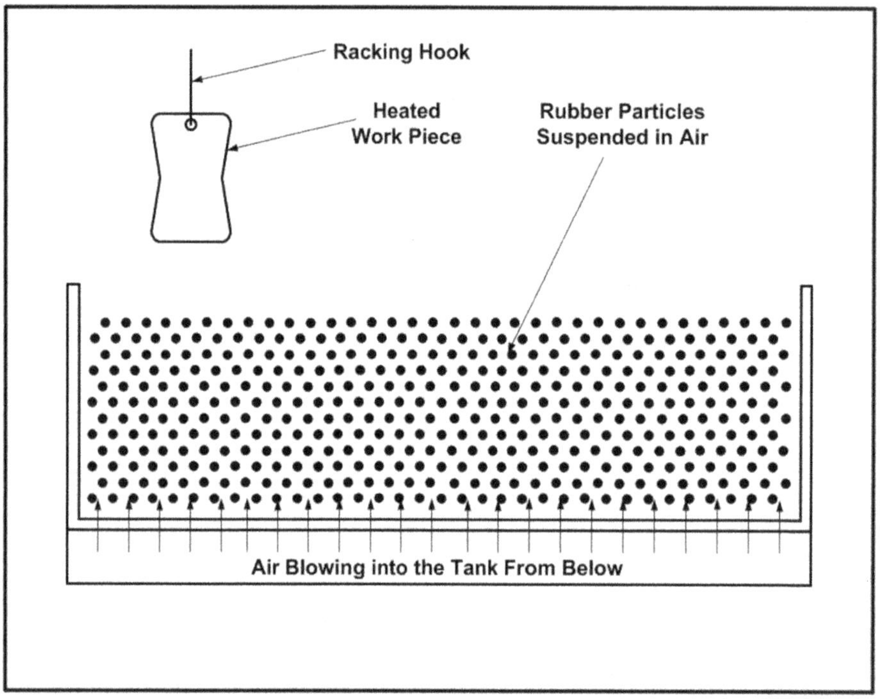

Figure 7.7 Rubber Particles Suspended in Air and Heated Part to Be Run Through

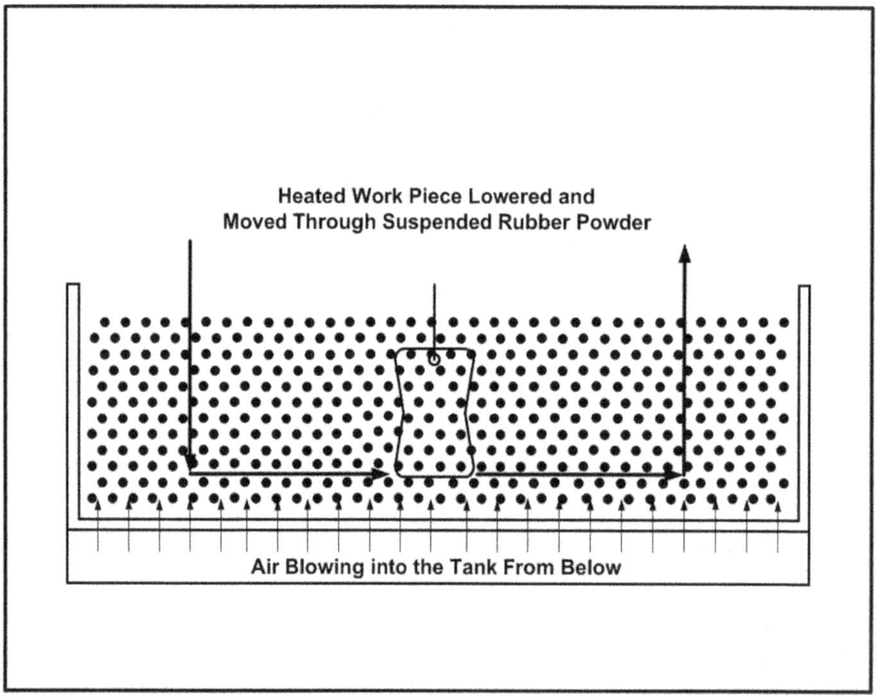

Figure 7.8 Heated Part Run through Rubber Particles, Melting them to the Surface of the Part

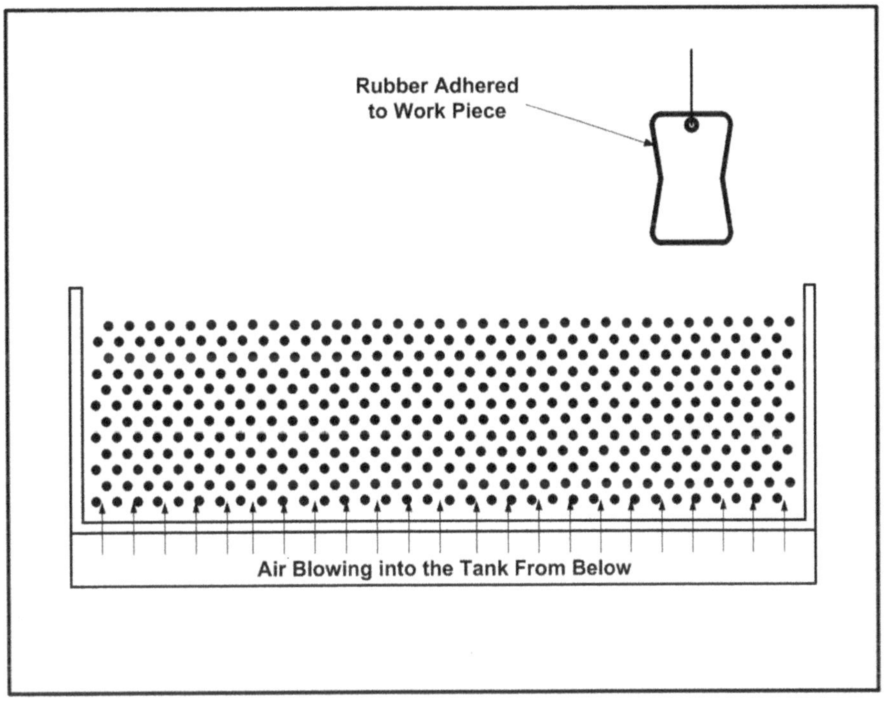

Figure 7.9 Work Piece Coated with Rubber

7.3.3.4 Injection Molding

Rubber can be injection molded the same manner that plastics can be. A permanent mold is constructed consisting of 1 to multiple cavities and placed into an injection molding machine. The molten rubber is injected under pressure into the cavities creating the configurations desired. Shrinkage is a factor as it is in plastic injection molding and is calculated in the same manner to ensure that the cavities are produced over sized at the proper dimensions. A very common injection molding application process is to rubber coat metal inserts used for seals. The inserts are loaded into the mold cavity and the rubber is injected to coat the meal insert. This produces a metal, rubber coated insert which is used for seals in applications such as hydraulic cylinders, toiled flushing assemblies, wheel bearings, etc.

7.3.3.5 Casting

Rubber casting is performed in the same manner as metal casting. The two main processes utilized for this is sand casting and investment casting (see Chapter 5). A mold is constructed and poured with molten rubber to produce the desired shapes. Shrinkage is allowed for in the construction of the patterns in the same manner as metal casting and utilized in the casting process.

7.3.3.6 Rolling

Producing rubber sheets and slabs requires the material to be rolled out to achieve the desired thickness. With the material heated to a given temperature, it is run through rollers over and over in order to reduce the thickness. Figure 7.10 illustrates the rolling process for producing rubber slabs and sheets. This process is identical to rolling plastic sheets, however, rubber is a material that has a memory and the viscoelasticity is extremely high, causing the material to expand greatly as it exits the rollers. Owing to the high viscoelasticity of rubber materials, it is extremely difficult to produce thin sheets of rubber in a rolling process. Keeping in mind that the volume never changed in a displacement process such as rolling, the starting volume is critical. Also, as the thickness of the material reduces, the exiting velocity will be increasing.

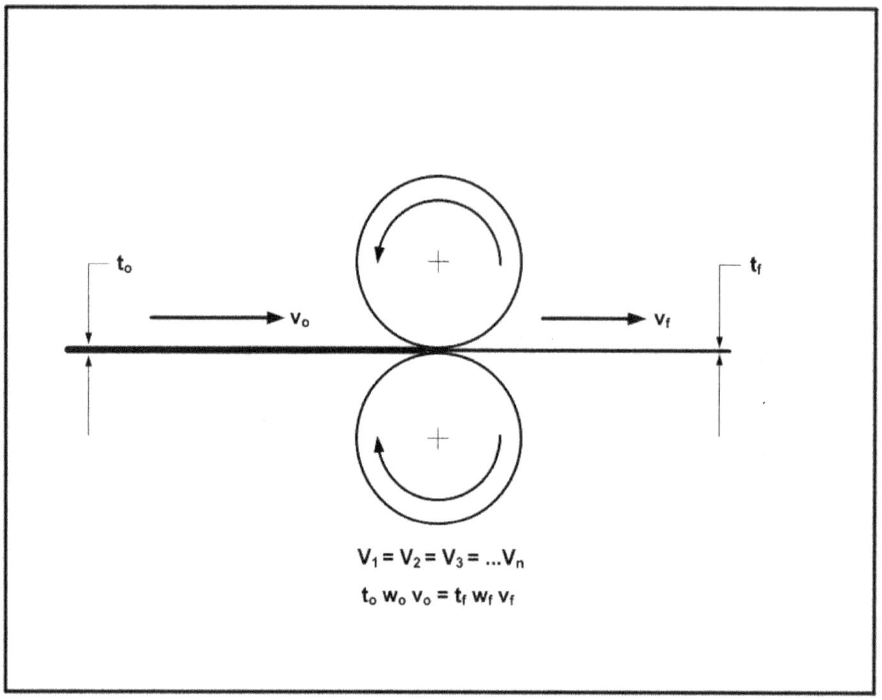

Figure 7.10 Rolling Process for Rubber Sheets

7.3.4 Vulcanizing

The final stage in rubber component production is vulcanization. The purpose of utilizing this process is to perform cross-linking of the elastomer molecules in the rubber. This is a critical stage in the rubber component production to add strength to the parts. During the vulcanization stage, the rubber material will obtain a different molecular structure as the molecules are cross-linked together as illustrated in Figure 7.11. Once this process has been performed, the rubber material cannot be re-melted for reuse. Vulcanized rubber is non-recyclable for reuse and needs to be disposed of in another manner. All tires (except one) go through vulcanization to add strength and rigidity for longer tire life. The only exception in the tire industry is drag racing slicks. These are non-vulcanized, leaving them very soft for more traction and track gripping. They wear down fast, but the objective for these tired is for maximum traction and no turning of the vehicle while competing.

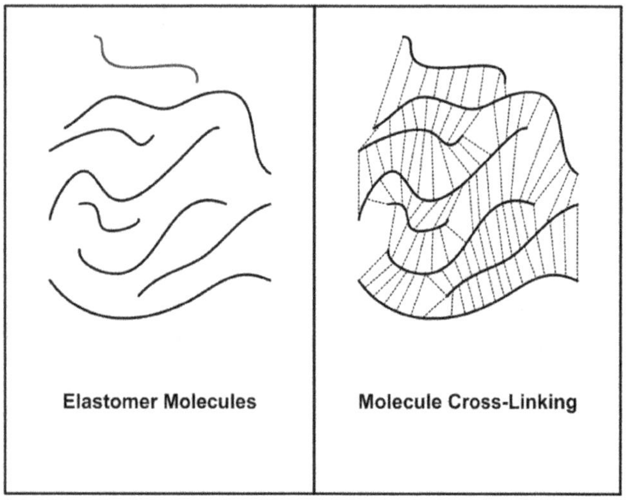

Figure 7.11 Elastomer Molecule Cross-Linking from Vulcanizing

Vulcanization required heat to accomplish the cross-linking of the molecules and can be performed in two ways. The first is to leave the component part in the molding cavity at the required temperature. This is very common in the injection molding process thus eliminating an operation in the overall process. The second method is to load the component part into a unit know as an *Autoclave* which is a steam-heated pressure vessel. This is charged with *super-heated steam* to apply the desired heat required. Once this is performed, the component parts need to be dried to remove the moisture from the steam which is performed in drying ovens in a controlled environment utilizing a heated inert gas such as nitrogen. If 100% of the moisture needs to be removed, a process call chlorinating is performed.

The parts are loaded into a large vat resembling a washing machine and tumbled as chlorine gas is injected into the process. This is utilized when rigged rubber products such as windshield wiper blades.

7.4 DESIGN CONSIDERATIONS FOR RUBBER

Rubber is a unique material and is used in products we come in contact with on a daily basis. As with any material or process, there are certain design considerations that need to be adhered to in order to produce acceptable components. These considerations are listed below.

Economic Production Quantity: With the long and complex processes required to produce rubber products, there needs to be a high enough quantity in order to make it cost effective. Besides compounding, mixing, and vulcanizing, the production tooling such as injection molds have a high cost attached to them. In order to be cost effective, a substantial production quantity is required.

Draft Angles: As with any solidification processes, draft angles are required for removal from the mold, or to remove the component part from the pattern.

Corner Radii and Chamfers: As with any component being produced in a molding process, the corners need to have either a radii or a chamfer on them to allow the component to be extracted from the mold and to allow the material to flow into those areas freely.

Holes: Holes are extremely and at time impossible to mold into rubber. Owing to the extremely high viscoelasticity of the material, as the core pin utilized to produce the hole is withdrawn from the material, the hole will reduce in size dramatically. The best method for inserting holes in rubber it to punch them in during a secondary operation. Punching and piercing holes will produce a hole to the desired size the best. Drilling a hole into rubber will produce a hole, however the rubber will again contract after drilling far more than it will from punching a hole and drilling will not produce a round hole in the material. Figure 7.12 shows how holes are punched into rubber with a punch and die assembly.

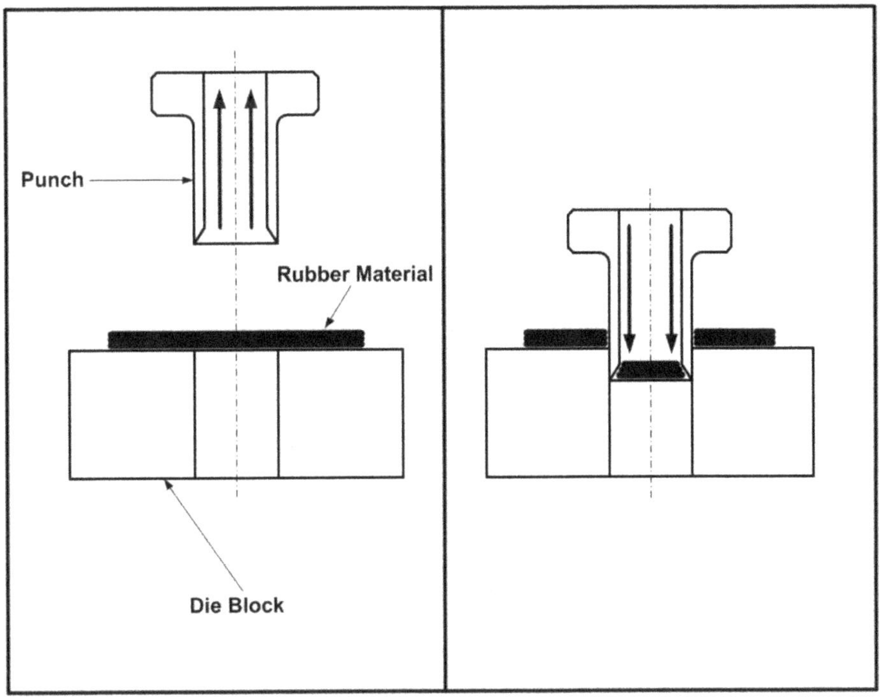

Figure 7.12 Holes Punching in Rubber Material

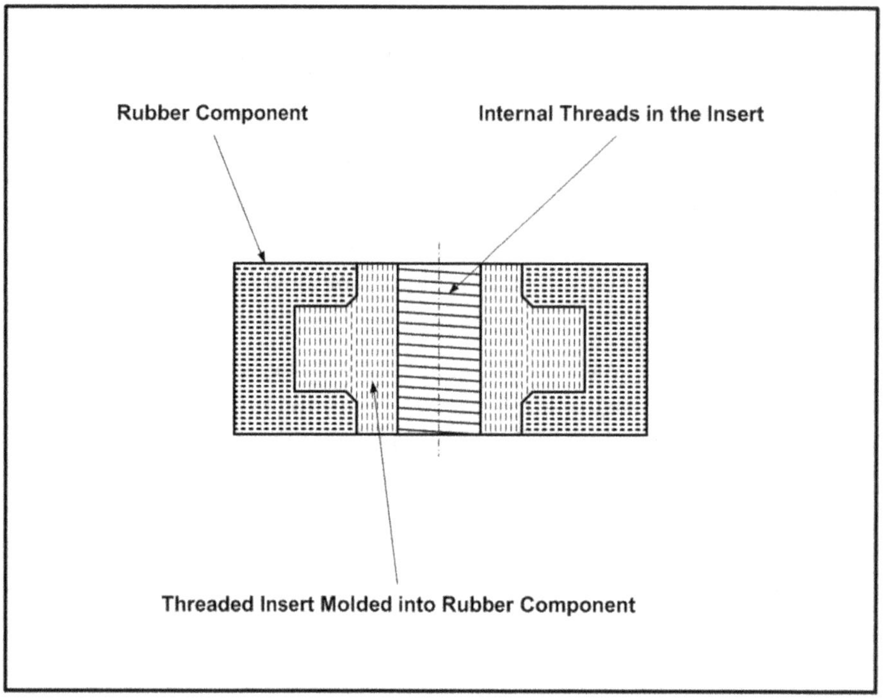

Figure 7.13 Threaded Insert Molded into a Rubber Component for Internal Threads

Screw Threads: Threads of any type cannot be implemented into a rubber component part. With rubber being extremely ductal, threads either internal to external will have no holding power whatsoever. If threads are required in a rubber component, a threaded insert would be molded into the rubber material at the shaping process. This is most commonly performed in injection molding of a rubber part and is shown in Figure 7.13.

Traction: Traction is one of the biggest redeeming features of rubber. If designing a component where traction is required, rubber is the best material that can be selected. Depending on the type of rubber, and the application where traction is desired, there is a rubber material that can satisfy the needs. Looking at rubber products that provide traction such as tires and drive belts, none over the years have replaced the rubber material that is being used.

Ductal: Rubber is very ductal and has elastic properties. Depending on how much and what type of strengthening elements are mixed into the material at compounding, it can be extremely elastic and stretching to 500–600% of its original size. Even rubber with a vast amount of strengthening elements mixed into it will display some elasticity. If a ridged material is required in the design, rubber may not be the best material selection to use.

Shelf Life: The shelf life of synthetic rubber is not a big factor, but with natural rubber it may be. Being that natural rubber is a fully organic material; it will have a limited shelf life as a raw material. If natural rubber is the material being utilized for producing the component parts, an inventory system needs to be in place ensuring that the inventory is turned by the first delivery in the first material that will be used.

7.5 COMPOSITE MATERIALS

Composites are materials that comprises a mixture of two or more materials together. The objective of composite materials is to create a superior material for high strength and light weight. They possess very high strength to weight ratios and modulus to weight ratios. These materials are widely used in the aerospace industry along with automotive applications and boats. The introduction of composite materials into the manufacturing and product arena has far expanded the capabilities of both. Component parts that were produced of metal in the past because of the strength required can now be produced from a composite rendering the same strength, or even higher strength with much less weight. This section will cover various composite materials and how they are shaped. Because a composite material is the mixing of two or more elements to create a new material they are not recyclable. Once the additives are combined with the parent material they cannot be separated in a recycling procedure.

7.5.1 Composite Matrix

As mentioned, composite materials is a mixture of two or more materials. Three of the main materials utilized in producing composites are metals, polymers, and ceramics/glass. The mixing of these produces materials possessing the strength of metals or ceramics, but the lighter weights of plastic. Figure 7.14 illustrates the composite matrix for metal/polymer/ceramic-glass mixtures. The metal/ceramic composites will be covered later in Chapter 10 with Cermets, and below is a description of the polymer composite mix materials.

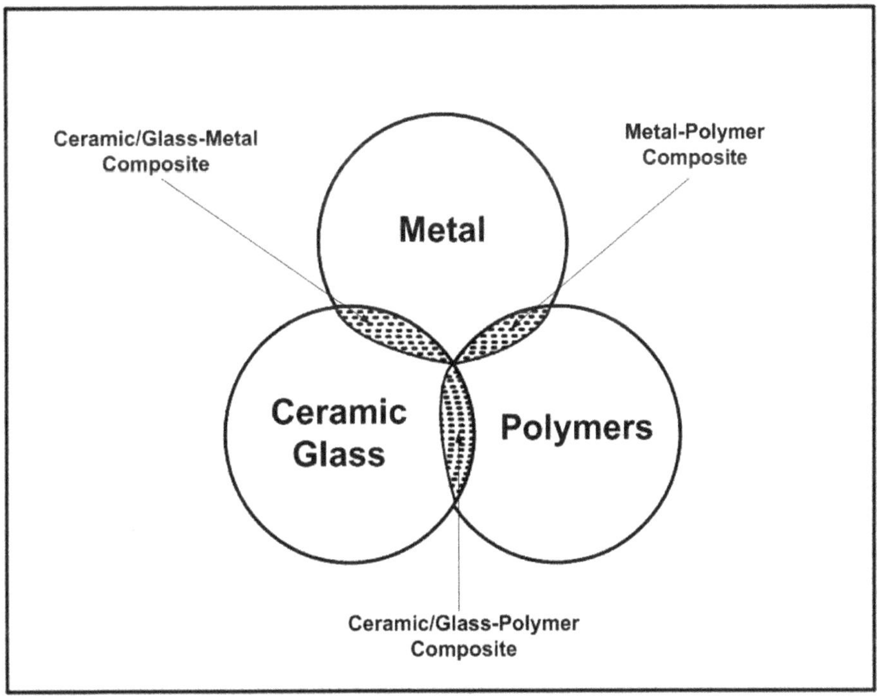

Figure 7.14 Metal/Glass/Polymer Composite Matrix

7.5.2 Polymer Matrix Composite

A polymer matrix composite is a material consisting of a polymer plastic material mixed with either a ceramic/glass or metal material. With the plastic materials being extremely light compared to metals and glass, the mixture of the two renders are extremely strong, yet light weight material. These composite materials have replaced several component parts that were historically fabricated from castings or machined from solid metals. One of the more popular composite mixtures is a glass/polymer material. This provides the light weight of plastic with the strength enhanced from the glass. The glass is mixed with the plastic and then injected into the mold. Glass particles can be in either the form of fibers of flakes, and depending on how much strength is required would determine the amount of glass being added in to the mixture. Figure 7.15 shows a cross-section of a plastic component mixed with either glass fibers or glass flakes.

Polymer matrix composites are divided into three categories being thermosets, thermoplastics, and elastomers. These are all similar in nature but the processes for producing composite components require different steps. A description of each of these along with the process required to produce components is detailed below.

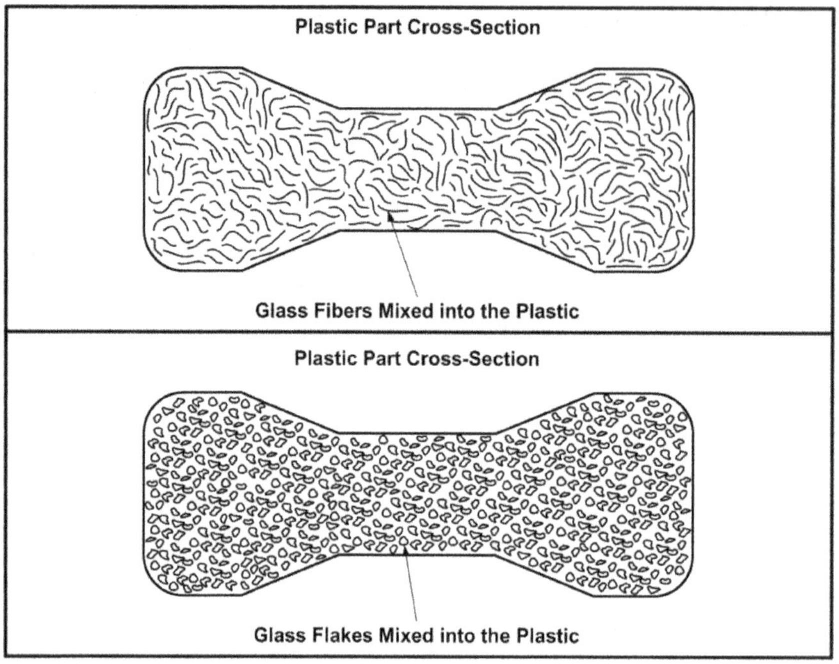

Figure 7.15 Glass/Plastic Composite Mix

7.5.2.1 Thermoset Composites

As described in Chapter 6, a thermoset is a plastic material created by mixing to elements together to form a third. Each of the elements alone will not harden by themselves and no heat is applied in the process. As the two elements are mixed together, there is a chemical reaction between them and the result is a third element that hardens due to the chemical reaction. For a thermoset composite material, the addition of other materials either metals or glass/ceramic is included in the mixing process. This is performed for each run of a part being fabricated due to the short timeframe in chemical reaction and hardening of the material.

7.5.2.2 Thermoplastic Composites

Thermoplastic materials are those which when heat is applied become molten and can flow as a thick liquid. For thermoplastic composite materials, the additives are mixed with the plastic material after it is heated and in a molten state. From there, the shaping process is performed resulting in a polymer composite component part.

7.5.2.3 Elastomer Composites

Elastomers are materials such as rubber, the reason why rubber and composite are in the same chapter. During the compounding and mixing stages of rubber processing, other elements are combined with the rubber creating a rubber-based composite. Once the vulcanization procedure is performed, the rubber takes on a new structure classifying it as a composite.

7.5.3 Reinforcing Agents

Several reinforcing agents are utilized in the creation of composites. These agents combined with the parent material create the mixture giving the new material its unique features. Each agent will perform different reinforcing functions depending on the material it is added to. They can be added as a mixture to a given material or be layered with a given material in the shaping process. Some of the more common reinforcing agents are listed below along with how they would be utilized in the composite shaping processes.

Glass: Glass is probably the most widely used reinforcing agent utilized in the composite arena. This agent is used as both an additive and a main material depending on the application for fabrication. For the production of fiberglass, the glass fibers are woven tightly into sheets that get layered with a resin holding them together.

Ceramics: Ceramics is mostly mixed with metals to produce a cermet composite. This renders cutting tools with far superior cutting abilities over standard high speed steel. It is also mixed with plastics in the same manner as glass to increase the wear resistance and add strength to the plastic materials.

Metals: Metals are very heavily used with ceramics for producing cermet composites. It is also mixed with plastics to provide some light conductivity to the plastic materials. If enough metal is added to the plastic material, plating is possible in a secondary finishing operation.

Carbon: Carbon is the main reinforcing agent additive used in rubber production. This was covered earlier in the chapter along with the other additives utilized in rubber component fabrication. Carbon is also mixed with glass to produce carbon fiber materials providing the strength of glass with some flexibility.

Kevlar: Kevlar is a unique agent that also gets added to rubber products. It has very extreme strength with very little elongation. This is also an additive for carbon fiber composite materials, increasing the strength of the glass-carbon mixtures.

7.6 COMPOSITE SHAPING PROCESSES

There are several production methods for producing composite materials. Some are utilized for special applications while others are much more generic and are used for a multitude of products. Three of the more common shaping processes for composite component fabrication are discussed below.

7.6.1 Open Molding

The open mold process involves a mold cavity that is used to lay-up the materials in it. This is mainly performed by hand and involves several layers of material(s) to produce the desired product. This is utilized for producing fiberglass product such as boat hulls. A layer of fiber glass sheet is laid into the mold cavity and then sprayed with a resin to hold it in shape. This is repeated several times until the desired thickness is obtained. Other sheet material such as Kevlar can be layered into the mold and also to obtain different material properties. Figure 7.16 illustrates how the materials are layered into the mold cavity and the resin applied between the layers of materials.

7.6.2 Closed Molding

Closed molding is very similar to the open molding process, except that there is a top to the mold that is lowered down by pressing the materials together. With this process, all the materials are laid up in the mold together with the resin, and the upper mold is lowered down pressing everything into the desired shape at once. Figures 7.17 and 7.18 show how the closed molding process for composite component fabrication operates.

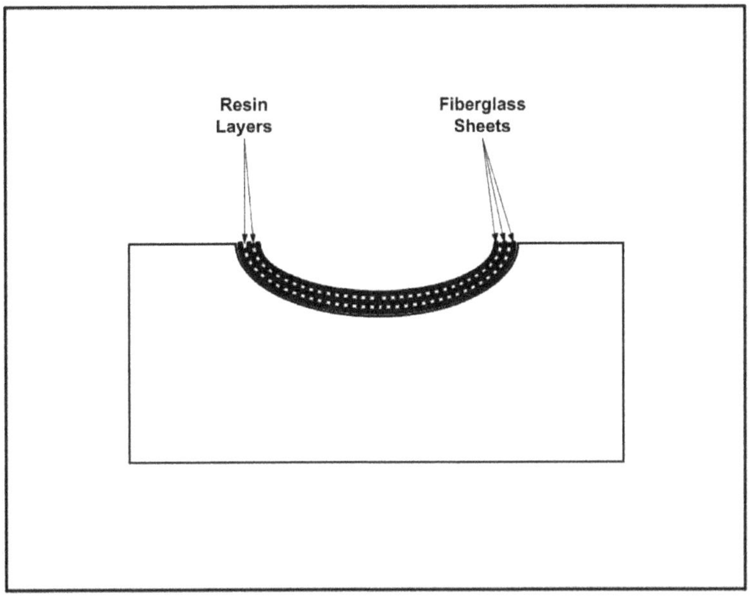

Figure 7.16 Open Mold Forming Process

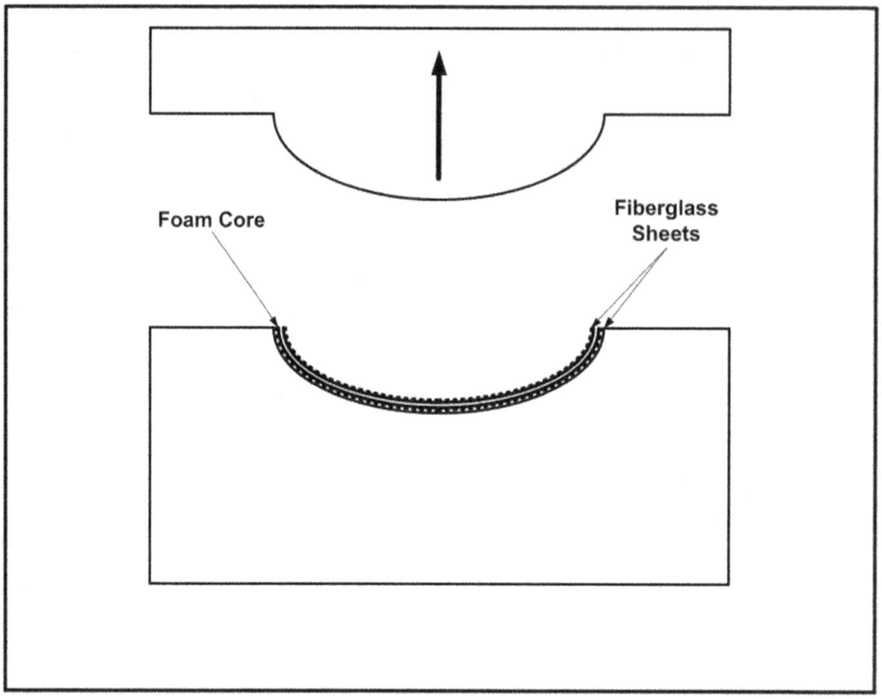

Figure 7.17 Material Loaded into a Closed Mold

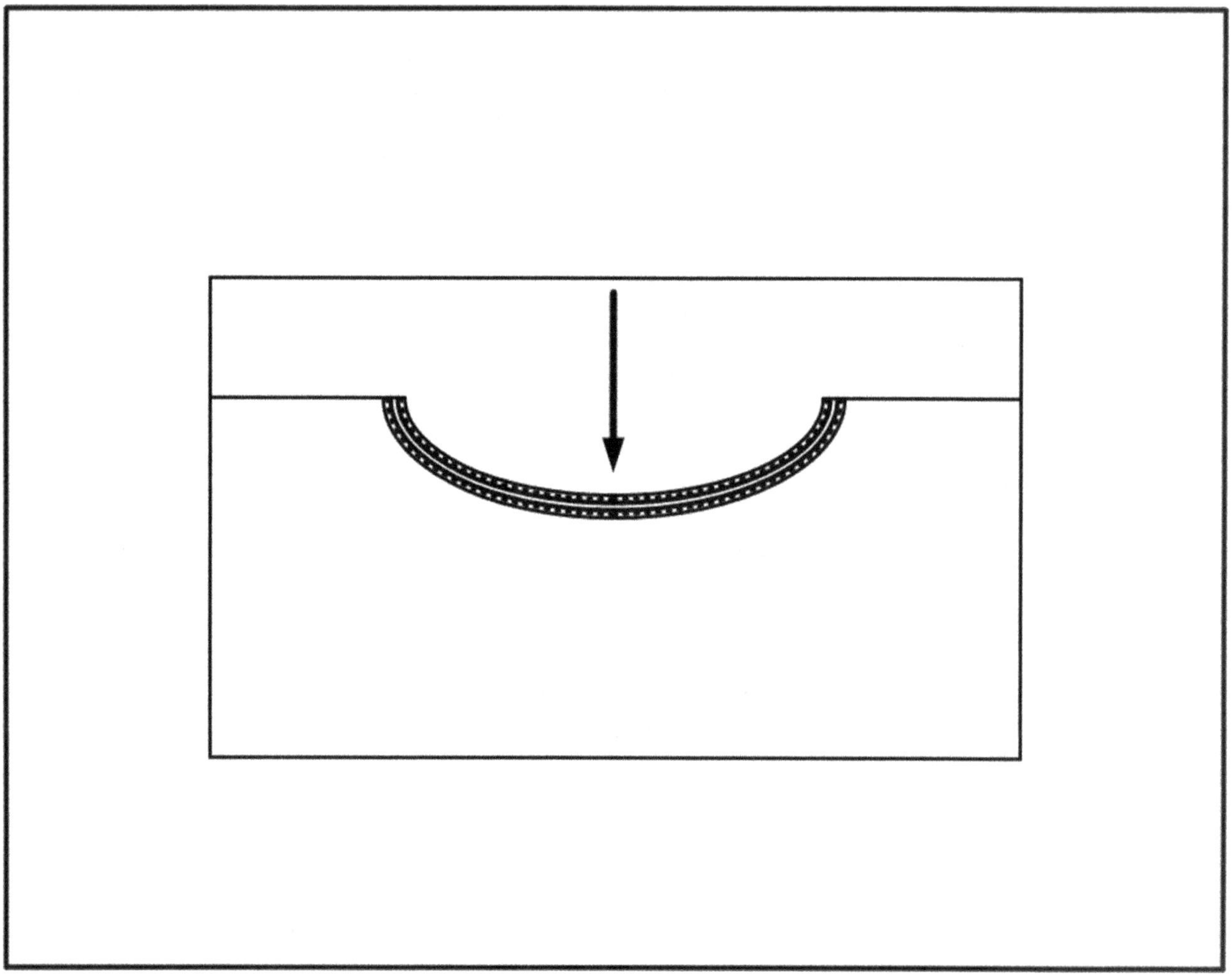

Figure 7.18 Closed Mold Compressing Material

7.6.3 Fiber Winding

Fiber winding is a process that is utilized for producing vessels which can withstand very high pressures and are extremely light weight. The process begins with fabricating a wax-based mandrel as shown in Figure 7.19. Once the mandrel has been produced, it is placed on a rotating system and sprayed with resin while glass fibers are run along it in a cross-hatch manner as shown in Figure 7.20. This winding process continues with a constant spraying of resin until the desired thickness if obtained. The final step in the process is to remove the mandrel

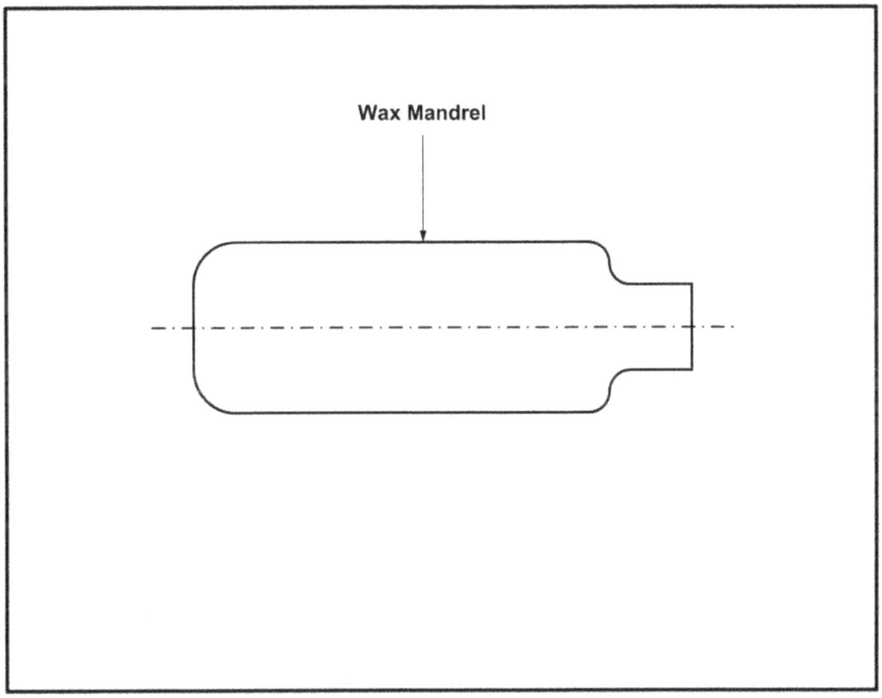

Figure 7.19 Fiber Winding Process; Wax Mandrel Developed

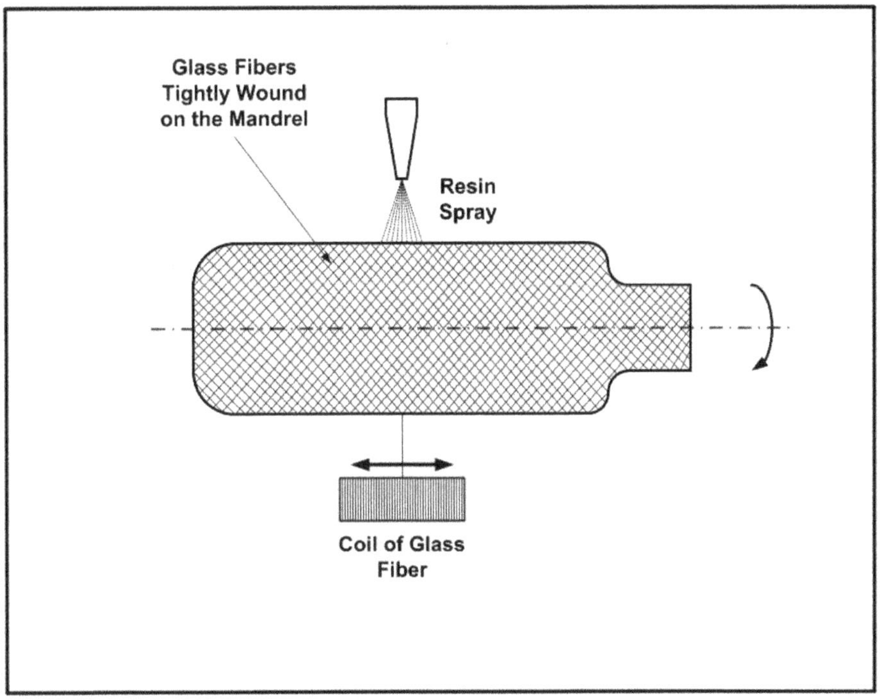

Figure 7.20 Fibers Winding on the Mandrel and Sprayed with Resin

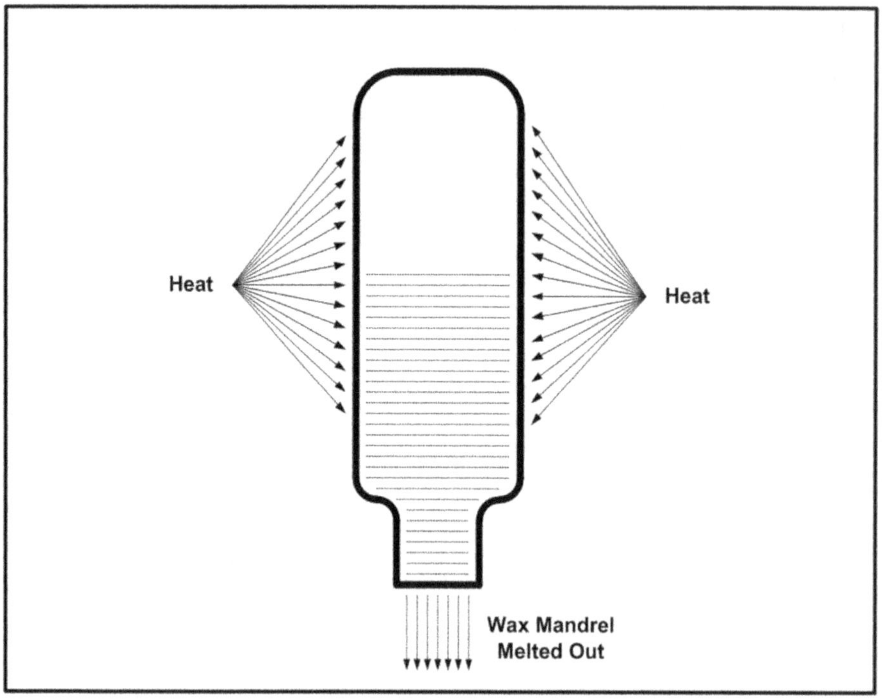

Figure 7.21 Wax Mandrel Melted out and Part Cured

and apply heat to it. The heat performs two functions. First, it will cure the resin bonding the fibers together. Second, the wax mandrel will be melted out leaving the vessel hollow inside. The advantage of this process is that a pressure vessel is produced with no seams unlike metal vessels. By having no seams in the vessel, there are no weak joints that can fail. Also, the finished product will be far lighter than a metal vessel that can hold the same pressures.

7.7 SUMMARY

The world of composite materials is ever changing as new materials are constantly being developed. This exciting and ever-changing field of materials has revolutionized the manufacturing field for being able to produce high strength and light-weight component parts. Without the advent of composites, the aerospace industry would never have been able to produce the exciting crafts that they have. The same can be said with the automotive industry. The vehicles that we drive today have several composite components, which were all metal in the past. As this field continues to expand, new materials will lead to products being produced in the future that are not possible today.

7.8 QUESTIONS

1. What are the four stages for producing natural rubber material?

2. Synthetic rubber is not a natural product. What is the main ingredient in producing it?

3. What is the compounding process for rubber and why is it performed?

4. After shaping rubber components, the final stage is vulcanization. What does this process do to the rubber material?

5. Why cannot screw threads, either internal or external be implemented in a rubber component? And if internal threads are required, how can they be put into a rubber component?

6. Why is the shelf life for natural rubber important to keep track of in the raw material inventory?

7. What is meant by a composite material?

8. What are the three main elements in the composite matrix and what does the mixing of them produce?

9. Define a thermoplastic composite, thermoset composite, and an elastomer composite.

10. In the fiber winding process, what is the big advantage of producing a component part utilizing this process over other processes?

CHAPTER

8

Glass Shaping Processes

8.1 INTRODUCTION

Glass is a material utilized in several ways that we see on a daily basis—from bottles and containers, to windows, to several other areas we come in contact with every day. It has properties not found with other materials, such as being clear and having optical properties. Glass is a material that has a molecular structure unlike other materials. In general, materials have three different molecular structures: a gas structure, a liquid structure, and a solid structure. As they pass from one to the other, the molecular structure of the material changes. Glass is a unique material where, when it is in liquid state, it hardens to what resembles a solid state; the molecular structure does not change. This classifies glass as a noncrystalline structure, which is referred to as a vitreous material. So, because glass does not change its structure, in layman's terms, glass, when it is in a solid state, is an extremely slow moving liquid. If you look at the windows in very old buildings, you will see that they are a little wavy. This is because the glass flows down over a long period of time because of the vitreous molecular structure. This chapter discusses the production of glass and how it is made along with various shaping processes as shown in Figure 8.1.

8.2 GLASS FABRICATION

The first stage in glass fabrication is to create the glass. This is performed by mixing the materials together required for producing glass. The interesting thing about glass is that even though it is clear in its natural state before any color is added to it, the materials utilized in producing it have no clear properties at all. Basically, glass is produced from solid materials, mainly sand. The mixing of these unique materials will produce what is

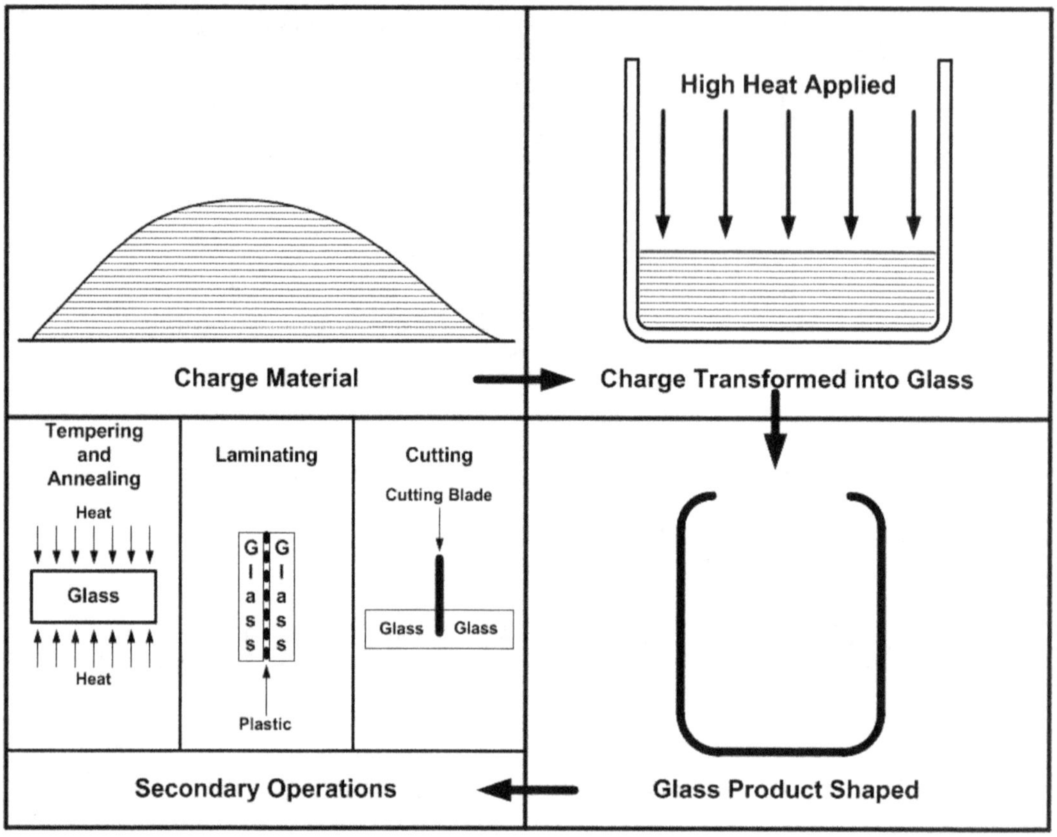

Figure 8.1 Four Steps in Glass Processing

known as the *Charge*, and it is the starting point of glass production. The main materials used to produce the charge are listed as follows.

Silica: Silica sand is the main ingredient in producing glass. It is one of the most common varieties of sand found in the world. It is used for a wide range of applications, from glass production, water filtration, industrial casting, to sand blasting.

Soda Ash: Soda ash is the popular name for the very common compound called sodium carbonate. It is used for manufacturing glass, and it is also contained in many cleaning agents and detergents, especially in areas where you have hard water and no water softener is available.

Limestone: This is a common sedimentary rock consisting mostly of calcium carbonate. Minor amounts of silica, feldspar, pyrite, and clay may also be present. Limestone can occur in many colors but is usually white, gray, or black.

Aluminum Oxide: This is a white or nearly colorless crystalline substance that is used as a starting material for the smelting of aluminum metal. It also serves as the raw material for a broad range of advanced ceramic products and as an active agent in chemical processing. Aluminum Oxide is also used as an abrasive for blasting and sanding.

Potash: Name used for various inorganic compounds of Potassium, chiefly the for Carbonate (KCO), a white crystalline material formerly obtained from wood ashes. They are used to make special types of glass, potassium silicate, and a dehydrating agent.

These materials need to be both washed and classified before they can be used to produce glass. Washing is done by taking the starting materials and hitting them with steam. This will remove any foreign substances on them to ensure that the materials are pure. Next, the materials need to be classified by size. This is critical in the process where having the starting materials to the proper size will create the type of glass being produced. The way the material is classified is by sifting it through a screen to ensure that the proper sizes are attained. Figure 8.2 shows a screen that is used to separate the particles. This is performed through a series of screens,

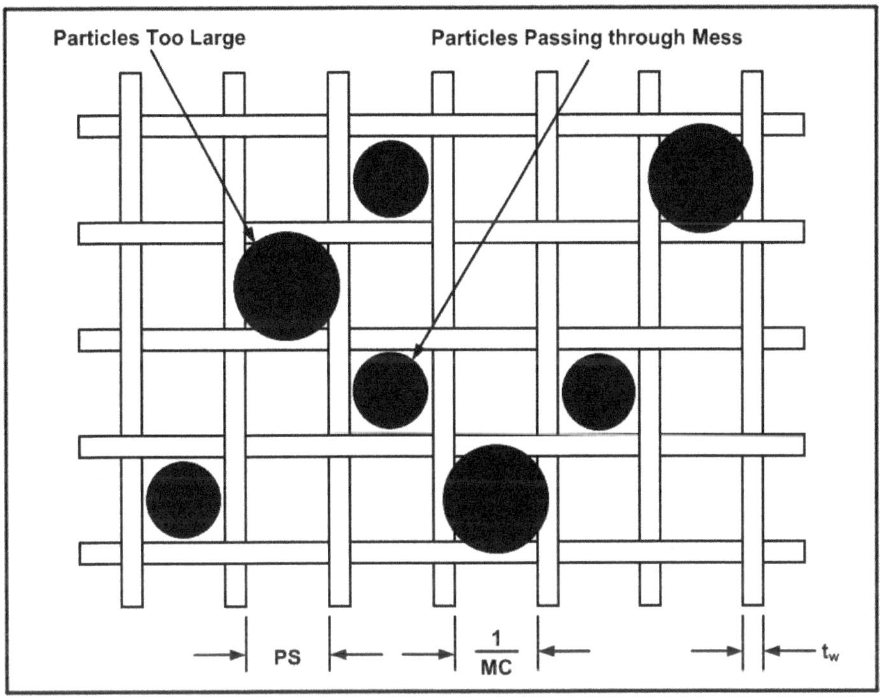

Figure 8.2 Classifying Grid to Separate Particles by Size

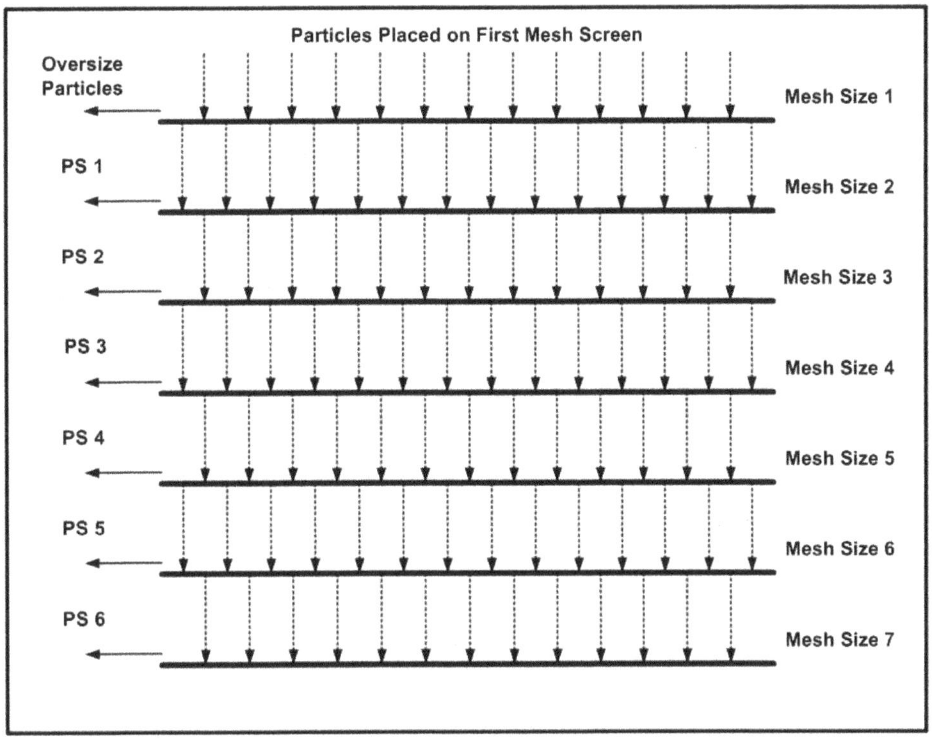

Figure 8.3 Several Classifying Grids Separating the Particles by Size

each having smaller and smaller openings in them. As the particles drop through a screen to the next, the ones that did not pass through are collected by their sizes. Figure 8.3 illustrates this process of screening to classify by size. From these materials all washed and classified, they are mixed to the proper proportions of the charge, which will be used to create the glass.

8.3 GLASS PRODUCING FURNACES

Once the charge has been mixed and prepared, it needs to be loaded into a furnace, which is called *Charging*. Depending on the volume of glass that is required to produce, three different furnace types can be used. These are *Pot Furnaces*, *Day Furnaces*, and *Continuous Furnaces*. Each of these furnaces is utilized to produce a given volume of glass and is defined as follows.

8.3.1 Pot Furnaces

A pot furnace is used for producing small volumes of glass. They have a limited capacity and are used for very small glass projects such as prototype work or glass hobby work. The way a pot furnace operates is to first load the charge into it. From there, heat is circulated around to heat up the ceramic walls of the furnace, which will become hot enough to transform the charge into glass. Note that the heat being applied does not actually heat the charge, but rather the internal walls of the furnace. Figure 8.4 shows how a pot furnace operates for producing glass.

8.3.2 Day Furnaces

Day furnaces are utilized for a larger volume of glass needed to be produced. The charge is loaded in the furnace and then heat is applied above the charge. Unlike the pot furnace, the heat is applied directly onto the charge to transform it into glass. These furnaces are used to produce a day's worth of glass for low production runs or for specialty fabrication methods. Figure 8.5 illustrates how a day furnace operates for producing glass.

8.3.3 Continuous Furnaces

Continuous furnaces are just what they state. Charge materials are continuously loaded into one end of the furnace, and glass is produced continuously out of the other end. These are used for mass production of glass for extremely high-volume production such as bottles, tubes, and fibers. Figure 8.6 shows the structure and operation of a continuous furnace producing glass.

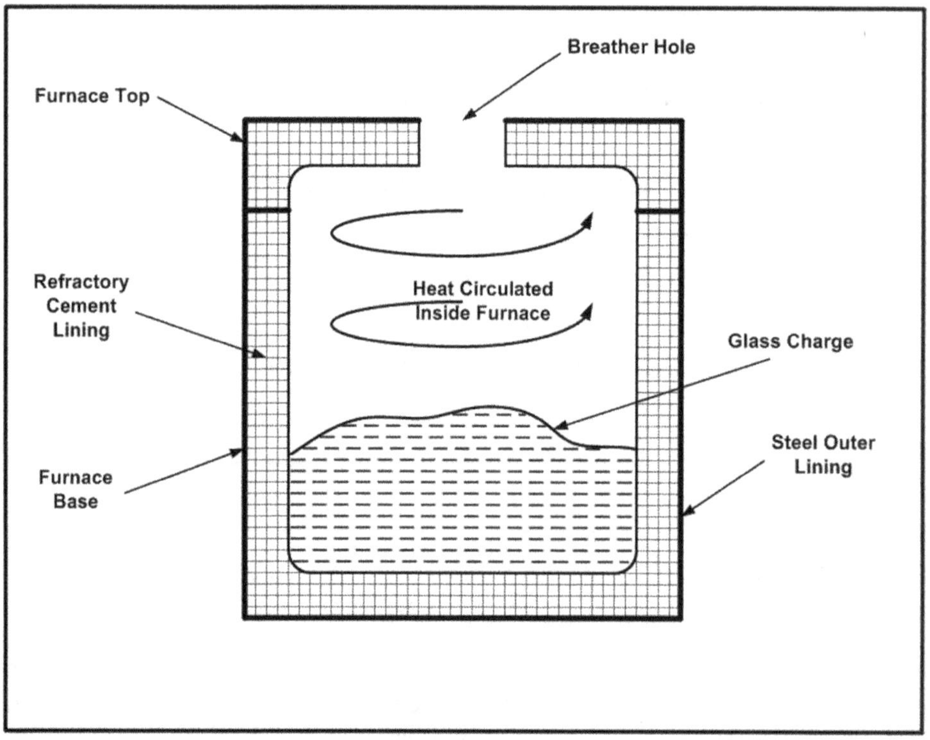

Figure 8.4 Pot Furnace for Producing Small Batches of Glass

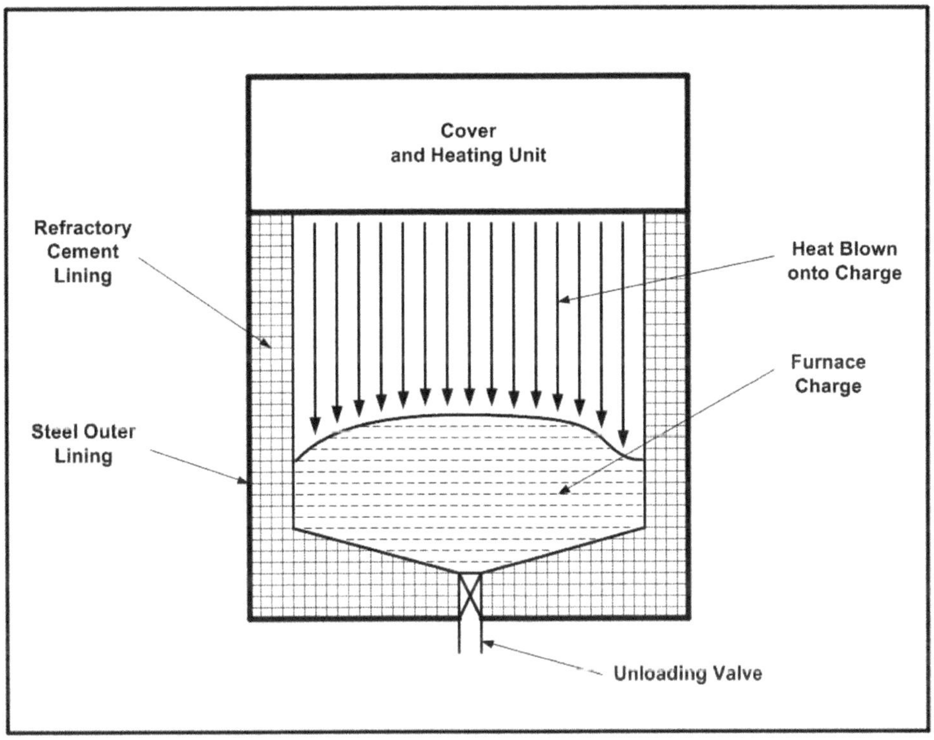

Figure 8.5 Day Furnace for Producing Larger Batches of Glass

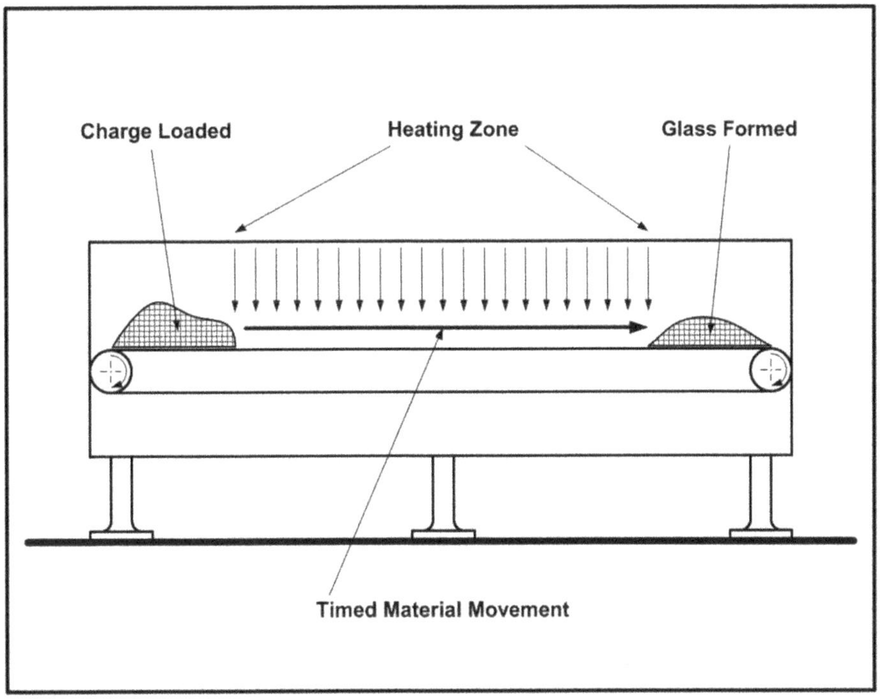

Figure 8.6 Through Furnace for Producing a Continuous Batch of Glass

8.4 SHAPING PROCESSES

Glass can be shaped into products in a variety of methods. Each of the processes will render a different shape and can be performed either on an individual part-by-part method or in a continuous manner. Each of the shaping processes used for producing glass products falls into one of three main categories, which are *Discrete Processes*, *Continuous Processes*, and *Fiber Processes*. These categories are defined as follows along with the main processes utilized.

8.4.1 Discrete Processes

Discrete processes are those that produce products of a unique shape and can be either automated or performed manually. This category of shaping processes is used for both production and hobbies. In a production environment, these would be automated, but will produce one product at a time, thus considering them to be discrete in nature. The more common discrete processes are detailed as follows.

8.4.1.1 Press Forming

Pressing glass into a certain shape is the most basic shaping process. A given amount of glass is loaded into the bottom of the die as shown in Figure 8.7. This load of the glass is called a *Gob* and will be to the consistency of soft clay. The volume of glass being loaded is very critical in order to produce an acceptable part. If there is too much glass, it will flow over the sides of the die when compressed. With too little glass, the shape will not be completely formed. Once the gob is loaded, the upper die is lowered as shown in Figure 8.8, forming the desired shape. After cooling, the upper die is raised and the glass product is retrieved as shown in Figure 8.9.

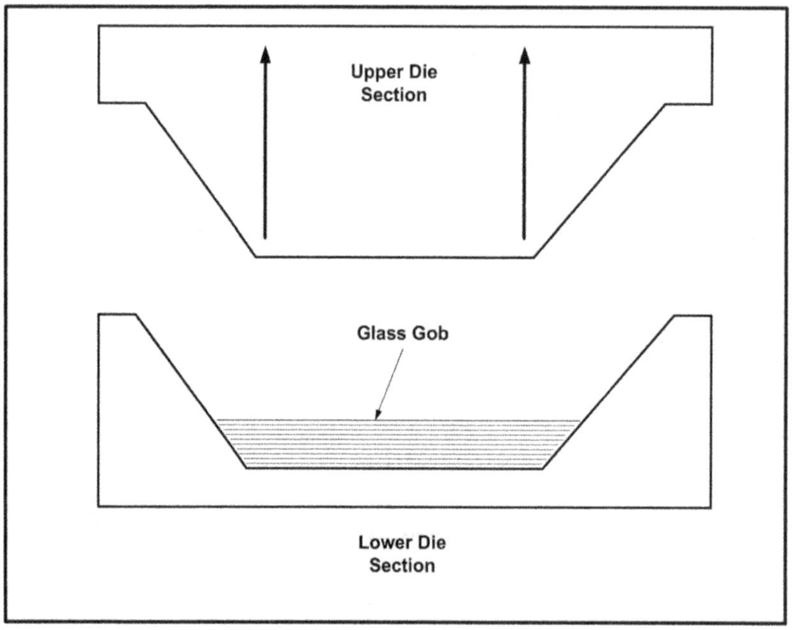

Figure 8.7 Pressing Glass Forming Process with Gob Loaded into the Lower Die Section

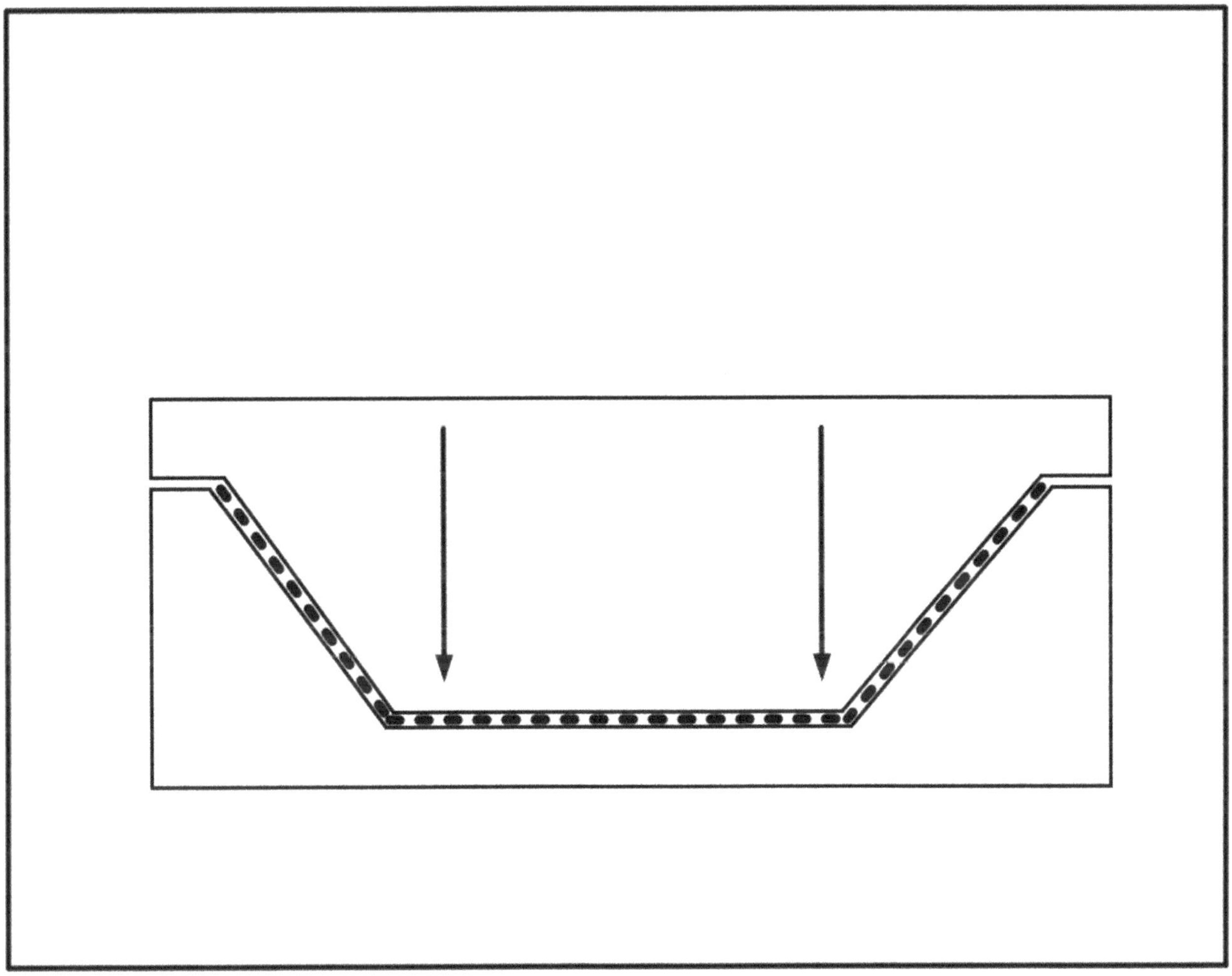

Figure 8.8 Glass Gob Pressed into Shape by Lowering the Upper Die Section

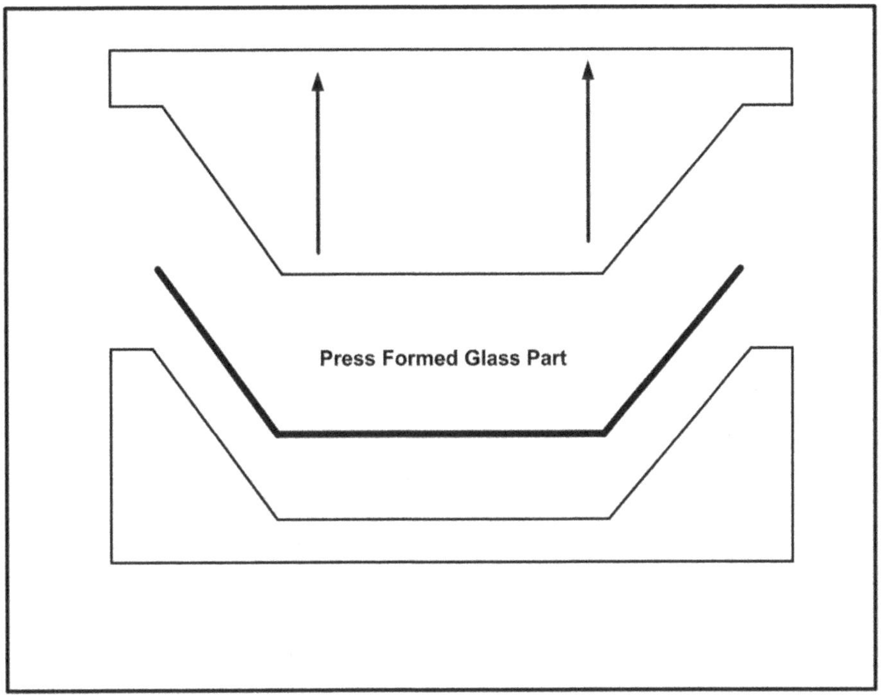

Figure 8.9 Upper Die Lifted and Glass Component Retrieved from Lower Die

8.4.1.2 Blow Forming

Blow forming is a process that utilizes air, which is injected into a mold to create a shape. This is the process used to create glass bottles and containers. A gob of a precise volume is loaded into the mold as shown in Figure 8.10 to begin the process. From there, air is injected into the mold, forcing the glass against the sides as shown in Figure 8.11. In most cases, the mold is rotated, so the centrifugal force will assist the air in pushing the glass to the sides of the mold. The consistency of the glass for this process is very fluid so it can be blown into the desired shape. After a cooling period, the mold is opened and the glass container is retrieved as shown in Figure 8.12. This process for producing bottles is automated, but because it renders one part at a time, it is considered discrete.

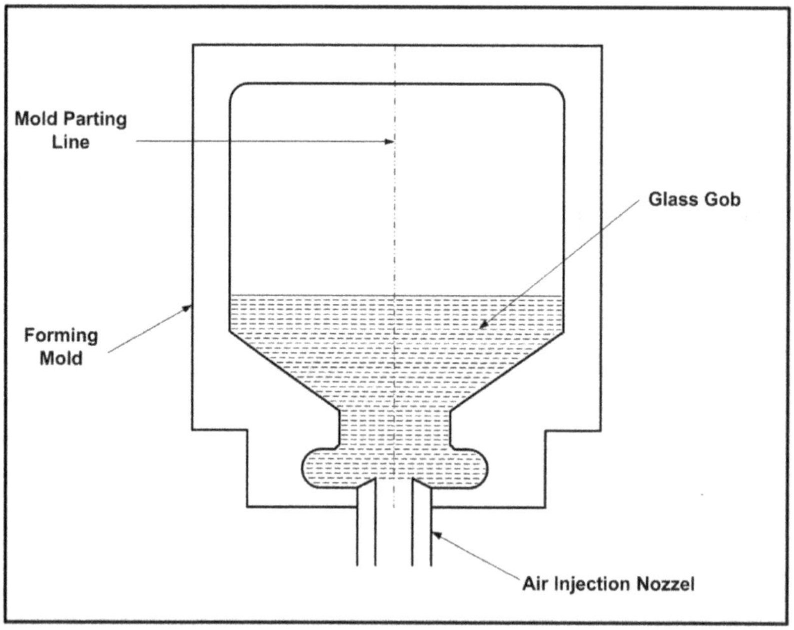

Figure 8.10 Glass Gob Loaded into Blow Molding Cavity

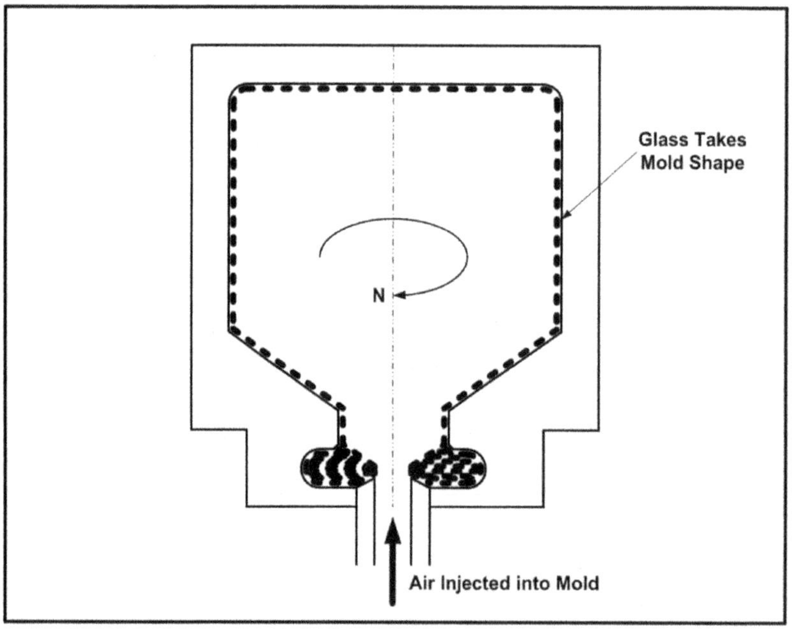

Figure 8.11 Air Injected into the Rotating Blow Mold to Create the Glass Shape

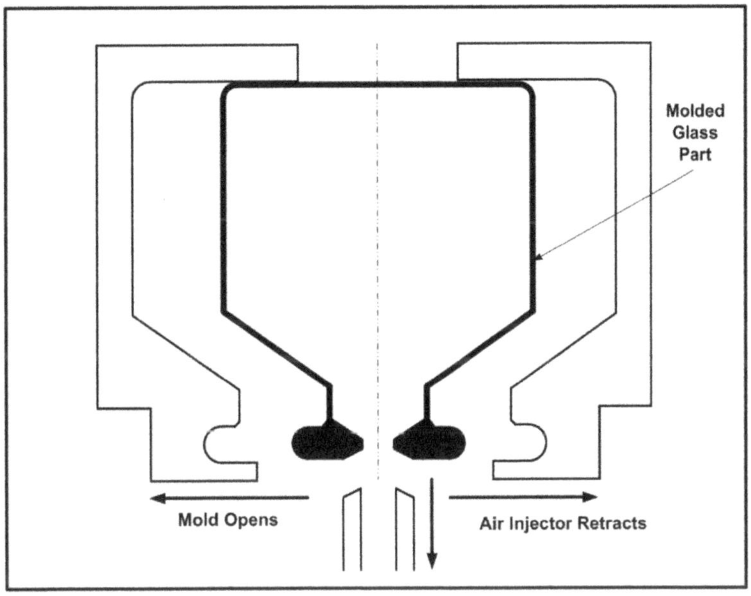

Figure 8.12 Mold Opens and Blow Molded Glass Component Retrieved

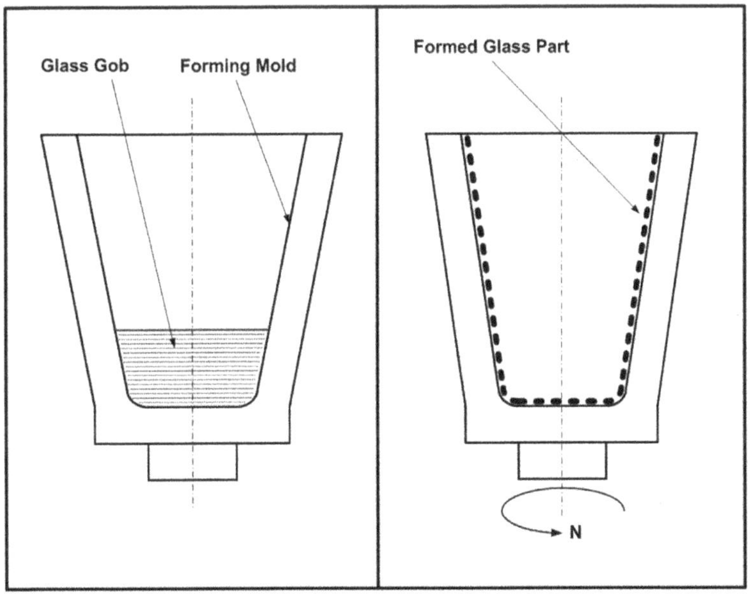

Figure 8.13 Glass Spin Forming Process Utilizing Centrifugal Force to Create the Shape

8.4.1.3 Spin Forming

Spin forming is a process that utilizes centrifugal force to create the desired shapes. The gob is loaded into the forming mold and then the mold is rotated at a given RPM to force the glass up the sides. For this process to work properly, the glass has to be at the right temperature and the rotation of the mold needs to be exact. If either of the two variables are not exact, the product will not shape properly. Figure 8.13 illustrates how the spin shaping process works.

8.4.2 Continuous Processes

The continuous process category is where the glass products are continually produced. This allows for the process to either produce glass products of a given size or produce them continually in whatever size desired. With the continuous processes, the glass is fed into whichever process is being used and the temperature must

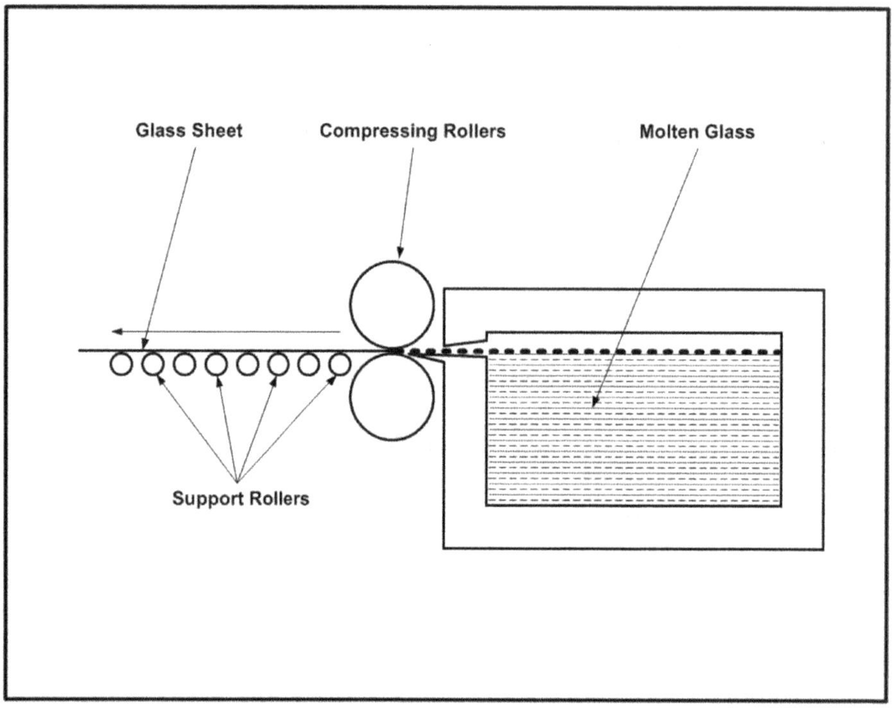

Figure 8.14 Glass Roll Forming Process

be maintained in order for proper shaping to take place. Following are some of the more common continuous shaping processes for glass.

8.4.2.1 Roll Forming

Rolling is a process used to create sheets of glass for a variety of uses. The molten glass is allowed to flow out of the melting container and is gripped by the rollers as it runs through them. The rollers are set to the desired thickness that is to be obtained, and as it exits the rollers, it is cooled and broken off at the desired lengths. The rollers can be adjusted to whatever thickness of glass that is desired, but very thin glass cannot be rolled. Figure 8.14 shows a typical sheet glass rolling process.

8.4.2.2 Float Forming

The float forming process is used to create thin sheets of glass. With this process, there is no rolling of the glass at all. The molten glass is caused to spill out onto a bath of molten tin. When this occurs, the glass will spread out over the liquid tin, creating a thin sheet of glass. The temperature of both the glass and the molten tin will determine how thin the glass will be formed. It moves across the tin into a cooling chamber and then exits the system of rollers, which carry it away. Figure 8.15 shows how the float process functions for producing thin glass sheets.

8.4.2.3 Tube Forming

Glass tube forming utilizes a process known as the Danner Process. Glass at the proper temperature and consistency is poured onto a hollow, rotating mandrill, which has an air stream blowing through it. With the mandrill pointing down at a given angle, the glass flows off the mandrill with a hollow center, thus creating the tubing. The newly created tubing is supported as it exits the mandrill and is broken off at the desired lengths. The temperature of the glass being poured onto the mandrill will determine the thickness of the tubing walls. Figure 8.16 illustrates the tube forming process.

8.4.2.4 Extrusion Forming

The extrusion process for glass is similar to that for plastics. The molten glass is pushed through a die opening that has the desired shape. As the glass extrusion exits the die opening, it will be supported on rollers and broken off at the desired lengths. Figure 8.17 shows the extrusion process with the glass shape exiting the die opening.

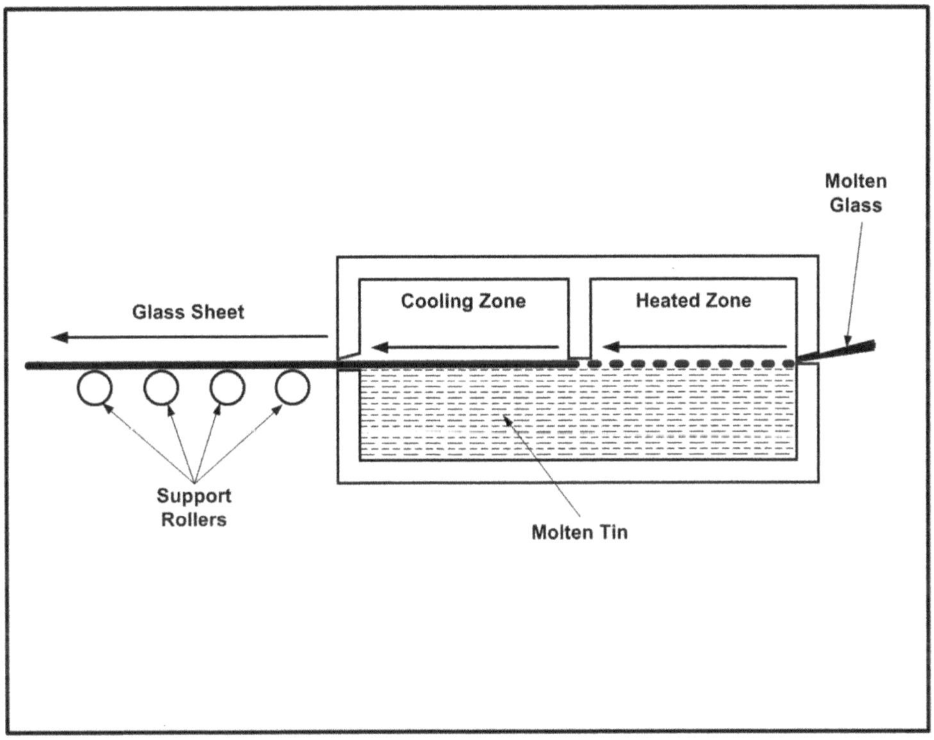

Figure 8.15 Glass Float Forming Process Utilizing Liquid Tin

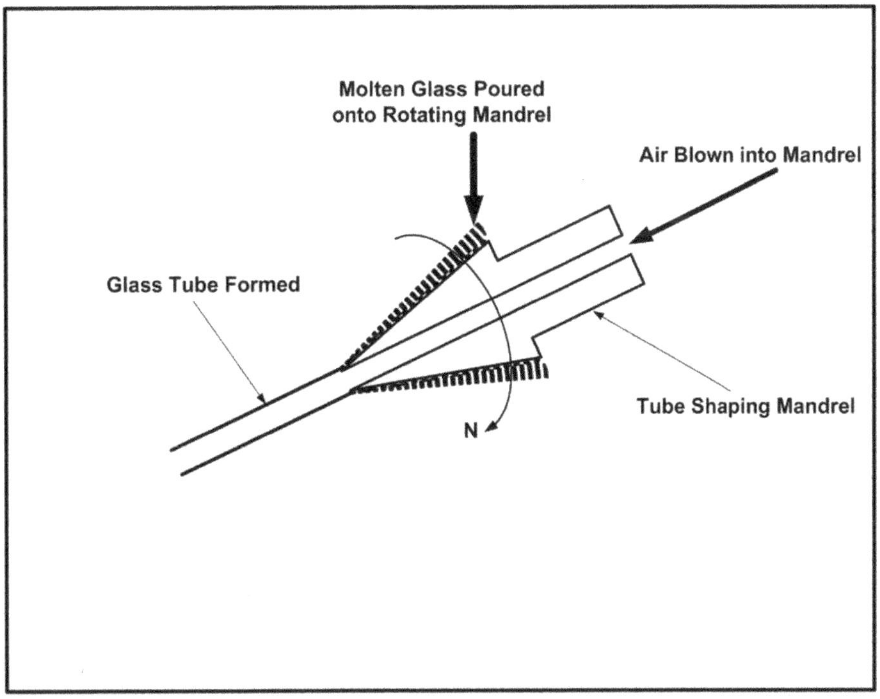

Figure 8.16 Glass Tube Forming Process over a Rotating Mandrel with Air Injected

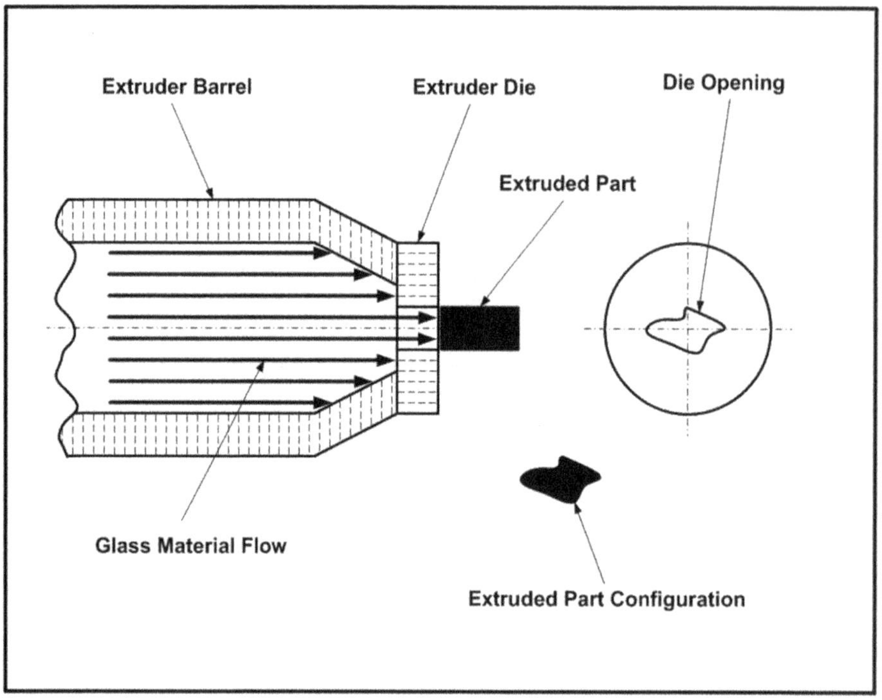

Figure 8.17 Glass Extrusion Process Creating a Two-Dimensional Shape to Unlimited Lengths

8.4.3 Fiber Processes

Glass fibers are produced in a continuous manner. These fibers can either be of a random length or be continuous, depending on the process utilized and the function of the finished fibers. The two main processes used for fiber fabrication are *Centrifugal Forming* and *Continuous Forming*. Although both produce glass fibers, they are completely different in how the fibers are formed. These two processes are explained as follows.

8.4.3.1 Centrifugal Fiber Forming

The centrifugal fiber forming process will produce glass fibers, but they will be at random lengths and not to any precise diameters. Fibers produced in this process are used in the production of fiberglass insulation and for reinforcing agents used in the production of composite materials. This process is simple in nature. Molten glass is injected into a rotating disk, which has holes along the outside. As the disk rotates and the glass is injected under pressure, glass will exit the holes, creating the fibers. As the glass cools, the fibers fall to the bottom of the unit and are collected. Figure 8.18 shows the centrifugal forming process where the rotating die ejects the created fibers. Depending on the temperature of the molten glass, the rotational speed of the forming die and the pressure that the glass is being injected under will determine the length of the fibers being produced.

8.4.3.2 Continuous Fiber Forming

The production of continuous fibers is basically an extrusion process. Molten glass is maintained at a set temperature in a containing vat and pressure is applied. The bottom of the vat has a die with an opening to the desired diameter, and as the pressure is applied, the glass fiber is extruded from the die opening. From the extrusion point of the die, the fiber is supported and guided through a cooling stage and then rolled up on a roller large enough to coil the glass fiber without breaking it. Figure 8.19 shows the continuous fiber production method for fibers of unlimited length. This production process is used for fabricating fibers that will be used for weaving fiberglass sheets that are used for a variety of applications and for producing glass fibers that are used for fiber optic applications.

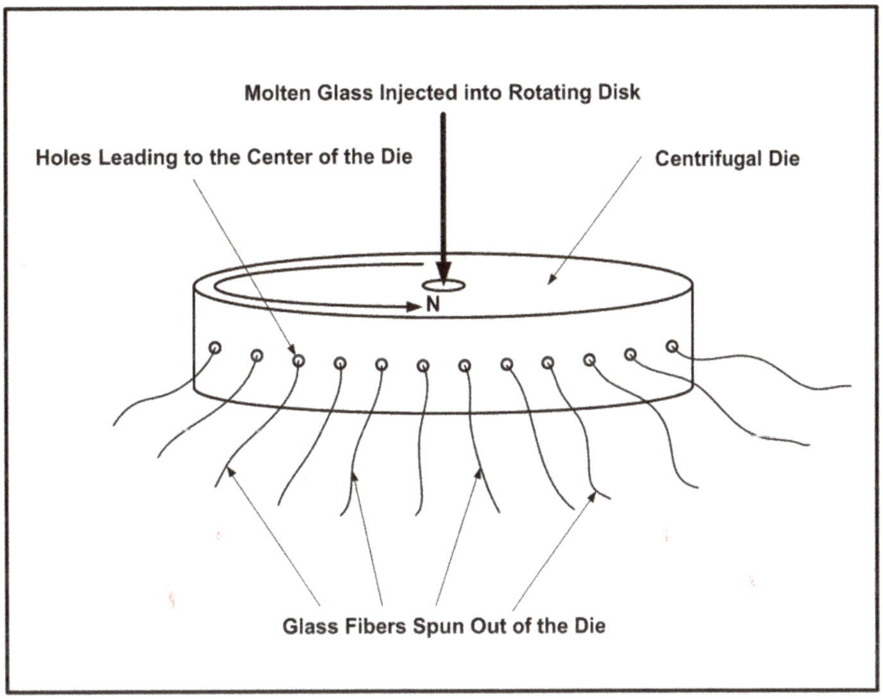

Figure 8.18 Centrifugal Fiber Glass Forming Process

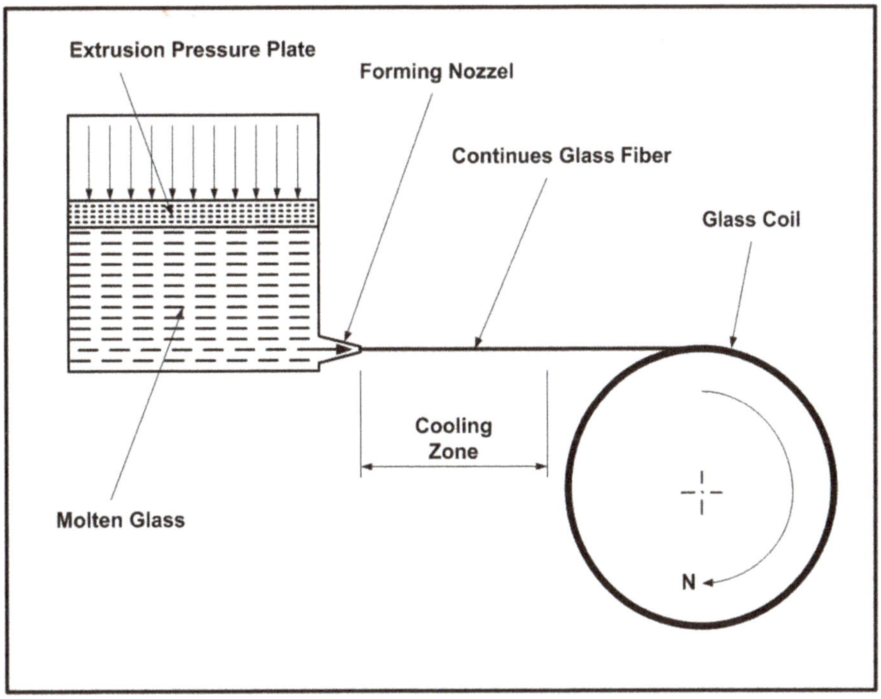

Figure 8.19 Continues Fiber Production Process

8.5 GLASS FINISHING

Once the glass product has been fabricated, there are some finishing operations that can be performed to either alter the internal structure of the glass or develop it into a different product structure. These include processes such as *Annealing, Tempering, Laminating,* and *Machining and Cutting.* Depending on the application of the glass product, one or more of these finishing operations may be required to produce a component that will function to the desired levels. Each of these finishing operations is discussed as follows.

8.5.1 Annealing

Annealing is a process utilized to relieve any internal stresses in the glass after the shaping process. When being shaped and cooled, the glass will cool rapidly from the outside of the component, thus causing internal stresses. To anneal a component, it is heated slowly to a given temperature, around 900°F and held there for a period of time. Once the time frame has been met, it is very slowly cooled down to room temperature. This reheating and slow cooling will relieve any internal stress in the component caused from the forming operations. Figure 8.20 shows the heating and cooling curve required for proper annealing.

8.5.2 Tempering

Tempering is performed very similarly to annealing, but the glass is raised to a higher temperature and then quenched with air for rapid cooling. The glass is heated to its "plastic range," meaning that it is close to becoming pliable before being cooled. What tempering performs on glass is to give it a hard outer surface resulting in a much stronger surface. This operation will render what is called safety glass that is used for all the vehicle windows except the windshield. This gives the glass the ability to withstand a harder impact without breaking. However, when it does break, it will shatter into countless small pieces rather than just cracking or breaking into large pieces. Figure 8.21 illustrates the hard case on the outer surface produced from tempering.

8.5.3 Laminating

Laminated glass is a process where two sheets of annealed glass are molded with a sheet of clear plastic between them. This is the process utilized for producing the glass used for vehicle windshields. With the plastic sheet laminated between the two sheets of glass, it will prevent anything from crashing through from the outside. When a crack occurs on the windshield of a vehicle, it is visible from the inside, but the crack is on the outer layer of glass. This reduces the chances of anything blowing into the vehicle causing danger to the driver. Figure 8.21 shows the two layers of glass with the plastic sheet laminated between them.

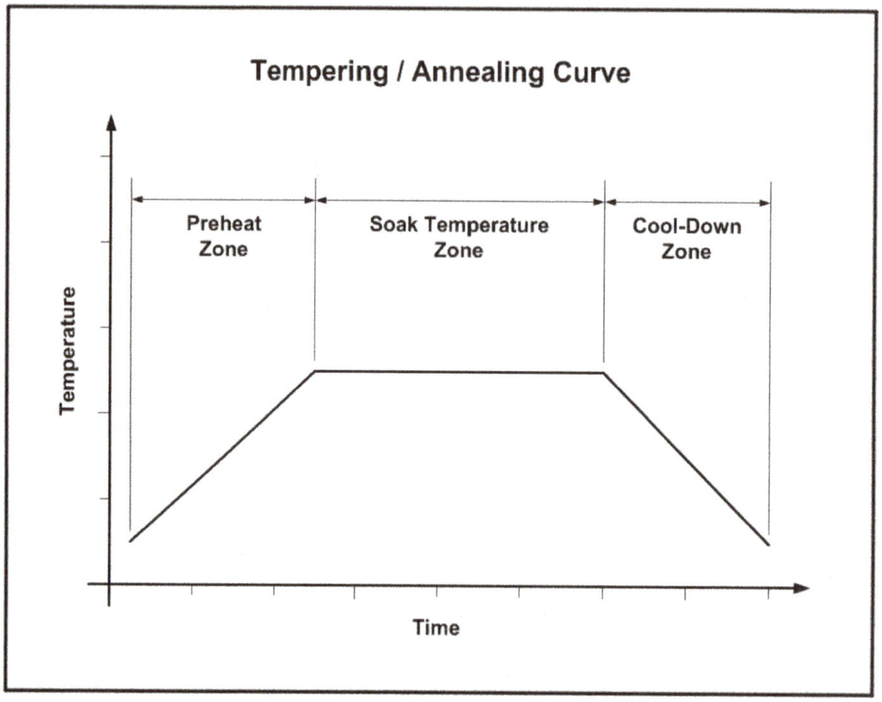

Figure 8.20 Glass Tempering Heating and Cooling Curve

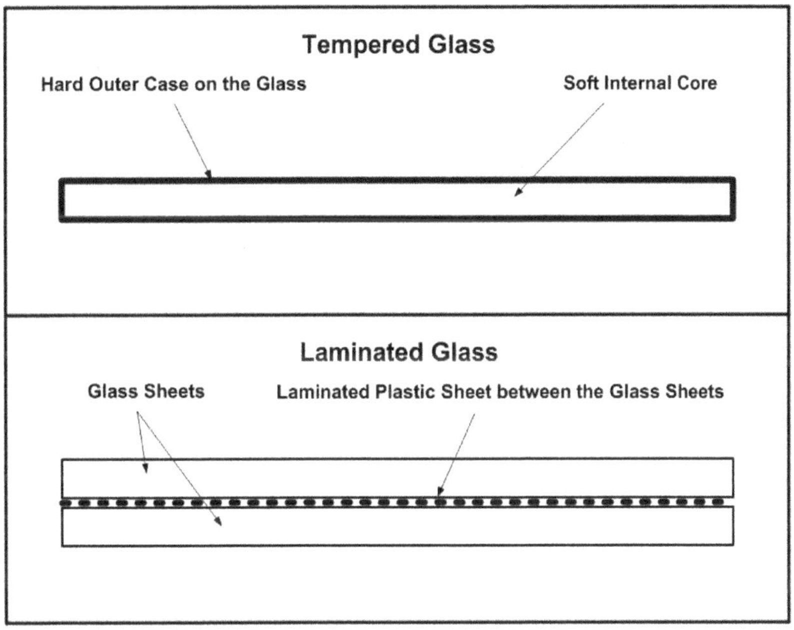

Figure 8.21 Laminated and Tempered Glass

8.5.4 Glass Cutting and Machining

Secondary operations on glass can be performed, but are very costly and time-consuming. Due to the fact that glass is extremely hard and brittle, it requires special tooling, mainly constructed from diamond and very slow cutting times to prevent any breaking during processing. For thicker glass, it can be ground to a final dimension, drilled to insert holes, or cut to a given length. These operations all require diamond tipped tools with a lot of coolant and slow cutting times. For sheet glass, it can be cut much easily. To perform cutting of sheet glass, a diamond style is dragged across the surface, causing a fracture crack through the glass. Once this has been performed, the sheet glass is tapped from the opposite side, causing the glass to crack along the fracture line. These cuts can be either straight or of a given configuration. Figure 8.22 shows these four secondary operations and how they are performed.

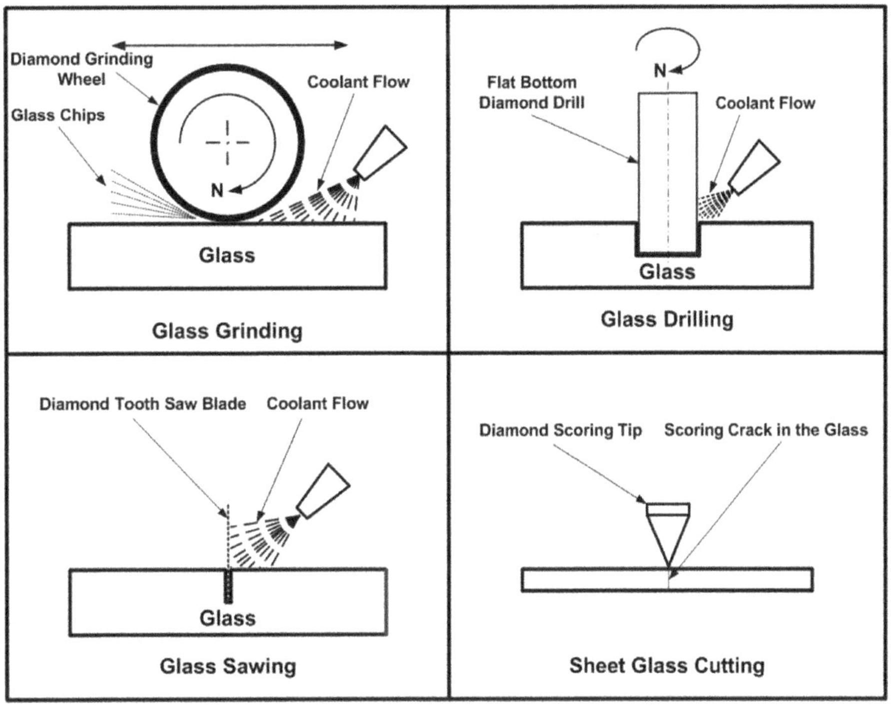

Figure 8.22 Secondary Glass Forming Operations

8.6 DESIGN CONSIDERATIONS

When designing a component that will be fabricated from glass, there are several design considerations that need to be understood. Glass does have its place in the manufacturing of parts, but by understanding what it will and, more importantly, will not do is critical in producing a functional component that will not fail in operation. Listed in the following are the major design guidelines that should be understood.

Transparent: Glass is a very transparent material, with clear glass being the most transparent. Even after adding color to the glass, it remains transparent and can be used in several applications requiring transparency in the products.

Optical Properties: When ground properly, glass will render optical properties. All the lenses for cameras, scopes, and eye glasses are ground glass, giving them the desired optical properties. These can also render both magnification and distance viewing.

Physical Properties: Glass possesses certain physical properties for components. It is much stronger under a compressive force than it is in tension. In a shear mode, it will fail almost instantly because of the extreme brittleness of the material.

Very Brittle: As mentioned before, glass is an extremely hard material, thus rendering it very brittle. If a glass component is to be applied in a situation where it may be struck with a shear force, another material should be considered.

Low Thermal Expansion: Because glass is a vitreous material and does not go through any transformation when changing from its liquid state to a solid state, it has extremely low thermal expansions. This is very advantageous for applications where the component part will be exposed to a wide range of temperatures during service.

Radii on Edges: During any of the shaping processes discussed earlier, radii on all the edges are required. This allows the glass to flow into those areas freely when a square corner will not fill. The larger the radii, the better the forming operation. If a square corner is required on the component, it will need to be ground in during a secondary operation after shaping.

High Wear Resistance: With glass being extremely hard, it will provide an extremely good wear surface. If the wear surface will be in compression only, glass slides are an excellent way to increase the wear resistance for the sliding actions. Several automated machines utilize glass slides to increase the wear resistance and maintain the precision of the equipment.

Secondary Operations Difficult: Even though secondary operations like grinding, drilling, and sawing are possible with glass, they should be avoided if possible. The time required to machine glass is extremely long, and even under the most perfect conditions, the glass can fail during processing.

Recycling: All glass materials are completely recyclable. At a recycling center, they would be separated by color and then remelted to be used over and over. It is important to recycle glass rather than disposing it in a land fill because glass is a material that will never decompose.

8.7 SUMMARY

Glass is a unique material possessing properties that other materials do not. It is utilized for a large variety of products that we see and use on a daily basis. Utilizing glass in a proper way will render an extremely useful component or product that is used in the proper application. By understanding the conditions that glass will perform under, it can be the best selection for components in any product being produced.

8.8 QUESTIONS

1. What is meant by a vitreous structure of glass?
2. When producing glass, what is the main ingredient along with other ingredients utilized?
3. How is the starting material classified for producing glass?
4. What is the difference between a Charge and Charging?
5. How are Pot Furnaces, Day Furnaces, and Continuous Furnaces different?
6. What is meant by a Discrete Process, and how does it differ from a Continuous Process?
7. Which process is utilized for producing glass fibers of varying lengths, and how is it different from continuous fiber fabrication?
8. What does Annealed glass do, and why is it important?
9. Define the difference between Laminated Glass and Tempered Glass.
10. Why is performing secondary operations on glass difficult?
11. Glass has very low thermal expansion. Why?
12. What are the physical properties of glass and which applications need to be avoided?

SECTION

2

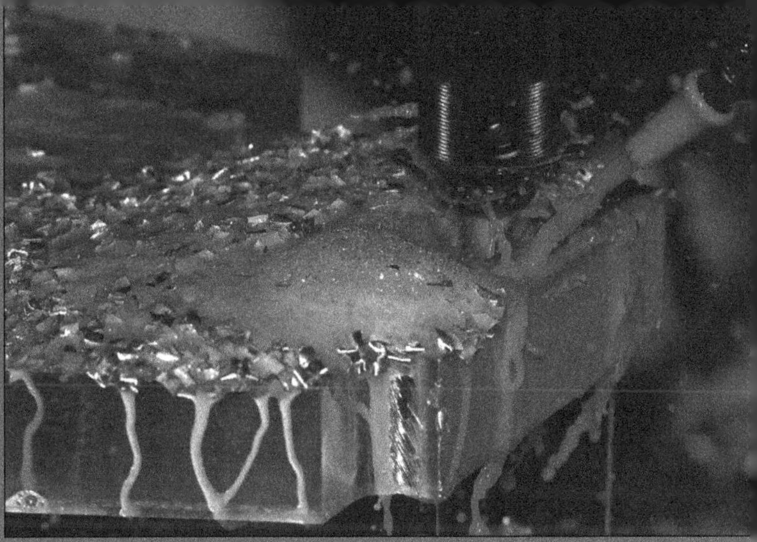

PARTICULATE FORMING PROCESSES

This section deals with processes that fabricate component parts utilizing material in a powder form. The material is not melted into a liquid matter, but rather the processes create the part shapes with the material in a solid, powdered form. These unique processes are used to create component parts that have unique properties, which no other processes are capable of performing. Because the powdered material is pressed together in its solid state, the finished parts will not be solid and have porosity within the internal structures. This porosity within the internal structure of the parts is a favorable trait and is utilized for strength structures and is filled with other materials if desired.

Particulate forming for component parts is performed using metal powders for metallic parts and ceramic powders to produce ceramic and ceramic composites. The powder particles are classified by both size and shape and then mixed with a binder material, which will hold the powders together during compression. After the components are pressed into the desired shape, heat is applied to them in what is called a sintering operation. During this stage, the binder may or may not in some unique applications burn out of the compressed part. Sintering will bond the particles together, creating what appears to be a solid metal or ceramic piece.

Secondary operations may or may not be performed on the component parts either to enhance the physical properties or to add features not possible in the pressing stage. These secondary operations include filling the porosity with an additional material or to machine other features into the pressed parts. Although the majority of the component part configurations are implemented during the pressing stage, certain features need to be added in separately and can bring features and properties not possible with solid part fabrication.

CHAPTER

9

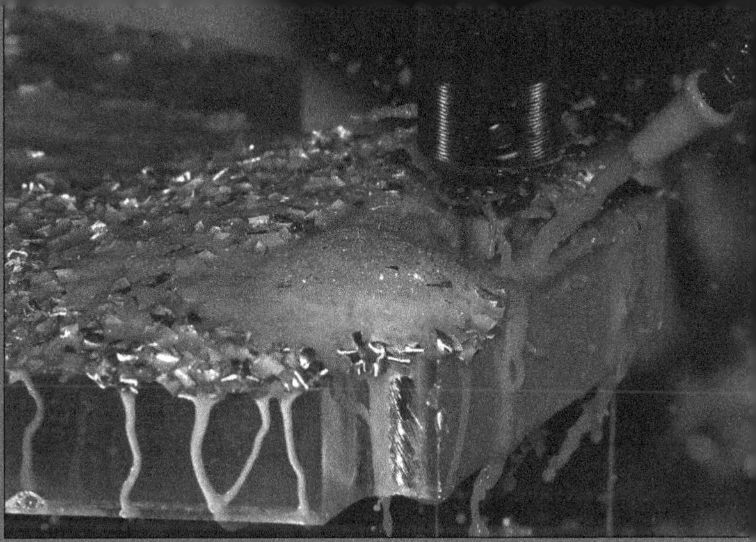

Powder Metallurgy Principles

9.1 INTRODUCTION

Powder metallurgy is a process that involves producing component parts from metallic powders. The material is small, solid metallic particles, which are compacted into a desired configuration utilizing different methods. The most common and widely used is the conventional *Press and Sinter* process. The powder is first produced using a method called atomization and is then classified by size and distribution. Before the shaping process, the powder is mixed with various additives to ensure proper shape processing. After shaping, a sintering process is performed to bond the powder particles together, rendering a component that resembles a solid part, but it will have an internal structure containing porosity. Unlike the casting process where porosity is not a desirable feature, in powder metallurgy, it is a feature that is both desirable and functional. After sintering, a variety of secondary processes can be performed either to add additional geometric features not possible in the shaping stage or to enhance the internal structure of the components. Figure 9.1 shows the steps from powder production through the finished components for the conventional Press and Sinter process. For the fabrication process, excluding powder production and powder preparation, there are three basic steps, which are:

- *Pressing*
- *Sintering*
- *Secondary Operations*

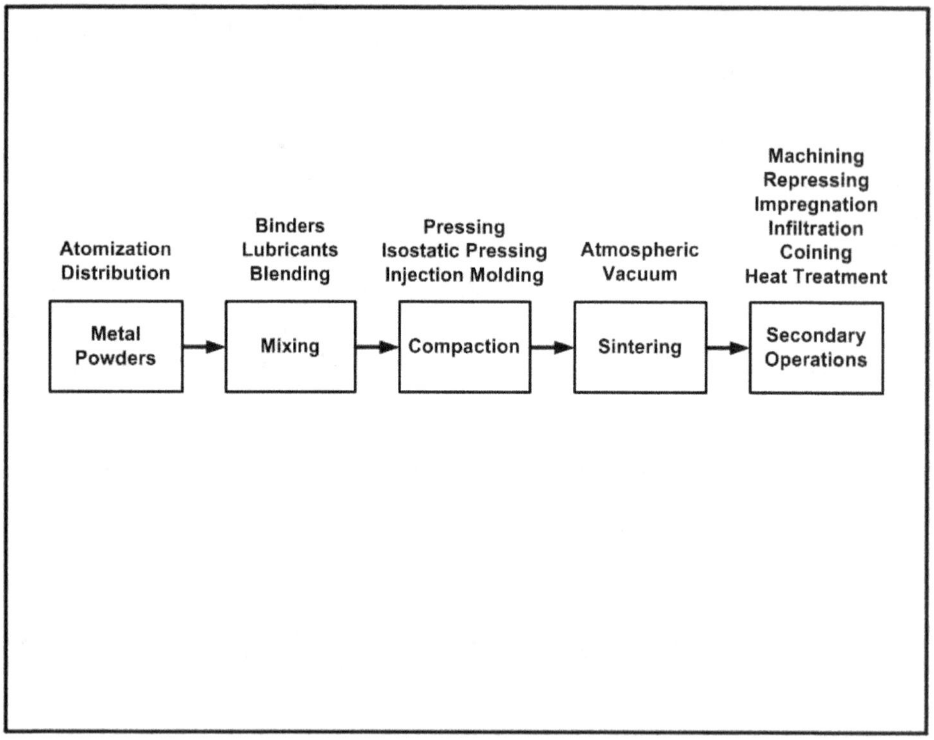

Figure 9.1 Conventional Press and Sinter Powder Metal Process Steps

This chapter covers these three steps in detail and also includes the production of powder metal particles. This process for producing metal parts is very effective for certain configurations and for materials that are extremely difficult to machine. As with any manufacturing process, there are both advantages and disadvantages, which are covered in detail.

9.2 POWDER PRODUCTION

The production of metallic powder is an involved process. It involves liquefying the metal that is to be transformed into powder, then performing an atomization process to create the powders. Various atomization processes exist, and the three most common are detailed as follows.

9.2.1 Gas Atomization

Gas atomization is a process where the metal to be transformed into powder is liquefied to a high temperature and then sprayed through a nozzle into a cold environment. As the liquid metal sprays out and is rapidly cooled, it solidifies into metallic particles, which fall to the bottom of the atomizer. The most common atmosphere that is used is cool nitrogen gas, but for certain applications, other inert gasses are used. By using an inert atmosphere, no changes are made to the material during the process. Figure 9.2 illustrates how the gas atomization process is performed for creating metal powders.

9.2.2 Water Atomization

Water atomization works in very much the same way gas atomization does, but rather than spraying the liquefied metal into a cold environment, cold water is sprayed onto the metal to cool the metal and create the powder. The water is pure, which means it is distilled and deionized to remove any foreign matter that is always present. These foreign matters can be minerals, bacteria, and other elements present in the water. The

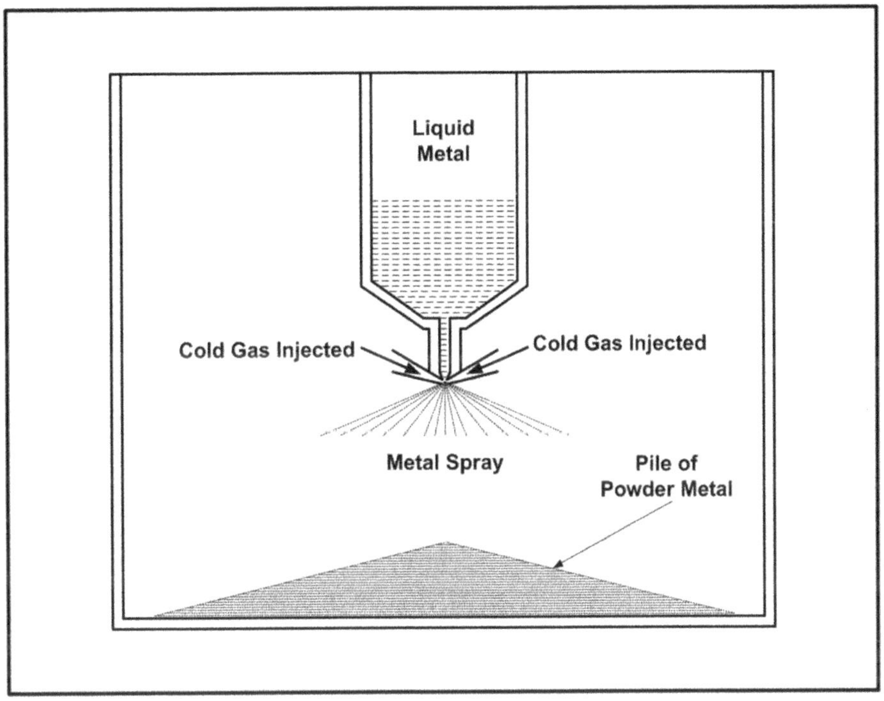

Figure 9.2 Gas Atomization Process

powder particles fall to the bottom of the atomizer along with the water. Once the process is completed, the water is drained off and the powder is collected. From there, the powder must be dried using a low heat to remove any moisture from the process. This process is used for nonferrous metals because ferrous metals such as irons will rust with the addition of water to it. Figure 9.3 shows how the water atomization process produces metallic powders.

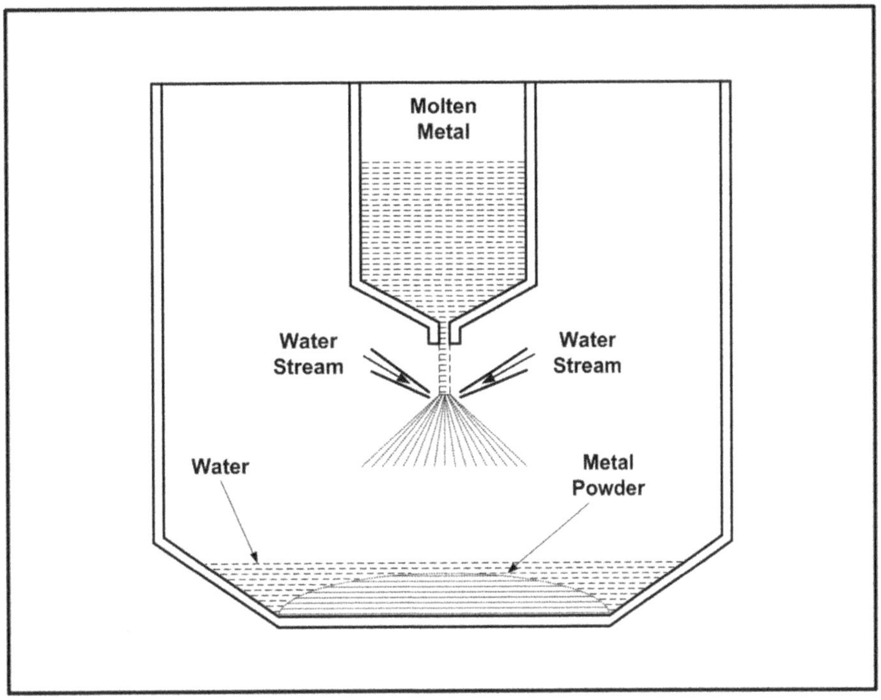

Figure 9.3 Water Atomization Process

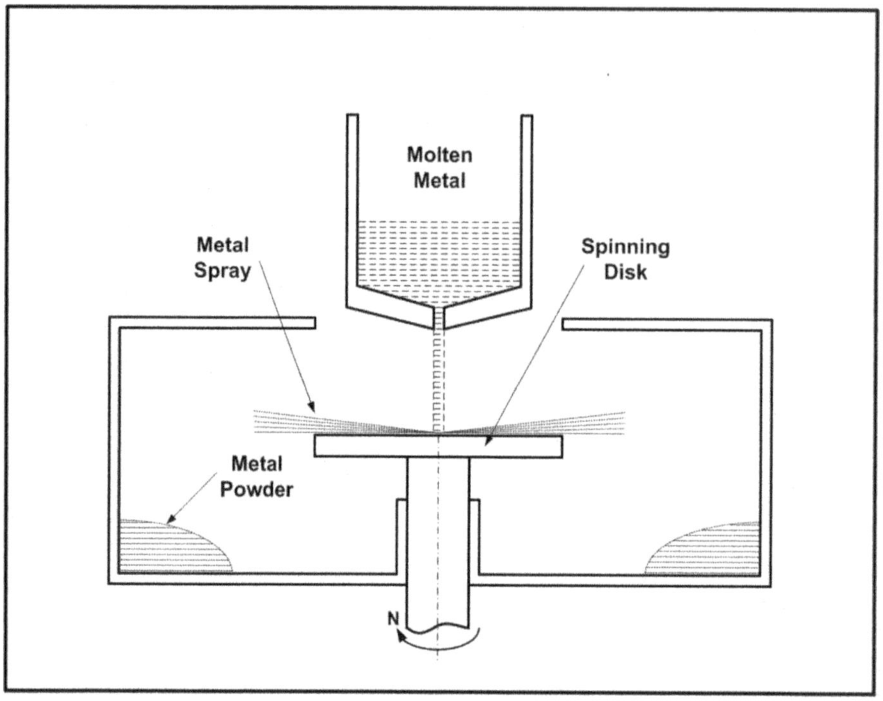

Figure 9.4 Centrifugal Atomizing Process

9.2.3 Spin Atomization

Spin or centrifugal atomization does not utilize water or a cold environment to turn the liquefied metal into powder. This process works by having a small stream of the molten metal fall down onto a rotating disk. When the stream of liquid metal hits the rotating disk, it is sprayed out into the atmosphere creating the powder particles. The atmosphere is neutral where an inert gas such as nitrogen is used. The powder is thrown to the sides of the atomizer and is then collected. Figure 9.4 shows how the spin atomization process operates to produce metallic powders.

9.3 POWDER PARTICLES AND SIZING

After the powder particles have been produced, they must be classified by size. Depending on the component parts to be produced, certain size particles are required to create the internal structures. The atomization process simply produces particles, but they will vary in size greatly. In order to be able to produce components that will function properly, particle size is critical to maintain the proper internal structures. Powder classification is a process that will separate the particles by size and is the second stage in powder production.

9.3.1 Powder Size Classification

Size classification is a simple yet time-consuming process for separating the metallic powder into uniform sizes. To do this, a series of screens are used to sift the metal particles through to separate the atomized powders by their sizes. Each screen has smaller openings, thus separating the particles. The screen openings are measured in what is called a Mesh Count, which denotes how many openings per inch a screen has. They are constructed with very thin and precise wires to separate the opening, thus allowing the particles to either pass through them or collect those of size that did not. Figure 9.5 illustrates the structure of a classifying screen showing the mesh count and the thickness of the wire in the mesh. The larger particles will not pass through the mesh, whereas the smaller particles will. As the particles fall through the screen, they will fall onto another screen with a smaller mesh to separate the next size. This process continues through several screens as shown in Figure 9.6, and the various particles that are smaller, as they continue down through the series of screens, are collected.

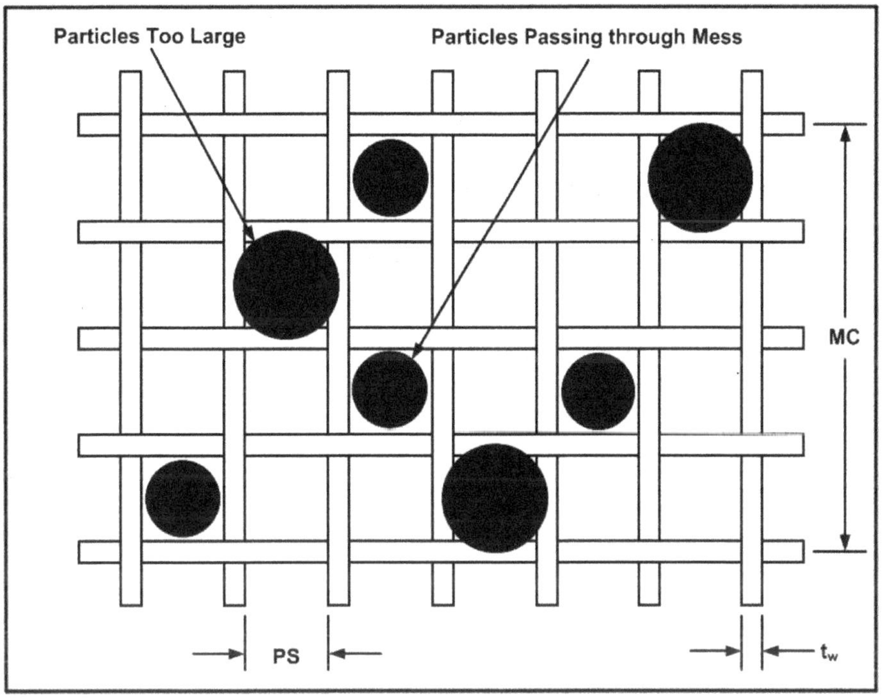

Figure 9.5 Particle Classification through a Sifting Screen

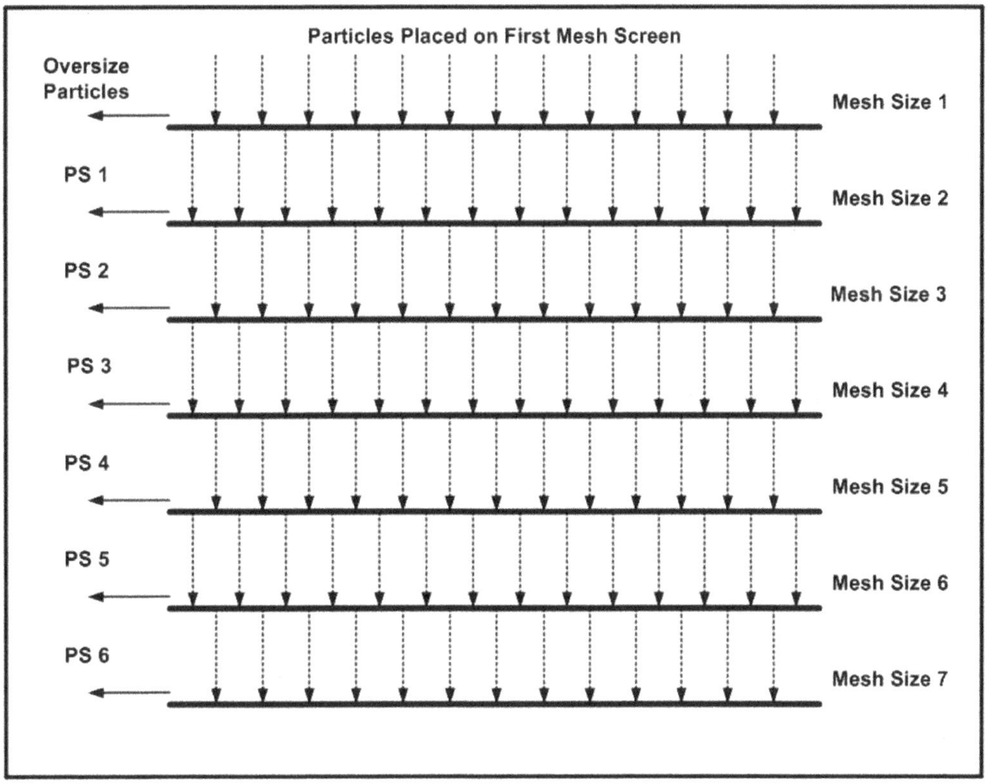

Figure 9.6 Particle Size Sifting Configuration for Separations

To calculate the size of the particles that a given screen will allow to pass through, the following equation is used:

$$PS = (1/MC) - t_w$$

PS = Particle Size

MC = Mesh Count: Openings/in. of the Screen

t_w = Thickness of the Wire

Example:

Data: The Mesh Count of the screen is 350 openings/in. and the wire thickness is 0.0015 in.. Determine the particle size that a screen of these dimensions will produce.

$$PS = (1/MC) - t_w$$
$$PS = (1/350) - 0.0015$$
$$PS = 0.0014 \text{ in.}$$

9.3.2 Powder Shape and Structure

Another important feature of the metal powder is its shape and structure. Depending on the atomization process utilized, along with the temperature of the molten metal and the type of cooling and atmosphere, the powder will take a different shape. Various shapes of the particles are possible to produce and will render different internal structures of the component parts. Figure 9.7 shows the various shapes of the powder particles that are produced in the atomization processes. For certain applications, a mixture of different shapes may be mixed together to obtain unique internal structures of the component parts.

Besides the shape of the particles, the pores of the material are also important for producing components. Pores are classified into two categories, *Open Pores* and *Closed Pores.* Open pores are the spaces between the particles in a pile. These are air gaps that separate the particles, which will produce the desired porosity during

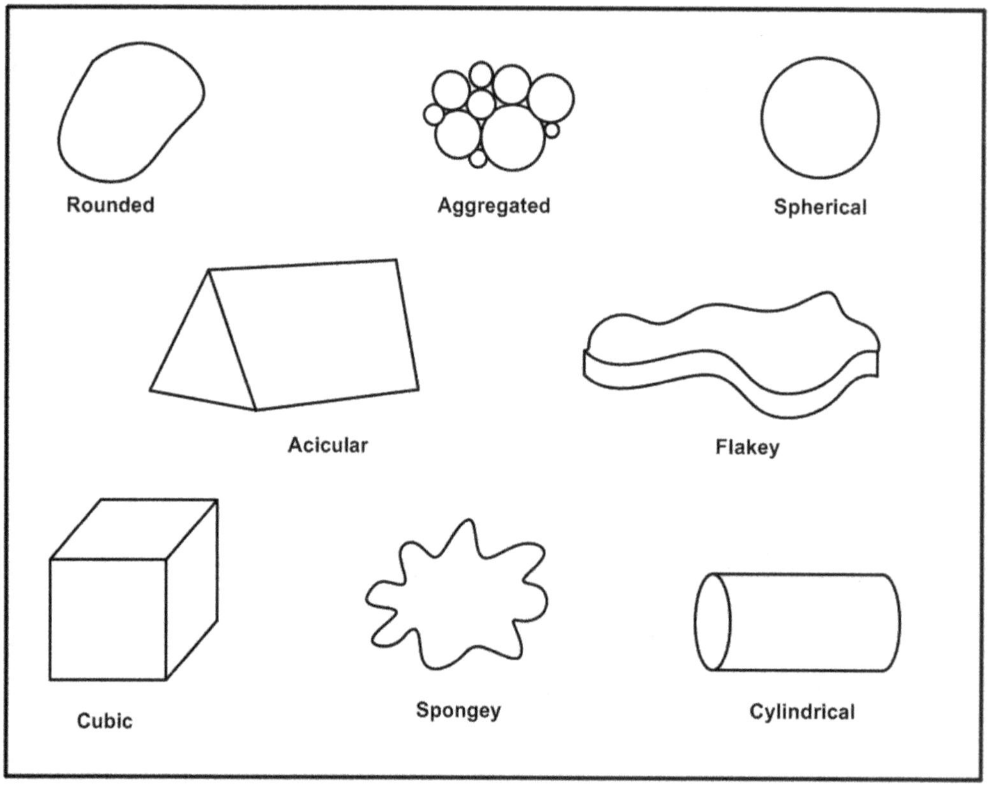

Figure 9.7 Particle Shapes Created in Atomization

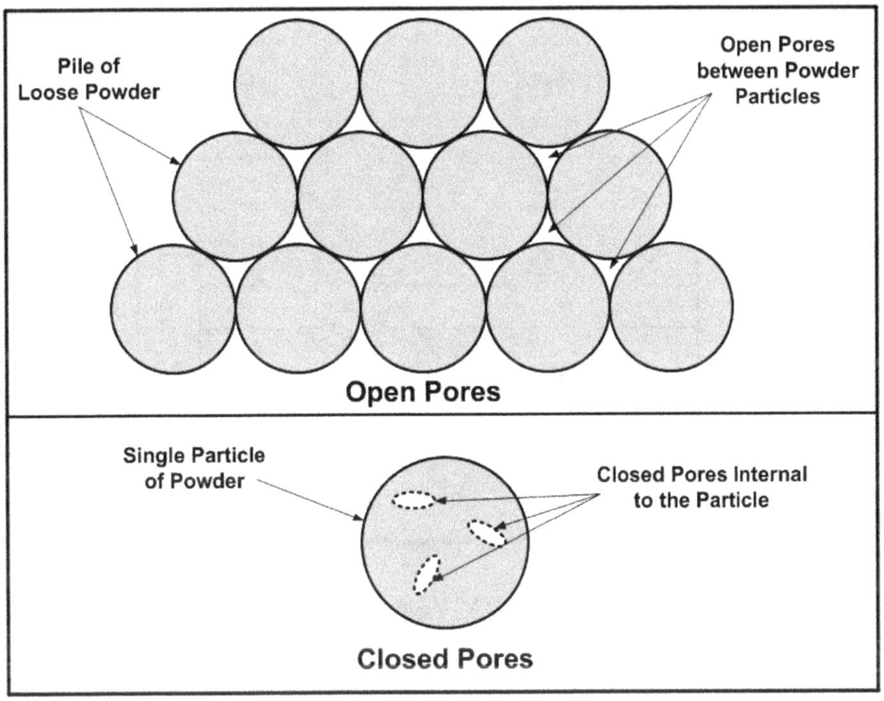

Figure 9.8 Open and Closed Pores in Metallic Powder

the compacting stage of the fabrication. Closed pores are voids internal within an individual particle. These are caused during atomization where air gets trapped in the particles. Closed pores are not a desirable feature and can only be detected by weighing each individual particle. Figure 9.8 shows the difference between open and closed pores.

9.3.3 Powder Preparation

Before the powder metal can be used in the shaping operations, it needs to be mixed with various additives along with different materials or particle sizes. This stage can be performed either during production of the powders or as an in-house process. The breakdown of this process is listed as follows.

Blending: Blending is when the same powder materials are used, but different size particles are combined. This will render different porosity structures internal to the component parts being fabricated.

Mixing: This process is combining different powder materials to produce a unique alloy. This is a major advantage of powder metallurgy as purchasing a unique material in solid form can be extremely costly, if possible to procure it at all. The only consideration for this is to ensure that the melting temperatures of the materials being mixed are very close. This is required so that during the sintering phase, they will bond together and not melt.

Lubricants: Lubricants are added to help reduce the friction generated during the pressing and compaction of the powder. As the powder is compressed under pressure, a lot of friction is generated in the pressing tools and the lubricants will reduce it greatly. This will reduce tooling ware and extend the life of the forming tools.

Binders: Binders are an extremely important additive to the mixture. As the powder is being pressed, the binder material will hold the powder together and not allow the particles to fall apart before they can be sintered. However, the percentage of binder is critical to maintain because during sintering, the binder material will burn off and the component part will shrink by the amount of binder added to the mixture. Depending on the powder that is being used, the most common binder material is either a thermoplastic or wax. These are added in fine powder form.

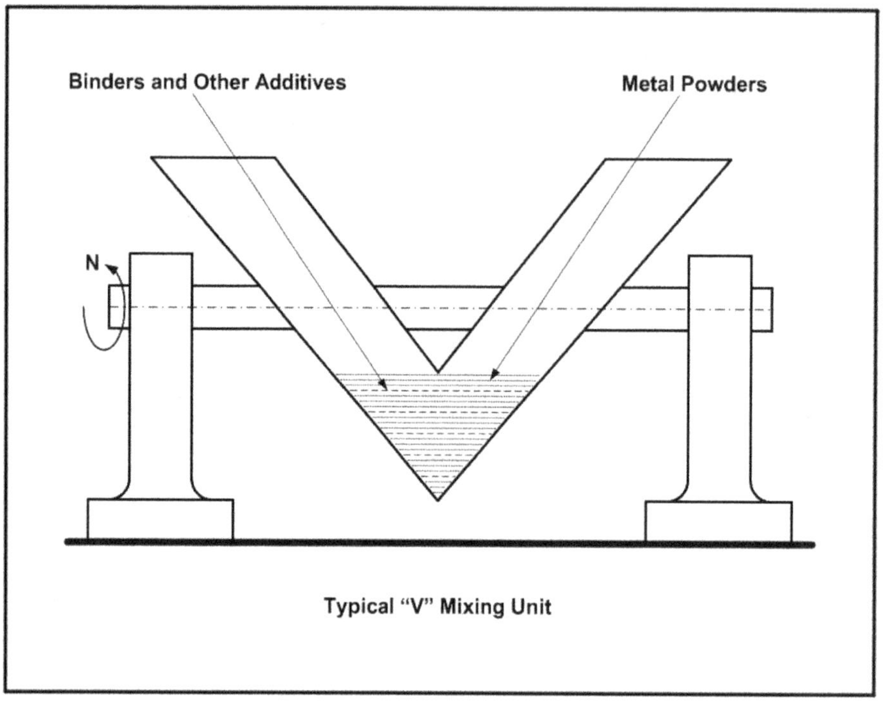

Figure 9.9 Typical "V" Style Powder Mixing Unit

Deflocculants: This additive is used to prevent clogging of the mixture to ensure an even mixture during the forming operations. It is very important that the powder mixture remains constant and does not clog together, ensuring an even mixture of all the materials at all times.

Once all the ingredients are selected, they need to be mixed thoroughly. There are several different types of mixing apparatus that are utilized. Figure 9.9 shows one of the simplest yet very effective mixers. This is a "V" style mixer where just two hollow tubes are joined together in a "V" shape. The materials are added into this and the unit is rotated, thus causing the various mixtures to blend together evenly.

9.4 PACKING, POROSITY, AND DENSITY

The amount that the powder material is compacted will determine what the porosity and density levels will be for the fabricated component. These three characteristics of a powder metal component are based on the combination of two things. First is the size of the particles being utilized in the compacting process to be pressed together. The size of the particles will determine how much open space there will be in the internal structure of the part, which is the porosity. With the porosity being the open space within the component part, the amount of metal will determine the density. The second factor is how tightly the particles are pressed together. This is known as the packing factor, and as more pressure is used, the particles will be compacted closer, thus reducing the amount of porosity and increasing the overall density of the component part being fabricated. It is this combination of the size of the particles and how much pressure is being used to compact the powder material that determines the amount of porosity and density the part will ultimately have when completed. These three factors are discussed as follows.

9.4.1 Porosity

As mentioned, the porosity is the amount of open pores that will be in the internal structure of the component part. To put this in simple terms, a powder metal part is not solid but rather resembles a structure like a pile of stone. Regardless of how much pressure is used to form the parts, it cannot and will not ever be a solid mass.

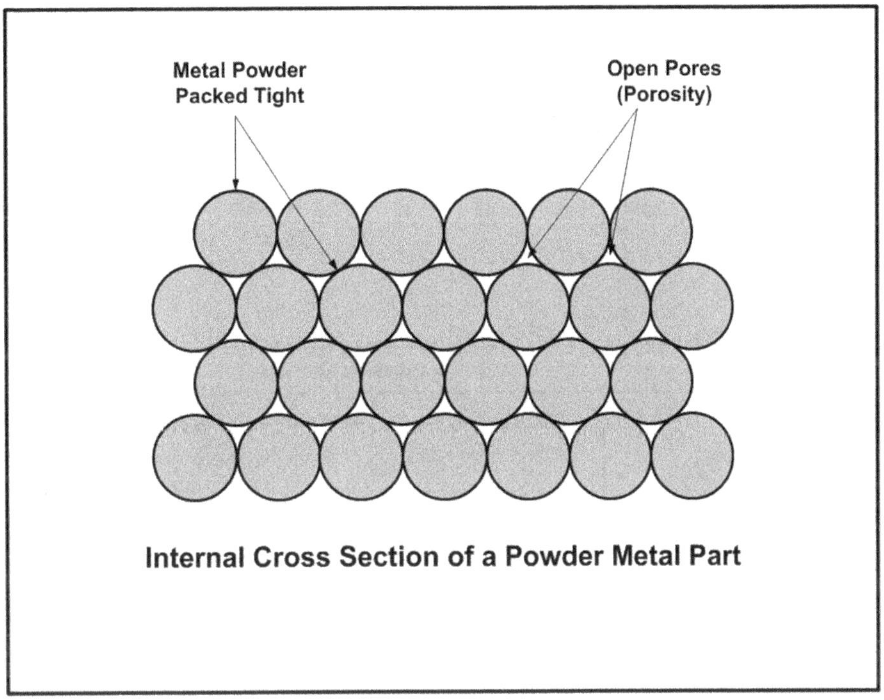

Figure 9.10 Cross Section of a Powder Metal Part Showing Porosity Present

The porosity within the internal structure performs two key functions for powder metal components. First, it will allow a metal part to have less weight than a solid metal part of the same configuration. Due to the pockets inside the part, it will always weigh less than a solid part of the same configuration. Second, the pores can be filled with another material during the secondary operations to give the component part a microstructure what no other fabrication process can render. Figure 9.10 illustrates the powder particles compacted and shows the open pores or porosity within the internal structure.

9.4.2 Density

The density of the component is related to the amount of metal material within the structure. As seen in Figure 9.10, the particles or solid mass in the part will determine the overall density. This density/porosity is a percentage combination that is closely held by the amount of pressure being exerted when compacting and the size of the particles being utilized. The smaller the particles being used, the completed component will have a higher density and the porosity will be lower. Depending on the application of the part, the porosity can be as high as 50%, thus leaving the density at 50%. This means that even though the component appears to be solid, it contains 50% air in the internal structure.

Density is measured in three ways. First there is the *True Density (TD)*, which is the density of a solid part. For solid components, the density will be 100% because there is no porosity within a solid mass of metal. The second measure is the *Pack Density (PD)*, which is the density found in a powder metal part. The pack density will always be less than the true density because as previously stated, powder metal components always contain porosity, so the air will always lower the density. Third, is the *Bulk Density (BD)*, which is the density of a loose pile of powder particles. This is the starting point for the compression process and is used to calculate how much powder will be required to press the component to the desired density.

9.4.3 Packing Factor

Packing factor (PF) relates to how tightly the powder particles are pressed together during the compacting operation. As the loose powder is loaded into the compressing die, it will be pressed with some calculated tonnage to create the part. The more pressure that is used, the tighter the powder particles will be packed together.

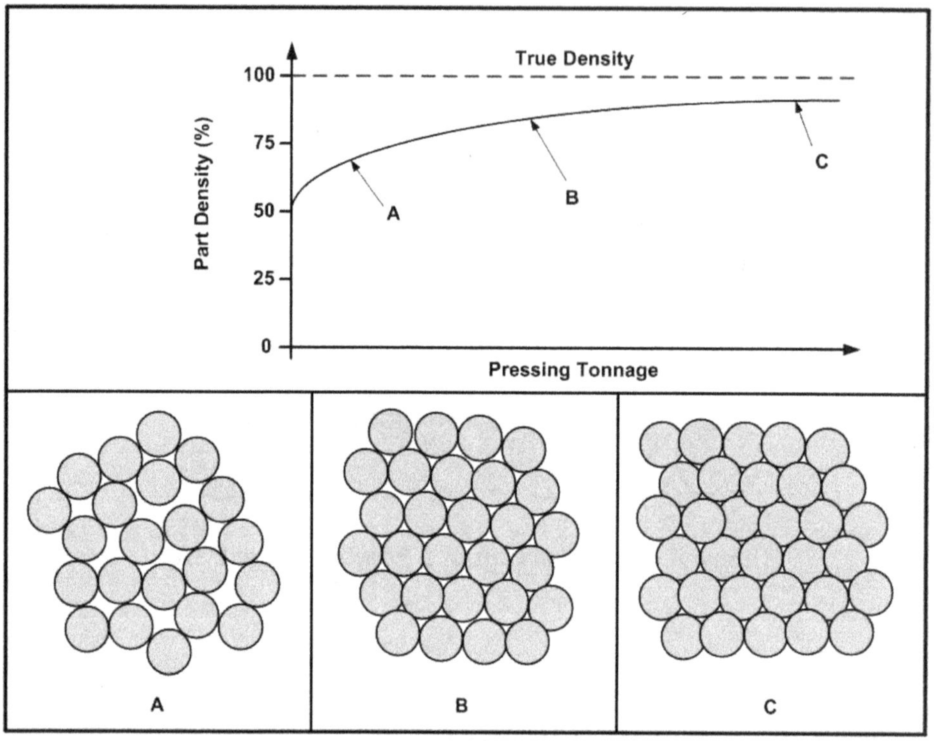

Figure 9.11 Relative Compacted Part Density vs. Pressing Tonnage

This is what will control the amount of porosity the component will have when the compression is completed. Figure 9.11 illustrates the packing factor forces with the particle compaction. The top line on the chart is the true density, which is that of a solid part and is 100%. As shown, the density curve is not linear and never reaches 100% regardless of the amount of pressure being used in the compacting operation.

Knowing that there is a relationship between porosity, density, and packing factor, each can be calculated in a variety of ways. The packing factor can be derived from the following three equations:

$$PF = PD/TD$$

PD = Packing Density of a Powder Metal Part
TD = True Density of a Solid Part

$$PF = TSV/BSV$$

BSV = Bulk Specific Volume of a Powder Metal Part
This will always be greater than 1
TSV = True Specific Volume of a Solid Part = 1

$$PF = PW/SW$$

PW = Weight of the PM Part
SW = Calculate Weight of a Solid Part
Of the same configuration as the powder metal part

Once the packing factor has been determined, the porosity can be calculated by the following equation. Both packing factor and porosity combined must equal 1 as they are given as a percentage of the amount of metal and air contained within the powder metal part.

$$1.0 = PF + Porosity$$

Applying these equations in the following three examples will demonstrate how the Packing Factor and the Porosity can be determined.

Example 1: PD = 1.743, TD = 2.829

PF = PD/TD

PF = 1.743/2.829

PF = 0.616

1.0 = PF + Porosity

1.0 = 0.616

Porosity = 0.384

Example 2: BSV = 1.973, TSV = 1

PF = TSV/BSV

PF = 1/1.973

PF = 0.507

1.0 = PF + Porosity

1.0 = 0.507 + Porosity

Porosity = 0.493

Example 3: PW = 2.150 lbs., SW = 3.412 lbs.

PF = PW/SW

PF = 2.150/3.412

PF = 0.630

1.0 = PF + Porosity

1.0 = 0.630 + Porosity

Porosity = 0.370

9.5 POWDER SHAPING PROCESSES

Shaping and forming the powder metal components are a simple but critical operation. The three main processes for performing this are the *Conventional Press and Sinter*, *Isostatic Pressing*, and *Powder Injection Molding*. Each of these processes will render different component configurations and have their positive and negative attributes. These three common forming processes are detailed as follows.

9.5.1 Conventional Press and Sinter

The conventional press and sinter process is the most common and most widely used to produce powder metal parts. The tooling used to fabricate part in this process is called a *Punch and Die Set* consisting of both an upper and a lower punch and a die. The die is the portion of the tooling set that contains the configuration of the part that is to be pressed. The upper punch will slide into the die to compress the powder from the top, while the lower punch will compress the powder from the bottom and also eject the part from the die once the pressing has been completed. Powder metal tooling is extremely expensive due to the fact that very complex configurations are usually required and that the punches need to fit into the die with a clearance of about 0.0002 in.. This extremely close fit is required so that the powder will be fully compressed and not escape from the die, causing a burr on the parts. This process is completely automated in the press and begins with the filling shaker moving over the die with the upper and lower punches both retracted. The shaker will fill the die with a closely calculated amount of powder and then retract off to the side, leaving the die cavity filled to the top with the powder material. Figure 9.12 shows the first two steps in the process where the shaker fills the die and moves off to the side. Next, the upper punch moves down into the die and the lower punch moves up, compacting the powder tightly to a calculated tonnage to the correct packing factor. After the powder has been compacted, the upper punch retracts and the lower punch moves up to the level of the die surface ejecting the completed part. Figure 9.13 shows these two final steps in the process where the part is compacted and ejected from the die. Then the shaker slides over and pushes the completed part aside, the lower punch retracts, and the process repeats itself.

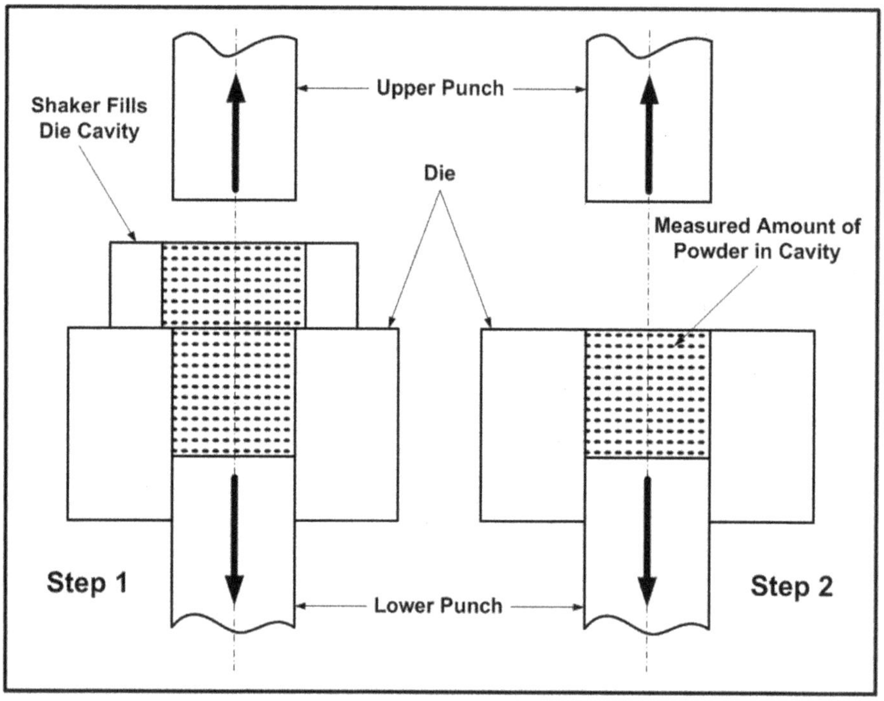

Figure 9.12 Pressing; Filling the Die Cavity

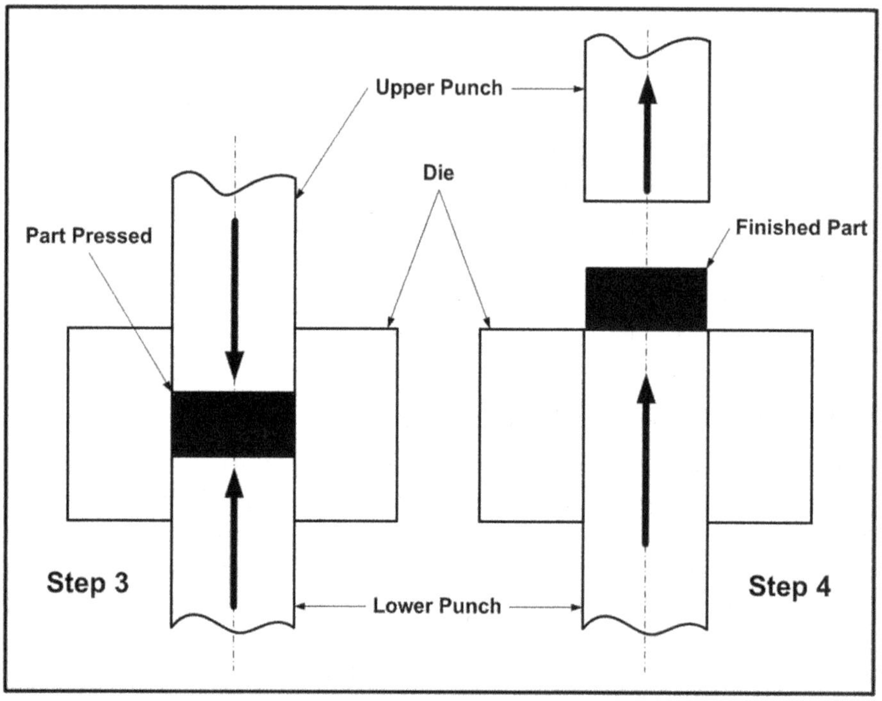

Figure 9.13 Pressing; Compacting the Powder and Ejection from the Die

Depending on the size of the part being fabricated and the pressure being utilized, this entire cycle can take between 2 and 5 s to complete. This shows the efficiency for producing components regardless of the complex geometry configuration of the part. With the process being automated, one operator can normally work on two presses and only has to remove the completed parts and place them on a tray to be ready for sintering. At this point, the compressed parts are very fragile because the only thing that is holding the powder particles together is the binder material. In this stage of the process, the component parts are known as a *Green Part* and will break very easily if dropped.

9.5.2 Isostatic Pressing

The isostatic forming process is one that can render parts of configurations that the press and sinter process cannot. The difference between the two is that the press and sinter process only applies pressure from two directions, the top and bottom. Isostatic will apply pressure from all sides equally. In order to accomplish this, a rubber mold is used, which will collapse as pressure is applied to it. The thickness and composition of the rubber are very closely calculated against the pressure that will be applied against the size of the component that is to be formed to attain the dimensions required. This process will not maintain the tolerances that press and sinter can, but is capable of producing complex geometric shapes, which may require secondary machining to render a finished part.

The first step is to load the rubber mold with the powder metal material and then seal it. The mold is placed in a chamber containing hydraulic fluid, which denotes how the pressure is going to be applied. Figure 9.14 shows the rubber mold loaded with the powder metal and placed in the pressure chamber. From there, the chamber is filled with hydraulic fluid and subjected to the desired pressure. This will apply the pressure against the rubber, collapsing it against the powder inside. Figure 9.15 illustrates the pressure being applied. Depending on the configuration of the part being formed, the thickness of the rubber may vary in order to adjust the pressure on the powder in different locations. The time that the mold is under pressure is calculated, and when the compacting stage is completed, the pressure chamber is drained and the completed component is retrieved from the mold. Figure 9.16 shows the completed part displaying a configuration that could not be attained in the press and sinter process.

Isostatic forming can be performed in two ways. *Cold Isostatic Pressing (CIP)* is performed with the hydraulic fluid at room temperature. The fluid will rise in temperature due to the pressure it will be under, but the process begins at room temperature. *Hot Isostatic Pressing (HIP)* will start with the hydraulic fluid at an elevated temperature. This is used for certain materials that require additional heat to form the component part correctly.

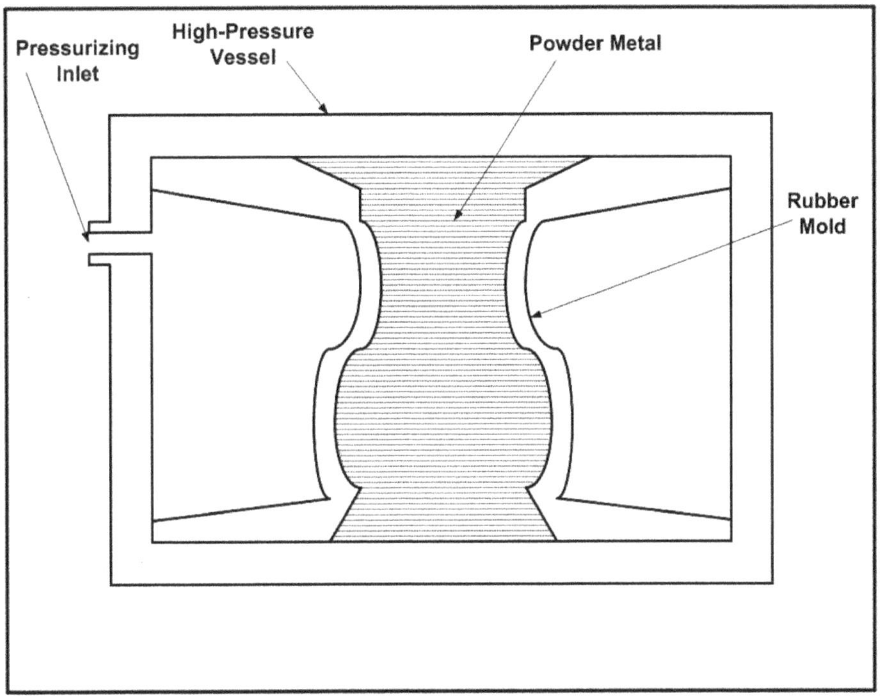

Figure 9.14 Powder Metal Loaded into an Isostatic Compacting Die

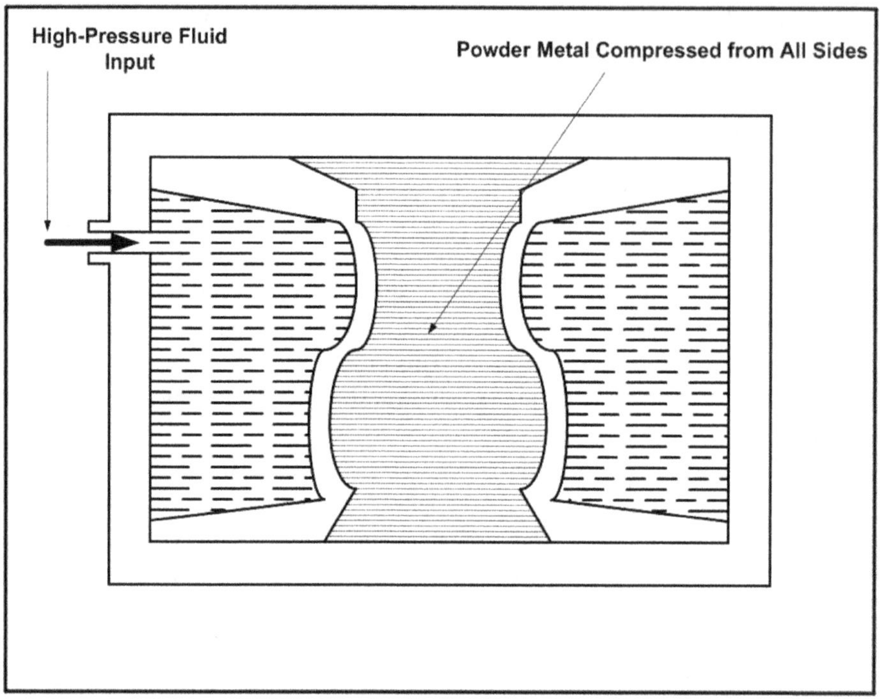

Figure 9.15 Compacting Die Placed under High Pressure in a Hydraulic Enclosure

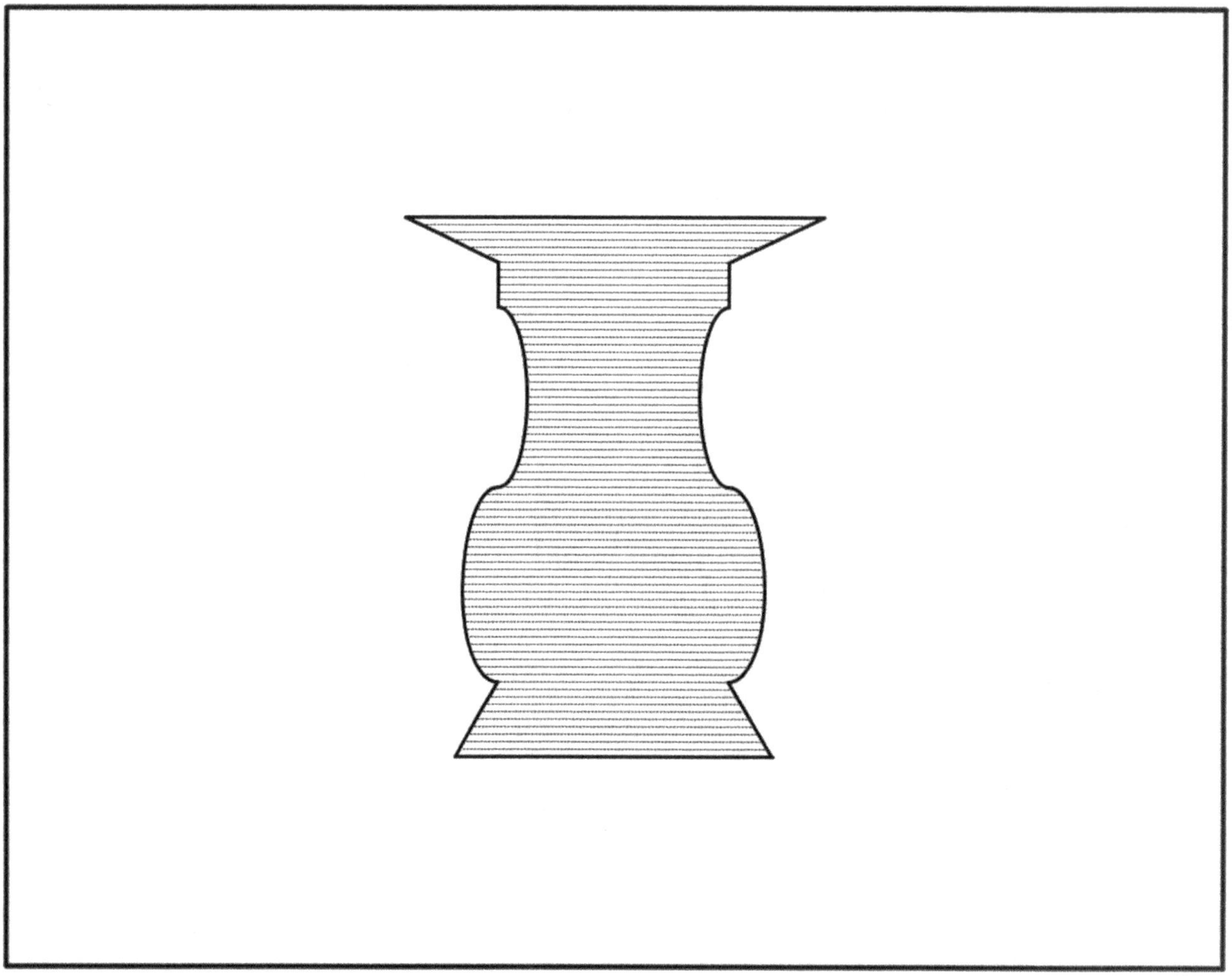

Figure 9.16 Isostatic Part Configuration Created in Pressing

9.5.3 Powder Injection Molding

Powder Injection Molding is the newest powder forming process. This began in the 1980s in an experimental manner and has now become a very viable process that will deliver part configurations and internal structures that no other powder forming process can produce. The powder injection molding process can be used for producing components from any solid particles, but is mainly used for producing metal parts with *Metal Injection Molding (MIM)* or ceramic part with *Ceramic Injection Molding (CIM).*

Unlike the press and sinter and the isostatic forming processes, MIM requires a different mixture of the powder before starting. This process requires very small metal powder at about 1–2 microns in size, and the particles must be spherical in shape. Binder mixing is next, and it is about 15–20% binder material that is mixed with the powder. This mixture needs to be very evenly mixed with the binder in a semimolten state. The mixture is then let to cool and chopped into small pellets of about 1/8 in. in size to be placed into an injection molding injection tube. To form the part configurations, the mold that is identical to the molds used for plastic injection molding is utilized, which is loaded into a molding machine. Figure 9.17 shows the starting setup for producing MIM parts.

Next, the powder/binder mixed pellets are loaded into the injection tube for melting. This stage of the process is extremely critical! The temperature of the injection material needs to be closely maintained in order for the mixture to remain equally distributed. If the binder material becomes a fluid, the metal particles will settle to the bottom and not be equally distributed. For this, the consistency of the melt should be about that of clay for this to be accomplished. Also, the injection tube is customized with mixing blades to keep the metal and binder mix uniform.

This melt is then injected into the mold under a controlled pressure and acceleration speed. If the acceleration is too fast, the binder may inject ahead of the metal particles and a uniform mixture will not be injected into the cavity. The cavity is filled in the same manner as it would be with plastic molding and allowed to cool as shown in Figure 9.18. Once the part is cooled, it is ejected from the mold as shown in Figure 9.19 and the green part is retrieved. The part at this stage will have rigidity similar to that of a crayon where it can be scratched or dented and must be handled carefully.

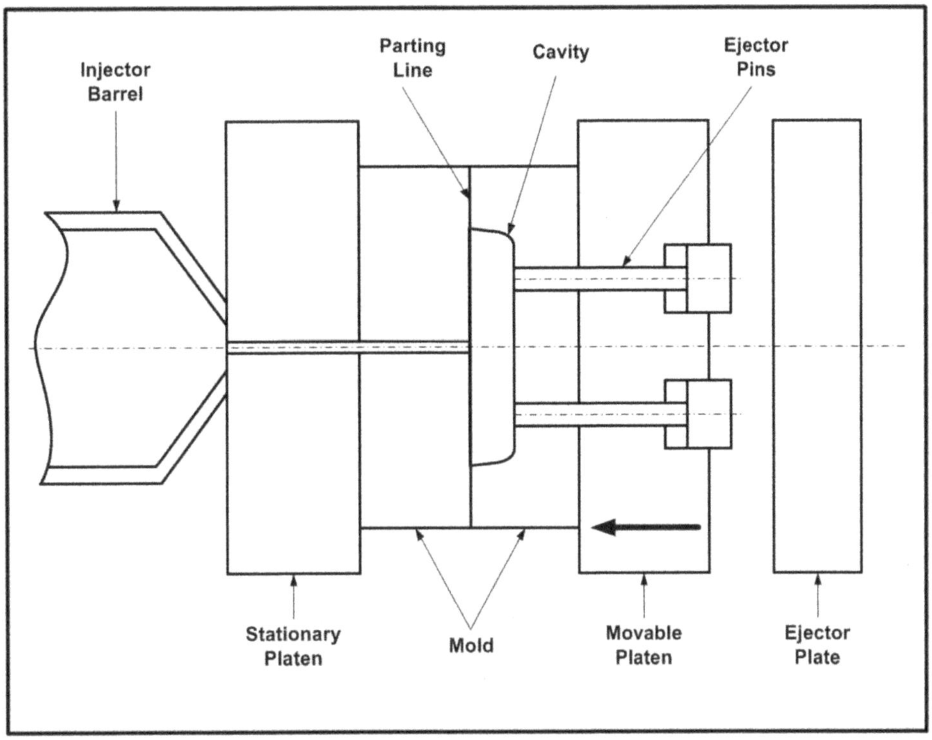

Figure 9.17 Injection Mold for Metal Injection Molding

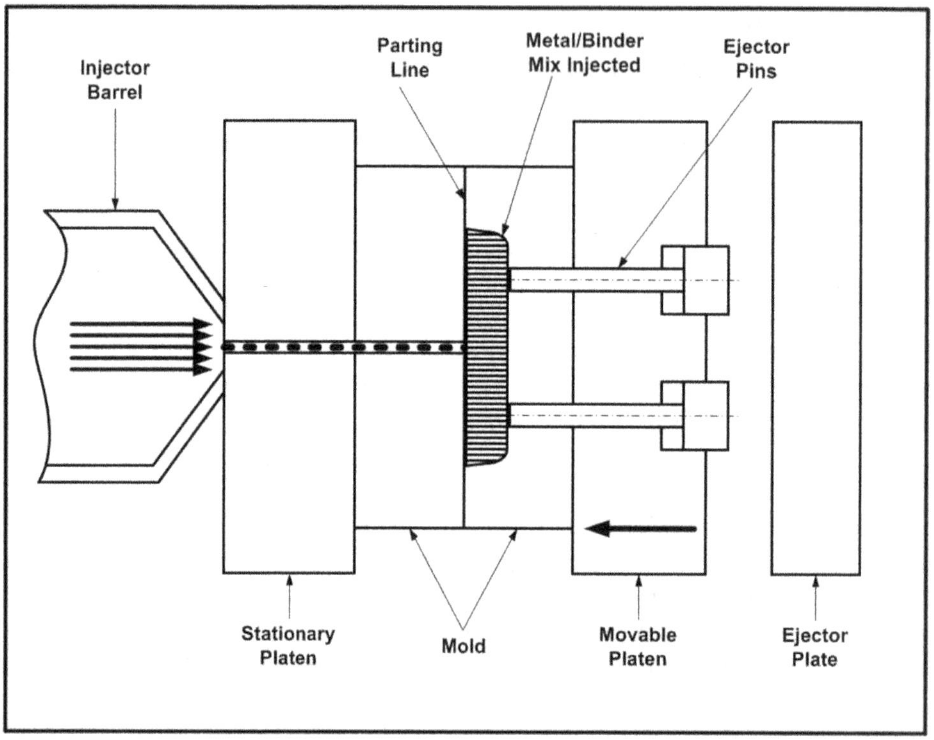

Figure 9.18 Powder and Binder Mixture Injected into the Mold Cavity

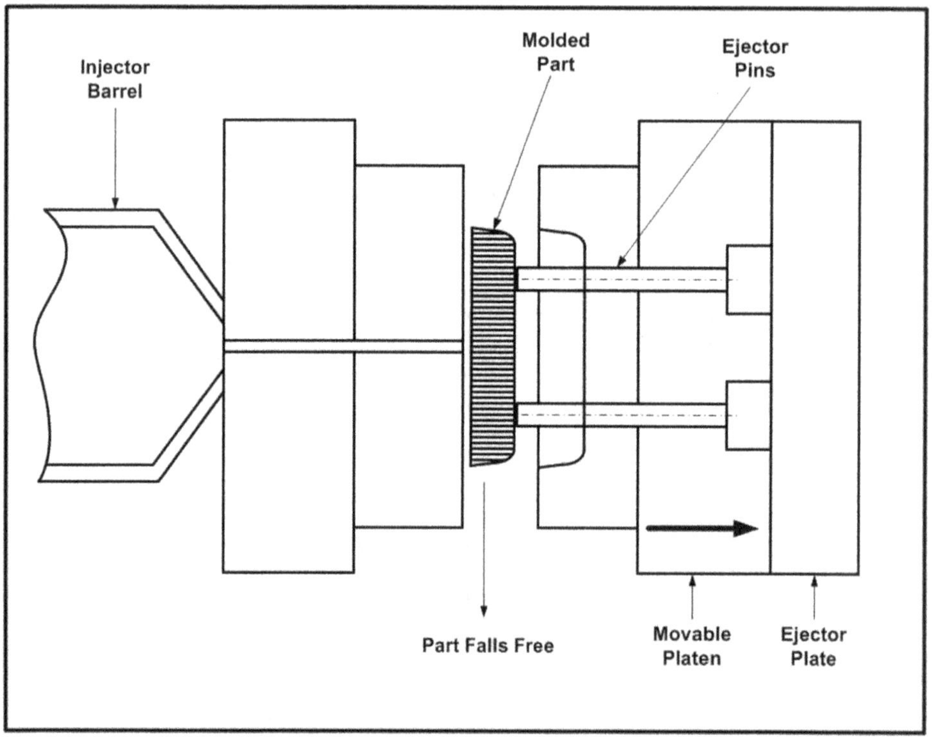

Figure 9.19 Injection Molded Green Part Ejected from the Mold Cavity

A major consideration with this process is the shrinkage that will occur during the next phase, which is sintering. The binder material is used to transport the powder metal into the mold and will be burned out of the part during sintering. If 15–20% binder is mixed with the powder, then 15–20% shrinkage will occur. Knowing this, the cavity of the mold needs to be calculated to allow for that percentage of shrinkage of the part. Also, the binder mixture must be EXACT every time. If say 17% binder is used in one mix and 19% is used in another mix, the shrinkage will be different for each run and the finished parts will not be of the same size.

Although the basic steps for powder injection molding are the same as those for the conventional processes, which are forming, sintering, and secondary operations, there are some adjustments required to the overall phases. Because there is far more binder material in the powder mixture, a debinding step is required. During this step, the binder material needs to be burned out of the part the same way as with the conventional processes, but requires much more time. This is performed in a vacuum furnace, rendering what is known as a brown part. Depending on the material that is being formed, this may have to be a separate step and not performed in line with sintering. Some materials can be processed in line with both debinding and sintering in the same furnace, but others require this to be separate.

Powder injection molding will deliver a component part with a completely different internal structure than the conventional processes will. Because the powder size is so small, the relative density of these parts will be between 98% and 99% of the relative density. There will be not linked porosity as with the conventional processes, but rather small, isolated pockets of porosity. For all intensive purposes, the powder injection molding process renders a solid component that will have the same properties as a solid metal part would have.

9.6 SINTERING

The second phase in the powder metal process is sintering. Here, the binder material is removed and the particles are bonded together. To accomplish this, the powder metal parts are heated up to between 75% and 85% of the melting temperature of the material. This will bond the particles together at their contact point, which, in simple terms, welds each of them together. Figure 9.20 shows the bonding point created from sintering.

There are three stages to successful sintering, which are *Preheating, Sintering (Soaking),* and *Cool-Down.* Neither of these steps can be avoided in order to sinter the parts correctly. Each step performs a different but critical function in the overall sintering process and is discussed as follows.

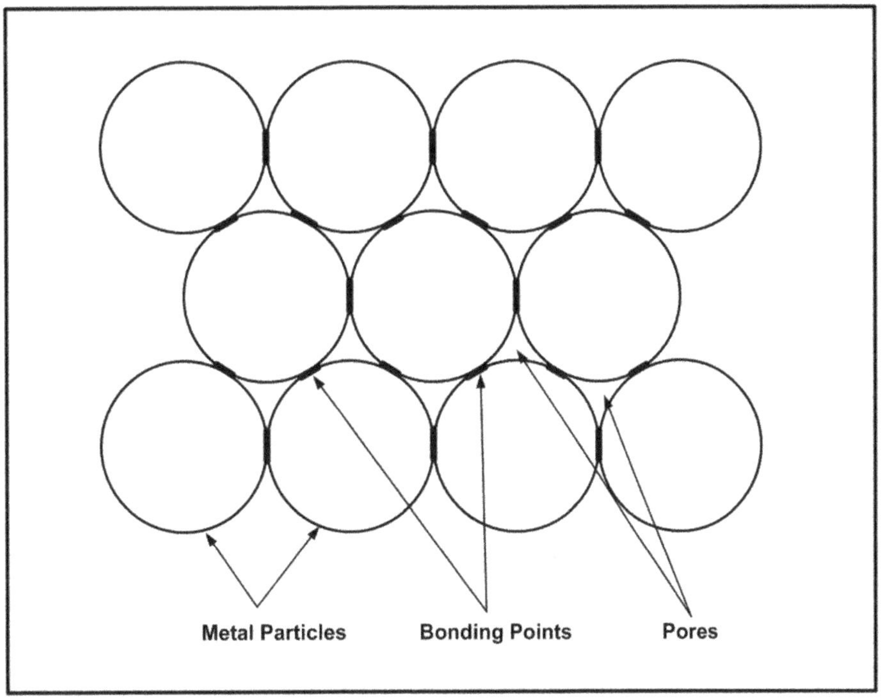

Figure 9.20 Particles Bonded Together at Their Contact Points from Sintering

9.6.1 Preheating

During the preheat stage, the parts are raised in temperature to the sintering temperature slowly. During this portion of the process, the binder material will be burned out of the part. By raising the temperature slowly, as the binder burns off, the part will hold its configuration and not crumble apart. There is a maximum rate at which the part temperature can be raised, but no minimum rate. For production purposes, the temperature will be raised as rapidly as allowed, without causing damage to the components; but for this stage, there is no minimum rate and performing this longer will not affect the component parts. If the temperature is raised too rapidly, the binder will burn off too fast and the parts will crumble and not maintain the desired geometry. Knowing this, it is not possible to skip this stage of the process and place the components into a furnace set at the sintering temperature.

9.6.2 Sintering

In the sintering stage, the particles will be bonded together as shown in Figure 9.20. The furnace will be set at the given temperature range for the material being sintered and the parts will be held there for a minimum amount of time. Figure 9.21 shows the sintering temperature and minimum amount of time to accomplish this. If the parts were to remain at that temperature longer, it will not affect them at all. Even if they remained there 2–3 times longer than the required minimum time, it would have no affect on the component parts or their internal structure.

9.6.3 Cool-Down

After the parts have gone through the sintering stage and the particles are properly bonded together, there needs to be a cool-down stage. With the components at 75–85% of their melting temperature from sintering, a slow cooling is required so as not to damage the parts. Noting that these are not solid parts, but rather particles bonded together, the thermal contraction needs to be performed slowly. If the cooling is performed too rapidly, the contraction will cause the bonded points to fail and the parts will crack and even break apart. By controlling the cooling rate, the part will contract slowly and the bonds will remain intact within the component.

Sintering is a simple yet complex process. The parts are heated up slowly, held at a given temperature, and then cooled down slowly. However, the rate of heating them up needs to be precisely calculated as does the cooling rate in order not to damage the part. Likewise, the temperature and time in the sintering stage are also critical. If the time is enough, the bonds will not be created. Also, too low or too high of a temperature will either not create the bonds or completely melt the parts. Figure 9.22 shows the sintering curve with the three stages of the process for the parts produced with conventional processes.

For component parts fabricated with the injection metal molding process, there are additional steps that need to be performed. The binder material needs to be burned off before the actual sintering can be performed as shown in Figure 9.23. This requires an additional process in the furnace. Once the binder has been removed, the work piece is called a brown part and is extremely fragile at this point. Once the binder is removed, the component part is sintered to bond the particles the same way as with the press and sinter components.

Figure 9.24 shows the sintering curve for parts produced with powder injection molding. Notice that debinding involves a soak time due to the large amount of binder that needs to be removed.

Sintering Temperatures & Times for Select Materials

Material	Sintering Temperatures		Minimum Sintering Time (Min)
	°C	°F	
Brass	850	1,600	24
Bronze	820	1,500	15
Copper	850	1,600	25
Iron	1,100	2,000	30
Stainless Steel	1,200	2,200	45
Tungsten	2,300	4,200	480

Source: Fundamentals of Modern Manufacturing; Second Edition. Groover, M.

Figure 9.21 General Sintering Temperatures for some Select Materials

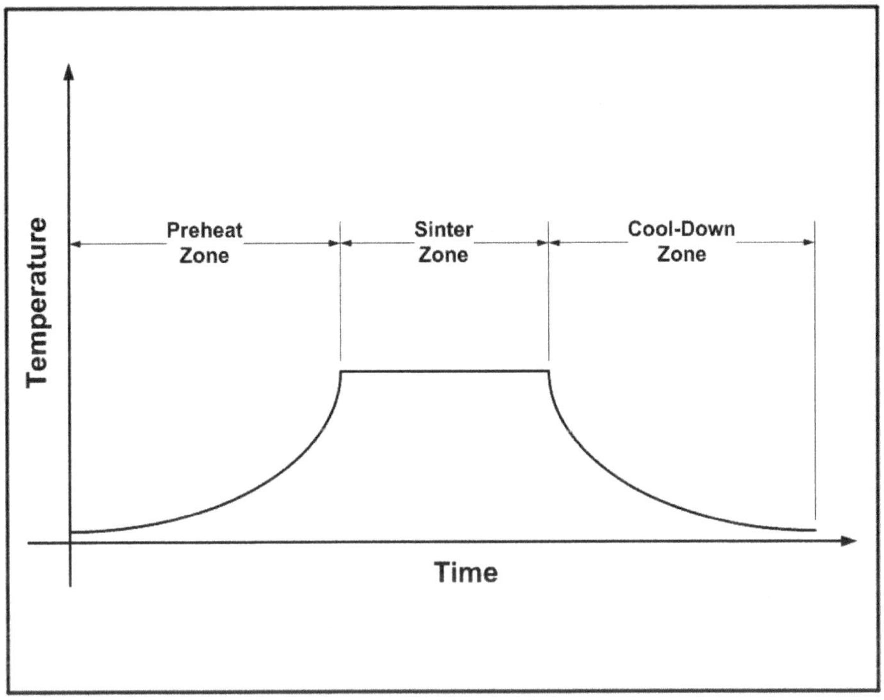

Figure 9.22 Conventional Powder Metal Part Sintering Curve

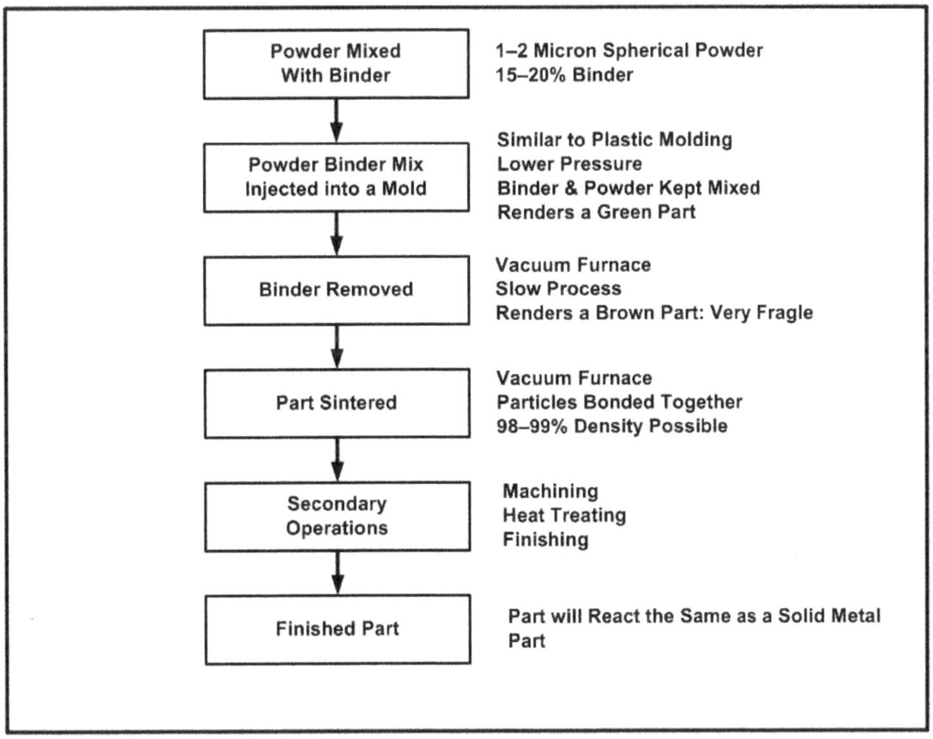

Figure 9.23 Process Steps for Metal Injection Molding of Powder Metal

In production, sintering is a batch, flow-through process performed in a continuous furnace. The furnace has stages for preheat, sintering, and cool-down with a controlled internal environment such as an inert gas, mainly nitrogen. Parts are loaded onto a slow-moving belt, which is timed for the three stages that they will pass through the furnace. As they move from stage to stage, each phase of the process is performed and when they exit the furnace, they simply fall into a container. Figure 9.25 shows the typical structure of a continuous sintering furnace.

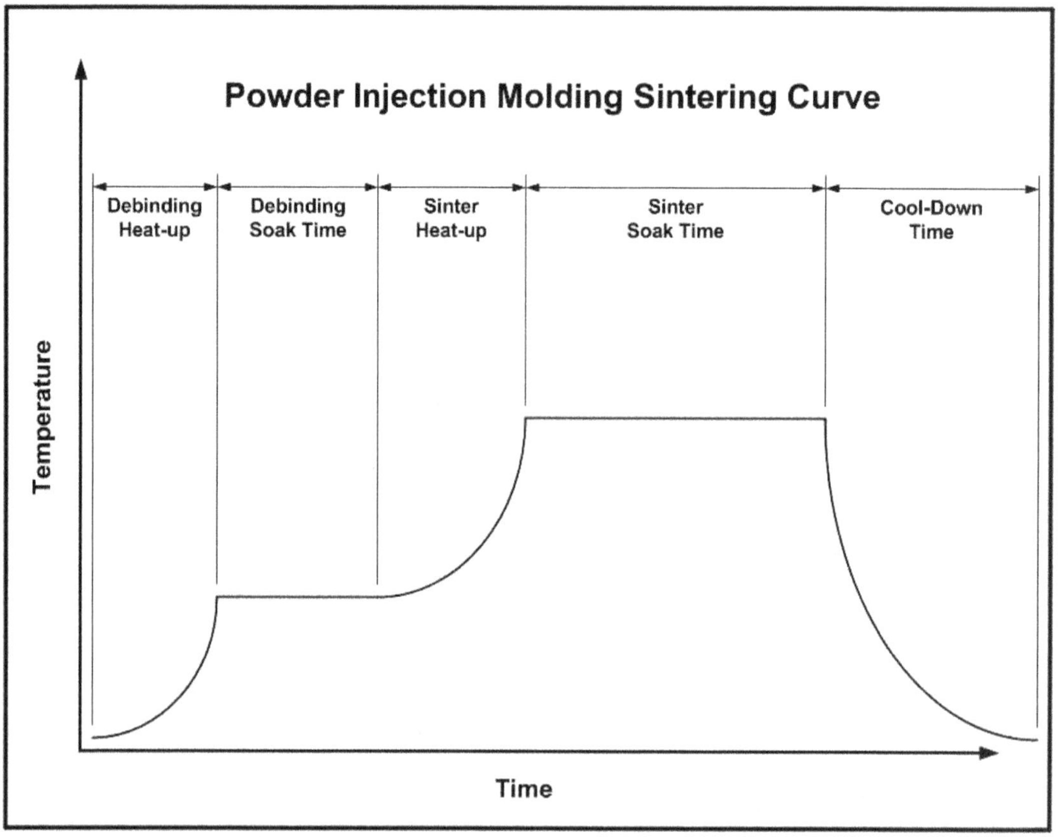

Figure 9.24 Metal Injection Molding Part Sintering Curve

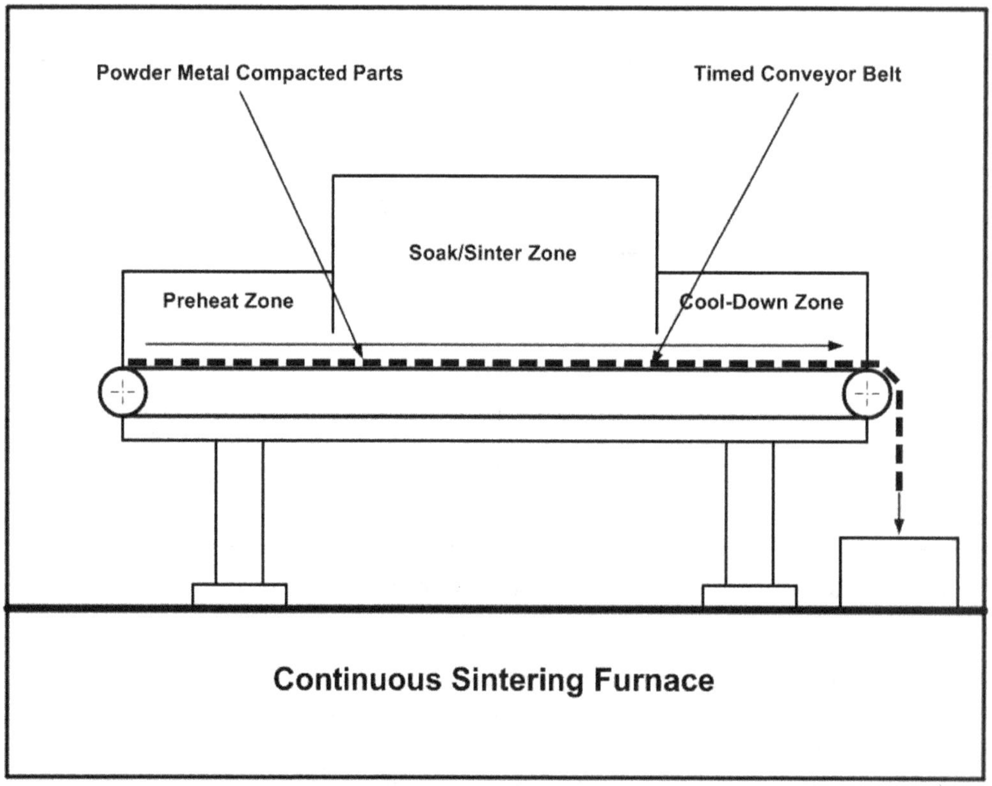

Figure 9.25 Continuous-Flow Powder Metal Sintering Furnace

9.7 SECONDARY OPERATIONS

Powder metal parts can undergo a variety of secondary operations. Some of these operations cannot be performed on solid metal components. Because of their unique internal structure, these operations will change and enhance the functionality of the parts to deliver unique characteristics not able to be found with solid metal parts. All of the secondary operations are performed after sintering and not before that. Some of the main secondary operations performed on powder metal parts are listed as follows.

9.7.1 Repressing

In repressing, the parts are reloaded into a die and pressed down. This is performed to increase the density by collapsing the porosity. High tonnage is used and the components are compressed to reduce the amount of porosity within the part. It is possible to compress the parts to very close to 100% density, but this will NOT make it a solid part because the particles are still only bonded from sintering.

9.7.2 Coining

Coining is identical to repressing, except that the upper and lower punches are of certain configuration. As the parts are repressed, the design in the two punches is imbedded into the top and bottom. This is also how the currency designs are applied to the coins we see every day.

9.7.3 Sizing

Sizing is used to bring either the outside or inside configuration to a closer tolerance. Unlike repressing, the parts are loaded into a die and then pushed through to squeeze the part to size. For the external shape, the part gets pushed through a die that has the internal geometry to bring the external configuration of the part closer. For internal sizing such as holes, a punch is pushed through the parts to bring the internal configuration to size. This is a common practice for all bearings that are produced with power metal.

9.7.4 Machining

Powder metal components can be machined to add configurations that are not possible in the pressing stage. Features such as side holes, groves, threaded holes, and undercuts, which are not possible in pressing, can be cut with the same effort as a solid part can. However, with the powder metal components not being solid in nature, there are certain precautions that need to be considered in order to accomplish machining and not cause damage to the part. One drawback with any conventionally formed powder metal part is that it cannot handle shear in any manner. When a part is being machined, say, a surface is being milled down to a dimension, as the cutter reaches the edge of the part, it is placed into a shear mode and will fracture off. Figure 9.26 shows how the edge of the component is placed into shear from the cutter and breaks off at an angle equal to the depth of the cut. To resolve this fractured edge, the part should have a 45° chamfer on the edge where the cutter will exit the machined surface. Figure 9.27 illustrates how the chamfer will eliminate this fracture edge on the machine surface. Also, cuts with light depth should be used vs. heavy deep cuts when machining powder metal parts. Powder injection molded parts can be machined exactly the same way as a solid metal part can.

9.7.5 Impregnation

Impregnation is a process that is unique to conventionally fabricated powder metal parts. This process is used to fill the open pores of the part with another substance to give it additional properties. To give the part lubrication properties, grease or wax can be impregnated into the internal structure of the part. The parts are loaded into a basket and lowered into a vat of the liquid grease or wax. The vat is sealed and a vacuum is pulled on the vat to force the liquid grease or wax into the part. Figure 9.28 shows these two steps in the process. What happens when vacuum is pulled on the vat such that the air is sucked from the pores in the part and the liquid in the vat is drawn inside of the part? Figure 9.29 illustrates how this step of

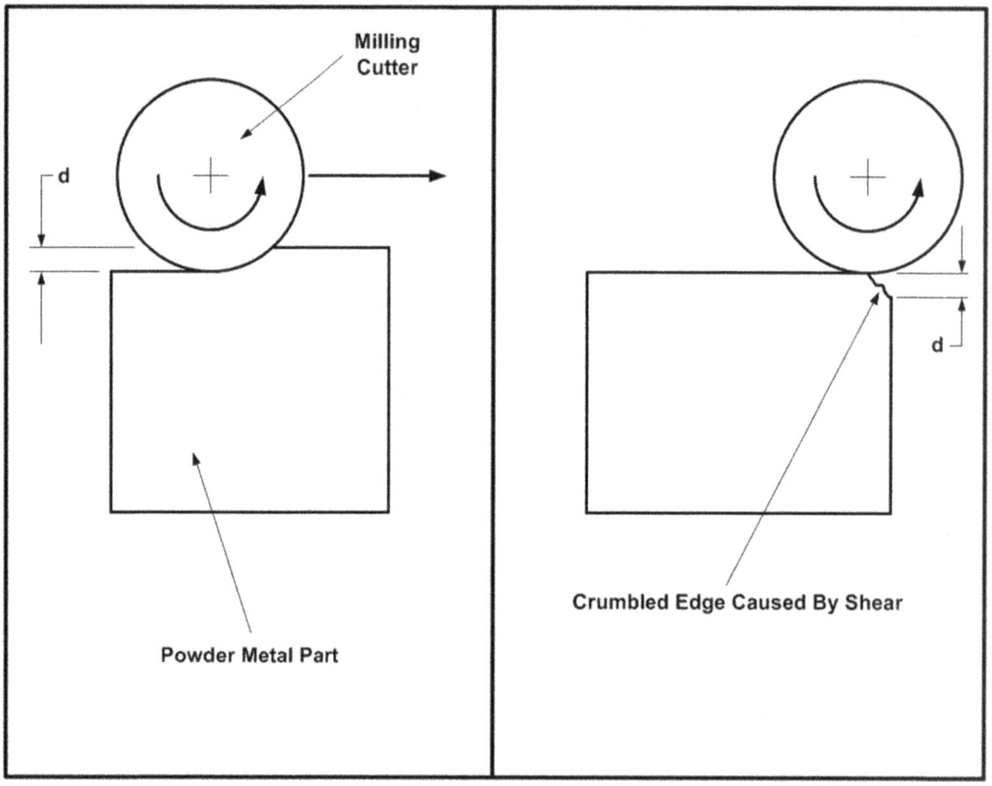

Figure 9.26 Crumbled Edge of Machined Powder Metal Part Caused by Shear

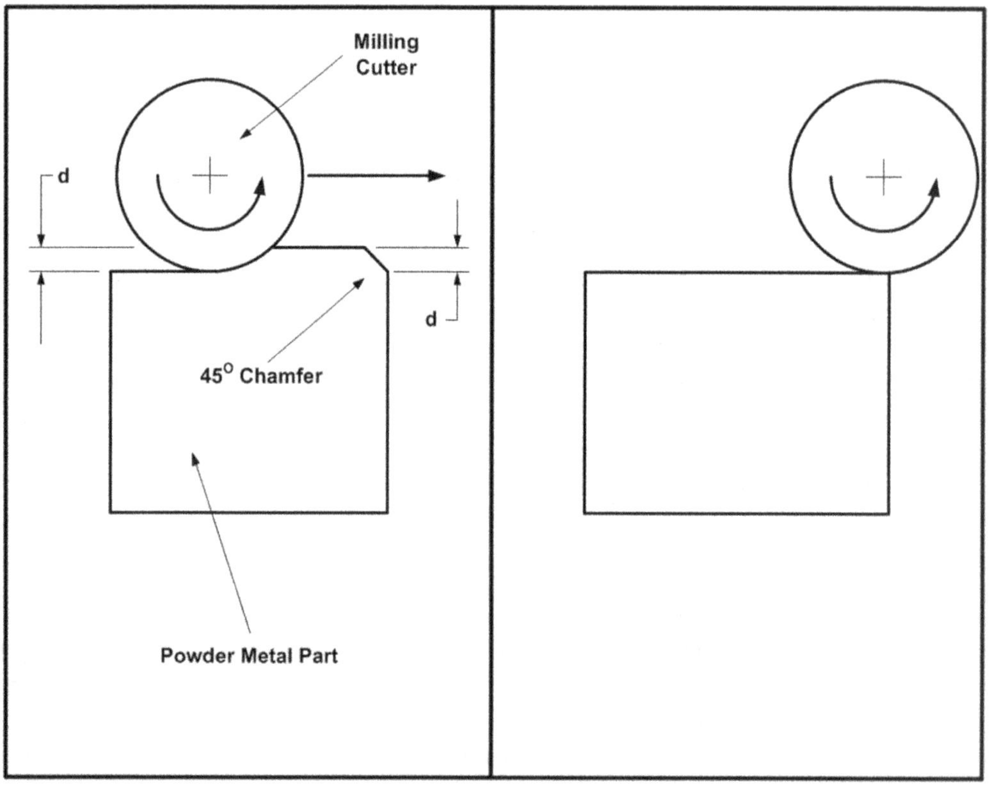

Figure 9.27 Crumbled Machining Edge Eliminated with a 45° Chamfer

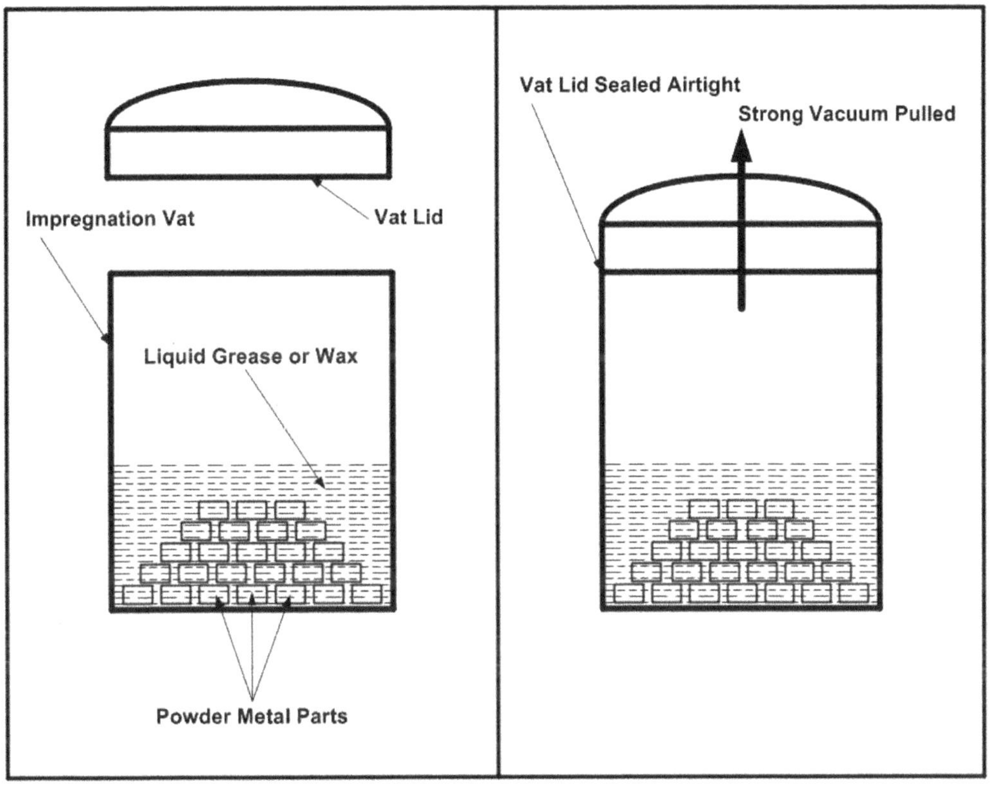

Figure 9.28 Powder Metal Impregnation Process

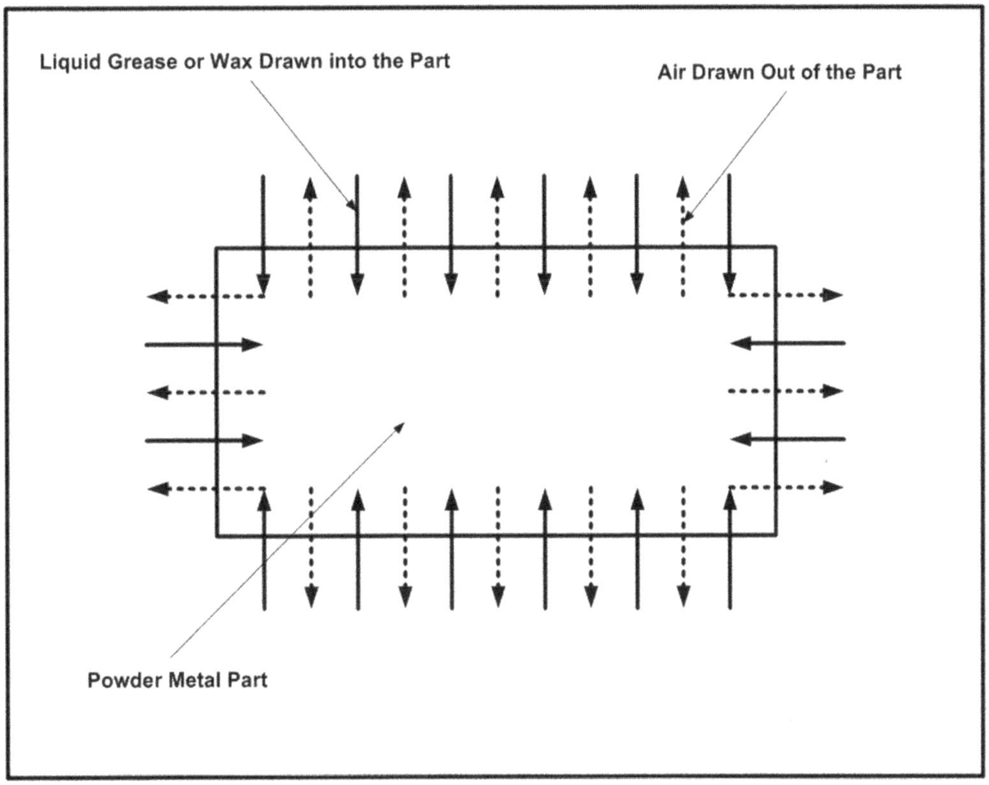

Figure 9.29 Impregnation Showing Air Drawn Out and Wax/Grease Drawn into the Part

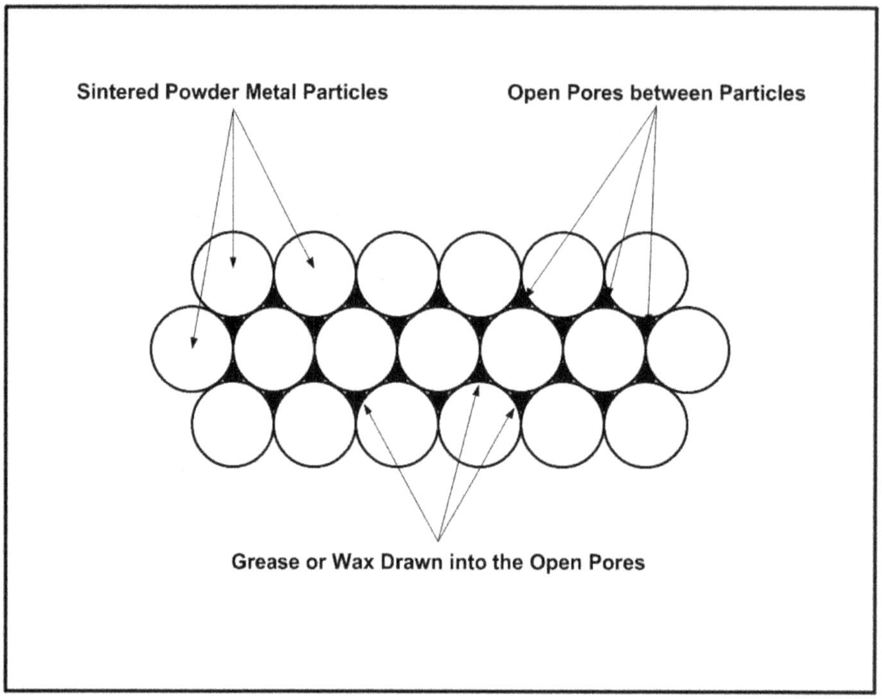

Figure 9.30 Powder Metal Part with Solidified Wax/Grease Impregnated into the Pores

the process works. After the parts have been impregnated with the liquid material they are removed from the vat and as the liquid cools, it resolidifies in the pores of the part as shown in Figure 9.30. This process is how all of the bronze bearings are produced providing them the lubrication properties. Impregnation is also used on iron parts that may not require lubrication, but the internal grease will prevent them from rusting over time.

For certain applications, plastic can be impregnated into the pores of the parts as well. This plastic material is impregnated into the component parts for two principle reasons. The first is if the open pores will have an effect on the part when functioning, such as acoustics or for vibration. The second reason to impregnate with plastic is if the part is going to be plated, it seals the pores so that the plating chemicals will not soak into the internal structure during the plating process.

9.7.6 Infiltration

The infiltration process is another process unique to powder metal. This also fills the open pores in the part, but in a different manner and with metal. The most common infiltration performed is with copper being infiltrated into iron parts. This fills the pores and gives the material different properties and structure for certain applications.

To perform infiltration, the same furnace is used as in sintering. A piece of carbon called a boat is placed on the slow moving belt. Next, a piece of copper is placed on the belt, followed by the powder metal parts on top of that, and then a second piece of copper on top of the parts. Sometimes, a weight is placed on the top piece of copper for thicker powder metal parts as shown in Figure 9.31. The assembly moves through the furnace at a temperature slightly above the melting temperature of the copper. A precaution is that the powder metal parts much have a melting temperature higher than that of the infiltration material. As this moves through the furnace, the copper material is drawn into the pores of the powder metal part through capillary action as shown in Figure 9.32, filling them completely.

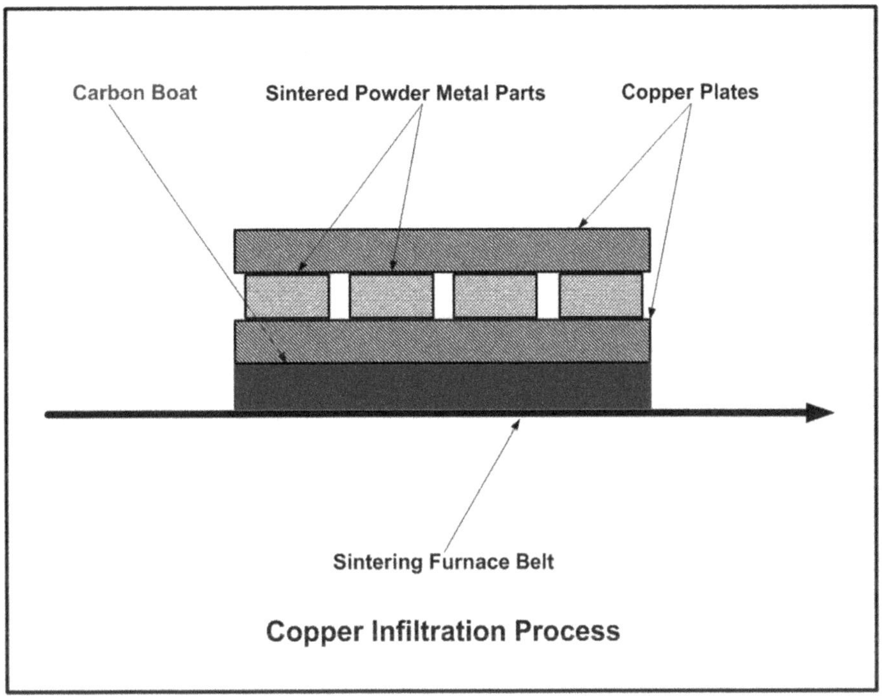

Figure 9.31 Example of Double Side Copper Infiltration of Powder Metal Parts

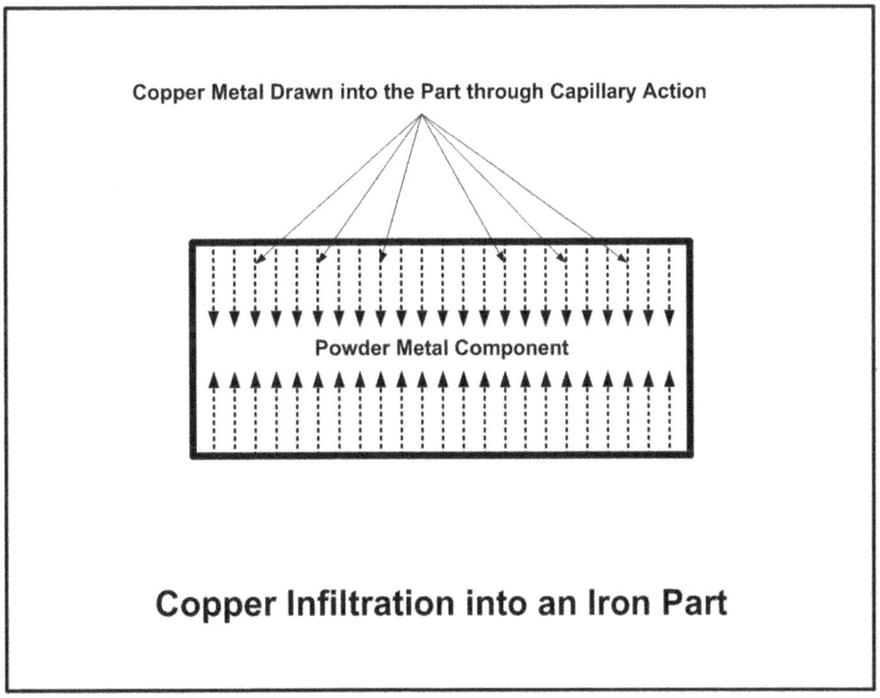

Figure 9.32 Copper Material Drawn into the Iron Part through Capillary Action

9.7.7 Hardening

For powder metal parts constructed from iron material, they can be heat-treated to increase the hardness. The parts are placed into a furnace that contains a high-carbon environment. What this does is let the carbon from the environment enter the surface of the part to a depth to turn just the outside into high-carbon steel. This creates what is called surface hardening and is covered in detail in Chapter 20.

9.8 DESIGN GUIDELINES

As with any manufacturing process, certain design guidelines are required in order to construct a component part that will be functional in operation. With the internal structure and the shaping process unique to the powder metal processes, the design for both fabrication and function needs to be understood. The guidelines for a successful powder metal component design are given in detail as follows.

Controlled Porosity: Possibly the biggest advantage to powder metal parts is being able to control the density. Depending on the overall size of the particles and the packing factor, 50% density is possible to attain with the conventional forming processes. Powder injection molding will always render component parts of extremely high density, similar to a solid piece.

Physical Properties: Powder metal parts, because of their nonsolid internal structure, will posses certain physical properties that need to be understood from a functional standpoint. Powder metal parts will be much stronger under compression than under tension. That is because they were created under compression. When they fail under tension, there will be no elongation of the part as with solid metal, but will reach the failure point and break. The worst situation for powder metal parts is to place them in shear. Because of their nonsolid structure, they cannot withstand a shear load at all and will fail very quickly when exposed to a situation like that. Powder injection molded components will perform in the same way as a solid metal part

Difficult Metals: Another major advantage with powder metal is being able to construct components from materials that are difficult to machine. Because the only thing that is used to create the parts is either pressure from the conventional processes or molding for injection parts, it does not matter if the components are made from soft aluminum or hard materials such as titanium.

Alloy Mixing: Another huge advantage with powder metal is being able to construct custom alloy material very easily. Because the material starts out in powder form, various powder materials can be mixed together to form a unique alloy. The only requirement for this alloy mixing is that all the materials being mixed together must have the same approximate sintering temperatures.

Shrinkage: Shrinkage will occur in all the powder metal processes. As the parts are being sintered, the binder material will burn out and the parts will shrink by the percentage of binder used. This needs to be considered when developing the tooling that will be used to create the components.

Die Ejection: With the conventional press and sinter forming process, the components need to be designed so that they can be ejected from the die. The parts may have certain steps in them, but there needs to be a straight line push from the bottom punch to eject the parts. This will be covered in more detail with the part geometry.

Screw Threads: Any screw threads, either internal or external, cannot be incorporated in the forming processes. If these features are required, they need to be machined into the part after forming and sintering.

Part Geometry: In conventional press and sinter part fabrication, there are certain geometric configurations that are required to be designed into the parts in order for proper pressing to be attained. Every corner needs to have either a chamfer or radii in order for the powder to be properly compressed and to allow for ejection out of the die without causing any damage. If a hole is going to be cored in the part during pressing, the minimum wall thickness of the part would be 0.0625 in.. Figure 9.33 illustrates these design features that are required to produce acceptable components during forming. As mentioned, the components being formed in the press and sinter process need to be able to eject from the die after pressing. Figure 9.34 shows part configurations that can and cannot be produced in this process. Parts A, B, and C can be produced and allow for easy die ejection and would easily produced. Parts D, E, and F have features not possible to produce during the pressing step in the process. These features would need to be machined during a secondary operation.

Powder Injection Molding: Parts produced with powder injection molding will follow the same design guidelines as those being plastic injection molded. These guidelines are covered in Chapter 6.

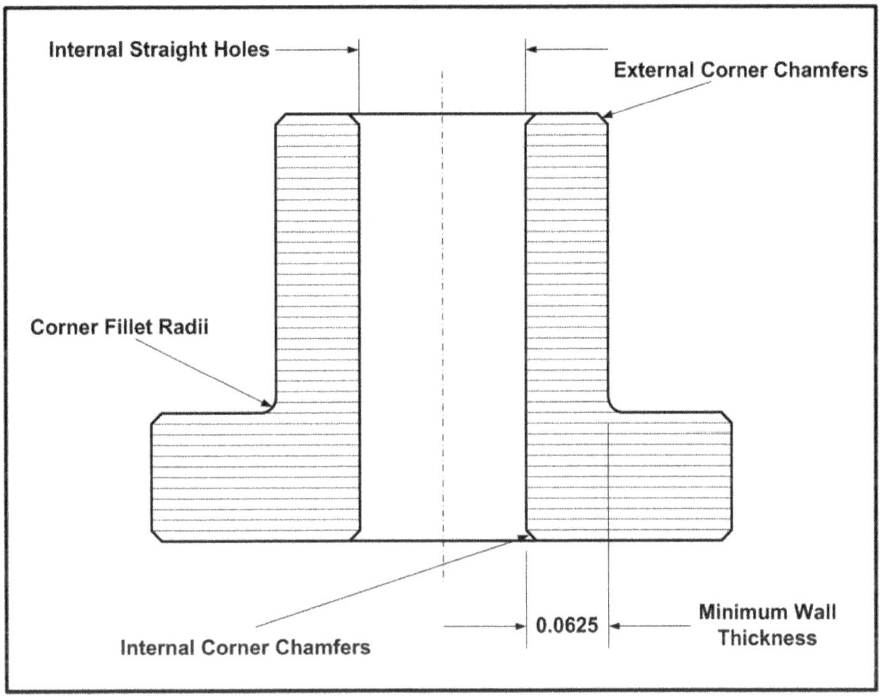

Figure 9.33 General Design Guidelines for Conventional Press and Sinter Powder Metal Process

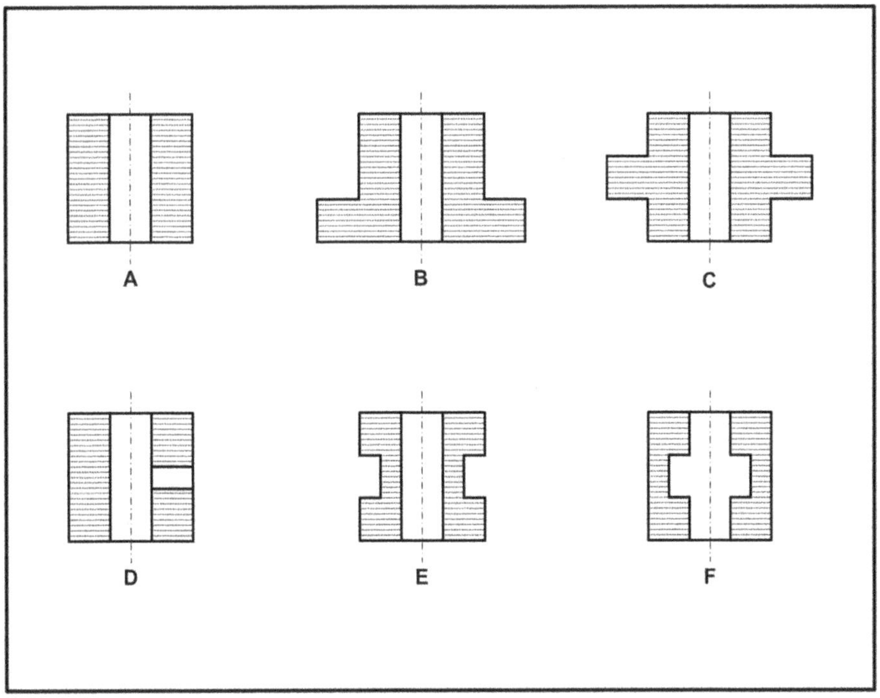

Figure 9.34 Part Configurations that Are Possible and Not Possible in Conventional Pressing

9.9 SUMMARY

Powder metallurgy is a very unique method for producing component parts in manufacturing. Very difficult geometric configurations can be formed in seconds as the pressing is fully automated along with working with materials that are extremely difficult to machine. Being able to control the internal porosity of the parts is advantageous for producing metal parts that have a lower weight than a solid metal part would, along with being able to impregnate and infiltrate these internal pores with a second material, providing the components features not possible to attain in a solid part. Understanding what the processes are capable and not capable of producing is important so that powder metal can be selected for the proper applications within a product and function perfectly. This interesting process has the ability to construct custom alloys of materials that may be able to be produced in a solid matter, but would be extremely expensive to procure. The ability to produce complex components in a very timely manner with unique physical properties is what establishes powder metallurgy as an extremely viable manufacturing process.

9.10 QUESTIONS AND PROBLEMS

Questions

1. What are the three basic steps in the powder metal part fabrication process?

2. In the production of powder metal particles, why is water atomization not utilized for the production of iron powders?

3. What is meant by powder classification, and how is it performed?

4. Powder metal particles can be produced in different shapes and geometric configurations. What determines how the shapes of the particles will be produced?

5. What is the difference between open and closed pores?

6. In preparing the powder for pressing, what is the difference between blending and mixing?

7. Binder materials are utilized in every powder metal forming process. What is the function of the binder, and what issues can arise if the percentage of binder varies from lot to lot?

8. Powder metal parts can be produced with a controlled porosity. What two factors determine how this can be controlled?

9. The sintering step has three stages. What are the three stages, and why is each of them critical to the process?

10. In performing secondary operations, what is the difference between impregnation and infiltration?

Problems

1. A screen is going to be used to separate the particles for classification. If the mesh count of the screen is 250 and the thickness of the wire in the screen is 0.0015 in., what is the largest size particle that will fall through the screen?

2. If a screen being used allows a particle size of 0.0012 in. to fall through and has a wire size of 0.0004 in., what would the mesh count of the screen be?

3. A particle size of 0.0008 in. is desired. The screen being used has a mesh count of 325. What wire size is required in the screen to deliver the desired particle size?

4. Component parts are being pressed to a packing density of 2.543 lbs/in.^{3}. If the true density of the material is 3.105 lbs/in.^{3}, what would the porosity and packing factor be?

5. A powder metal part has a bulk specific volume of 1.328. What would the porosity of the part be, and what packing factor was used?

6. If a powder metal part has a weight of 2.517 lbs. and a solid part of the same configuration would weigh 2.894 lbs, what would the porosity of the part be and what packing factor was used to produce it?

7. A given powder metal part has a porosity of 0.3 and the true density of the material is 3.709 lbs/in.^{3}, what would the packing density of the part be and the packing factor used to form it?

8. A rectangular powder metal part is pressed to the following dimensions:
   Length = 1.7616 in., Width = 1.1384 in., Height = 0.6306 in.. After sintering, the dimensions of the same part are: Length = 1.7500 in., Width = 1.125 in., Height = 0.625 in.. What would the percentage of binder need to be for this to happen?

9. If a rectangular powder metal part is pressed to a Length = 0.7763 in., Width = 0.5693 in., and Height = 0.2588 in. and the binder percentage is 3.5%, what would the dimensions of the part be after sintering?

10. A disk is being produced in powder metal and has a diameter of 1.750 in. with a height of 0.250 in.. If the packing factor is 0.71, what would the porosity be? The same disk was produced using the same packing factor, but this time the height came out to be 0.235 in. What would the porosity of this part be?

CHAPTER 10

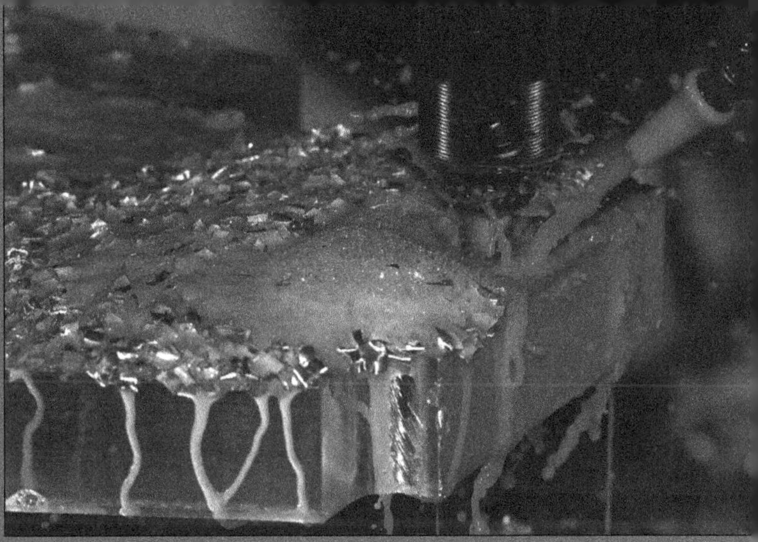

Ceramic and Cermet Principles

10.1 INTRODUCTION

Ceramics are widely used both in products and for producing products in manufacturing. Several products we use daily are fabricated from ceramic materials such as dishes, bowls, and vases. In other manufacturing processes, ceramics are utilized in the fabrication such as in investment casting where ceramic is used in the construction of the actual mold. The craft industry is a major user of ceramics for the hobbies and producing unique items.

There are two categories of ceramics, which are Traditional Ceramics and New Ceramics. Traditional ceramics are what are used for producing products and in some fabrication processes where new ceramics that are unique and possess extremely high strength are required, making them excellent for long-life cutting tools for hard materials. Both types of ceramics have their place in both products and manufacturing processes.

This section covers the traditional and new ceramics along with cermets, which are a mixture of ceramic and metal to produce a composite material. The production of these materials and their shaping processes along with the uses and design considerations will be discussed in detail.

10.2 TRADITIONAL CERAMICS

Traditional ceramic is a natural material that is extracted from the ground. This material is used for producing ceramic products such as plates and bowls. In the craft and hobby arena, traditional ceramics are what are used for the various processes that are utilized to handcraft their unique items. This material can be used in a variety of ways for shaping and creating the geometric configurations depending on the process being used.

Traditional ceramics are also utilized in fabrication process as a support material. In the casting and molding processes, some expendable molds are created in ceramic to attain configurations that would not be possible with other mold constructions. Because this type of ceramic material can be used in a consistency from very liquid to a paste like clay, it makes it very versatile for a variety of uses.

Traditional ceramics go through four steps to produce components. These are:

1. Material Preparation
2. Shaping
3. Drying
4. Firing

The preparation of the material and the process steps for shaping and processing traditional ceramic components are covered as follows.

10.2.1 Raw Material Preparation Process

Traditional ceramic is a natural material that is mined from the ground. It is extracted in large clumps and needs to go through various steps before it can be utilized for processing components. After it is extracted, the first step is *Comminution.* Here, the large clumps of ceramic material are reduced in size and broken down into smaller pieces. From there, the smaller pieces are *Crushed* to further reduce the size. During crushing, the ceramic material is reduced to the size of a powder, which prepares it for the shaping process operations. The material can either remain in the dry state from crushing or be mixed with water to the desired consistency. Figure 10.1 illustrates the processing stages from the raw material to the finished ceramic material ready for shaping.

Water is the ingredient that is used to change the consistency of the dry ceramic. The shaping process being used will determine the amount of water to be mixed with the dry powder. About 50% water mixed to the powder will allow it to flow in a liquid state, and less water will give it a firm consistency like clay. Note that the mixing of water to the particles does not dissolve the ceramic material. It is used either as a binder to hold the particles together during shaping or as a transport vehicle when in a liquid state into various molding processes.

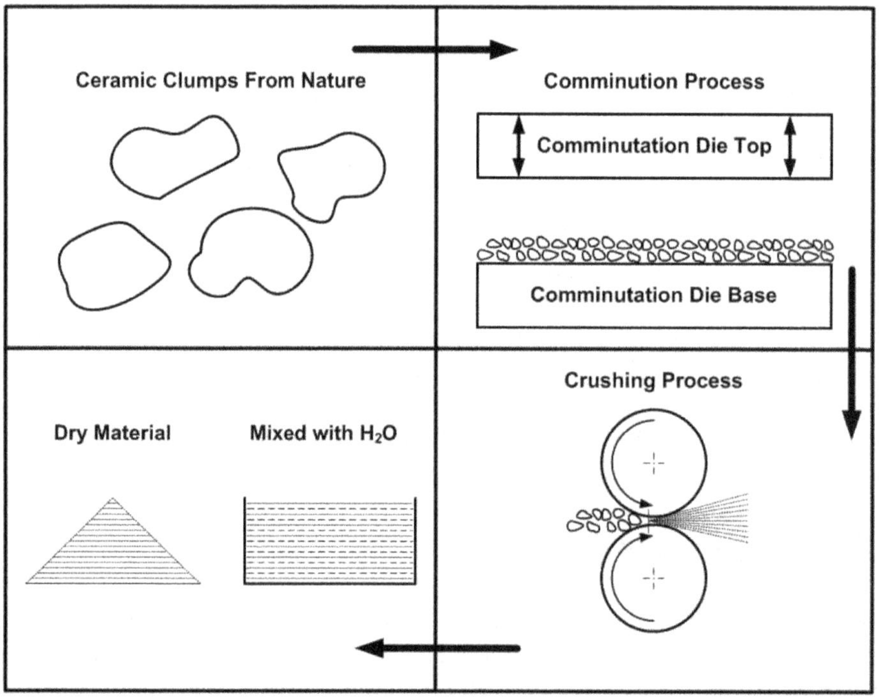

Figure 10.1 Traditional Ceramic Preparation Stages

10.2.2 Shaping

A variety of methods can be utilized to shape traditional ceramics. Some of these processes are similar to those used in other material fabrication, but with some distinct differences. As with any material being shaped, there are similarities and differences that need to be accommodated in order to obtain an acceptable component. Following are some of the more common methods for shaping traditional ceramics and how they differ from other materials that use similar processes.

10.2.2.1 Slip Casting

The slip casting process is used to create shapes that are hollow inside, such as vases. It involves a two-section mold, which is closed and poured with a liquid mixture of ceramic and water. The mixture is allowed to solidify for a given amount of time and will solidify from the outside of the pour as shown in Figure 10.2. After the outside has solidified to the desired thickness, the mold unit is turned upside down and the nonsolidified material mixture is poured out. From there, the mold is returned upright and separated. The slip cast component is removed as shown in Figure 10.3.

This process can be time-consuming depending on the wall thickness desired for the component being fabricated. The production of components that require slip casting requires several molds, at all the various stages of solidification in order to maintain the production requirements.

10.2.2.2 Molding

Molding is performed in the same manner as slip casting. The mold is closed and filled with a ceramic/water mixture. But here, the mixture is allowed to completely solidify before the mold is opened as shown in Figure 10.4. Depending on the overall thickness, the solidification times for molding can be 24–38 h or more. As with slip casting, several molds are required in order to keep up with the production requirements.

10.2.2.3 Plastic Forming

Plastic forming is more of a craft process and produces a single part by hand. This involves working with a water–ceramic mixture to attain the consistency of clay. From there, the material can be modeled by hand and sculptured into the desired shape required. Another method is called *Throwing*. This is performed on a

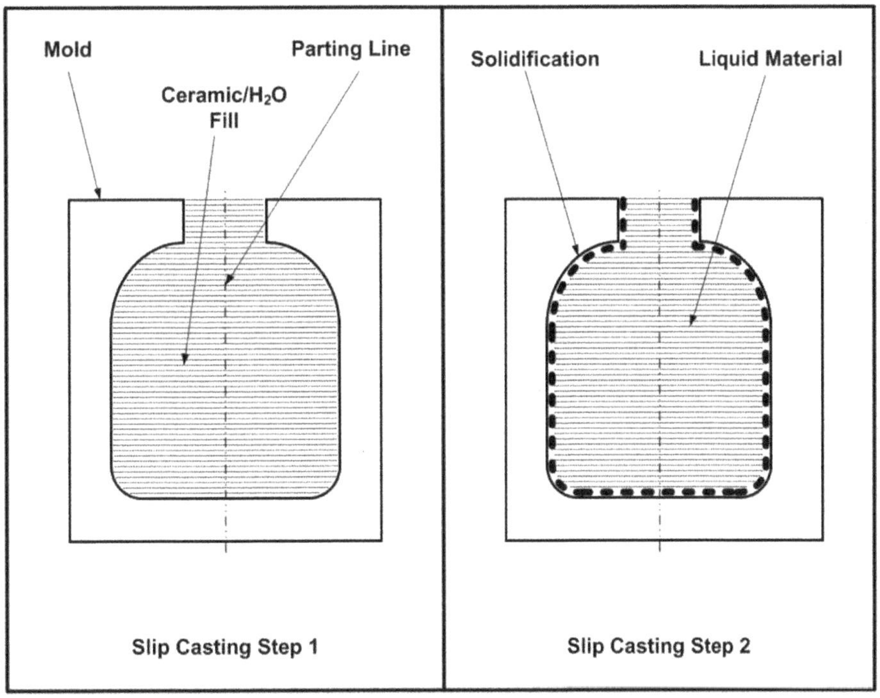

Figure 10.2 First Two Stages of the Slip Casting Process

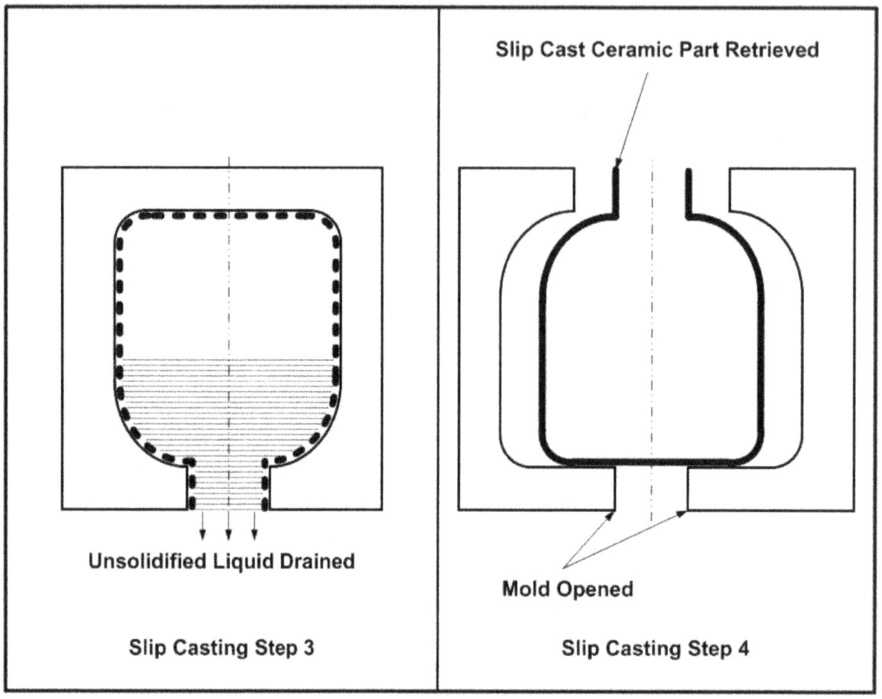

Figure 10.3 Last Two Stages of the Slip Casting Process

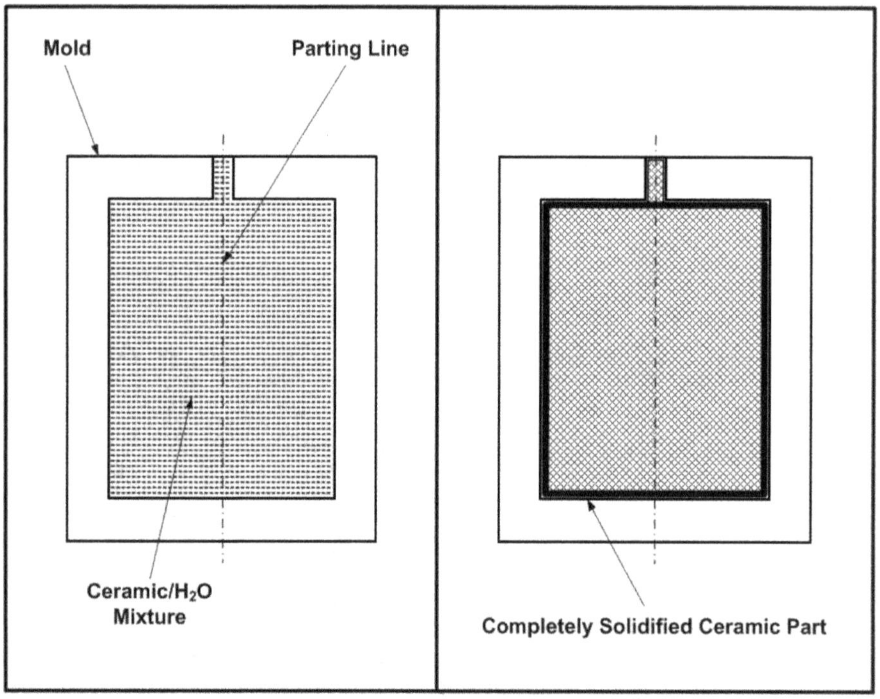

Figure 10.4 Ceramic Molding Process

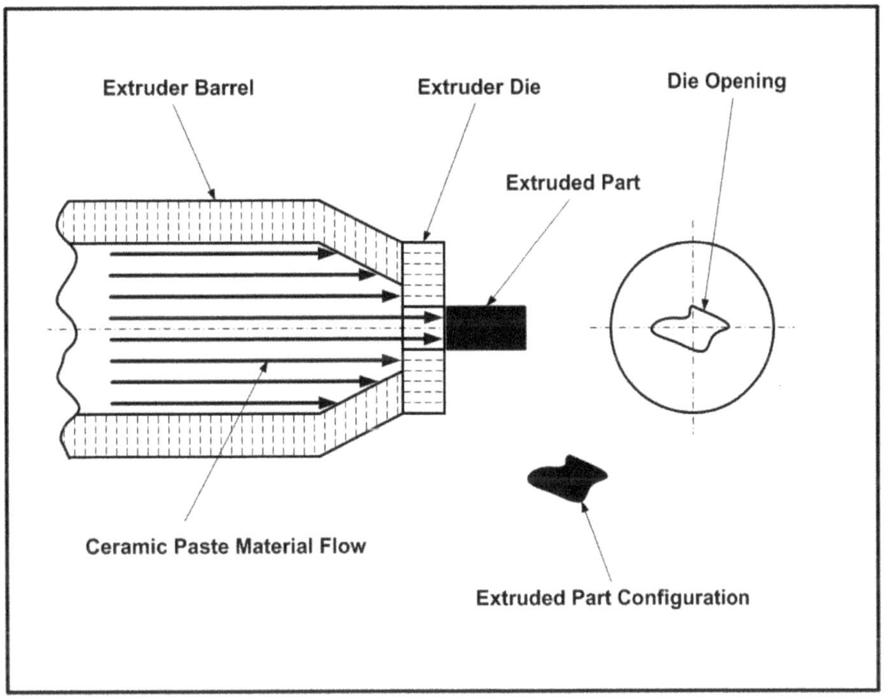

Figure 10.5 Ceramic Extrusion Process

potter's wheel in the same manner that clay pots are made. The ceramic mixture is rotated on the wheel and then shaped by hand.

This hand shaping of ceramic is also used to create prototype parts that will eventually be produced from other processes. From the prototype part, the molds can be designed and constructed for mass production. Prototyping by hand requires a very artistic individual to generate the original shape that will become a production component.

10.2.2.4 Extruding

Extruding is performed in the same manner as with plastics or other materials. A paste containing mixture of ceramic and water is loaded into the extrusion barrel and pushed through an opening. The shape of the die on the end of the barrel determines the configuration of the extruded material exiting. Figure 10.5 shows the material being forced through the die opening creating the two-dimensional shape of the ceramic material.

Extrusion of ceramic material is performed extremely slowly, and the extruded section needs to be well supported. This is required because unlike other materials that are extruded, ceramic material can bend easily due to the fact that it is still wet and can be deformed very easily.

10.2.2.5 Pressing

Pressing is performed using an upper and a lower die section. The ceramic–water mixture, which is in a paste consistency, is loaded into the lower die section as shown in Figure 10.6. The upper die is lowered, pressing the mixture into the desired shape, which Figure 10.7 depicts. After the material is pressed into the desired shape, the upper die retracts and the completed component is removed, which Figure 10.8 illustrates. This process is used for producing ceramic plates and bowls and does not take the time that molding or slip casting requires.

An important factor in the pressing process is the starting volume of the material being loaded into the lower die section. If too little material is loaded, the component will not fully form. Likewise, if too much material is loaded, it will bleed out of the sides when the upper die is lowered. The starting volume is extremely critical in a process of this nature and needs to be accurately measured each time.

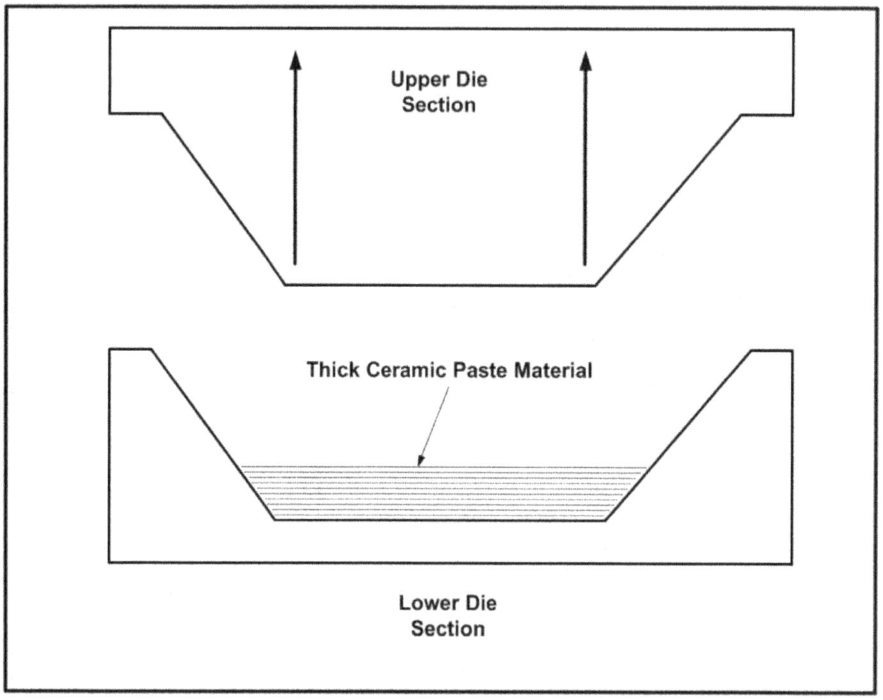

Figure 10.6 Ceramic Paste Loaded into the Press

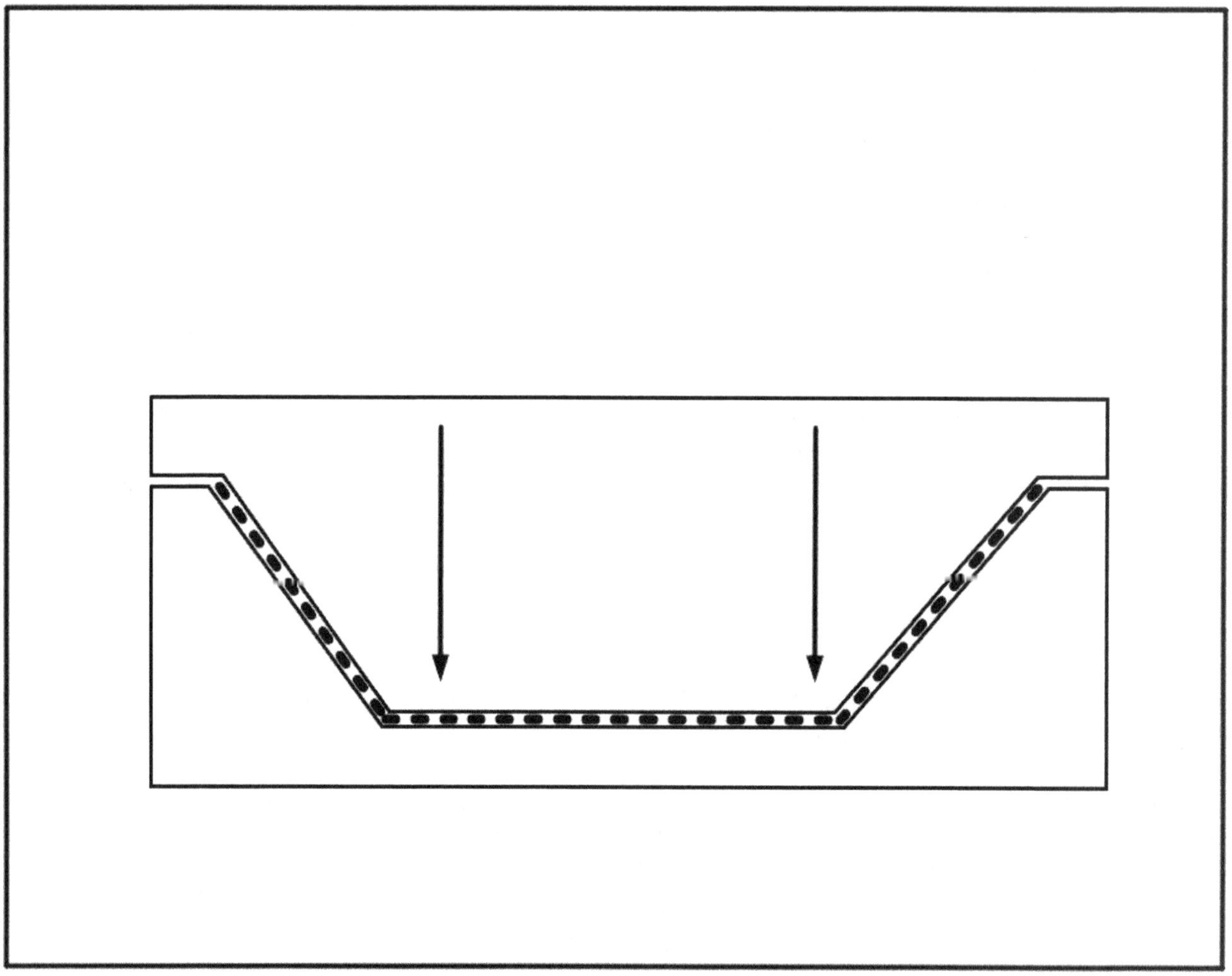

Figure 10.7 Ceramic Paste Compressed into the Desired Shape

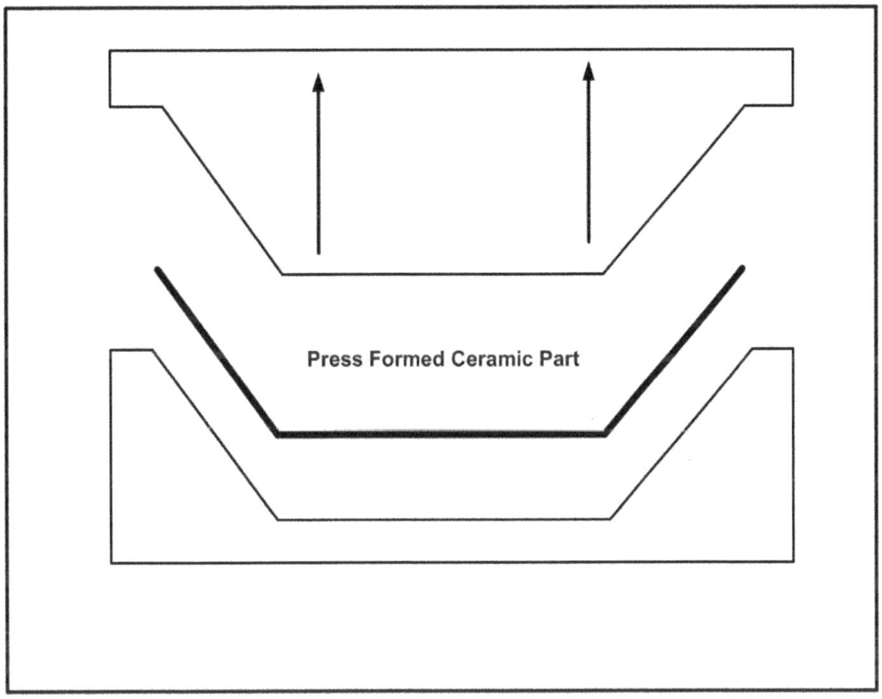

Figure 10.8 Ceramic Press Formed Part Retrieved

10.2.3 Drying

Due to the fact that the traditional ceramic is mixed with water as a binder and a transport vehicle, it needs to be dried to remove the water in the mixture. Drying will render what is known as a *Green Part* and is the next step in the overall processing. This process is performed in two stages in order to control the rate at which the water is being extracted from the ceramic component. A component that has just been shaped has water in the pores between the ceramic particles. Each stage is required in order to successfully evacuate the water from the part without causing any defects. This is performed in a drying chamber with controlled heat and humidity for each stage of the drying process.

Stage 1: During the first stage of drying, is accomplished rapidly and the water is removed from the surfaces of the component. The capillary action moves the water inside the component to the surface and the water is removed. This is performed in the drying chamber in a dry atmosphere at a temperature between 100 and 120°F. During this stage, shrinkage will occur as the amount of water in the mixture is being evacuated from the part. In stage 1 of drying, 90–95% of the water is removed, which explains the dramatic shrinkage of the component.

Stage 2: During the second stage of drying, the remainder of the water is removed and the ceramic particles are in contact with each other. In the drying chamber, the temperature is raised to between 140 and 160°F to ensure that all of the moisture is removed. The shrinkage during stage 2 of the drying process is far less than in stage 1, and when completed, it produces what is known as a *Brown Part.*

Drying can be a long process depending on the overall thickness of the component. This can range from several hours for thin components to several days for thick components. The parts must have 100% of the moisture removed before moving on to the next step in the processing. If there is any trace of water remaining in the part, it will cause a major issue in the following step, which is firing.

10.2.4 Firing

The final step in the process is firing. Firing will bond or sinter the ceramic particles together similar to that of powder metal. The parts are loaded into a kiln, which will heat them to a high temperature. Figure 10.9 illustrates the component part in a kiln with the heat applied evenly. During this phase, the porosity in the part that contained water is altered where there are no gaps between them, thus rendering a solid structure. Prior

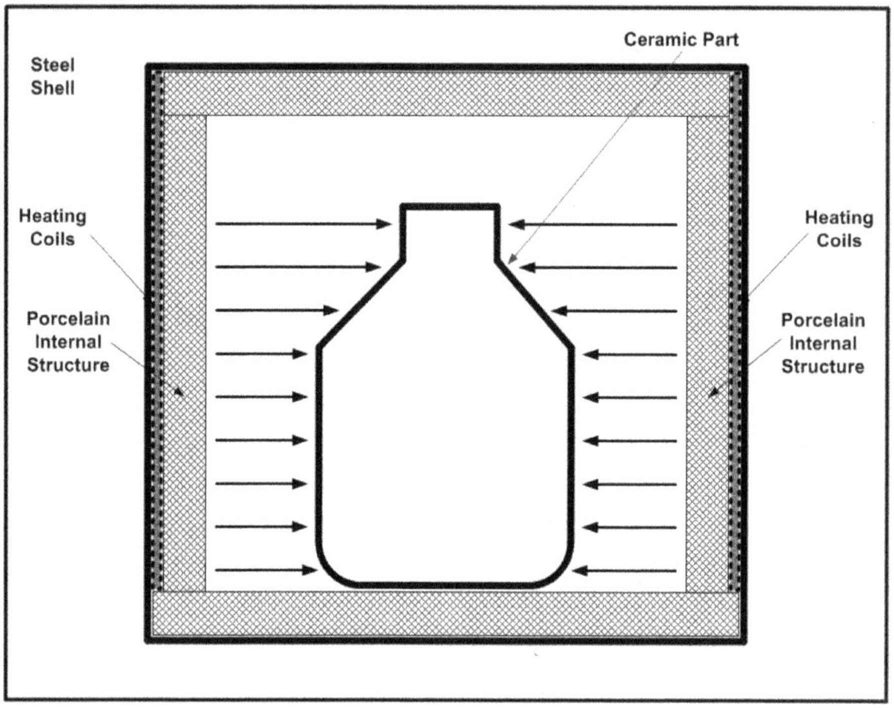

Figure 10.9 Firing Kiln for Traditional Ceramic Parts

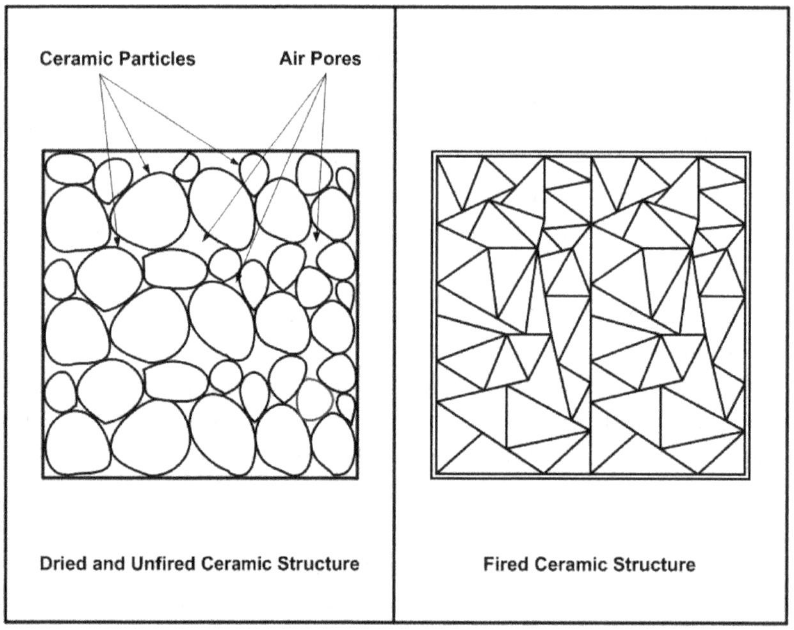

Figure 10.10 Traditional Ceramic Grain Structure Change from Firing

to firing, if water were to be added to the part, the ceramic would return to the paste or liquid consistency. Post firing, the vessel is capable of holding water and will not deform. Figure 10.10 shows the altered microstructure of the ceramic material from pre to post firing.

If all of the water is not completely dried out of the component before it is placed in the kiln, it will start to boil in the internal pores of the part and cause it to crack. If there is too much water remaining in the part, it can literally explode in the kiln. Ensuring that the drying has been 100% accomplished prior to firing will allow for this step to be performed properly.

An optional process for ceramic parts is called *Glazing*. Glazing is a coating that is applied on the surface of a fired part to give it the desired color and shine. After the parts have been fired, the glazing is applied and the part is refired to harden the glaze. This is not required to make the component functional, but is always performed to enhance the appearance of ceramic products.

10.3 NEW CERAMICS

New ceramics are much different than the traditional ceramics. They are not derived from natural sources, but rather from a chemical mixture. This type of ceramic renders a very high strength and a hard material, which has several uses in manufacturing and in products. In some products, it is used in areas of high heat due to the resistance it posses to such environments. The process for making new ceramic components requires four steps, which are:

1. Material Preparation
2. Shaping
3. Sintering
4. Secondary Operations

These steps for fabricating new ceramics are covered as follows.

10.3.1 Material Preparation

The preparation of the starting material is similar to that of producing powder metal particles in many ways and unique in others. The creation of the ceramic starts with the mixing of the chemical salts, which are dissolved in water. The mixture is then sprayed out of a nozzle into an extremely cold environment and frozen rapidly. This is similar to the atomization process utilized for creating powder metal, but rather than just cooling the spray, it is frozen to form the ceramic particles. The newly created ceramic particles are then loaded into a vacuum furnace to remove the water that was used to dissolve the creating salts, resulting in dry ceramic powder.

As with powder metal, the powder particles must be classified by size. This process is performed exactly the same way as with powder metal particles. A screen is produced, which will sift the particles, allowing larger particles to be trapped on top and smaller particles to fall through as shown in Figure 10.11. To classify the particles to exact sizes, a series of screens are utilized and the ceramic powder is screened multiple times, with each screen capturing the sizes. Figure 10.12 demonstrates this screening process utilizing multiple screens.

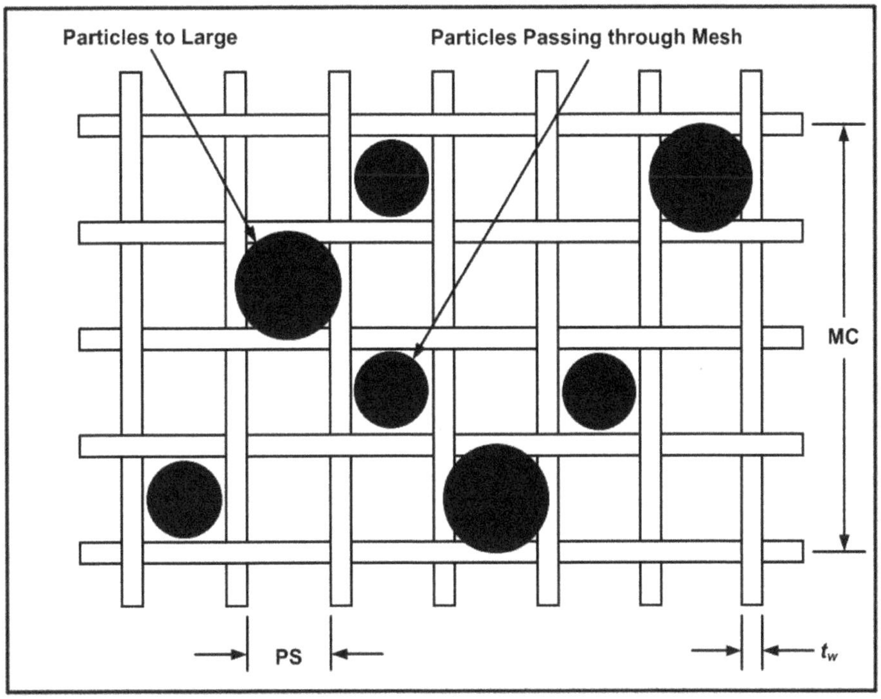

Figure 10.11 Particle Screen for Size Classification

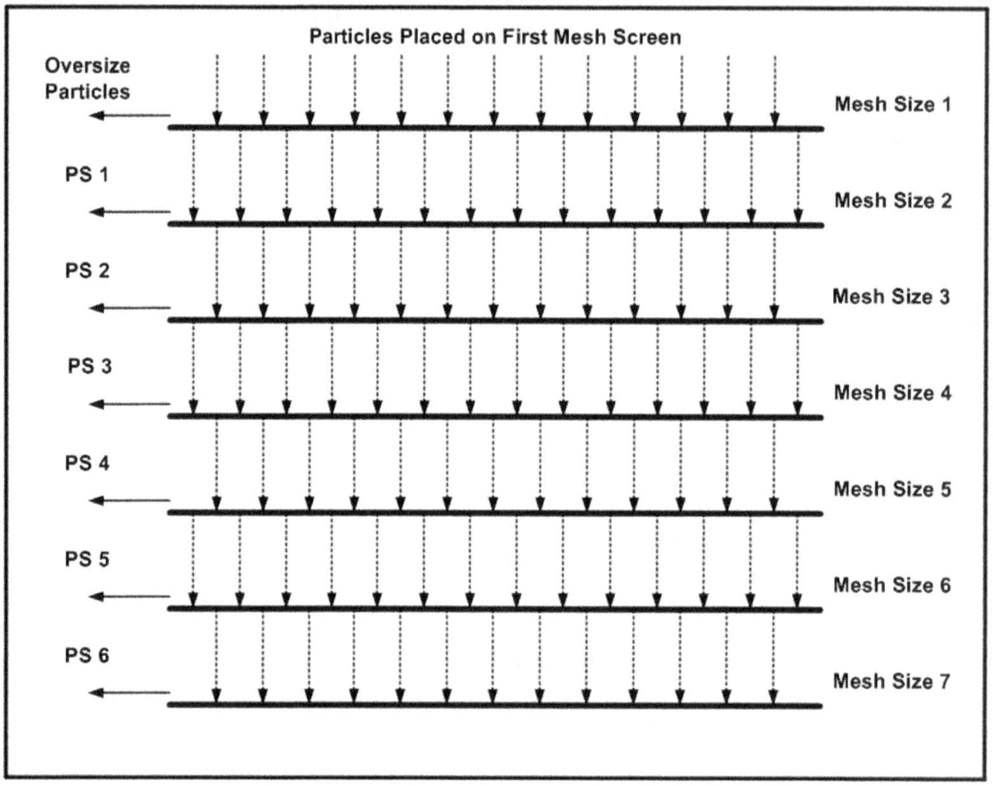

Figure 10.12 Multiscreen Particle Size Classification System

To determine the openings in the screen that separates the particles by size, which is known as the mesh count, the following equation is used:

$$PS = (1/MC) - t_w$$

PS = Particle Size
MC = Mesh Count (openings/in.)
t_w = Thickness of the Wire

(Refer to Chapter 9 on Powder Metal for examples of how this is applied.)

The powder ceramic particles are then mixed with various additives the same way as with powder metal for shaping. These additives are:

1. *Binders*: To hold the particles together during shaping
2. *Lubricants*: To reduce the friction in pressing
3. *Deflocculants*: To prevent clumping of the powder particles.

(Refer to Chapter 9 on Powder Metal for details on these additives.)

10.3.2 Shaping

Shaping of powder new ceramics is almost identical to that of powder metals, which are covered in Chapter 9. The main processes for shaping are *Hot Pressing, Isostatic Pressing*, and *Powder Injection Molding*. These are covered as follows.

10.3.2.1 Hot Pressing

Hot pressing is the same as cold pressing except that the die and punches are heated during compaction. The die cavity is loaded with the ceramic powder mix as shown in Figure 10.13. Next, the upper and lower heated punches compact the powder to the desired packing factor, then the upper punch is raised, and the lower punch ejects the compacted component. Figure 10.14 illustrates this process.

(Refer to Chapter 9 on Powder Metal for more details on the pressing process).

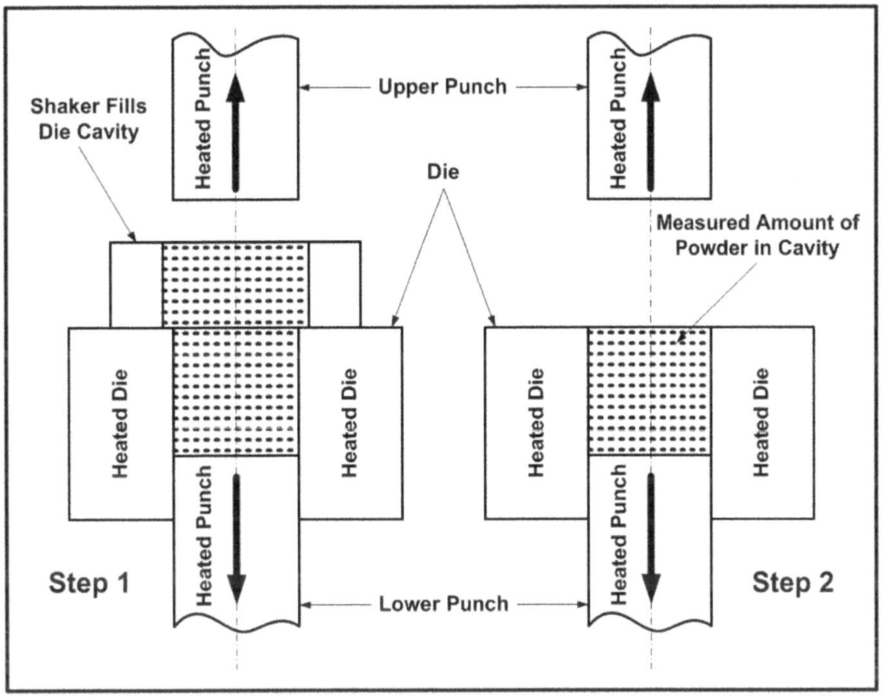

Figure 10.13 Hot Pressing Die Loading with Ceramic Powder

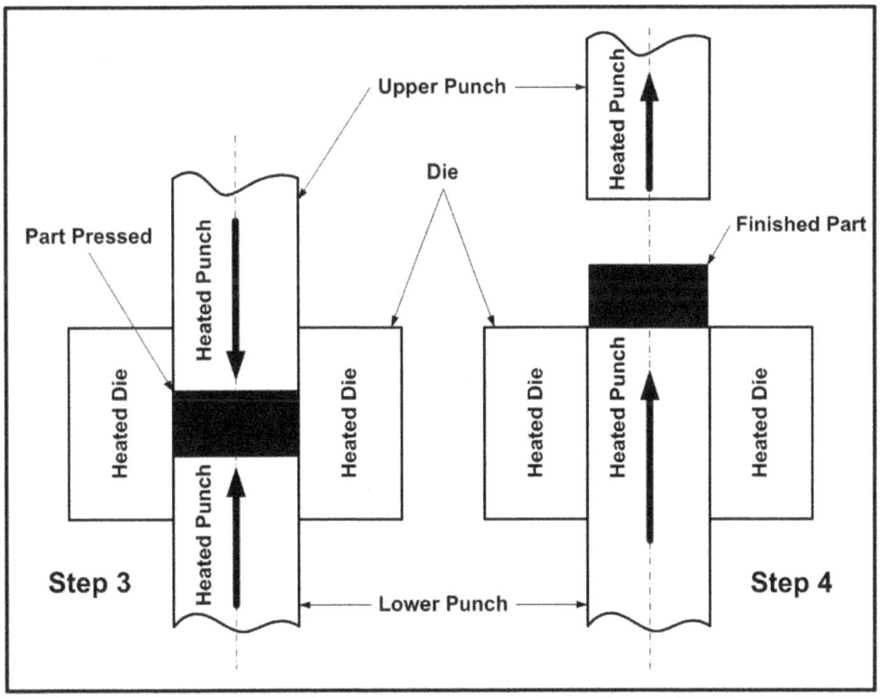

Figure 10.14 Hot Pressing Compaction and Ejection of Ceramic Component

10.3.2.2 Isostatic Pressing

Isostatic pressing is performed similarly to that for powder metal, but with ceramics, it is always performed hot. A rubber collapsible mold is used, but the oil in the hydraulic chamber is heated. The heat from the hydraulic oil performs the same function as the heated die and punches as in hot pressing.

(Refer to Chapter 9 on powder metal for more details on isostatic pressing.)

10.3.2.3 Powder Injection Molding

New ceramics can be produced utilizing the powder injection molding process the same way as with powder metal. An injection molding machine and a mold are utilized for the shaping of the component parts. The ceramic and binder mixture is injected into a mold, which performs the shaping of the parts, and then the mold is opened and the completed part is ejected.

(Refer to Chapter 9 on powder metal for more details on powder injection molding.)

10.3.3 Sintering

The sintering phase for the fabrication of new ceramics bonds the particles together. This is performed similarly to that for powder metal, except that it is performed in a vacuum furnace. During sintering, the binder materials are burned off and shrinkage will occur. As with powder metal, the components will shrink by the percentage of the binder in the powder mixture. With new ceramics, the sintering temperatures are between 80 and 90% of the melting temperature of the material, which is between 3,000 and 3,500°F.

(Refer to Chapter 9 on powder metal for more details on sintering.)

10.3.4 Secondary Operations

New ceramics can have limited secondary operations performed on them due to the extreme hardness and brittleness of the material. If additional shaping is required, it needs to be performed with diamond cutters and grinding wheels. With new ceramics being one of the hardest man-made materials, only diamond cutting tooling is able to cut and remove the material during shaping.

10.4 CERMETS

A cermet is a composite material consisting of a mixture of metal and new ceramic materials. This mixture can be seen in Figure 10.15, within a composite matrix, which renders a unique new material having the hardness of the ceramic with the ductility of the metal. Depending on the mixture, the cermet is classified as:

1. Metal Matrix Composite (MMC), having a higher percentage of metal than ceramic
2. Ceramic Matrix Composite (CMC), having a higher percentage of ceramic than metal.

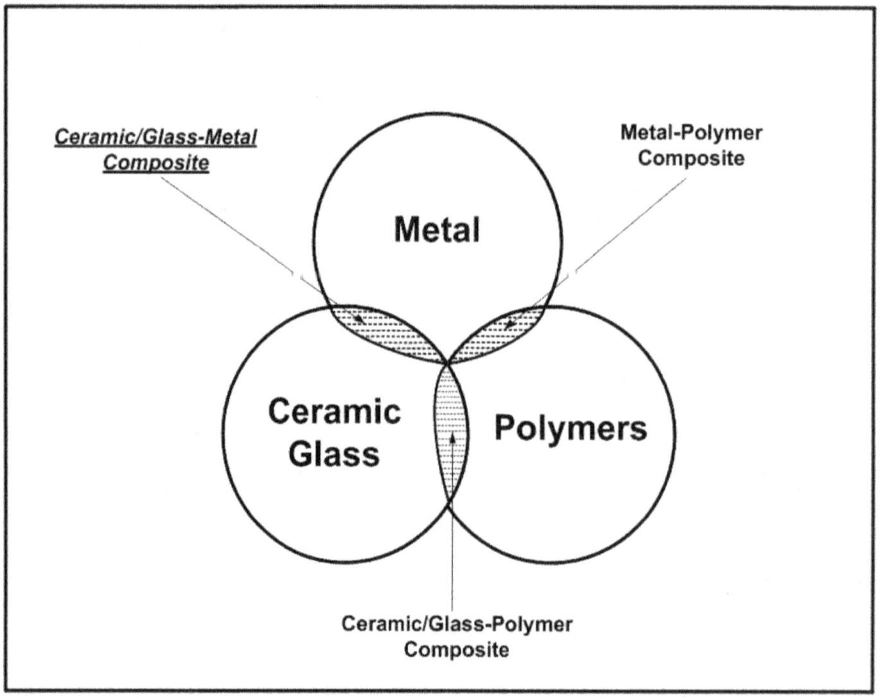

Figure 10.15 Metal–Ceramic Composite Composition

One of the most common is *Tungsten Carbide*, where carbide ceramic is mixed with cobalt metal. This composite material is widely used for the cutting tool in the machining arena. Because of the ceramic material, the cutting tools will last far longer than those produced from high-carbon steel. The cobalt metal mixed in will absorb the shock when the tool makes contact with the workpiece being machined. If a solid ceramic tool were to be used in the same application, it may shatter when making contact with the workpiece due to the high brittleness of the ceramic.

Tungsten carbide is comprised of between 80 and 90% ceramic and 20–10% cobalt. The particles are mixed together with a wet binder material to hold them together during forming. This mixture is worked into a sludge and then dried in a vacuum furnace to prevent oxidation of the metal particles. Once the particle and binder mixture is dried, it can be shaped and sintered.

10.4.1 Shaping and Sintering

Shaping is primarily performed by pressing in the same way powder metal is pressed. (See Chapter 9 on powder metal for more details on pressing) Once the component has been pressed into the desired shape, sintering is the next step in the process. This is performed in two stages. The first stage is where the binder material, which was used to hold the ceramic and metal particles together during pressing, is burned out of the part. The second stage is critical. Unlike other materials, ceramic has a much higher melting temperature than do metals. During this stage, the cobalt particles are melted within the components and fill up the open pores between the ceramic particles. This finalizes the composite structure, and this can be seen in Figure 10.16 in a before and after sintering structure. Note the percentage of cobalt particles being melted and distributed within the pores of the ceramic particles.

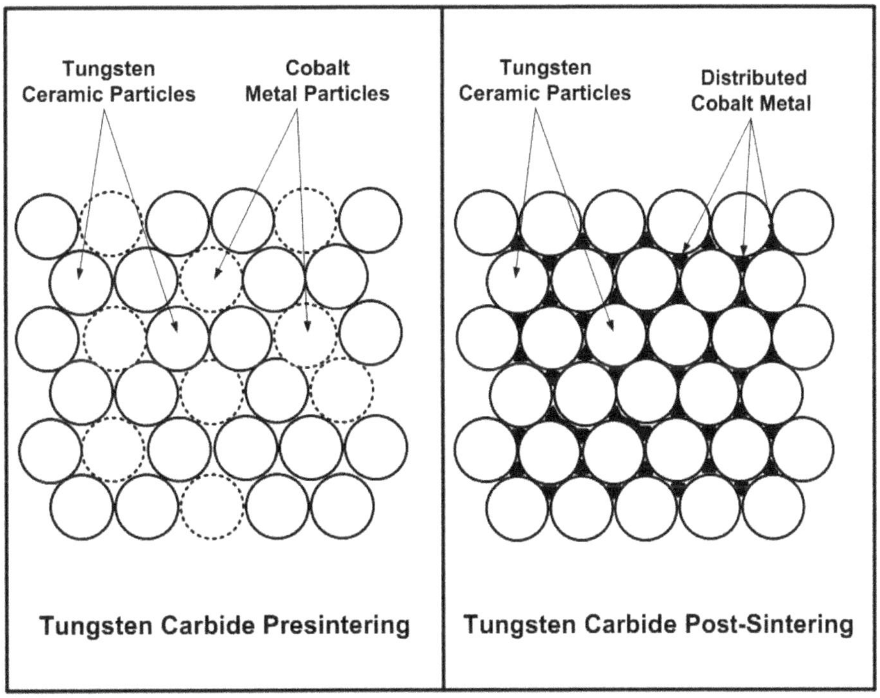

Figure 10.16 Tungsten Carbide Pre- and Postsintering Structure

10.5 DESIGN CONSIDERATIONS

Working with ceramic materials, either traditional or new, and understanding what performance criteria are required for both component fabrication and function. Following is a list of areas that need to be considered when designing components that will be manufactured from ceramic materials.

Physical Properties: Ceramic materials will be stronger in compression than in tension. Because of the extreme brittleness of the material, they cannot take any amount of shear load without failing. Due to the brittleness of the material, there will be no deflection and no ductility in the component part.

Part Geometry: As a rule of thumb for designing any component regardless of the process that will be used to fabricate it, the geometry of the design should always be a simple as possible. The more complex the design of a component is, the higher the cost will be to manufacture it.

Chamfers and Radii: All corners on ceramic components need to contain either a chamfer or radii. As with any particle forming process, the powder needs to be able to flow into the corners and be able to be removed from the shaping process without causing any damage to the newly formed part. The larger the chamfers and radii are, the better the shaping process will work.

Shrinkage: Shrinkage is a factor as it is with powder metal. For new ceramics, the percentage of binder will burn out of the component, and it will shrink by that amount. With traditional ceramics, the shrinkage can be very large, depending on how much water was used in the mixture prior to the shaping process.

Screw Threads: Screw threads, either internal or external, cannot be incorporated into a ceramic component of any material. Due to the extreme brittleness of the material, the slightest amount of torque on the threads will cause them to shear off and strip. If the component needs to be fastened to something, a hole should be incorporated into the part where a screw can pass through to be fastened into the mating part.

Secondary Operations: Secondary operations on components fabricated from traditional ceramic materials are not possible. The material is extremely brittle and does not have the strength to withstand any material removal whatsoever. With new ceramic materials, secondary operations are very limited. The removal of material requires the use of diamond tooling and is costly and very time-consuming. If possible, design the features into the component so they can be incorporated in the shaping process and avoid any secondary processing.

Wear Resistance: A worthy feature of new ceramics is the extremely good wear resistance that they posses. Because these materials are extremely hard, they can be used for wear surfaces in the same manner as glass can. Just keep in mind that if there is going to be any impact to the surface, or if it will be placed into a shear mode during function these should be avoided.

10.6 SUMMARY

Ceramic materials are very widely used in both products and in manufacturing processes. Due to the unique structure of these materials, it can be used to fabricate components that posses properties unlike other materials. In the traditional ceramic arena, the fabrication of dishes and other ceramic goods is far easier than manufacturing them from other materials such as glass. Also, this ceramic is used in other manufacturing processes for constructing expendable molds as in the investment casting process. Because of its high resistance to heat, traditional ceramic is used in the linings of furnaces where extreme heat is used for heat-treating operations.

New ceramic, being a chemical mixture, renders an extremely hard material. This has become a major ingredient in the production of cermet composites along with being used in its ceramic state. In applications where high strength and high wear resistance are required, new ceramics have become a lead player. When mixed with metal to produce either an MMC or CMC composite, it delivers the hardness and wear resistance of the ceramic with some ductility of the metal. There are new materials being invented in this arena, and this will only expand the arena of usage for this critical material throughout the manufacturing world.

10.7 QUESTIONS

1. What is the difference between traditional ceramics and new ceramics?

2. What are the four steps required when working with a traditional ceramic material?

3. What is the difference between comminution and crushing?

4. How does the slip casting process work?

5. What does firing do to a part shaped with a traditional ceramic material?

6. What is the process for producing a new ceramic material?

7. When performing secondary operations on a new ceramic component, what type of tooling is required and why?

8. What is a cermet, and where is it mainly used?

9. How is the internal structure of a cermet different from that of a solid new ceramic component?

SECTION

3

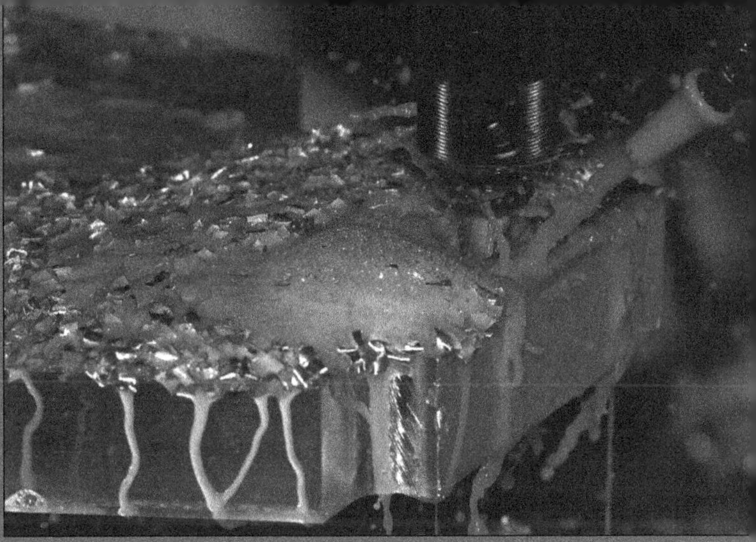

METAL FORMING PROCESSES

Metal forming processes work with metal in a solid state. The material is formed by bending and shearing it into a desired shape through various operations and processes. This can be performed either at room temperature or at a raised temperature but not at a temperature at which the metal becomes liquid. Some metal forming processes date back to the old blacksmith days where the metal was heated and then hammered into the shape that was required. Although this has been automated and refined over the years, the concept for shaping metal in this manner has not changed much.

Another metal forming process deals with fabricating components from thin sheet of metal where the configuration will be stamped out of the material and then shaped by bending or forming operations. Very complex geometric configurations can be attained with the stamping process, and for high-volume manufacturing, it can be almost fully automated. Some simple products that are seen daily fabricated utilizing this process are door hinges. They are stamped out of a sheet of metal and then bent where the hinge pin holds them together.

The metal forming processes are a very viable and effective method for producing components that would take a vast amount of machining or not be able to be produced at all. This section deals with these processes along with the restrictions and precautions for fabrication. Each of the metal forming processes is different in nature and is used for producing components with different properties and geometries.

CHAPTER

11

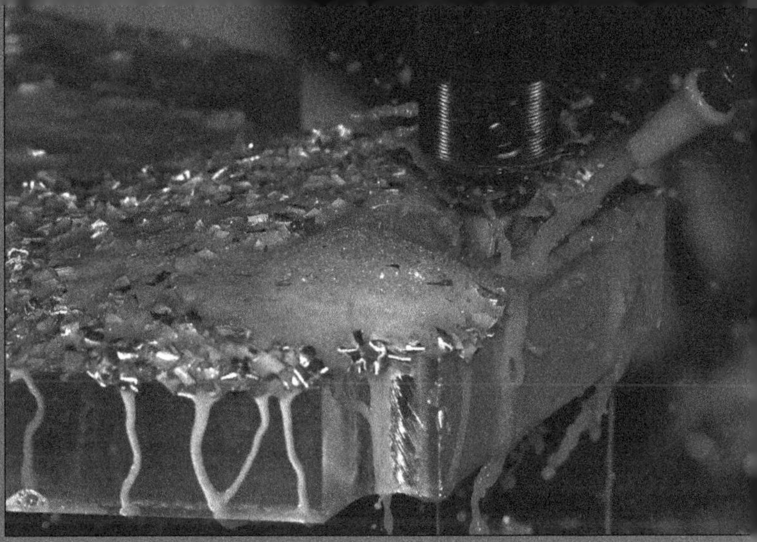

Metal Forming Principles

11.1 INTRODUCTION

Metal forming consists of processes that take metals in either sheet or bulk shapes and form them into a finished configuration. These processes do not melt the starting material as does the solidification processes but rather work with the material either in its cold state or being heated up to a given temperature. The starting material is run through various shaping operations depending on which process is selected, and for the most part, little to no material is removed while working it to the final configuration. Metal forming is a very widely utilized process for delivering metal components for products or for producing the starting metals that are utilized for other fabrication processes. This chapter discusses the principles for the shaping of metal within the metal forming processes.

11.2 METAL FORMING PROCESSES

The metal forming processes are divided into two main categories. Each of these categories shapes the metal but is extremely different in how it performs the actual shaping. These process categories are as follows and will be covered in detail in Chapters 12 and 13 independently.

Bulk Deformation: The bulk deformation starts with a block of metal and performs operations that create significant deformation and massive shape changes of the metal. With these processes, no material is removed during shaping as it is formed into the desired configuration. A volume of metal is selected for starting and is formed through various processes. Figure 11.1 shows how a metal piece is transformed into

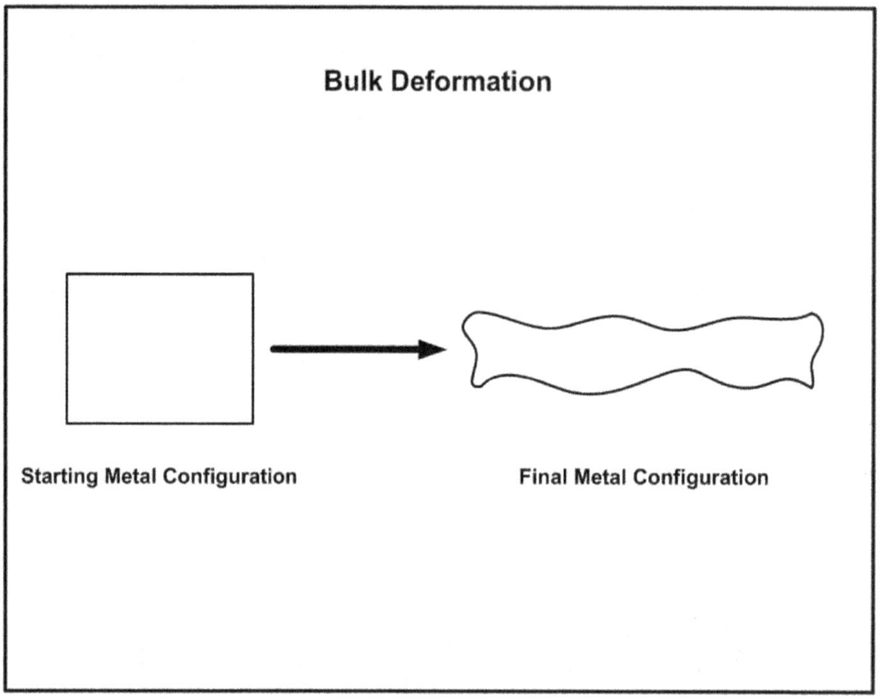

Figure 11.1 Bulk Deformation Creating a Metal Configuration

a new shape through a deformation process. The main bulk deformation processes are as follows and will be covered in Chapter 12

- Rolling
- Forging
- Wire Drawing
- Bar Drawing

Sheet Metal Forming: Sheet metal forming processes are those that are performed on thin sheets of metal, cutting and shaping them into the desired configuration. These processes are performed with tooling know as a punch and die setup. Normally, it begins with the punch stamping out a configuration from a sheet of metal, similar to a cookie cutter. Figure 11.2 shows the punch and die setup punching a part from a metal strip, while Figure 11.3 illustrates the strip where the components have been punched from. Very complex geometries can be produced with this process rapidly and to close tolerances. Secondary operations can be performed, such as bending where the stamped out part can be formed into various angular configurations. Sheet metal forming operations are mainly performed at room temperature, with a few exceptions. The main sheet metal processes are as follows and will be covered in Chapter 13

- Blanking
- Piercing
- Shearing
- Bending
- Drawing

11.3 MATERIAL BEHAVIOR IN FORMING

In order to perform metal forming process correctly, there needs to be an understanding of how the physical properties of the metal will perform. As the metal starts to get deformed, it will undergo both stress and shear. As force is applied to the material, it will begin to deform. Once it reaches the yield strength, it will undergo a

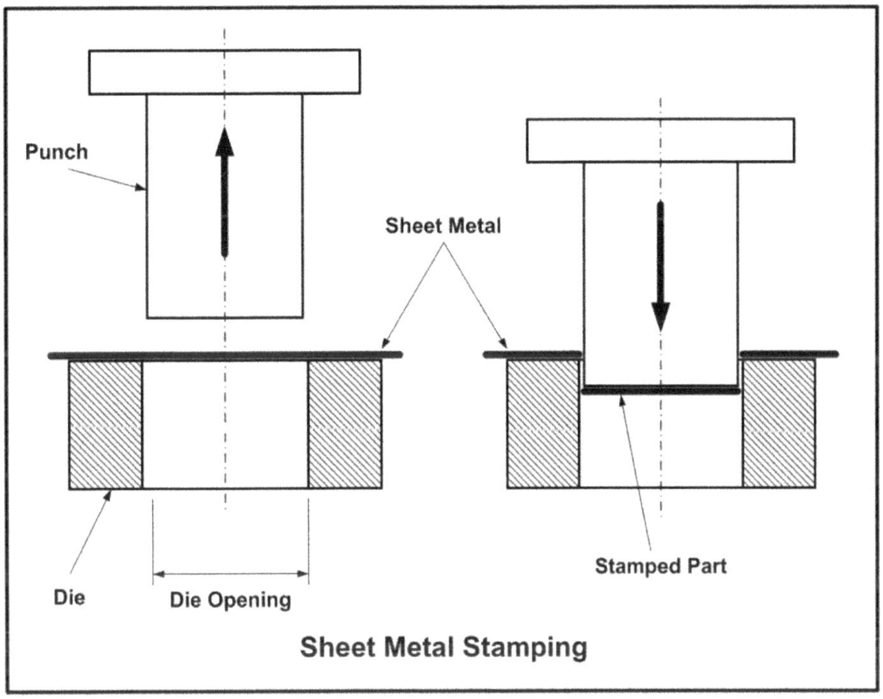

Figure 11.2 Sheet Metal Creating Stamped-Out Configurations

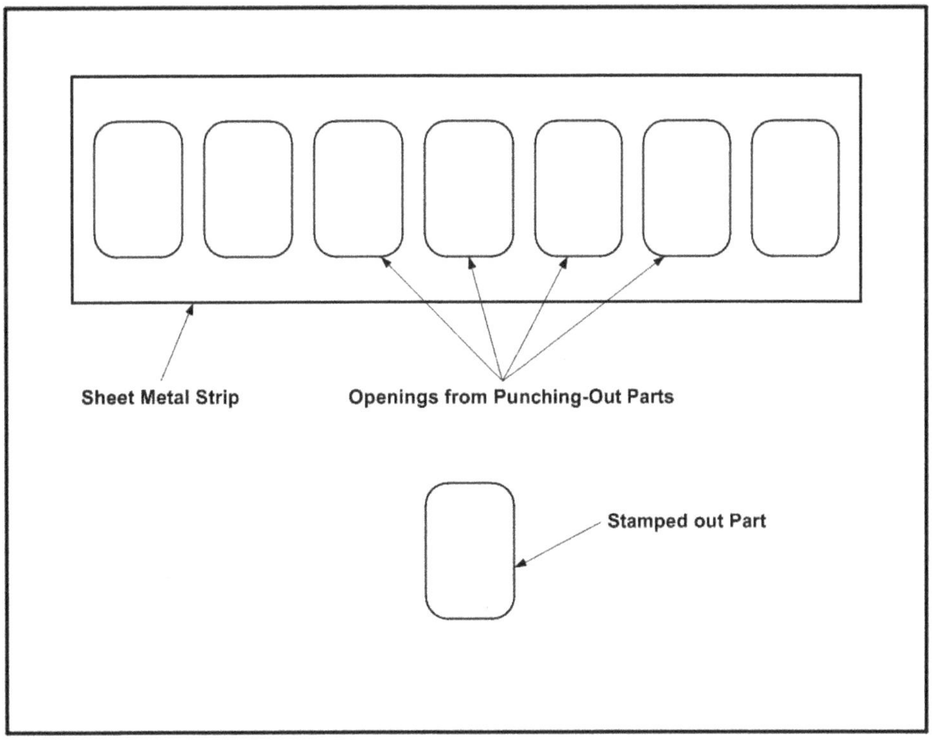

Figure 11.3 Multiple Parts Stamped Out of a Metal Strip

permanent deformation in the material. Prior to reaching the yield strength, if the force is removed, the material will return to its original shape with no deformities. As the pressure on the material increases, deformation will occur until it reaches its ultimate strength and at that point it will shear off and break. Figure 11.4 shows a typical shear/stress curve for metals.

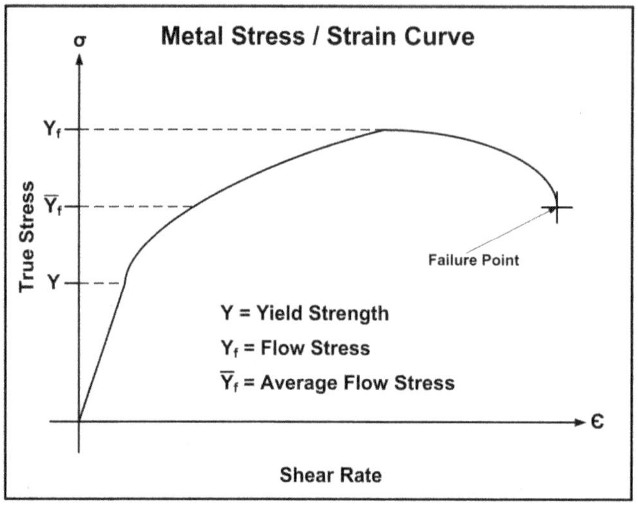

Figure 11.4 Stress/Strain Curve for the Metal Being Formed

Another factor in the shaping of metal occurs during bending. As the material is bent into a given angle, it will produce internal stresses on both sides. These stresses will place the metal in either compression or tension alternatively from the bend. As this happens, the stresses will increase as they moves from the center of the metal, which is the neutral axis, toward the outer surfaces of the part. These stresses will cause strain hardening in the metal, which may be or may not be desired. Figure 11.5 illustrates these compressive and tension stresses being produced in the formed area of the metal.

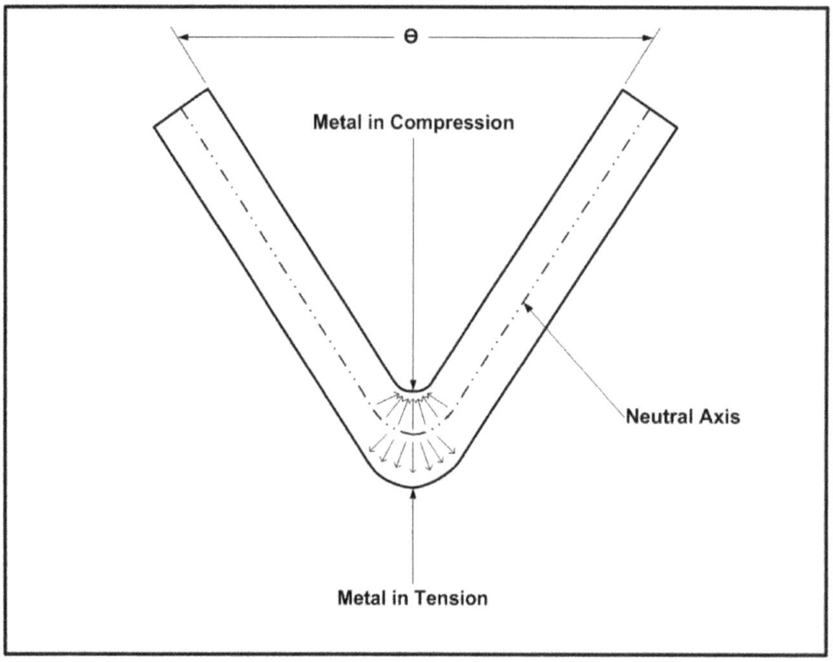

Figure 11.5 Tension and Compressive Stresses on a Sheet Metal Part from Bending

11.4 TEMPERATURES IN METAL FORMING

Metal forming can be performed either at room temperature or at elevated temperatures depending on the processes being used. Each will produce both positive and negative attributes for both the components being produced and the process forming them. The three categories of temperatures utilized in metal forming processes are *Cold Working, Warm Working,* and *Hot Working.* These are defined as follows along with their positive and negative attributes.

11.4.1 Cold Working

Cold working refers to processing the material at room temperature. It does not mean that the metal is cooled down at all. This is almost always used in sheet metal forming with a few exceptions for special processes. The advantages and disadvantages of cold working are listed as follows.

Accuracy: Because the metal is not heated when performing the shaping operations, it will have the highest accuracy of the three temperature ranges. With no thermal expansion during the processing and no thermal contraction when the components are cooled down, the accuracy from the forming will be maintained.

Surface Finishes: For the same reasons as accuracy, the surface finishes with cold working will be the best of the three temperature ranges. As the metal is heated, it will produce a scale on the outer surfaces. Cold working will not produce a scale and will render the best surface finish on the component parts being fabricated.

Less Energy: Without having to heat the metal and maintain the temperature during processing, cold working will use much less energy than warm or hot working will.

Strain Hardening: Straining hardening can be either a positive or a negative feature of cold working. When the metal is bent, it will produce stresses in the bent area of the component as shown in Figure 11.5. If added strength is required in the bent section, then the strain hardening in the area would be an added advantage. However, it the metal is required to have equal strength, the components will need to be annealed later to eliminate the straining hardening in those areas.

Higher Forces: Due to the fact that the metal is not heated, there is no malleability of the material, so the forces required to shape the material are higher than the other temperature ranges. These higher forces will be evident for stamping out the component shapes from a strip of metal to bending them into the desired configurations.

11.4.2 Warm Working

Warm working in metal forming is performed with the metal heated to a temperature of about 0.3 of its melting temperature. This gives the material more malleability allowing it to be formed into tighter configurations than would cold working. This is used for special sheet metal forming processes requiring the metal to be shaped into geometric shapes that would fail in cold working, along with some bulk deformation processes. The advantages and disadvantages of warm working are listed as follows.

Lower Forces: With the metal having an elevated temperature, there is less forces required for the shaping processes. The increased malleability of the metal allows it to flow easier than a cold material would.

Intricate Geometries: Once again, with the increased malleability of the metal, it can be formed into more intricate configurations than with cold working. This is very evident for extremely tight angles being bent into the metal, which would crack and fail in cold working.

Annealing Reduced: With the elevated temperature, any straining hardening that would be present during forming will require less annealing to eliminate it. This will be required on very tight bends as some strain hardening will be present, but the amount of annealing will be far less than with cold working.

Strain Hardening: The straining hardening will be greatly reduced because of the raised temperature of the metal. As with cold working, this can be either a positive or a negative feature. If additional strength is required in the formed areas, it will not be present. However, if the material required uniform strength, it will be more present in the metal throughout. For extremely tight formed areas, strain hardening will be present and annealing will be required to eliminate it.

Accuracy: With the starting material being at an elevated temperature, there is thermal expansion during the forming operations. As the material cools back to room temperature, thermal contraction will occur causing warm working to not have the accuracy that cold working will.

Surface Finishes: Warm working will produce some scale on the metal at the elevated temperature and produce a surface finish not as good as cold working would. If required, this scale would need to be removed in a secondary operation.

Higher Energy: With having to maintain the metal at the elevated temperature during the shaping processes, more energy is required. Depending on how many operations are required to attain the final configuration, this may require heating the material several times using much more energy than cold working would.

11.4.3 Hot Working

Hot working in metal forming is performed with the metal heated to a temperature of between 0.5 and 0.7 of its melting temperature. This gives the material complete malleability and an orange glow during the shaping processes. Hot working is never performed with sheet metal forming but is almost always performed with the bulk deformation processes. Because bulk deformation changes the shape of the starting material drastically, having the starting material close to the melting temperature allows for it to be shaped easily where cold or warm working would only create cracks and defects in the metal. The advantages and disadvantages of warm working are listed as follows.

Major Deformation: With the metal being extremely close to the melting temperature, it can withstand the most shape change of the three temperature-range working environments. With the malleability close to the flowing point of the metal, extreme shaping can be performed without causing damage to the component part.

Fracturing: As with being able to perform major deformation due to the extremely high temperature of the material, fractures caused from shaping are greatly reduced. When shaping at the higher temperature range (0.7 T_m), they will be eliminated completely. The material can flow during the shaping in a way that is not possible with cold or warm working temperature ranges.

Strain Hardening: Working in a temperature range of 0.5–0.7 of the material's melting temperature, there is no strain hardening in the component being shaped. Regardless of how dramatic the shape change is, with the temperature elevated to the annealing range during the fabrication process, the material is annealed during processing.

Isotropic Strength: A very positive feature with hot working is the internal structure of the component part being produced. With the temperature of the metal being elevated to an extremely high range, the strength of the component is isotropic. This means that the strength of the completed part will be equal in all directions.

Lowest Forces: With the metal being close to the melting temperature, the shaping will require less force than either cold or warm working. Having the highest malleability of the three working temperature ranges allows the metal to be shaped the easiest, requiring the least amount of force.

Accuracy: Of the three working temperature ranges, hot working will have the least accuracy. The metal will be at its highest thermal expansion during shaping and will contract the most when cooled down, resulting in a difficult situation for controlling accuracy. In many cases, the final dimensions would be dialed in during a secondary operation.

Surface Finishes: Very heavy scaling on the outer surface will occur in hot working with the working temperatures being highly elevated. In several instances, this scale will be removed with a secondary operation after shaping has been completed.

High Energy: Having to maintain the working temperature between 0.5 and 0.7 of the melting temperature of the metal, extremely high energy is required. During processing, the material is constantly reheated in order to perform the dramatic shaping required in hot working, and the most energy, of the three working temperature ranges, is expelled.

11.5 SUMMARY

Metal forming is a process utilized for working with either sheet metal or blocks of metal to create shapes that either are not possible with other processes or would be extremely difficult and costly to produce. By shaping or stamping configurations out of metal, an understanding of how the material acts during these operations is critical to creating the geometric shapes that are desired. All metals have a certain yield strength that allows it to be either shaped or cut, and these physical properties need to be understood while shaping and working during processing. Depending on the process being utilized, the working temperature can be elevated in order to shape the metal properly and not cause any damage to the component parts. Factors during processing such as accuracy, surface finishes, strain hardening, and the amount of shaping forces are major considerations in order to select the proper process. Both bulk deformation and sheet metal forming processes will be covered in detail in Chapters 12 and 13, respectively.

11.6 QUESTIONS

1. How are the metal forming processes different from solidification processing?

2. What is the difference between bulk deformation and sheet metal working?

3. What is the yield strength of a material, and why does it have to be forced past that point in order to form a component part?

4. When forming a bend in metal, there will be both compressive and tension stresses in the bend. Where would both of these stresses be located in the bent areas?

5. What are the three working temperature ranges for metal forming, and what are the temperatures that they are performed at?

6. Strain hardening can occur in forming metal. What is strain hardening, and can it be both an advantage and disadvantage and why?

7. Why is accuracy difficult to control during some of the working temperature ranges during metal forming?

8. What is meant by isotropic strength in a formed component, and with which of the three temperature ranges for working metal can it be accomplished?

CHAPTER 12

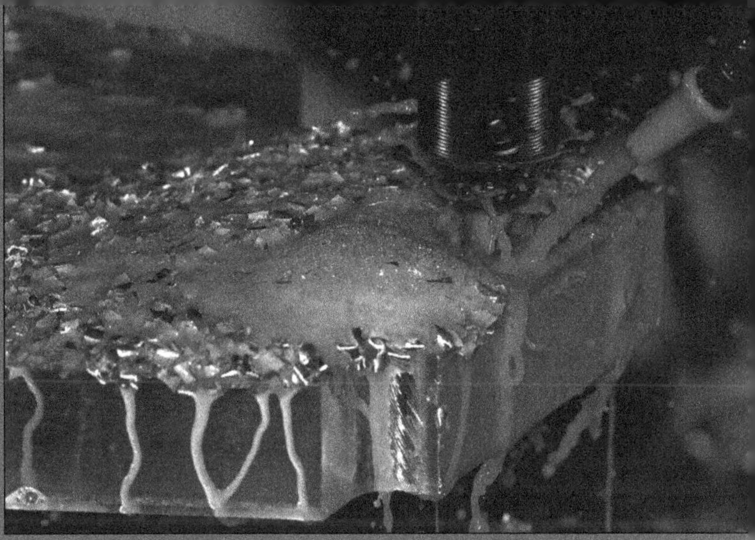

Bulk Deformation Principles

12.1 INTRODUCTION

Bulk Deformation by definition is significant displacement and shape change of the metal being worked. The starting configuration of the metal after processing is completed will not resemble the finished component at all. This process is utilized either for producing raw materials for other fabrication processes or for creating rough shapes that can be finished with secondary operations. Utilizing a deformation process can be the most efficient method for creating complex shapes in an efficient manner. By being able to accomplish dramatic configuration changes without machining and removing the material in the process allows the bulk deformation processes to be very efficient and not wasteful of material during processing. This chapter covers the concepts of bulk deformation along with some of the major processes utilized for creating geometric shapes.

12.2 DEFORMATION CONCEPTS

Deforming metal to create various geometric shapes is an old concept. Blacksmiths have utilized the deformation process by starting with a piece of iron, heating it, and then hammering it into the desired shapes to create horse shoes, shackles, and other iron products. Although the processes have been well refined where the majority of the human intervention and craftsmanship has been replaced by technology, the concept has remained virtually untouched. It begins with a selected volume of metal being heated and then either rolled or pounded into the desired shape. With bulk deformation, the starting metal is always heated to perform the processing. Most of the shaping processes are performed utilizing *Hot Working* in order to accomplish the dramatic configuration changes without creating any defects. A select few processes will be performed utilizing *Warm Working*, but none will be performed with *Cold Working*.

A very critical factor in the bulk deformation processes is the starting volume of the material. The deformation processes are shaping processes and will utilize the volume from the beginning to the end to create the desired shape. If the volume is too little, the shape will not be fully created, and likewise, if the volume is too high, the shape will not be correct either. The volume never changes during processing, so it needs to be exact from the beginning to accomplish a correct geometric shape at the end.

$$V_1 = V_2 = V_3 = \ldots V_n$$

12.3 DEFORMATION PROCESSES

Deforming metal involves several processes that are utilized to reshape the starting material into a desired shape. Although each method is different in nature to create different geometric shapes, the concept is virtually the same with each. The metal is reshaped and none of the starting material is removed during processing. This section covers four of the more common bulk deformation processes, which are *Rolling*, *Drop Forging*, *Hammer Forging*, and *Bar/Wire Drawing*. These are detailed as follows.

12.3.1 Rolling

Rolling is a method utilized for producing flat materials, which are utilized as raw materials for other processing operations like sheet metal working and machining. To accomplish this, the starting metal is heated into the hot working range and sent through a series of rollers to reduce the thickness. To control the width of the material being rolled, side rollers are used to contain the width during the thickness reduction. This process is repeated several times, reducing the material thickness more and more until the final desired thickness is achieved. Figure 12.1 illustrates the rolling process and the factors that are required in order to accomplish this in a satisfactory manner. As the material is reduced in thickness (or width also, if required), the volume will remain constant as shown in the following equation:

$$t_o W_o L_o = t_f W_f L_f$$

t_o = Original Thickness
W_o = Original Width
L_o = Original Length
t_f = Final Thickness
W_o = Final Width
L_f = Final Length

As the material passes through the rollers as shown in Figure 12.1, when it exits the rollers, the velocity of the material will increase. This velocity increase will become more and more intense as the material reduction gets smaller. If, say, the starting thickness is 1.000 in. and reduced by 0.020 to 0.980 in., there will be a 2% increase in the exiting velocity. However, if the starting thickness was 0.150 in. and reduced by the same 0.020 in., the exiting velocity would be increased 13.3%. The following equation shows the velocity increase based on width and thickness reductions:

$$t_o W_o v_o = t_f W_f v_f$$

t_o = Original Thickness
W_o = Original Width
v_o = Original Velocity
T_f = Final Thickness
W_o = Final Width
v_f = Final Velocity

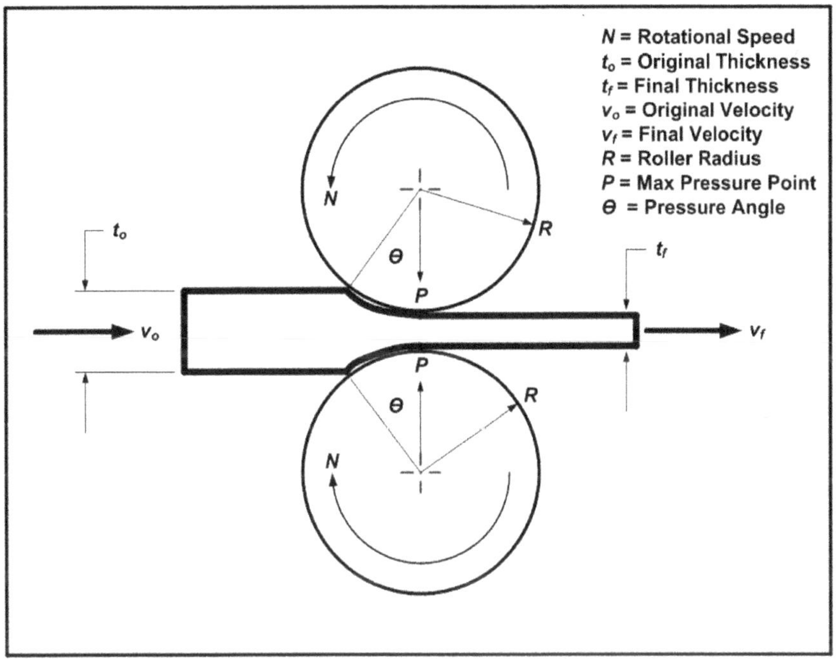

Figure 12.1 Deformation from the Rolling Process

The amount of material that is being reduced is measured in two different manners. The first is called the draft (d) and is expressed by the following equation:

$$d = t_o - t_f$$

d = Draft
t_o = Original Thickness
t_f = Final Thickness

Another measure for the delta in the material thickness is known as the reduction (r). This gives the amount of material that has been reduced as a percentage. The reduction is expressed by the following equation:

$$r = \frac{d}{t_o}$$

r = Reduction
d = Draft
t_o = Original Thickness

As the material is passed through the rollers, it will be in contact based on the amount the material that is being reduced as shown in Figure 12.1. The contact angle (Θ) along with the roller radius (R) defines the contact length (L) at which the material is subjected to reduction. As the material is passed through the roller length, the pressure (P) increases until it reaches the maximum pressure at which the reduction is the greatest. As the rollers rotate at a given RPM (N), it generates a roller velocity (v_r). During this, the material will increase in velocity along with reduction in thickness and friction will occur. The material will experience some slip as it passes between the rollers, which is known as the forward slip (s). This forward slip measurement is expressed by the following equation:

$$s = \frac{v_f - v_r}{v_r}$$

s = Forward Slip
v_f = Final Velocity
v_r = Roller Velocity

The true strain (∈) on the material is a factor of the original thickness and the final thickness and is expressed by the following equation:

$$\in = \ln \frac{to}{tf}$$

t_o = Original Thickness

t_f = Final Thickness

When reducing the material between the rollers, there is a maximum that can be accomplished in each pass. This is predicated in terms of both the roller radius (R) and the friction constant (μ) of the material, which is shown in the following equation:

$$d_{max} = \mu^2 R$$

d_{max} = Maximum Draft

μ = Friction Constant of the Material

R = Roller Radius

When performing rolling of material, all of these factors need to be considered in order to accomplish the process correctly. An example of how these factors work together is illustrated as follows.

Example Problem

A rolling operation is to be performed using hot working. The following information is available for the process:

Original Material Thickness: 0.750 in.; Original Material Width: 4.000 in.; Original Material Length: 12.000 in.; Original Material Velocity: 3.75 in./s. The roller has a radius of 3.750 in. and a rotational velocity of 3.25 in./s with a friction constant between the material and the rollers of 0.097.

The material will be reduced from 0.750 in. to 0.720 in. while the width remains at 4.000 in.

Step 1: Final Length

$$t_o W_o L_o = t_f W_f L_f$$

$$(0.750)(4.000)(12.000) = (0.720)(4.000)L_f$$

$$L_f = 12.500 \text{ in.}$$

Step 2: Final Velocity

$$t_o W_o v_o = t_f W_f v_f$$

$$(0.750)(4.000)(3.75) = (0.720)(4.000)v_f$$

$$v_f = 3.906 \text{ in./s}$$

Step 3: Draft

$$d = t_o - t_f$$

$$d = 0.750 - 0.720$$

$$d = 0.030 \text{ in.}$$

Step 4: Reduction

$$r = \frac{d}{t_o}$$

$$r = \frac{0.030}{0.750}$$

$$r = 0.040 \text{ } or \text{ } 4.0\%$$

Step 5: Forward Slip

$$s = \frac{v_f - v_r}{v_r}$$

$$s = \frac{3.906 - 3.250}{3.250}$$

$$s = 0.219$$

Step 6: True Strain

$$\epsilon = \ln \frac{t_o}{t_f}$$

$$\epsilon = \ln \frac{0.750}{0.720}$$

$$\epsilon = 0.410$$

Step 7: Maximum Draft

$$d_{max} = \mu^2 R$$

$$d_{max} = (0.097^2)(3.750)$$

$$d_{max} = 0.035 \text{ in.}$$

12.3.2 Drop Forging

The drop forging process works by placing a calculated volume of heated metal into a lower die section and having the upper die slam down on it. Both the upper and lower dies have a configuration with the desired geometry cut into them. The upper die section is slammed down hard with a vast amount of force, causing the heated metal to take the shape of the dies. This may require several strikes for the desired shape to fully form depending on how severe the displacement is. Figure 12.2 illustrates the drop forging process starting with an ingot of heated metal and ending with the shape pounded into it from the upper and lower die section. It is evident that the starting volume of the metal is critical in order for the geometric shape to be properly formed in this operation.

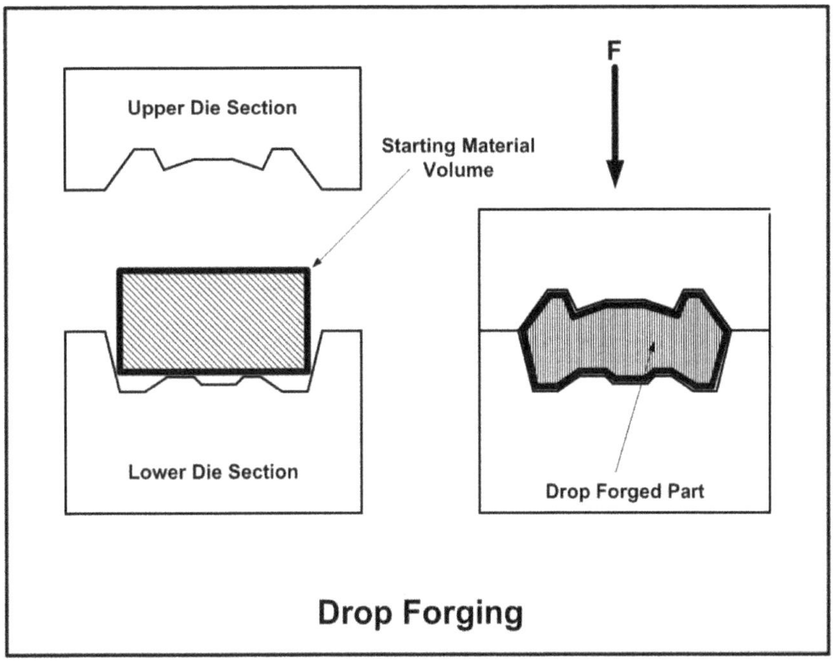

Figure 12.2 Component Shaping from the Drop Forging Process

12.3.3 Hammer Forging

Hammer forging is similar to drop forging as it utilizes a slamming force to displace the materials. The difference is that, with drop forging, the starting material remains in one location while the upper and lower dies pound it into shape. With hammer forging, the upper ram will move up and down hammering the heated material into size. Also, the material being hammered will move under the ram as it pounds the metal into its new thickness. This is the exact process utilized by the old blacksmiths, which would shape the iron into the various shapes such as horse shoes and shackles. Figure 12.3 shows the hammer forge process pounding the metal into a shape as it is moved under the hammering ram.

A manufacturing part fabrication that utilizes the hammer forging process is the making of Damascus steel. The way this is performed is by starting with a given volume of heated metal, which is pounded out very thin. From there, the metal is reheated and folded in half and rehammered to become thin again. This is repeated over and over, creating steel that is comprised of several thin layers hammered together. Figure 12.4 shows how the fabrication of Damascus steel is performed. If this were to be repeated 10 times, the steel would be comprised of over 1,000 layers, giving it the ability to hold a sharp edge better than any other steel. This type of steel was invented by the ancient Japanese, and this is how the Samurai swords were made.

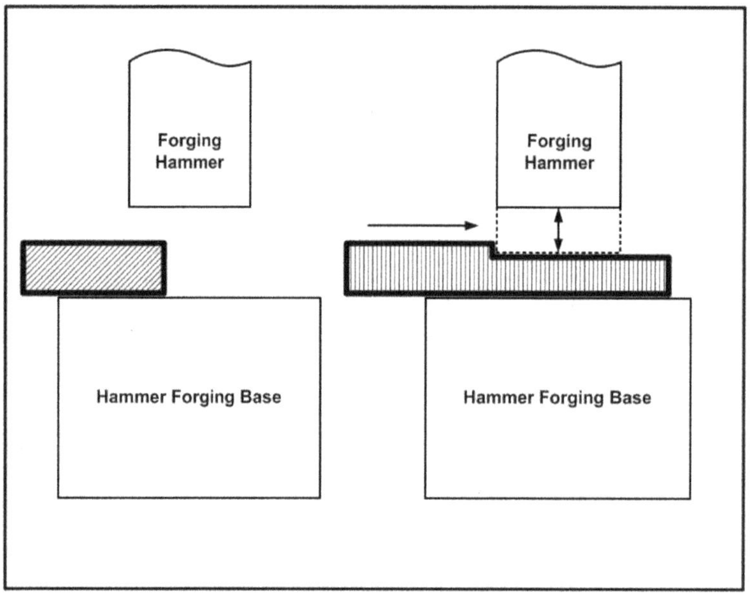

Figure 12.3 Hammer Forging Shaping Component Parts

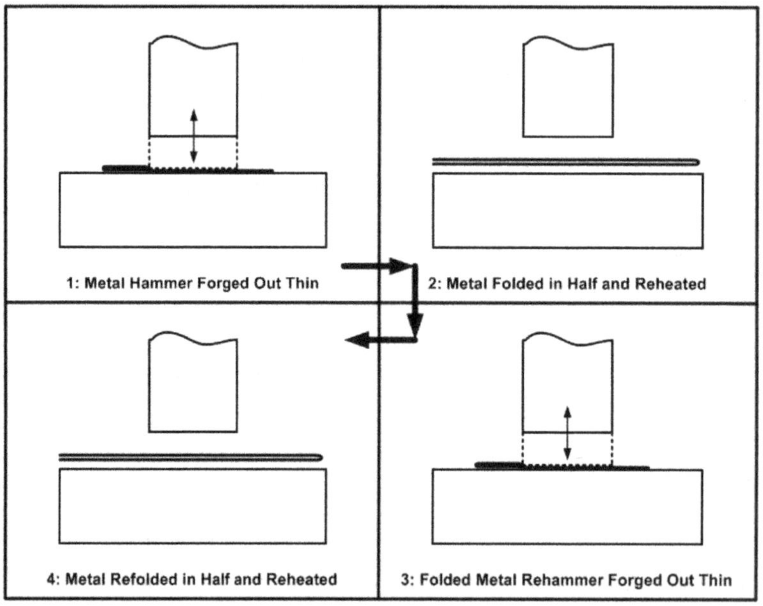

Figure 12.4 Hammer Forging Use for Producing Damascus Steel

12.3.4 Wire and Bar Drawing

Drawing is a process where the material is pulled through a die opening to create a shape. This shape can be either round as for wire or of another configuration such as that obtained with the extrusion process. Although this process resembles extrusion in many ways, the difference is that the material is not pushed through the die opening but rather pulled through. The metal is heated into the warm working range and then fed through the die opening. A clamp is attached to the portion that is protruding through the die and the metal is pulled through, creating the shape desired. If a close tolerance is required, after the metal is worked in the warm working state, it can be pulled through cold to bring the tolerances closer. The cold drawing, however, will produce external stresses on the bar or wire. Figure 12.5 shows how a bar or wire drawing process is set up and performed. All bar stock that is utilized for the raw materials utilized in machining is produced with this process as are all the wire products.

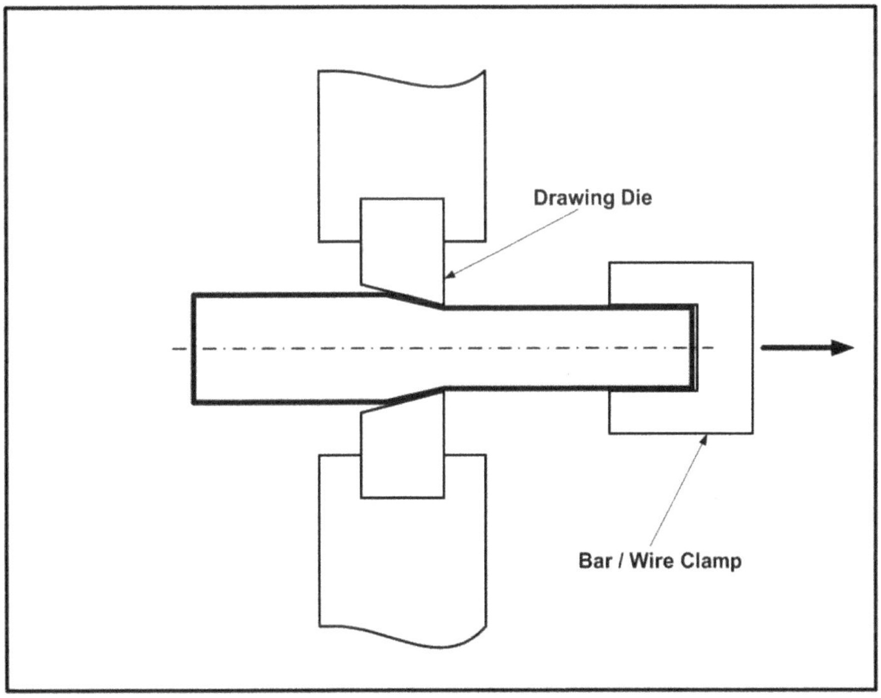

Figure 12.5 Bar or Wire Being Drawn through a Die for Shaping

12.4 SUMMARY

Bulk deformation of metal is a viable process for producing massive shape changes to metals in an efficient and effective manner. Machining a component to the geometric configurations that can easily be produced with deformation would be very time-consuming and would produce waste from removing the unwanted material to accomplish the same configurations. Also, to fabricate a material such as Damascus steel is not possible by any other means. One critical factor that needs to be closely calculated for any deformation process is the starting volume of the material. Bulk deformation is a process where no excess material is removed, simply reshaped. If the volume at the beginning of the operation is not exact, the final configuration of the component will not be correct. Because this process is conducted in the warm-to-hot working ranges, thermal expansion and contraction of the metal will occur, rendering a scaly surface finish and making it difficult to maintain close tolerances. In many cases, secondary operations may be required to improve the surface finish and bring the dimensions into a closer tolerance.

12.5 QUESTIONS AND PROBLEMS

Questions

1. Define bulk deformation.

2. What working temperature ranges is bulk deformation performed at and why?

3. Why is the starting volume of the material being used in bulk deformation critical to the process?

4. During the rolling process, why is the exiting velocity important to understand and monitor?

5. For the forging processes, what is the difference between drop forging and hammer forging?

6. Explain how Damascus steel is produced.

7. During the drawing process, if it were to be performed cold, what issues can occur with the material being drawn through the shaping die?

Problems

1. A piece of steel is to be heated and rolled out. If the starting thickness is 1.250 in. and starts at a width of 3.750 in. and a length of 15 ft 6 in., what would the final width of the material be after rolling?

2. An aluminum ingot that starts at a thickness of 0.375 in. and a width of 1.125 in. is to be rolled with a starting velocity of 5.750 in./s. If the width remains the same but the thickness is reduced to 0.325 in., what would the exiting velocity of the metal be?

3. Several rolling operations are to be performed on a brass ingot. It starts out at a thickness of 1.000 in. and is 3.500 in. wide. Each rolling operation will reduce it by a draft of 0.060 in. and six rolling operations will be performed. If the width remains constant and the starting velocity is 8.250 In/sec, what would the velocity be for each operation and what would be the percentage increase from the starting operation to the final operation?

4. A rolling operation involves rollers that are 2 ft 8 in. in diameter. If the material thickness is to be rolled from 0.900 in. to 0.825 in., what would the reduction and friction constant be if this reduction is the maximum draft that the process will allow?

5. A copper ingot that starts with a width of 6.750 in., a length of 10 ft 5 in., and a thickness of 1.750 in. is desired to be rolled to a thickness of 1.250 in. and maintain a constant width. What would the final length

of the part be, and if the rollers on the machine have a diameter of 3 ft 7 in. and a friction constant of 0.095, is this possible to be performed in one roll?

6. A rolling operation will be performed on an ingot of steel with a starting thickness of 1.500 in., a width of 2.000 in., and a starting velocity of 8.932 in./s. With the width remaining constant and the material being reduced to 1.450 in., with a 16 in. diameter roller rotating at 0.15 RPM, what would the true stress and forward slip be?

7. During a rolling operation, a piece of bronze will start at a thickness of 0.750 in., with a length of 12' and width of 1.875 in. After rolling, the width expands to 1.925 in. and the length increases to 12' 9 in. With a starting velocity of 3.10 in./s, the machine has a 2 in. diameter roller, which is turning at 0.008 RPM, what would the final thickness of the material be? Also, if this reduction is the maximum draft possible, what is the friction constant along with the forward slip and true stress?

8. Rolling of an iron ingot is required to increase the length and decrease the thickness. The ingot starts with a thickness of 5.500 in. and is 14.000 in. long with a width of 7.500 in. and will be reduced by the maximum draft each time. The rollers in the machine are 48 in. in diameter and are rotating at 0.005 RPM with the original velocity of the material being 12.75 in./s. If the friction constant is 0.158, and after running through seven passes, what would the final thickness, length, width, velocity, and forward slip be if the width increased by 1.015% with each pass?

CHAPTER

13

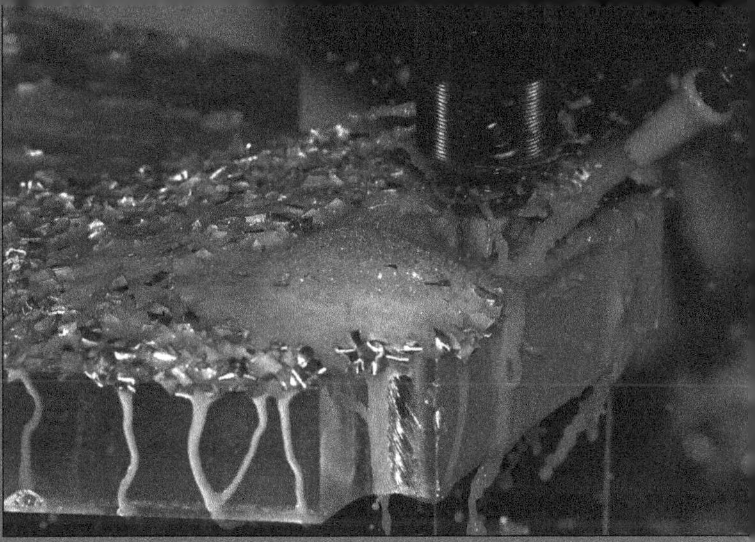

Sheet Metal Working Principles

13.1 INTRODUCTION

The concept of sheet metal forming is simple yet very complex. Sheets of metal are used and through various processes, component parts are stamped out and shaped into the desired configurations. These processes are performed in thin sheets of metal and are extremely efficient for producing complex configurations rapidly. Automation, both programmable and fixed, ensures that the components will be repeated consistently. To fabricate the parts that can be produced easily through sheet metal working by other means would be extremely difficult and very time-consuming. An understanding of how the metal will react during forming and shaping is critical to successful sheet metal working as described in Chapter 11. What forces are required for stamping and bending the metals along with the stresses that will be produced to render a process capable of delivering sheet metal components of consistent quality. This chapter covers the principles and concepts of sheet metal working along with design guidelines for component parts.

13.2 SHEET METAL CONCEPTS

Sheet metal working deals with producing component parts from thin sheets of metal. Configurations are stamped out of strips of metal and from there can be bent and formed into various desired geometric shapes. For the majority of the sheet metal working processes, they are performed utilizing cold working or at room temperature. These processes can be performed either on a part-by-part basis or fully automated, depending on the desired configurations and production quantity required.

Performing sheet metal operations requires an understanding of how the metal will react during processing. When metal configurations are being stamped out, the metal will go through three stages of deformation. Each of these stages will produce a certain level of deformation to the metal that cannot be reversed. These three stages in sheet metal working are *Plastic Deformation*, *Penetration*, and *Fracture*, which are defined as follows.

Plastic Deformation: During this stage, the metal is deformed as pressure is applied. The punch begins to penetrate the metal and deforms the surface. For some applications, plastic deformation is the only stage performed to insert a strengthening rib in the sheet metal component.

Penetration: During the penetration stage, the punch starts to shear the metal as it begins to pass through the plastic deformation stage. When this occurs, the metal will be sheared and have a smooth surface finish as it is sheared clean.

Fracture: After the material has been penetrated, it will eventually fracture and the component being stamped out will break when it reaches the ultimate strength. When it fractures, it will render a rough, fractured surface finish below the shear surface obtained during penetration.

Figure 13.1 illustrates these three stages that occur when metal is being stamped out from a strip. All three of these stages will occur rapidly as when the sheet metal is being stamped out, the punch does not move slowly, but rather very fast and slams into the metal forcing it through the die opening. Cycles of 0.10–0.50 s are not uncommon for performing stamping operations with sheet metal.

When a component is stamped out of sheet metal, the surface finish on the outside of the part will have a given structure. First, there will be a rollover caused from the punch entering the material during the plastic deformation stage. Second, there will be a smooth shear surface of about 1/3 the thickness of the material as a result of the penetration stage. Finally, there will be a fracture surface where the material breaks free during the fracture stage of about 2/3 the material thickness. A heavy burr will also be produced from the fracture stage. Figure 13.2 shows blowup of the various conditions that the side of a stamped part will render.

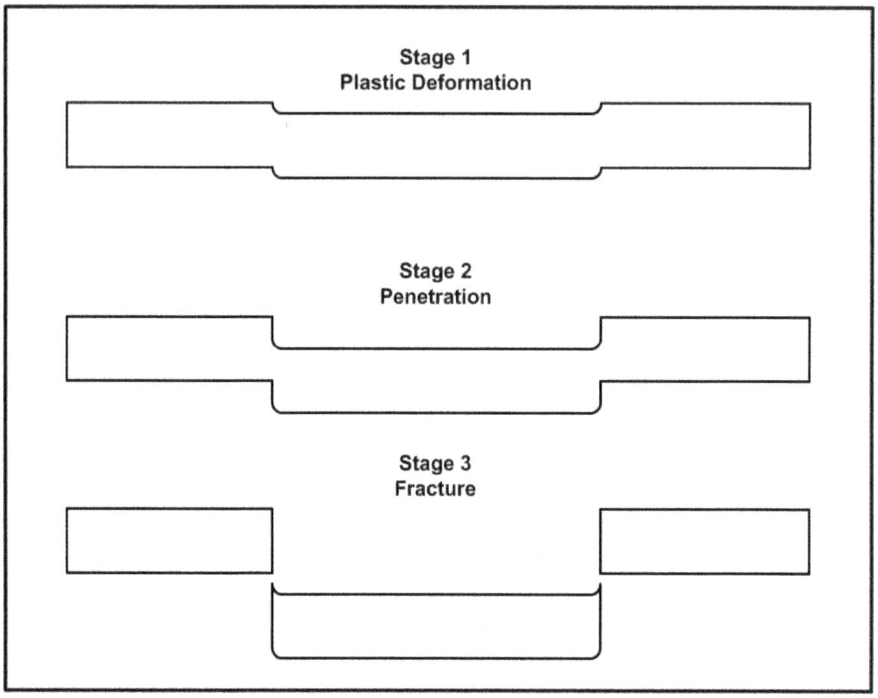

Figure 13.1 Three Stages of Sheet Metal Deformation

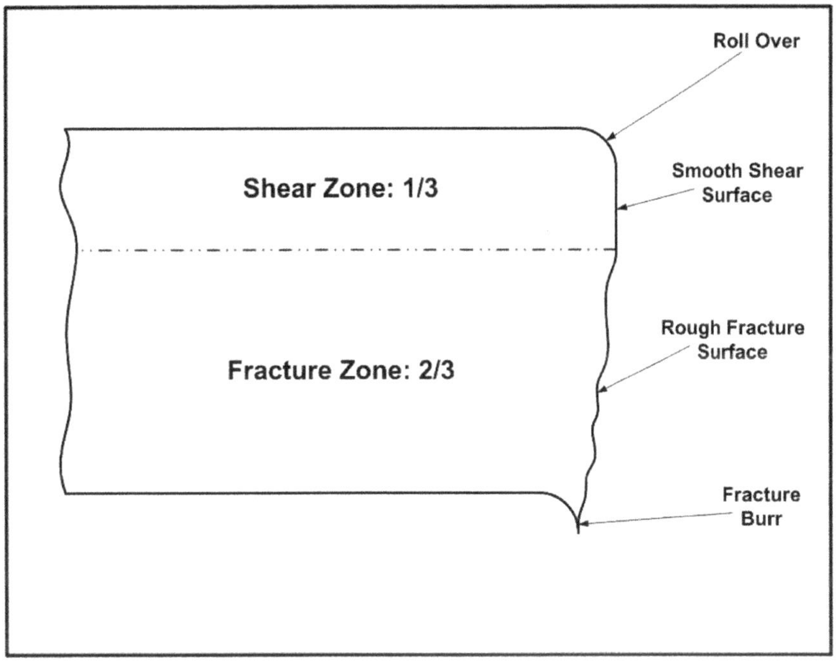

Figure 13.2 Shear/Fracture Structure of a Blanked Sheet Metal Component

13.3 SHEET METAL PROCESSES

The operations performed on sheet metal are divided into three categories, which are *Shearing and Cutting*, *Blanking and Piercing*, and *Bending and Forming*. Each of these three categories will perform completely different functions on the material and depending on the geometric shapes desired, may require two of more of them to accomplish the final configuration. Each of these categories is discussed as follows and will cover the main processes utilized in sheet metal working.

13.3.1 Shearing and Cutting

Both shearing and cutting perform the same function, which is to take the sheet metal material and bring it to the desired length. This can be done on large sheets of metal to produce the widths that will be utilized in blanking and piercing operations or to cut a formed component from a strip. The only difference between shearing and cutting is the type of tooling that is used. For shearing, the punch that is used to cut the metal has two angles on it. Looking at this from a front view, the cutting edge of the punch has an angle raking to one side, allowing the punch to not enter the metal all at once, but rather in a manner similar to how scissors cut a piece of paper. The second angle is seen when viewing the punch from the side where an angle is raking away from the die, allowing for a sharp edge to make contact with the material. Also, the punch or shear will come down with no clearance between it and the die creating a sharp cutting action. Figure 13.3 illustrates how a shearing process is performed with the dual angle punch.

Cutting is performed in very much the same manner as blanking and piercing, which is explained during the process. Here, the punch has a flat surface and is run down into the material to cut it off. There are no angles on the punch, allowing it to fracture off the material. A major difference besides the angles on the punch is that cutting requires a calculated clearance between the punch and die. This clearance calculation is covered in the section on blanking and piercing. Figure 13.4 shows how the cutting process is performed along with the tooling configurations required.

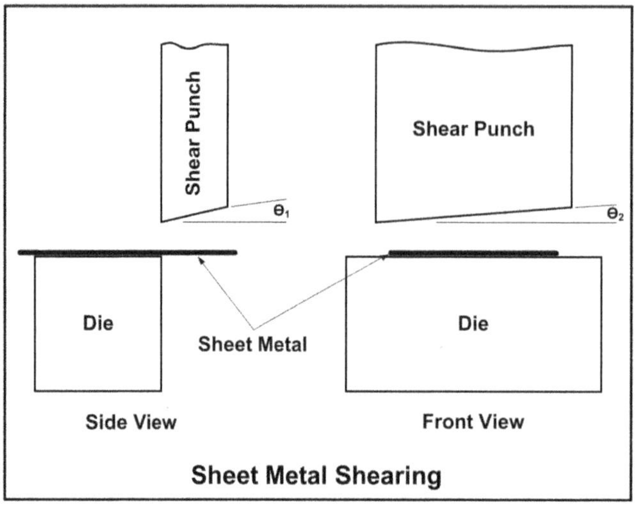

Figure 13.3 Sheet Metal Shearing Process

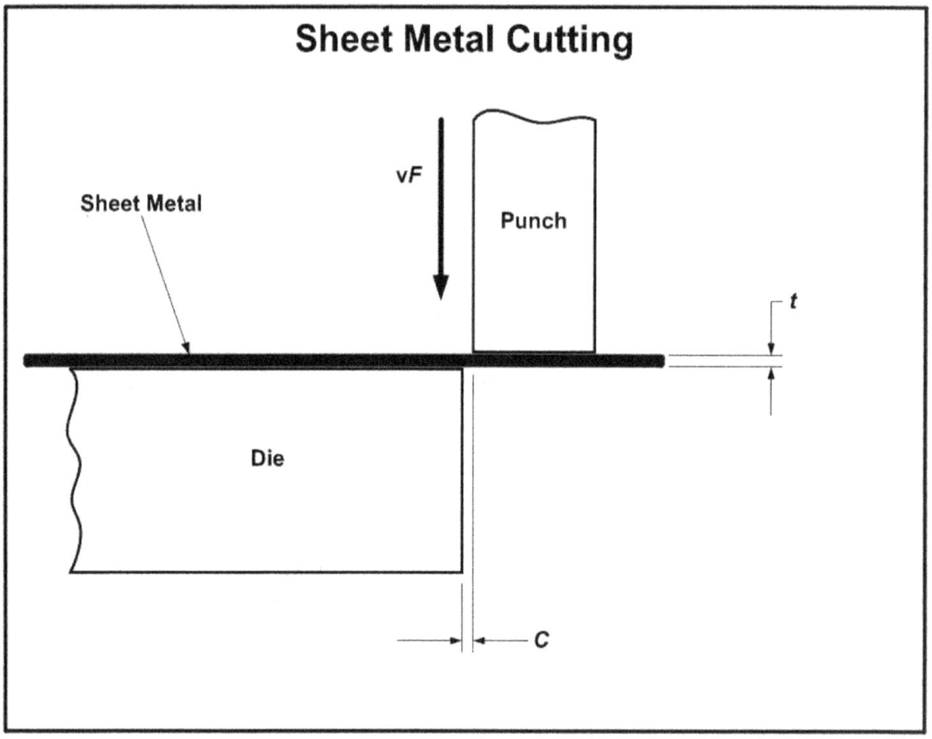

Figure 13.4 Sheet Metal Cutting Process

13.3.2 Blanking and Piercing

Possibly the most commonly utilized process in sheet metal working is blanking and piercing. This is performed utilizing tooling know as a punch and die setup and is commonly referred to as presswork. The die will contain the configuration desired for the component as will the punch. Sheet metal material is placed between the die and punch, and the punch is slammed down, sending the material through the three stages, rapidly fracturing it and creating the geometric shape desired. Figure 13.5 shows a basic punch and die tooling setup and how it creates the component part.

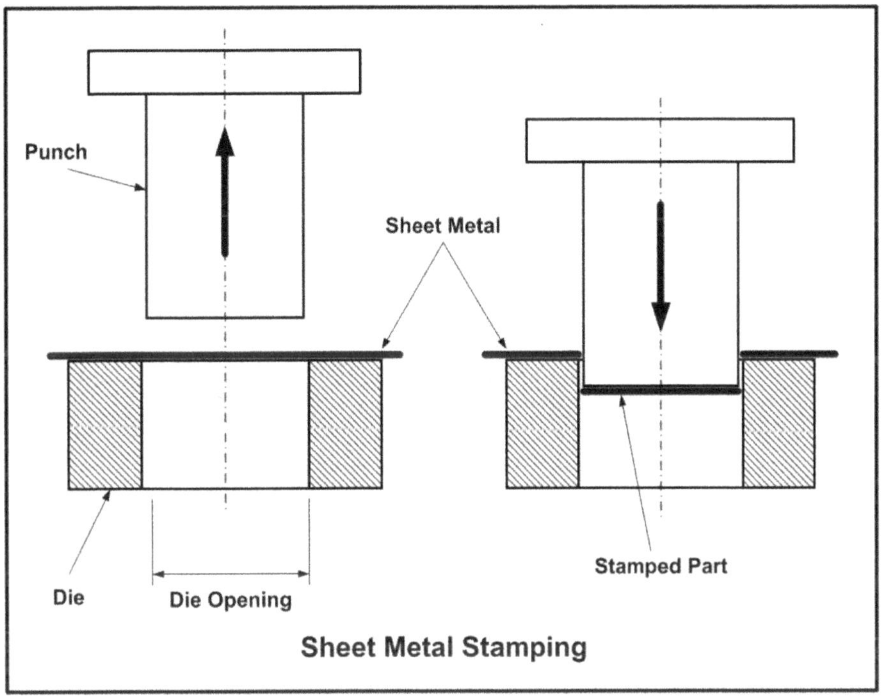

Figure 13.5 Typical Punch and Die Setup for Stamping Components

In order to create the geometric shapes with blanking and piercing, the tooling is constructed with certain dimensions all based on the thickness and type of material that will be worked with. The die section, which has the opening for the desired shape, will have the exact dimensions required on the part drawing. The punch, however, will have a clearance, reducing the dimensions by a calculated amount. If there was no clearance between the punch and die, when the punch slams into the sheet metal, it would either break off completely or seize up in the die. Figure 13.6 illustrates the clearance on both sides of the punch from the die opening.

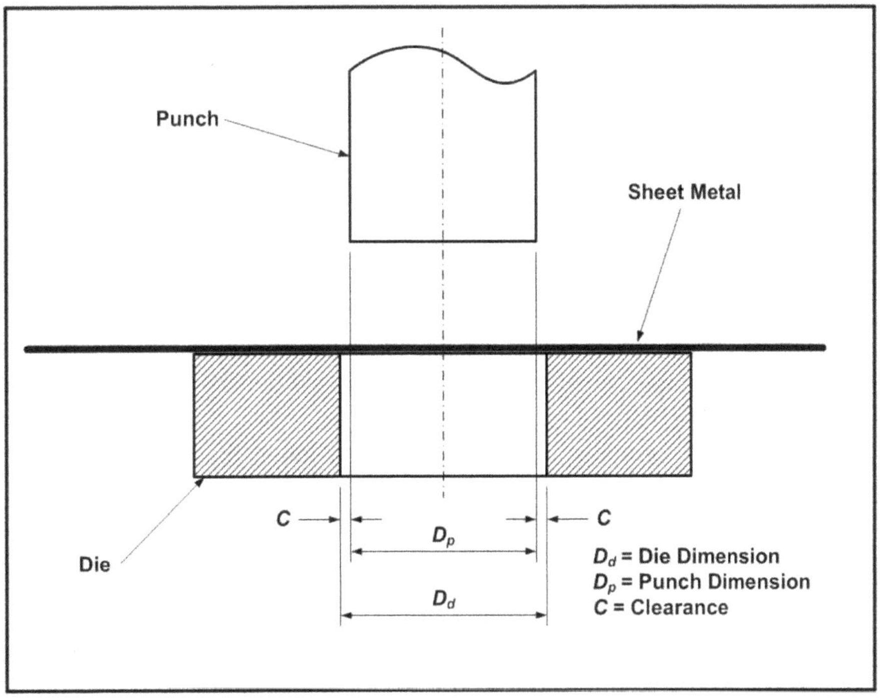

Figure 13.6 Punch and Die Setup with Established Clearances

Clearance Factors for Select Metals	
Material	**A_c Value**
1100S and 5052S Aluminum Alloys, All Tempers	0.045
2024ST and 6061ST Aluminum Alloys; Brass, All Tempers; Soft Cold Rolled Steel, Soft Stainless Steel	0.060
Cold Rolled Steel. Half Hard; Stainless Steel, Half Hard and Full Hard	0.075

Source: Groover; Fundamentals of Modern Manufacturing. 2nd Edition

Figure 13.7 Clearance Factors for Select Metals

To calculate the clearance for the punch, the following equation is used:

$C = A_c t$

C = Clearance

A_c = Clearance Factor (See Figure 13.7)

t = Material Thickness

Once the clearance has been calculated and the punch and die have been fabricated, the amount of force that will be required in order to stamp out the components needs to be calculated. This is normally given in tonnage to select the proper press to adequately perform the stamping process. There are two equations used to calculate the amount of force required depending whether you are working with the shear strength or tensile strength of the material. These equations are displayed as follows:

$F = S\, t\, L_b$

or

$F = 0.7\ \mathrm{TS}\ t\, L_b$

F = Blanking or Piercing Force

S = Shear Strength

TS = Tensile Strength

t = Material Thickness

L_b = Length of Blank (Perimeter of the Part)

The length of the blank is a critical factor in the equations. This is the perimeter of the component being stamped out, not just the longest length of the part. Figure 13.8 shows how the length of the blank is calculated to obtain the perimeter of some simple configurations. But regardless of how complex the configuration is, the perimeter will always be the length of the blank.

The efficiency of stamping out components in sheet metal working is due to the speed it can be performed at. For manual operations, strips of metal are hand-fed through the die and the punch is activated and stamped as produced. For fully automated operations, the strip material in a large coil is fed through the punch and die setup automatically. The remainder of the coil exiting the die setup is cut into smaller pieces to be discarded. For the manual strips, they are discarded in the strip form. Figure 13.9 shows a strip of material where configurations have been stamped out. A part that is stamped out of the strip is called a blank, and it is the configuration that is desired to be obtained. If a secondary operation is to be performed on a blank to produce open configurations, it is called piercing operation. The discarded material from a piercing are called slugs. Calculations for piercing, whether for clearances or forces, are performed exactly the same way as for blanking.

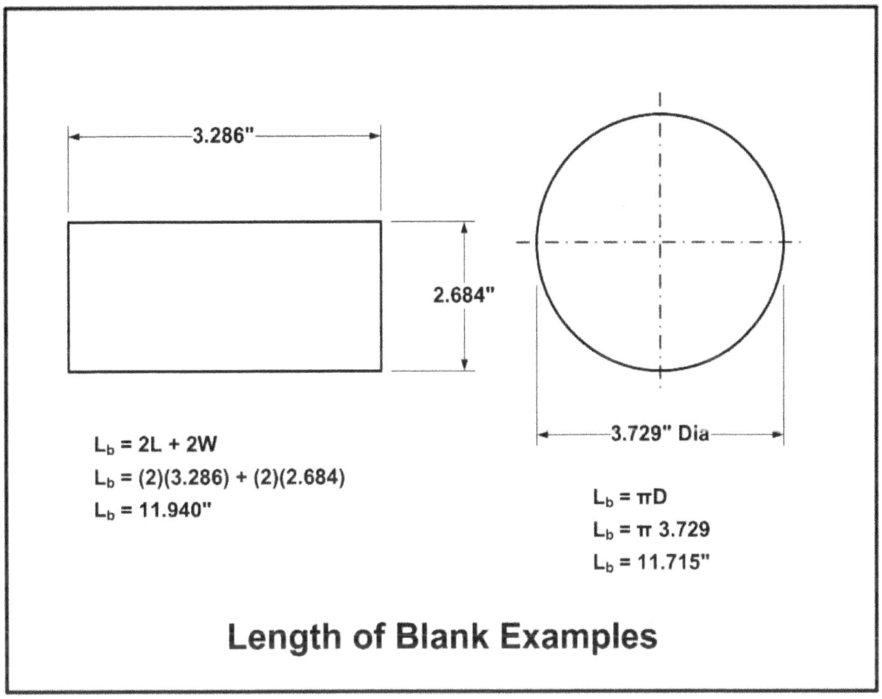

Figure 13.8 Blanking Length Calculation Examples

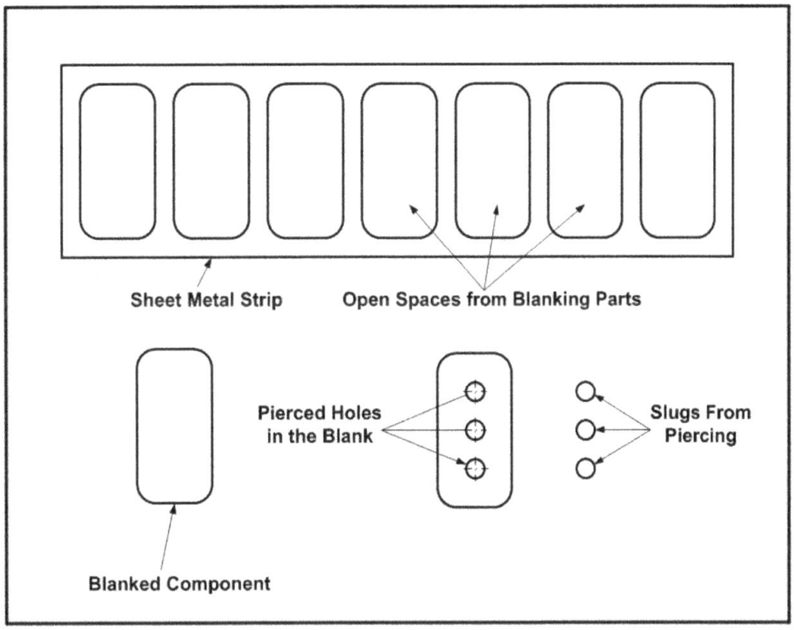

Figure 13.9 Blanked Parts from a Strip and Pierced Holes in a Blank

Some manual punch and die setups will perform a single operation at a time. For high-production operations, a tool called progressive die will be utilized. With this type of die, it will perform multiple operations as the strip of metal is fed through. Figure 13.10 shows how a progressive die setup works where the material will get various operations performed on it in set stations. Looking at this example, Station 1 is stamping out location notches that the following operations will locate from in order to locate the strip properly. Stations 2 through 5 are stamping out various configurations in the strip that will be part of the finished part. Station 6 blanks out the final configuration, rendering a completed part.

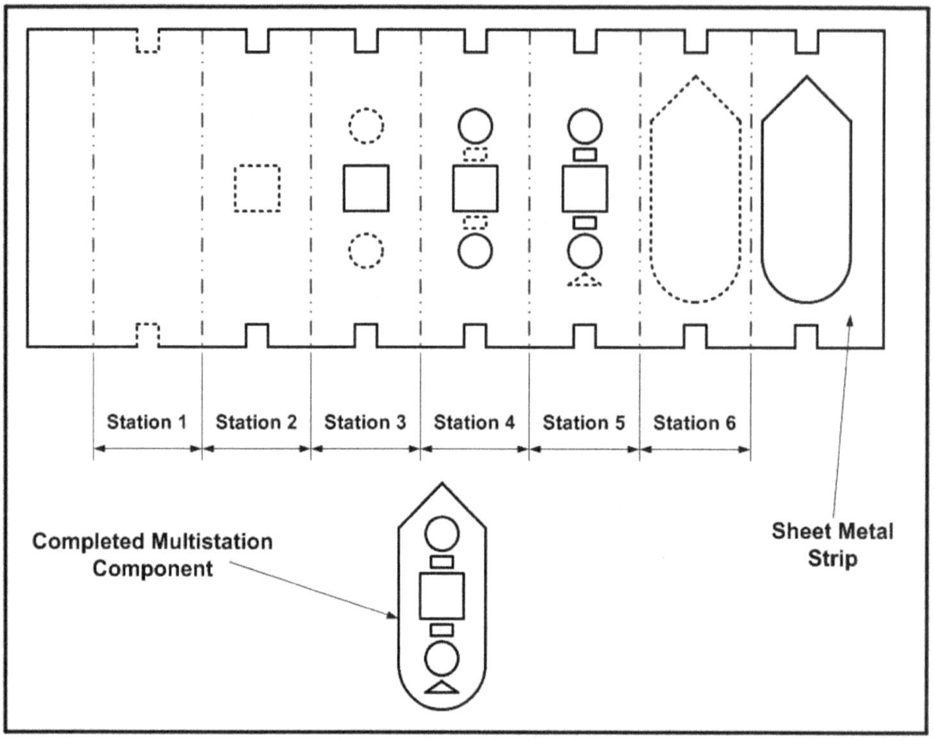

Figure 13.10 Multistation Pierce and Blank Progressive Die

Example Problem

A part is to be blanked out of soft stainless steel that is 0.0625 in. thick. The shape of the part is a rectangle measuring 4.275 in. long by 2.750 in. wide. The shear strength of the material is 150,000 psi and the tinsel strength is 214,285.7 psi.

Step 1: Determine the proper clearance factor (Figure 13.7).

Soft Stainless Steel = 0.060

Step 2: Calculate the clearance.

$$C = A_c t$$
$$C = (0.060)\ (0.0625)$$
$$C = 0.00375 \text{ in.}$$

Step 3: Determine the punch and die sizes.

Punch Sizes

$$L_p = L - (2C)$$
$$L_p = 4.275 - (2 \times 0.00375)$$
$$L_p = 4.4600 \text{ in.}$$
$$W_p = W - (2C)$$
$$W_p = 2.750 - (2 \times 0.00375)$$
$$W_p = 2.7350 \text{ in.}$$

Die Sizes

$$L_d = 4.275$$
$$W_d = 2.750 \text{ in.}$$

Step 4: Calculate the length of the blank.

$$L_b = 2W + 2L$$
$$L_b = (2)\,(2.750) + (2)\,(4.250)$$
$$L_b = 14.0500 \text{ in.}$$

Step 5: Calculate the blanking force.

$$F = S\,t\,L_b$$
$$F = (150{,}000)\,(0.0625)\,(14.0500)$$
$$F = 263{,}438 \text{ pounds or } 131.72 \text{ tons}$$

or

$$F = 0.7\ \text{TS}\ t\,L_b$$
$$F = 0.7\,(214{,}285.7)\,(0.0625)\,(14.0500)$$
$$F = 263{,}438 \text{ pounds or } 131.72 \text{ tons}$$

13.3.3 Bending and Forming

Bending and forming of sheet metal are widely performed to create shapes other than flat. This can be accomplished in a variety of methods with the most common being either an edge die or a "V" die setup. An edge die will bend the material into a 90° angle only and is shown in Figure 13.11, whereas a "V" die can produce different angles based on the angle cut into the punch and die. A "V" die setup is shown in Figure 13.12.

Remembering from Chapter 11, when metal is formed into a shape, there will be both compressive and tension forces where the material has been deformed as shown in Figure 13.13. The material will be formed into a configuration, most commonly as angles. When bending a metal, there are some factors that need to be

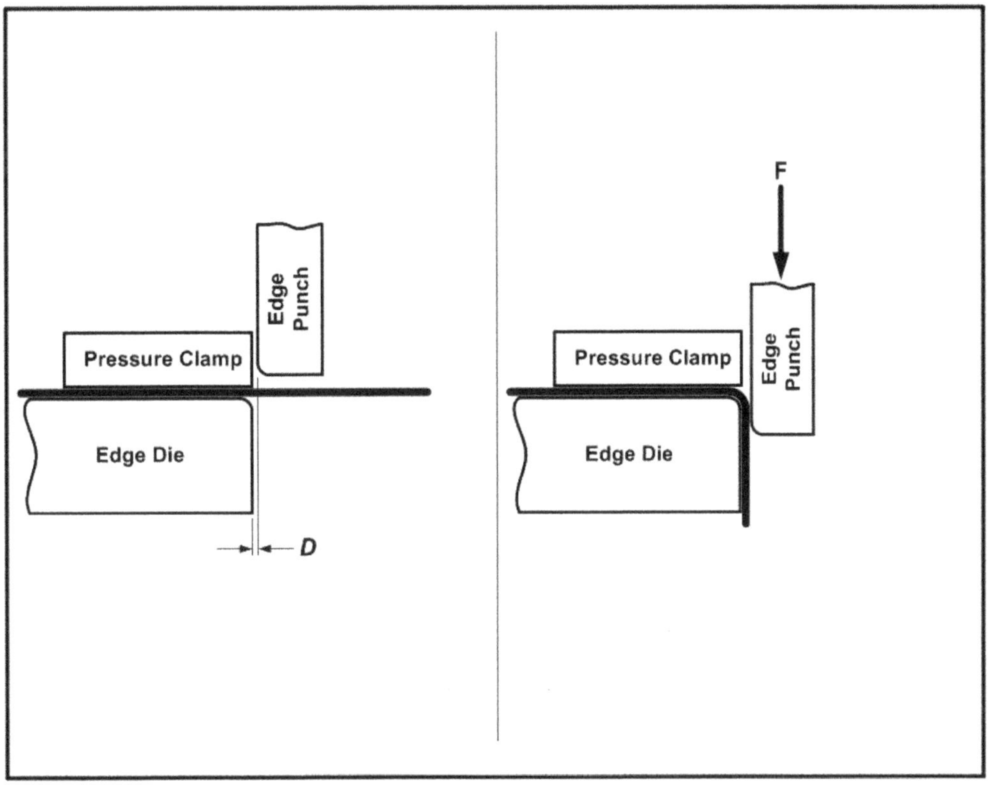

Figure 13.11 Sheet Metal Edge Bending Setup

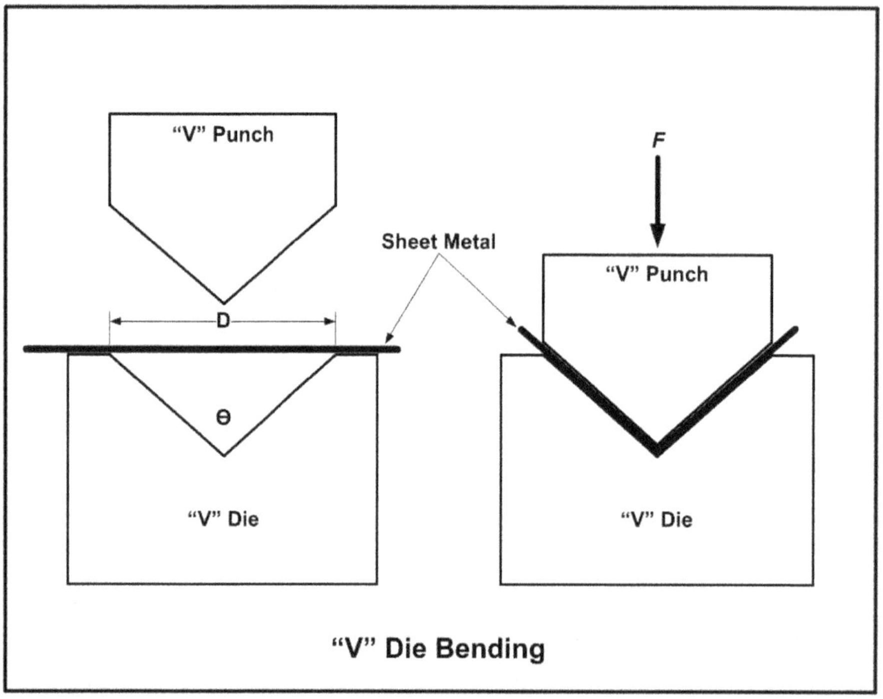

Figure 13.12 Sheet Metal "V" Die Bending Setup

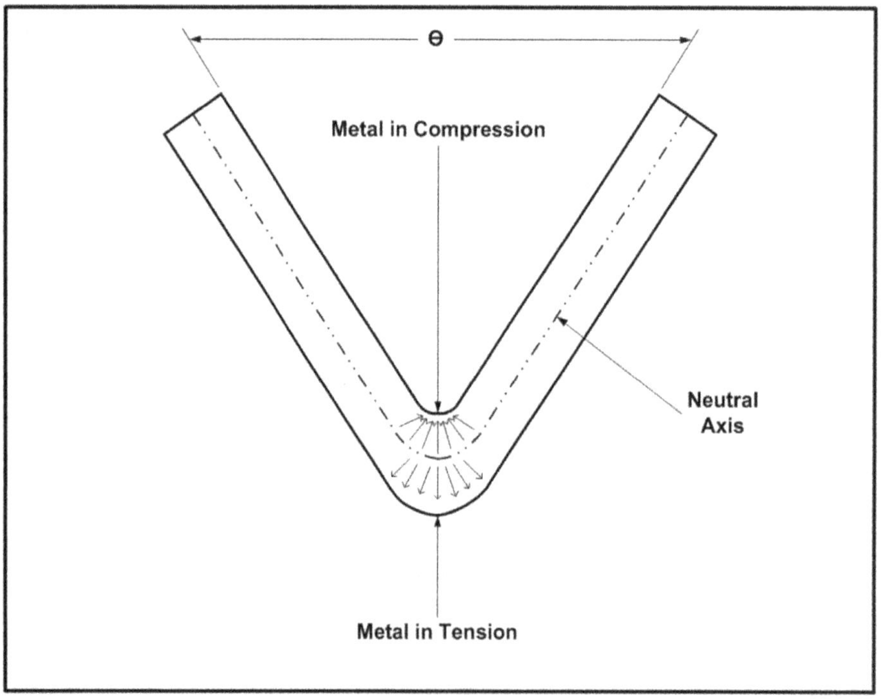

Figure 13.13 Tension and Compressive Forces on a Formed Sheet Metal Component

understood in order to perform the forming correctly. Figure 13.14 shows a typical simple bend and the factors requiring attention. The first is what is called the Bend Allowance (BA). This is the factor that needs to be accounted for when calculating how large the blank of material needs to be prior to forming. A certain radius

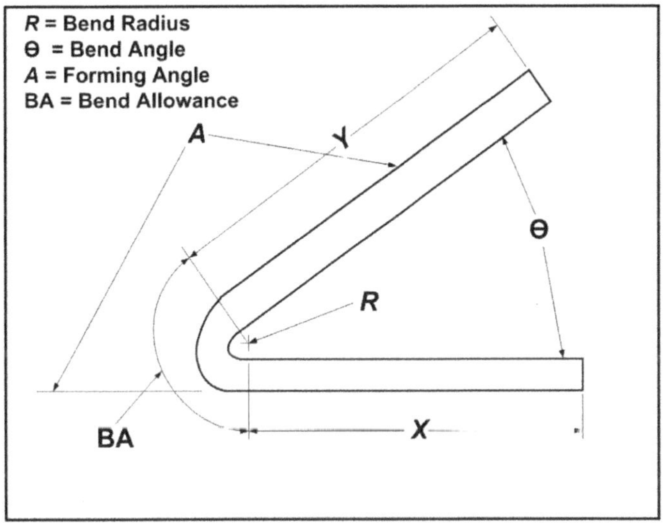

Figure 13.14 Sheet Metal Bending Factors

is always required in a formed part and is measured from the inside of the bend. No sheet metal can be bent to a sharp corner, and if the radius is too small, the material will have extremely high stress concentrated in the bend area which can cause a failure. The angle that the form needs to take (Θ) is what the part print calls for where the forming angle (A) is how much the material needs to be formed in order to obtain Θ.

The first factor that needs to be calculated is the bend allowance (BA). This is the additional factor that needs to be added when a flat blank is being produced prior to the actual bending. The equation for calculating the bend allowance is as follows:

$BA = 2\pi A/360 (R + (K_{ba} t))$

BA= Bend Allowance

R = Bend Radius

t = Material Thickness

Θ = Bend Angle

A = Forming Angle

K_{ba} = Stretching Estimate Factor

If $R < 2t$; $K_{ba} = 0.33$

If $R \geq 2t$; $K_{ba} = 0.50$

Forming Angle: $A = 180 - \Theta$

After the bend allowance has been calculated, the amount of force required to perform the bend is next. The amount of force required is critical for selecting the proper size press for forming the parts. The following equation is used for calculating bending forces of sheet metal:

$F = K_{bf} TS\, w\, t^2 / D$

F = Bending Force

TS = Tinsel Strength

w = Width of the material

t = Material Thickness

D = Die Opening

K_{bf} = Bending Constant

For Edge Die: $K_{bf} = 0.33$

For "V" Die; $K_{bf} = 1.33$

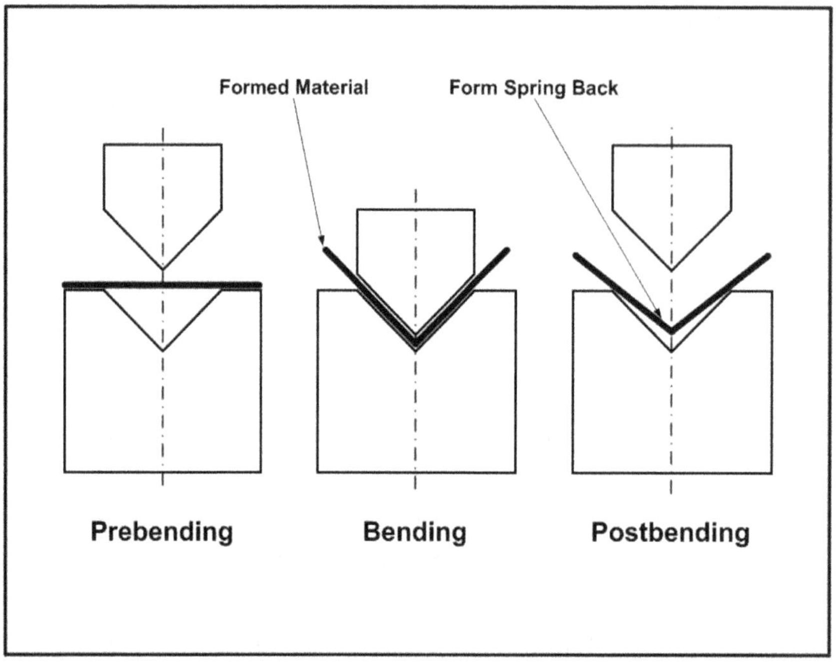

Figure 13.15 Sheet Metal Spring Back during Forming and Bending

The die opening is the amount that the material requires in order to perform the bending properly. This is the same for both edge die and "V" die setups, and the equation for calculating this is as follows:

$D = 2(R + t) \sin \Theta/2$

D = Die Opening

R = Bend Radius

t = Material Thickness

Θ = Desired Angle

An important factor that will always occur during any sheet metal forming is something called spring back. This is where the metal is formed, but will try to return back after the forming is completed. Spring back is a very difficult factor to account for as it is dependent on several conditions. These being the material type, the thickness, the angle it is being formed to, and whether it is being formed in cold, warm, or hot working. Figure 13.15 shows how the material will spring back after being formed.

Example Problem

A sheet metal part shown in Figure 13.16 will be bent to the angle shown in a "V" die. The material width is 2.125 in. with a tinsel strength of 175,500 psi. What would the size of the blank need to be prior to bending, and how much force will be required?

Step 1: Determine the forming angle.

$A = 180 - \Theta$

$A = 180 - 34.5$

$A = 145.5^\circ$

Step 2: Calculate the bend allowance.

$BA = 2\pi\, A/360\, (R + (K_{ba}\, t))$

$BA = 2\pi\, 145.5/360\, (0.250 + (0.50 \times 0.125))$: $2t = 0.500$: $K_{ba} = 0.50$

$BA = 0.7936$ in.

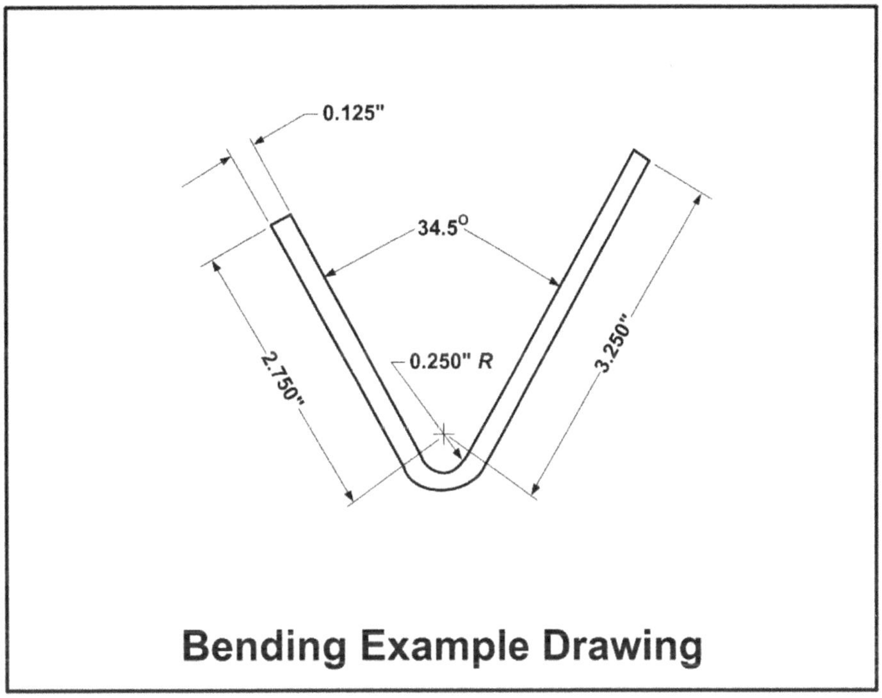

Figure 13.16 Bending Example Part Print

Step 3: Determine the size of the blank.

$L_b = L_1 + L_2 + \text{BA}$
$L_b = 3.250 + 2.750 + 0.7936$
$L_b = 6.794$ in.
$W_b = 2.125$ in.

Step 4: Calculate the die opening.

$D = 2\,(R + t)\sin \Theta/2$
$D = 2\,(0.250 + 0.125)\sin 34.5/2$
$D = 0.2224$ in.

Step 5: Calculate the bending force.

$F = \text{K}_{\text{bf}}\,\text{TS}\,w\,t^2/D$
$F = (1.33)\,(175{,}500)\,(2.125)\,(0.250^2)/0.2224$: "V" Die; $\text{K}_{\text{bf}} = 1.33$
$F = 34{,}846.59$ pounds or 17.42 tons

13.3.4 Deep Drawing

A deep drawing process is one that takes sheet metal and shapes into a deep configuration. Products such as metal cups are all fabricated in this manner, and the process is performed utilizing warm working rather than cold working. Depending on how deep a drawing is required to create the geometric shape, it may require several stages and operations as to not stretch the material too much in one drawing. Also, a lot of lubrication is used to allow the sheet metal to be able to exit the die after being shaped. Another important factor that needs to be considered is the material that will be drawn. The material needs to be ductile, allowing it to be stretched into the desired shape when drawing. If a metal that is brittle or hard is used, it will tear and crack when it is drawn out. Figure 13.17 illustrates how the deep drawing process operates, and normally several die sets are used as the drawing increases from one die to the next.

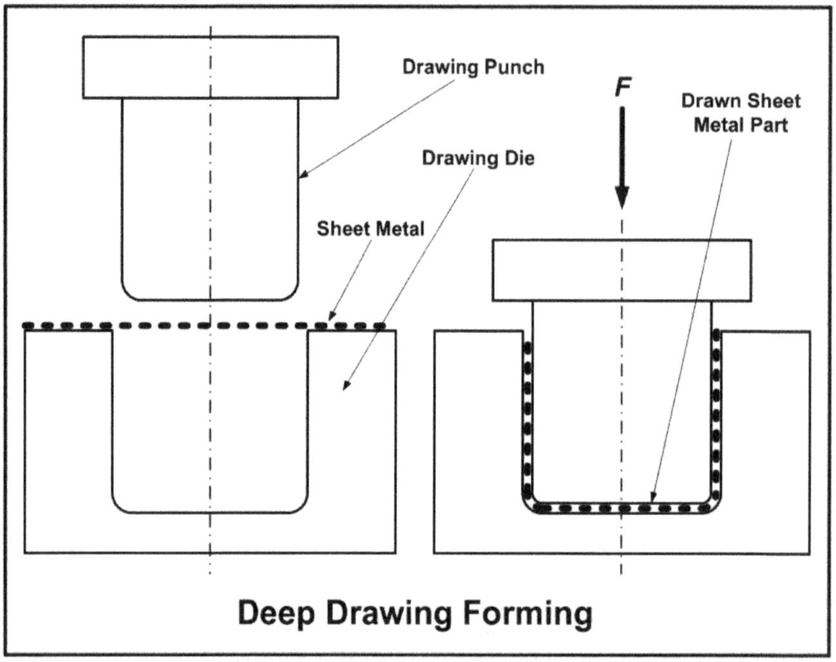

Figure 13.17 Sheet Metal Deep Drawing Setup

13.3.5 Stretch Forming

Stretch forming is a deformation process that is performed on sheet metal. The material is clamped at both ends and heated into the warm working range. From there, it is stretched in both directions as a die is pushed into the center and the metal takes a form. As with drawing, the material being utilized in stretch forming needs to be ductile in order for the forming to take place without cracking. Figure 13.18 shows how the stretch forming process is performed.

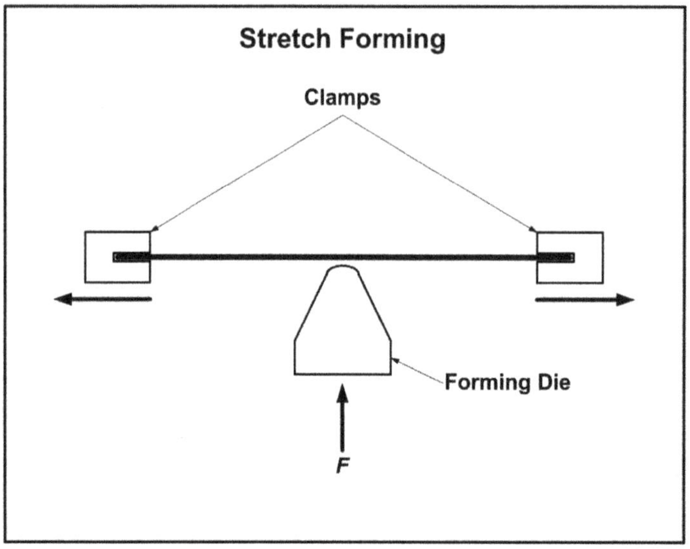

Figure 13.18 Sheet Metal Stretch Forming

13.3.6 Spin Forming

Spin forming is yet another sheet metal deformation process that is performed in the warm working range. A round piece of sheet metal with a hole in the center is clamped to a mandrel, which is also heated and rotated. Next, a forming tool is run across the mandrel shaping the sheet metal disk against it, allowing the sheet metal

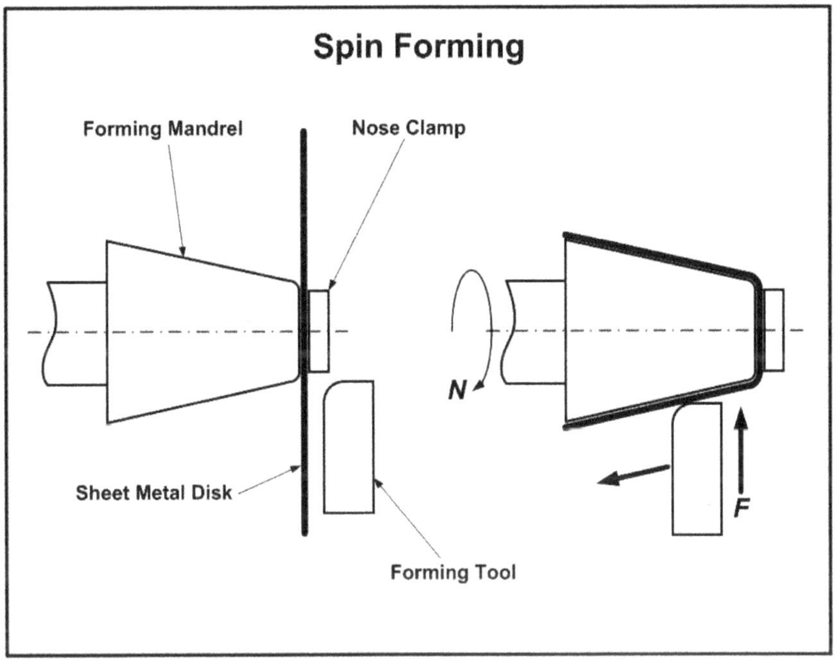

Figure 13.19 Spin Forming of Sheet Metal Components

disk to take the desired shape. With any type of deformation of sheet metal, the material requires ductility in order to take the geometric shape without any damage. Another consideration for spin forming is that the finished component will have a hole in the center, and this needs to be acceptable if the spin forming process is to be utilized. Figure 13.19 illustrates how the spin forming process is performed.

13.4 DESIGN CONSIDERATIONS

Sheet metal working is very versatile and there only a few items that need to be considered to perform the processes successfully. These considerations for properly designing sheet metal components are covered as follows.

Geometry: Sheet metal components can be stamped out with sharp 90° corners. Because there is no material flowing performed during the process as with solidification processes, the parts can be produced with sharp corners. Also, there is no limits on the two-dimensional geometry of the parts. The punch and die can be fabricated to any required configurations.

Heavy Burrs: As explained earlier, a stamped out component will always have a heavy burr caused from the fracture that occurs when being stamped. If these burrs are not acceptable on the completed part, then it will require deburring in a secondary operation.

Bending: For bending a sheet metal, it requires a certain radius where the bend will be implemented. A sharp forming die will not bend the material, but rather cut into it. There is a minimum radius that can be used, but the larger the radius, the less stresses that will be present in the bend. Also, if the material is going to be bent into an extremely tight angle, failure will occur. If extremely tight bends are required, a warm or even hot working process should be selected.

Spring Back: When a sheet metal component is bent or formed into any configuration other than flat, there will be spring back from the shape it was formed to. Knowing this, the material will have to be overformed so that when it springs back, it will be to the desired geometric shape desired.

Work Hardening: When forming sheet metal, especially when bending, there will be stresses in the formed areas. This causes work hardening, which was explained in Chapter 11. Work hardening can be either a positive or a negative trait in the formed components. Work hardening will increase the strength of the formed area, but will also cause it to be more brittle. If this is a negative feature, the component parts will require an annealing operation to eliminate the stresses and work hardening.

Material Selection: Material selection is important for forming operations. There needs to be a level of ductility in order for forming to be accomplished successfully. These is especially critical for deep drawing, stretch forming, and spin forming where the configurations will be dramatically altered.

Production Quantity: Sheet metal working tooling is very expensive. To make this a viable process, there needs to be a sufficient production quantity to make it cost-effective. Also, a stable design is required so that the tooling does not require alterations or replacement.

13.5 SUMMARY

Sheet metal working can produce very complex geometric shapes in a timely manner. With production rates capable of stamping out 4 to 1,500 parts per hour, it can be the most effective and efficient method for thin sheet metal parts. Almost all sheet metal parts are fabricated in this manner, and if the production quantities are high enough, the process can be almost completely automated. Understanding what the processes will yield during designing will render successful fabrication results.

13.6 QUESTIONS AND PROBLEMS

Questions

1. What are the three stages that sheet metal will go through when it is being stamped out of a strip?

2. Explain how the side of a stamped out component will look, and what causes this type of surface finish?

3. What is the difference between cutting and shearing sheet metal?

4. When stamping out sheet metal, what is the difference between blanking and piercing?

5. What is a clearance in blanking and piercing and which is it implemented on, the punch or die?

6. When calculating the force required for stamping out sheet metal, what is the blanking length and why is this important?

7. Why is the opening in the die of the exact dimensions that are required from the part drawing?

8. When bending sheet metal, what stresses will be present in the bent area?

9. What is the difference between the bend angle and the forming angle?

10. What is spring back, and what causes it?

11. What is the difference between edge bending and "V" die bending?

12. For bending sheet metal, what is the die opening?

13. What is meant by a bend allowance, and why is it important to calculate properly?

14. If a part is going to be formed utilizing the deep drawing process, what heat range is it performed in and why?

15. Can a part requiring a very deep draw be fabricated in one die? If not, why?

16. What material considerations need to be taken into account when performing deep drawing, stretch forming, and spin forming?

17. Strain hardening will occur when forming sheet metal. How can this be a positive trait, and if it is not desired, how can it be eliminated?

Problems

1. A sheet metal component is going to be blanked out of soft cold-rolled steel that is 0.03125 in. thick. What would the proper clearance need to be?

2. A blanking operation is going to be performed on 1100S aluminum alloy that is 0.145 in. thick. The desired shape is a rectangle measuring 3.583 in. long by 2.628 in. wide. What would the proper punch and die sizes need to be?

3. For the data in problem #2, if the material has a tinsel strength of 128,750 psi, what force would be required to blank it out?

4. A part has been blanked out of full hard stainless steel that is 0.0625 in. thick with a shear strength of 195,600 psi. Two holes with a diameter of 0.4375 need to be pierced into it and both holes will be pierced at the same time. What would the punch and die sizes be, and how much force would be required?

5. A progressive die is going to be used to create a tamped component. This is a five-station die where each station is performing a given operation to the sheet metal strip being fed through it. Figure 13.20 shows the strip layout and what is being performed during each operation, and Figure 13.21 shows the details of each station. If the material thickness is 0.1093 in. having a tinsel strength of 124,750 psi, what would be the required force for each station and the total force needed?

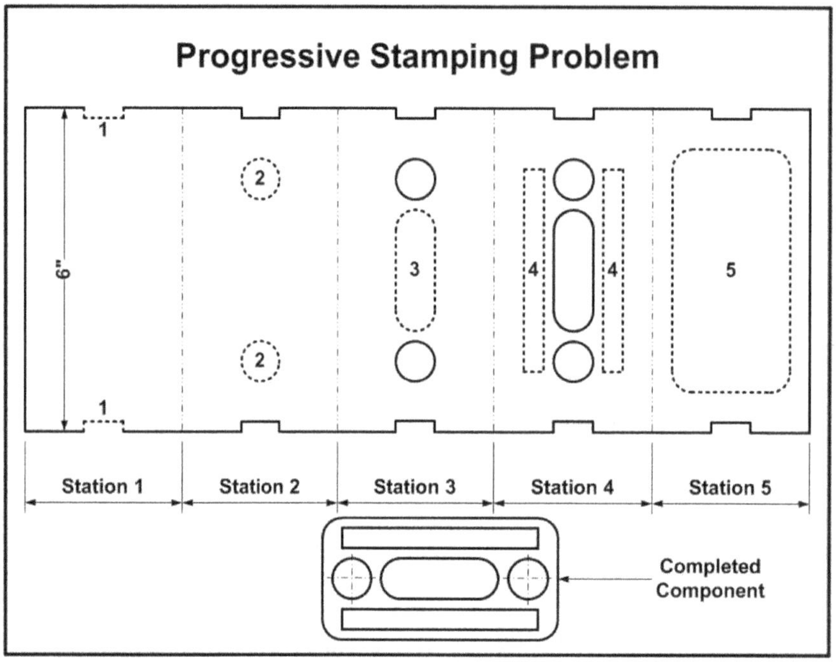

Figure 13.20 Progressive Stamping Problem Number 5

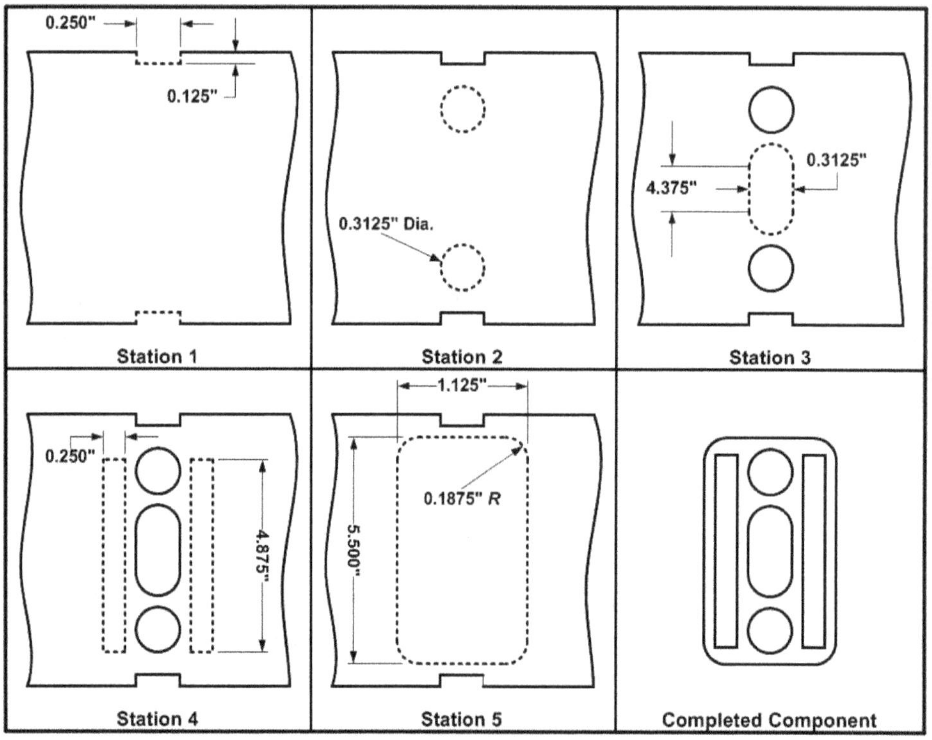

Figure 13.21 Progressive Stamping Problem Number 5 Details

6. A 62° bend is going to be placed in the center of a blank. The thickness of the material is 0.125 in. and the width is 2.125 in. with a tinsel strength of the material of 135,625 psi. A "V" die will be used to form the angle and from the center of the 0.1875" radius, each end of the formed part needs to be 3.750 in.. What size should the blank need to be, and how much force is required to bend the blank?

7. A blanked component will be formed with two angle bends. Figure 13.22 shows the final configuration required. The blank is 3.500 in. wide with a tinsel strength of 173,750 psi. Each bend will be performed in a separate operation and will be performed in a "V" die tool. What would the size of the blank be, and how much force is required for each bend?

8. A finished component is shown in Figure 13.23. There are three separate operations to be performed. The first is stamping out the blank, the second is to pierce the four holes at the same time, and the third is to bend it to the desired angle. The material is brass and has a tinsel strength of 115,225 psi. Determine everything required to perform the three operations: pinch and die sizes, blank sizes, blanking and piercing forces, and bending force utilizing a "V" die tool.

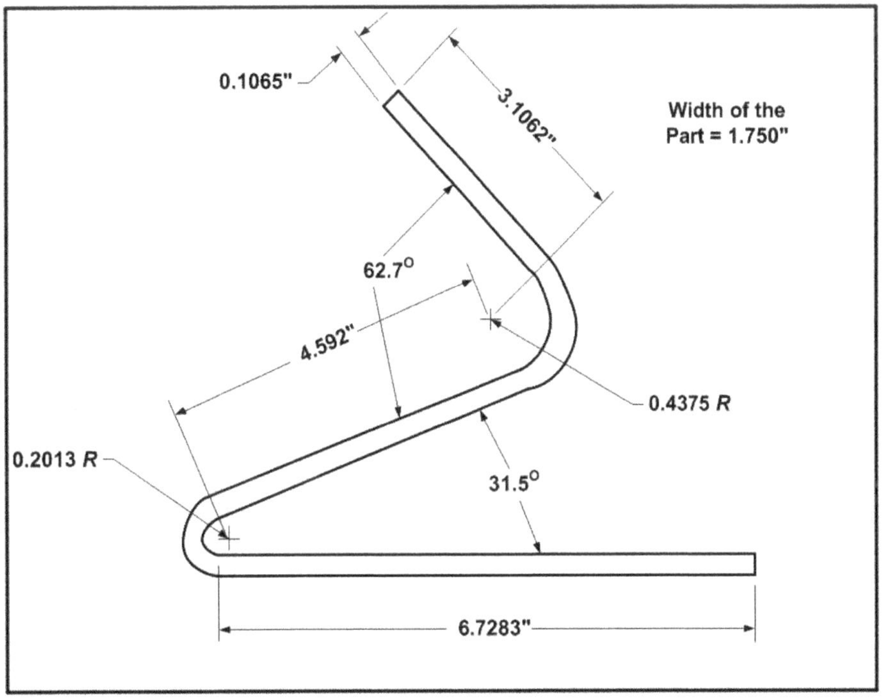

Figure 13.22 Bending Drawing for Problem 7

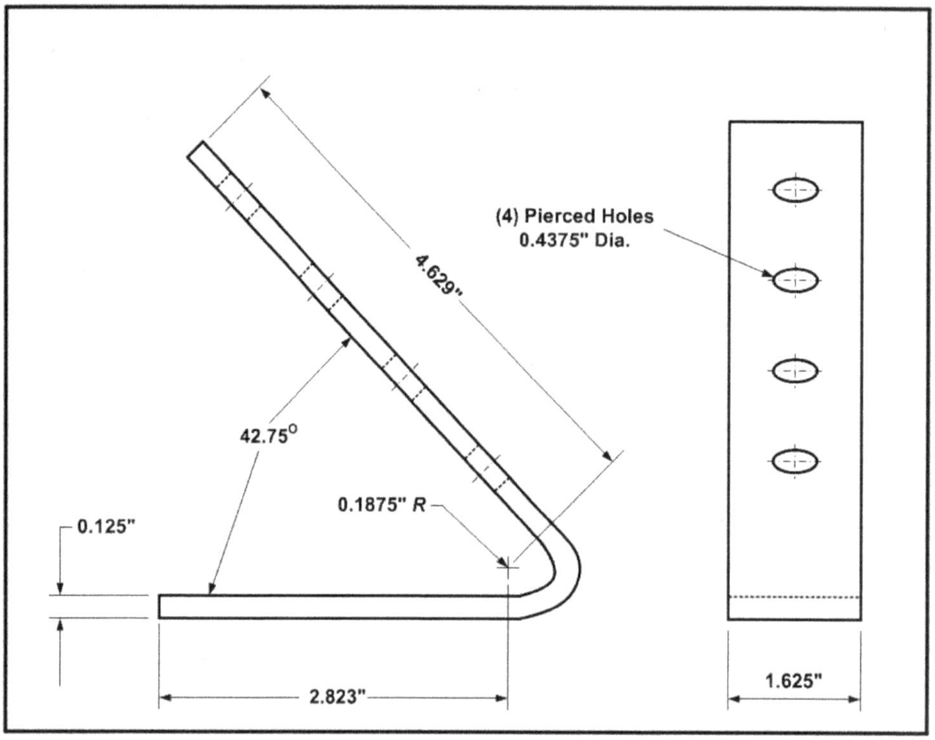

Figure 13.23 Drawing for Combination Problem 8

SECTION 4

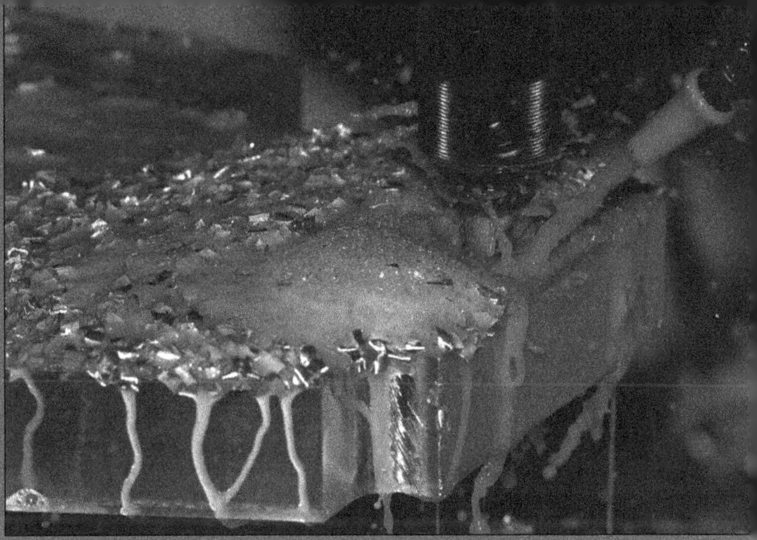

MATERIAL REMOVAL PROCESSES

The material removal processes create geometric configurations by beginning with a piece of material and then extract from the block to create the desired shape. As the material is removed from various processes, it begins to take shape until it reaches the final geometric shape. Depending on the process selected, the material can be removed by machining with metal cutters where the material is cut away, or by the use of abrasive machine where instead of a metal cutter, hard abrasive particles will remove the material. Some of the nontraditional processes will remove the material either by burning, or vaporizing, or dissolving. Regardless of how the material is removed, these processes always begin with a piece of material larger than the finished size.

A major advantage of the material removal processes is that they are capable of holding the closest tolerances and can provide the smoothest surface finishes of any other part fabrication process. Tolerances of ± one millionth of an inch and surface finishes smoother than glass are attainable. However, the tighter the tolerances and the smoother the surface finishes, the more time-consuming the processing will be.

A drawback besides being time-consuming is that the material removal processes are extremely wasteful of material. Unlike the solidification processes covered earlier where an exact amount of material is consumed in the fabrication, material removal will start with a larger volume and then extraction of the material as the component is being shaped.

This section covers a variety of material removal processes, from the conventional to the nonconventional. The positive and negative factors of these processes are discussed along with how they are a major role-player in manufacturing. Also included in the section are the production economics and how all the fabrication processes can be analyzed for efficiency and cost-effectiveness.

CHAPTER

14

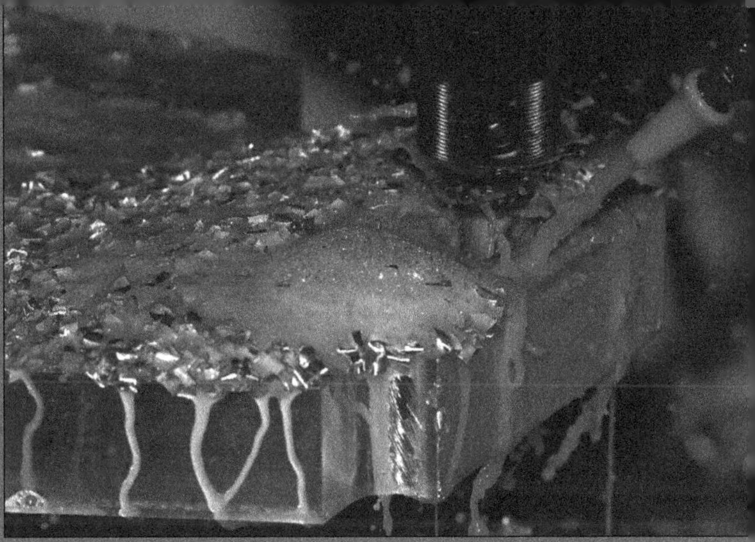

Material Removal Principles

14.1 INTRODUCTION

The material removal processes create component parts by extracting the material, rendering the desired geometric configurations. A piece of material is selected and areas of it are removed through various processes during the creation of the desired shape. Figure 14.1 illustrates how a block of material is transformed into a component part with the removal of material from it. Depending on the process being utilized, two terms remain constant to define them. The first is the *Machine Tool*, which is the unit that will hold the workpiece and deliver the power required to remove the material. The second term is the *Cutting Tool*, which is the implement that actually removes the material during processing. It is the combination of these two items that is required in order to perform material removal in manufacturing.

Material removal is the most versatile process group for the fabrication of parts. With a few exceptions, most geometric configurations can be attained through a wide variety of processes within this group. As with any of the manufacturing processes, there are both advantages and disadvantage that encompass them. These are described as follows.

Material Removal Advantages

- *Wide Variety of Materials*: Just about any solid material can be shaped with material removal processes. Even extremely hard materials like hardened steel and ceramic can be processed and have the material removed. However, the harder the material, the longer the process will take.

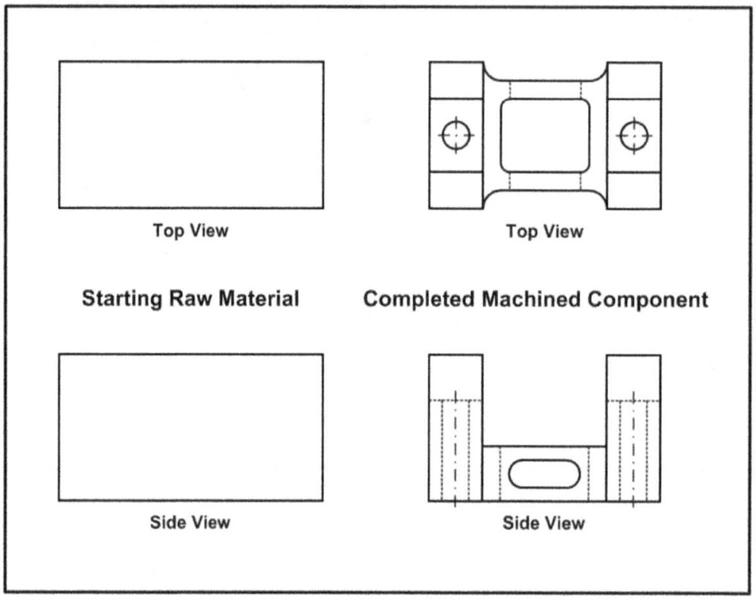

Figure 14.1 Component Part Created from the Extraction of Material

- *Geometric Shapes*: With very few exceptions, almost any geometric shape can be produced with material removal. The more complex the geometry of the part being shaped, the more time it will take to accomplish the end result.
- *Dimensional Accuracy*: A major advantage of material removal is its ability to maintain very close tolerances. With just general machining, ±0.001 in. is attainable, and with other processes, it is not impossible to produce component parts of ±1 millionth of an inch.
- *Surface Finishes*: The best surface finishes possible can be attained with the removal processes. It is not uncommon to produce smooth surface finishes that are smoother than glass.

Material Removal Drawbacks

- *Wasteful of Material*: As these processes involve extracting material from a larger workpiece, a vast amount of waste is produced. For the majority of the processes, this is in the form of material chips, which at the most, can only be recycled but not reused.
- *Time-Consuming*: Material removal is the most time consuming of any component fabrication process. Even simple geometric configurations take more time than it would with other processes. The more complex the configuration, the more the time required for fabrication. Also, the tighter the tolerances and the smoother the surface finish requirement, the longer it will take to accomplish them.

14.2 MATERIAL REMOVAL CATEGORIES

The material removal processes are divided into two main categories, determined on how they remove the material from the workpiece. These categories are *Traditional Processes* and *Nontraditional Processes* and are explained as follows.

- *Traditional Processes:* A traditional process is one in which, when the material is being removed, chips are created during the shaping process. These processes are further segmented into two subcategories, which are *Machining Processes* and *Abrasive Processes* and are explained as follows.

- *Machining Processes*: These processes remove the material from the workpiece utilizing cutters made from a variety of materials. As the cutter removes the material, chips are formed and directed away from the cutting surfaces. Machining can remove massive amounts of material and is used both for roughing out a component part and to finish shaping it, depending on the tolerances required. The types of processes included in this group are *Milling, Turning, Drilling, Broaching, Sawing,* and *Shaping,* which are covered in Chapter 15.
- *Abrasive Processes*: Abrasive processes utilize hard abrasive materials for removing the material. Although difficult to see, chips are formed as the abrasive matter removes the material from the workpiece. This process will remove very small amounts of material and is utilized to attain both extremely close tolerances and extremely fine surface finishes. The types of processes included in this group are *Grinding, Honing, Lapping, Polishing,* and *Superfinishing,* which are covered in Chapter 16.
- *Nontraditional Processes*: The nontraditional processes are those that remove the material in unique ways. During the removal of the material, no chips are formed, but rather the material is either dissolved or vaporized. They are capable of producing geometric configurations that the traditional processes either cannot or would take substantially longer to produce. The types of processes included in this group are *Ultrasonic Machining, Electro-Discharge Machining, Water Jet Cutting, Laser Cutting, Plasma Cutting,* and *Chemical Etching,* which are covered in Chapter 17.

14.2.1 Types of Material Removal Shaping

The removal of material during either a traditional or a nontraditional fabrication process is performed in one of two methods. Each will produce the desired configurations but the methods in which the geometric shapes are produced are very different in nature. These two methods for creating shapes in material removal are *Generating* and *Forming* and are explained as follows.

Generating: Generating of a geometric shape on a workpiece is performed by the trajectory of a standard cutting tool. The tool is moved along the workpiece, and depending on how the tool is moved, its trajectory will determine the configuration of the workpiece being shaped. Figure 14.2 illustrates how the standard tool trajectory shapes the workpiece.

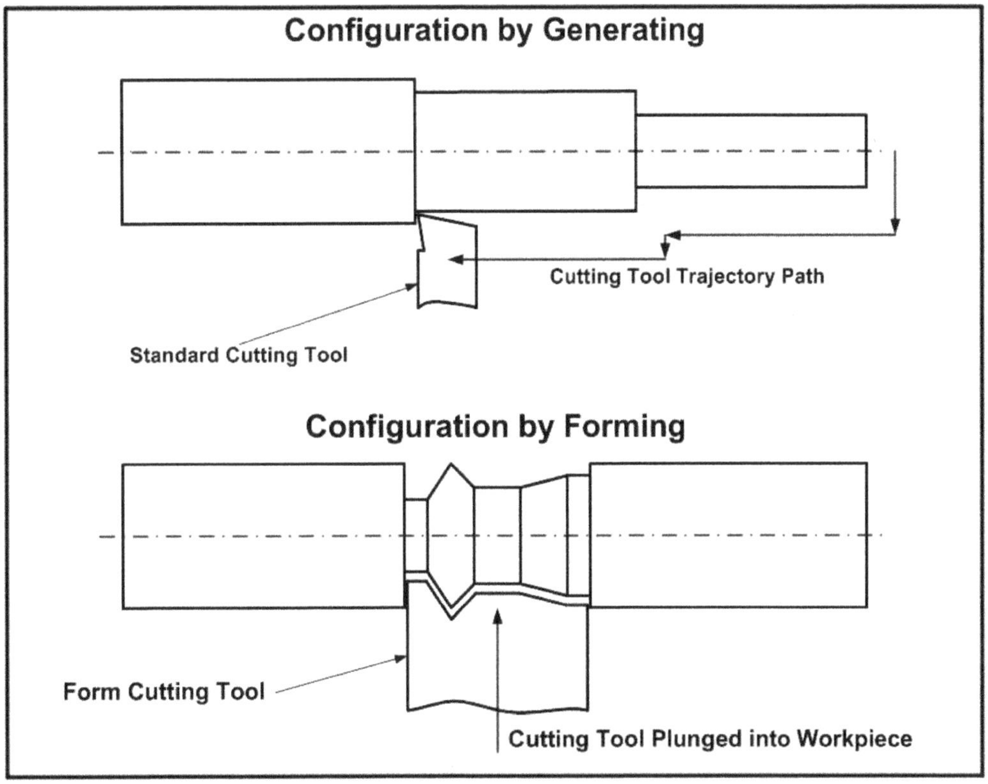

Figure 14.2 Part Configurations Created from Generating and Forming Operations

Forming: A geometric shape created by forming is totally dependent on the configuration of the cutting tool being utilized. Here, the cutting tool is created to the exact shape that is desired and rather than moving alone the workpiece, it is plunged into the workpiece to produce the geometric shape desired. Figure 14.2 illustrates how the form cutter is fed into the workpiece to shape the desired configuration.

14.3 CUTTING TOOL PRINCIPLES AND TECHNOLOGY

Within the material removal arena, cutting tools are the most important factor for creating part configurations. Several considerations need to be analyzed in order to select the proper cutting tool based on the process being used, the material being processed, and the cutting conditions of the operation. The geometric shape and the material the cutting tools are made from all play a major role in a successful selection removal process. These factors are described in detail as follows.

14.3.1 Cutting Tool Geometric Features

The cutting tools utilized in machining all share four common features. These are the *Lead Angle*, *Rake Angle*, *Clearance Angle*, and the *Nose Radius*. Depending on the material being machined, they may vary greatly, but in order for the tool to remove the material, all four of these features will be present. Each of these features is described as follows.

Lead Angle: The lead angle is that angle in which the cutting tool enters the workpiece during cutting. In the majority of the cutting applications, this angle will be positive where the tip of the tool makes contact with the workpiece first, but in some cases, depending on the material being removed, it can be negative. Figure 14.3 illustrates the lead angle of the cutting tool entering the workpiece in the top view.

Rake Angle: The rake angle is that angle in which the cutting tool is flowing the chip away from the workpiece during cutting. This angle can vary greatly from a positive or a negative angle, depending on the material being processed and the overall cutting conditions. Figure 14.3 illustrates this angle of the cutting tool in the side view.

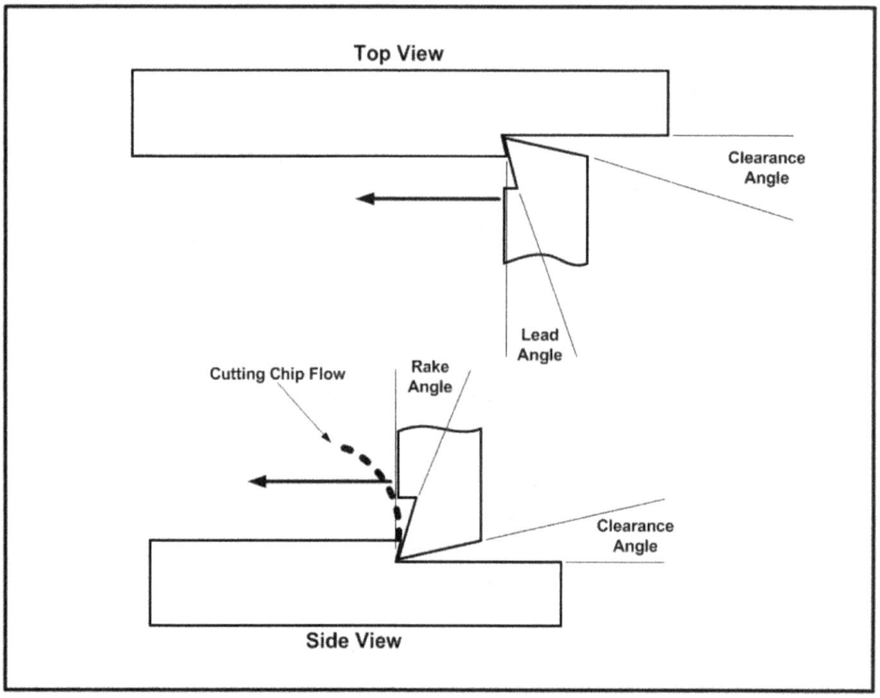

Figure 14.3 Cutting Tool Angles and Their Functions in Machining

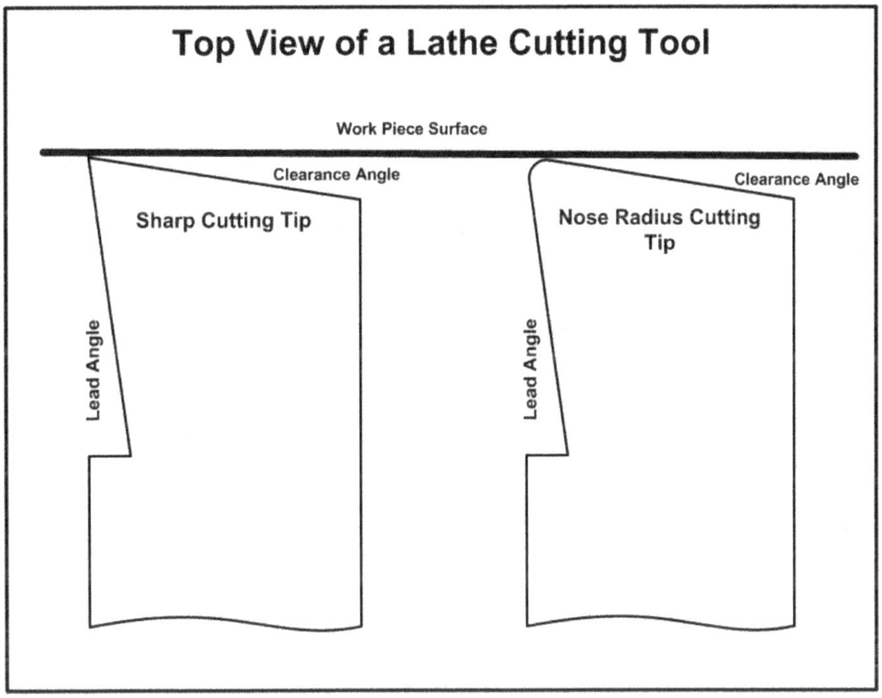

Figure 14.4 Cutting Tool Nose Radius

Clearance Angle: The clearance angle is that angle in which the cutting tool is clearing the workpiece to avoid and scratching of the material and to allow the rake angle and the lead angle to function properly. It is implemented for both the rake and lead angles and is shown in Figure 14.3 in both the top and side views.

Nose Radius: The nose radius is a feature that removes the sharp point on the end of the cutting tool to increase both the strength of the cutting tool and the tool life. Depending on the application and material being processed, this radius can be extremely small or large, ranging from a radius of 0.005–0.125 in. Figure 14.4 shows a typical lathe cutting tool with both a sharp point and a nose radius.

14.3.2 Cutting Tool Materials

Cutting tools for machining can be constructed from a variety of materials. Depending on the material being machined and the process for machining being utilized, selection of the proper cutting tool is critical to the operation. Some materials are extremely hard and brittle, offering little to no ductility. Other materials will posses less hardness but will be more ductile in nature. When selecting a cutting tool, the selection is a balancing act in order for the cutting tool to posses the proper properties. The three criteria that need to be evaluated for cutting tools are Toughness, Hot Hardness, and Wear Resistance. Figure 14.5 shows these three criteria and how they need to be balanced in order to optimize the machining operation.

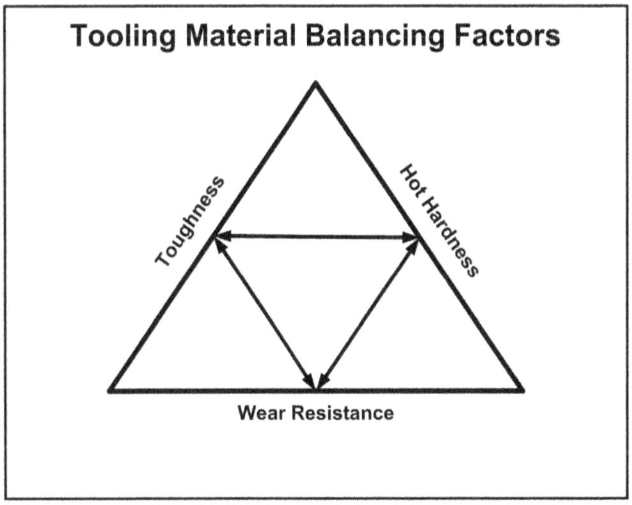

Figure 14.5 Tooling Material Balancing Factors of Toughness, Wear Resistance, and Hot Hardness

Toughness: The toughness of a material depicts how much energy it can absorb while cutting

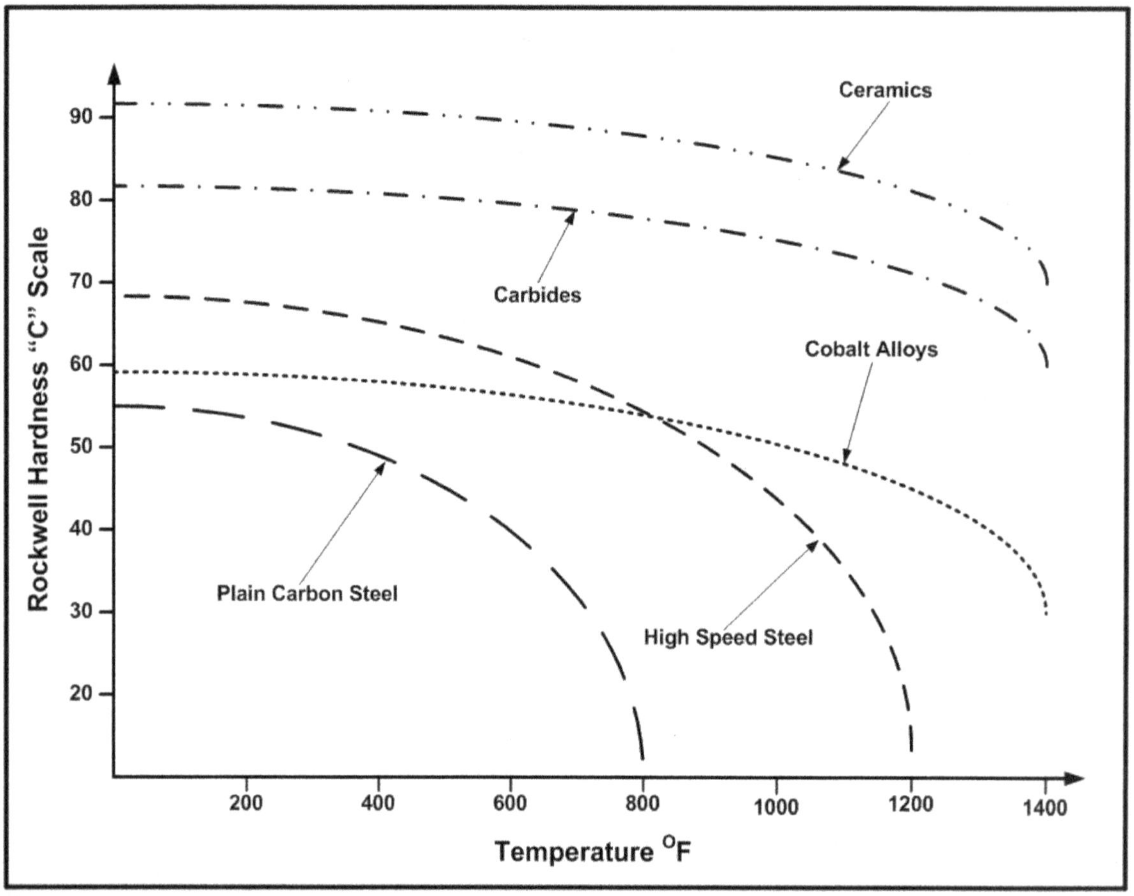

Figure 14.6 Tooling Material Hardness vs. Temperature Chart

the workpiece. As the cutting tool makes contact with the workpiece, the toughness will absorb the shock, causing it not to break. Toughness is in direct contrast with wear resistance and, in some cases, in direct contrast to the hot hardness.

Hot Hardness: The hot hardness of a material is its ability to maintain the hardness of the cutting tool under heat. As friction and heat are generated at the cutting edge of the tool, the tooling material will start to anneal causing it to lose the hardness required to perform the cutting of the workpiece. Figure 14.6 shows a chart of various cutting tool materials with the starting hardness and the loss of hardness as the temperature rises.

Wear Resistance: Wear resistance is determined by the hardness of the cutting tool material. All materials will possess a certain hardness, which allows them to be utilized as a cutting tool. However, as the hardness of the material increases along with the wear resistance, the toughness of the material decreases. Looking at some materials, ceramics will be extremely hard and provide excellent wear resistance but will not be able to absorb energy very well when they come in contact with the workpiece. Likewise, common carbon steel will have very low hardness and wear resistance but can absorb energy extremely well.

The balancing act of toughness, hot hardness, and wear resistance needs to be analyzed in order to select the proper cutting tool for the operation. Each of these factors is evaluated against the material being machined, the operation process being performed, and the geometry of the cutting tool.

14.3.3 Cutting Tool Failure

Failures can and will occur with all cutting tools. These failures in some instances can be avoided, prolonging the life of the cutting tool. There are three categories of cutting tool failures and some of the probable causes are listed as follows.

Fracture Failure: A fracture failure is where the cutting tool breaks during processing. There are several conditions that can cause a breakage of the cutting tool with the main conditions being: *Heavy Cuts, Heavy Feeds, Improper Cutting Tool Configurations*, and *Wrong Cutting Tool Material.*

Temperature Failure: A temperature failure is caused by applying extreme heat to the cutting tool, causing it to lose the properties of hardness. This is mainly caused by the friction created by the cutting tool as it removes the material during processing. Conditions that can result in a temperature failure are: *Lack of Coolant or Lubricant, Heavy Cuts, Heavy Feeds, Fast Rotational Speeds*, and *Wrong Cutting Tool Material.*

Gradual Wear: Gradual wear is the desired way for a cutting tool to fail. Regardless of the material that the cutting tool is constructed from, it will eventually wear out and need to be replaced. In any material removal process, if all the cutting conditions are established correctly, the wearing down of the cutting tool will be optimized and the tool life will be extended to the limit.

Expanding the failures of cutting tools, they are further divided into various categories, which are *Abrasion, Adhesion, Diffusion, Chemical Reaction*, and *Plastic Deformation.* Each of these can be avoided or greatly minimized by proper cutting tool selection and proper cutting conditions for the process. Each of these tool failures is covered as follows along with their probable causes.

Abrasion: If the workpiece, which is being cut, contains particles within it that are harder than the material, the cutting tool will experience abrasion at the cutting edge. An example that can cause this would be machining an aluminum sand casting. If during the casting process some of the casting sand got mixed in with the aluminum, during machining, the tool will come in contact with it, which can cause an abrasion failure during cutting.

Adhesion: Adhesion is where the cutting material actually adheres or welds to the cutting edge of the cutting tool. This is caused by improper cutting conditions with the main causes being: *Heavy Cuts, Heavy Feeds, Extreme Temperatures, Lack or Coolants or Lubricants.*

Diffusion: With a diffusion failure, the atoms responsible for the hardness of the cutting tool are being lost. Basically, the cutting tool is being annealed during the cutting process from extreme heating of the tool.

Chemical Reaction: A chemical reaction failure is one where the material of the cutting tool and the material being cut will have a reaction. A failure of this nature is normally rapid and is simply a result of a wrong cutting tool material being utilized in the process.

Plastic Deformation: Plastic Deformation is a failure mode where the cutting edge of the tool is deformed or melted during the cutting operation. This is caused by the same factors as an adhesion failure, but also can be caused by the cutting tool being constructed of the wrong material for the material being machined.

14.3.4 Cutting Fluids

When performing material removal, cutting fluids are utilized to aid in the physical cutting of the materials. These are used in both traditional and nontraditional processes and are broken down into two main categories, which are *Lubricants* and *Coolants.* Most cutting fluids will perform as either lubrication or a coolant with a few exceptions that will perform both tasks. These cutting fluids will perform three tasks during the processing of the workpiece, which are as follows.

Lubricate the Cutting Tool: By lubricating the cutting tool, the friction is greatly reduced, which will maintain the cutting edge longer. Also, friction reduction will reduce the amount of heat that is generated during the cutting operations.

Cool the Cutting Tool and Workpiece: Cooling the cutting tool will aid greatly in reducing the amount of heat, thus reducing the hardness that will be lost from temperature failures. Cooling the workpiece is also an important factor. If extremely close tolerances are to be maintained and the workpiece is heated during processing, it will experience thermal expansion. Measuring a heated workpiece while being heated can lead to a false reading when the workpiece cools down.

Flush Away the Chips: In some automated machine tools, the coolant is applied in a flood manner in order to maximize the cooling of the workpiece and the cutting tool. This will flush the chips away from the work surface to avoid any scratching during machining and maintain the surface finish of the component part.

14.3.4.1 Types of Cutting Fluids

Cutting fluids come in a vast number of varieties depending on both the material being machined and the process being utilized. Some are natural fluids, while others are synthetic compositions. Each of the cutting fluids will fall into one of three categories, which are *Cutting Oils*, *Emulsified Oils*, and *Chemical Fluids*. Each of these is described as follows.

Cutting Oils: Oils are available in a multitude of varieties and consistencies. Depending on the material being machined, straight oils can be applied to lubricate the cutting tools during processing. Cutting oils will perform mainly lubrication and minimal cooling as oils are not very effective at dissipating heat.

Emulsified Oils: Emulsified oils are special oils that can actually mix with water. These are mixed with H_2O in the ratio of 30:1 (30 counts of H_2O to 1 count of oil.) The consistency is that of water and will perform both cooling and lubrication to the cutting tool and workpiece during machining.

Chemical Fluids: Chemical fluids are synthetic liquids that are mixed with water to a given count. This type of cutting fluid is mainly utilized in automated production equipment in a flooding manner where a vast amount is applied to the cutting tool and workpiece. The main function of the chemical fluid is to cool the cutting tool and workpiece with little to no lubrication properties.

14.3.4.2 Application Methods

Application of cutting fluids during processing can be performed by three basic methods. Depending on the process being utilized, each will perform the required function and is effective for the given operation. These methods are described as follows.

Manual Application: Manual application is performed either with a brush or with a spray bottle. The cutting fluid is applied with a brush on the surface of the workpiece before a cut is to be made or sprayed on the surface while the machining is being performed. This technique is mainly utilized for single part fabrication in manual machining operations. The cutting fluid is not recycled during the operation for manual application methods.

Misting Application: Misting is used for special processes that require very little cutting fluid and the process is removing extremely light amounts of material. The cutting fluid is misted out of a nozzle on to the work surface during the removing of the material. This allows the workpiece to remain at a low temperature during processing in order to maintain extremely close tolerances. Depending on the process utilizing a misting application, the cutting fluid can be either recycled through a closed loop system on the machine tool or discarded.

Flooding Application: Flooding is utilized in automated equipment where the cutting fluid is applied in vast amounts during the processing. Chemical fluids mixed with H_2O is mainly used for flooding and will cool the cutting tool and workpiece, along with flushing the chip away to maintain the desired surface finish. The amount of fluid being applied during flooding is measured in Gallons Per Minute (GPM) and is recycled through a filter system and reused in a closed loop system on the machine tool.

14.4 MATERIAL REMOVAL CUTTING CONDITIONS

For all machining processes, maintaining proper cutting conditions is critical for an efficient and effective operation. These conditions are all dependent upon three main criteria, which are *Material Being Processed*, *Machine Tool Utilized*, and the *Cutting Tool Selection*. The cutting conditions are dependent upon each other and will determine the optimum processing settings for a given machining operation. These cutting conditions are described as follows.

Cutting Speed (v): Cutting speed is defined as either how fast the material is passing in front of the cutting tool as in a turning operation or how fast the cutter is passing in front of the material as in a milling or drilling operation. This is given in units of Surface Feet Per Minute (SFPM) and is dependent on the material being cut.

Rotational Speed (N): Rotational speed is either how fast the drill or cutter is rotating or in a turning operation or how fast the material being turned is rotating. This is given in units of Revolutions Per Minute (RPM) and is dependent on the material being cut or the setup holding the workpiece.

Feed (f): Feed is a measure of how much material is being removed in a given condition. This is either how much material is being removed for each revolution of a drill or a turning part or how much is being removed by each tooth of a milling cutter. It is given in units of Distance/Revolution or Distance/Tooth. As with the aforementioned conditions, this is also material dependent.

Feed Rate (f_r): The feed rate is a factor of the feed and the rotational speed. This renders a value that shows how much the cutting tool is removing the material over a unit of time. The unit of measure for this is in./min.

Machining Time (T_m): Machining time is the time required to perform the material removal over a given distance. This is a factor of the length of the cut being performed and the feed rate, and the unit of measure is minutes.

Material Removal Rate (MRR): Material removal rate is a measure of the volumetric removal of material over a given time frame. This shows the efficiency of the operation and is given in units of Cu.in./min.

14.5 SUMMARY

Material removal is an extremely viable manufacturing process family and possibly one of the most utilized. Although it is time-consuming and wasteful of material, it is also capable of maintaining the closest tolerances and highest surface finishes of any manufacturing process. The combination of both traditional and nontraditional material removal processes allows this group to fabricate component parts to almost any geometric shape required. By selecting the proper cutting tools for any given process and operation and maintaining the cutting conditions, machining can be performed in a very effective manner and can render some of the highest quality component parts of any manufacturing process today.

14.6 QUESTIONS

1. What is the difference between a machine tool and a cutting tool?

2. What is the difference between generating and forming a geometric shape on a workpiece?

3. Traditional processes are divided into two categories. What are they?

4. What are the four common features of a cutting tool, and describe their functions during machining.

5. What are the three categories of tool failure?

6. Describe failure of a cutting tool by diffusion.

7. Cutting fluids perform three main functions. What are they?

8. What is the difference between cutting oils and emulsified oils?

9. What are the three criteria for determining the cutting conditions for a given operation?

10. In determining the cutting conditions, describe cutting speed and how is it measured?

CHAPTER

15

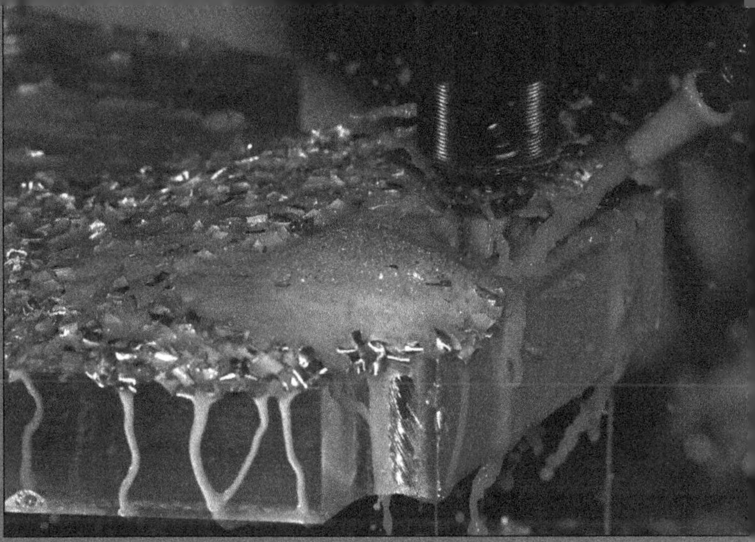

Machining Operation Principles

15.1 INTRODUCTION

Within the material removal category of manufacturing processes, machining is one of the most versatile and widely used processes. Component configurations of a wide variety can be produced to very close tolerances. Machining is capable of producing finished parts or can be utilized for rough cutting of material to be finished in more precision-based material removal processing. These processes can be performed either manually, semiautomated, or fully automated, depending on the configurations and production quantities.

The three main categories for machining segment of the component being fabricated based on the configurations required. These categories are *Turning*, *Drilling*, and *Milling*, which are covered in detail further on in this chapter. Other machining processes exist and are capable of rendering massive material removal or will provide special configurations to the workpiece. Regardless of the process being utilized, each will be performed with a cutting tool and on a particular machine tool and requires calculations in order to maximize the efficiency of the operation. In performing these critical calculations, three factors are selected based on the process being performed and the material being machined. These selected factors are *Cutting Speed*, *Feed*, and *Diameter of the Cutter or Material*. Each of these is described as follows.

Cutting Speed (v): The cutting speed is very material and cutter size dependent in machining. This is a measure of how much of the cutter is passing in front of the material for milling and drilling operations or how much material is passing in front of the cutter for turning operations. This is measured in units of *Surface Feet Per Minute (SFPM)*, and Figure 15.1 shows a table of cutting speed ranges for selected materials being machined.

Recommended Cutting Speed Ranges for Various Materials			
Material	**Speed (SFPM)**	**Material**	**Speed (SFPM)**
Alumimum	900 - 1,500	Plastics	900 - 1,800
Brass / Bronze	400 - 750	Plastics, Glass Filled	400 - 1,000
Copper	450 - 850	Refractory Alloys	125 - 350
Cast Iron, Soft	300 - 750	Steel, Low Carbon	300 - 600
Cast Iron, Hard	125 - 350	Steel, Medium Carbon	150 - 450
Ductile Iron	150 - 450	Steel, Hardened	50 - 225
Malleable Iron	300 - 550	Stainless Steel, Soft	250 - 500
Magnesium	1,100 - 1,600	Stainless Steel, Hard	75 - 275
High Nickel Steel	225 - 350	Titanium, Soft	175 - 400
Nickel Base Hi-Temp	50 - 125	Titanium, Hard	75 - 250

Figure 15.1 Cutting Speed Ranges for Select Material Groups

Feed (f): Feed is selected based on the material being processed or the overall process and setup being utilized. This factor renders the value of the distance per revolution that the cutting tool will move for turning and drilling operations or the distance per tooth for milling operations. The higher the feed for removal, the more aggressive the cut will be, and likewise, the lower the feed, the better the surface finish the process will render.

Diameter (D, D_o): It can be either the diameter of the milling cutter or drill being utilized, or for turning operations, it is the original diameter of the material being cut. This will be used during calculation of the values for determining the rotational speed of the cutter or the rotational speed of the material for turning operations.

Once the three machining factors have been determined, calculated values need to be obtained in order to correctly perform the operation in the most efficient manner. These calculated values are the *Rotational Speed*, *Feed Rate*, *Machining Time*, and *Material Removal Rate*. Each of these is described as follows.

Rotational Speed (N): Rotational speed refers to at what speed the milling cutter or drill will be rotating or at what speed the material will rotate during a turning operation. This is determined by both the cutting speed and the diameter of the drill/cutter or the original diameter of the material for turning operations. The value calculated here will determine what revolution per minute is optimal for the operation being performed.

Feed Rate (fr): The feed rate calculation determines how much the cutter will move over a given amount of time. This is a factor of the rotational speed and the feed rendering the value of distance/min for the operation being performed.

Machining Time (Tm): Machining time tells us how long a particular operation will take in minutes. This is determined by the calculated feed rate and the length of cut that is being performed.

Material Removal Rate (MRR): The material removal rate is a volumetric measure of the amount of material that is being removed per minute. Depending on the operation being performed (turning, milling, or drilling), it is calculated using the cutting speed, diameter, feed, or feed rate and renders the value of $in.^3/min$. This is explained in detail for the various operations later in this chapter.

15.2 TURNING PROCESSES

Turning is a process that will produce round component parts in manufacturing. It is performed on a machine tool known as an *Engine Lathe*, which is available in several different sizes and configurations. The workpiece is loaded into the lathe, and depending on the desired shape and configuration, several operations can be performed to render the completed component part. Turning operations involve the movement of two major axes, which are the "X" axis and the "Z" axis. The "X" axis moves the cutting tool in and out of the workpiece, while the "Z" axis moves the cutting tool along the workpiece. Figure 15.2 illustrates the tool movement along these two axes for a turning operation. There are several unique operations (see Figure 15.3) that can be performed within the turning arena, which are described as follows.

15.2.1 Turning Operations

Facing: A facing operation deals with cutting the end of the workpiece to remove the saw cuts, square the face to the workpiece diameter, and bring the part to the desired length. The cutting tool is moved into the workpiece along the "X" axis while the "Z" axis remains stationary. Figure 15.4 illustrates a facing operation where the cutting tool is moved from the outside of the workpiece to the center. One area that is extremely difficult to control during a facing operation is the cutting speed (v). If a constant cutting speed is to be maintained, as the tool moves inward on the workpiece, the diameter constantly changes, thus changing the parameters of the cut. To aid in this, the feed or the rotational speed needs to be constantly adjusted to compensate for the change in the workpiece diameter.

Contouring: Contouring is creating a workpiece from a standard cutting tool, which is called generating. The tool is moved along the workpiece and the geometry is created based on the trajectory of the tool. Figure 15.5 shows the standard single-point cutting tool contouring the workpiece along a given trajectory.

Forming: Forming is performed with a special cutter where the desired geometry is contained in the cutting tool itself. Here, the tool is not moved along the workpiece but rather being plunged in while the workpiece is rotating. Figure 15.5 shows a forming tool being plunged into the workpiece to create the desired geometric shape.

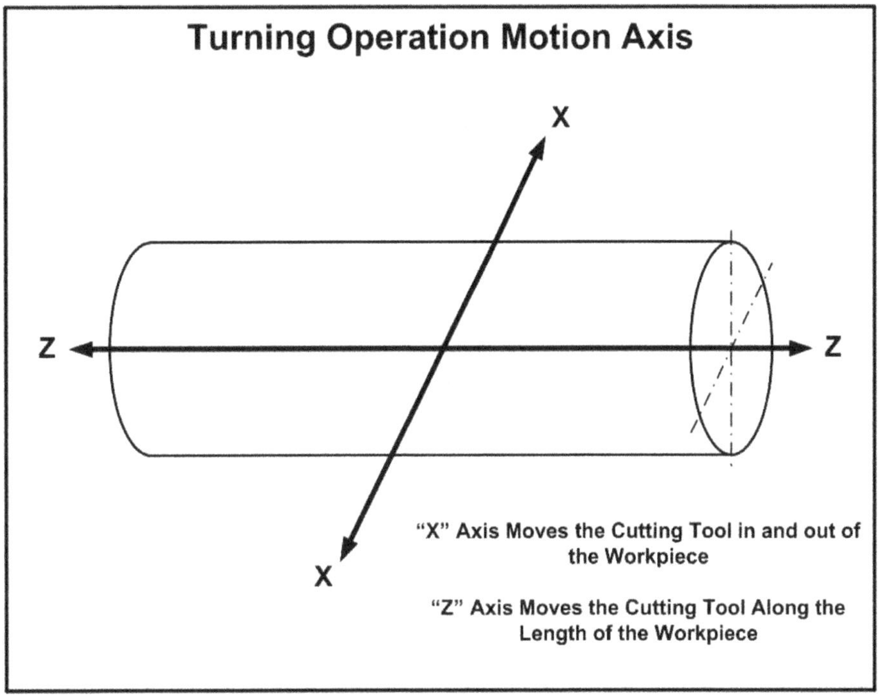

Figure 15.2 Axis of Motion for Turning Operations

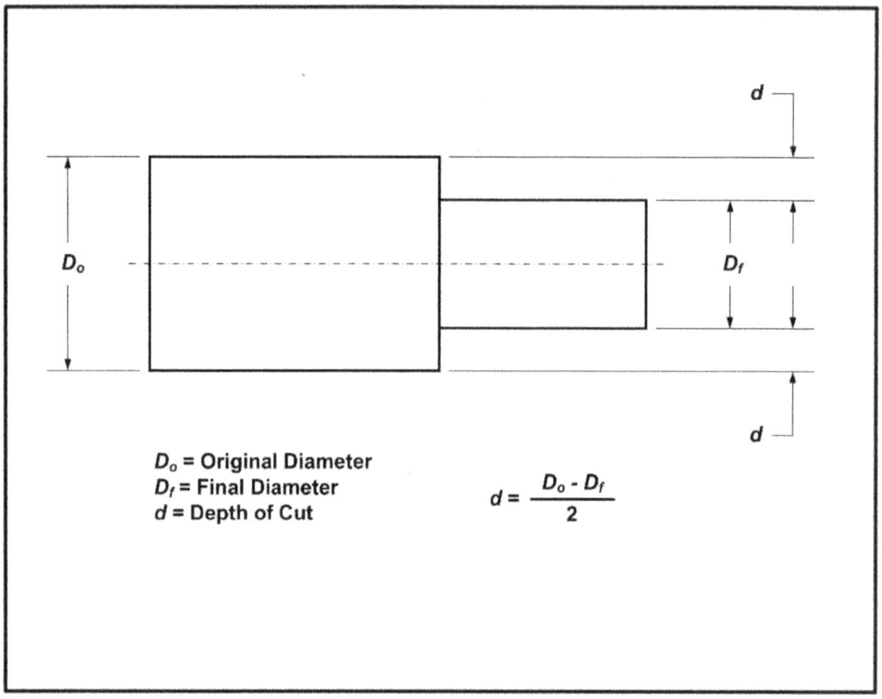

Figure 15.3 Turning Diameter Reduction and Depth of Cut

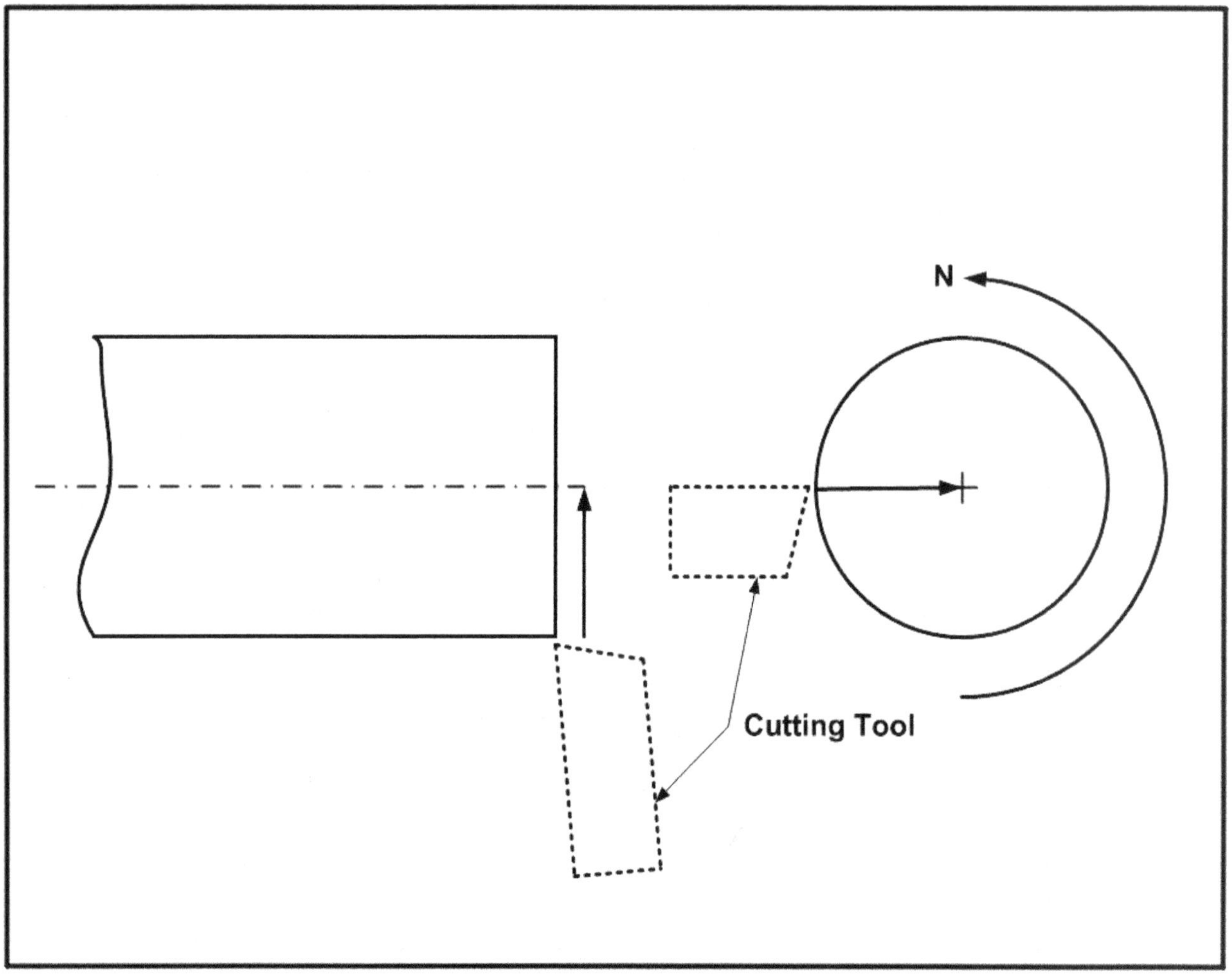

Figure 15.4 Turning Facing Operation

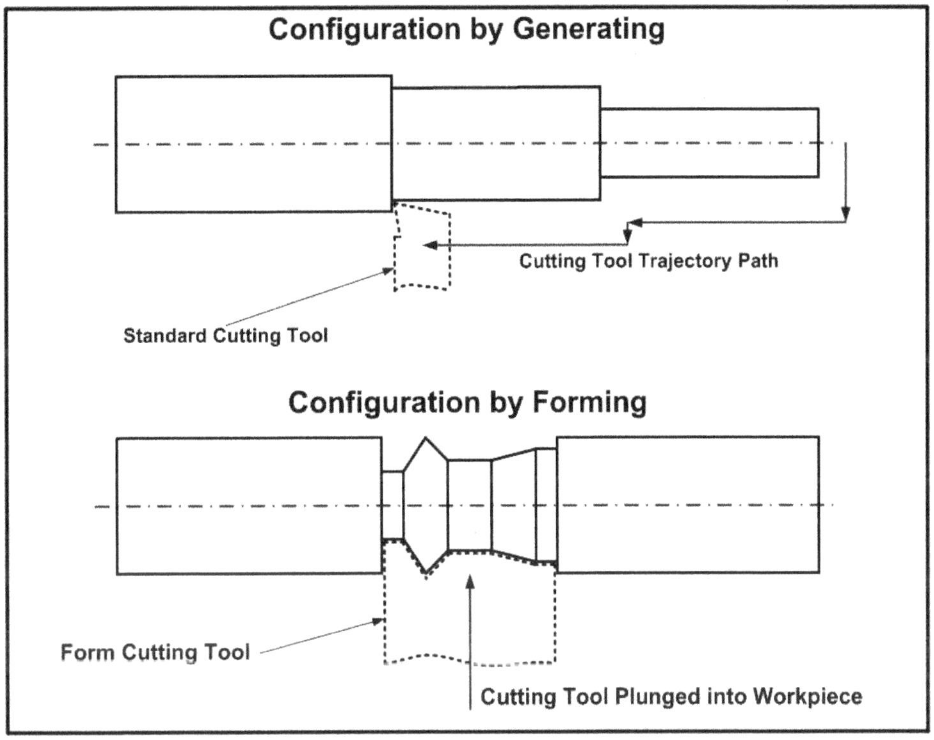

Figure 15.5 Turning Contouring and Forming Operations

Chamfering: Chamfering is creating an angle on the edge of the workpiece. These angles are normally 45°, but other angles are utilized as well. Chamfers can be created by a form tool being plunged into the edge of the workpiece or by moving the "X" and "Z" axes in unison. Figure 15.6 illustrates chamfers of various sizes on a workpiece along with the proper part print notations for them.

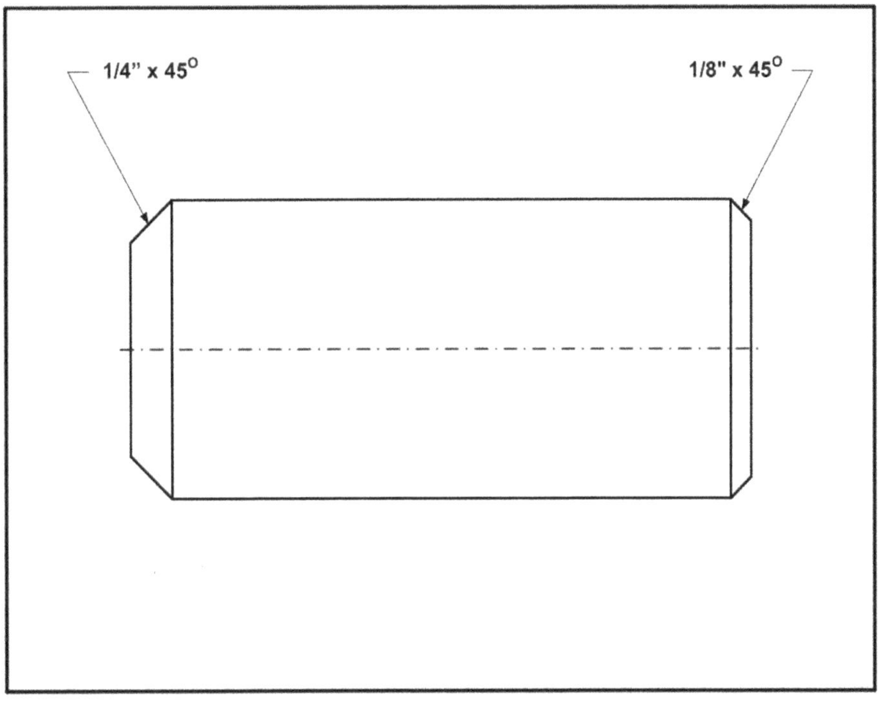

Figure 15.6 Proper Chamfer Designations on a Part Drawing

Tapering: Tapering involves cutting a diameter which will either get larger or smaller. Tapers can be created by a form tool being plunged into the edge of the workpiece, or by moving the "X" and "Z" axis in unison. Figure 15.7 displays the two methods for specifying a taper on a workpiece.

Grooving: Grooving involves cutting a configuration into the diameter of the workpiece. This is performed by moving the cutting tool into the workpiece with the "X" axis while the "Z" axis remains stationary. A groove is cut to a determined depth as shown in Figure 15.8.

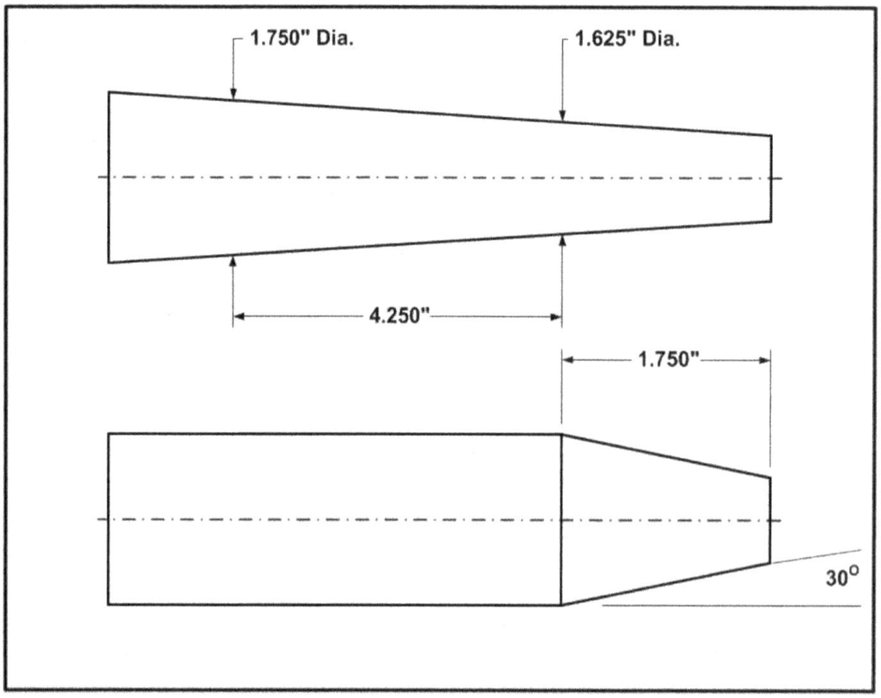

Figure 15.7 Proper Taper Designations on a Part Drawing

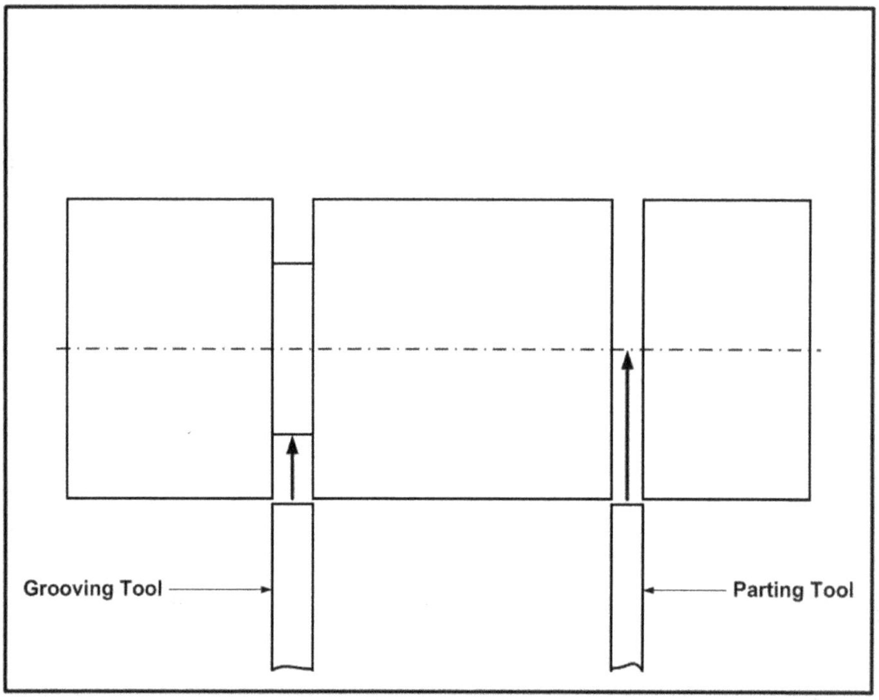

Figure 15.8 Turning Grooving and Cutoff Operations

Cutoff: Cutoff is performed identical to grooving, except that the cutting tool is moved into the workpiece to the center and the end of the workpiece is cut off. This is used for removing either excess material from the workpiece or a completed component from the end of the raw material. Figure 15.8 illustrates a cutoff operation.

Threading: Threading is performed with a single-point form tool, which is timed for the desired number of threads being cut. The tool is moved into the workpiece and the threads are formed over several passes. Each pass will be timed so the cutting tool will pick up the same path each time. For standard threads, the form tool angle is 60°, but for other specialty threads, a special cutting tool will be required. Figure 15.9 shows a 60° cutting tool being used to form standard threads on a workpiece.

Boring: A boring operation is performed identical to a contouring operation except that is it performed internally on the workpiece. An existing hole is required, and the boring tool has a single-point cutting tool, which will form the desired hole diameter over several passes. Boring will produce a straight hole to any required diameter. Figure 15.10 illustrates a boring operation producing an internal hole in the workpiece.

Drilling: Drilling inserts a hole into the workpiece, and in a turning operation, it will only be in the center of the workpiece. Here, the workpiece will be rotating at the proper speed while the drill remains stationary and is fed into the workpiece. This is covered in detail later in this chapter, in section on drilling.

Knurling: Knurling is not a material removal process but rather a material displacement operation. This is performed by moving hardened rollers into the workpiece while it is rotating and several passes are made to displace the material into the knurling configuration desired. This is utilized on a workpiece when a gripping surface is required. Knurls can be either of a diamond shape, which requires two rollers, or of a straight shape requiring only one roller. Figure 15.11 shows a diamond knurling pattern being cut into a workpiece.

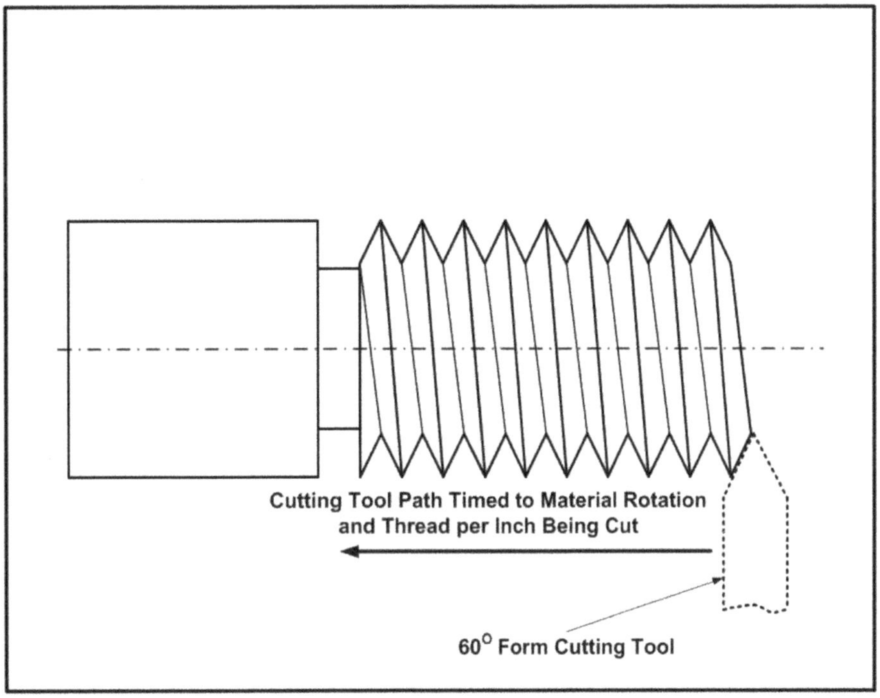

Figure 15.9 Turning Single-Point Threading Operation

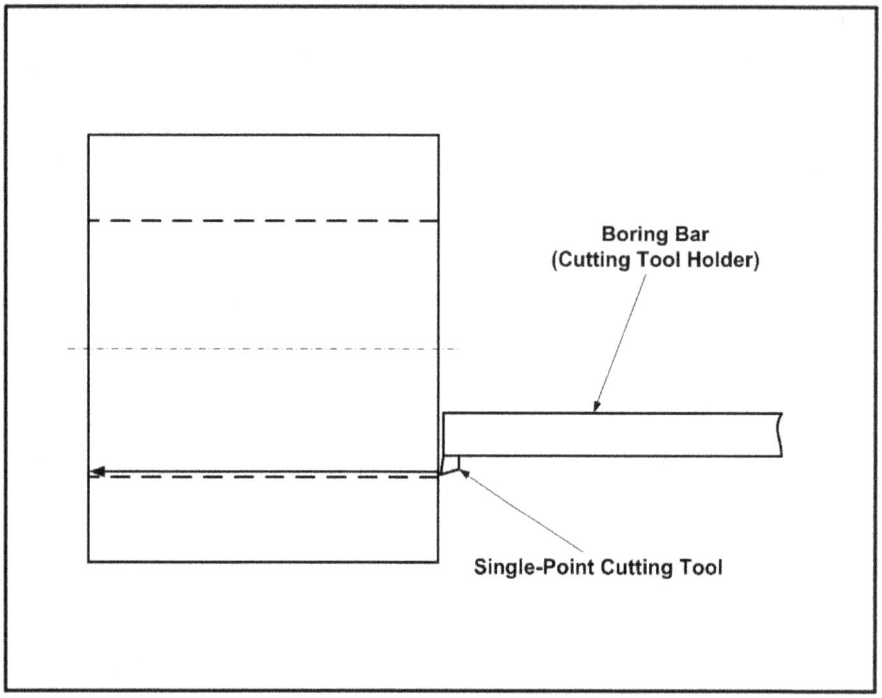

Figure 15.10 Turning Single-Point Internal Boring Operation

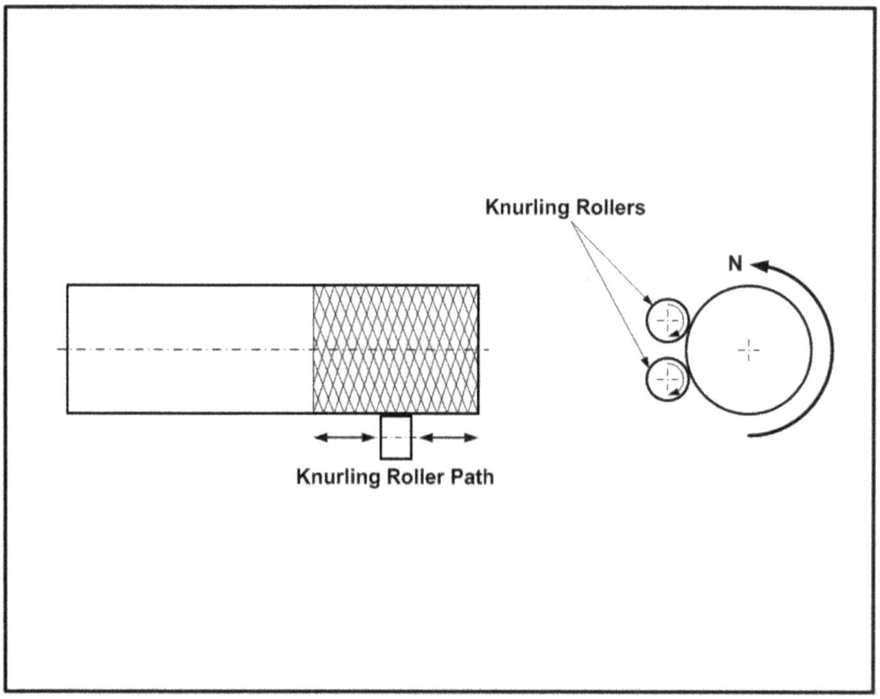

Figure 15.11 Turning Cross-Hatch Knurling Operation

15.2.2 Turning Calculations

To perform a turning operation as efficiently as possible, certain values must be calculated. By optimizing the operation, it will enable the best removal of the material and prolong the cutting tool life, allowing for uninterrupted machining. The values that need to be calculated are *Depth of Cut*, *Rotational Speed*, *Feed Rate*, *Machining Time*, and *Material Removal Rate*. These are explained as follows for a turning operation.

Depth of Cut: The depth of cut determines how much the diameter of the workpiece will be reduced with each pass of the cutting tool. As shown in Figure 15.3, the diameter will be reduced by two times the depth of cut due to the fact that in a turning operation, it is one long continuous cut and whatever is being removed from the front of the workpiece is also being removed from the back.

$$d = \frac{D_o - D_f}{2}$$

d = Depth of Cut
D_o = Original Diameter
D_f = Final Diameter

Rotational Speed: The rotational speed determines the optimum revolutions per minute that the material is to have. This is material-dependent along with the starting diameter of the workpiece. To optimize this, a new rotational speed needs to be calculated for each pass of the cutting tool because the final diameter (D_f) of one pass will be the original diameter (D_o) of the next pass.

$$N = \frac{v}{\pi D_o}$$

N = Rotational Speed (RPM's)
v = Cutting Speed (SFPM)
D_o = Original Diameter

Feed Rate: The feed rate renders the value of the distance the cutting tool is moving along the workpiece over a given unit of time. This value is dependent upon the rotational speed of the material and the feed, which is how far the cutting tool is advancing with each revolution of the workpiece.

$$f_r = Nf$$

f_r = Feed Rate (distance/min)
N = Rotational Speed (RPM)
f = Feed (distance/revolution)

Machining Time: The machining time will relinquish how long each cutting pass will take. This is used when calculating the efficiency and determining what the cost of the operation will be for component pricing.

$$T_m = \frac{L}{f_r}$$

T_m = Machining Time (min)
L = Length of Cut
f_r = Feed Rate (distance/min)

Material Removal Rate: The material removal rate renders the volumetric removal rate of material over a given unit of time. This measures both the efficiency and effectiveness of the operation being performed.

$$MRR = v f d$$

MRR = Material Removal Rate (in.^{3}/min)
v = Cutting Speed (SFPM)
f = Feed (distance/revolution)
d = Depth of Cut

Turning Example Problem

A piece of brass material is to be turned from an original diameter of 3.500 in. to a diameter of 3.250 in. A feed of 0.005 in./revolution will be used and the length of the cut is 5.750 in. For the cutting speed, we will use the middle of the range for the material as shown in Figure 15.1, which will be 575 SFPM.

Step 1: Determine the depth of cut for the operation.

$$d = \frac{D_o - D_f}{2}$$

$$d = \frac{3.500 - 3.250}{2}$$

$$d = 0.125 \text{ in.}$$

Step 2: Determine the rotation speed of the material for the operation.

$$N = \frac{v}{\pi D_o}$$

$$N = \frac{(575)(12)}{\pi 3.500}$$

$$N = 627.53 \text{ RPM}$$

Note: v is given in Surface Feet/Minute and must be converted to Surface Inches/Minute.

Step 3: Determine the feed rate for the operation.

$$f_r = Nf$$

$$f_r = (627.53)(0.005)$$

$$f_r = 3.138 \text{ in./min}$$

Step 4: Determine the machining time for the operation.

$$T_m = \frac{L}{f_r}$$

$$T_m = \frac{5.750}{3.138}$$

$$T_m = 1.833 \text{ Min}$$

Step 5: Determine the material removal rate for the operation.

$$\text{MRR} = v\,f\,d$$

$$\text{MRR} = (575 \text{ x } 12)(0.005)(0.125)$$

$$\text{MRR} = 4.312 \text{ in.}^3/\text{min}$$

15.3 DRILLING PROCESSES

Drilling provides the opportunity to insert holes into the workpiece for a variety of reasons and applications. The machine tools that are utilized for drilling are *Drill Presses*, *Milling Machines*, and *Engine Lathes*, but holes can also be inserted by hand drilling operations. A drill will insert a hole into a workpiece, but the drill may render a hole that is slightly oversized along with the potential of the hole not being perfectly straight. Both of these conditions can be caused either by the drill not being sharpened properly or by using too much force and not having the chips exiting the hole, causing burnishing inside. Also, very deep holes can also produce a nonstraight hole in the workpiece. Another factor of drilled holes is that they will never be perfectly round. This nonroundness may not be a factor in many applications, but if a perfectly round hole is required, it will require reaming. This is covered in the section on reaming.

Holes can be drilled into a workpiece in two different manners, which are *Through Holes* and *Blind Holes*. A through hole is one that is drilled completely through the workpiece, while a blind hole is drilled to a desired depth. Figure 15.12 shows a through hole and a blind hole along with the proper part drawing dimensioning for both. To obtain the desired depth of a blind hole, or to know the exact distance the drill must travel for

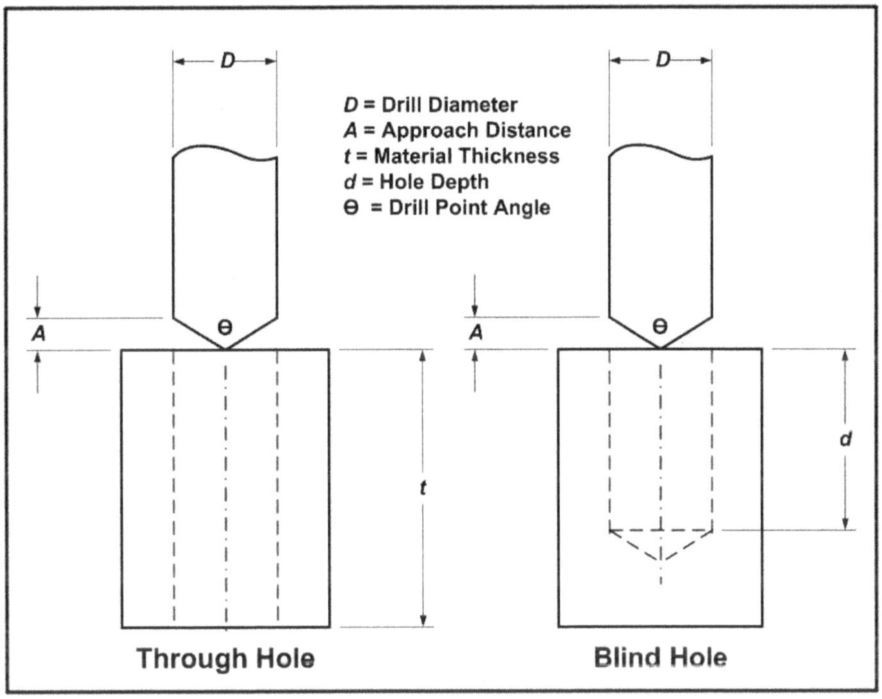

Figure 15.12 Through and Blind Hole Drilling

a through hole, there is an *Approach Distance* that needs to be accounted for within the operation. This is explained in the calculation section on drilling.

For inserting a large hole in a workpiece, drilling needs to be performed in stages starting with a smaller hole and then progressing up in size to the desired diameter. If a large hole is attempted to be drilled into the workpiece in one pass, the drill will not be able to remove the material properly, causing either damage to the drill or a hole that is not acceptable. A rule of thumb is that the large size that should be drilled into a workpiece for the first hole is ¼", but this is also dependent upon the material being drilled and the setup of the operation. Figure 15.13 illustrates how large holes are inserted into a workpiece in stages starting with a center drilled hole.

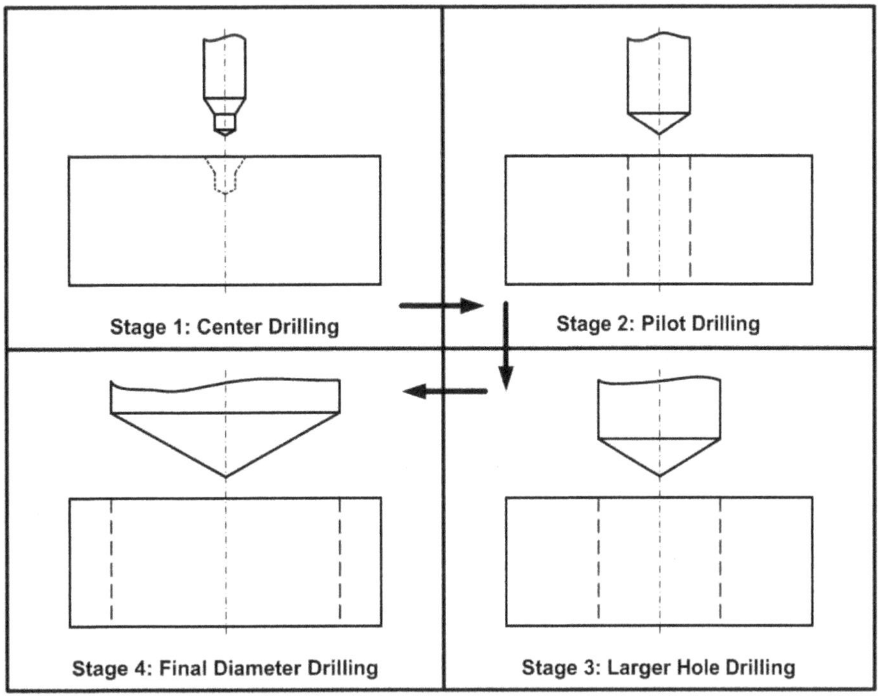

Figure 15.13 Step Drilling Stages for Large Holes

15.3.1 Drill Sizes and Types

Drills are available in a multitude of sizes and configurations depending on the material being processed or the type of hole that is desired. The standard drill point, which is the norm and can be used on just about any material, is 118°, but different drill points are available with either a steeper or a shallower angle for more effective drilling on various materials. Figure 15.14 shows various drill point angles utilized for different materials and/or configurations.

Standard Drills: Standard drills are utilized for inserting holes of various sizes into a workpiece. With 118° being the standard drill point, they can be purchased with custom drill points or resharpened to a special angle. A wide variety of sizes are available to choose from with a breakdown of the English system of drill sizes listed as follows.

English System Drill Sizes

Number Drills: 80–1 (Small to Large) 0.0135–0.2280 in.

Letter Drills: A–Z (Small to Large) 0.2346–0.4130 in.

Fraction Drills: Starting at 1/64 in. From 1/64 in. to 1/2 in. by 1/64 in. Steps.

Table 15.1 lists the drill sizes for both English system drills and metric drills.

The number and letter drills offer a variety of sizes between the fractional sizes to expand the hole diameter selection. These are available up to 1/2 in. because after 1/2 in., a hole can be bored to any given size desired.

Center Drills and Step Drills: Center drills and step drills share common geometric features as they both have more than one cutting diameter. A center drill, or countersink drill, as it is also referred to, is used primarily as a starting point for drilling holes. This is a short, ridged drill, which will not walk when it comes in contact with the surface of the workpiece. The center drill has a short drill point with a 118° angle, which starts the drilling and then adopts a 60° angle to enlarge the hole so that a standard drill can pick it up to maintain the hole's location without walking off. Another function that the center drill performs is to insert a hole in the end of a round piece of material during turning operations. Here, the configuration of the hole is functional as it will act as a support means. The 60° angle on the center drill is identical to the 60° center, which will support the material from the end while turning a diameter.

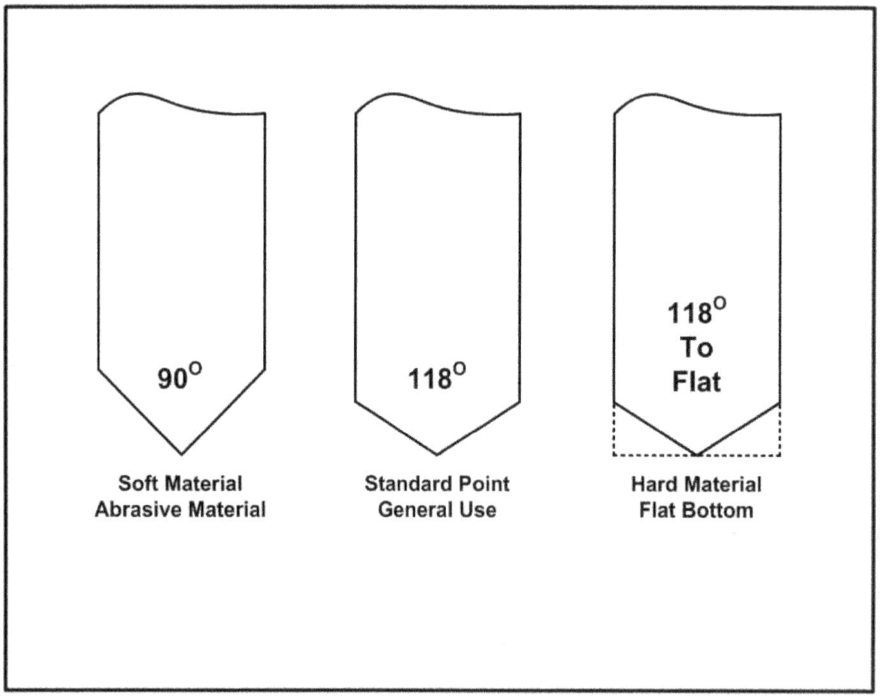

Figure 15.14 Drill Point Angles for Various Materials and Applications

INCH/METRIC TAP DRILL SIZES & DECIMAL EQUIVALENTS

DRILL SIZE	DECIMAL EQUIVALENT	TAP SIZE
80	.0135	
79	.0145	
1/64	.0156	
78	.0160	
77	.0180	
76	.0200	
75	.0210	
74	.0225	
73	.0240	
72	.0250	
71	.0260	
70	.0280	
69	.0292	
68	.0310	
1/32	.0312	
67	.0320	
66	.0330	
65	.0350	
64	.0360	
63	.0370	
62	.0380	
61	.0390	
60	.0400	
59	.0410	
58	.0420	
57	.0430	
56	.0465	
3/64	.0469	0 - 80
55	.0520	
54	.0550	
53	.0595	1 - 64, 72
1/16	.0625	
52	.0635	
51	.0670	
50	.0700	2 - 56, 64
49	.0730	
48	.0760	
5/64	.0781	
47	.0785	3 - 48
46	.0810	
45	.0820	3 - 56
44	.0860	
43	.0890	4 - 40
42	.0935	4 - 48
3/32	.0938	
41	.0960	
40	.0980	
39	.0995	
38	.1015	5 - 40
37	.1040	5 - 44
36	.1065	6 - 32
7/64	.1094	
35	.1100	
34	.1110	
33	.1130	6 - 40
32	.1160	
31	.1200	
1/8	.1250	
30	.1285	
29	.1360	8 - 32, 36
28	.1405	
9/64	.1406	
27	.1440	
26	.1470	
25	.1495	10 - 24
24	.1520	
23	.1540	
5/32	.1562	
22	.1570	
21	.1590	10 - 32
20	.1610	
19	.1660	
18	.1695	
11/64	.1719	
17	.1730	
16	.1770	12 - 24
15	.1800	
14	.1820	12 - 28
13	.1850	
3/16	.1875	
12	.1890	
11	.1910	

DRILL SIZE	DECIMAL EQUIVALENT	TAP SIZE
10	.1935	
9	.1960	
8	.1990	
7	.2010	1/4 - 20
13/64	.2031	
6	.2040	
5	.2055	
4	.2090	
3	.2130	1/4 - 28
7/32	.2188	
2	.2210	
1	.2280	
A	.2340	
15/64	.2344	
B	.2380	
C	.2420	
D	.2460	
1/4 E	.2500	
F	.2570	5/16 - 18
G	.2610	
17/64	.2656	
H	.2660	
I	.2720	5/16 - 24
J	.2770	
K	.2810	
9/32	.2812	
L	.2900	
M	.2950	
19/64	.2969	
N	.3020	
5/16	.3125	3/8 - 16
O	.3160	
P	.3230	
21/64	.3281	
Q	.3320	3/8 - 24
R	.3390	
11/32	.3438	
S	.3480	
T	.3580	
23/64	.3594	
U	.3680	7/16 - 14
3/8	.3750	
V	.3770	
W	.3860	
25/64	.3906	7/16 - 20
X	.3970	
Y	.4040	
13/32	.4062	
Z	.4130	
27/64	.4219	1/2 - 13
7/16	.4375	
29/64	.4531	1/2 - 20
15/32	.4688	
31/64	.4844	9/16 - 12
1/2	.5000	
33/64	.5156	9/16 - 18
17/32	.5312	5/8 - 11
35/64	.5469	
9/16	.5625	
37/64	.5781	5/8 - 18
19/32	.5938	
39/64	.6094	
5/8	.6250	
41/64	.6406	
21/32	.6562	3/4 - 10
43/64	.6719	
11/16	.6875	3/4 - 16
45/64	.7031	
23/32	.7188	
47/64	.7344	
3/4	.7500	
49/64	.7656	7/8 - 9
25/32	.7812	
51/64	.7969	
13/16	.8125	7/8 - 14
53/64	.8281	
27/32	.8438	
55/64	.8594	
7/8	.8750	1 - 8
57/64	.8906	
29/32	.9062	

DRILL SIZE	DECIMAL EQUIVALENT	TAP SIZE
59/64	.9219	1 - 12
15/16	.9375	1 - 14
61/64	.9531	
31/32	.9688	
63/64	.9844	1 1/8 - 7
1	1.0000	
1 3/64	1.0469	1 1/8 - 12
1 7/64	1.1094	1 1/4 - 7
1 1/8	1.1250	
1 11/64	1.1719	1 1/4 - 12
1 7/32	1.2188	1 3/8 - 6
1 1/4	1.2500	
1 19/64	1.2969	1 3/8 - 12
1 11/32	1.3438	1 1/2 - 6
1 3/8	1.3750	
1 27/64	1.4219	1 1/2 - 12
1 1/2	1.5000	

METRIC TAP DRILL SIZES

METRIC TAP	TAP DRILL (mm)	DECIMAL (Inch)
M1.6 x 0.35	1.25	.0492
M1.8 x 0.35	1.45	.0571
M2 x 0.4	1.60	.0630
M2.2 x 0.45	1.75	.0689
M2.5 x 0.45	2.05	.0807
M3 x 0.5	2.50	.0984
M3.5 x 0.6	2.90	.1142
M4 x 0.7	3.30	.1299
M4.5 x 0.75	3.70	.1457
M5 x 0.8	4.20	.1654
M6 x 1	5.00	.1968
M7 x 1	6.00	.2362
M8 x 1.25	6.70	.2638
M8 x 1	7.00	.2756
M10 x 1.5	8.50	.3346
M10 x 1.25	8.70	.3425
M12 x 1.75	10.20	.4016
M12 x 1.25	10.80	.4252
M14 x 2	12.00	.4724
M14 x 1.5	12.50	.4921
M16 x 2	14.00	.5512
M16 x 1.5	14.50	.5709
M18 x 2.5	15.50	.6102
M18 x 1.5	16.50	.6496
M20 x 2.5	17.50	.6890
M20 x 1.5	18.50	.7283
M22 x 2.5	19.50	.7677
M22 x 1.5	20.50	.8071
M24 x 3	21.00	.8268
M24 x 2	22.00	.8661
M27 x 3	24.00	.9449
M27 x 2	25.00	.9843
M30 x 3.5	26.50	1.0433
M30 x 2	28.00	1.1024
M33 x 3.5	29.50	1.1614
M33 x 2	31.00	1.2205
M36 x 4	32.00	1.2598
M36 x 3	33.00	1.2992
M39 x 4	35.00	1.3780
M39 x 3	36.00	1.4173

PIPE THREAD SIZES (NPSC)

THREAD	DRILL	THREAD	DRILL
1/8 – 27	11/32	1 1/2 – 11 1/2	1 3/4
1/4 – 18	7/16	2 – 11 1/2	2 7/32
3/8 – 18	37/64	2 1/2 – 8	2 21/32
1/2 – 14	23/32	3 – 8	3 1/4
3/4 – 14	59/64	3 1/2 – 8	3 3/4
1 – 11 1/2	1 5/32	4 – 8	4 1/4
1 1/4 – 11 1/2	1 1/2		

Table 15.1 Drill Size Chart for English System and Metric System Drills

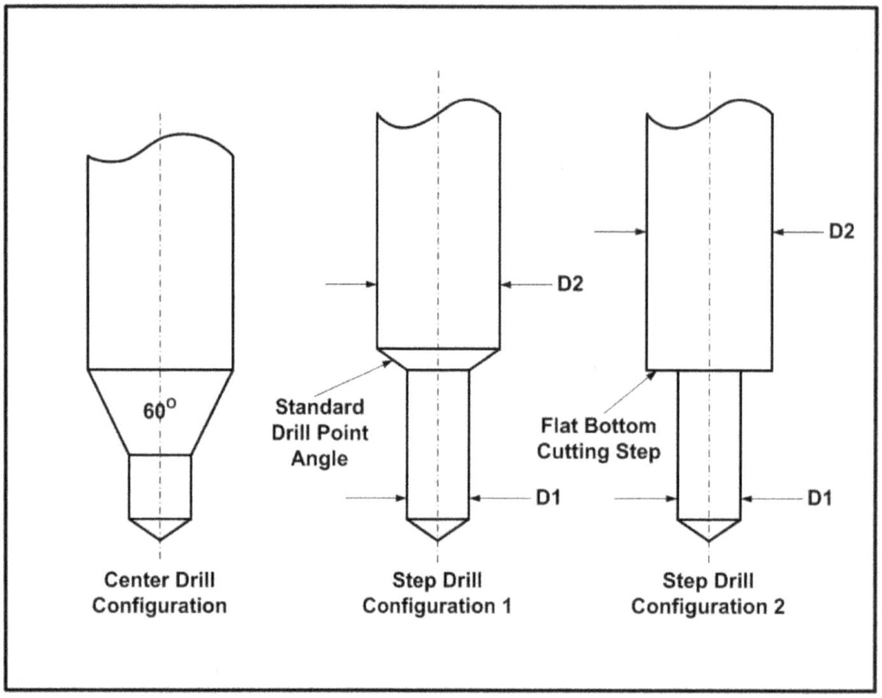

Figure 15.15 Center Drill and Step Drill Configurations

Step drills are drills that have multiple cutting diameters on one drill. These can be used in operations where a hole will have two or more diameters allowing for this operation to be performed with one drill rather than having to change the standard drill to obtain the same configuration. Also, when drilling a larger hole, a step drill can open the hole in the same way as using two or more standard drills as shown in Figure 15.14. Figure 15.15 shows the configuration of a center drill and two different styles of step drills.

15.3.2 Counterboring and Countersinking

Countersinking and counterboring are performed primarily to insert clearance for threaded fasteners. An existing hole is required and then the countersink or counterbore is lowered to form the configuration to a determined depth. For standard flat-head threaded fasteners, the angle of the countersink is 82°, but then can be purchased at a wide range of angles also. Counterboring performs the same functions but will render a square bottom hole, which is used for cap screws. There is a pilot on the nose of the counterbore, which will guide it into the hole to maintain alignment. Figure 15.16 illustrates a countersinking and a counterboring operation.

15.3.3 Reaming

Reaming is an operation that will create a hole to an extremely close tolerance as opposed to a drill, which will have a wider tolerance range when creating a hole. A reamer will open the hole to a tolerance range of about ±0.0002 in. allowing for more precision. Also, a reamer will create a perfectly round hole unlike a drill. These are available in all standard sizes the same as drills, and custom sizes are available as well. Standard reamers are available with both under- and oversized diameters besides the standard sizes. These are used if an object needs to be pressed into a hole where a 0.001–0.002 in. interference is required or if an object needs to slide into a hole in the same manner.

Reaming removes a very small amount of material and requires an slightly undersized hole from the finished size the reamer will create. Depending on the size of the reamer, the starting hole will be drilled, leaving the amount that the reamer will remove. Following are the guidelines to drill an undersized hole that is required prior to reaming.

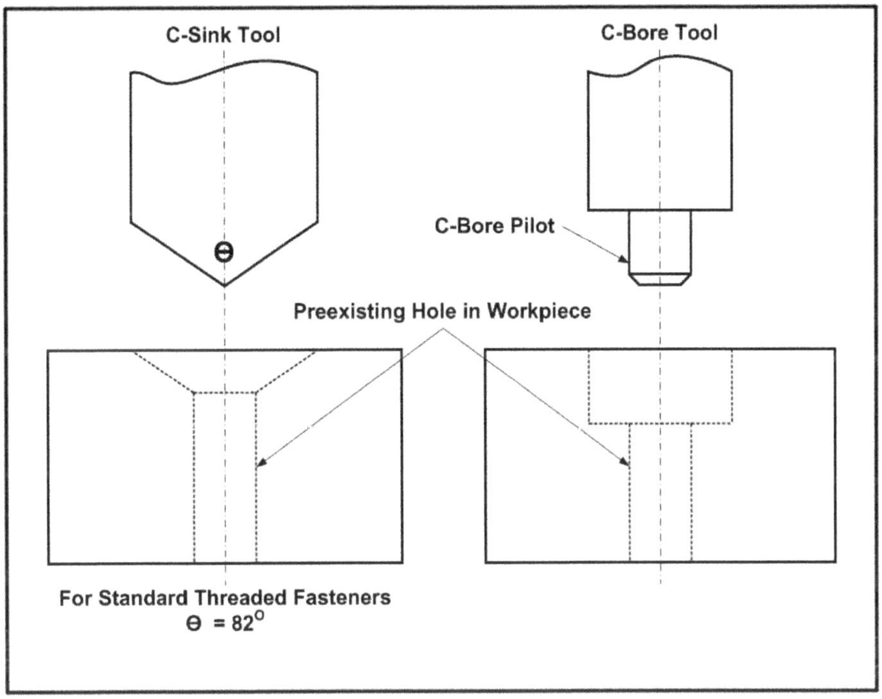

Figure 15.16 Countersinking and Counterboring Hole Configurations

Reamer Size: 80–1/4 in.; Hole Drilled 1/64 in. Undersize
Reamer Size: 1/4–3/4 in.; Hole Drilled 1/32 in. Undersize
Reamer Size: 3/4 in. and up; Hole Drilled 1/16 in. Undersize

Figure 15.17 shows an undersized hole to be reamed along with the guidelines for the hole's diameter prior to reaming.

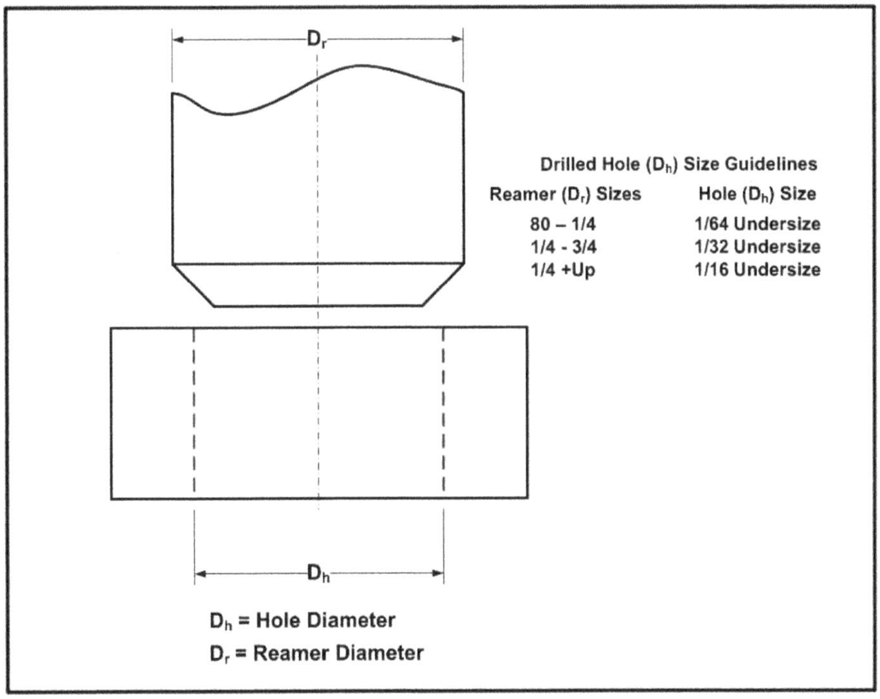

Figure 15.17 Hole Drilling for Reaming Operations Guidelines

15.3.4 Drilling Calculations

To perform a drilling operation as efficiently as possible, certain values must be calculated. By optimizing the operation, it will enable the best removal of the material and prolong the cutting tool life, allowing for uninterrupted machining. The values that need to be calculated are *Rotational Speed, Feed Rate, Approach Distance, Machining Time*, and *Material Removal Rate*. These are explained as follows for a drilling operation.

Rotational Speed: The rotational speed determines the optimum revolutions per minute that the drill is to have. This is material-dependent along with the diameter of the drill.

$$N = \frac{v}{\pi D}$$

N = Rotational Speed (RPM)

v = Cutting Speed (SFPM)

D = Drill Diameter

Feed Rate: The feed rate renders the value of the distance the drill is moving into the workpiece over a given unit of time. This value is dependent upon the rotational speed of the material and the feed, which is how far the drill is advancing with each revolution of the workpiece.

$$f_r = N f$$

f_r = Feed Rate (distance/min)

N = Rotational Speed (RPM)

f = Feed (distance/revolution)

Approach Distance: The approach distance is that distance along which the drill has to travel based on the angle of the point before the actual diameter begins to be formed (See Figure 15.12). This is a factor of the diameter of the drill and the angle of the drill point.

$$A = 0.5\ D \tan 90 - \frac{\theta}{2}$$

A = Approach Distance

D = Drill Diameter

Θ = Drill Point Angle

Machining Time: The machining time will relinquish how long the drilling operation will take. This is used when calculating the efficiency and determining what the cost of the operation will be for component pricing. There are two equations for machining time depending on whether the holes being drilled are through holes or blind holes.

Through Hole

$$T_m = \frac{t + A}{f_r}$$

T_m = Machining Time (min)

t = Material Thickness

A = Approach Distance

Blind Hole

$$T_m = \frac{d + A}{f_r}$$

T_m = Machining Time (min)

d = Hole Depth

A = Approach Distance

Material Removal Rate: The material removal rate renders the volumetric removal rate of material over a given unit of time. This measures both the efficiency and effectiveness of the operation being performed.

$MRR = \pi D^2 f_r/4$

MRR = Material Removal Rate (in.³/min)

D = Drill Diameter

f_r = Feed Rate (distance/min)

Drilling Example Problem

A 0.375 in. diameter hole is to be drilled in soft stainless steel. The drill has a point angle of 118° and will be a through hole with the thickness of the material being 1.750 in. A feed of 0.0005 in./revolution was selected along with a cutting speed in the middle of the range for the material as shown in Figure 15.1, which will be 375 SFPM.

Step 1: Determine the rotational speed of the drill for the operation.

$$N = \frac{v}{\pi D}$$

$$N = \frac{(375)(12)}{\pi 0.375}$$

$$N = 3819.72 \text{ RPM's}$$

Note: v is given in Surface Feet/Minute and must be converted to Surface Inches/Minute.

Step 2: Determine the feed rate for the operation.

$$f_r = n f$$

$$f_r = (3819.72)(0.0005)$$

$$f_r = 1.910 \text{ in./min}$$

Step 3: Determine the approach distance of the drill for the operation.

$$A = 0.5 \text{ D Tan } 90 - \frac{\theta}{2}$$

$$A = (0.5)(0.375) \text{ Tan } 90 - \frac{118}{2}$$

$$A = 0.1127 \text{ in.}$$

Step 4: Determine the machining time for the operation.

$$T_m = \frac{t+A}{fr}$$

$$T_m = \frac{1.750+0.1127}{1.910}$$

$$T_m = 0.975 \text{ Min}$$

Step 5: Determine the material removal rate for the operation.

$$MRR = \pi D^2 f_r / 4$$

$$MRR = \pi (0.375^2)(1.910)/4$$

$$MRR = 0.211 \text{ in.}^3/\text{min}$$

15.4 MILLING PROCESSES

Milling operations involve removing material from a workpiece that is predominantly square. Various cutting tools are utilized for the generating and forming of the workpiece as it is moved along the workpiece. Milling processes have three major axes of motion, which are the "X" axis, the "Y" axis, and the "Z" axis. The "X" axis moves the cutting tool left and right along the workpiece, the "Y" axis moves the cutting tool in and out of the workpiece, and the "Z" axis moves the cutting tool up and down along the workpiece. Figure 15.18 shows the three major axes for milling operations. For certain operations, it may be required to move two or even all three of the axes simultaneously to shape the workpiece into the desired geometric configurations. The machine tool utilized for milling is either a *Horizontal Milling Machine* or a *Vertical Milling Machine*. Horizontal milling is where the cutter is held horizontal to the workpiece, which is shown in Figure 15.19; while in vertical milling, the cutter is held at 90° to the workpiece, and it only utilizes the face of the cutter to perform the material removal.

When cutting material in a milling operation utilizing the side of the cutter, it can be performed in two manners. Depending on which direction the side of the milling cutter is rotating when it is removing the material, the cutting action is referred to as either *Conventional Milling* or *Climb Milling*. With conventional milling, the cutter rotates in the opposite direction of the material being fed into it. This style of milling is capable of delivering heavier depth of cuts and is safer because if the material becomes loose while being cut, it will be pushed away from the cutter. However, conventional milling will relinquish a bad surface finish. This is a result of the chips that are caught in the cutter teeth and then being imbedded into the surface of the workpiece, which has already been machined. With climb milling, the cutter rotates in the same direction of the material being fed into it. Lighter cuts are taken with climb milling due to the fact that it is not as safe as conventional milling. If the workpiece becomes loose, the cutter will pull it in and can cause a dangerous situation. Climb milling will, however, render a much better surface finish because any chips that are caught in the teeth of the cutter will get imbedded into the surface of the workpiece, which has not been machined yet. Figure 15.19 illustrates both climb and conventional milling along with the advantages and disadvantages of each. For most milling operations, the majority of the material will be removed with conventional milling, then a light finishing cut will be performed with climb milling to render a good surface finish.

Along with the conventional and climb milling styles, milling operations can be performed with *Peripheral Milling* or *Face Milling*. Both can accomplish most of the same geometric configurations on the workpiece, but they are accomplished in a different manner. Both peripheral and face milling are discussed in detail as follows.

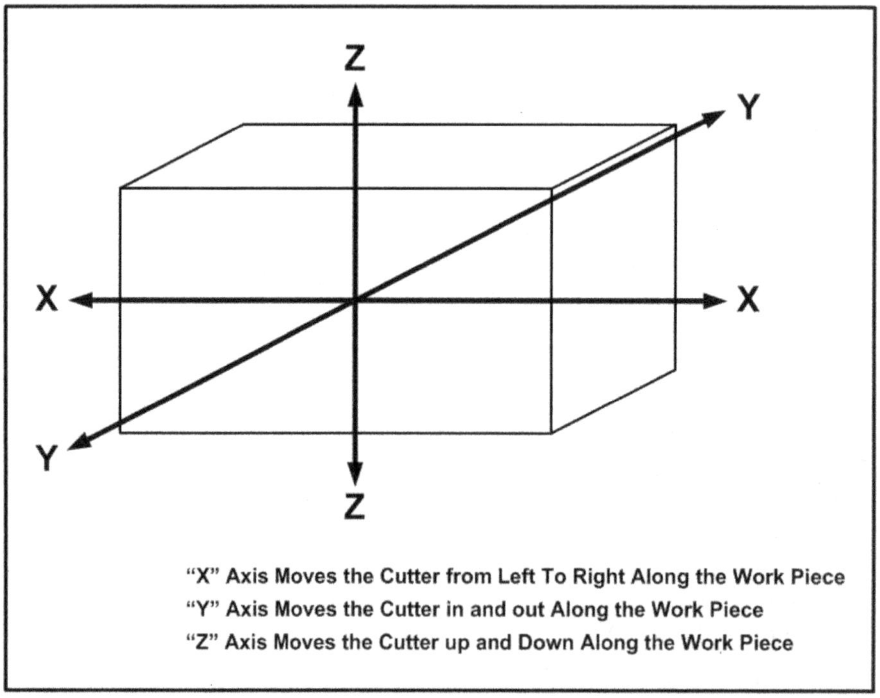

Figure 15.18 Axes of Motion for Milling Operations

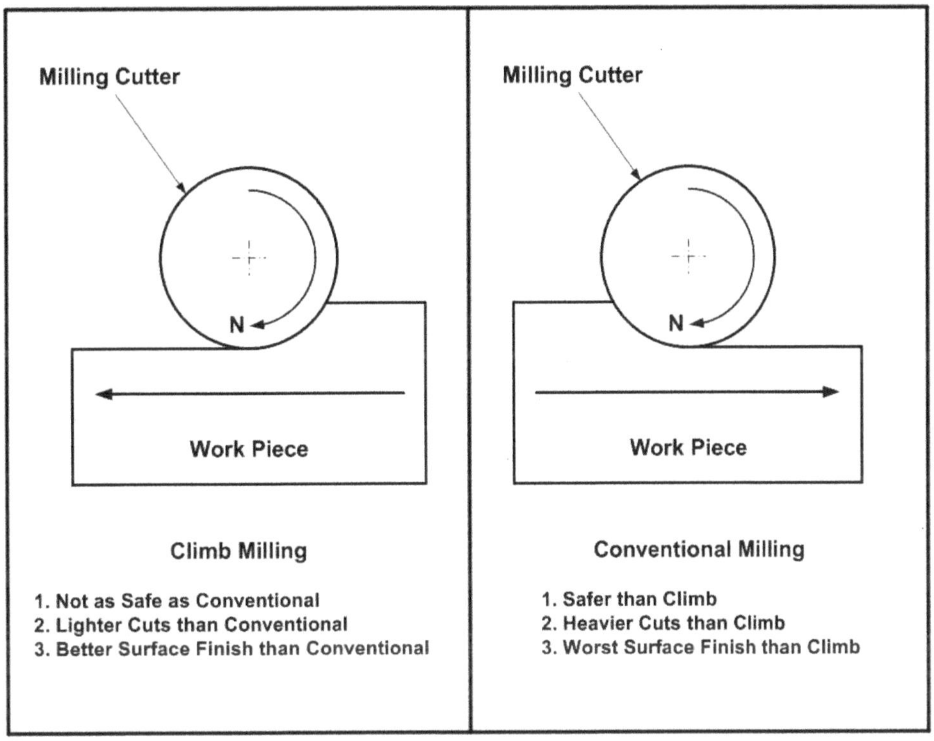

Figure 15.19 Climb and Conventional Milling Operations

15.4.1 Peripheral Milling

Peripheral milling utilizes the side of the cutting tool to perform the material removal of the workpiece. This can be performed on either a horizontal or a vertical milling machine depending on the application. The cutting teeth of the cutter engage into the workpiece in either conventional or climb milling and the cutter is moved along the component part as the material is removed. The cutter will be set to the desired depth of cut and then moved into the workpiece to remove the material. Figure 15.20 shows a side view of peripheral milling along with the approach distance for the cutter.

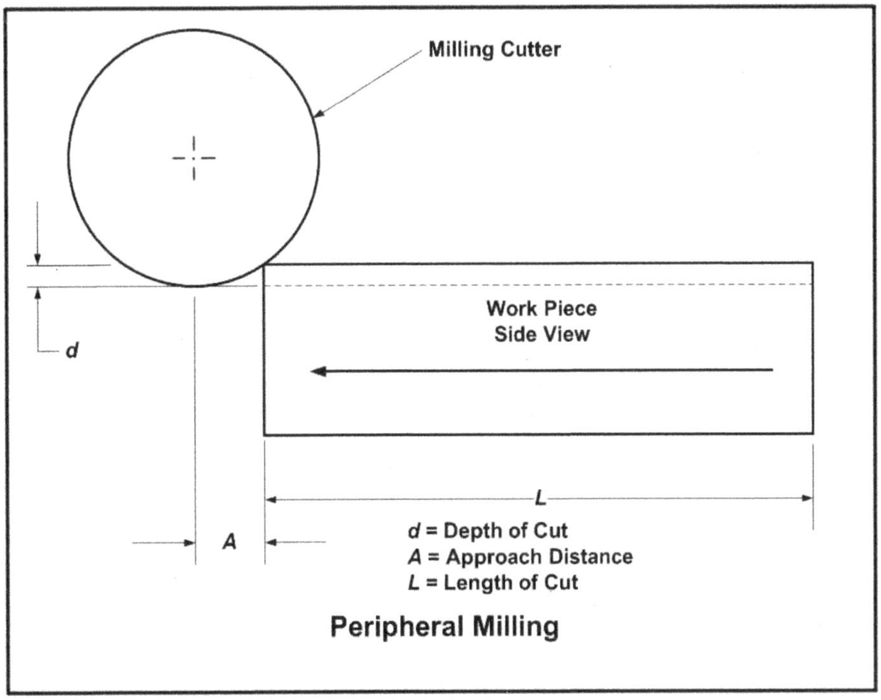

Figure 15.20 Peripheral Milling Operation

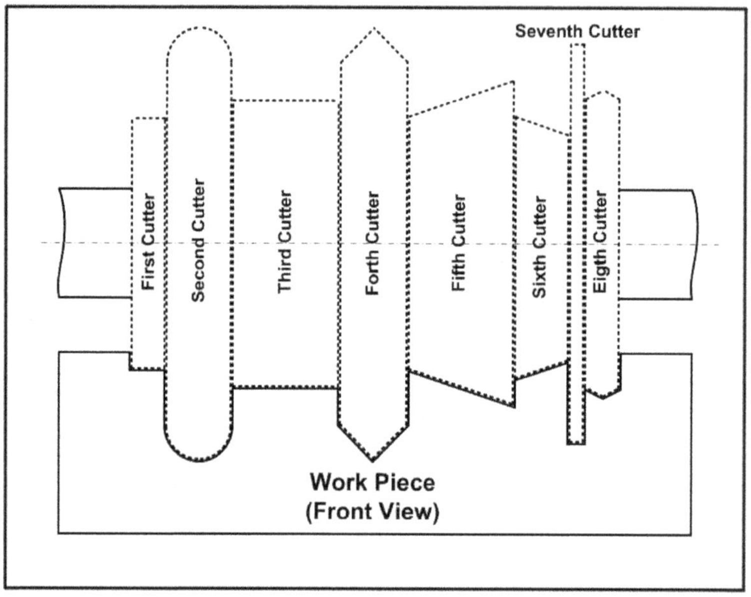

Figure 15.21 Multicutter (Gang Milling) Peripheral Milling Operation

For some applications, multiple cutters can be stacked together to perform a forming operation on the workpiece rather than a generating operation, which is more common with a single cutter. Figure 15.21 illustrates a front view of a multimilling cutter setup for forming a workpiece.

15.4.2 Face Milling

Face milling is performed by utilizing the face of the milling cutter to remove the material from the workpiece. This is only performed on a vertical milling machine where the cutting tool is held at 90° from the surface of the component part. In a face milling operation, there is no conventional or climb milling difference as the cutting tool is only engaged with the face and not the side for cutting. Figure 15.22 shows a top view of face milling along with the approach and runout distances for the cutter.

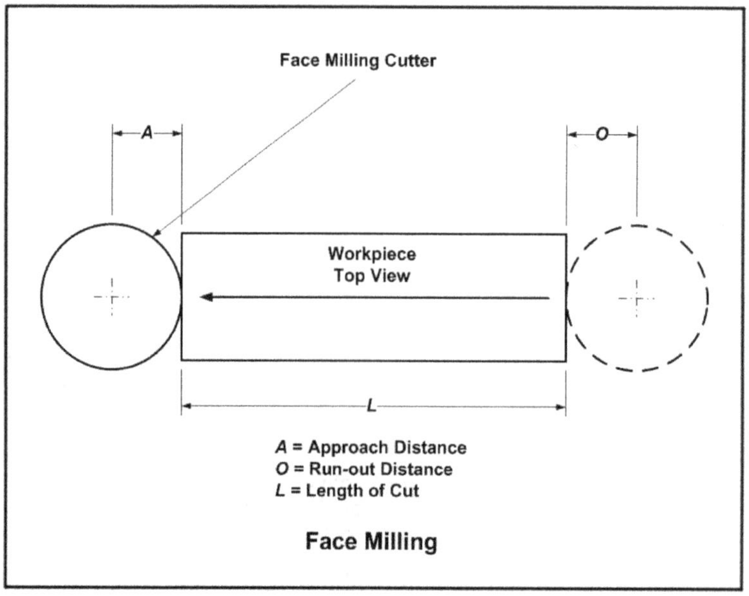

Figure 15.22 Face Milling Operation

15.4.3 Milling Calculations

To perform a milling operation as efficiently as possible, certain values must be calculated. By optimizing the operation, it will enable the best removal of the material and prolong the cutting tool life, allowing for uninterrupted machining. The values that need to be calculated are *Rotational Speed, Feed Rate, Approach Distance, Machining Time,* and *Material Removal Rate.* These are explained as follows for a milling operation.

Rotational Speed: The rotational speed determines the optimum revolutions per minute that the milling cutter is to have. This is material-dependent along with the diameter of the milling cutter.

$$\mathrm{N} = \frac{v}{\pi D}$$

N = Rotational Speed (RPM)
v = Cutting Speed (SFPM)
D = Cutter Diameter

Feed Rate: The feed rate renders the value of the distance the milling cutter is moving into the workpiece over a given unit of time. This value is dependent upon the rotational speed of the material and the feed, which is how far the drill is advancing with each revolution of the workpiece along with how many cutting teeth are on the milling cutter.

$$f_r = \mathrm{N\,n\,f}$$

f_r = Feed Rate (distance/min)
N = Rotational Speed (RPM)
n = Number of Teeth
f = Feed (distance/Tooth)

Approach Distance: The approach distance is that distance along which the milling cutter has to travel before it starts cutting to the desired depth, or in face milling, before it is fully engaged into the workpiece (See Figure 15.20 for peripheral milling and Figure 15.21 for face milling). This is a factor of the diameter of the milling cutter and the depth of cut being performed.

Face Milling

$$A = O = \frac{D}{2}$$

A = Approach Distance
O = Runout Distance
D = Cutter Diameter

Peripheral Milling

$$A = \sqrt{d(D-d)}$$

A = Approach Distance
d = Depth of Cut
D = Cutter Diameter

Machining Time: The machining time will relinquish how long the milling operation will take. This is used when calculating the efficiency and determining what the cost of the operation will be for component pricing. There are two equations for machining time depending on whether it is peripheral or face milling operation.

Face Milling

$$T_m = \frac{L+A+O}{fr}$$

T_m = Machining Time (min)
L = Length of Cut

A = Approach Distance
O = Runout Distance
f_r = Feed Rate (distance/min)

Peripheral Milling

$$T_m = \frac{L + A}{fr}$$

T_m = Machining Time (min)
L = Length of Cut
A = Approach Distance
f_r = Feed Rate (distance/min)

Material Removal Rate: The material removal rate renders the volumetric removal rate of material over a given unit of time. This measures both the efficiency and effectiveness of the operation being performed.

$$\text{MRR} = W\,d\,f_r$$

MRR = Material Removal Rate (in.^3/min)
W = Width of Cut
d = Depth of Cut
f_r = Feed Rate (distance/min)

Peripheral Milling Example Problem

A copper workpiece that is 8.000 in. in length and 1.750 in. wide is to be peripheral-milled to a depth of 1.125 in. The cutter is 6.500 in. in diameter and has 24 teeth. The feed is 0.0003 in./tooth and a lower end of the cutting speed for copper as shown in Figure 15.1, which is 450 SFPM, is used.

Step 1: Determine the rotational speed of the cutter for the milling operation.

$$N = \frac{v}{\pi D}$$

$$N = \frac{(450)(12)}{\pi 6.500}$$

$$N = 264.44 \text{ RPM}$$

Step 2: Determine the feed rate for the milling operation.

$$f_r = N\,n_t\,f$$

$$f_r = (264.44)(24)(0.0003)$$

$$f_r = 1.904 \text{ in./min}$$

Step 3: Determine the approach distance for the milling operation.

$$A = \sqrt{\text{d}(\text{D} - \text{d})}$$

$$A = \sqrt{(1.125)(6.500 - 1.125)}$$

$$A = 2.459 \text{ in.}$$

Step 4: Determine the machining time for the milling operation.

$$T_m = \frac{L + A}{\text{fr}}$$

$$T_m = \frac{8.000 + 2.459}{1.904}$$

$$T_m = 5.493 \text{ min}$$

Step 5: Determine the material removal rate of the milling operation.

$$\text{MRR} = W\,d\,f_r$$
$$\text{MRR} = (1.750)(1.125)(1.904)$$
$$\text{MRR} = 3.748 \text{ in.}^3/\text{min}$$

Face Milling Example Problem

A bronze workpiece is to be face-milled with a 1.500 in. diameter cutter. The length of the cut is 5.750 in. and the width will be the diameter of the cutter. With a feed of 0.0006 in/tooth and a depth of cut of 0.125 in., the cutting speed was selected from the top of the range as shown in Figure 15.1, which is 750 SFPM.

Step 1: Determine the rotational speed of the cutter for the milling operation.

$$N = \frac{v}{\pi D}$$

$$N = \frac{(750)(12)}{\pi 1.500}$$

$$N = 1909.86 \text{ RPM}$$

Step 2: Determine the feed rate for the milling operation.

$$f_r = N\,n_t\,f$$
$$f_r = (1909.86)(4)(0.0006)$$
$$f_r = 1.904 \text{ in.3/min}$$

Step 3: Determine the approach and runout distance for the milling operation.

$$A = O = \frac{D}{2}$$

$$A = O = \frac{1.500}{2}$$

$$A = O = 0.750 \text{ in.}$$

Step 4: Determine the machining time for the milling operation.

$$T_m = \frac{L + A + O}{fr}$$

$$T_m = \frac{5.750 + 0.750 + 0.750}{4.584}$$

$$T_m = 1.582 \text{ min}$$

Step 5: Determine the material removal rate for the milling operation.

$$MRR = W\,d\,f_r$$
$$MRR = (1.500)(0.125)(4.584)$$
$$MRR = 0.859\ \text{in.}^3/\text{min}$$

15.5 SHAPING PROCESSES

The shaping process is one that is used for massive material removal from the workpiece. For shaping, the machine tool being used is called a *Shaper* and the process functions by utilizing a single-point cutting tool, which is moved across the surface of the workpiece that is being held stationary. Depending on the size of the shaper, the depth of cut can be up to 1.000 in. or more, removing a huge amount of material. Once the tool has completed the stroke removing the material, it retracts and will index over for the next cut. Figure 15.23 illustrates a single-point cutting tool being moved across the workpiece to remove the material.

Plaining is a very similar process except the cutting tool remains stationary and the table holding the workpiece moves back and forth. The machine tool used for planning is called a *Plainer* and is capable of removing even more material on extremely large workpiece. This process is still used for squaring up very large blocks such as mold casings and more cavities. Both shaping and planning are mainly utilized for roughing a workpiece close to the finished sizes, then completed utilizing milling or other removal processes.

Both shaping and plaining have been greatly replaced by the introduction of CNC equipment, which can perform both the roughing and finishing of a workpiece. This automation requires only the setup of the material and writing of the program that will inform the machine what to do and can operate unattended. These processes will be covered in Chapter 27.

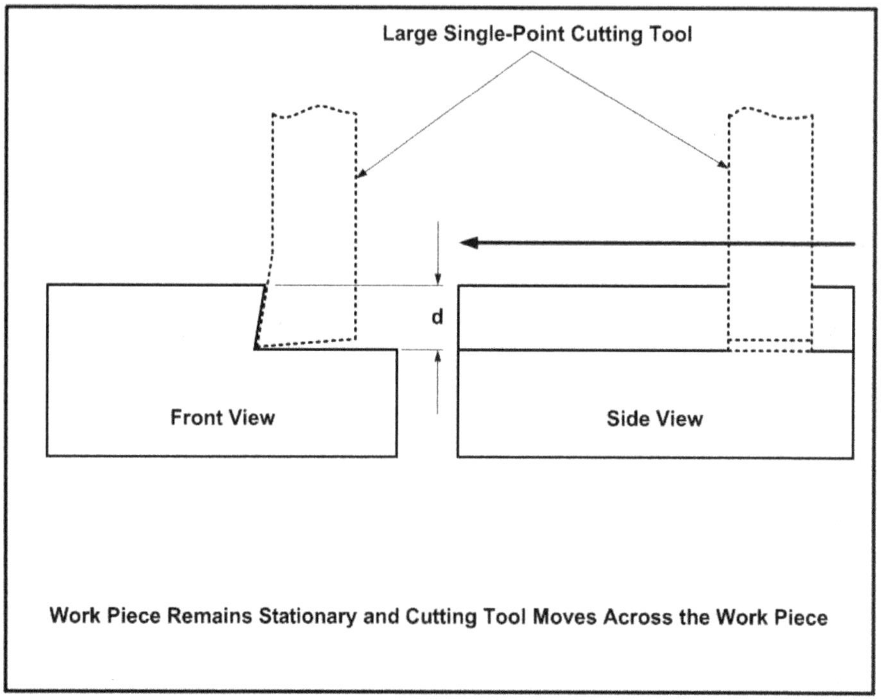

Figure 15.23 Shaping Operation

15.6 BROACHING PROCESSES

Broaching is a process that will render an internal shape not capable with most other processes. Shapes such as a square hole or other configurations can be easily obtained through this process. This requires a starting hole in the workpiece and a tapered cutting tool to the configuration desired. The cutting tool is inserted into the starting hole in the workpiece, then either pushed or pulled through depending on the machine and setup being used. The tool cuts the configuration of desired shape as it moves through the workpiece. Depending on how much material removal is required to create the desired shape, several cutting tools may be required to create the shape in stages. Figure 15.24 illustrates a typical broaching process where a starting hole is in the component and the broach is pushed through to render an internal shape not possible from drilling or milling operations.

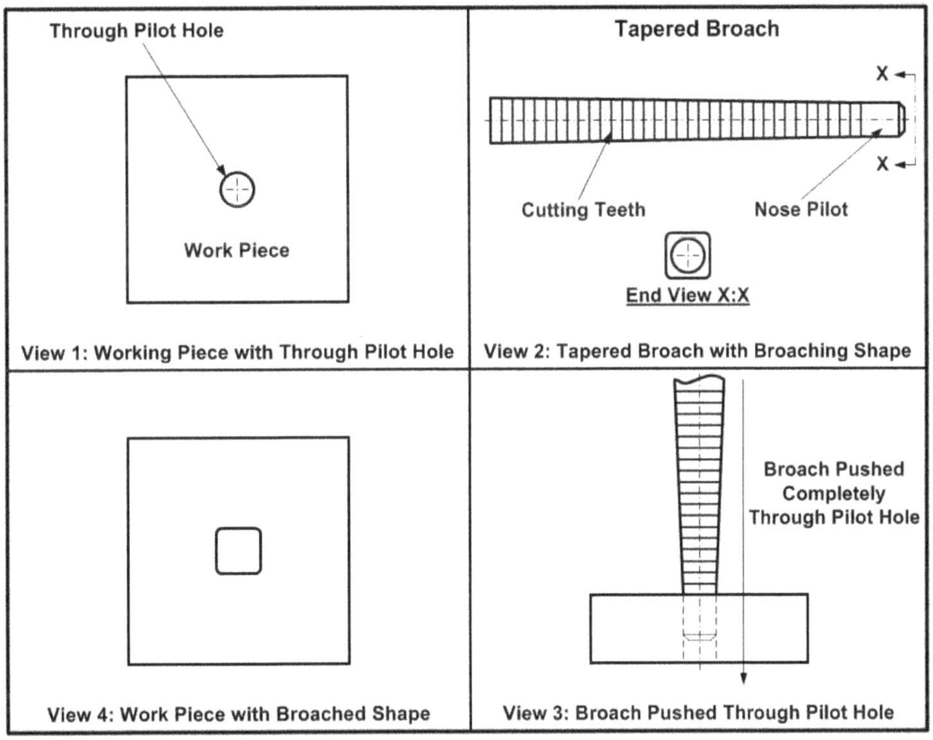

Figure 15.24 Broaching Operation

15.7 SAWING PROCESSES

Sawing is the simplest process of material removal. This is normally the first process used to cut the raw material that will be used to fabricate the desired component parts. This process can be performed in several ways utilizing a horizontal band saw, a vertical band saw, or a handheld hack saw. This process only has two rules to make it effective. The first is to let the saw do the cutting, without forcing it. Forcing the saw will either damage the blade or create a cut that is not straight. The second rule is to, at all the times, have a minimum of three cutting teeth in contact with the workpiece while cutting. If less than three teeth are in contact with the workpiece, the blade will bounce as it moves across and cause the teeth to break. Figure 15.25 shows a saw blade in contact with the material maintaining a minimum of three teeth in contact with the workpiece.

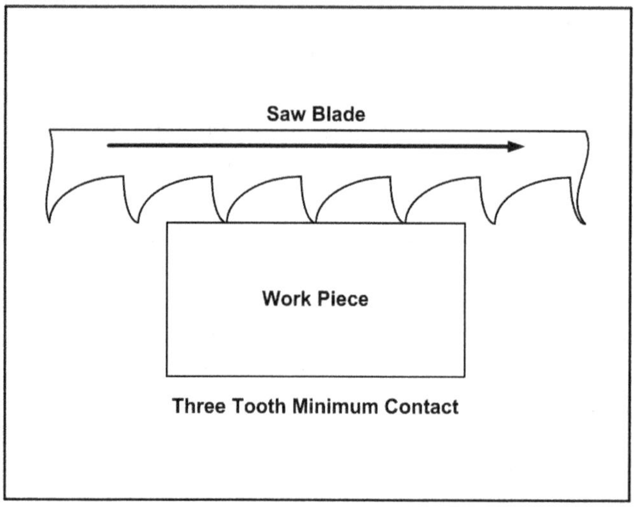

Figure 15.25 Sawing Operation Showing Minimum Three Teeth Contact

15.8 SUMMARY

Machining is possibly the most widely utilized material removal process. With a wide variety of processes within this arena, configurations of almost any shape can be created to very close tolerances. With the turning process, round components are formed or generated along with tapers, holes in the center, and machine threads. Milling renders square parts with steps, pockets, slots, and groves, depending on the part print requirements. Drilling inserts holes into the workpiece, and in combination with reaming, it is capable of producing holes that are perfectly round to extremely close tolerances. Countersinking and counterboring will allow for screws of different configurations to be able to sit recessed into the workpiece. Lastly, broaching offers the capability of producing internal configurations that are not possible with turning, milling, or drilling. This combination of processes offers the manufacturing world a wide variety of options for the fabrication of component parts.

15.9 QUESTIONS AND PROBLEMS

Questions

1. Explain what the cutting speed is in material removal.

2. What is the difference between feed and feed rate?

3. What does the material removal rate tell you in a machining operation?

4. What are the two axes that a turning process utilizes?

5. Explain the difference between contouring and forming in material removal.

6. How is threading performed in the turning arena for standard machine threads?

7. In turning, does the depth of cut reduce the diameter of the workpiece by that amount? If not, why?

8. For drilling, what is the standard drill point angle?

9. Reaming requires a starting hole that is undersized. What is the amount that the hole needs to be undersized based on the reamer size that will be used?

10. Milling utilizes three major axes of motion. What are they, and what movement does each perform?

11. What is the difference between climb milling and conventional milling, and what are the advantages and disadvantages of each?

12. Define the difference between face milling and peripheral milling.

13. What is meant by an approach distance, and why is it different for face and peripheral milling?

14. What is the difference between shaping and planning?

15. When sawing material, there are two rules that need to be followed. What are they?

Problems

1. A component part made of malleable iron material is to be created in a turning operation. The material starts out with a diameter of 2.250 in. and will be turned down to 2.143 in. with a length of cut of 4.750 in. Utilizing a feed of 0.0025 in./revolution and a cutting speed in the middle of the recommended range, what would the cutting conditions be for the operation?

2. Three turning cuts are to be made on a workpiece made of aluminum that is 4.750 in. long and starting out at a diameter of 1.500 in. The first pass has a depth of cut of 0.120 in. and a feed of 0.005 in./revolution. The second pass has a depth of cut of 0.090 in. and a feed of 0.008 in./revolution. The final finishing pass has a depth of cut of 0.020 in. with a feed of 0.002 in./revolution. The first two passes will utilize the cutting speed at the bottom of the range and the final cut will use the cutting speed at the top of the range. What are the cutting conditions for each pass, and what would the final diameter of the workpiece be?

3. If a turned workpiece with a diameter of 3.500 in. and a length of 11.250 in. made of refractory alloys is taking 17.361 min to process. If the cutting speed being used is in the middle of the recommended range, what would the feed need to be?

4. A component part made of hard stainless steel that is 1.125 in. thick will have a 0.1875 in. diameter hole drilled through it. Utilizing a feed of 0.0015 in./revolution and a drill with a tip of 118°, what would the cutting conditions be?

5. A workpiece constructed of glass-filled plastic that is 3.750 in. thick requires two holes to be drilled in it. The first is a through hole with a diameter of 0.375 in., and the second hole is a blind hole with a diameter of .4375 in. at a depth of 2.250 in. The 0.375 in. diameter drill has a point of 118° and will utilize a feed of 0.0025 In/Rev and the lower end of the cutting speed range. The 0.4375 in. diameter drill has a point of 125° and will utilize a feed of 0.0005 In/Rev and the upper end of the cutting speed range. What are the cutting conditions for both holes?

6. A large hole is to be drilled through into a workpiece made of medium-carbon steel that is 3.875 in. thick. This needs to be step-drilled, so it will be performed in three stages. The first drill is 0.250 in. in diameter using a feed of 0.0008 in./revolution and a cutting speed at the bottom of the material's recommended range. The second drill is 0.625 in. in diameter using a feed of 0.0035 in./revolution and a cutting speed in the middle of the material's recommended range. The final drill is 1.250 in. in diameter using a feed of 0.0015 in./revolution and a cutting speed at the top of the materials recommended range. All three drills have a point angle of 118°. What are the cutting conditions for each hole and the total machining time for drilling all three holes?

7. Eight holes are to be reamed into a workpiece. The reamed hole diameters are as follows:
   (1): #14, (2): "G", (3): 7/16, (4): 27/32, (5): #47, (6) "A", (7): 3/16, (8): 47/64
   What sizes do the holes need to be drilled in order to properly ream them to the desired sizes?
   Note: Give the closest *English DRILL Size*, not just the decimal value. (No Metric Drills are available.)

8. A milling operation is to be performed in two stages on a magnesium workpiece that is 9.750 in. long and 3.125 in. wide. The first stage will be accomplished with peripheral milling utilizing a cutter that is 8:000 in. in diameter and having 48 cutting teeth across the entire part to a depth of 1.625 in. This will use a feed of 0.0001 in./tooth and a cutting speed in the middle of the material's recommended range. The second stage will be performed with face milling to cut a slot down the center of the workpiece. A 1.500 in. diameter cutter with four cutting teeth will be used along with a feed of 0.0005 in./tooth and a cutting speed at the bottom of the material's recommended range. What are the cutting conditions for both stages of the operation?

9. A face milling operation is to be performed on a workpiece of hard titanium that is 7.750 in. long by 4.250 in. wide. The cutter being used has a diameter of 0.750 in. with six cutting teeth. A feed of 0.0008 in./tooth will be used along with a cutting speed at the bottom of the material's recommended range. If the entire surface of the workpiece needs to machined down to a depth of 0.095 in. and only 2/3 of the cutter width per pass can be used, what are the cutting conditions for the operation and what would the total machining time be?

10. A component part will be processed through three separate operations. It is made of hard cast iron and the starting size is 3.75 in. wide × 14.500 in. long × 2.250 in. thick. The first operation is peripheral milling operation, which will remove 0.625 in. from the top surface using a 4.750 in. diameter cutter with 12 cutting teeth. A feed of 0.0003 in./tooth will be used along with a cutting speed at the bottom of the material's recommended range. The second operation will be face milling inserting a slot down the center of the workpiece 0.110 in. deep using a 0.500 in. diameter cutter with two cutting teeth. A feed of 0.0007 in./tooth will be used along with a cutting speed in the middle of the material's recommended range. The third operation will be to drill a hole through the part that will be reamed later. The reamer size that will be used is a size "M." The drill has a drill point of 135° and a feed of 0.0012 in./revolution will be used along with a cutting speed at the top of the material's recommended range. What are the cutting conditions for each operation?

CHAPTER

16

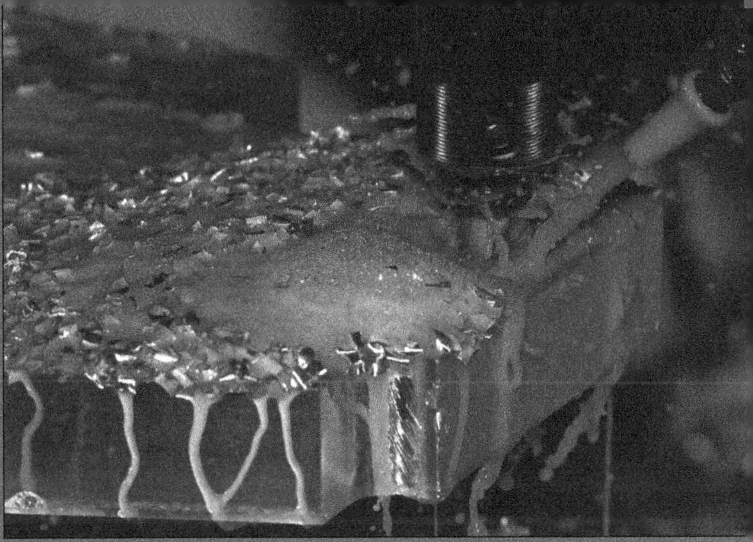

Abrasive Machining Principles

16.1 INTRODUCTION

Abrasive machining deals with the removal of material from the actions of hard, abrasive particles. As the particles rub across the surface of the work piece, chips of material are produced resulting in the removal process. Because chips of material are produced abrasive machining falls into the arena of traditional machining. For the most part, the chips being produced may not be visible to the naked eye, but are produced and can be evaluated under microscopic power.

There are several categories of abrasive machining with each being utilized either for attaining close tolerances, extremely fine surface finishes, or a combination of both. The most basic abrasive material removal would be the use of sand paper or emery cloth to smoothen the surface finish of the material. As this is rubbed across the surface, the abrasive particles of the paper or cloth will remove a light amount of material to either improve the surface finish or to reduce the size by a small amount. As with all manufacturing processes, there are both advantages and disadvantages that need to be understood. These are described below.

Abrasive Machining Advantages

- *Materials*: The abrasive machining process can be performed on all types of materials. Even materials that are difficult or not possible with machining can be performed with abrasive material removal. Glass, ceramics, carbides, and tool hard steel can all be machined with abrasive machining and can be performed easily.
- *Tolerances*: The closest tolerances possible are attainable with the abrasive machining processes. Depending on the process being utilized, tolerances from ±0.0001 in. to ±1 millionth of an inch are possible. When extreme tolerances are required the measurement and inspection will require special equipment and in some cases a very controlled environment to ensure a precise reading.

- *Surface Finishes*: Abrasive machining can render the smoothest surface finishes of any manufacturing process. Depending on the process being used and the particle size of the abrasive material, surface finishes smoother than glass can be attained. The roughness of a surface finish is measured by R_a readings which is measuring the micro inches that the roughness is. Figure 3.11 in Chapter 3 displays R_a readings of various materials and process yields. Looking at these readings, superfinished gage blocks have a smoother reading than glass. This was obtained through an abrasive machining process.

Abrasive Machining Disadvantages

- *Material Removal*: Unlike standard machining, very light material removal is accomplished with abrasive machining. The abrasive particles will remove the material based on the particle size and process being utilized. As surface finish and tolerance requirements become more critical, less material will be removed to accomplish the desired requirements.
- *Time Consuming*: All of the abrasive machining processes are extremely time consuming and costly. Owing to the fact that extremely small amounts of material are being removed and along with obtaining close tolerances and extreme surface finishes, long processing times are required. For component parts requiring this, they will go through several operations and will be extremely costly and time consuming.

There are several manufacturing processes utilized in the abrasive material removal arena. The most common and widely used are *grinding*, *honing*, *lapping*, *superfinishing*, *buffing*, and *polishing*. These are described below.

16.2 GRINDING PROCESSES

Grinding is a process that utilizes a wheel constructed of bonded abrasive particles that is rotated as it is moves along the surface of the work piece. This process is the least precise being capable of rendering tolerances of ±0.0001 easily. Grinding can be performed on flat surfaces, round configurations, and internally. These are described later in this section in more detail.

16.2.1 Grinding Wheel Construction and Wear

Wheels utilized for grinding are constructed of hard, abrasive particles which are bonded together. There is a hole located in the center of the wheel to allow for mounting on an arbor of a machine that will provide the power and rotational requirements. These wheels that rotate are at very high RPMs to perform the material removal and require balance and concentricity. As the wheels wear down they will be re-faced to maintain the cutting ability and surface finish requirements.

To produce grinding wheels, the hard abrasive particles are mixed with a binder material that is loaded into a die and pressed. The wheels are then placed in a furnace to cure the bonder material bonding the particles together. Unlike the components produced in powder metal, the binder does not leave the finished product. The wheels will consist of three elements contained within them which are *Grains*, *Binder*, and *Pores*. The combination of these elements is shown in the equation below: with the cross section of the structure illustrated in Figure 16.1.

$$\mathbf{P_g + P_b + P_p = 1.0}$$

P_g = Portion of Grains

P_b = Portion of Binder

P_p = Portion of Pored

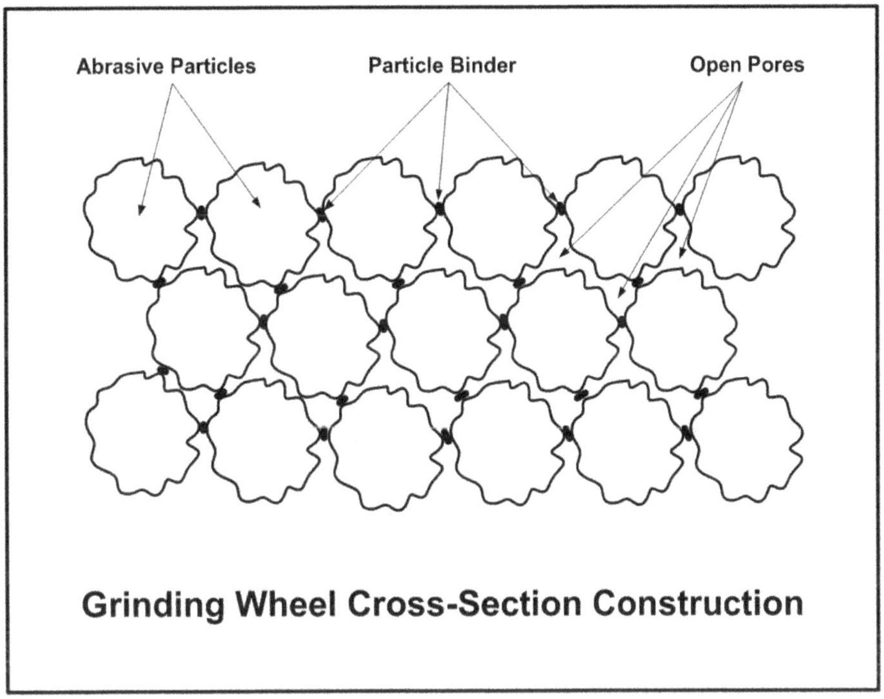

Figure 16.1 Typical Grinding Wheel Internal Structure

Wheels are available in a multitude of materials, sizes, and configurations. Four of the most common utilized are listed below:

Grinding Wheel Types

- *Aluminum Oxide* Al_2O_3: These are white grinding wheels and the most commonly utilized. They are used for all ferrous metals and materials within the steel arena.
- *Silicon Carbide* S_iC: These are green grinding wheels and are harder than the aluminum oxide wheels. They are used on nonferrous materials such as aluminum, brass, bronze, and copper. Also, they perform very well for brittle material such as cast iron and on hard materials such as stainless steels.
- *Cubic Boron Nitride cBN*: These are gray grinding wheels and are used when grinding hardened tool steels.
- *Diamond*: A diamond grinding wheel is not constructed like the other wheels are. These are made by mixing diamond particles with rubber that is molded onto a steel wheel as shown in Figure 16.2. Diamond wheels are used for extremely hard materials such as carbides, ceramics, and glass. With the diamond particles contained in the rubber, the energy from coming into contact with the work piece is absorbed by the rubber which the diamond grains perform the cutting.

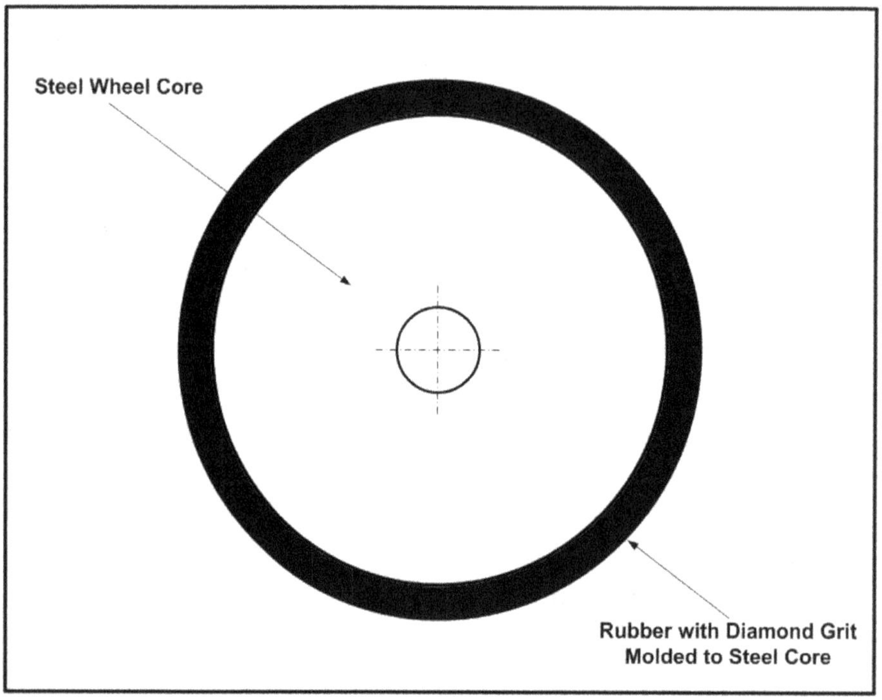

Figure 16.2 Diamond Grinding Wheel Construction

As the grinding wheel is moved across the material surface removing the material it will wear down in the process. As with any manufacturing process the operation needs to be optimized to accomplish the removal in the most efficient manner. There are four considerations in grinding that need to be accounted for in order to obtain the most effective material removal, which is listed below.

Grinding Operation Considerations

- *Optimize Surface Finish*: With grinding being a finishing process for the component part, the surface finish being attained is important. This is critical on the last grinding pass of the wheel along the work piece.
- *Maximize Material Removal Rate*: Material removal rate determines how much material is being removed from the work piece. Maximizing this factor will allow the operation to be performed in the minimum amount of time.
- *Minimize Heat Damage*: The grinding process is capable of generating a good amount of heat that can damage the work piece. Coolant needs to be applied in order to minimize the damage to the work piece surface and to maintain a low work piece temperature to attain extremely close tolerances.
- *Select Proper Wheel*: Wheel selection is important in order to maximize the operation. If the wrong wheel is selected for the material being ground, it will either wear down rapidly or not cut the material properly resulting in an inefficient operation.

The chart in Figure 16.3 illustrates the three possible zones while grinding. Zone 1 is used for a finishing cut where very little material is being removed to bring the work piece into the final dimension and maximizing the overall surface finish. Zone 2 is where the operation should be to maximize the material removal rate while having the least amount of wheel wear during machining. Zone 3 should be avoided due to the fact that the wheel is wearing down faster, but the material removal rate is not much more than Zone 2.

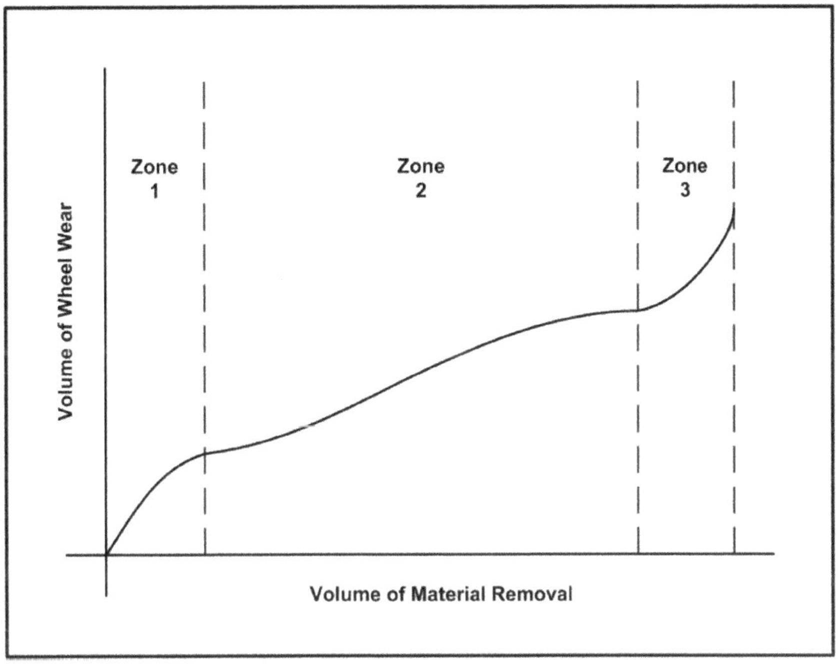

Figure 16.3 Wheel Wear vs. Material Removal Chart

16.2.2 Surface Grinding

Surface grinding is a process that deals with grinding flat surfaces. A grinding wheel is rotating and is moved across the surface of the work piece in a straight line removing the depth of cut that is desired. As a general rule of thumb, the depth of cut for most ferrous materials is about ±0.005 in. maximum. After the wheel has completed the cutting path, it is indexed over a specified amount and another pass is made. The cutting can be performed in both directions due to the high surface speed of the grinding wheel. In almost all cases, coolant is applied in a flood manner to reduce the amount of heat being generated on both the work piece and the grinding wheel. Under very controlled conditions, surface grinding can produce a tolerance of ±0.0001 in. depending on the depth of cut being considered and the grinding wheel selected for the operation. Figure 16.4 illustrates a typical surface grinding setup.

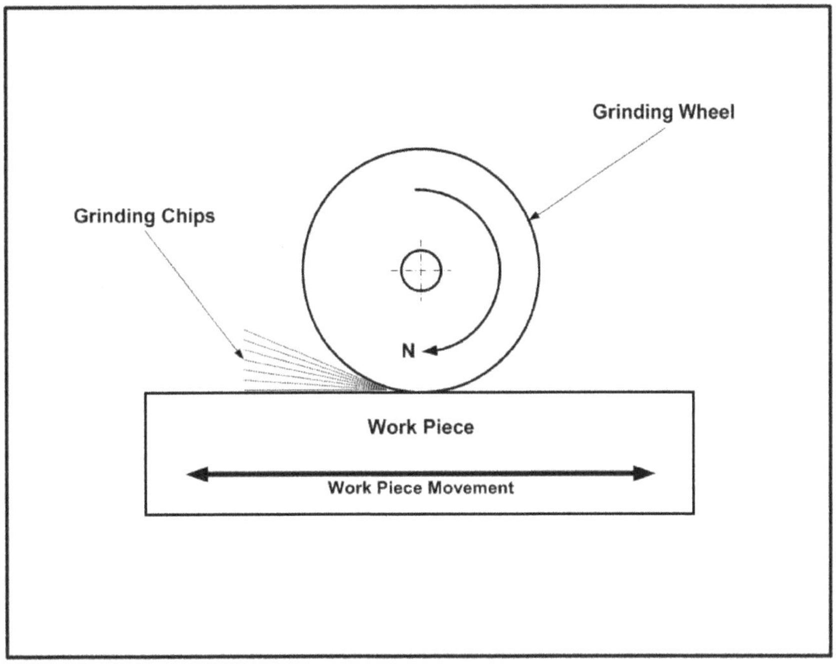

Figure 16.4 Typical Surface Grinding Operation

16.2.3 Cylindrical Grinding

This grinding process deals with grinding the outside diameter of a round work piece. This is accomplished by rotating the work piece and the grinding wheel so that they are rotating in unison. Figure 16.5 shows an end view of how the work piece and wheel rotation is set for this type of grinding operation. As the wheel is in contact with the work piece, it is moved slowly across from one end to the other to remove the material. Figure 16.6 shows a top view of the wheel movement along the work piece for a cylindrical grinding operation.

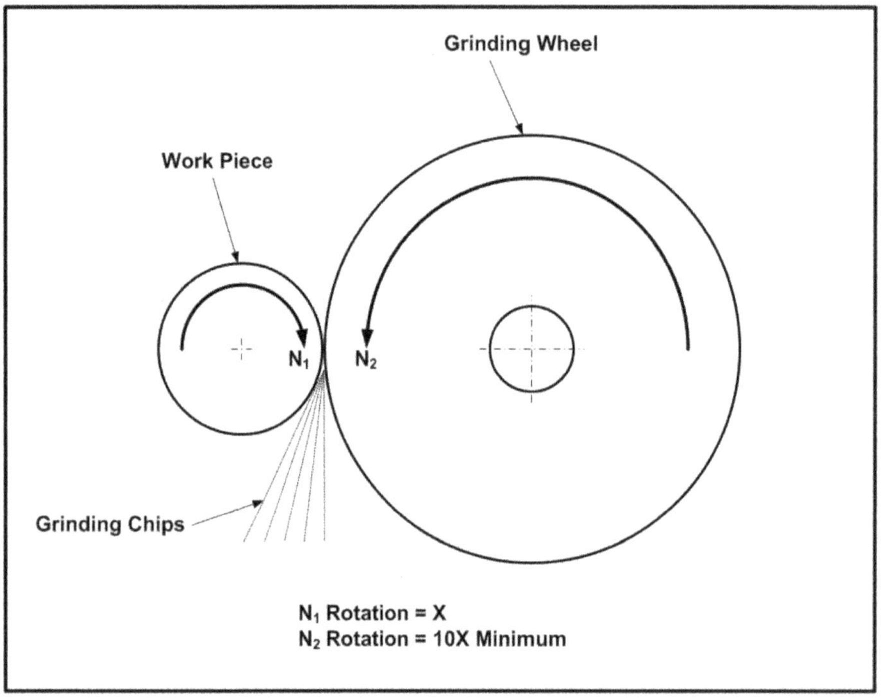

Figure 16.5 Cylindrical Grinding Operation Side View

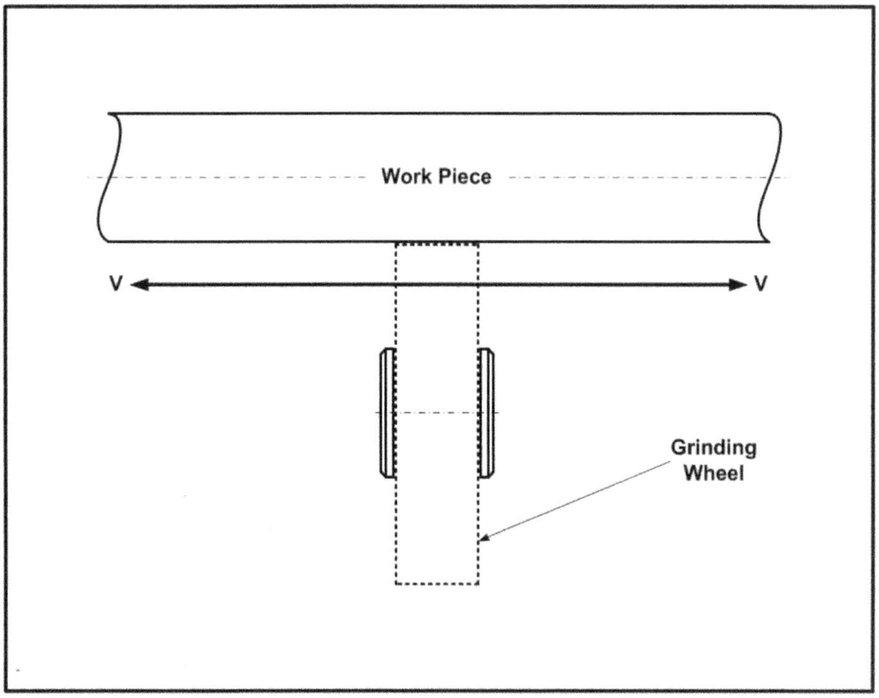

Figure 16.6 Cylindrical Grinding Operation Top View

16.2.4 Internal Grinding

Internal grinding operations are performed very similar to cylindrical grinding but done internal on the work piece. The wheel and work piece rotation is identical to cylindrical grinding as shown in Figure 16.7 displaying an end view of an internal grinding operation. The grinding wheel is held on a mandrel that moves in and out of the work piece illustrated in Figure 16.8, which shows a section view of a work piece and grinding wheel working internal on the part.

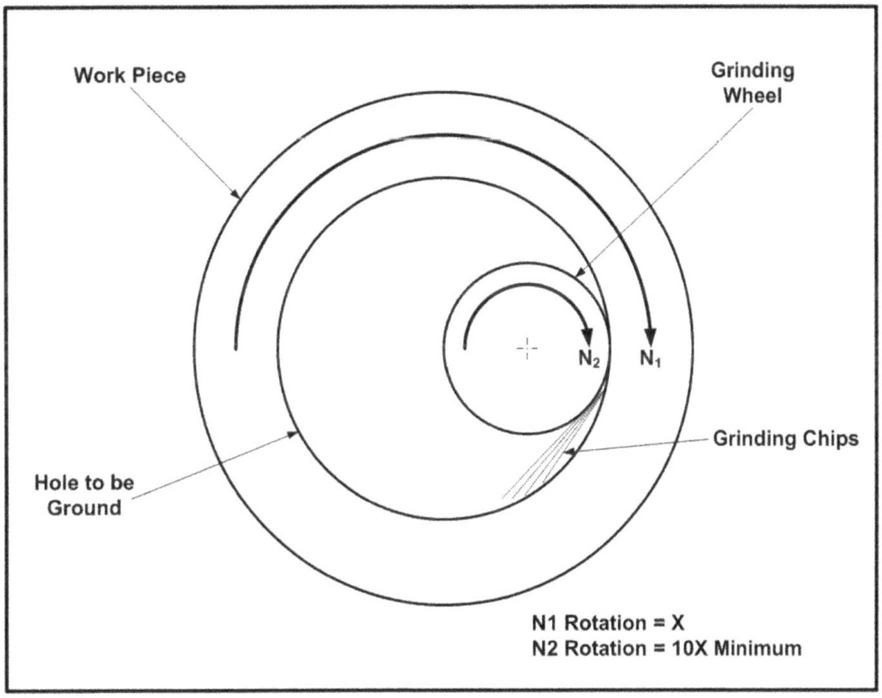

Figure 16.7 Internal Grinding Operation End View

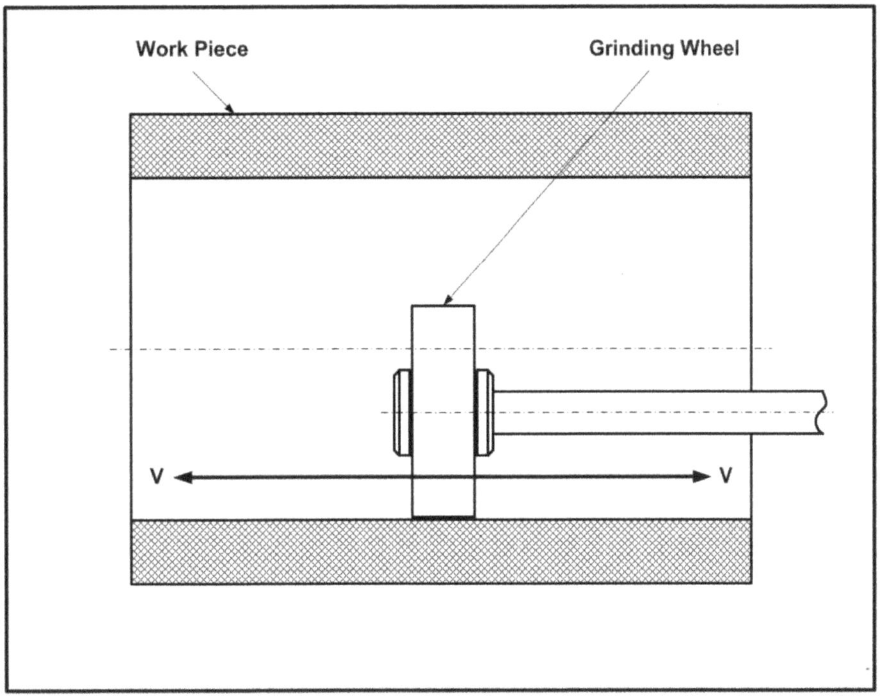

Figure 16.8 Internal Grinding Operation Section View

16.2.5 Grinding Calculations

Calculations for grinding are required to perform the operation as effectively as possible. The factors that need to be considered are *wheel surface speed, material removal rate, length of the chip,* and *number of chips being generated.* These equations are listed below:

Wheel Surface Speed: This is the cutting speed of the grinding wheel as it moves across the work piece removing the material

$$v = \pi D N$$

v = Wheel Surface Speed (SIPM)
D = Wheel Diameter
N = Rotational Speed of the Wheel (RPMs)

Material Removal Rate: This is the volumetric measure of how much material is being removed over a given timeframe

$$\mathbf{MRR} = V_w W d$$

MRR = Material Removal Rate (in.^{3}/min)
V_w = Work Piece Speed (SIPM)
W = Width of Cut
d = Depth of Cut

Length of Chip: This determines how long the chip being removed in the operation will be

$$L_c = \sqrt{Dd}$$

L_c = Length of Chip
D = Wheel Diameter
d = Depth of Cut

Number of Chips: This determines how many chips will be generated over a given timeframe

$$n_c = V W C$$

n_c = Number of Chips (chips/min)
V = Wheel Surface Speed (SIPM)
W = Width of Cut
C = Grit per Unit Area (particles/in.)

Grinding Example

A work piece is to be flat ground with a 12.000 in. aluminum oxide grinding wheel with a grit count of 450 particles/in. The wheel is rotating at 3750 RPMs and is 0.750 in. wide. The depth of the cut considered will be 0.0005 in. and the work piece is moving under the wheel at a speed of 16 SIPM.

Step 1: Determine the cutting speed of the wheel

$V = \pi D N$
$V = \pi\,(12.000)(3750)$
$V = 141{,}371.7$ SIPM

Step 2: Determine the material removal rate

$\mathrm{MRR} = V_w W d$
$\mathrm{MRR} = (16)(0.750)(0.0015)$
$\mathrm{MRR} = 0.018$ in.^{3}/min

Step 3: Determine the length of the chip

$$L_c = \sqrt{Dd}$$
$$L_c = \sqrt{(12.000).0005}$$
$$L_c = 0.0775 \text{ in.}$$

Step 4: Determine the number of chips

$$n_c = V\,W\,C$$
$$n_c = (141371.7)(0.750)(450)$$
$$n_c = 47{,}712{,}938 \text{ chips/min}$$

16.3 HONING PROCESSES

Honing is an internal abrasive process that is normally used after internal grinding, but not all the time. This process is used to bring the tolerances for an internal diameter in closer, or to obtain a given surface finish. Honing is done with a honing tool that has spring-loaded abrasive sticks that run in and out of the hole. Figure 16.9 illustrates an end view of the hole with the honing tool inserted into it and the abrasive sticks in contact with the work piece. The tool is rotated, but slowly and not like the rotation required for grinding. Figure 16.10 displays a section view of the honing tool moving in and out of the hole while rotating. This process removes very little material and the amount is dependent on the spring pressure exerted against the abrasive sticks. The surface finish is controlled by the grit count in the abrasive sticks that are removing the material from the work piece. Honing can accomplish internal tolerances of ±0.00005 in. or better depending on the spring pressure and grit count being utilized in the operation.

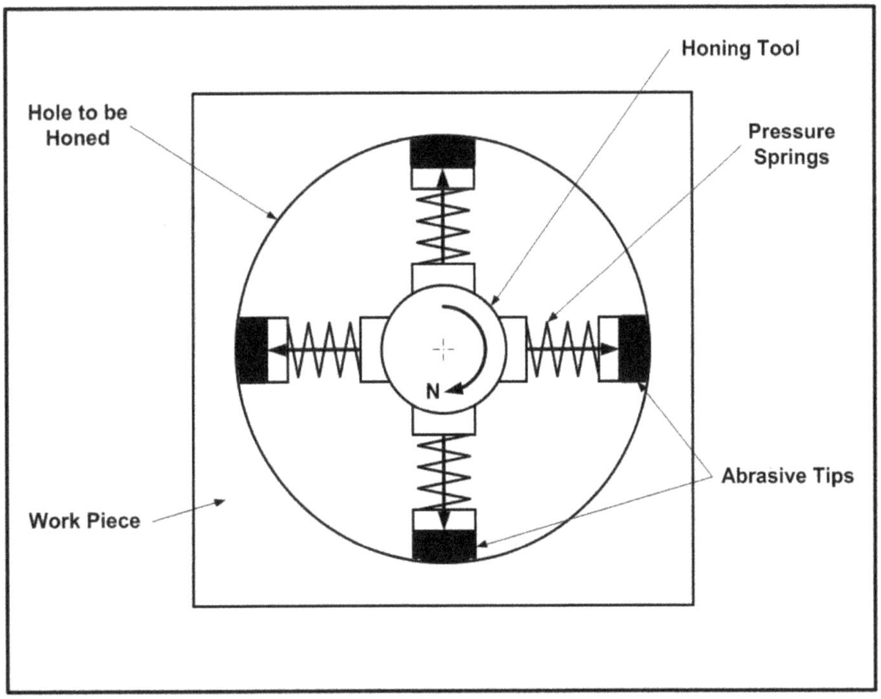

Figure 16.9 Honing Operation End View Showing Tool Set-up

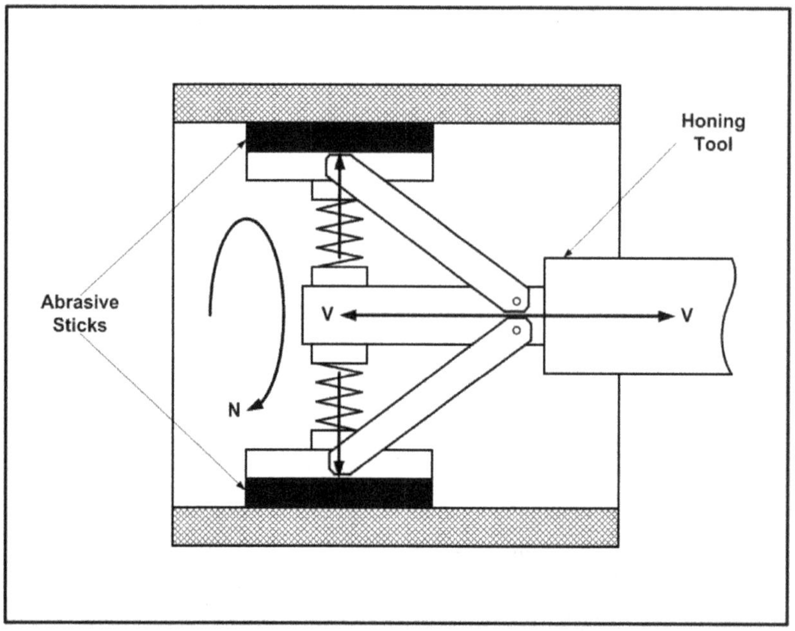

Figure 16.10 Honing Operation Section View

16.4 LAPPING AND SUPERFINISHING PROCESSES

The lapping process works with flat surfaces to improve the surface finish and bring the work piece into a closer tolerance. A lapping block, which is a flat cast iron block with small slots cut into it, is coated with a lapping compound. Lapping compound consists of a fine grit that is mixed with grease. The compound is spread on the lapping block and then the work piece is placed on the grit coated block and moved back and forth at a given frequency with a downward force. Lapping is capable of producing extremely fine surface finishes and tolerances to ±0.00001 in.

Superfinishing is performed exactly the same as lapping but with a much finer grit and at a higher frequency with a lighter downward force. The process is the most precise material removal process that is capable of producing the smoothest surface finishes and the closest tolerances. When performed under extremely controlled conditions, superfinishing can produce component parts to within ±0.000001 in. and surface finishes smoother than glass. Figure 3.11 in Chapter 3 shows various surface finish readings with the smoothest being superfinished gage blocks. Figure 16.11 illustrates a typical lapping/superfinishing operation.

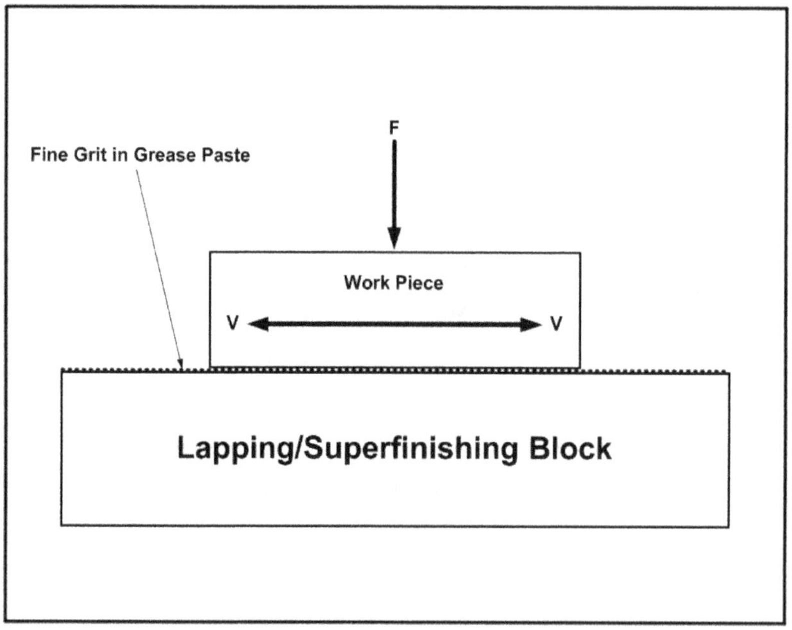

Figure 16.11 Lapping/Super Finishing Operation Set-up

16.5 POLISHING AND BUFFING PROCESSES

Polishing and buffing are primarily utilized for improving appearance or for preparation required for additional processing. This process in most instances is performed by hand and utilizes what is called a *rag wheel.* This is a wheel that is constructed of several layers of cloth material, which is held together in the center. The wheel rotates on an arbor and a polishing compound is applied to the wheel. The work piece is then run along the rotating wheel to buff and polish it to the desired surface finish. Several component finishing operations such as plating and anodizing require the component parts to be buffed and polished in order to produce a given appearance.

16.6 SUMMARY

Abrasive material removal processes are very widely utilized to produce component parts that require extremely close tolerances and very smooth surface finishes. From grinding processes to superfinishing processes, the capability of producing components of close tolerances and smooth surface finishes cannot be attained with any other manufacturing processes. With grinding capable of producing tolerances of ±0.0001 in. and superfinishing producing tolerances of ±0.000001 in., a range of extreme tolerance capability exists within the process arena. However, as extreme tolerances and surface finish smoothness is required, the cost and time to attain them goes up dramatically. With all the abrasive material removal processes removing very small amounts of material, the time to accomplish the tolerance and surface finish requirements is long. With time equating to money, the longer the time the higher the component part cost will be.

16.7 QUESTIONS AND PROBLEMS

Questions

1. How does the abrasive machining process remove the material from the work piece?

2. What are the advantages of abrasive machining over other component fabrication processes?

3. A grinding wheel has three components that make up the construction of the wheel. What are they and explain what each does in the grinding process.

4. A diamond grinding wheel is constructed differently than other wheels. Explain how a diamond wheel is constructed and why.

5. In grinding, what is the difference between wheel surface speed and work piece surface speed?

6. Explain how honing is performed and what is it used for?

7. Of all the abrasive machining processes, which is the most precise and what tolerances and surface finishes are possible to attain?

Problems

1. A grinding operation is being performed on a steel work piece. A 9.000 in. diameter wheel rotating at 2000 RPMs will remove 0.001 in. from the parts surface. With the wheel being 0.500 in. wide and having a grit count of 900 particles/in., it will be moving across the work piece at a speed of 10.000 SIPM. What are the cutting conditions for the operation?

2. A cylindrical grinding operation is going to be performed on a brass work piece. A grinding wheel which is 12.000 in. in diameter will be rotated at 3,500 RPMs and move across the work piece at 5.500 SIPM and will produce a cut which is 0.250 in. wide. If the process requires the length of the chips to be 0.060 in. long and the number of chips to be 364, what would the depth of cut and the grit count need to be in order to accomplish the operation correctly?

3. An internal grinding process is being performed to bring a hole into the final dimension. A 1.500 in. diameter wheel with a grit count of 1,200 particles/in., which is 0.375 in. wide, is being used. If the wheel surface speed is 4,250 SIPM and the material removal rate is 0.0021 in.^{3}/min, what would the rotational speed and depth of cut need to be along with the length of the chip and the number of chips produced in the operation?

4. Two grinding operations will be performed on a component work piece. The first is a roughing grinding cut using a 10.500 in. diameter wheel which is 0.625 in. wide. The grit count for this cut is 650 particles/in. and will be rotating at 2,500 RPMs. For the roughing cut the work piece surface speed will be 14.500 SIPS and the depth of cut of 0.0012 in. will be considered. The second step is a finishing cut still using a wheel that is 10.500 in. in diameter which is 0.625 in. wide. The grit count for this cut is 2,500 particles/in. and will be rotating at 3750 RPMs. For the finishing cut, the work piece speed will be 7.75 SIPS and a depth if cut of 0.0003 in. will be considered. What would the cutting conditions be for each of the grinding cuts being taken into account?

5. A cylindrical grinding operation is to be performed on a stainless steel work piece. The grinding wheel being used is 0.750 in. wide and will be rotating at 3,750 RPMs. A depth of cut on the work piece will be 0.0011 in. and the grinding wheel will have a surface speed of 125,173 SIPM. If the operation has a material removal rate of 0.010 in.^{3}/min and is producing 7,040,980 chips/min, what would the diameter of the wheel, the grit count of the wheel, the surface speed of the work piece, and the length of the chip be?

CHAPTER

17

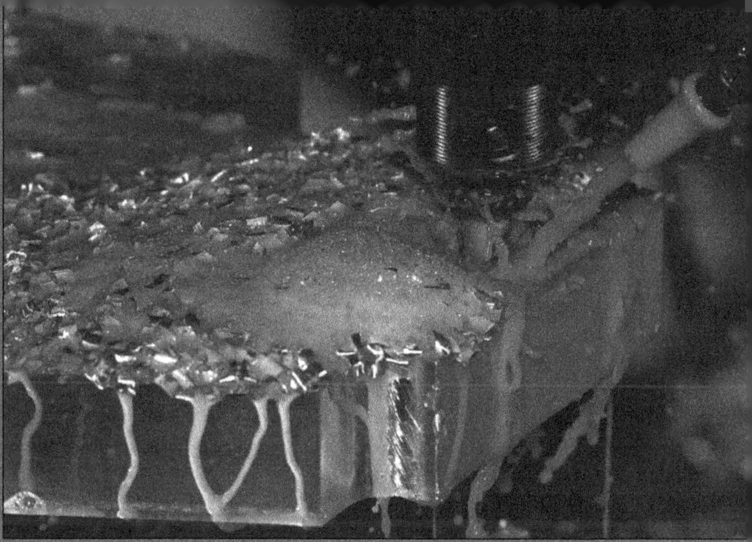

Non-Traditional Machining Concepts and Theory

17.1 INTRODUCTION

The non-traditional machining category of processes is under the material removal umbrella, but do not perform the same as the machining and abrasive processes. During the removal of the material in this category of process, for the most part no chips of material are produced during removal. With this category of processes not performing in a traditional manner, the material is vaporized, pulverized, or dissolved rather than producing chips that can be accounted for. Depending on the process being utilized and the material being processed, it can be either rapid or slow, but for the most part this category can hold tolerances equal to the machining group. This chapter will discuss some of the more common non-traditional manufacturing processes and where they would be utilized.

17.2 NON-TRADITIONAL PROCESSES

The non-traditional arena of material removal processes encompasses a variety of methods each of which utilizes a unique method to perform its tasks. Depending on the operation being performed, the material can be removed using electricity, water, vibration, laser beams, or chemicals. This section will discuss some of the more common non-traditional processes being utilized in manufacturing that are *electro discharge machining, ultrasonic machining, hydrodynamic machining, laser machining, plasma machining,* and *chemical etching.*

17.2.1 Electro Discharge Machining (EDM)

Electro discharge machining (EDM) as it is more commonly referred to, is a process that utilizes electrical current to remove the material from the work piece. Here, the material is burned away from the work piece as it is being removed. The work piece is positively charged and the EDM tool is negatively charged. As the tool comes into contact with the work piece, it creates a circuit and arcs burning the material away from the part. A solution called *dielectric fluid* is used to enhance the electrical contact between the cutting tool and the work piece, and to also allow for a more constant contact surface. The concept of EDM machining is simple, yet very complex to obtain the configurations and tolerances desired. As with all manufacturing processes, there are some guidelines that need to be understood which are explained below.

Conductive Material: EDM can only be performed on materials that are electrically conductive. With the work piece being charged positive and the cutting tool charged negative, the material from the work piece gets the material burned away during the process.

Material Hardness: The EDM process can be performed on soft materials or materials that are tool hard such as drills or machining cutters. The burn rate for the operation would be almost the same for soft or harder materials.

Process Precision: This process is capable of producing tolerances to within ±0.0005" under very controlled conditions and multiple cuts. This precision is required for constructing components that will be used in production tooling to produce manufactured parts in production.

Process Speed: EDM is an extremely slow process. As the electrical discharge removes the material, it does so in very light removals. For the most part, EDM equipment is automated and once setup and activated can run unattended. Operations using EDM can have several different machines and can set them up during the day, and then allow them to run until night or for extremely long jobs over a weekend.

EDM can be performed in two methods, which are sinking EDM and wire EDM. Both utilize the same process for removing the material, but accomplish it in different manners rendering different configurations. These two processes are explained below.

17.2.1.1 Sinking EDM Machining

The sinking EDM process operates utilizing a cutting tool which has the configuration desired that is moved up and down on the work piece to burn the material away. These machines have three main axis, which are "X," "Y," and "Z." The position of the tool is set with the "X" and "Y" axis while the "Z" axis moves the tool up and down on the work piece to burn the material. The desired configuration is contained in the tool and not from moving the "X" and "Y" axis. Some sinking EDM machines have a forth rotational axis that will allow the tool to rotate as it is plunged into the work piece for unique configurations.

With the material being conductive, the cutting tools require the same. A main material that is utilized for producing cutting tools for sinking EDM is graphite. These are produced with conventional machining to the required configurations allowing for the overburning that will occur during material removal. Depending on the size of the configuration and the tolerances required, it may require several burning tools starting with rough material removal down to final configuration cutting.

Sinking EDM is the main process utilized in producing cavities for injection molds. The cavity material will be hardened to the desired level and then the cavity configuration is burned into the mold. From there all that needs to be done to the cavity is to polish it out to the desired smoothness required. Prior to the introduction of sinking EDM, the mold cavities had to be hand machined into the soft metal with the allowances for material shrinkage and hardening. After hardening, the cavities would be hand ground and polished to the final dimensions. This was extremely labor intensive and with EDM running unattended, the mold building process has been reduced dramatically. Figure 17.1 shows how the sinking EDM process operates with the cutting tool plunging up and down on the work piece causing a burning arc.

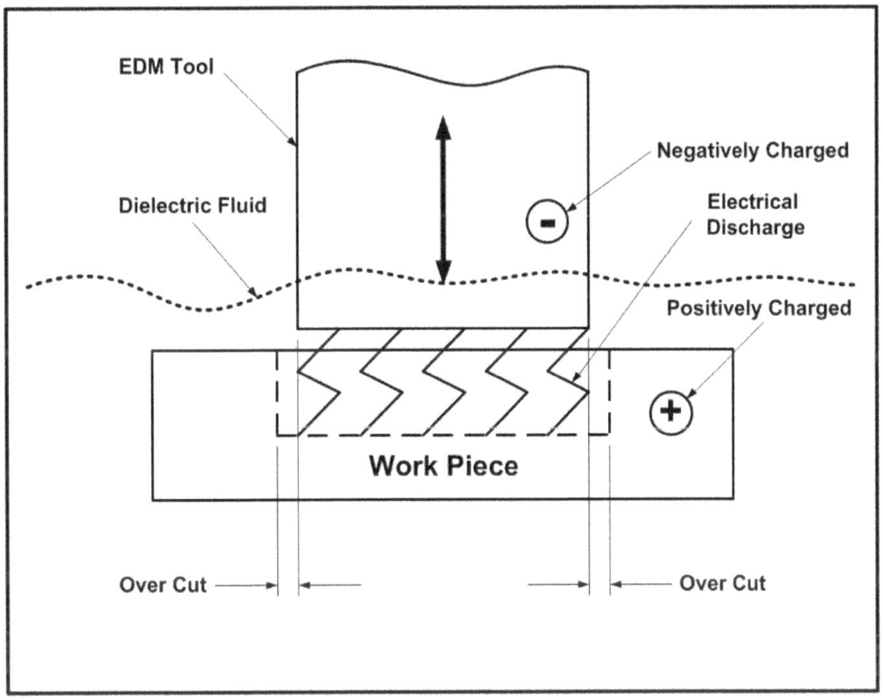

Figure 17.1 Sinking EDM Concept Showing Discharge Removal of Material

17.2.1.2 Wire EDM Machining

Wire EDM machining operates with the same electrical discharge process but utilizes a wire that runs through the work piece. This requires a starting hole in the work piece to thread the wire through to get the process started and the machine moves "X" and "Y" to the programmed pattern. The wire is charged negative that allows it to burn through the material as the machine moves slowly through the programmed pattern. Figure 17.2 illustrates how a coil of charged wire is supported on the top and bottom and is pulled through at a given rate. Once the wire has burned the material, it is not reusable and will get discarded.

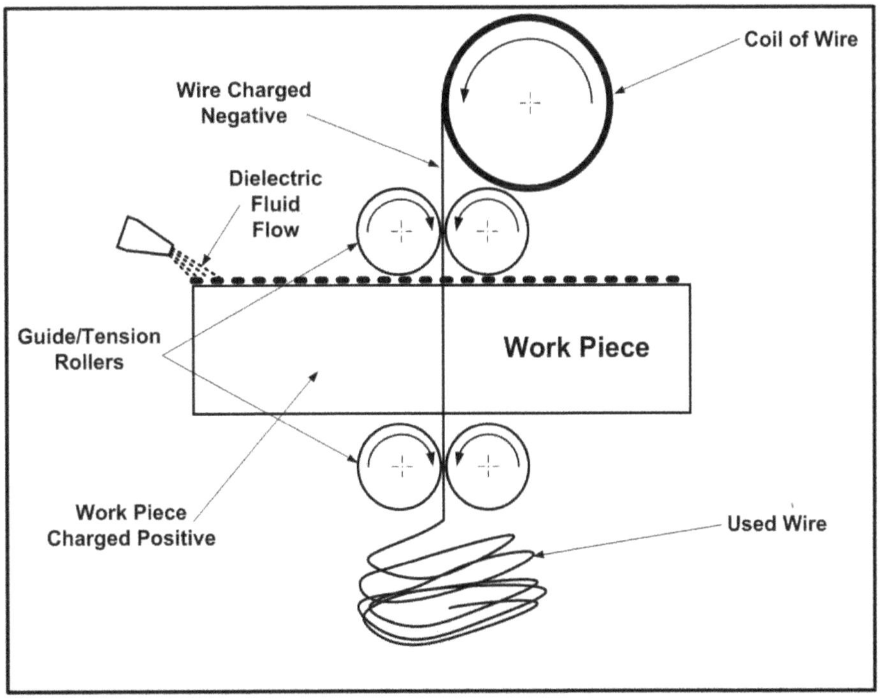

Figure 17.2 Wire EDM Concept Displaying Charged Wire Removing the Material

When the process is burning the material, there will be a cut which is as wide as the wire diameter plus the overburn on both sides of the wire that needs to be compensated for. Figure 17.3 shows the wire diameter and the overburn on both sides of the material while burning out the configuration. This process is used very similar to a band saw cutting out a pattern in a piece of material, but can maintain a tolerance of ±0.0005" under controlled conditions. Figure 17.4 displays a configuration that was burned out of a piece of hardened material with the wire EDM process.

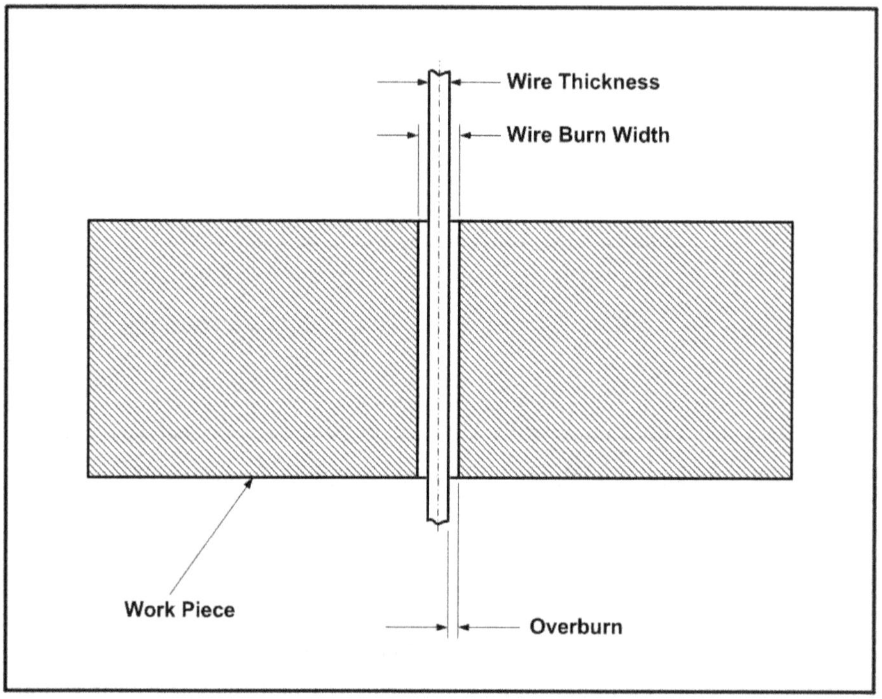

Figure 17.3 Wire EDM Overburn during Cutting

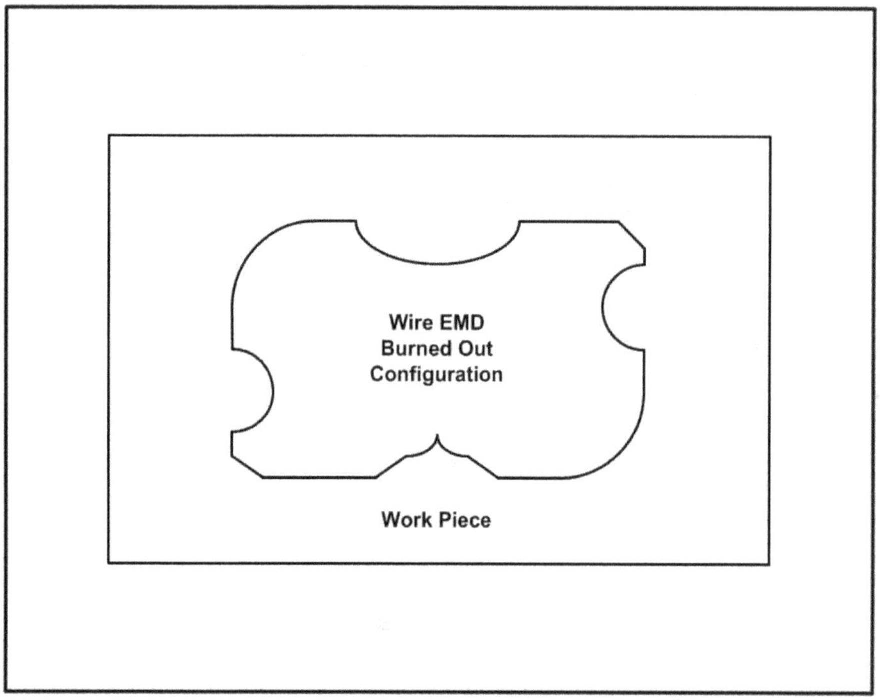

Figure 17.4 Illustration of a Configuration Burned Out Utilizing Wire EDM

With this process burning out a configuration through a work piece, it makes it a perfect application for producing a punch and die setup for sheet metal working. We recall from Chapter 13 that a clearance is required for the punch to perform correctly, the slug being burned out of the work piece and becoming the punch. By calculating the overburn on both sides of the wire, a wire diameter can be selected so that the total burn width will equal the required clearance for the punch. Figure 17.5 shows how the burned out slug from a work piece can be used for the punch of a die with the total wire burn width equaling the required clearance of the punch.

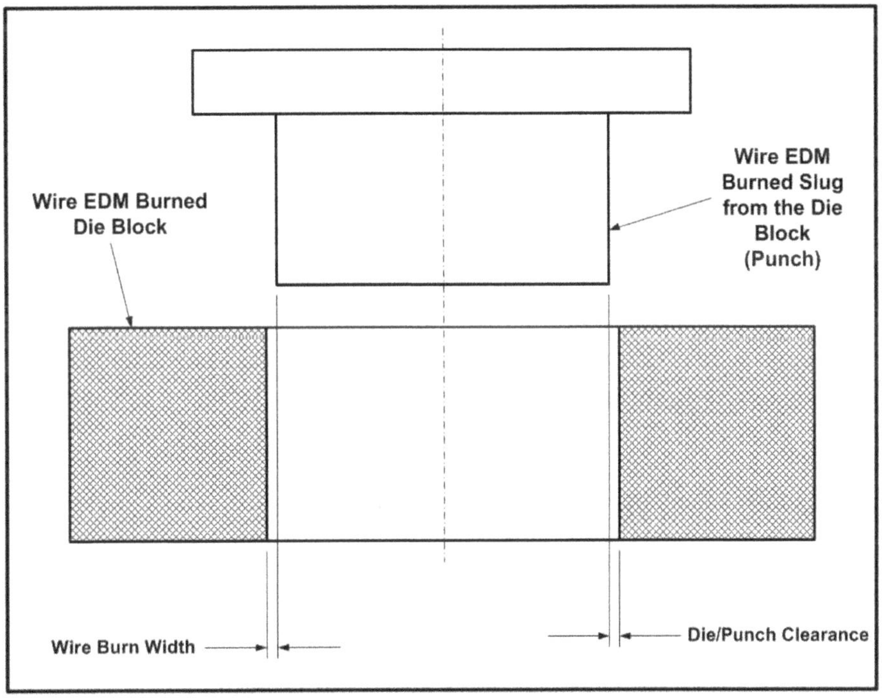

Figure 17.5 Wire EDM Produced Sheet Metal Punch and Die

17.2.2 Ultrasonic Machining

The ultrasonic machining process operates very similar to the sinking EDM process. Where EDM utilizes an electrical discharge, ultrasonic machining uses a high frequency vibration and a slurry grit to cut the material. A cutting tool known as a horn is fabricated from hardened steel and is moved up and down against the work piece. A slurry consisting of a grit and H_2O of between 20 and 60% grit is utilized and rubs against the horn and work piece as it makes contact. The horn is vibrated at a frequency of between 20,000 and 30,000 Hz which provides the cutting motion of the material. Figure 17.6 illustrates an ultrasonic machining operation where the grit slurry is rubbed between the horn causing the configuration to be cut along with the over cutting on the sides.

This process is not as precise as the sink EDM process and cannot operate on hard materials very well. With the cutting horn being constructed of hardened metal, a work piece of equally hard material would only produce the same wear on the tool as on the work piece. The advantage with ultrasonic machining is that it can be performed on materials that are non-conductive. Plastic materials are easily machined with this process to shape configurations into the material.

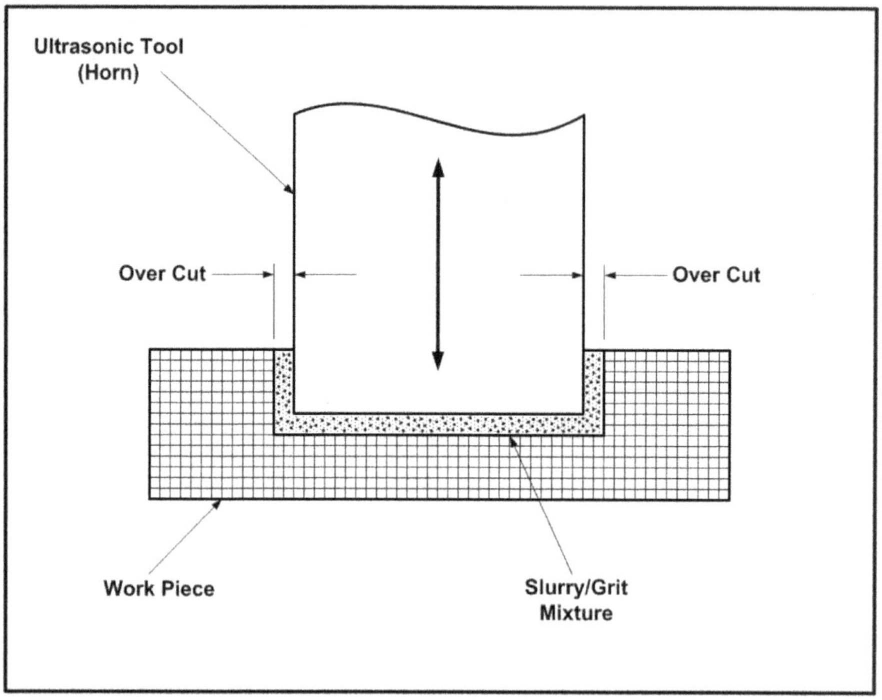

Figure 17.6 Ultrasonic Machining Concept Illustration

17.2.3 Hydrodynamic Machining (Water Jet Cutting)

Hydrodynamic machining, more commonly known as water jet cutting, utilizes a high pressure beam of water to cut the work piece material. Water is run through a nozzle that has an opening of between 0.004" and 0.016" at a pressure of 60,000–150,000 PSI and a velocity of 3,000–5,000$^{ft}/_{sec}$. This process is utilized to cut through material either as a cut-off process or to produce a cut-out shape in it similar to how wire EDM produces an internal void.

To setup and start the process, the water jet nozzle is set at a given stand-off distance from the surface of the work piece. This distance is similar to focusing a lens in order to get the optimum efficiency. The water jet is activated and is left stationary to cut completely through the work piece, producing a starting point. From there the table is moved in both the "X" and "Y" axis to produce the desired configuration. The desired configuration can be either the shape of the cut-out in the material or the material being cut-out. Depending on the material being cut and the thickness of the material, the cutting rate can range from 12 $^{in.}/_{min}$ to 1,200 $^{in.}/_{min}$.

There is a limitation of material types and thickness for this process. As the process has been refined since its conception, it is capable of cutting through 0.500" thick aluminum now. Some of the considerations that need to be addressed in hydrodynamic machining are nozzle wear and the blow-through angle. With the process utilizing extremely high pressure and velocity, the internal diameter of the nozzle gets worn down producing a larger diameter that will either deliver an oversized cut or require the stand-off distance to be adjusted. The blow-through angle is an angle that is produced during the cutting action. This process will not produce a straight-line cut, but will rather have an angular cut with the bottom cut of the work piece being wider than the top cut. Figure 17.7 illustrates the hydrodynamic machining process which is very simple when compared to laser and plasma machining.

17.2.4 Laser Machining

With laser machining, a concentrated beam of light is focused on the work piece to perform the material removal. The power is set for the laser and a stand-off distance is established to focus the beam to the work piece. Power is given to the laser and it starts to cut into the work piece either to a given depth or completely through. Depth cutting is used in applications such as engraving into the surface of the material while through cutting can produce flat blank components identical to sheet metal stampings. Laser cutting is used frequently for short-run manufacturing to produce flat blank sheet metal components when the cost of building a stamping die cannot be justified. Figure 17.7 shows the similarity between hydrodynamic and laser machining.

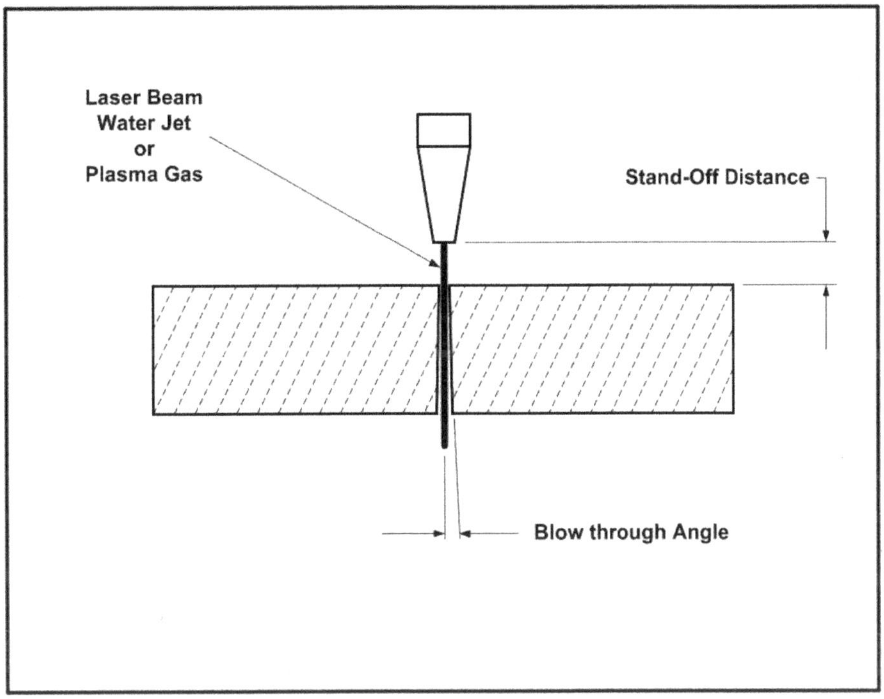

Figure 17.7 Laser/Plasma/Hydrodynamic Machining Concept and Theory

17.2.5 Plasma Machining

Plasma machining is very similar to laser machining but utilizes superheated, electrically ionized gas to perform the cutting. An inert gas such as argon is charged and heated to a temperature of 18,000–25,000 °F and blown through a nozzle. This heated gas literally vaporizes the work piece material as it cuts through. Plasma cutting cannot cut to a depth the way laser machining can but can cut through the material to produce flat blank sheet metal parts. As with both laser machining and hydrodynamic machining, a given stand-off distance is required to focus the energy of the plasma beam on the work piece material. Figure 17.7 again illustrates the plasma machining operation and the similarities to both laser and hydrodynamic machining.

17.2.6 Chemical Etching

Chemical etching is a material removal process that requires four basic steps to produce the desired component. These steps are *cleaning and masking*, *selected unmasking*, *etching*, and *de-masking*. First the work piece must be cleaned to remove any substances that would interfere with either the masking or etching processes. Once cleaned, a masking material is applied over the entire component. This material is one that the chemical utilized to etch the work piece and will not affect it at all. Once the masking material has been applied and allowed to dry, the second step is to strategically cut away the masking material in the areas of the work piece where etching is desired. The masking material that will remain on the work piece will prevent the etching chemical from affecting those areas. After the masking material has been removed in the areas where etching is desired, the third step is the physical etching. The work piece is submerged into a chemical that will dissolve the work piece in the areas where the masking has been removed. The depth of the etching is dependent on two variables which is the concentration of the etching chemical and the time that the work piece is submerged into the chemical. The last step of the process is to remove all of the masking material to expose the desired etched configuration on the component.

Etching can be very closely controlled for the depth with the chemical concentration and time allowed to be submerged. However, the etched will always eat under the masking material and if left for long can completely dissolve the component. Figure 17.8 displays the four stages of chemical etching and shows how the chemical will dissolve under the masking to a certain level.

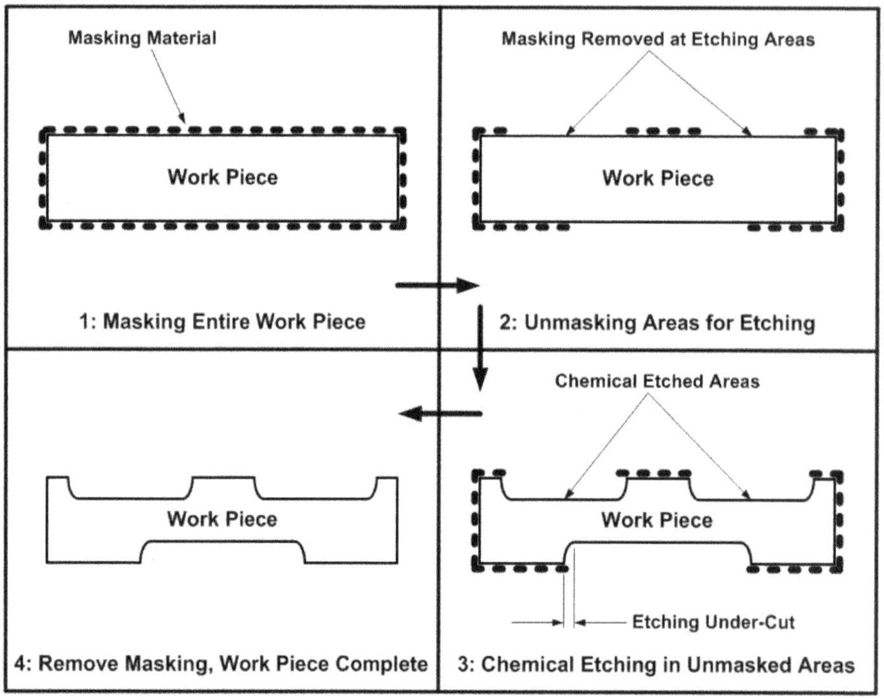

Figure 17.8 Chemical Etching Process Stages

17.3 SUMMARY

The non-traditional material removal arena is a very viable process set for manufacturing. Configurations that are not possible to construct with the traditional material removal processes are possible utilizing the non-traditional processes. Although some of these processes are extremely slow in the removal of the material, they can work with materials that are tooling hard where the traditional processes cannot easily. By including these unique processes in the manufacturing portfolio, the fabrication capabilities of the operation will be greatly expanded.

17.4 QUESTIONS

1. With the non-traditional material removal processes not producing chip like the traditional processes, what happens to the material being removed?

2. In EDM, what is the charge on the cutting tool and the work piece?

3. What does the dielectric solution in EDM machining do during the operation?

4. Explain the difference between sinking and wire EDM machining.

5. What is meant by a slurry mixture in ultrasonic machining and what is it comprised of?

6. Why cannot tool hard materials be cut with the ultrasonic machining process?

7. In hydrodynamic machining, what is the pressure and velocity range that is normally utilized for cutting the material?

8. For hydrodynamic machining, laser and plasma machining, what is meant by a stand-off distance and why is this critical to the operation?

9. What are the four stages required for chemical etching?

10. To control the depth of etching, what are the two variables that need to be controlled?

CHAPTER

18

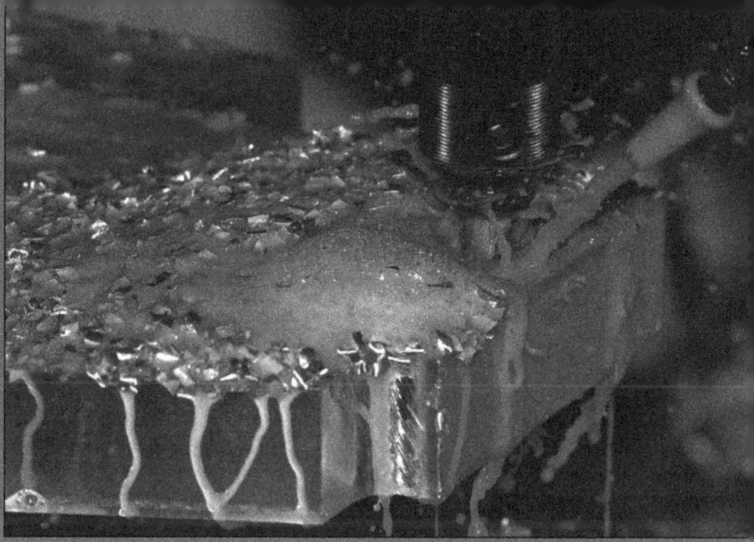

Machining Economic and Design Principles

18.1 INTRODUCTION

In any business manufacturing included, operating as economically as possible is critical to its overall success. By very closely monitoring the time that is required to produce the component parts and working to keep this as optimized as possible, the efficiency of the operation will be maximized and the cost minimized. Cost is, without a doubt, the most critical factor in any business. Every business will strive to keep their operating costs as low as possible in order to maximize the profits of the company. No business, including nonprofit businesses, can survive by losing money.

When performing an economic analysis in manufacturing, it is done from the bottom up. The first analysis is a microanalysis where each individual operation for a given component part is evaluated to determine the exact time and cost. Each of these operations is then rolled up for a total time and cost for that particular component part. Figure 18.1 illustrates a typical microanalysis and rollup for a given component part. Once all the component parts have had a microanalysis constructed, the second stage is to construct a macroanalysis. For this, the time and cost for each component are rolled up along with any assembly time and cost and other factors that are included in the complete product fabrication. Figure 18.2 shows a typical macro rollup of a product.

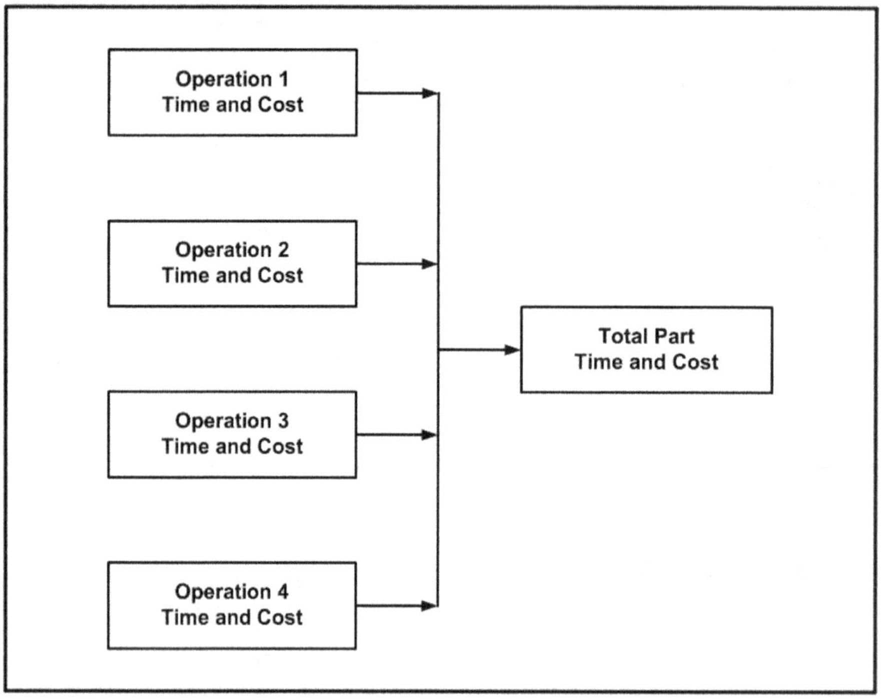

Figure 18.1 Typical Micro Time and Cost Development for a Single Component Part

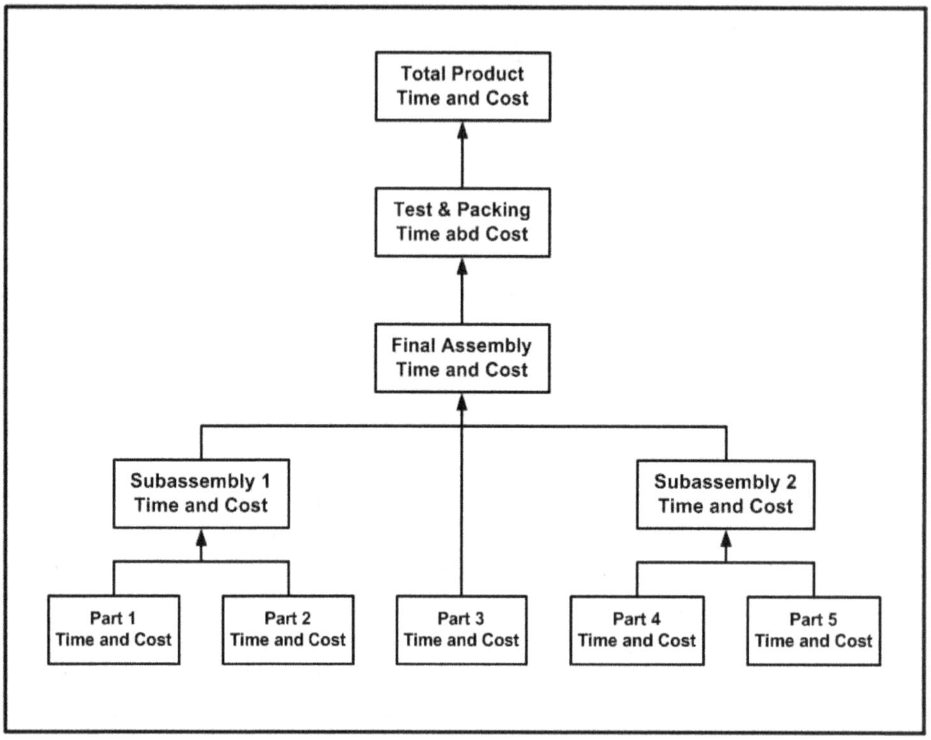

Figure 18.2 Typical Macro Time and Cost Structure for a Product Rollup

This chapter looks at how machining operations are evaluated economically on a microscale. It discusses the economic criteria for machining and the cutting conditions that are used to develop the overall operation. Along with these, the optimization of an operation will be presented along with what portions of an operation add value to the component and how costs are derived.

18.2 ECONOMIC MACHINABILITY CRITERIA

From an economic viewpoint, the machinability of an operation is based on four criteria and considerations. Each of these criteria will encompass several elements, which either have been covered or will be covered from a micro viewpoint. These areas that are considered for economic machinability are *Tool Life*, *Force and Power*, *Surface Finish*, and *Chip Disposal*, which are discussed as follows.

Tool Life Considerations: The life of the cutting tool is dependent upon several factors within the operation. By extending the life of the tool, the operation will be more efficient because the time to replace the tool will be greatly reduced as will be the cost of the tool. The four main considerations for extending the life of the cutting tool are listed as follows:

- *Machine*: The selection of the machine tool that will be performing the operation can affect the life of the cutting tool. If the machine is not capable of delivering the required force or supporting the workpiece properly, the cutting tool will experience faster wear and will require replacement more often.
- *Material*: The material being machined will greatly affect the life of the cutting tool. The harder the material, the more rapidly the cutting tool will wear down and require replacement. Even under perfect machining conditions, harder material will wear a cutting tool faster than softer material.
- *Geometry*: Having the proper geometry of the cutting tool for a given material and setup will increase the tool life. For any operation, the rake, lead, and clearance angles along with the nose radius all need to be precisely calculated for the material being machined as discussed in Chapter 14.
- *Feed Rate*: The rate that the cutting tool will be moving at across the workpiece can affect the life of the cutting tool. If an extremely high feed rate is being utilized, then a stronger force will be exerted on the cutting tool. Even if the machine tool is capable of handling a deeper cut and an aggressive feed rate, the cutting tool will wear down much faster than if the parameters were reduced.

Force and Power Considerations: The amount of force and power required to perform the material removal will affect the economics of the operation. The more the force that is required to perform the operation, then more power is required. Power needs to be generated or purchased, so the higher the power requirement, the higher the cost to supply it to the operation.

Surface Finish Considerations: Controlling the surface finish on a workpiece will require closer monitoring of the operation. From an economic viewpoint, a smooth surface finish will be more costly than a surface that does not have to be controlled. When controlling a surface finish on a workpiece, there are four main considerations that need to be addressed, which are listed as follows:

- *Cutting Tool*: The geometry of the cutting tool is critical to maintaining a good surface finish. In some instances, the same cutting tool can be used, which is being utilized for the general material removal, but will need to be in excellent condition. In other cases, a different cutting tool for the finishing cut will be required. Either way, there will be time required to replace the cutting tools along with the cost of the tools.
- *Material*: The material being processes will affect the surface finish. Every material has an internal structure, which affects the machining properties. In order to control the surface finish, along with the cutting tool, the feed and speed will need to be adjusted down to accomplish the required R_a finish.
- *Feed/Speed*: As mentioned, obtaining a smooth surface finish will require the feed and speed of the cutter to be greatly reduced. A much lower feed rate will produce a smoother surface finish on the workpiece. However, the slower the feed rate in the operation, the longer the cutting time will be, thus creating a higher cost to the operation.
- *Vibration*: Vibration in the setup of a machining operation will affect the surface finish greatly. This will cause the workpiece to oscillate during the cutting, creating a rough finish on the workpiece. Vibration control is performed in the manner the workpiece is being held in the machine tool. The addition of supports, damping, and additional clamps may be required to reduce the vibration created during machining. Also, reducing the feed and speed along with the depth of cut being taken will assist in controlling the harmonic generated during the material removal.

Chip Disposal Considerations: Disposal of the chips being generated during a machining operation can be viewed in two different manners. The first is to get the chips away from the workpiece during the actual machining to avoid scratching of the machined surface. This removal is also required after the completed workpiece is removed from the machine prior to loading the next one. The chips need to be cleaned and removed from the holding fixture to ensure that the workpiece will be located and clamped properly for processing.

The second area is the physical disposal of the chips themselves. Machining generates a vast amount of chips during the process, which eventually will have to be disposed of. In today's environment, chips will be recycled rather than being brought to a land fill. Proper recycling requires that the chips be kept separate based on the material being machined. Along with the recycling, some metals are considered hazardous materials because of the alloys contained in them. Stainless steel, e.g., contains a high alloy percentage of chromium, which is classified as a hazardous material. These need to be kept separate regardless of recycling or not.

18.3 ECONOMIC CUTTING CONDITIONS

The cutting conditions from an economic viewpoint will be primarily dealing with time. As the time increases, so does the cost of the component workpiece. Several of the elements that are considered in the cutting conditions have been covered in detail in the section on machinability criteria. These conditions for machining from an economic viewpoint are listed as follows:

Feed First, Speed Second: When calculating the cutting conditions for a material removal operation, the first consideration is the feed. This will determine the rate at which the material will be removed along with the associated time. The feed will also be a major factor in determining the surface finish that will result from the cut. The slower the feed, the smoother the surface finish. Once the feed has been determined, the cutting speed will be applied. As shown in Figure 15.1 in Chapter 15, the cutting speed is very material dependent with a range to apply based on the material and any extraneous elements in the operation.

Cutting Tooling: The cutting tool will be performing all the removal of the material during the operation. Ensuring that the tool material and geometry are correct for the material being machined and the setup of the operation is critical. Also, the cost of the cutting tool needs to be a consideration as well. The tools will wear out eventually, even under perfect machining conditions and require replacement. So, every time the cutting tool is replaced, the cost of the tool will be absorbed into the processing of the workpieces. Section 18.4.2 on operational time and cost discusses how the tool cost is added to the cost of the component workpieces.

Material Constraints: As mentioned in the section on machinability criteria, the material being machined will have to be considered. The hardness, internal structure, and setup of the machine need to be evaluated so that the machining time can be reduced without affecting the removal operation.

Depth of Cut: How deep of a cut the process requires is important for time considerations. The more material that can be removed without affecting the operation will allow for a shorter machining time in the operation. If the material can be removed in three passes vs. five passes, the processing time will be less along with the cost. This is covered in Section 18.4.2 in more detail.

Surface Finish: Maintaining a surface finish will require adjusting the cutting conditions in order to render the required R_a finish. This normally will need a lower feed and cutting speed along with either a new cutting tool or a different cutting tool. With these alterations to the process, the machining time will increase and the cost of the cutting tool will need to be included in the operation.

Vibration: Any vibration in the setup of the material being machined will require alterations to the machining conditions. If the workpiece will vibrate during machining, then the feed, speed, and depth of cut will be greatly reduced to compensate for the vibration occurring during the material removal. By adjusting these parameters, the machining time will be increased along with the cost of the operation.

18.4 PROCESS OPTIMIZATION

The process for fabricating component parts and assembling the final product requires each step to be optimized. What this means is that both the time and cost should be at a minimum for each stage of the overall manufacturing process. By reducing the processing times, the operation can increase their capacity for additional work. Also, with time equaling cost, the less time it takes to perform the operation, the less the component part or the assembly will cost. Time and cost are connected to each other in manufacturing and form a closed loop system. Both the processing time and cost are critical for the operation along with a constant evaluation of whether the time of operation will add value or not to the process. These elements for manufacturing economics are described in detail in the next two sections.

18.4.1 Value Add and Non-Value Add Elements

A value add element in manufacturing is one that alters or improves *Form*, *Fit*, or *Function* of the component part or product. By altering any of the three mentioned attributes of the component part or assembled product, the operation increases the value of what it is producing. The value add tasks within the operation are what physically produce the component parts and the completed product for the company. Although they are few in number, they constitute full attention to ensure that they are functioning as efficiently as possible. Following is a list of the value add elements within a manufacturing operation:

Value Add Elements

Machine Time: The machine time is the time during which the material is processed. This can be the time for the cutter to remove the material, for the die to stamp out a sheet metal part, or for the mold to be filled to produce an injection molded component. What is not included in the machine time is the time required to load the part into the machine and the unload time to remove the part after the processing has been completed.

Assemble Time: Assemble time is when two or more component parts are being joined in some manner to produce either a subassembly or a completed assembly of a product. The time that is required to orientate the parts that will be joined, along with the time required to join them together either by welding, screws, or rivets, falls under the assembly time umbrella. This assembly time can be extremely long depending on how many component parts are required in the finished product. If we were to look at the automotive industry and a completely disassembled vehicle, the number of parts required to produce a finished product is astounding.

Function Adjustment Time: Adjustment time does add value to the product but is a category, which, through close monitoring of the manufacturing operations of the component parts, should be at a minimum. Here, final adjustments of the product are performed in order to ensure that it will perform properly and to the customer's needs. For standard products, this stage should be kept to a minimum and be controlled throughout the fabrication process by maintaining dimensions and tolerances. However, for custom products being fabricated, the adjustment time is critical in order to have the product perform to the customers' requirements.

Non-value add elements in manufacturing are those tasks that need to be performed, but do not alter the *Form*, *Fit*, or *Function* of the component part or product. Looking at the three areas listed, which constitute the value add times in manufacturing, there are several other areas that need to be performed but add no value to the component part or to the product. In all manufacturing operations, the non-value add times are always reviewed to attempt to minimize the times that are required but do not contribute directly to the value in any manner. Following is a list of the non-value add elements within a manufacturing operation:

Non-Value Add Elements

Machine Loading Time: This time is required for loading the component part in the machine for processing. This includes picking up the part from the bin it is located in, assembling it into the fixture in the machine, locking the part into the fixture with clamps or in a vice, closing any safety doors, and activating the machine cycle.

Machine Unloading Time: This time starts directly after the machine time has been completed. The safety door is opened, all clamping devices are loosened, and the component part is removed and placed aside. Once the part has been removed, the holding fixture may need to be cleaned before the next piece can be loaded into it for processing.

Tool Changing Time: In all the material removal processes, the cutting tool being utilized will require changing at some point of time. When this is performed, the machine tool will be stopped and the old cutting tool will be removed and a new one installed. After the new tool is installed, it will require setting it up to establish the "X," "Y," and "Z" axes to the machine. This time for replacing the cutting tool will start after a given number of component parts have been produced and a small portion of the overall time will be granted to the cycle time based on the number of parts a given tool will produce.

Part Inspection Time: Although many feel that inspection times are a value add element, they are not. Inspection is a task that only confirms that the operation that was just performed truly did produce the component to the required dimensions and tolerances. During inspection, the Form, Fit, and Function of the component or product are not altered in any manner, merely confirmed for confirmation.

Part Packing Time: After the component parts have been machined and/or inspected, they need to be packed in some type of container to be moved either to the next operation or to the assembly area. If the operation is in business to only produce component parts, then they will require packaging in order to be shipped to the customer. Depending on the requirements, this stage can be lengthy if the finish and surface finish need to be protected. For instances like this, the packing many include placing the part in a protective bag and be packed in special containers in order to not damage the critical areas.

Material Movement Time: Material movement time is when the component parts or the product needs to be transported from one location to another. For parts, it could be the movement from one machine tool where an operation has been completed to another for additional processing. During this moving time, no value is added to the parts or product in any manner.

System Testing Time: System testing time falls into the same category as inspection. This is normally performed on the finished product to confirm that the functions are operating correctly. No value is added to the product; only a confirmation of the functionality is being performed.

Both the value add and non-value add elements are required in a manufacturing operation, but the balance between them requires constant monitoring. The objective is to minimize the non-value add times and maximize the value add times. It is impossible to remove 100% of the non-value add time within an operation, but by reducing it, the operation will be able to maximize the amount of time being utilized to produce products.

For example, let us look at a football game. The actual playing time is four quarters for 15 min per quarter. This equates to a value add time of 1 h. However, with the clock stopping between plays, team time outs, and half-time, the game runs a total time of about 3 h. If we view this as a percentage, we would have 33.3% value add time in the game.

By constant evaluation of methods to reduce non-value add times in the operation, the overall efficiency will greatly increase. Figure 18.3 shows an operation being performed in two different ways. Both produce the exact same results, but the total time is greatly different. The original structure of the operation has all the tasks being performed in order, resulting in a 9-min total cycle time. When the operation was modified, the inspection and packing were moved to be performed internal to the machining time, resulting in a 6-min total cycle. Viewing this from a value add/non-value add percentage, the original operation had a 38.9% value add time and the modified operation increased to 58.3% value add time. Nothing changed in the actual machining time, just simply moving the elements around increased the value add time and reduced the total cycle time, allowing for additional processing for the operation.

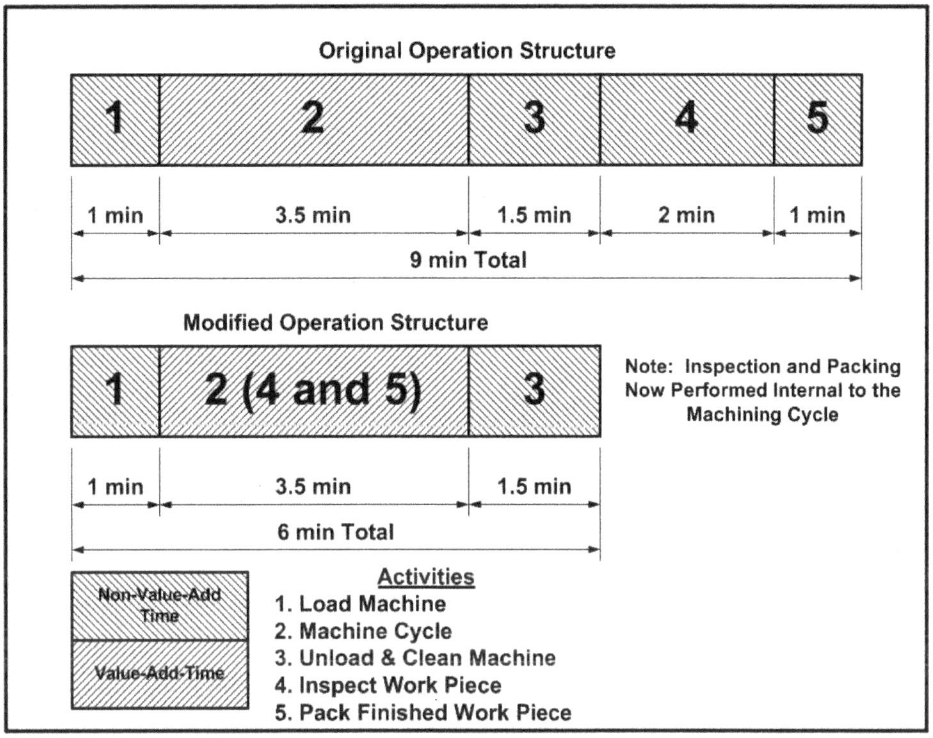

Figure 18.3 Value Add and Non-Value Add Times

18.4.2 Operational Time and Cost

As mentioned earlier, time equals money in a manufacturing operation. The longer it takes to process a workpiece or to perform an assembly operation, the more that particular process will cost. Calculating the time for any given processing step will provide the operation important information for calculating the capacity required based on the individual time per piece and the annual requirements.

Calculating the time required on a per part basis will include both value add and non-value add times. This combination of times will equate to the total time required to perform the given operation and be utilized for determining the production schedule and overall capacity of the business. The following equation shows the elements that are included in a typical processing cycle time calculation. For every operation, not all the elements may be included, but the possible ones are included in the equation along with whether they are a value add or non-value add task.

$$T_c = T_l + T_m + T_u + T_i + T_p + T_t/n_p$$

T_c = Cycle Time per Part (min)

T_l = Part Loading Time (min); Non-Value Add

T_m = Part Machining Time (min); Value Add

T_u = Part Unloading Time (min); Non-Value Add

T_i = Part Inspection Time (min); Non-Value Add

T_p = Part Packing Time (min); Non-Value Add

T_t = Tool Change Time (min/Number of Pieces); Non-Value Add

n_p = Number of Pieces per Tool (min)

As you can see, the majority of the elements in the equation are non-value add to the overall cycle time. This is why, as mentioned in Section 18.4.1, the non-value add elements need to be constantly evaluated to minimize them when and wherever possible. By reducing the times, combining them with value add times or just eliminating them will increase the capacity and reduce the cost of the operation.

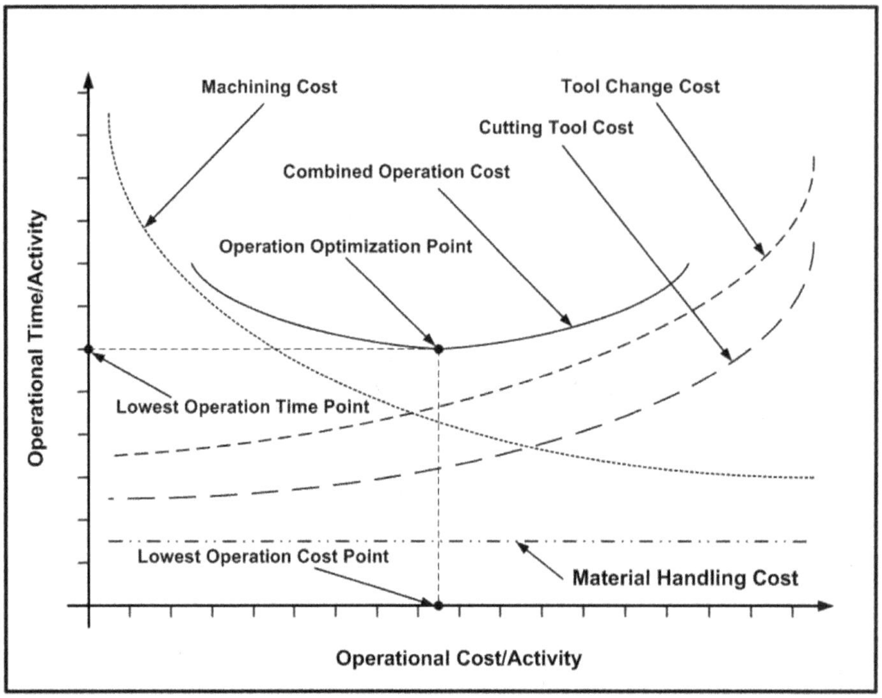

Figure 18.4 Operational Optimization Curve

The cost of the operation will utilize the times in the aforementioned equation along with two other elements. The first is the operational cost. Every operation will incur a cost that is developed from all the costs it takes to run the operation. This is always given in dollars per hour and is referred to as the *Fully Absorbed Cost*. The second element is the cost of the perishable tooling that is used in the operation. As the cutting tool wears down, it will need to be replaced at a given frequency, which is normally after a certain number of parts have been fabricated. The following equation is used to calculate the cost for a given operation utilizing the operational cost and the cutting tool cost with the individual times for the operation.

$$C_p = T_l C_o + T_m C_o + T_u C_o + T_i C_o + T_p C_o + T_t/n_p C_o + C_t/n_p$$

C_o = Operational Cost \$/h.

C_t = Tool Cost

When optimizing an operation, the objective is to minimize both the cost and time to the optimum levels. This becomes a balancing issue with the machining time and other elements that affect the process. By simply reducing the machining time as low as it can possibly get may not be the most economical method for processing the component part. Figure 18.4 illustrates an optimization curve, which shows that as the machining time and cost decrease, the time and cost for replacing the cutting tool increase. At some point, the processing time may be less, but the cost will begin to increase. Finding the optimization point for each operation is critical to balance both the processing time and cost within the operation.

Following are two example problems showing how both the processing time and costs are calculated for a microlevel economic analysis.

Example Problem 1

A given machining operation is being performed. The time to load the machine is 1.152 min with a machining time of 8.528 min. It takes 1.079 min to unload the part from the machine and 2.857 min to inspect it. The final task is to pack the finished part, which takes 1.325 min. If the cutting tool requires changing every 300 workpieces and it takes 17.528 min to change it, what is the total time per workpiece for the operation?

$T_c = T_l + T_m + T_u + T_i + T_p + T_t/n_p$

$T_c = 2.251 + 8.528 + 1.079 + 2.857 + 1.325 + 13.719/300$

$T_c = 14.987$ min/unit

Example Problem 2

With the times given in example problem 1, the additional data is provided. The cutting tool that is being used has a cost of $110.00 and the cost for the operation is $125.00 per hour. How much would the cost per part be?

$C_p = T_lC_o + T_mC_o T_uC_o + T_iC_o + T_pC_o + Tt/n_pC_o + Ct/n_p$

$C_p = (2{,}251)(125/60) + (8.5288)(125/60) + (1.079)(125/60) + (2.857)(125/60) + (13.713/300)(125/60) + (110/300)$

Cp = $31.589/unit

18.5 SUMMARY

Economic analysis on both a micro and a macro level is critical for a manufacturing facility. By performing both and constantly working to perfect each level to reduce the time and cost, a higher profit margin can be realized for the business. Understanding the criteria required in manufacturing for favorable economic results will allow the business to benefit economically and increase both revenue and capacity. This constant evaluation of the manufacturing operation is necessary to be able to control and monitor the cost of the products being produced. Without this analysis, the business will not fully know what the product is costing to fabricate and how high their profit is. More than one business has failed by not understanding the importance of strong economic analysis from the bottom up.

18.6 QUESTIONS AND ANSWERS

Questions

1. What is the difference between a micro and a macro economic analysis?

2. What are the four elements for economic machinability criteria, and what is their importance?

3. There are two considerations for chip disposal. What are they?

4. What are the factors that can affect the surface finish of a workpiece in machining?

5. When establishing the cutting conditions, why is the feed established before the speed?

6. What are the three criteria that determine a task is value add to the process?

7. Inspection is an important area to ensure that the product meets the design requirements. Is this a value add or a non-value add task, and explain your answer.

8. When a process is optimized, will the machining time always be as low as it can possibly be? Explain your answer.

Problems

1. A component part will be machined in a given operation. It will require 1.184 min to load the part into the machine and 6.418 min to perform the actual machining. Once the machining is completed, the unloading time is 1.068 min and it needs to be inspected, which takes an additional 2.118 min. After inspection, the part will be packed requiring 0.624 min. The tool being used costs $135.00 and will require replacement every 250 pieces, which will take 12.727 min. If the cost to run the operation is $105.00 per hour, what will the processing time and cost per part be?

2. An operation needs to produce 35,000 units of a given component. The times required to perform the operation are as follows:

   Loading Time: 1.375 min

   Machining Time: 12.428 min

   Unloading Time: 1.052 min

   Inspection Time: 3.946 min

   Packing Time: 1,629 min

   Tool Change Time: 19.824 min

   Tool Change Frequency: 400 pieces

   Tool Cost: $115.69

   Operation Cost: $107.82perhour

   What is the time and cost per piece along with the annual cost to perform the operation?

3. The process in problem #2 had some process improvements so that the machining time is now 10.083 min. However, by increasing the machining time, the cutting tool needs to be replaced every 225 pieces now. What is the new time and cost per piece along with the annual cost, and would this be savings for the operation or not?

4. Still working to improve the operation from problem #3, a new "super tool" has been implemented, which costs $325.50. This has reduced the machining time to 9.827 min and can last for 325 pieces before requiring it to be replaced. The new tool takes longer to replace and the tool change time is now 28.276 min. What is the new time and cost per unit, and was this change profitable for the operation?

5. A given operation works 2,000 h/ year. It produces 13,000 component parts during that time but does not know what the actual machining time is. With the following information, determine the machining time along with the time and cost for the component and the annual cost to the operation.

   Loading Time: 1.068 min

   Unloading Time: 0.925 min

Inspection Time: 1.873 min

Packing Time: 1.399 min

Tool Change Time: 13.892 min

Tool Change Frequency: 450 pieces

Tool Cost: $135.75

Operation Cost: $94.71 per hour

6. To improve the process in problem #5, after determining what the machining time was, the inspection and packing times were evaluated to see if they could be performed internal to the machining cycle. If they can, what would the new time and cost per part be, and how much of a savings would this equate to for the operation?

7. A piece of hard cast iron is 14.750 in. long and will have a peripheral milling process performed on it to a depth of 0.625 in.. The cutting speed will be at the low end of the spectrum for that material (refer to Figure 15.1) utilizing a cutter that is 8.750 in. in diameter with 60 teeth and a feed of 0.00015 in./tooth. The cutter costs $273.29 and needs to be replaced every 675 pieces requiring 14.825 min. With an operational cost of $121.73 per hour, and if it requires 1.015 min to load the machine and 0.947 min to unload and no inspection of packing is required, what would the time and cost per unit be?

8. A component part that is 9.375 in. long will have two turning operations performed on it. The first is a roughing cut, which will take the diameter from 2.750 in. down to 2.500 in. utilizing a feed of 0.007 in./revolution. The second cut will be a finishing cut reducing the diameter to 2.450 in. using a feed of 0.0025 in./revolution. The material is High Nickel Steel, and the roughing cut will use a cutting speed at the top of the range while the finishing cut will use a cutting speed at the bottom of the range. The operation incurs a fully absorbed cost of $102.93 per hour, and the part requires 2.183 min to load into the machine and 1.836 min to unload. If inspection takes 8.527 min, but only has to be performed every 25 pieces and the cutting tool needs to be replaced every 350 pieces and takes 10.529 min to be replaced and costs $46.32, what would the time and cost per part be?

SECTION

5

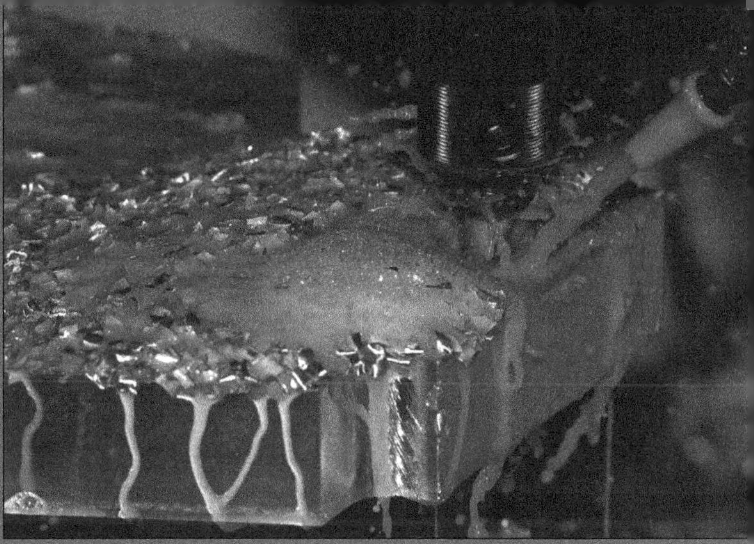

PHYSICAL ENHANCEMENT PROCESSES

Physical enhancement processes are categorized as *processes that alter the structure of the component part but do not intentionally alter the geometry*. In some cases there may be some slight changes in the geometry from the process, but it is not a desired outcome. Reflecting on the typical path for a process plan, the physical enhancement operations can be preformed either after the last secondary operation or during them where further processing is required after the enhancement has been performed. Figure 1 displays where the physical enhancement processes are situated in the overall fabrication process. As Figure 1 illustrates, not all component parts will require this processing, while some will require several enhancement operations. These unique processes are used to alter the microstructure of the component's material in order to improve the functionality and increase the components life.

This section covers some general metallurgy principles and discusses the microstructure of materials along with how they are affected by tension, compression, and shear forces. Understanding how various materials will react when placed under non-static conditions will allow the selection of proper material for a given function. Some component parts require enhancement operations as a result of previous processing to remove any stresses within the component that were inherited as a result of shaping, bending, forming, etc. Other parts require their internal structure to be altered in order to function properly during the assembly of the finished product. Heat treating is the main physical enhancement process that is utilized to perform the required alterations of the material of the component parts and is covered within this section of physical enhancements operations.

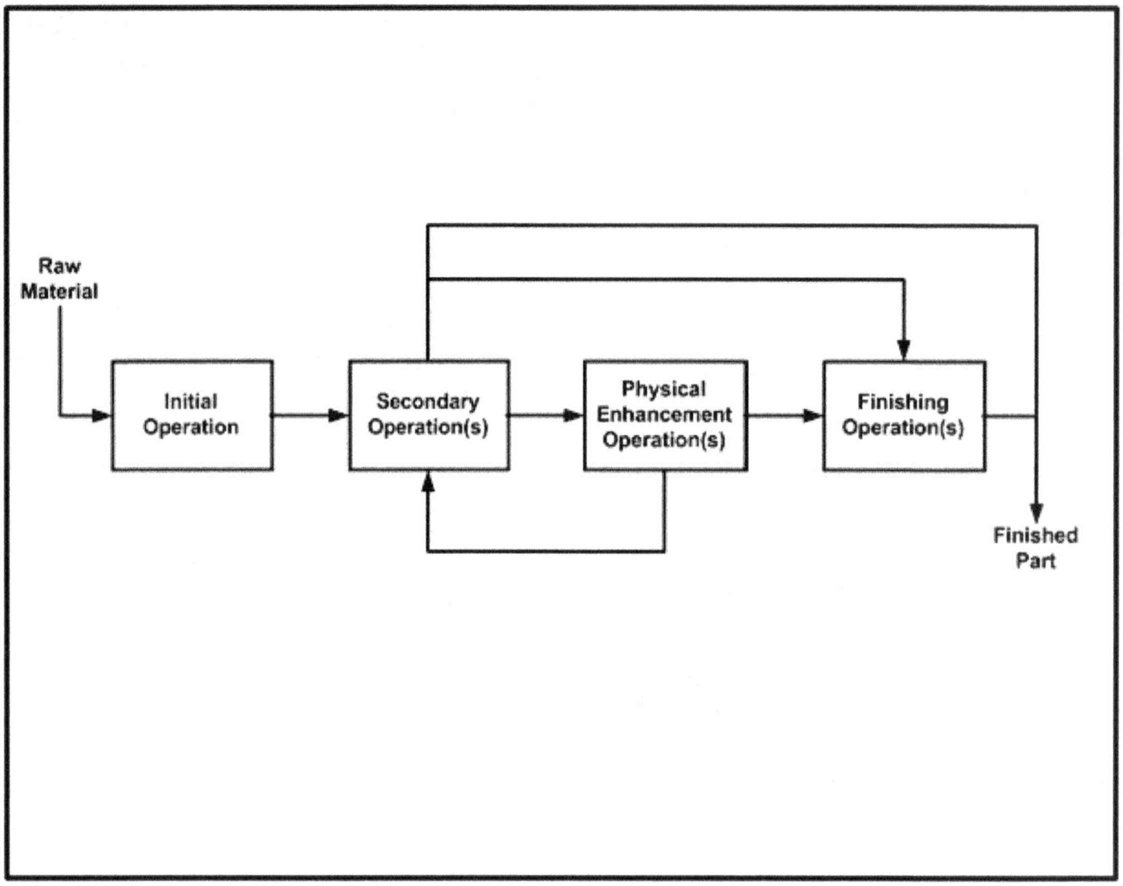

Figure 1 Physical Enhancement Operations in the Overall Process Flow

CHAPTER

19

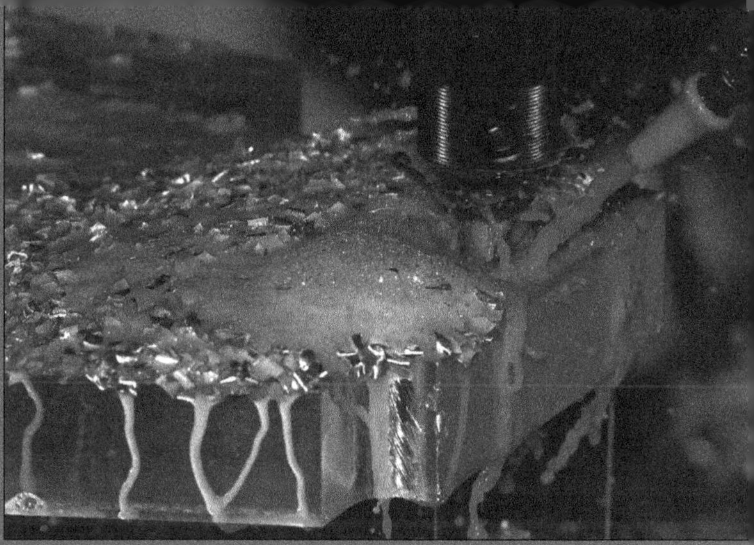

Principles of Metallurgy

19.1 INTRODUCTION

Metallurgy is the science of metal behaviors and structures under stressed conditions. As various stresses are applied to metals, they undergo changes in their structures both internal and external. These changes can affect the functionality of the component part during its service. By understanding the physical attributes of metals, errors can be avoided by selecting the wrong material for a given application.

This section will discuss the structures of metals, mainly steels, and the identification compositions of various steel alloys along with the effect that various stresses have on them. Various testing methods for tension and hardness and what measurements are taken to determine the materials failure points as well as their performance will be discussed.

19.2 METAL STRUCTURE

Metals have a crystalline structure with the molecules arranged in an orderly manner. This orderly structure is what gives the material its strength when placed under stress conditions. Figure 19.1 shows the organized molecular structure of metals when not under any stresses. However, when metals are melted into a liquid state, the structure becomes semi-unorganized. The molecules start to move and the crystalline structure is altered causing the material volume to expand. Figure 19.2 shows the semi-unorganized structure of the metal in a liquid state.

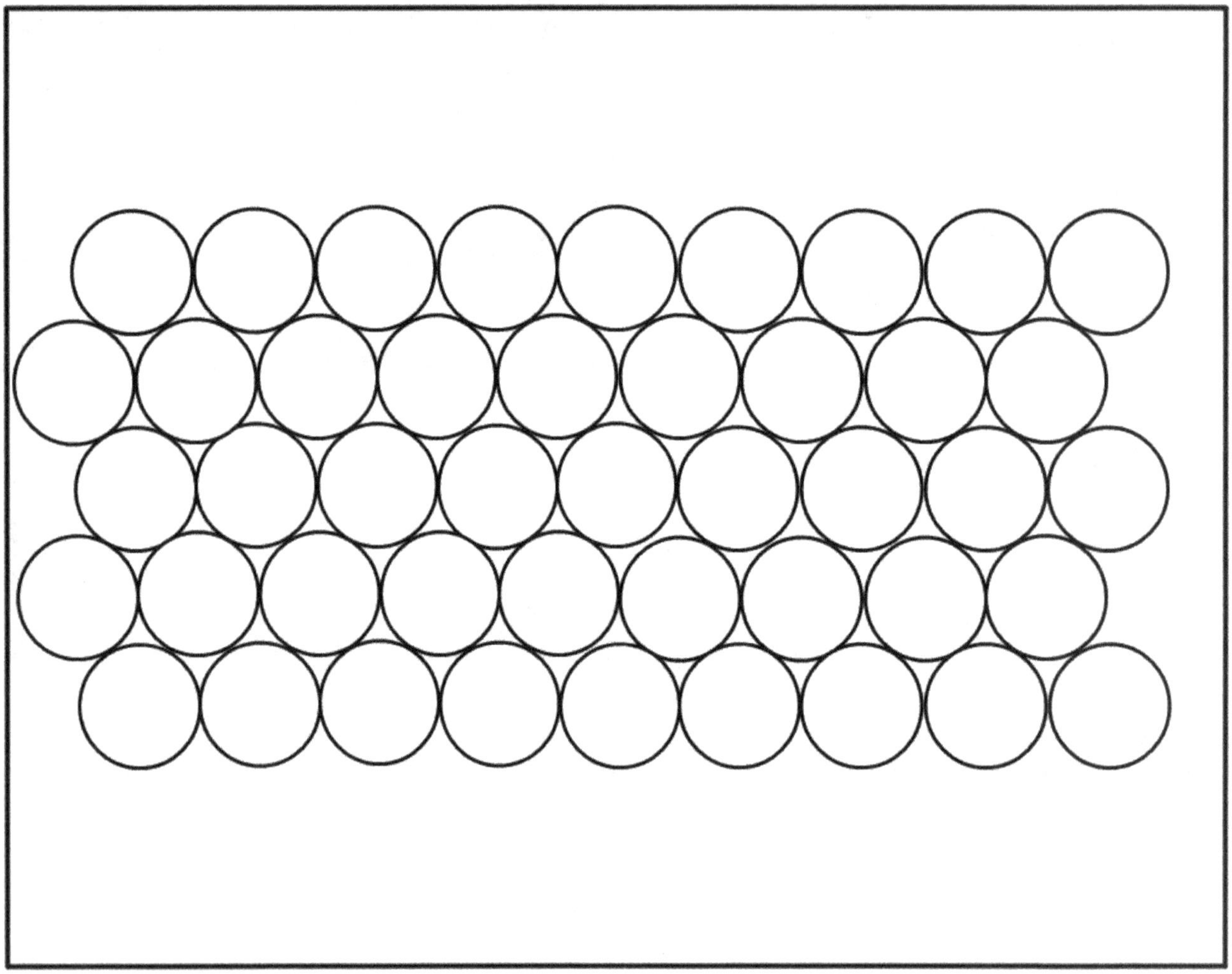

Figure 19.1 Molecular Structure of Solid Metals

As metals go from their liquid state to their solid state, shrinkage will occur. This is known as *solid solidification shrinkage* when the molecules move back to their organized structure. At this point however, the metal is still just slightly below the melting temperature and as it cools down, there will be additional shrinkage as it cools to room temperature. This is known as *solid thermal contraction shrinkage* as the material cools and contracts. All materials will expand and contract with heat and Figure 19.3 illustrates the transition from liquid to a solid along with the contraction occurring.

Every metal will go through this transition from a liquid to a solid and the percentages of solidification shrinkage and solid thermal contraction will differ from material to material. See Figure 5.7 in Chapter 5 on castings for percentage of shrinkage values of various materials.

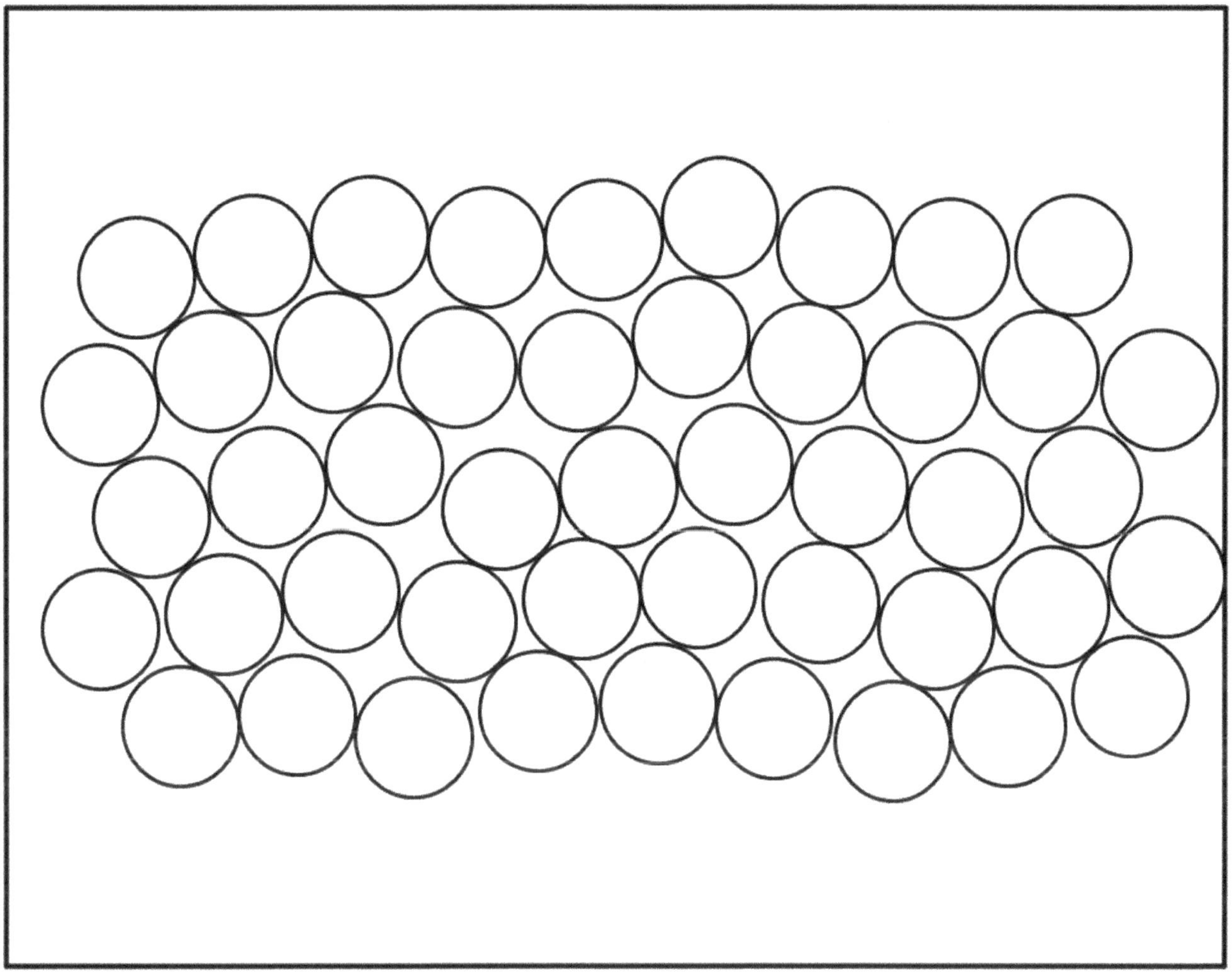

Figure 19.2 Molecular Structure of Liquid Metals

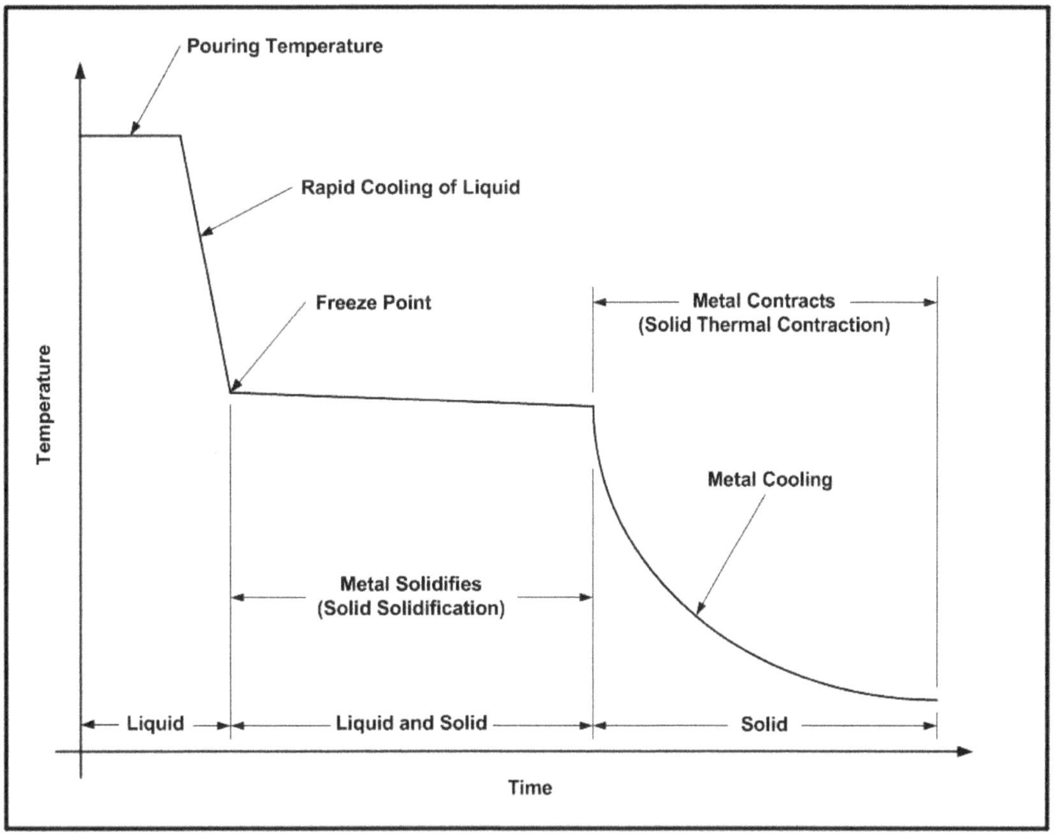

Figure 19.3 Solidification and Contraction of Metals from Liquid to Solid Structures

19.2.1 Steel Identification

Steel has a numerical identification scale based on the main ingredient and the alloys and carbon percentages mixed with it. The Society of Automotive Engineers (SAE) developed an identification scale that is utilized for steels and nonferrous materials that has been accepted throughout the Unites States. An example of some of the steel classifications and an example of how the numerical values explain the material and the mixtures are listed below:

Steel Classification Examples:

- 1XXX: Carbon steel family
- 2XXX: Nickle steel family
- 3XXX: Nickle chromium steel family
- 4XXX: Molybdenum steel family
- 5XXX: Chromium steel family
- 6XXX: Chromium-vanadium steel family
- 9XXX: Silicon-Manganese steel family

Number Code Example 3540 Steel

- 5% Chromium
- 0.35–0.45% Carbon

Steel is produced in a low carbon, medium carbon, or high carbon alloy. The amount of carbon in the steel composition will determine what hardness it will achieve when heat treating. The hardening of steels will be covered in Chapter 20 with heat treating and will explain the process for achieving hardness in the material. The three categories of steel carbon content are outlined below and what hardness they can achieve when heat treating:

- *Low carbon steel:* Less than 0.2% carbon (will not harden)
- *Medium carbon steel:* 0.2–0.5% carbon (will achieve 1/2 hardness)
- *High carbon steel:* <0.5% carbon (will achieve full hardness)

19.3 METAL BEHAVIOR

As metals are exposed to various forces, the internal molecular structure begins to get stressed and becomes altered in the process. These stresses that get applied will change the material strength until it will eventually fail as shown in the stress/strain curve in Figure 19.4. The major stresses that can be introduced into the metal are *compressive stresses*, *shear stresses*, *tension stresses*, and *torsion stresses*. These are explained below along with the type of failures that can occur.

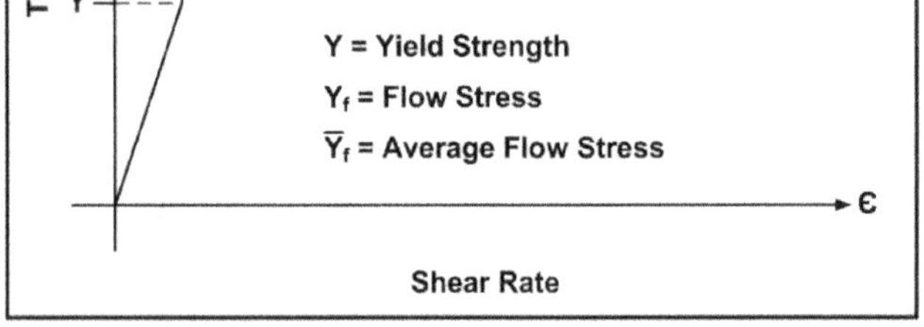

Figure 19.4 Stress/Strain Curve of Metals

19.3.1 Compressive Stress

Compressive stresses in a material is caused by forces being applied from the ends and pushing straight in toward the center. As the forces increase, the material will compress and the center of the material will begin to expand. All materials can handle some amount of compressive forces without any permanent deformation until the forces reach the yield point of that particular material. As the forces continue to increase, the material will compress and deformation will occur, expanding the width and

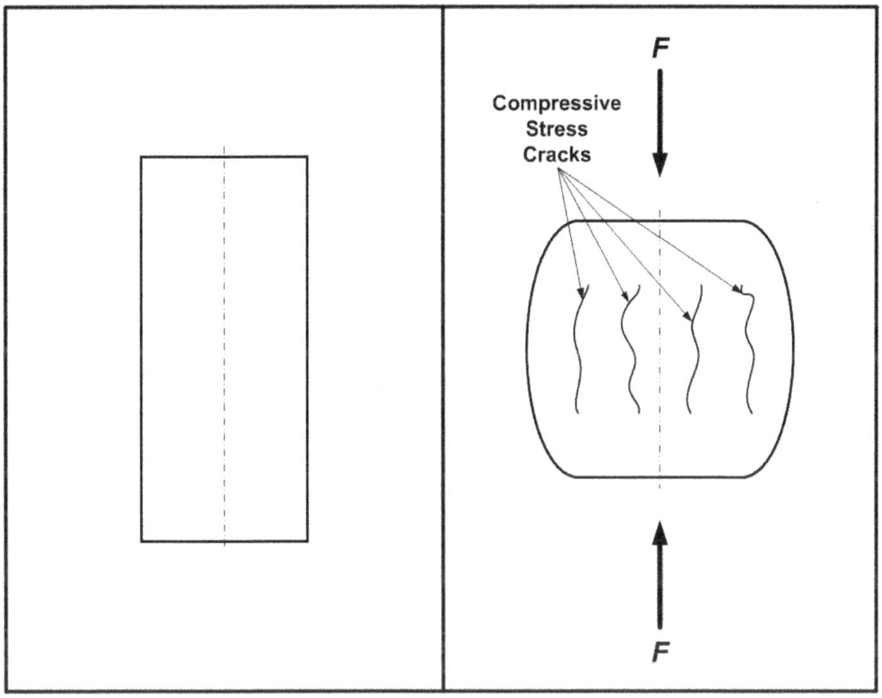

Figure 19.5 Compressive Stress Cracks from a Side View

the length becomes shorter. If the compression continues, failure of the material will result in stress cracks present on the outer surface of the component. Figure 19.5 shows a side view of the material in compression and how the cracks will radiate from the center toward the ends of the component as the material begins to fracture. These cracks will starts from the outer layer of the material and move toward the center. Figure 19.6 illustrates a top view, showing the cracks forming on the outer layer of the material and radiating toward the center.

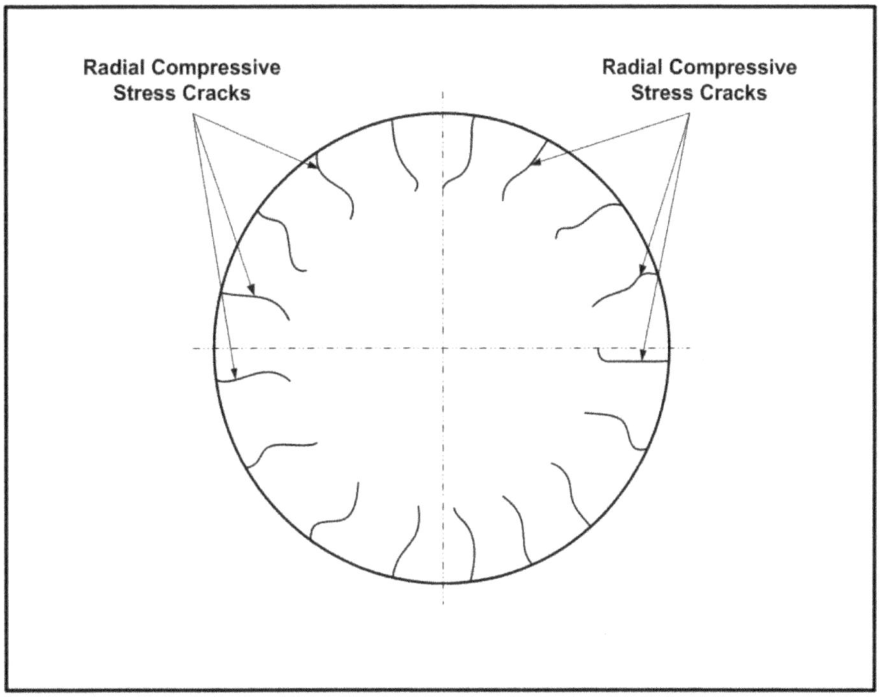

Figure 19.6 Compressive Stress Cracks from a Top View. Cracks are Radial from the Outside toward the Center

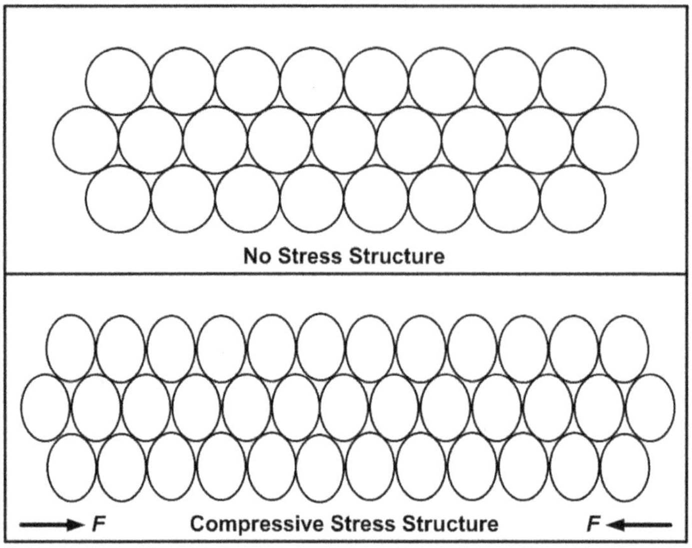

Figure 19.7 Compressive Molecular Structure of Metals

As compressive forces are exerted on the material, the molecules of the material will go from their steady-state structure to a compressed structure. As the forces increase, the outer layer will fail first owing to the fact that there is no structure to hold them in place. Figure 19.7 shows the molecular structure change of the material as the compressive forces are applied.

19.3.2 Shear Stress

A shear stress is caused by the material having opposite forces applied on the surfaces of the material. In a radial situation, the component will experience forces on both the top and bottom in opposing directions simultaneously. As the material begins to reach the yield point, there will be stretch lines appearing on the sides of the material as shown in Figure 19.8. Many of the ductal materials can experience a vast amount of shear stress before failure, but it will be greatly deformed in the process. Likewise, a brittle material may show very little to no deformation before a failure happens. This failure will be almost instantaneous with no visible deformation.

As shear forces are exerted on the material, the molecules of the material will go from their steady state to an elongated structure as shown in Figure 19.9. Failure from stress forces for ductal material will show stretch

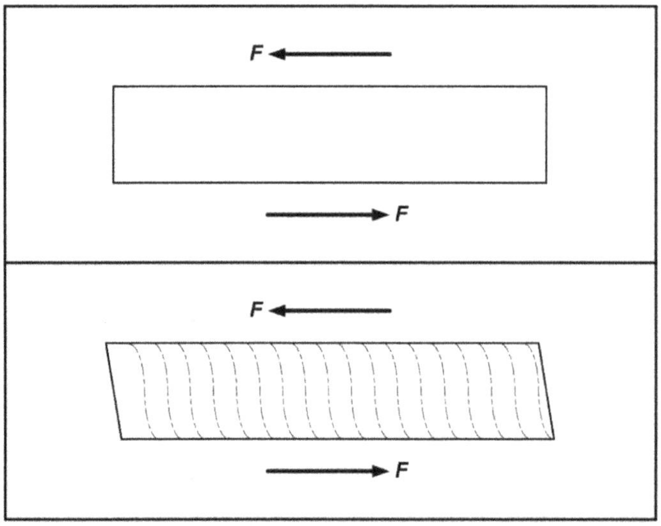

Figure 19.8 Shear Force Example on a Metal Component

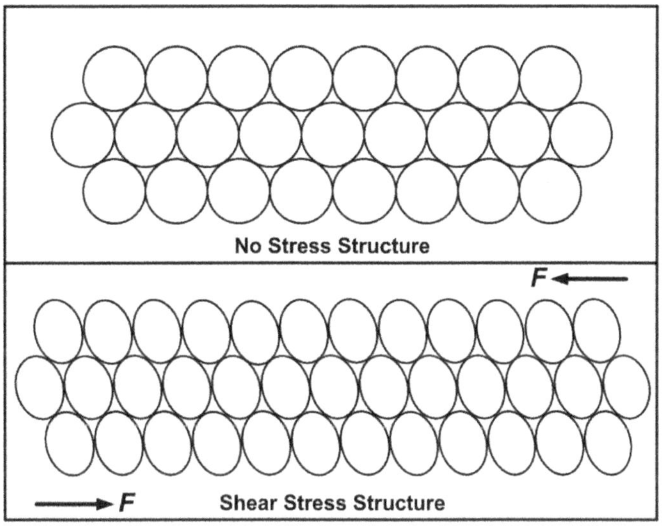

Figure 19.9 Shear Molecular Structure of Metals

lines that will be more obvious near the surfaces where the forces are being applied on; however, brittle material may show no visible deformation line before a failure occurs.

19.3.3 Tension Stress

Tension stresses are the complete opposite of a compressive stress. In a tension situation, the component will have opposing forces pulling on the ends. This causes the material to have a straight elongated molecular structure as shown in Figure 19.10 where the molecules are stretched from their static structure. As the forces on the component increase, the length will increase and the thickness will decrease. Tension tests are commonly performed on sample material pieces where the starting length and diameter are the same. The test piece is placed in a tension testing machine and forces are applied until the yield point is reached. As the forces are increased, the length of the material will increase which is illustrated in Figure 19.11 and measured. This continues until the ultimate strength is achieved and the sample breaks. Along with the length, the change in diameter is also measured, which is shown in Figure 19.12. The more ductal the material, the more elongation and diameter reduction will be observed. But with extremely brittle materials, the elongation of length may be unmeasurable along with the reduction in diameter and the test piece will just instantly fail.

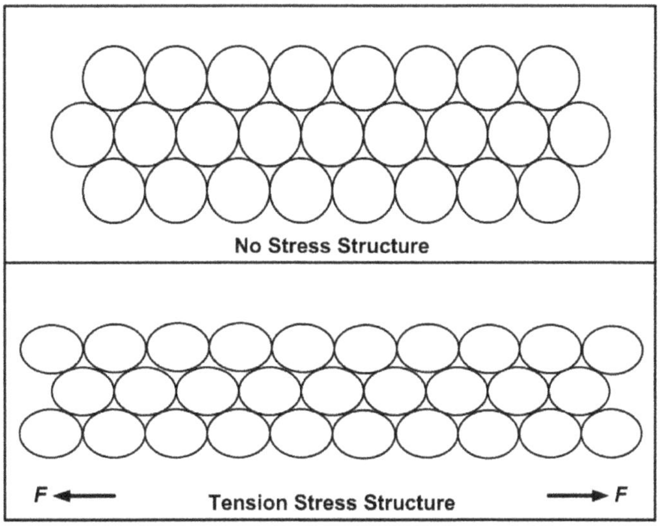

Figure 19.10 Tension Molecular Structure of Metals

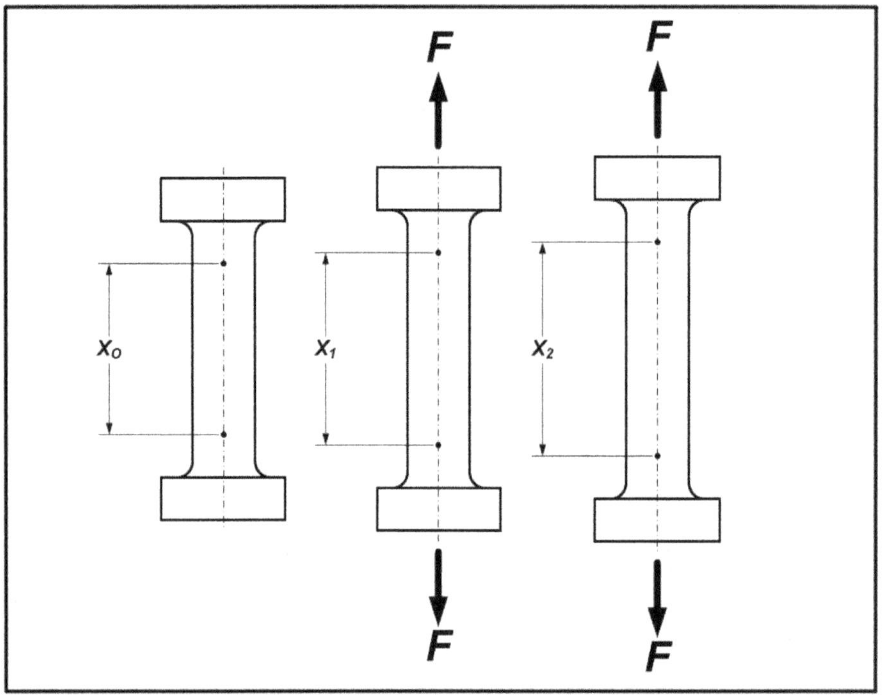

Figure 19.11 Tension Testing of Metals and Length Change

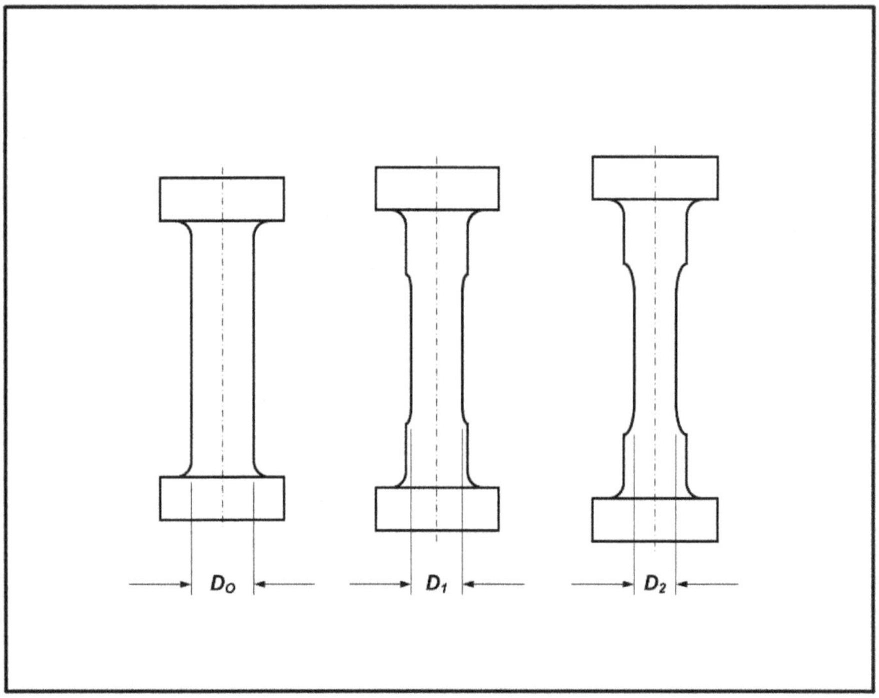

Figure 19.12 Tension Testing of Metals and Diameter Change

19.3.4 Torsion Stress

Torsion stresses occur when two opposing radial forces are applied on the ends. This is mostly prevalent in round shafts that are being held at one end and a radial force that is being applied on the other end. As the forces increase, the shaft will witness a stress in the center and as the forces decrease, the yield point will cause it to be twisted until a failure occurs. Figure 19.13 shows how a torsion stress gets applied to a shaft and how the break will be radial in nature. All materials regardless of whether it is ductal or brittle will fail with the same radial pattern caused by the twisting nature of the stresses. Figure 19.14 illustrates the molecular structure of a torsion stress where the molecules are more highly stressed in the center of the shaft and less stressed at the ends. This is caused by the amount of twisting that is felt in the center rather than at the ends.

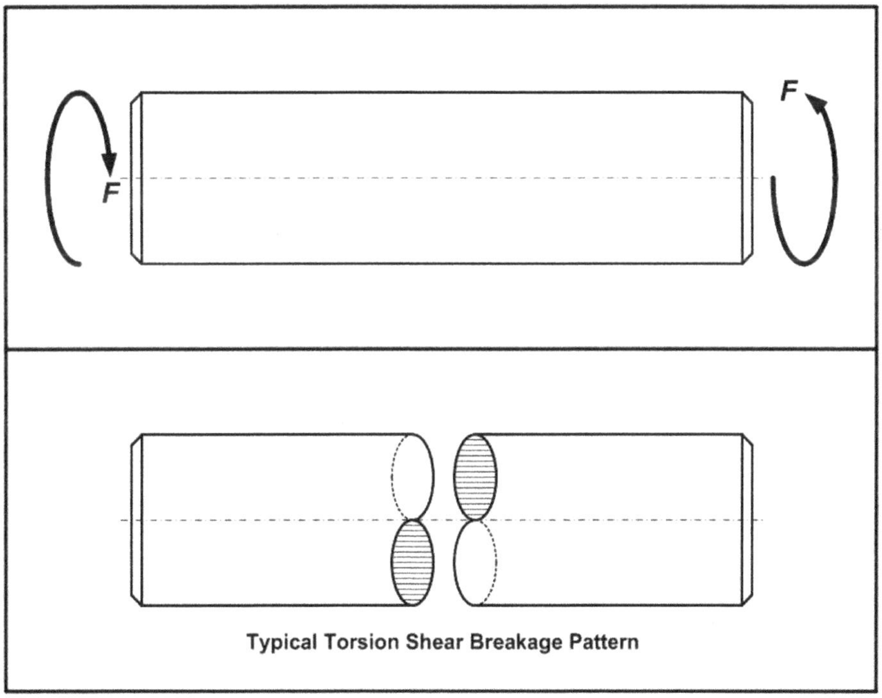

Figure 19.13 Torsion Forces on a Shaft and the Typical Torsion Breaking Pattern

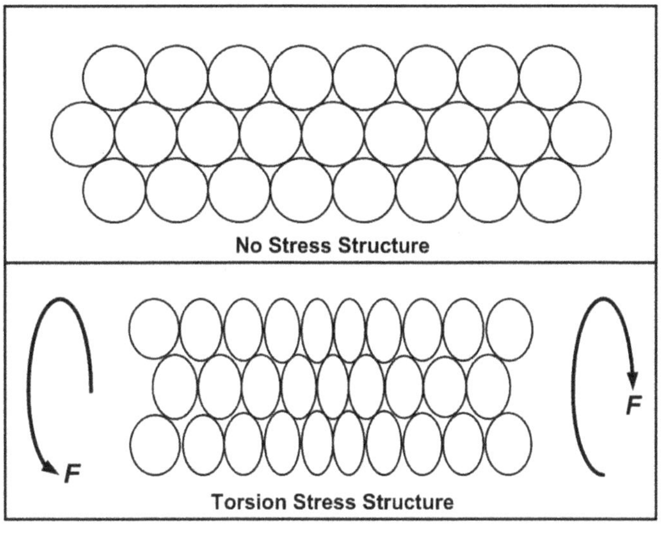

Figure 19.14 Torsion Molecular Structure of Metals

19.4 SUMMARY

The study of metallurgy is a very in-depth and complex field. This chapter only brushed the surface of the study showing the effects of various forces that are applied materials. Understanding the effects that various forces will have on metal components and what type of failures can occur is critical when selecting a specific material for a component part in the design stages. As forces get applied to metals, the molecular structure will be altered depending on what type of stresses are being applied. By knowing how a specific material will either react or hold up to the stresses being induced can make the difference between a successful or unsuccessful application. Before blindly selecting a material for a component part that will experience any type of stress applied to it during operation, research the material and evaluate if the structure is capable of withstanding the forces before proceeding any further in the design stage of a project.

CHAPTER

20

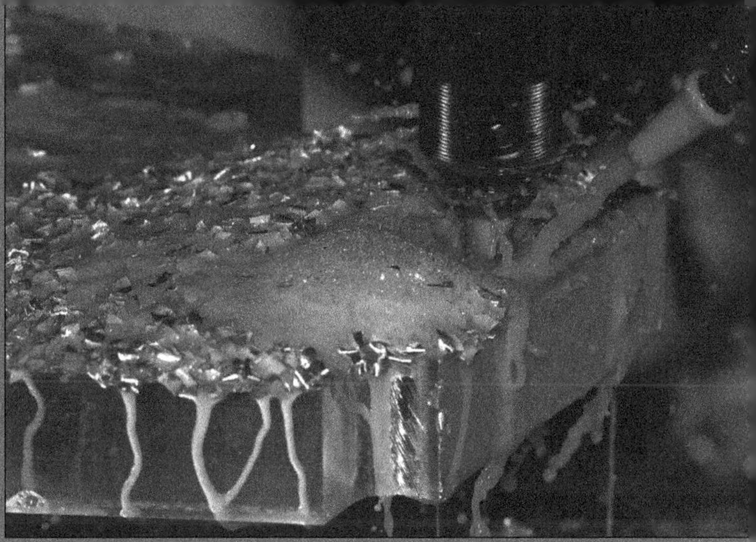

Principles of Heat Treatment for Metals

20.1 INTRODUCTION

Heat treatment for metals involves various heating and cooling operations to alter the internal microstructure of the materials. These operations can be utilized for reducing internal stresses that have been induced from previous operations or to either increase or decrease the hardness of certain steels. What allows certain steels to be able to obtain increased hardness when certain heat treating operations are performed is dependent on the level of carbon that the material contains.

Besides hardening of steels, heat treating operations are utilized for both reducing stresses of components and reducing induced hardness from hardening to various levels. Internal and external stresses can be present in materials that have not been hardened, but have been induced from previous operations such as bending and forming. These stresses can cause failures in certain functional applications and reducing them allows the components to perform as designed.

This section will discuss some of the more common heat treating operations such as metal hardening, annealing, tempering, and surface hardening. It will also cover testing the hardness of metals along with the testing of heat reduction vs. time for carbon steels. By understanding what heat treating operations can be performed on materials to obtain various internal structure changes is important for producing functional component parts for the products being designed.

20.2 HARDENING OF METALS

Steel containing a certain level of carbon in the mixture can be hardened. The carbon content in the steel reacts with heat to produce a new microstructure producing a work piece far harder than the original structure. Not all steel materials can achieve a new hardness level. The carbon content needs to be at a certain level in order for the microstructure change to occur. Low, medium, and high carbon steels displaying the carbon content and the hardness that can be realized are listed as follows:

- Low carbon steel: less than 0.2% carbon (will not harden)
- Medium carbon steel: 0.2%–0.5% carbon (will achieve 1/2 hardness)
- High carbon steel: <0.5% carbon (will achieve full hardness)

The procedure for hardening requires heating the steel to a certain level, and then cooling it down to room temperature. The material is heated into the austenite zone, and then rapidly cooled into the martensite zone. This causes the molecular structure of the material to become much harder than the original structure. During cooling, the time required is critical in order to avoid the material entering either the pearlite or the bainite zones that will induce unwanted properties in the metal. Figure 20.1 illustrates the heating and cooling curve for hardening steels, noticing how the cooling trajectory avoids entering the pearlite and bainite zone which is referred to as the "nose." This process will bring the steel to its maximum level of hardness depending on the carbon levels within it. When hardening steel, it is always hardened to the maximum, and then if less hardness is desired it will be drawn down in a process known as tempering, which is covered later in the chapter.

Steel is a unique material that displays a vast color spectrum as it goes through a heating range. As it warms up, the steel will first go through a rainbow of colors until it reaches a certain temperature, and then it will display a wide variety of colors from dull red up to yellow and even white when it becomes liquid. This is extremely helpful when viewing the material in the furnace to get an idea if the desired temperature has been achieved for the hardening process. Figure 20.2 shows the color spectrum for hardening steels ranging from 1,000–2,500 °F. Even though a production furnace will raise the temperature exactly to the desired level, if a

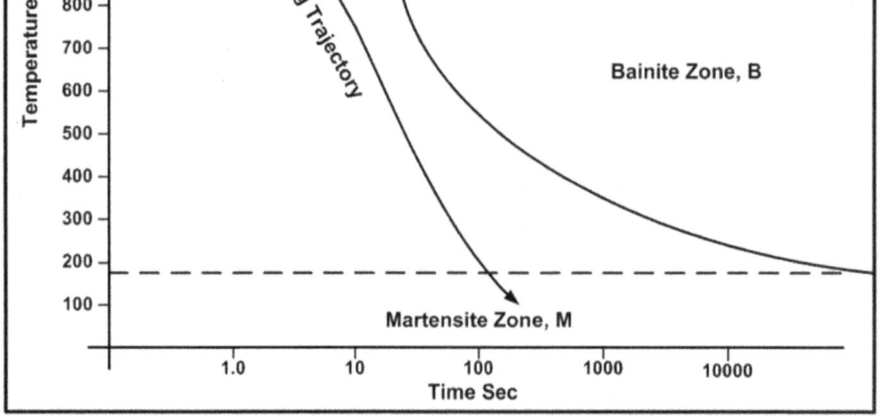

Figure 20.1 Heating and Cooling Curve for Hardening Medium and High Carbon Steel

Steel Hardening Color Spectrum	
Steel Color	**Temperature**
Dull Red	1,100 °F
Bright Red	1,300 °F
Orange	1,450 °F
Bright Orange	1,650 °F
White Orange	1,800 °F

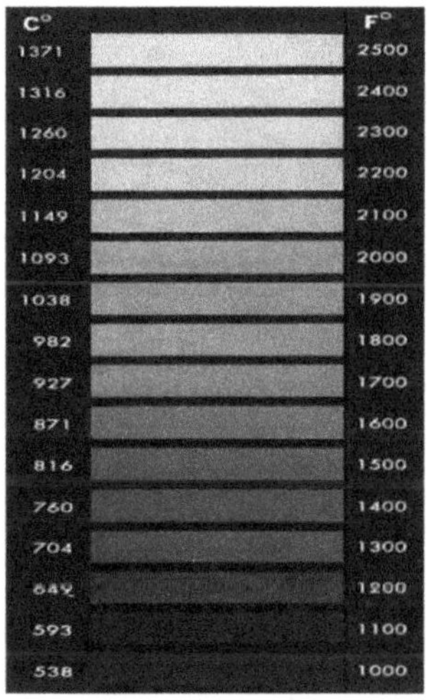

Figure 20.2 Steel Color Spectrum for Hardening Operations (see inside back cover for color version of Figure 20.2)

manual heat treating operation were to be performed the color of the material being heated can be determined fairly close from the color of the steel.

For production hardening, the component parts are loaded into a furnace and heated to the desired level before being cooled down. The cooling process is known as quenching and is covered later. Most hardening furnaces maintain a controlled environment where the ambulant air is replaced with a neutral environment such as nitrogen to eliminate the foreign elements in the furnace. The common furnaces operate using either electricity or gas to produce the heat and render hardened components easily. Some unique furnaces operate in the same way, but have additional features. Figure 20.3 shows a vacuum furnace where after the atmosphere is neutralized, a vacuum is pulled to either reduce the internal pressure or eliminate it completely. Certain

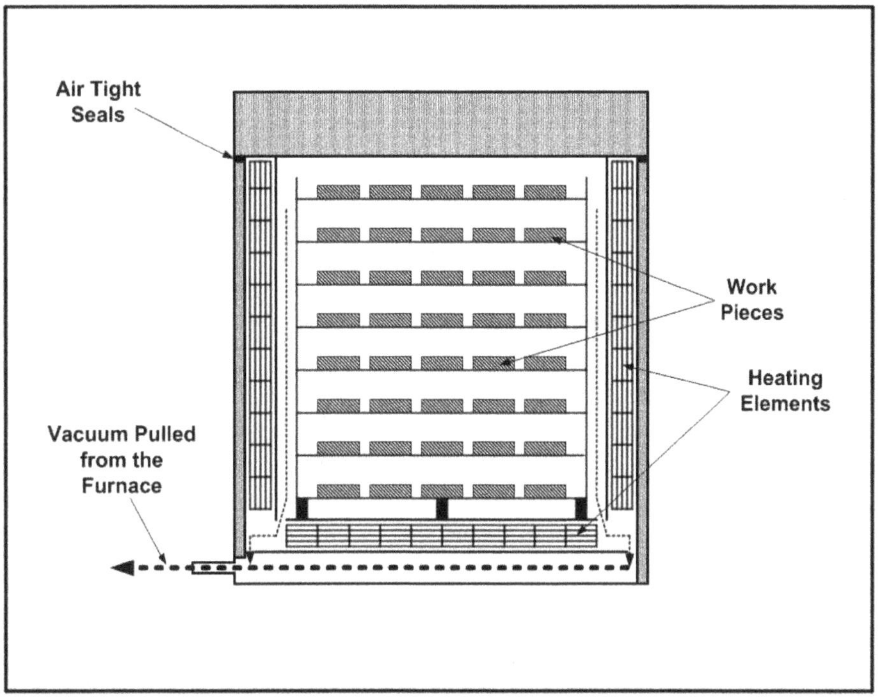

Figure 20.3 Vacuum Furnace Where the Atmosphere Is Reduced or Eliminated

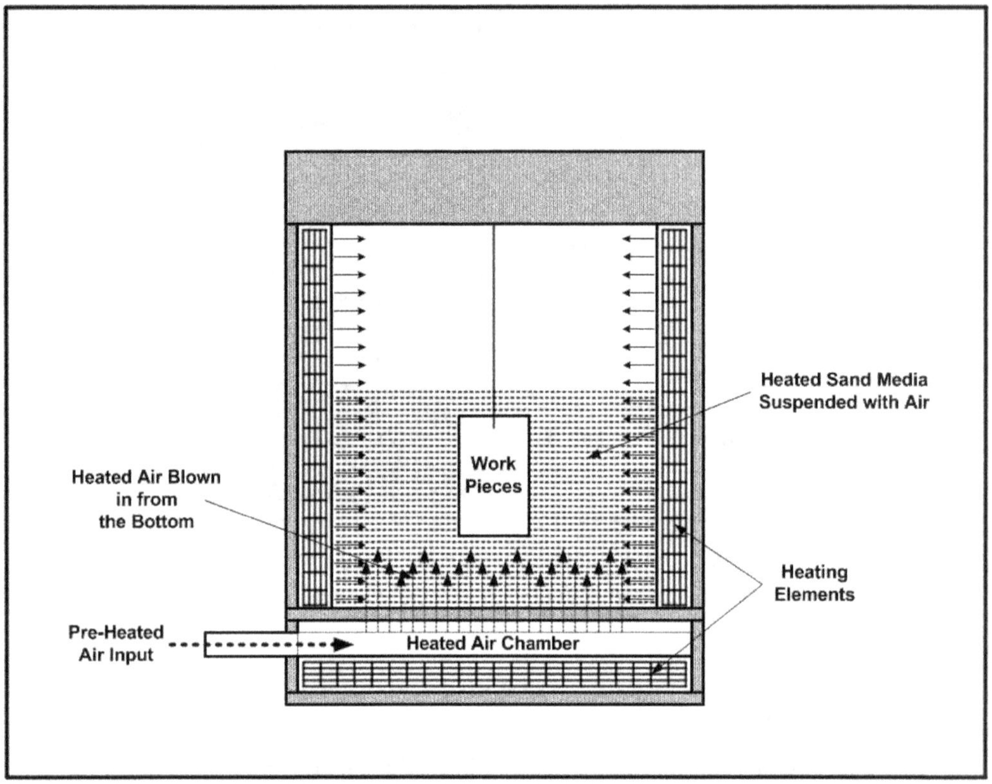

Figure 20.4 Fluidized Bed Furnace Where the Heat Is Contained in a Sand Media

materials require a reduced atmospheric pressure and attain better hardening results. Another unique furnace is shown in Figure 20.4. This is a fluidized bed furnace where a sand type media is heated and then it is lightly suspended with air from below. The furnace heats the media and the parts are loaded into it either on a hook or in a basket. The media heats the parts and with it being a type of sand (aluminum oxide the most common), the scale on the outside of the parts is greatly eliminated owing to the abrasive action. Also, the media holds the heat for a long time so when the furnace isn't in use, it remains warm thus requiring less energy to bring it back up to the required temperature.

20.2.1 Quenching Methods

Quenching is the term for lowering the temperature of a heated component for hardening. Depending on the type of steel being quenched, the cooling times can vary greatly along with the media the component is being quenched in. As illustrated in Figure 20.1, the cooling time needs to avoid the nose curve, but the time can be as low as second to minutes depending on the type of steel. Figure 20.5 illustrates some examples of quenching media and the materials that would be quenched in. Each type of steel will specify what media is required to produce the desired hardening results.

Two methods are utilized for quenching, which are *submersion quenching* and *air quenching.* With submersion quenching, the heated work piece is removed from the furnace and immediately immersed into the quenching media. This is utilized for a variety of steels that require either a fast or medium quenching time in order to avoid the nose of the cooling curve. Figure 20.6 illustrates a submersion operation where the work piece is completely immersed into the media until the cooling has completely occurred. With air quenching, the heated work piece is placed in a chamber where the air is pulled away from the part. By pulling the heat away from the work piece, it aids in more uniform cooling. If the air were to be blown onto the work piece, the outside of the part would cool faster than the interior and the heat would be retained. Figure 20.7 shows the process of removing the heat from the component by evacuating the heat from the surface rather than pushing cool air onto it.

Quenching Media with Quenching Duration and Material Type			
Quench Symbol	Quenching Media	Quenching Duration	Material Being Hardened
A	Room Temperature Air	Slow Quenching Time	Low Air Tooling Steel
D	Oil/Salt Mixture	Medium Quenching Time	Special High Carbon Tooling Steel
M	Melted Salts	Slow Quench Time	Special High Carbon Tooling Steel
MT	Melted Salts Special Mixture	Slow Quench Time	High Carbon Tungsten Cutting Tool Steel
W	Still Water	Fast Quenching Time	High Carbon Steel (Drill Rod)
O	Still Oil	Fast Quenching Time	General Purpose High Carbon Steel

Figure 20.5 Quenching Symbols and Media for Various Steel Hardening

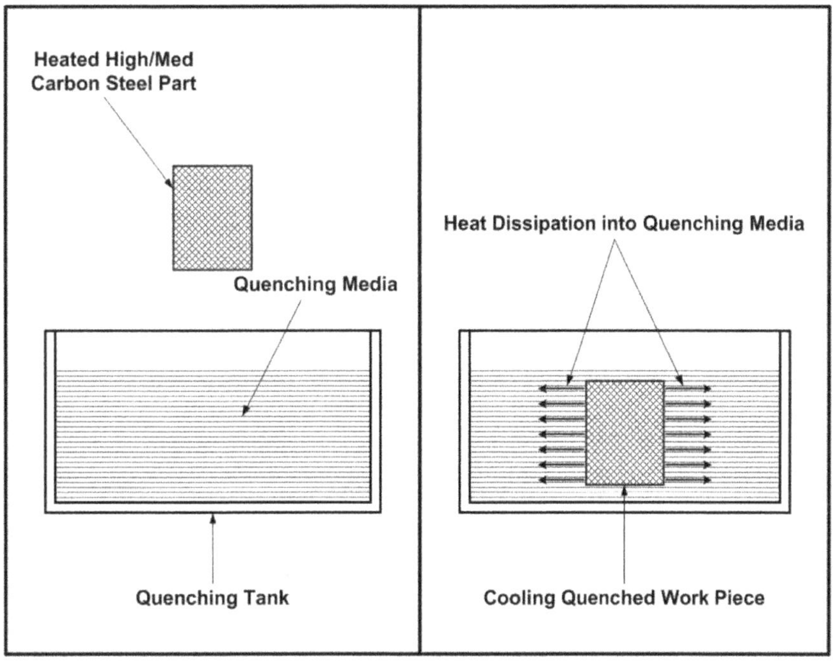

Figure 20.6 General Submersion Quenching Method for All Media Except Air

This rapid cooling operation will drop the temperature of the heated medium or high carbon steel to alter the microstructure and induce the maximum hardness of the part. As shown in Figure 20.5, the type of quenching media is dependent on the type of steel being hardened. It is critical for the hardening operation to utilize the proper quenching media to achieve the optimum hardening results.

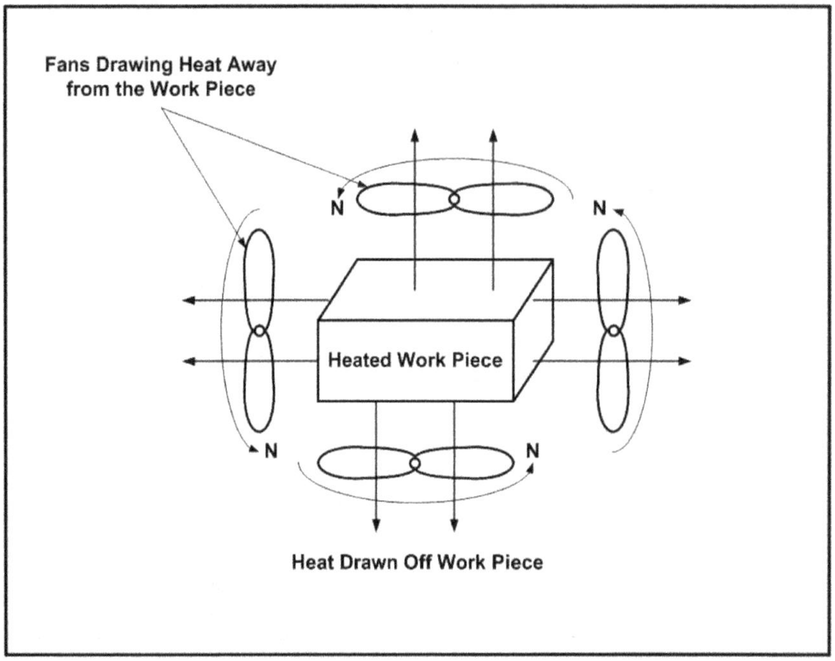

Figure 20.7 Air Quenching Method with the Heat Being Drawn Away from the Hardened WorkPiece

20.3 TEMPERING AND ANNEALING

Both tempering and annealing are processes that perform relatively the same operations to the component work piece. The major difference is that tempering will lower the hardness of a hardened component to a desired level, where annealing will remove all the induced hardness and return the metal to the original hardness. Annealing is often performed on components that have induced stresses resulting from previous shaping to relieve the stresses that may affect the functionality of the work piece. The annealing operation will eliminate any strain hardening that is caused from bending of the material or from surface stresses built up from machining. The functions that tempering and annealing perform on the component work piece are as follows:

- Reduces hardness and brittleness
- Alter microstructure of the material
- Soften the material
- Recrystallize cold worked metals
- Relieve residual stress from prior shaping

Tempering and annealing is performed by slowly heating the material up, letting it soak at a given temperature and time, and then slowly allowing it to cool down. Figure 20.8 illustrates the heating and cooling curve for tempering and annealing. With production furnaces, the tempering can be performed immediately following hardening. The work pieces are heated then quenched under the furnace. From there, the parts are raised back up into the furnace and tempered to the desired hardness. Figure 20.9 shows the heating and cooling curve of a hardening and tempering operation performed together in a production furnace.

As mentioned earlier in the chapter, steel has a wide color spectrum when exposed to heat. Figure 20.10 shows the color spectrum of steel at lower than temperature that turns the material into various shades of red. You will notice that there is a very small temperature increment between the various colors of the material. This color change is extremely useful for tempering and annealing to help identify the temperature of the material during the process. An example of various tempering hardness results is illustrated in Figure 20.11 where component work pieces has some of the hardness removed in order to establish the desired hardness for a given function.

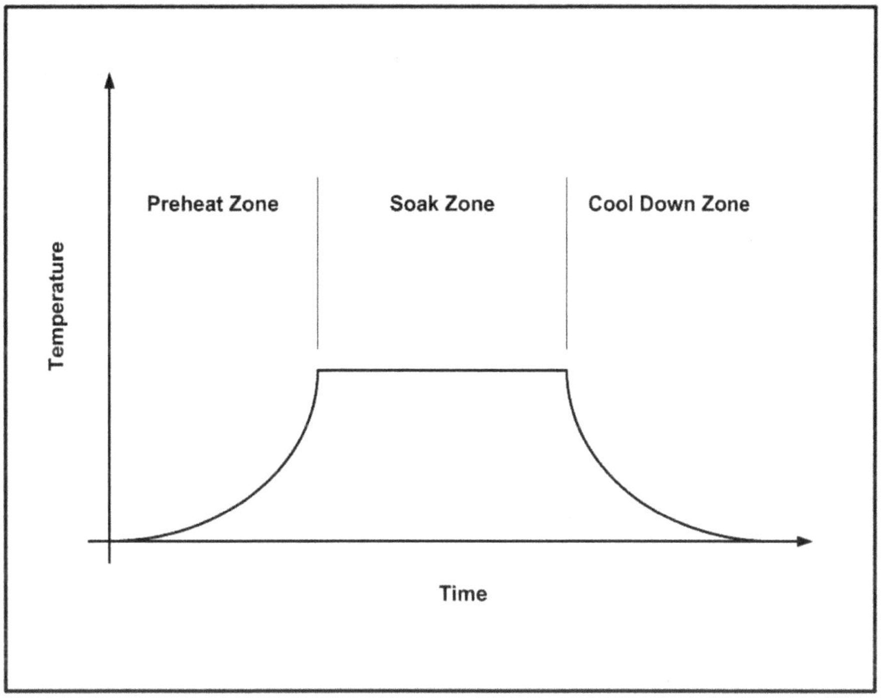

Figure 20.8 General Tempering and Annealing Heating and Cooling Curve

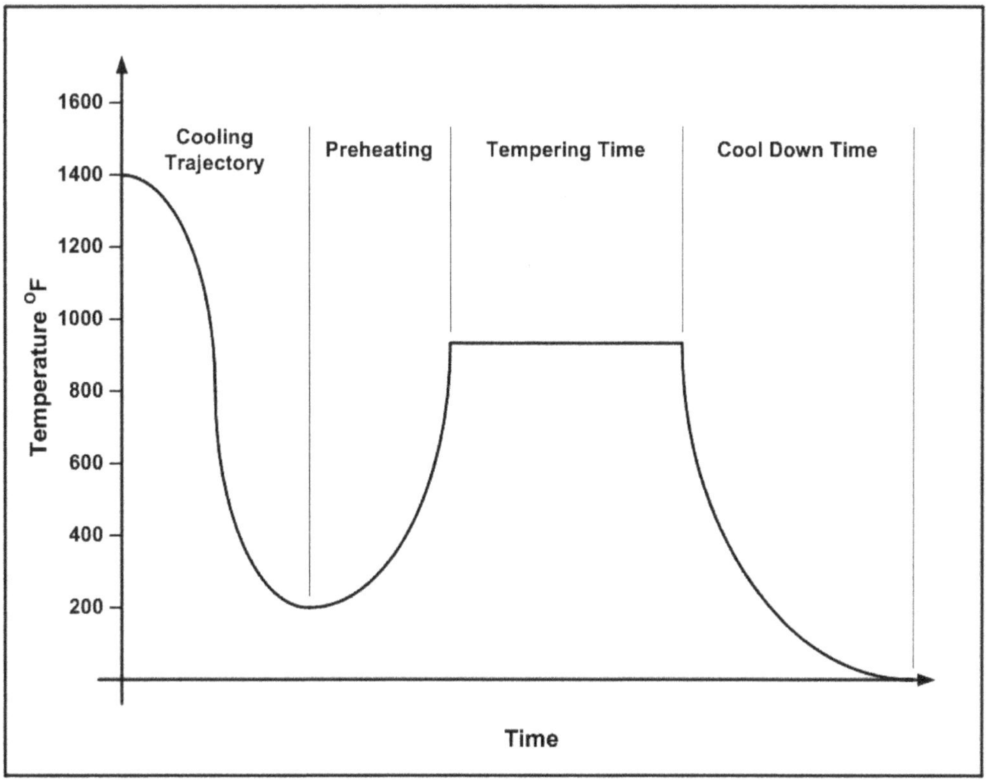

Figure 20.9 Combination Hardening Followed by Tempering Immediately

Steel Tempering Color Spectrum	
Steel Color	**Temperature**
Pale Yellow	430 °F
Straw Yellow	450 - 490 °F
Dark Yellow	480 - 490 °F
Brown Yellow	500 °F
Brown Purple	520 °F
Light Purple	530 °F
Purple	540 °F
Dark Purple	550 °F
Blue	560 °F
Dark Blue	570 °F
Light Blue	600 & Up °F

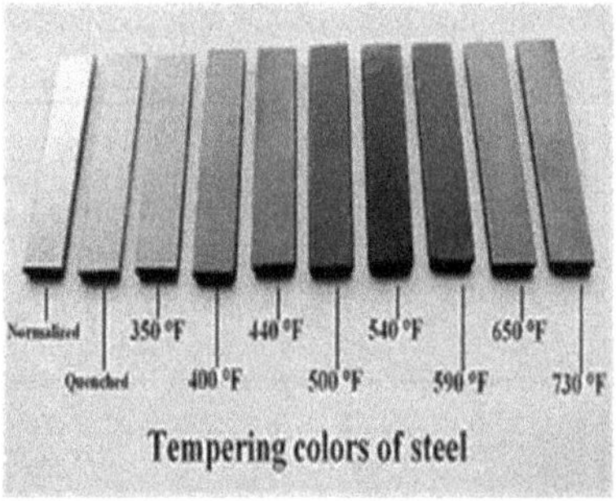

Figure 20.10 Steel Color Spectrum for Tempering and Annealing Operations (see inside back cover for color version of Figure 20.10)

Tempering/Annealing Levels and Achieved Hardness for Various High Carbon Steel Components		
Rockwell Hardness	**Examples of Applied Tempering & Annealing**	**Temperature**
68 C	Full Hardness of High Carbon Steel	No Tempering
58-62 C	High Speed Steel Machining Cutters	Temper @ 375 – 400 °F
56-58 C	Saw Blades: For Cutting Most Materials	Temper @ 450 °F
54-56 C	Punches, Chisels, Hammers, etc.	Temper @ 500 °F
52-54 C	Wear Plates, Wear Shafts, Slides, etc.	Temper @ 550 °F
48-52 C	Guide Bushings, Liner Bushings, etc.	Temper @ 600 °F
Full Soft	All Induced Hardness Removed	Anneal @ 1,400 °F

Figure 20.11 Examples of Various Tempering Hardness and Applications

20.3.1 Hardness Testing

The hardness of a material needs to be tested to ensure that either the material has reached full hardness, or after tempering or annealing it has been reduced to the desired level. There are several methods for hardness testing with the most precise being a penetration test. This is performed by placing the component work piece under a special machine that has a penetration stylist which is pushed into the surface of the work piece under a weighted pressure. The amount of resistance is recorded with a numerical value. A common hardness scale

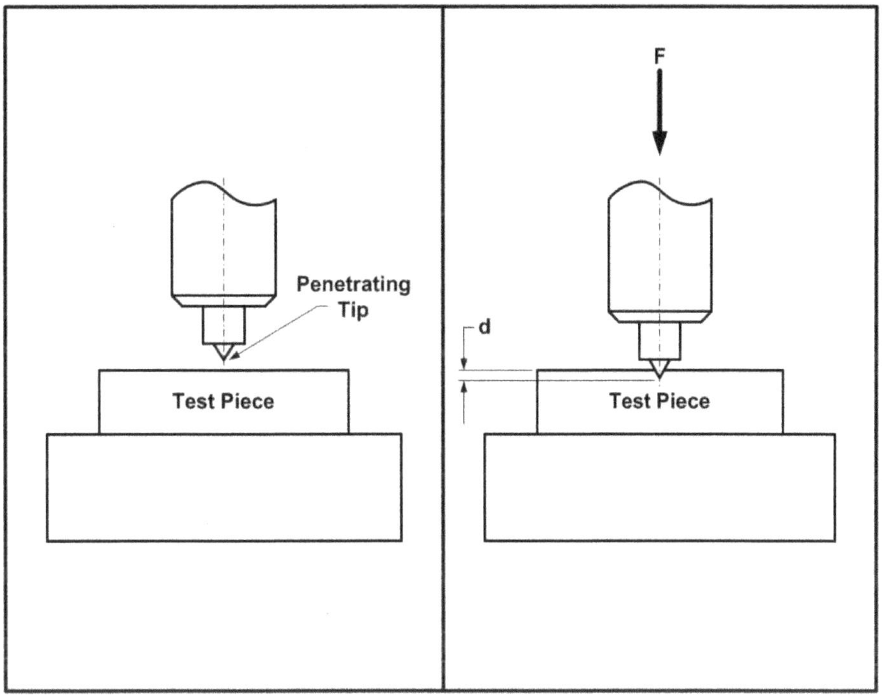

Figure 20.12 Hardness Penetration Testing (Rockwell Hardness Testing)

is the Rockwell scale that will render a value of the hardness based on the amount of resistance present in the component being tested. For high carbon steel at full hardness, the maximum value on the "C" scale is 68. As shown in Figure 20.11 previously, tempering the material can have the hardness lowered and the Rockwell "C" reading will be less than 68. This scale is all relative and a comparison to the maximum hardness can be realized. Figure 20.12 illustrates how the penetration testing for hardness is performed on a work piece. The only drawback to performing a penetration hardness test is that the component will have a small indentation on the surface, which may in some instances be unacceptable for the function of a given work piece.

Another test that is widely utilized to test the cooling time for a given material is the *Jominy End Quench Test*. Performing this test requires fabrication of a test specimen of the material being evaluated that will be hardened to the temperature required to achieve full hardness. The specimen is 1.000 in. in diameter with a testing length of 4.000 in. A flat is ground on the surface of the 1.000 in. diameter, so hardness testing can be easily performed with a head on the non-testing end for supporting the specimen during testing. Once the test specimen is at the required heat, it is placed in a hole and water is sprayed to the bottom of the test piece. This will perform a cooling curve where the end closest to the water will cool more rapidly, and slower moving up to the head. Figure 20.13 shows how the Jominy end quench test is performed with a test specimen. Once the test specimen has been completely cooled, penetration tests are performed starting at the end which was exposed to the water cooling stream and working up to the 4.000 in. length in 1/8 in. increments. The hardness at each test point is recorded and charted to determine the effect of the cooling time vs. hardness. Figure 20.14 illustrates a test performed on a given material and the hardness decline from the end. This can be used to determine the time lag to achieve a determined hardness of a material from hardening.

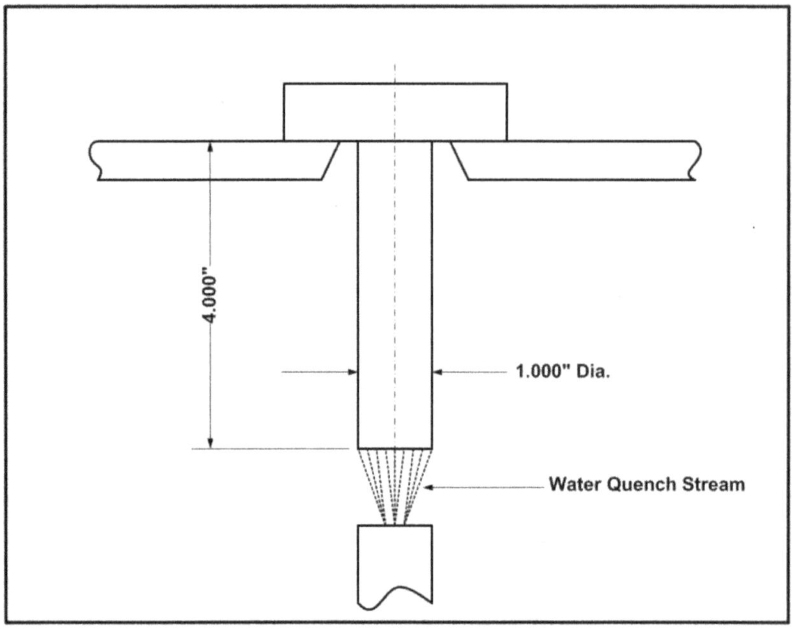

Figure 20.13 Jominy End Quench Testing Procedure

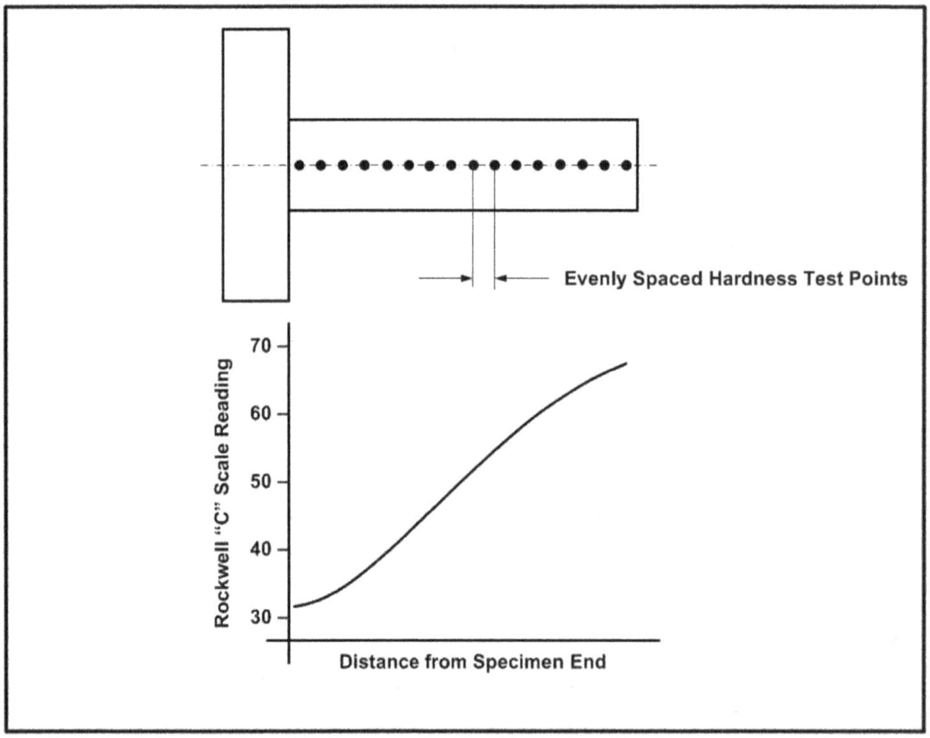

Figure 20.14 Jominy End Quench Hardness vs. Quench Time Testing and Hardness Results Chart

20.4 SURFACE HARDENING

If low carbon steel is attempted to be hardened in the same manner as medium and high carbon steel, it will achieve no hardness and remain the same. A method that will put a case of hardness on the outer layer of the material is surface hardening. The process will deliver a hard case on the component work piece to a determined depth depending on which process is utilized. The core of the material will remain at its original hardness with only a hardened case on the outside. Figure 20.15 illustrates a surface hardened component with a soft internal core and a hardened surface to a desired depth on the outer layer.

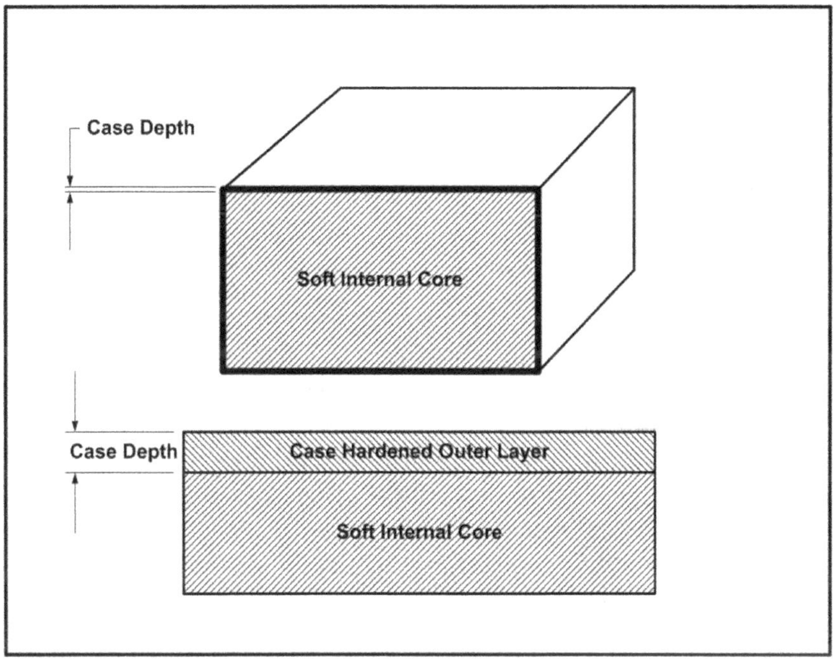

Figure 20.15 Surface Hardening Examples of the Depth of the Case on Low Carbon Steel

This process is widely used if the outer surface requires some wear resistance thus saving the cost of utilizing either medium or high carbon steel that will require some tempering after hardening. However, it the component part will experience any high forces or a repeated hammering during its function, the case will be penetrated and the part may fail. The three processes that utilized for surface hardening are *carburizing*, *nitriding*, and *carbonitriding*, which are explained below.

20.4.1 Carburizing

Carburizing is a process that embeds carbon into the surface of the work piece while it is being heated in the furnace. The components are loaded into a furnace and heated up to between 1,600 and 1,700 °F in a high carbon atmosphere. During the process, the carbon will infiltrate the surface of the work piece to a determined depth dependent on the amount of time the parts are in the furnace. This infiltration of carbon into the surface of the part will produce a hard case on the outside in the same manner as it performs in a medium and high carbon steel material. This will only penetrate to a depth while the core remains soft, and when quenched produces the hardened skin on the work piece. Two methods are used in carburizing which are *pack carburizing* and *gas carburizing*, which are described below.

Pack carburizing: This is performed in a furnace under pressure that will drive the carbon into the component work piece deeper. A case depth of between 0.025 and 0.150 in. can be attained.

Gas carburizing: The component work pieces are heated under normal pressure allowing the carbon to infiltrate to less of a depth than with pack carburizing. A case depth of between 0.005 and 0.030 in. can be obtained with this process.

20.4.2 Nitriding

Nitriding works very similar to carburizing with a high nitrogen atmosphere in the furnace. The nitrogen infiltrates into the surface of the component work piece that will also harden when quenched. The temperature for this process is lower with the parts being heated to about 950 °F before quenching. Depending on how long the parts remain in the furnace, a case depth of between 0.001 and 0.020 in. can be attained.

20.4.3 CarboNitriding

With both carburizing and nitriding allowing a hardened case to be infused into the surface of the work piece, carbonitriding is a combination of both. Here, and atmosphere consisting of a mixture of both carbon and nitrogen is used to infiltrate into the component work piece surface. The temperature similar to nitriding of 950 °F is used and a case depth of between 0.003 and 0.020 in. can be attained.

20.5 SUMMARY

Heat treatment of metals is a physical enhancement process that alters the internal structure of the material. Hardening which is performed on medium and high carbon steels will increase the over through hardness of the material dependent on the amount of carbon in the steel. The material is raised to a given temperature and then quenched to induce the hardness to avoid both pearlite and bainite entering the material. The quenching time will vary from rapid to slow depending on the material being hardened.

Tempering and annealing can be performed to reduce the hardness and internal stresses in the component work piece. Both tempering and annealing will perform the same microstructure changes with the difference being that annealing will remove all the hardness in the work piece while tempering will reduce it to a desired level. Annealing is often utilized to eliminate strain hardening and surface stress in non-hardened components to return the internal structure back to the original state.

The hardness of a material can be tested through the penetration method where a stylist is pushed into the surface under a given load to determine the hardness. Using the Rockwell scale, a reading of Rockwell C68 is full hardness of a high carbon steel and through tempering, the reduction in hardness can be measured to the desired level. This test will however leave an impression in the part being tested which may not be acceptable.

For low carbon steel, surface hardening can produce a hard case skin on the work piece. The three processes that are utilized for surface hardening are carburizing, nitriding, and carbonitriding. The work pieces are exposed to a given atmosphere in a furnace that induces the hardness to a given depth on the parts. With this process, the core of the material will remain soft with only the outer surface obtaining hardness.

20.6 QUESTIONS

1. Of the three types of carbon steel, what is the percentage of carbon contained in each of the following and how much hardness can be attained?

   Low carbon steel
   Medium carbon steel
   High carbon steel

2. In the hardening curve, what is meant by the "nose" and what elements are trying to be avoided during the quenching process?

3. What is the difference between a vacuum furnace and a fluidized bed furnace and how do they operate?

4. If a work piece were to be hardened and required heating with a torch or by some means other than a temperature controlled furnace, how would you know that the required temperature has been attained?

5. All steel work pieces that are to be hardened require quenching to obtain the hardness. What is the difference between submersion quenching and air quenching?

6. What is the difference between tempering and annealing?

7. Tempering and annealing both perform the same function on a work piece. What functions do they perform and what is the difference between tempering and annealing?

8. When utilizing a penetration hardness test, how is it performed and what is the maximum hardness that can be attained on high carbon steel?

9. What is a Jominy end quench test and what does it show?

10. What is surface hardening and what type of material is it performed on and why?

11. There are three processes for surface hardening. What are they and what is the difference between them?

SECTION

6

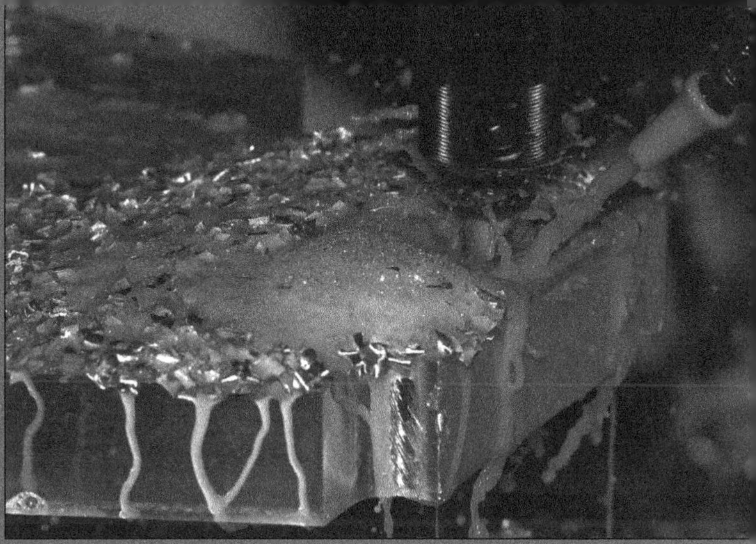

SURFACE PREPARATION AND FINISHING

During the processing of the component workpiece, surface preparation needs to be performed to ensure that the following operation can be performed correctly. Depending on the proceeding operation, the amount of surface preparation may be minimal to extreme. Moving from one machining to another normally requires very minimal surface preparation where completed components requiring some form of finishing operation will require a far more in-depth preparation before the operation can be performed. Several types of cleaning and surface treatments are utilized and are covered in this section along with the usage during processing of the component workpieces.

The final process in the fabrication of workpieces is finishing. There are several reasons that component finishing is performed, from protection from the elements to decorative coating, to enhance the overall appearance. For the most part, all components fabricated from materials that have the capability of corrosion will require some form of finishing protecting them during their functions in the completed product. This finishing can vary from complete corrosion protection to a corrosion and decorative coating, depending on the application of the component workpiece. There are some processes that require no finishing such as plastic parts, but in some cases, they may receive a finishing operation to enhance the appearance for cosmetic purposes.

This section covers a variety of cleaning and finishing processes and where each would be utilized in the fabrication of component workpieces. The combination of these operations is dependent on the function of the parts, the amount of cleaning or protection required, and their appearance in the finished product. Each process will perform a different function on the components, and the selection of the correct order and process will be discussed.

CHAPTER

21

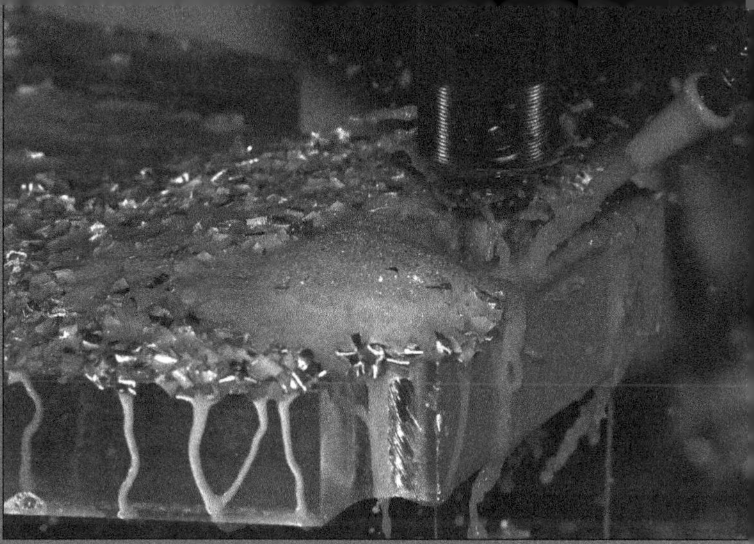

Cleaning and Surface Treatment Principles

21.1 INTRODUCTION

Cleaning and a variety of surface treatments are highly required during the manufacturing process and are performed during the various operations as the component work part moves through the operations from start to finish. This is performed for a variety of reasons and is performed after the completion of one operation before moving the workpiece on to the next. Figure 21.1 illustrates the cleaning or surface treatments as the component workpiece moves through the processing. As the components are assembled into the completed finished product, the last operation performed is cleaning. The visual appeal of a finished product is important to the customer, so efforts need to be made to ensure that the product is acceptable to the general public. No one wants to purchase something new that does not look appealing, regardless of whether or not the appearance affects the function in any manner. If you look at the vehicles on a dealer's lot, the engine under the hood is spotless as the customer inspects it for purchase. How clean the vehicle is under the hood has no effect whatsoever on the performance of the vehicle, but it does to the potential customer.

This chapter covers a variety of cleaning and surface treatment operations and how they affect the component workpieces. The reasons for performing these operations and the considerations for selecting various processes will be discussed along with the amount of cleaning required prior to the next operation that will be performed. Some operations that will be performed may require very minimal cleaning whereas others require cleaning and surface treatments that, after being performed, the component workpieces cannot even be touched with bare hands. All of the processes do not intentionally alter the geometry; however, some can either induce surface etching or cause surface stresses on the materials.

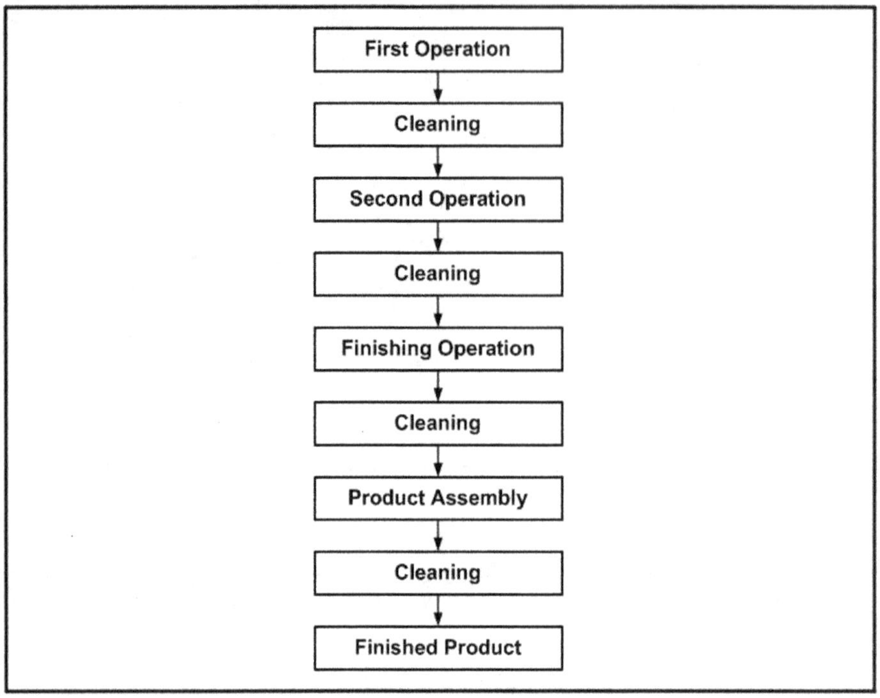

Figure 21.1 Surface Treatment and Cleaning during Processing

21.2 REASONS AND CONSIDERATIONS

When cleaning component workpieces or finished products, there are reasons and considerations for performing the cleaning that need to be understood. There are several processes for performing cleaning and surface treatments and no one process can do it all depending on the reasons and considerations. Following is a list of these factors and how they can affect the materials being cleaned along with determining just how clean the material is required to be in the process.

Reasons

1. *Remove Contaminants from the Workpiece*: The main purpose of cleaning is to remove any contaminants and debris from the workpieces. As the component workpieces move through their operations, cleaning is important in order to not cross-contaminate the following operation from debris and fluids that may remain on them. Cross-contamination can either require changing the cutting fluids in the following operation or, in extreme cases, require the operation to be shut down and cleaned.
2. *Prepare Workpiece Surface for More Processing*: Depending on the operation to follow, some require very critical cleaning or treatments to the component workpiece surfaces. Parts that are moving into finishing receiving a coating or deposition on them always require certain surface treatments and always require extreme cleaning. In some cases, the cleaning needs to be so extreme that the workpieces cannot even be touched after cleaning.
3. *Improve Worker Hygiene Conditions*: By removing the contaminants from the workpieces, a cleaner environment for the workers handling them will be realized. Some fabrication processes may utilize cutting fluids that could be toxic and removing them will protect the operators from spreading the contaminant around to areas where it could be harmful.
4. *Increase Workpiece or Finished Product Appearance*: Appearance is everything to a finished product. No customer wants to purchase a product that is not clean. Whether the product being purchased is a complete assembly or a component part, which will be assembled into a product, cleanliness is important to the customer receiving it.

Considerations

1. *Purpose of Performing the Cleaning*: The question to ask is: why are we cleaning the workpieces? Is it just to remove the majority of the contaminants so as to not cross-contaminate the following operation or is it to produce a finished product? The reason for cleaning the workpieces or product will determine which process will be utilized to perform the operation.
2. *Degree of Cleaning Required on the Workpiece*: How clean does the component workpiece need to be? If the workpieces are just moving from one machining operation to another, just whipping them off with a rage might be sufficient. However, if the parts are moving into a finishing process, then the surfaces will need to be completely clean and free of any contaminants. Understanding how clean they would need to be will determine the process that will be utilized. If the parts are being cleaned more than they need to be, there is a waste of cost, and in the same manner, if they need to be extremely cleaned and are not, the operation they are moving into will fail.
3. *Contaminant to be Removed from the Workpiece*: What needs to be removed from the component workpieces? Depending on the contaminant on the parts, the proper process will be required from very aggressive to extremely mild. Also, if the contaminant on the work-pieces causes a reaction with the cleaning chemical, an undesirable and possibly dangerous situation may occur.
4. *Workpiece Material to be Cleaned*: What is the material that is to be cleaned? With several of the cleaning processes utilizing chemicals, it is critical to know if there will be any reactions between the cleaning solution and the material being cleaned. In some instances, the reaction can change the physical properties of the material being cleaned, and in extreme instances, the material can be completely destroyed in the process.
5. *Environment Considerations of the Cleaning Process*: What environment considerations need to be understood? The majority of the cleaning solutions being utilized can cause damage to the environment if not properly disposed of. Very caustic and acidic solutions will need to be neutralized or containerized and sent to a facility for proper disposal. Just dumping these chemicals without proper disposal can cause extreme damage to the environment and could result in very steep fines from the government.

21.3 SURFACE TREATMENT AND CLEANING PROCESSES

All cleaning and surface treatment processes fall into one of two categories. These are chemical processes and mechanical processes. The process category to be selected is dependent on the material being treated and the intensity of the treatment required. Each of these treatment areas will render a different treatment, and in some cases, a combination of both will be required to achieve the desired result. The chemical and mechanical treatment processes are explained as follows with a description of the more common processes in each arena.

21.3.1 Chemical Processes

Chemical treatment processes are performed utilizing solutions to treat and clean the materials. These can range from mild solutions such as common household cleaners to extremely harsh solutions like acids and caustic solutions. The amount of treatment can range from a simple cleaning to enhance the overall appearance or to remove minor contaminants to prepare the material for the next operation to actually performing surface etching into the material to remove any and all contaminants. The processes described as follows are a few of the more common processes that are utilized within the manufacturing arena during operations.

21.3.1.1 Alkaline Cleaning

The alkaline cleaning process is one of the most widely utilized in the manufacturing world. This process utilizes a caustic cleaning solution, which is heated to a given temperature, normally between 120 and 200°F. This process is used to remove contaminants such as oils, grease, waxes, and other contaminants like metal chip and

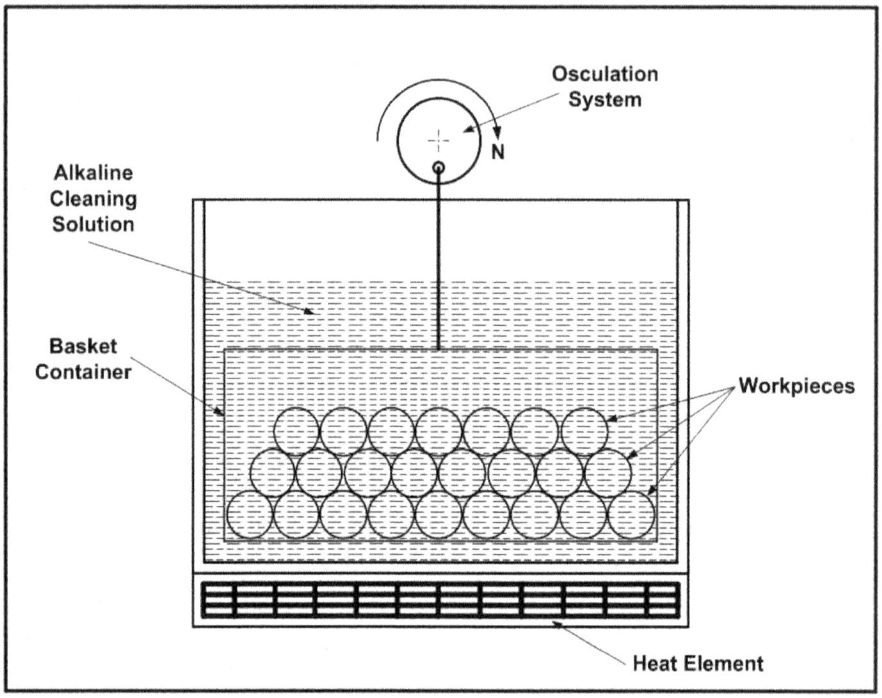

Figure 21.2 Alkaline Cleaning System

residue from previous manufacturing operations. The workpieces are placed in a basket, which is lowered into the solution and agitated slightly to allow the solution to come in contact with the workpieces being cleaned. Figure 21.2 illustrates an alkaline cleaning system utilized in a production environment

The solution used in this cleaning process can vary depending on the material requiring to be cleaned. Solutions like sodium carbonate, borax, potassium hydroxide, and various phosphates can be rather diluted with water or used directly. Alkaline cleaning of the component parts is performed before the plating operations, and the cleaning system in high-production facilities is integrated into the plating line.

These are extremely caustic solutions, and when the solution needs to be changed, disposal is an issue. The cleaning solution needs to be treated by first removing the contaminants that were cleaned from the workpieces and then neutralized to an acceptable pH level before it can be disposed.

21.3.1.2 Electrolyte Cleaning

The electrolyte cleaning process utilizes the same cleaning solutions as in alkaline cleaning but induces a current in the process to assist in the removal of the contaminants. An electrical current is alternated between positive and negative between the workpiece and the cathodes. Figure 21.3 shows the electrolyte cleaning process with the component workpiece being charged from positive to negative. A light direct current of between 20 to 60 V is used, and as it is applied and switched between positive and negative, it assists in dislodging the contaminants on the workpiece surfaces. Figure 21.4 illustrates how the current, as it is alternated, dislodges the contaminants from the workpiece surfaces.

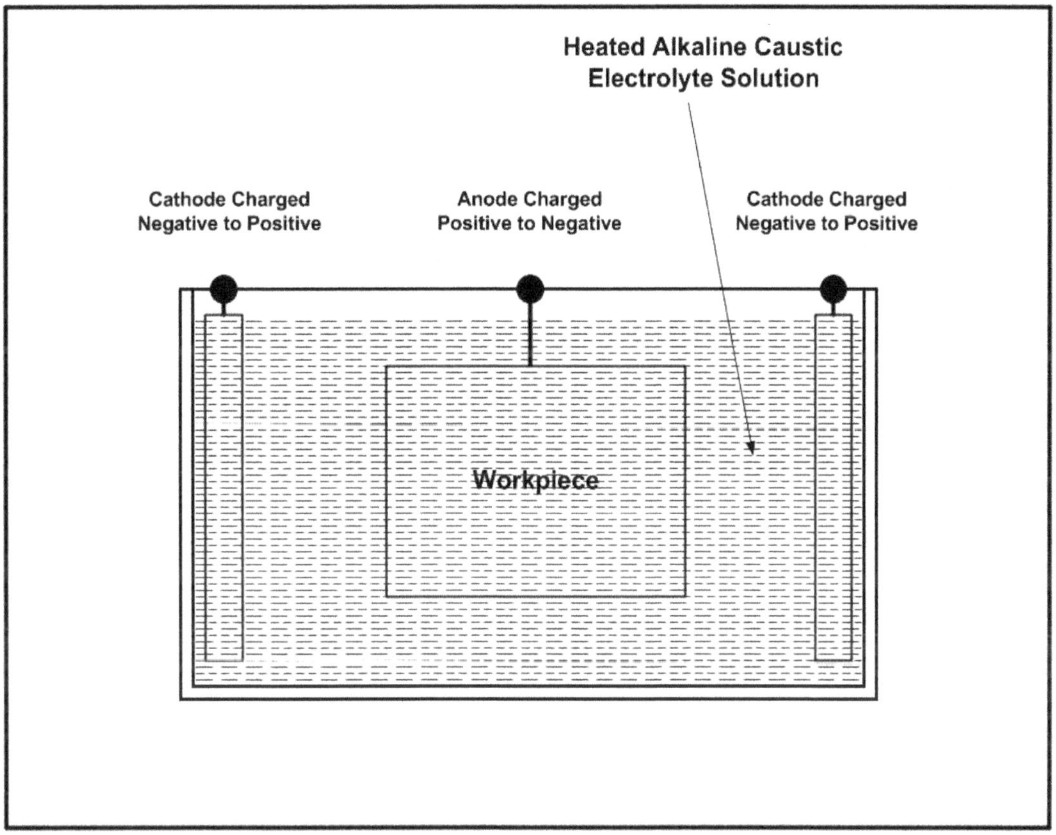

Figure 21.3 Electrolyte Cleaning System

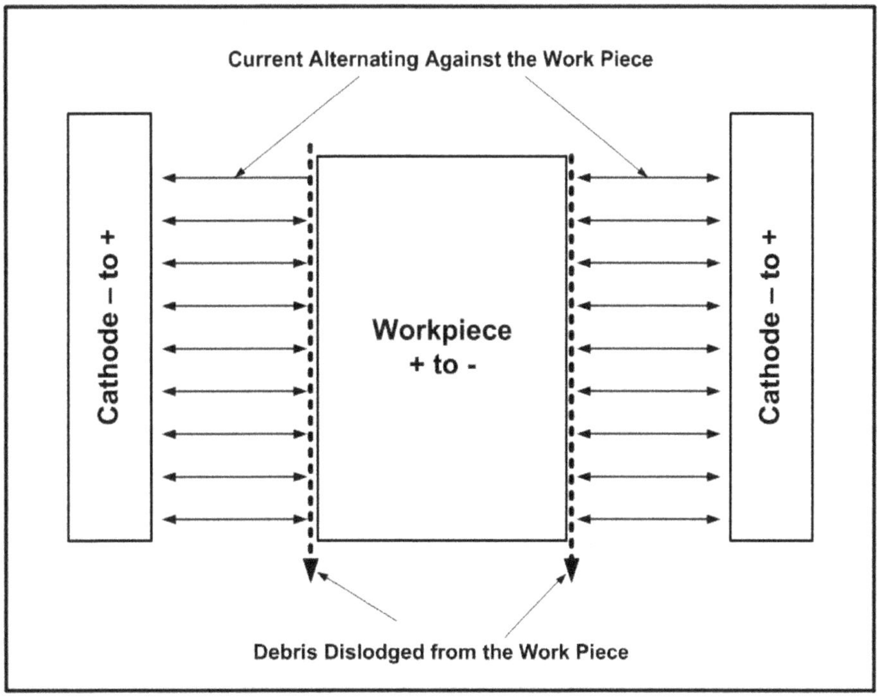

Figure 21.4 Electrolyte Cleaning Concept

21.3.1.3 Emulsion Cleaning

The emulsion cleaning process is one that is much more environmentally friendly. This process utilizes an organic solution either diluted in water or used straight. A common clean solution used in emulsion cleaning is a citrus cleaner, which will emulsify the oils and grease on the workpieces. With the solution being organic, the treatment required when it needs to be changed is only to separate the contaminants from the solution. Being an organic solution, the workpieces cleaned in this process need to be followed by alkaline cleaning to remove any of the organic solution remaining on the surface of the components. Even though this requires the operation of the less environmentally friendly alkaline process, it greatly reduces the amount of the caustic solution that needs to be changed, thus reducing the cost and time for neutralizing the alkaline solutions. The process greatly resembles alkaline cleaning where the workpieces are loaded into a basket and submerged into the cleaning solution. Figure 21.5 illustrates the emulsion cleaning process followed by the alkaline cleaning process.

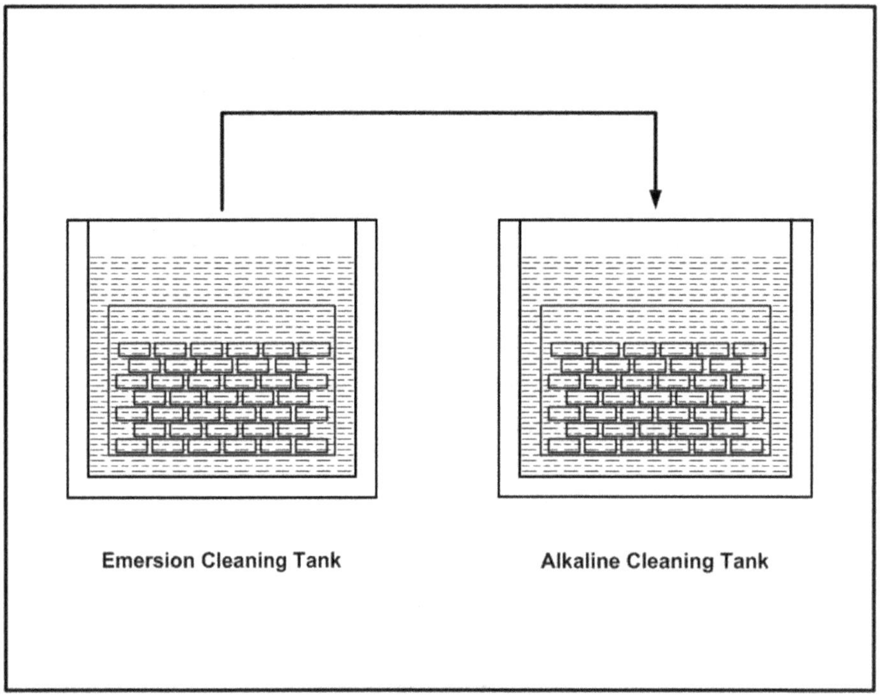

Figure 21.5 Emersion Cleaning Followed By Alkaline Cleaning

21.3.1.4 Solvent Cleaning

Solvent cleaning is performed by submerging the component workpieces into a solution that will dissolve the oil and grease on the surfaces. For most submerging solvent cleaning, the solutions are extremely mild and can be applied by hand for short terms without damaging the skin. For extended solvent cleaning, rubber gloves are required to avoid damage to the skin, and hand washing is required prior to exposure to avoid further removal of the skin oils.

Another solvent cleaning process is vapor degreasing. In this process, the solvent solution is much more aggressive and should not come in contact with the hands. Here, the solution is heated to a vapor and the workpieces are submerged into the vapor. The vapor dissolves the oil and grease on the workpieces and can penetrate into very tight areas where a liquid solution may not be able to. Solutions such as perchloroethylene, methylene, and trichloroethylene are used because of their low boiling temperatures. The solution is heated to between 150 and 200°F causing it to vaporize. The vapor rises up in the tank and the cooling/condensing coils at the top of the tank cause it to liquefy again and fall to the bottom to be revaporized. Figure 21.6 demonstrates the vapor degreasing process with the solution being heated to a vapor and reliquefied as it comes in contact with the cooling coils.

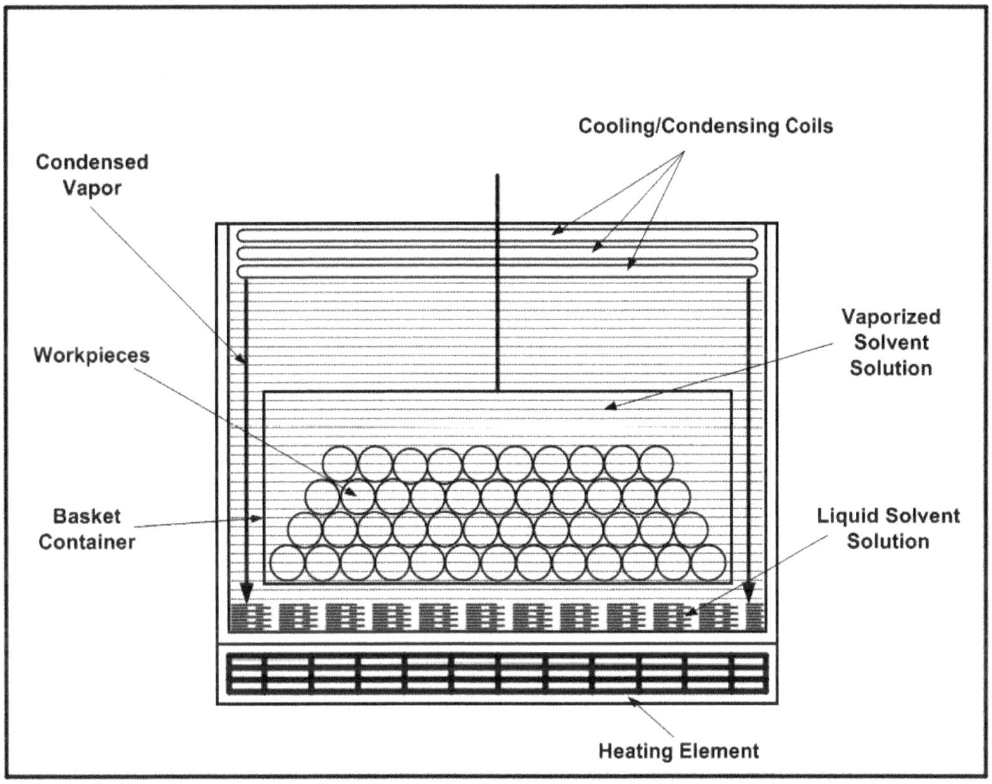

Figure 21.6 Vapor Solvent Cleaning System

21.3.1.5 Acid Cleaning and Pickling

Acid cleaning is the most aggressive of all the cleaning and surface treatment processes. This process involves submerging the component workpieces into a concentrated acid, which will remove all the contaminants from the surfaces. Heavy oils, grease, wax, and particle debris must be removed prior to acid cleaning through one of the previously mentioned methods. The acid will literally eat away at the surface of the workpiece, thus removing any remaining light oils and residue from alkaline or solvent cleaning. Also, due to the fact that acid is being used, light corrosion can be removed from the surfaces. The acids that are most commonly utilized in this process are hydrochloric, sulfuric, and phosphoric, depending upon the metal that is being submerged. Some materials will have a violent reaction such as zinc when placed in sulfuric acid. Pickling is performed similarly to acid cleaning, except that the time frame the component workpieces are left in the acid bath is longer and causes surface etching. Both acid cleaning and pickling are the final step in cleaning before plating, and once the parts have been cleaned, they cannot be handled by hand. The parts will be acid-cleaned, given a water rinse, and moved directly into plating to ensure that the surfaces are as clean as possible.

There are two processes for performing acid cleaning and pickling. The first is tank cleaning. Here the component workpieces are racked and placed into the tank. The racks being used are the same that the parts will be on for plating and moved directly to the next operation. Figure 21.7 illustrates a tank acid cleaning process with the workpieces individually racked and submerged. The second method is acid barrel cleaning. It utilizes a barrel fabricated from a material that the acids will not attack and is perforated with slam holes. The parts are loaded into the barrel and are submerged into the acid while the barrel rotates slowly. The acid enters through the holes in the barrel and, as the parts rotate, cleans the surface. A caution for barrel cleaning is to ensure that the rotation of the parts that will cause them to collide with each other will not cause damage to the geometric features. An example would be to have threaded components, but tumbling of the components together may cause damage to the threads while they are being cleaned. Figure 21.8 shows the barrel cleaning process with the unit submerged and rotated in the acid bath.

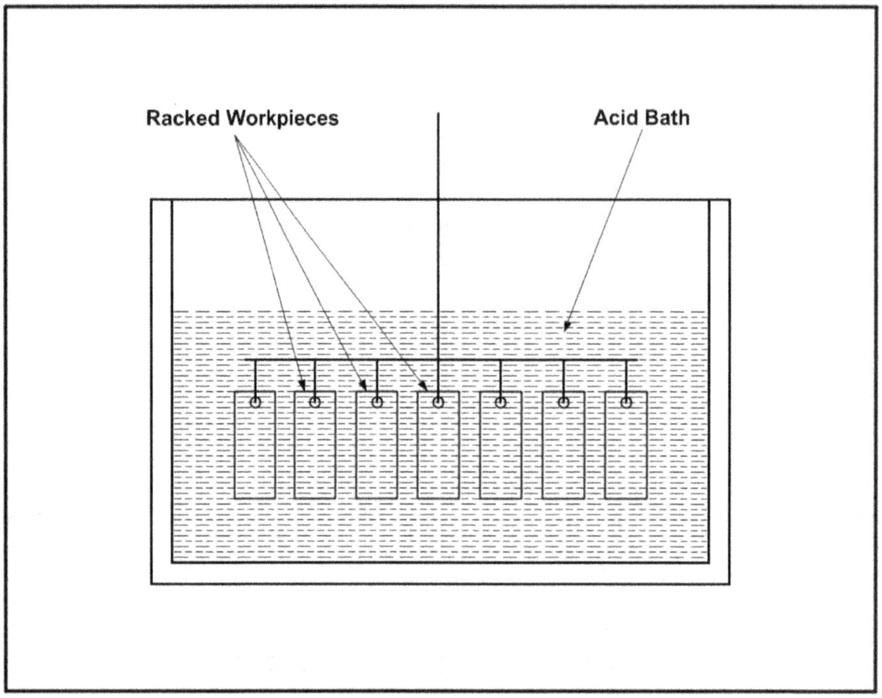

Figure 21.7 Tank Acid Cleaning System

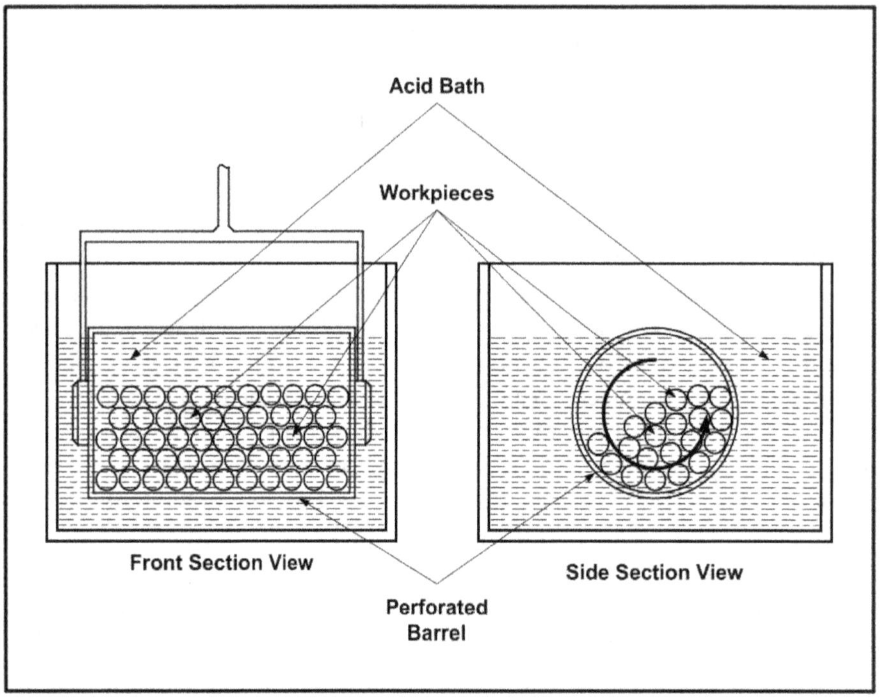

Figure 21.8 Acid Barrel Cleaning System

21.3.2 Mechanical Processes

The second category of cleaning and surface treatments is mechanical processing. Here, various mechanical means are utilized to remove contaminants that can be light or heavy. Some mechanical processes are mild and are used to perform light cleaning, while others are extremely aggressive and are utilized to remove heavy corrosion and even burrs on the workpieces. Some of the more widely used mechanical cleaning processes are described as follows.

21.3.2.1 Ultrasonic Cleaning

The ultrasonic cleaning process utilizes a high-frequency vibration to dislodge the debris on the workpiece. A tank is filled with a water solution that contains some mild cleaning solution and heated to between 150 and 190°F. The component workpieces are places in the solution and the internal container of the tank is vibrated at an extremely high frequency of between 20 and 45 kHz. With the liquid being noncompressible, the vibration of the cleaning fluid will cause it to create high-pressure implosions on the surface of the workpiece and, for all-intensive purposes, blast the debris from the piece. While the component workpieces are in the tank, the frequency will be altered from high to low, causing different implosions on the surface of the workpiece for better dislodging of the debris. Figure 21.9 shows an ultrasonic cleaning tank construction with the vibration system and heating elements.

This process is very highly utilized for jewelry cleaning because it is capable of dislodging dirt and contaminants from extremely tight areas. If we were to use a very tightly constructed neck chain as an example, the dirt gets imbedded into the weaves of the chain and through ultrasonic cleaning; the implosions will dislodge the debris from within the tightly woven chain. Figure 21.10 illustrates how the high-pressure implosions hit the workpiece and dislodge the debris.

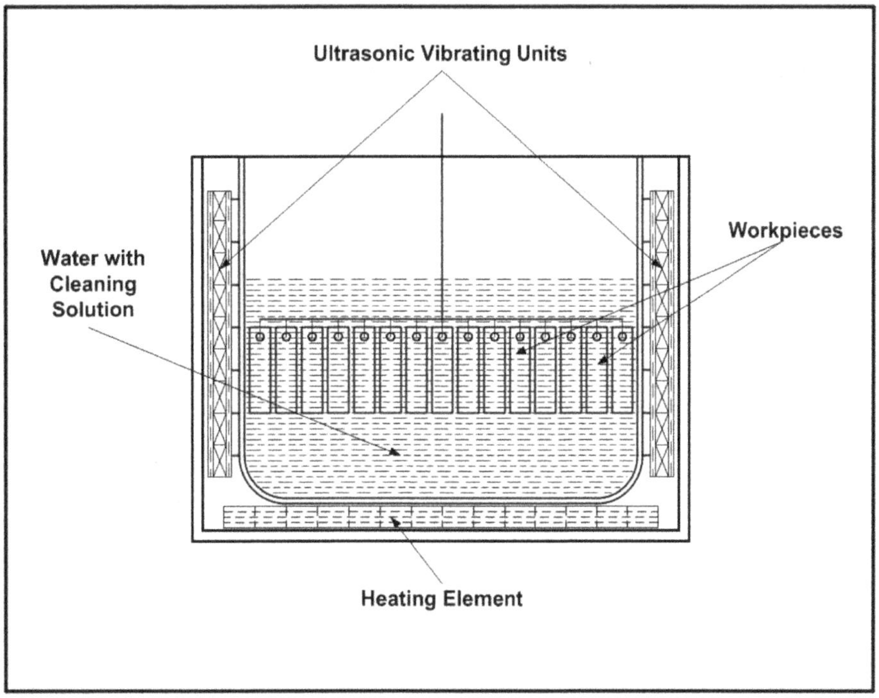

Figure 21.9 Ultrasonic Cleaning System

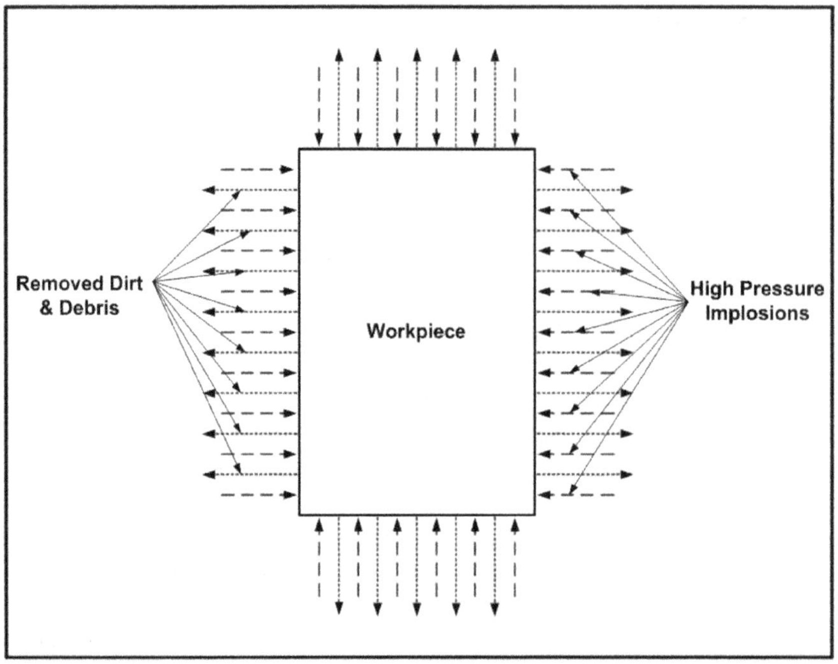

Figure 21.10 Ultrasonic Implosion and Debris Removal

21.3.2.2 Blasting and Shot Peening

Possible the most aggressive cleaning process is blasting and shot peening. This is performed by utilizing high-pressure air, which projects a hard medium at the workpieces. Blasting is used to remove heavy corrosion and heavy contaminants like scale, paint, and rust. High-pressure air between 100 and 160 psi is mixed with an abrasive medium and shot at the surface of the workpiece. The medium can vary from mild, which is used to basically polish the surface, to extremely aggressive, which will cut into the surface for heavy removal. Commonly used media are crushed walnut shells (for polishing), sand, aluminum oxide, silica, and glass. Sand, aluminum oxide, and silica will cut and etch the surfaces and remove heavy contaminants and are used for etching glass. Glass beads are used to produce a satin finish on metal surfaces, and when this finish is plated, it will render a satin-chrome appearance. Figure 21.11 illustrates how the blasting and shot peening process is performed.

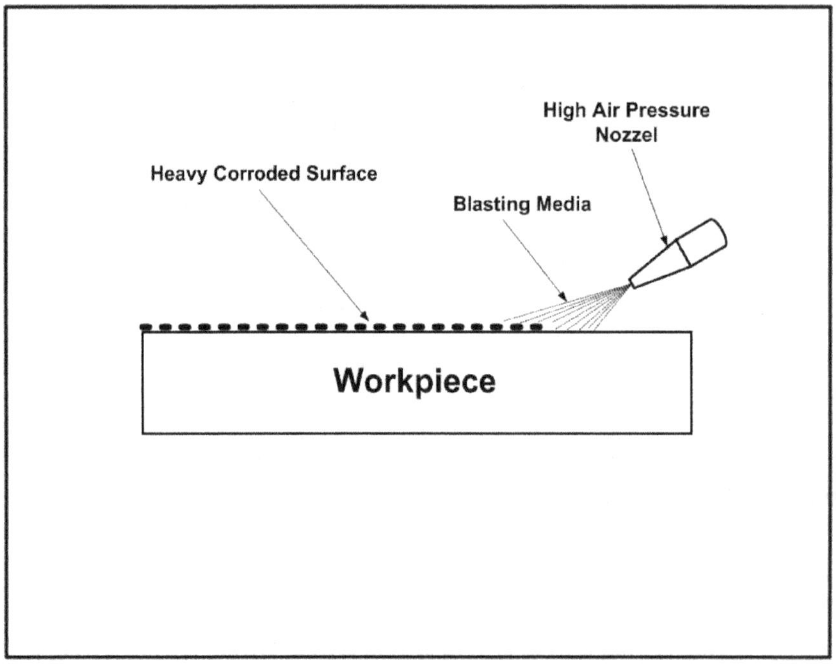

Figure 21.11 Blasting and Shot Peening Operation

Shot peening is performed exactly in the same manner as blasting but utilizes small cast-iron pellets. This process is used to induce compressive stresses and cold-work the surfaces of the workpiece to increase the fatigue strength. This will also remove lighter debris from the surface like blasting will as a by-product, but the main objective is to perform a surface treatment rather than cleaning. Blasting will also induce surface stresses on the workpiece, and if this is an undesirable feature, annealing operation will be required to remove the stresses that have been induced.

21.3.2.3 Tumbling

Tumbling is an aggressive process for both cleaning and removal of burrs on the component workpieces. It is performed in a barrel where the workpieces are loaded in along with an abrasive medium and tumbled together. The abrasive medium can be either dry or more commonly mixed with water. What performs the cleaning and deburring is the part-to-part contact as the workpieces are hit together in the rotating barrel. Figure 21.12 shows the tumbling operation with the pieces hitting together as they tumble with the media. A caution for this operation is to ensure that the part-to-part contact will not alter or destroy the geometric configuration of the part. Threaded components should never be tumbled due to the fact that the threads will be heavily damaged from the tumbling actions. But for parts produced in a sheet metal process where a heavy burr is produced from stamping, tumbling can be a very viable option for removing the burrs. However, if the components are left in the tumbler too long, the geometry can be either damaged or destroyed.

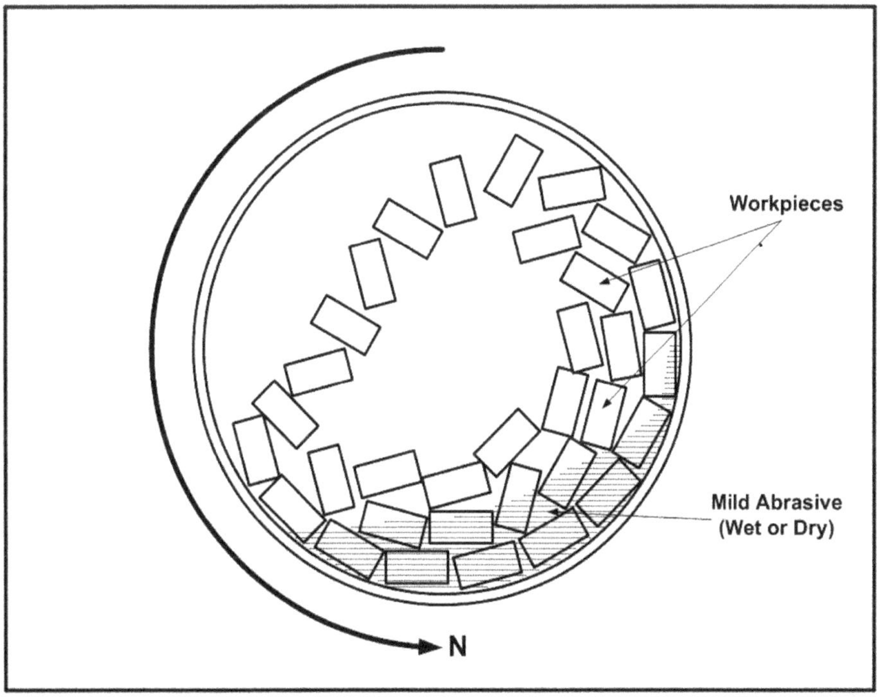

Figure 21.12 Tumbling System Showing Part-to-Part Contact during the Process

21.3.2.4 Vibrating

The vibrating process is similar in nature to tumbling but less aggressive to the workpieces. While tumbling relies on part-to-part contact to perform the cleaning and deburring, vibrating relies on media-to-part contact. A large container, normally round but can also be square, is loaded with the desired media along with the component workpieces. Care is taken to not overload the workpieces so that during the process, there will be maximum part-to-part contact. What is desired is for the workpieces to come in contact with the media being utilized and not with the other component workpieces. The tank is oscillated, causing the media and parts to rotate together in two different directions. Figure 21.13 shows the top view of the tank and the circular rotation of the parts and media, and Figure 21.14 shows the side view of the rotation of the parts and media. This duel rotation of the workpieces and media will both clean and deburr them with very minimal part-to-part contact.

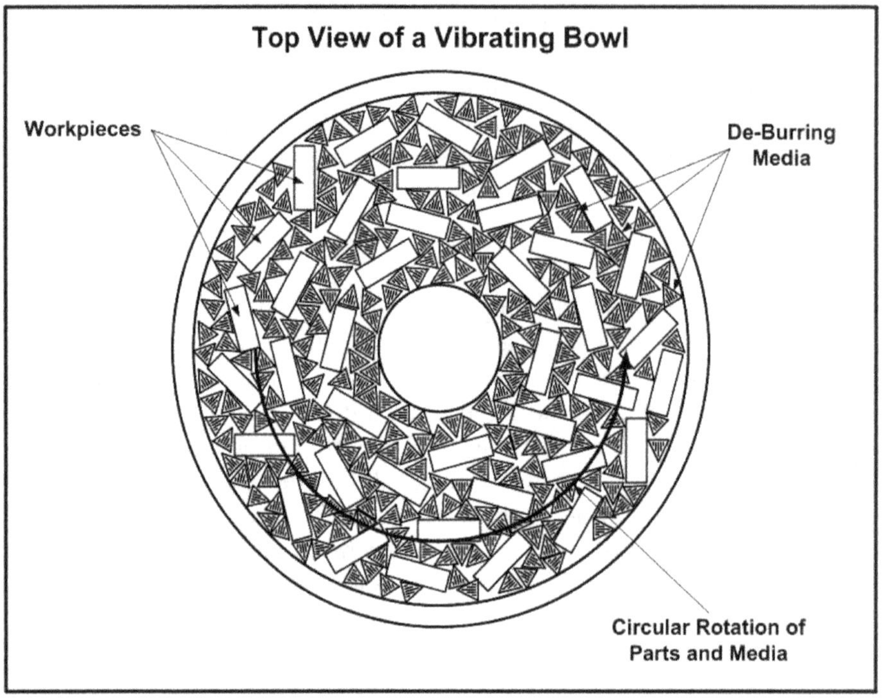

Figure 21.13 Top View of a Vibrating Tank Showing Part to Media Contact and Rotational Action

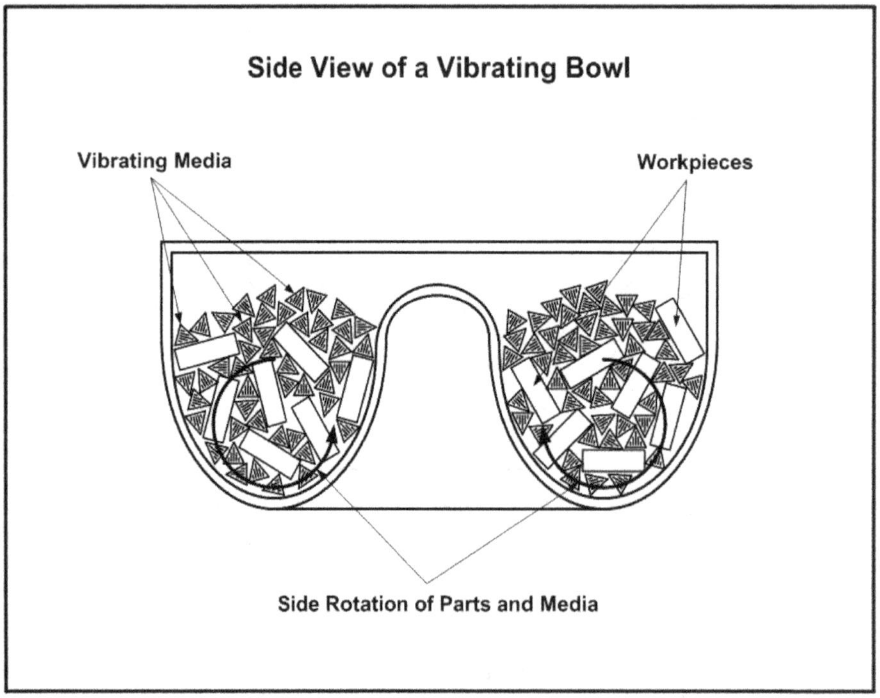

Figure 21.14 Side View of a Vibrating Tank Showing Part-to-Media Contact and Rotational Action

The media utilized in vibrating can range from extremely mild to very aggressive, depending on the desired outcome. Also, the media can be used dry or mixed with water depending on the media selected. For a vibrating operation where the purpose is to clean and polish the workpieces, the media can be materials such as ground corncobs, ground walnut shells, and dry sand. These are very mild and will produce a highly polished finish while light debris from the workpieces. For more aggressive vibrating, medium made from glass-filled plastics to ceramic in an almost unlimited geometric configuration is used and should be mixed with a small amount of water. This aggressive medium will both clean and deburr the workpieces, and as with tumbling, if they are left in the vibrator too long, the geometry of the components can be damaged or destroyed.

21.4 SUMMARY

Cleaning and surface treatments of component workpieces are both important and critical in the manufacturing arena. There are a multitude of reasons and considerations to evaluate before selection of a process and are important to understand. As mentioned, cleaning is not a final operation but rather performed between each operation as the workpieces are being fabricated. The degree of cleaning required is dependent on the next operation to be performed, and care should be taken to not overclean as this will require more time and thus induce a higher cost. At times, a simple cleaning with a rag may be sufficient before moving to the next operation, but for parts moving into a finishing operation, the cleaning required will be extreme to prepare the surface as clean as possible.

21.5 QUESTIONS

1. What are the reasons that cleaning and surface treatment operations are performed on component workpieces?

2. What are the considerations that need to be evaluated before selecting a cleaning or surface treatment operation?

3. What is the difference between chemical and mechanical cleaning?

4. Of the chemical cleaning processes, which is the most common and widely utilized and what disadvantages does it pose?

5. Emulsion cleaning utilizes what type of cleaning solution, and why does it need to be followed by alkaline cleaning?

6. How does electrolyte cleaning work, and what solution is it performed in?

7. What are the two types of solvent cleaning used, and explain how each is performed.

8. In the mechanical cleaning arena, what is ultrasonic cleaning and how does the process work?

9. Both blasting and shot peening will induce what into the surface of the workpieces and what can be done to eliminate it?

10. Explain the difference between tumbling and vibrating, and what precautions need to be considered to avoid damage to the component workpieces?

CHAPTER 22

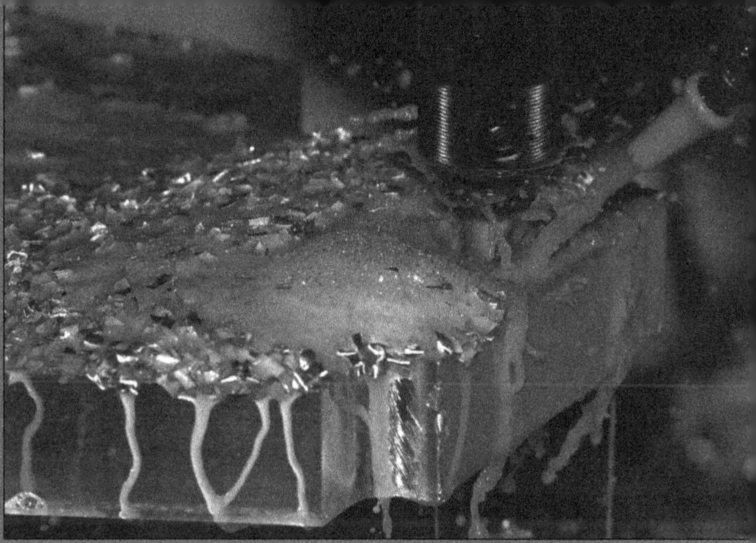

Surface Coating and Deposition Principles

22.1 INTRODUCTION

Coatings and depositions are applied to the surfaces of component workpieces for a variety of reasons. What we see on the products that we purchase are coatings that, to us, are there to enhance the appearance, but the purpose for applying them is much more than just esthetics. There are several reasons that depositions are applied to the workpieces, which are listed as follows:

1. *Corrosion Protection*: One of the main reasons that a coating or deposition is applied on component workpieces is to protect them from corrosion while in service. Steel is a material that is extremely susceptible to corrosion, and even when a steel component is assembled into a product that will be used in an office environment, the normal humidity in the atmosphere can cause corrosion.
2. *Appearance*: The overall appearance and esthetics of a product are important to the customer purchasing it. By applying a coating on the components, even if the material is not susceptible to corrosion, its appearance is enhanced, and thus it becomes more appealing to the potential customer.
3. *Wear Resistance*: Some deposition coatings are very hard and can offer wear resistance to the applied surfaces. These can be utilized on components that will have rubbing on them during the function in the product and can offer an alternative to using more expensive materials for the applications.
4. *Worn Surfaces*: In some applications, the surfaces of component parts can become worn down. As with wear resistance, some coatings that can be applied are extremely hard and can build up a worn surface rather than replacing the component. This is very useful for expensive pieces and can save money by building up the surfaces rather than replacing them.

5. *Further Processing*: Some depositions are used as a base coating on component workpieces that will receive further processing. These coating can be utilized for surface treatments such as painting or in the plating arena, and some coatings require layers of plating for the process to be performed correctly.
6. *Electrical Properties*: Various coatings and depositions can alter the electrical properties of the component workpieces. Depending on the material it is being applied to, and the deposition being applied, the electrical conductivity or resistance can be increased for a given application.

There are several processes for applying coatings and depositions to component workpieces, and depending on what is required, a proper process needs to be selected. What is required of the coating or deposition being applied and the material it is being applied to will determine which process is the most viable. The chapter discusses some of the more common deposition processes and where they are most widely utilized in manufacturing.

22.2 ELECTROPLATING

The electroplating process is one of the most widely utilized for coatings and depositions on component workpieces. This is because of the variety of plating materials that can be applied and the uses of these materials. What makes electroplating so versatile is that it can be used as a corrosion protection, a decorative coating, or a combination of both. A variety of materials can be plated on metallic components, and following is a list of the more common materials that are used in electroplating and their uses.

Zinc: This deposition is primarily used as a corrosion protection. This plating material is available in a multitude of colors with yellow being the most common. Zinc is a very soft material and offers no wear resistance, but is an excellent protection from the elements on steel parts.

Nickel: Nickel is used as both a corrosion protection and a decorative coating. It is much harder than zinc, but not enough to offer wear resistance. This deposition is also utilized as a base coating for other plating processes that require this material for proper processing.

Copper: The deposition has several uses when applied. It is very decorative on components when the appearance is important and will reduce the cost dramatically as opposed to fabricating the part from copper. It offers some amount of corrosion protection, but copper will corrode in extreme conditions. Being a good electrical conductor, copper plating can increase the conductivity of the material it is applied to. Copper is also utilized as a base coating for other plating processes to aid in the adherence of the next material being applied.

Chrome: Chrome is possibly the most known plating material. It is used for decorative applications and offers some corrosion protection. It is also an extremely hard material and is used for wear resistance applications and for rebuilding worn surfaces. A proper chrome plating process will be performed over a nickel plating where the nickel offers the majority of the corrosion protection as a base coating.

Silver: Silver has two distinct applications in plating. First, it is decorative and is used in costume jewelry to give it the appearance of being made of silver. Second, silver is one of the best conductors of electricity and can increase the conductivity of the material it is applied to.

Gold: Very similar to silver, gold plating is widely used for costume jewelry. Another main application of gold plating is on electrical connectors. Although gold does not have the electrical conductivity of silver or copper, the main feature is that gold will never corrode, leaving the connectors clean. This is evident from seeing gold retrieved from ships sunk over two hundred years ago in the ocean and still free from any corrosion.

Brass: This deposition will offer both decorative and corrosion resistance properties. It can be plated and offers a polished brass surface while offering mild corrosion properties in nonextreme environments.

The principle of electroplating is the transfer of the metal ions onto the workpiece through current flow, which is based on Faraday's two physical laws. The workpiece that is the cathode is submerged into an electrolyte

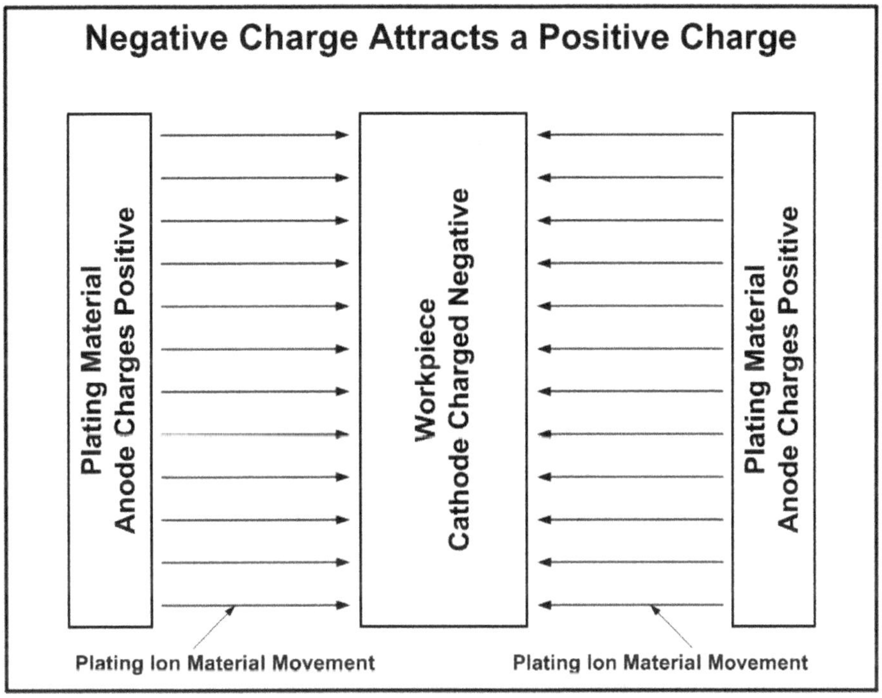

Figure 22.1 Negative Attraction of Positive Ions in an Electroplating Process

solution and is charged negative, while the material to be deposited onto the workpiece, which is the anode, is charged positive. As the current is applied and transfers through the electrolyte solution and negative attracts positive, the ions from the cathode are transferred from the material to be deposited to the anode, which is the workpiece. Figure 22.1 illustrates the theory of electroplating, showing the transfer of ions from the cathode to the anode.

With electroplating utilizing current to perform the operation, there are some areas that need to be understood and considered in order to produce quality components. One of the main issues causing electroplating to not properly adhere to the workpiece surfaces is insufficient cleaning. The cleaning process utilized on an electroplating line is acid pickling. This will clean and etch the surfaces rendering them as clean as possible, and the workpieces cannot be handled afterward. Even the oils in the human skin will leave residue on the surface, causing the plating material to not adhere properly. Another consideration is how the plating material will build up on the workpieces. How the component workpieces are attached to receive the negative charge will greatly affect the deposition thickness. The plating ions will be attracted where the current is the strongest, thus applying a heavier deposition in that area. Likewise, as the current moves less further away from the racking point, the deposition will be lighter. Electroplating will never render an even deposition, and the uneven deposition will not be extreme, but it will be present. Figure 22.2 demonstrates various racking points and how the deposition can vary on the component workpieces as the current is applied. Another area that needs to be considered is the surface finish from electroplating. During the process, the ions will be more attracted to the high points on the surface, thus building them up greater than the lower points. With all surface finishes having a certain roughness, it will be magnified through plating. The rumor that plating will render a smoother surface finish is false, and if a certain roughness is required for the plated component, it must be smoother prior to plating. An example would be if the final surface finishes on a workpiece are required to have a reading of R_a 32, the component prior to plating should be roughly R_a 26. Figure 22.3 illustrates how the plating depositions will build up on the high points of the surface roughness, rendering a higher roughness than prior to plating.

The electroplating process can be performed in two ways. Rack plating will plate each component individually, and barrel plating will plate them in a batch mode. Depending on the size, configuration, and requirements, a proper method needs to be selected. Both rack plating and barrel plating are explained as follows.

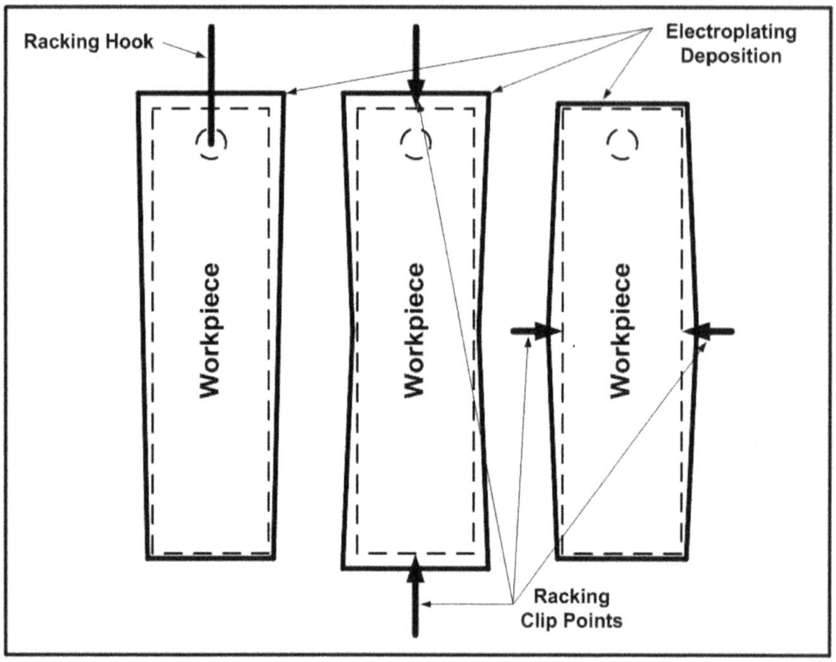

Figure 22.2 Uneven Deposition Buildup from Racking Positions

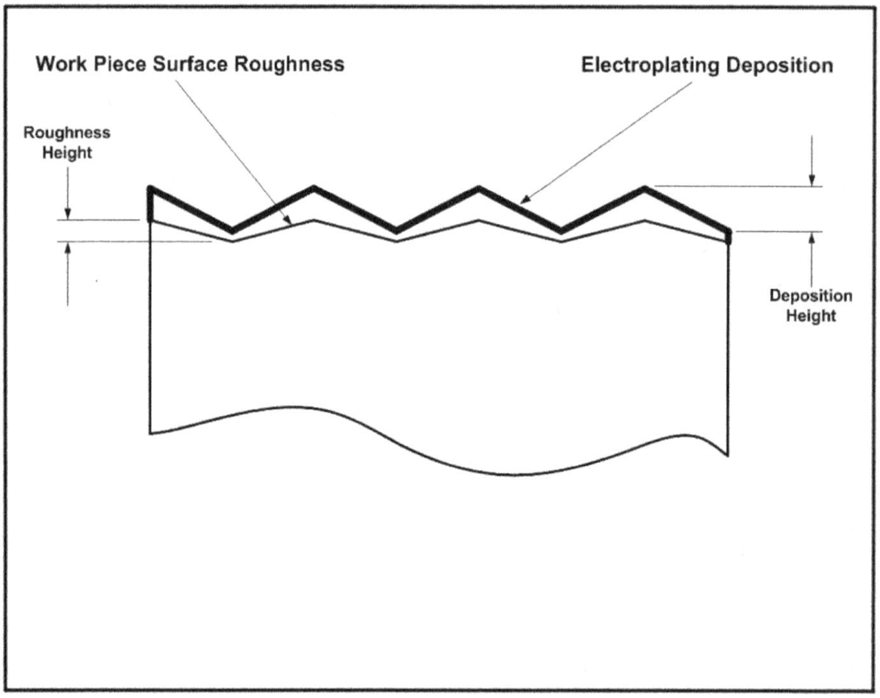

Figure 22.3 Surface Finish Roughness Buildup on Workpieces

22.2.1 Rack Plating

With rack plating, each individual component will be racked separately and placed into the plating tank. The racked parts will be run through the acid cleaning operation first and moved directly to the plating tank. Each rack is placed on a bar, which will be charged negative sending the current through the rack and to the component workpieces. The current is applied to the tank and the ions from the anodes on each side of the tank receive the positive charge and begin to transfer to the workpieces. Figure 22.4 shows how the workpieces are individually racked with the positive and negative charges applied, allowing the ions to transfer.

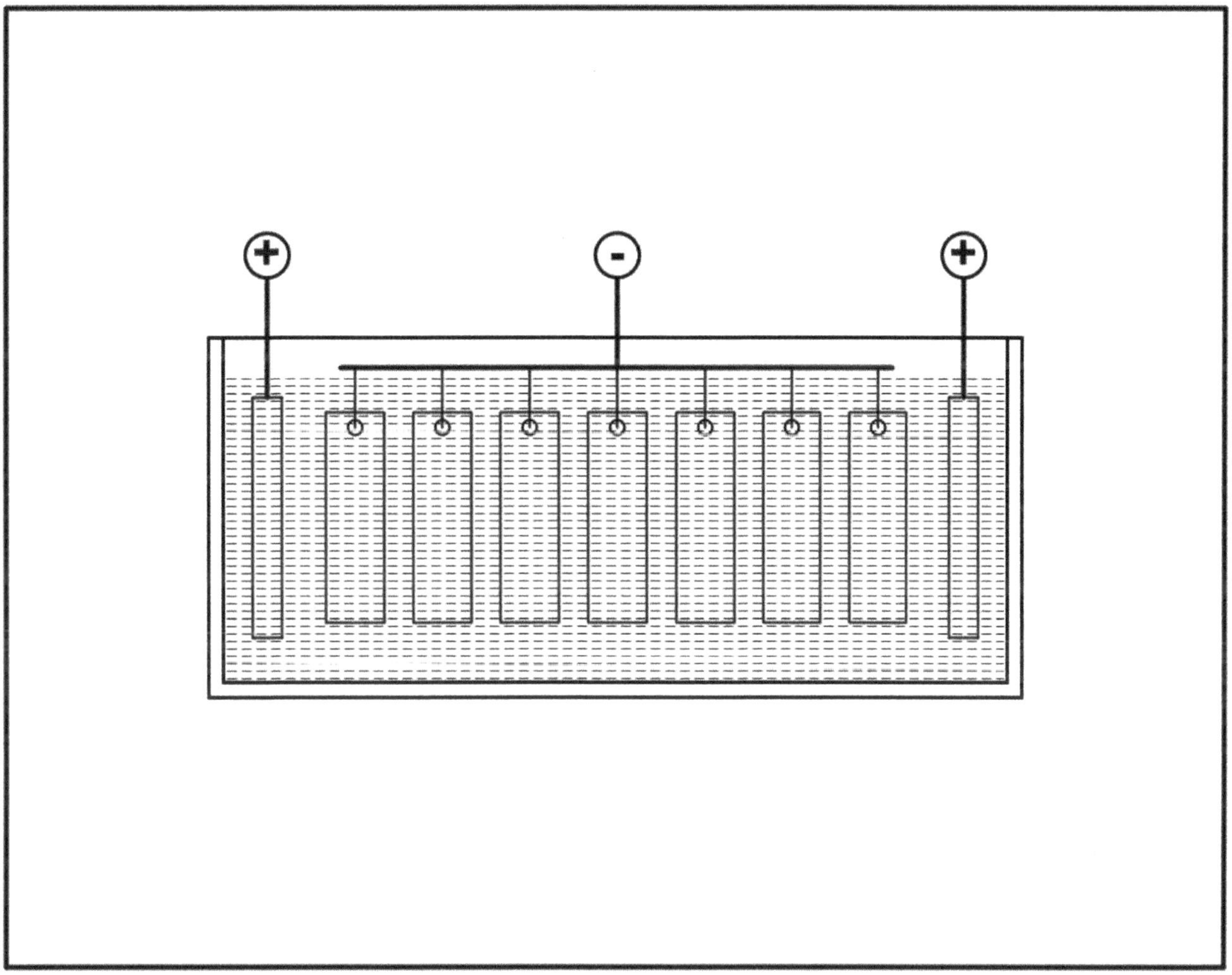

Figure 22.4 Rack Plating Tank System

22.2.2 Barrel Plating

Barrel plating is a batch process where the components are not plated individually but rather in a batch mode. A perforated barrel is used and the parts are loaded in to about 1/3 to 1/2 full. It is important to not overfill or underfill the barrel as this will not allow for proper movement during the process. The barrel has flexible anode rods on both sides, which will run the charge through all the components in the barrel to allow the plating process to activate. Figure 22.5 shows the configuration of the barrel with the anode rods from each side and the level of components for plating. A lid is placed on the barrel and submerged into the plating tank as shown in Figure 22.6. The barrel is rotated, which is shown in Figure 22.7, and the current is applied. As the workpieces tumble in the barrel, the flexible cathode rod transmits the current through the components, which are always in contact with each other. The ions are transferred through the perforations in the barrel and deposited on the workpieces.

There are some precautions for barrel plating that need to be considered before selecting this process for plating. If the components are threaded, the threads will get damaged during the tumbling in the barrel and may render them useless. Another consideration is the surface finish requirements. As the parts tumble together, the surface finish will get damaged as the threads would.

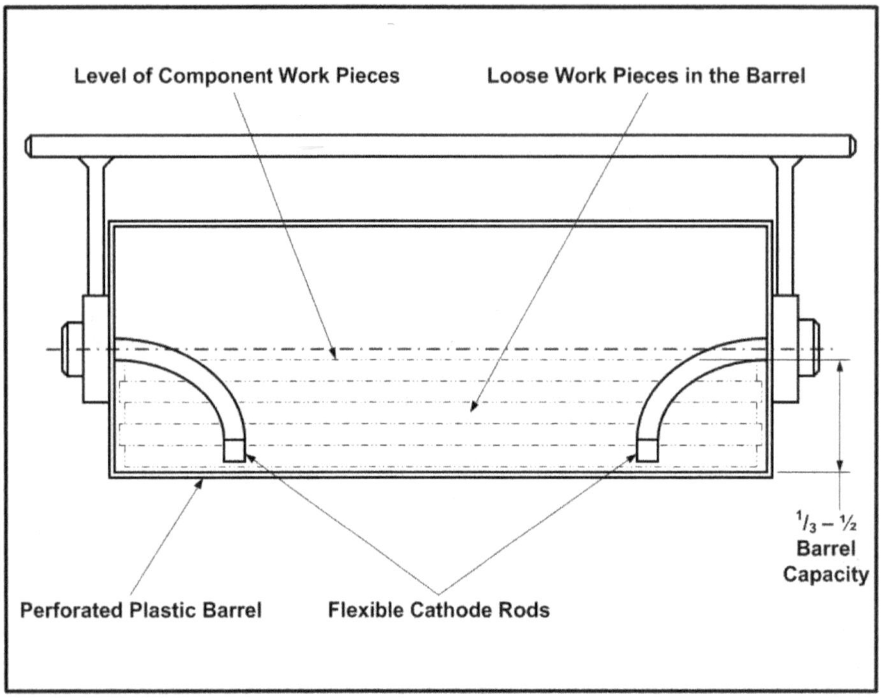

Figure 22.5 Barrel Plating System

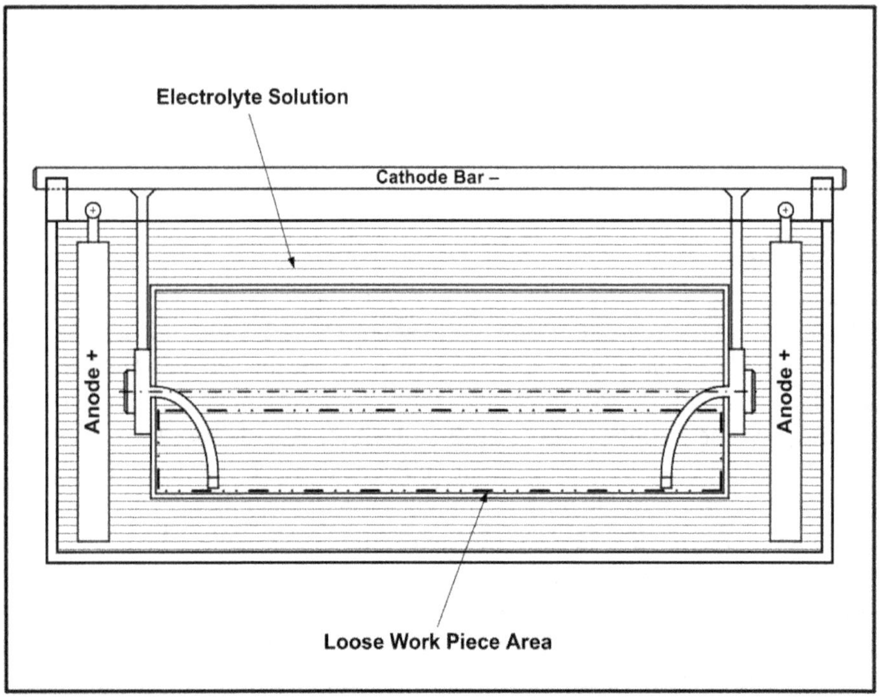

Figure 22.6 Front View of Barrel Plating System with Anodes and Cathodes on the Workpieces

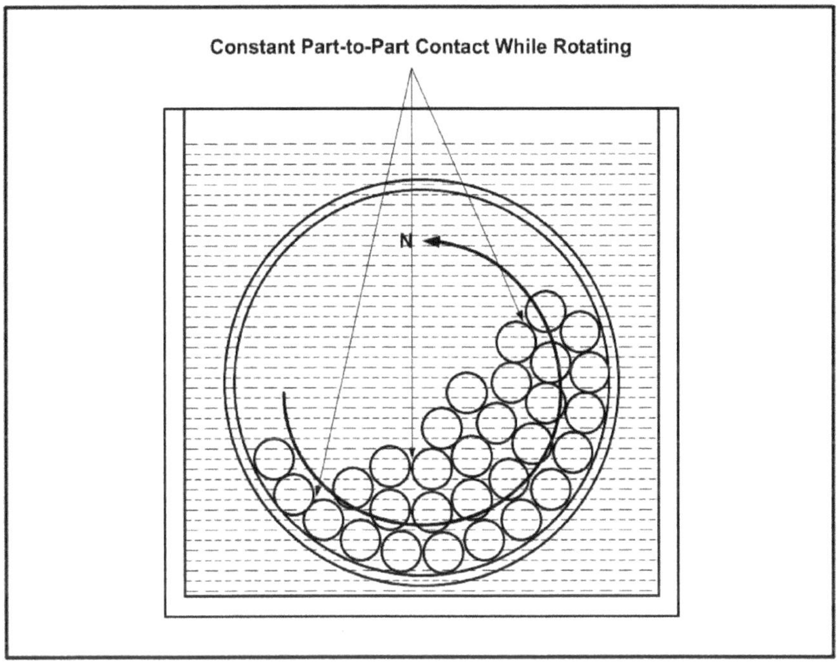

Figure 22.7 Side View of Barrel Plating System Showing Part-to-Part Contact for Current Flow

22.3 ELECTROLESS PLATING

This process deposits metallic plating on the component workpieces but utilizes no electricity whatsoever. Electroless plating is a chemical process where the solution the workpieces are submerged in reacts with metal, causing a deposition to occur. The thickness of the deposition is dependent on two factors. The first is how long the parts are left in the plating tank, and the second is the concentration of the plating chemical. As parts are plated, the metallic ion in the chemical solution diminishes and constant monitoring of the concentration is required. A major advantage to electroless plating over electroplating is that there will be an even deposition on the components. With no current to determine the ion movement, the coating will be uniform throughout the surfaces of the workpieces. Figure 22.8 illustrates the electroless plating process with the chemical reaction creating the deposition on the workpieces.

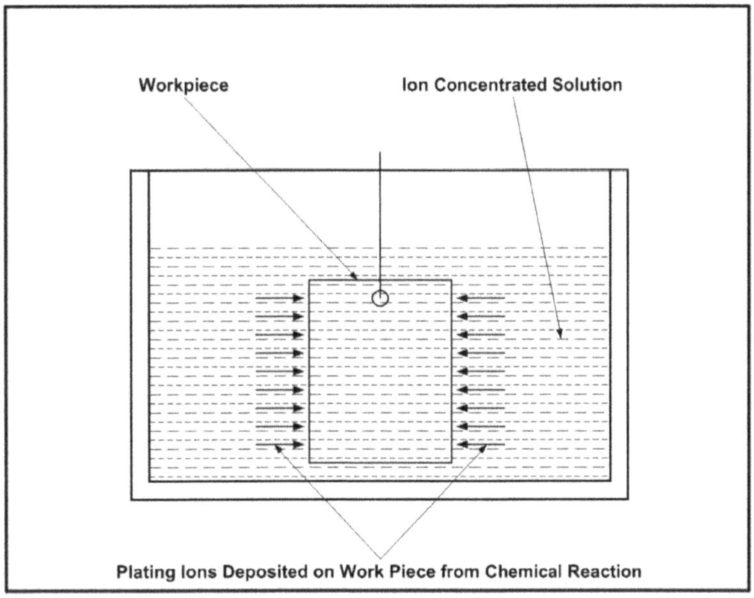

Figure 22.8 Electroless Plating System; Chemical Process Only

22.4 HOT DIPPING

The hot dipping process or more commonly referred to as galvanizing is used exclusively for heavy corrosion protection. There are no decorative features as a result of this process. For this process, a tank of molten zinc or aluminum is used into which the component workpieces are dipped, providing them with a thick coating on the surfaces. When performing hot dipping, there is one very important feature to consider and that is that the melting temperature of the component being dipped into the molten bath must have a higher melting temperature than the molten bath. If not, the component workpiece may be either deformed or melted during the process.

Hot dipping can be performed on individual components or on a continuous length part. Figure 22.9 shows the hot dipping process being performed on a single component part. For a continuous length part, the coil is pulled through the molten bath as illustrated in Figure 22.10. An example of a continuous part that is hot dipped would be a chain link fence. A large coil of the fence material is pulled through the hot dipping tank and coiled up on the opposite end.

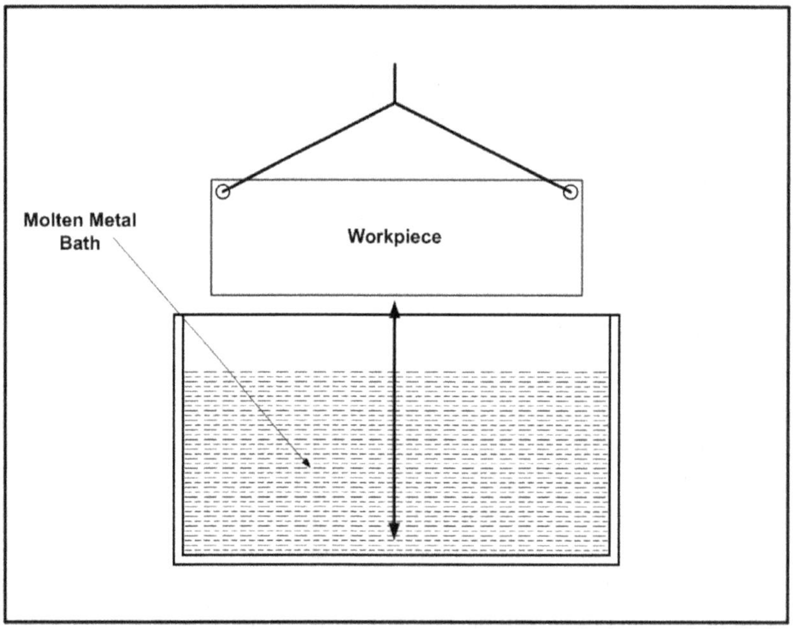

Figure 22.9 Single-Component Hot Dipping System

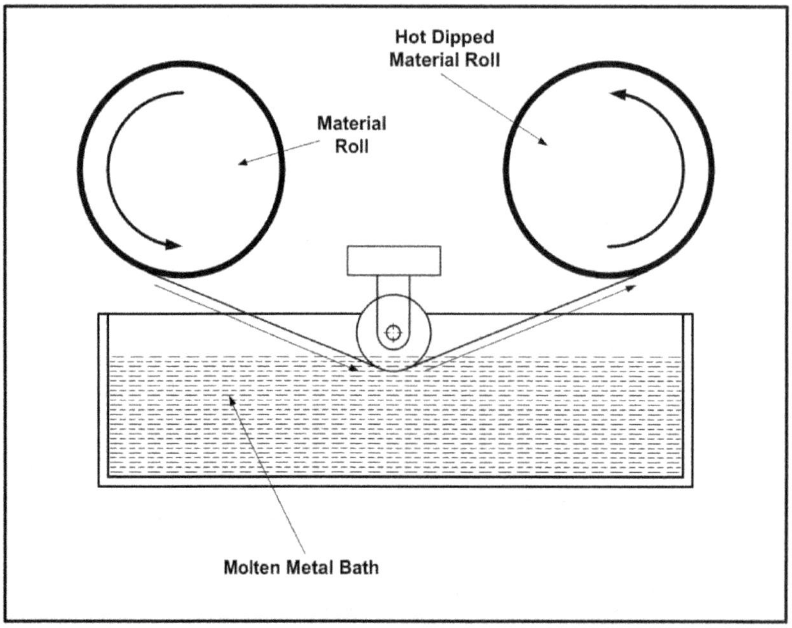

Figure 22.10 Continuous-Component Hot Dipping System

22.5 CONVERSION COATINGS

Conversion coatings are used for a variety of reasons on the surfaces of workpieces. The solution used in conversion coating induces a reaction on the metallic surface being applied to. Either these coating are sprayed onto the parts or the parts are submerged into the solution in production, rendering a deposition of about 0.0001 in. or less. Some of the reasons for applying conversion coating are to apply a base coating for painting and to provide light corrosion protection. The materials that these coatings are most commonly applied to are steel, zinc, and aluminum. Coatings of either phosphate or chromate are the two most common that are applied to the workpiece surfaces.

22.6 OXIDE COATINGS

This category of surface treatments actually involves forming a light layer of corrosion on the component workpieces. This involves submerging them into an extremely caustic solution to start to lightly corrode the surfaces. From there, various chemical treatments are done to apply the coating to protect the parts. The two major oxide coatings utilized in manufacturing are anodizing and black oxide, which are explained as follows.

22.6.1 Anodizing

Anodizing is primarily a decorative coating, but it does provide some corrosion protection as well. This process utilizes an electrolyte procedure to induce oxidation on the surfaces of the component workpieces. This process is mainly used on aluminum and zinc, but other nonferrous materials are also anodized. In some special cases, steel parts are anodized, but this is the exception rather than the rule for this process. Once the oxidation layer has been induced on the surfaces of the workpieces, they are subjected to coloring or dye tank next. Anodizing is available in almost any color desired and the oxidation layer will obtain the desired coloring. This coating process will render a deposition thickness of between 0.0001 and 0.0003 in. on the workpieces, which is extremely minimal. With a special anodizing process called *hard anodizing*, a heavier deposition of between 0.005 and 0.010 in. can be obtained, which will provide heavier corrosion protection and increase the wear resistance on the soft aluminum and zinc materials.

22.6.2 Black Oxide

The black oxide process is utilized mainly as a corrosion protection but can also be decorative. Like anodizing, this process starts with a light deposition of corrosion on the workpieces but does not utilize the electrolyte procedure. The component workpieces are submerged into an extremely caustic solution to create the oxidation layer. Another difference between anodizing and black oxide is that black oxide can only be applied to steel components. Once the components have obtained a light oxidation layer, they are placed in the color tank, which dyes the parts black. Once the components have been colored black, the final stage in the process is to give them an oil coating. With the parts being steel, this is a critical step to prevent further corrosion. Steel being highly prone to rusting, the oil layer prevents further corrosion. The drawback to this process is that the components must remain oil-coated in order for the protection to be maintained.

22.7 POWDER COATING

Powder coatings have become extremely popular due to the permanent nature of the process and the extended protection that it offers. Depending on the powder coating process selected, it can also be decorative in nature, but each process will render corrosion protection. There are several methods to apply powder coating, but the two main categories are the electrostatic method and the heated part process. These are described as follows.

22.7.1 Electrostatic Powder Coating

Powder coating utilizing the electrostatic procedure is performed by charging the component workpiece negatively and the coating powder positively. As with the electroplating process, the negatively charge workpiece will attract the positively charged powder. The workpiece is placed on a rack, which delivers the negative charge while the powder applicator is charged positively. As the powder is sprayed onto the workpiece, it is attracted to the component. At this point, the powder coating the part is only being adhered from the electrical charges and is fragile to the touch. After the powder application, the component workpiece still charged is placed into an oven to bake the powder onto the workpiece. The temperature utilized for this stage can vary greatly depending on the powder that has been applied. The amount of heat that is required is roughly between 75 and 80% of the melting temperature of the powder material applied to the part. If too little heat is used, the powder will not cure properly, and likewise, too much heat will cause it to run off the workpiece. Figure 22.11 illustrates the electrostatic powder coating process from application to heat curing. This process will render both a corrosive and a decorative coating on the component workpieces with the deposition thickness varying depending on the powder material and amount being applied.

The powder materials used in electrostatic powder coating vary from acrylics to enamels to metallic. When plastic-based materials are being utilized, they need to be able to receive a charge in order to adhere to the workpiece. Metallic powders are naturally able to be charged, but depending on the material of the workpiece, the curing temperature in the oven can cause damage. It is critical when using metallic powders to understand the curing temperatures required along with the temperatures that can cause damage to the component being coated.

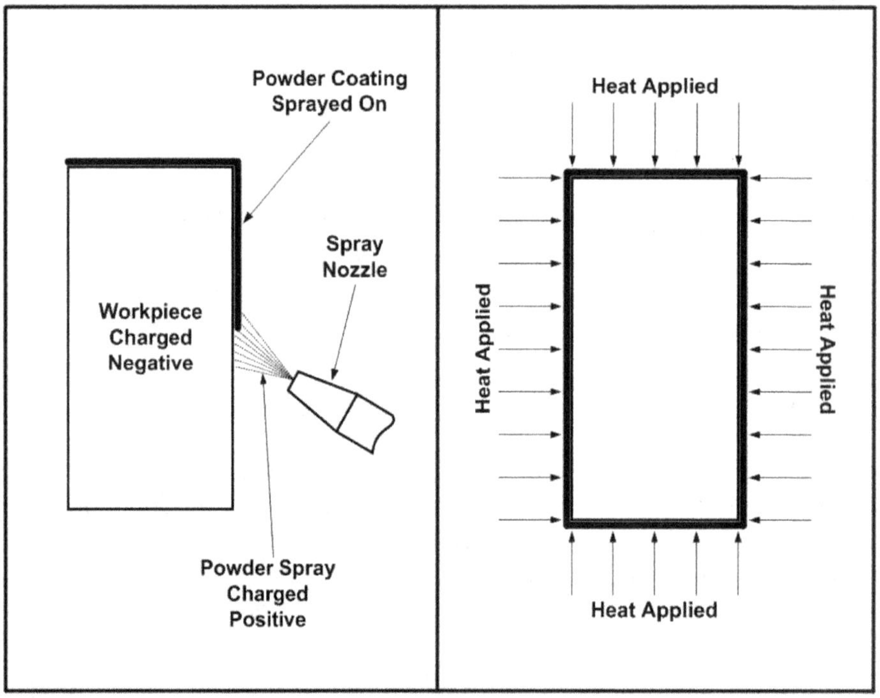

Figure 22.11 Electrostatic Powder Coating Concept

22.7.2 Heat Part Powder Coating

Deposition of a coating on workpieces along with the heated component process is utilized for corrosion protection only. This is a nondecorative process and is not used for any applications requiring enhanced appearance. The powder material that is to be applied is placed in a tank where air is pumped from the bottom, suspending the particles in the tank. What is required is a fog of the powder in the tank into which the component workpieces will be suspended. The workpieces are heated to a given temperature depending on the powder material being used and are suspended in the powder fog in the tank. The heat on the workpiece will cause the powder in the tank to adhere to the surfaces, depositing a heavy coating. With this being a completely corrosion coating, runs and drips on the workpiece are inevitable and not a defect from the process. Figures 22.12 through 22.14 shows how this process is performed. The main deposition materials used in heated component powder coating are plastics and rubber, which render a heavy corrosion coating for protection from the elements. Depending on the application of the component workpieces in the assembly into the product, this type of deposition can also render a cushioning and sealing effect as a secondary benefit.

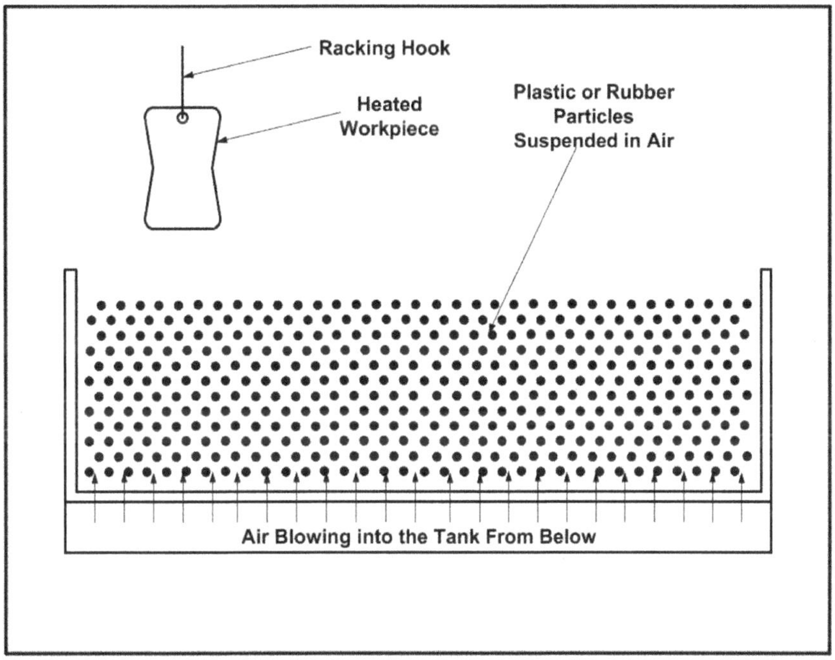

Figure 22.12 Heated Part Powder Coating 1

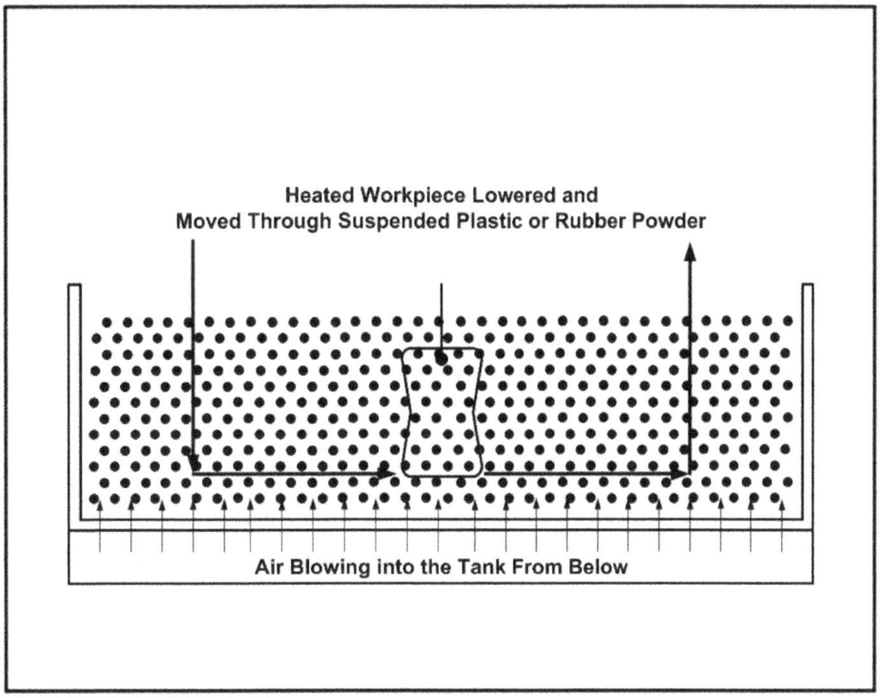

Figure 22.13 Heated Part Powder Coating 2

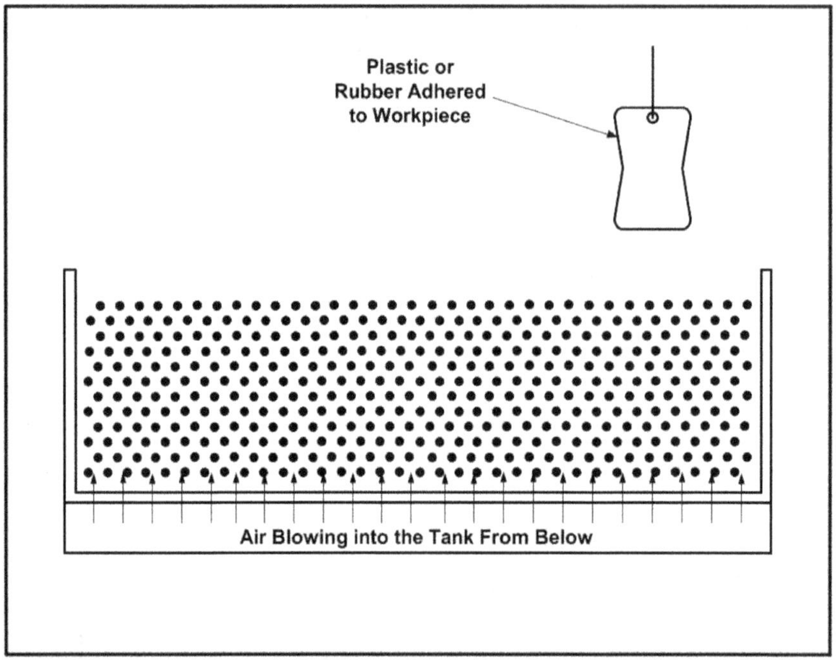

Figure 22.14 Heated Part Powder Coating 3

22.8 PAINTING

Painting is the most basic and universal coating process utilized. The application methods range from brushing, spraying, and dipping, depending on the finished requirements and appearance features. Both corrosion protection and appearance can be obtained from the painting process and the final finish will be determined by the process being utilized. Depending on the material that is to be painted, certain presurface treatments may be required for proper adhesion of the paint. As mentioned earlier in conversion coatings, aluminum and zinc material require a presurface treatment in order for the paint to properly coat the workpiece.

The main process used for appearance is spraying. This can be performed either with or without electrostatic applications. For electrostatic painting, the workpiece and paint are charged exactly the same way as in powder coating. Component dipping is used for corrosion protection only where the parts are dipped into a vat of paint and allowed to drain and dry. This process will render runs and drip areas on the part, but is one of the fast methods for applying paint when appearance is not required. In a production environment, the painted components are normally baked to speed up the curing process. Temperatures of between 120 and 160°F are used to reduce the drying times and aid in the adhesion of the paint to the surfaces.

22.9 SUMMARY

Coatings and depositions are applied to component workpieces for a variety of reasons. Each process utilized will perform different functions with the coating being applied, along with providing corrosion protection and some will render improved appearance. Some processes require several deposition processes to be performed correctly, while others may require only the initial process. In every deposition process, surface cleaning is critical for proper adhesion of the coating being applied. But whatever the reason, just about all metallic components will have some type of coating or deposition applied to them before being assembled into the final product.

22.10 QUESTIONS

1. What are the reasons that coatings and depositions are applied to workpiece surfaces?
2. In the electroplating process, what are the charges applied to the workpiece and the material that is to be plated?
3. Electroplating does not form an even deposition on the workpieces. Why does this happen?
4. Why does electroplating not improve the surface finish of the workpieces being plated?
5. When selecting barrel electroplating, what precautions need to be taken so as to not damage the component workpieces during the process?
6. Define how the electroless plating process work, and what advantage does it have over electroplating?
7. For hot dipping, what are the two materials that are the most commonly used for coating the workpieces?
8. What is the main consideration for the hot dipping process so that the workpieces do not get damaged during the process?
9. For conversion coatings, what are the main materials it is applied to and what are the reasons they are applied to the workpieces?
10. What are the two methods that conversions coatings are applied along with the two coating solutions that are used?
11. In the oxidation processes, what is the first stage for the surface treatment?
12. For anodizing, what are the main materials that this process is performed on?
13. What are the main differences between the anodizing and black oxide process?
14. Black oxide is only applied to one material type. Which material is that?
15. For electrostatic powder coating, what is the charge applied to the workpiece and powder?

16. Why is it critical to cure an electrostatic powder-coated component and what precautions need to be understood?

17. How does the heated component powder coating process work?

18. Besides corrosion protection, what are the secondary benefits that can be realized from heated component powder coating?

19. What are the three application methods for painting and which one is utilized for appearance?

SECTION 7

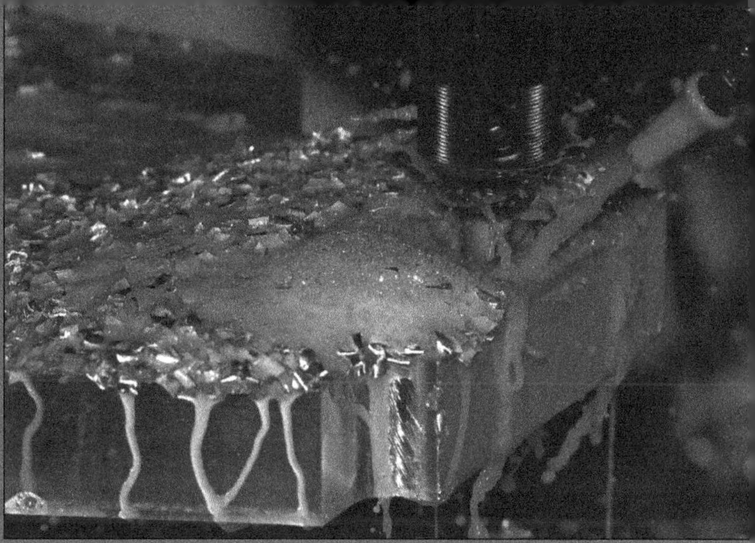

ASSEMBLY PROCESSES

The final stage in the manufacturing arena is assembly. Once all the components are fabricated and have had any coatings or depositions applied to them, the finished product requires assembly and testing. The majority of the assembly processes are performed after the finishing has been applied to the component workpieces, but some are performed prefinishing and the parts are coated later. There are several assembly processes where some are nonpermanent, allowing for disassembly of the joined components, whereas others are permanent in nature and intended for a nondisassembly situation. An evaluation of the final product requirements needs to be performed in order to select the proper assembly for the permanent and nonpermanent areas to assure final functionality. Areas that need to be accessed for service and adjustment need to be nonpermanent, while others may not require that function and can be permanent, rendering a stronger assembly overall.

This section covers a variety of assembly processes utilized in manufacturing. From the permanent processes such as welding and brazing along with some mechanical applications to the nonpermanent like various fasteners are be covered. Each of the assembly processes will have both advantages and disadvantages, and an evaluation of the products' function will determine the best process for any given application.

CHAPTER

23

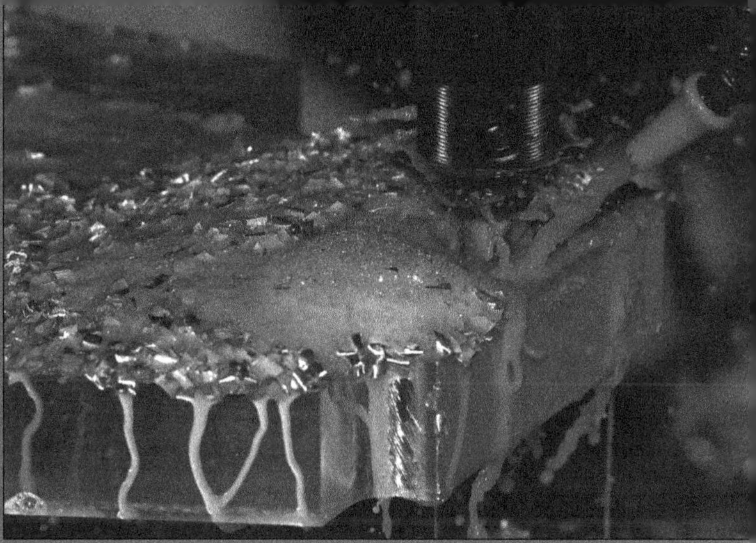

Welding Processes and Principles

23.1 INTRODUCTION

Welding is a widely utilized process for joining components in an assembly. This procedure will produce the strongest assembly of any of the joining processes available. The assembled component workpieces will be permanent in nature and not able to be disassembled after welding is completed. The concept behind the welding process is to have two components coalesced together at their contact points. This is accomplished by the melting of the two components together forming the permanent assembly. Figure 23.1 illustrates two component workpieces welded together creating an assembly. There are a variety of welding processes, all of which will provide the joining of the component workpieces through coalescence. If the base materials of the workpieces are not melted together, then a weld has not been accomplished. This chapter covers the principles of welding and various processes utilized in manufacturing assembly.

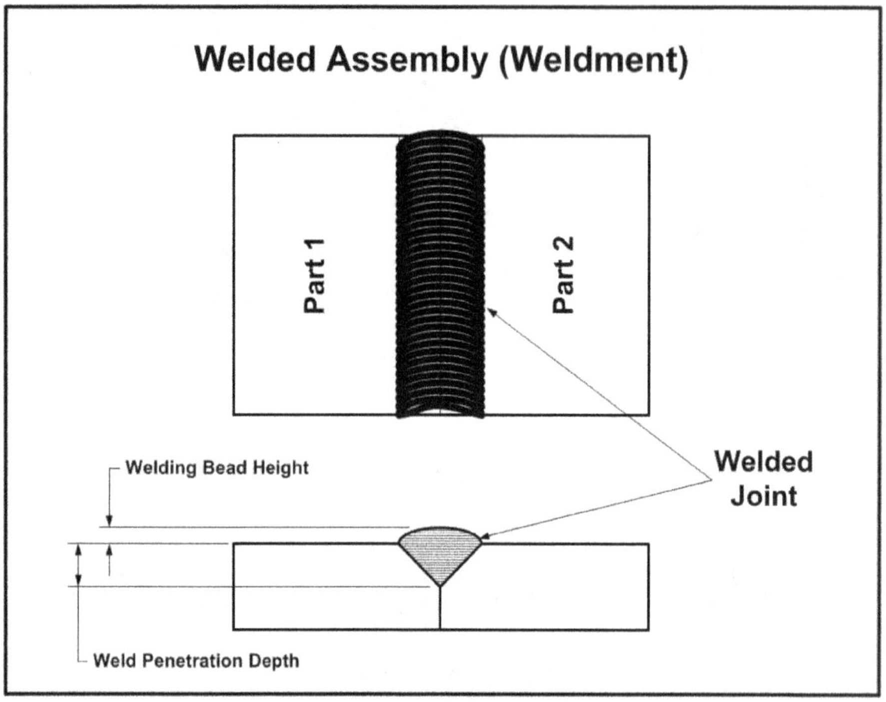

Figure 23.1 Welding Bead Height and Penetration

23.2 WELDING PRINCIPLES

Welding is a joining process where two workpieces are connected through the coalescence of their base materials. This is accomplished by raising the temperature of the two materials to be joined far above their melting point, thus allowing them to flow together creating the assembly desired. The terminology for a welded assembly is called a weldment and refers to the components that have been coalesced together. Figure 23.2 shows the melting temperature ranges for some select materials. Welding processes are categorized in two process families, which are Fusion Welding and Solid-State Welding. Each of these categories contains a variety of processes, which are defined as follows.

Fusion Welding: The coalescence is achieved by applying heat only to the concentrated area where the joining is required. The processes that fall into this category are:

— Arc Welding
— Oxyacetylene Welding
— Resistance Welding
— Laser Beam Welding
— Electron Beam Welding

Solid-State Welding: The coalescence is achieved by applying pressure, or a combination of both heat and pressure, to the concentrated area where the joining is required. The processes that fall into this category are:

— Forge Welding
— Diffusion Welding
— Friction Welding
— Ultrasonic Welding
— Explosion Welding

Melting Temperatures of Select Metals

Material	Degrees F	Material	Degrees F
Aluminum	900 – 1,220	Molybdenum	1,800 – 4,750
Brass	1,490 – 1,880	Nickel	2.650 – 2,750
Bronze	599 – 1,970	Stainless Steel	2,250 – 2,750
Cast Iron	2,050 – 2,390	Steels	2,550 – 2,700
Copper	599 – 2,260	Titanium	2,820 – 3,180
Magnesium	880 – 1,200	Zinc	707 – 1,470

Note: Temperatures Ranges are Approximate and not Alloy Specific

Source: www.matweb.com

Figure 23.2 Selected Materials' Melting Temperature Ranges

Creation of weldments can be accomplished in a multitude of welding types. Depending on the process being utilized, filler materials may or may not be used. Each process and procedure needs to be evaluated in order to select the proper method to create the best weldment for a given assembly and function. Figures 23.3 and 23.4 show a variety of weldment joining techniques used in welded assemblies. When performing weld on thick materials, it is common to apply more than a single bead of weld to the joint. This will also provide a much stronger connecting joint on the weldment especially when a side force will be applied during the desired functionality. Figure 23.5 illustrates a multi bead build-up on a welded assembly.

As with any manufacturing process, there are both advantages and disadvantages that need to be considered. Welding does produce the strongest and most permanent assembly, but understanding both the positive attributes and drawbacks along with the safety precautions of the process will allow for a proper selection of the required assembly. These factors are described as follows.

Welding Advantages

— *Permanent Assembly Joint:* Welding provides the most permanent joining of any assembly process. By melting the two workpieces together at their contact surfaces, the assembly for all-intensive purposes becomes one solid piece.

— *Stronger Assembly than Base Materials:* With some welding processes, the welded joint can actually be stronger than the base materials of the components joined. This can happen in two ways. First, if the process is utilizing filler material that is stronger than the base metal, the joint will subsequently be stronger. Second, if a buildup of filler material is applied to the welded joint, it will be thicker than the base material and, again, render a stronger joint than the base material.

— *Economical Process:* The welding process for some processes can be a very economical method for joining parts. Depending on the process being used, it may be able to be automated requiring manual labor. Other processes require a minimal amount of skill and training and once the process is set up, almost any operator can perform the operations.

— *Not Factory Restricted:* A big positive factor for welding is that it is not restricted to a factory environment as other manufacturing processes are. Welding is performed daily outside of a manufacturing facility with portable equipment, either electric or gas.

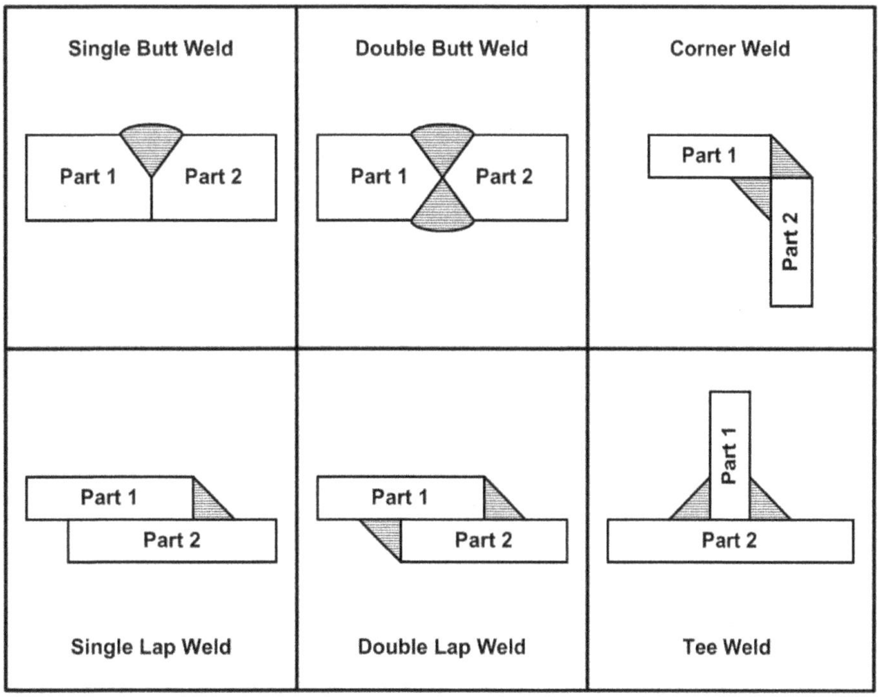

Figure 23.3 Various Welding Assemblies 1

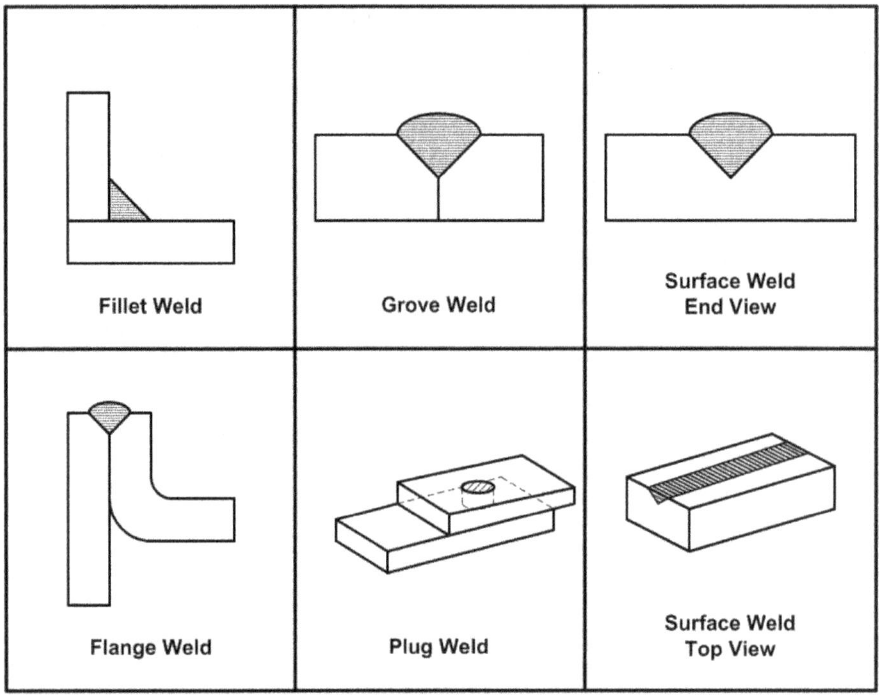

Figure 23.4 Various Welding Assemblies 2

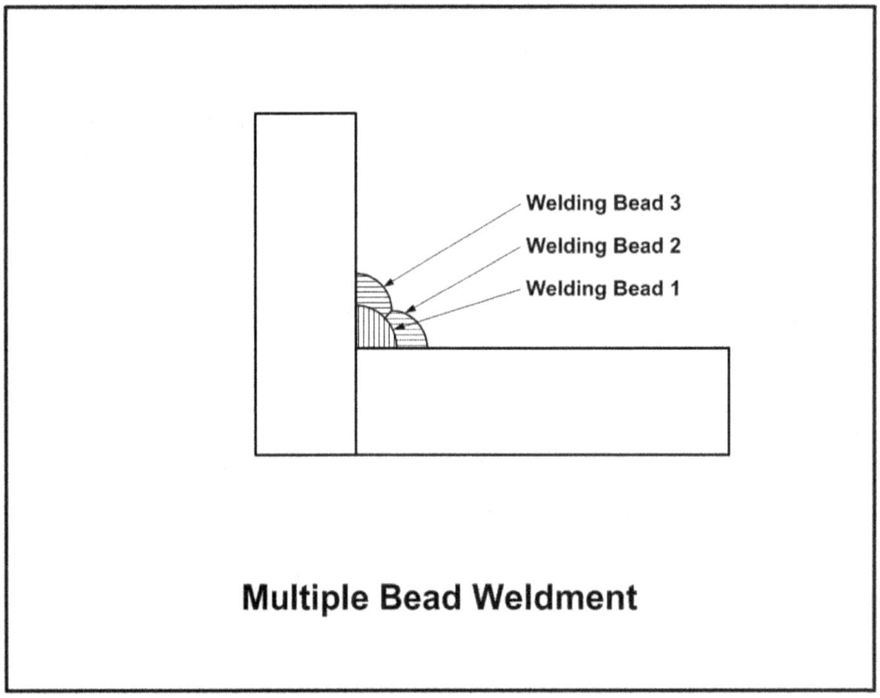

Figure 23.5 Examples of Multiple Beads on a Welded Joint

Welding Disadvantages

— *High-Energy Usage:* Welding requires the usage of either extremely high amperage. It is not uncommon to weld metals together with amperage of between 120 and 160 A, and in extreme cases, amperage of over 10,000 A will be used.

— *Dangerous Process:* As mentioned, welding utilizes high amperage to perform the coalescence of the materials. Other processes will utilize combustible gasses to produce the required heat. An error with either of these can result in serious injuries or even death.

— *Trained Operators:* For fully manual welding operations, a skilled craftsman is required. Welding of this nature is a skilled trade and the tradesmen performing the work are highly paid.

— *Disassembly Not Possible:* As mentioned, welding renders the most permanent assembly of any joining process. If an error is made when the components are welded together, disassembly is without destroying the parts. They would have to be discarded and new components fabricated and welded.

— *Not Possible to Join Dissimilar Materials:* With welding being the result of flowing the base materials together, joining of unlike materials is not possible. In order to weld properly, the materials being joined need to be of the same type and have the same melting temperature. To try to weld a piece of aluminum to a piece of steel could never happen due to the face that aluminum has a much lower melting temperature than does steel and would begin to flow long before the steel piece would.

— *Hidden Defects:* A welded joint can have internal defects, which may not be visible on the surface. These defects can vary and being hidden to the naked eyes can cause a failure in the assembly while operating.

Safety Considerations

— *High Temperatures*: The temperatures generated during welding are extremely high. With welding having to melt the metals being joined almost instantly, temperatures can be upward to 10,000°F.

— *High Currents*: In order to produce the heat required to melt the metals, high amperage is utilized in the process. As mentioned earlier, a current of between 120 and 160 A is normal. In some extreme welding applications of thicker materials, it can be as high as 10,000 A. Keep in mind that 1/4 of an amp is enough to kill a person.

— *Combustible Gasses*: For welding processes that utilize gasses to produce the heat, the gasses used are extremely combustible and flammable. Leaks from the tanks, nozzle, or the hoses can result in an explosion, resulting in a catastrophic situation.

— *Molten Metals*: Welding produces molten metal during the process of coalescence. Any drips or spattering of this molten material can cause extreme burns even through heavy clothing. This becomes extremely dangerous especially when welding upside down where the drips and spatter are capable of landing on the person performing the operation.

23.3 FUSION WELDING

The fusion welding category performs the welding of the materials by utilizing only heat to flow the materials together. By definition: *Fusion welding accomplishes the coalescence of the base materials by the application of heat to a concentrated area.* High heat is applied to the area where the welding is to be performed, allowing the base materials to flow together creating the weldment bond. The methods for applying the heat for the process are either electricity or with combustible gasses. This category of welding processes can be performed utilizing filler materials, which will be deposited into the welded joint, or can be performed with no filler materials. Depending on the material characteristics and geometry, the proper fusion welding process can be selected to perform the operations required. The various and more common fusion welding processes are explained as follows.

23.3.1 Arc Welding

The arc welding processes utilize an electrical current to generate the required heat for the weldment being assembled. This is performed using an electrode, which charges with one polarity while the workpiece is charged with the opposite. As the electrode comes in close proximity to the workpiece, the circuit is completed causing an electrical arc, which produces the required heat. Depending on the material and the electrode being used, the power source providing the current can be either Alternating Current (AC) or Direct Current (DC). An AC power source is used when welding ferrous materials, while DC can be used for all materials being welded. Also, process and electrode type will determine whether the workpiece is charged positive or negative.

Electrodes utilized in the arc welding processes can be either consumable or nonconsumable. The difference being that a consumable electrode will deposit the filler material into the welded joint thus being consumed in the process. A nonconsumable electrode will remain intact and only deliver the heat from the arc it creates. When utilizing a nonconsumable electrode, if filler material is required, it needs to be added separately by hand during the arcing process.

Welded joints are very prone to the atmosphere and require protection during the process. As the material is liquefied almost instantly and then rapidly cooled, the elements contained in the atmosphere can affect the overall quality of the weld. To assist in the protection of the weld, flux is utilized during the process. This can be either a chemical material, which is deposited over the weld, or an inert gas that shields the welded area. Chemical flux is a substance that coats some consumable electrodes and coats the welded area while the electrode deposits the filler material. For other arc welding processes, and inert gas, most commonly argon is blown onto the welded area during the process to shield the weld. Besides shielding the weld, flux will perform three beneficial functions to the weldment, which are as follows:

Flux Functions

— *Protect the Weld from the Atmosphere:* As mentioned, the welded joint is very prone to the atmosphere and the contaminants that are contained in it. By either coating the welded area or utilizing an inert gas to blow the atmospheric air away from the weld, it will be protected and create a stronger and higher quality weld.

— *Stabilize the Welding Arc:* Another function that the flux performs is to aid in stabilizing the electric arc. As the arc gets created from the circuit completion, the flux, whether it is chemical or gas, will aid in maintaining a smooth and stable arc, which will increase the quality of the weld being performed. Without the flux, the arc will spread and be as concentrated in the areas that the weld is being made.

— *Reduce Welding Spatter:* Spatter occurs when the liquefied welding material is splashed out of the weld and onto the sides. Although this is not a major issue regarding the strength of the weldment, it does affect the appearance of the welded joint and would require the spatter drops to be ground off.

The following welding processes fall into the arc welding arena and although they are basically the same, there are differences in how they are performed and where they are utilized. Each of the arc welding processes will utilize electrical current to produce the required heat that melts the base materials, thus joining them together. Some utilize filler, while in others, the filler is either not utilized or added separately.

23.3.1.1 Shielded Metal Arc Welding (SMAW)

Shielded Metal Arc Welding (SMAW) is more commonly referred to as stick welding. This process utilizes a consumable electrode that is coated with a chemical flux. It can be performed with either alternating or direct current and the polarity is dependent on the electrode being used. Figure 23.6 shows the construction of the electrode with the flux coating around the outer layer. As the electrode comes in close proximity to the workpiece, the circuit connection is completed and the arc is formed. The heat generated will melt both the base materials and the electrode, which will be the added filler material. For best results, the electrode should be maintained at a distance of approximately 1/8 of an inch from the materials being welded. If the electrode comes in contact with the material, it will stick and require removal before continuing. Also, an angle from the workpieces of between 30 and 45° needs to be maintained in order for proper penetration and deposition of the filler material. Figure 23.7 illustrates how the SMAW process is performed and the direction of the welding flow.

The flux material on the outer layer of the electrode will melt and create a hard coating on the welded joint as the filler material is being added. For a general explanation of the chemical flux coating on the electrode, it is similar to a sand-type substance. As the extreme heat created from the arc melts the base materials and the

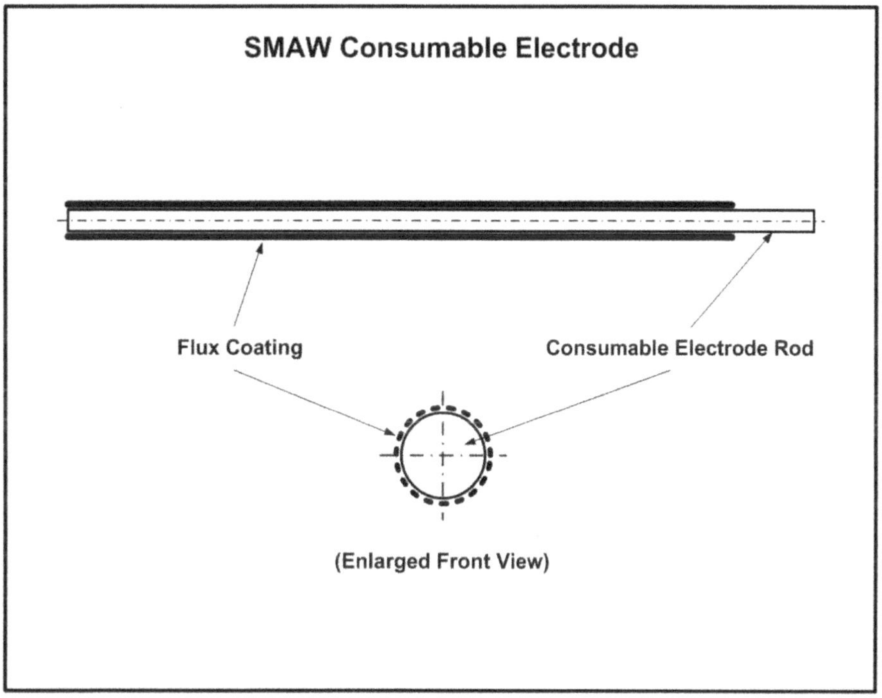

Figure 23.6 SMAW Flux Coated Consumable Electrode

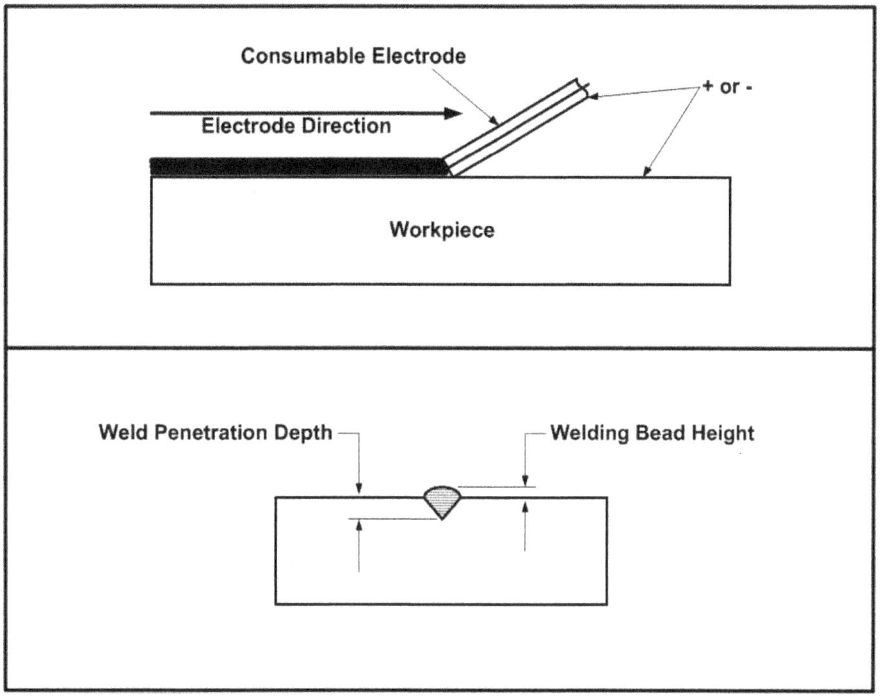

Figure 23.7 SMAW Process and Direction of Weld Flow

filler of the consumable electrode, the flux will turn into a glass-like substance and render a hard coating over the top of the weld. Figure 23.8 shows a sectional view of the weld filler material with the hard flux coating over it. This provides the required protection for the weld but will require it to be chipped away once the weld has cooled down.

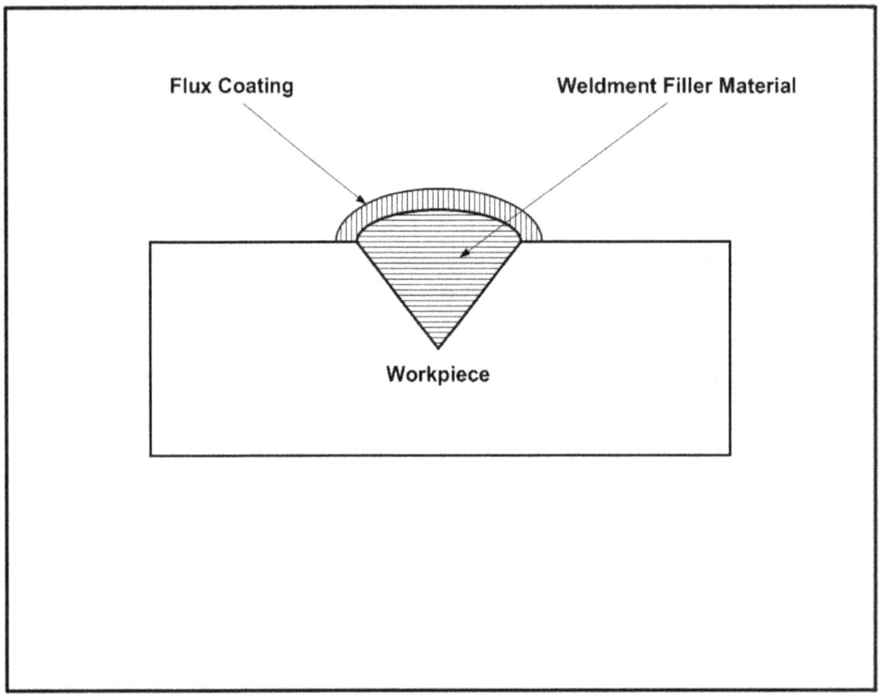

Figure 23.8 SMAW Hard Flux Coating

23.3.1.2 Gas Metal Arc Welding (GMAW)

Gas Metal Arc Welding (GMAW) is more commonly referred to as wire feed welding or MIG (Metal Inert Gas) welding. This process also utilizes a consumable electrode, but unlike the SMAW process, the electrode is not a stick but rather a coil of wire. The wire coil is fed through a nozzle, which comes in contact with the workpieces creating the required arc. Depending on the weldment being produced, the wire is available in various thicknesses and is plated with copper. The copper plating is applied to protect the wire from corrosion from the elements. To engage the movement of the wire through the nozzle, a foot peddle is pressed and the wire moves through the nozzle toward the workpieces. Depending on the weldment being produced, the speed of the wire is set and a continuous rate is established.

This process requires a flux, but rather than a chemical flux, and inert gas is utilized. Argon is the most commonly used, but in special applications, nitrogen or helium may be applied. As the feed peddle is depressed, the gas is blown through the nozzle protecting the weldment during the process. An advantage to utilizing a gas flux rather than a chemical flux is that there is no hard coating that needs to be chipped away. This reduces the overall time to complete the weldment and the weld can be viewed instantly rather than after the hard flux is chipped away.

GMAW also has the advantage of being easier to perform and faster than SMAW. With a coil of wire being fed through the nozzle, there is no time lost in changing electrodes and restarting the weld as with SMAW. Also, once the arc has been established, the nozzle can easily be maintained at the proper distance from the workpieces for the weld, whereas with SMAW, the electrode is constantly being consumed and the operator needs to move it closer to the weld as they are performing the operation. Figure 23.9 illustrates the gas metal arc welding process displaying the wire feed, inert gas jet, and direction of the weld.

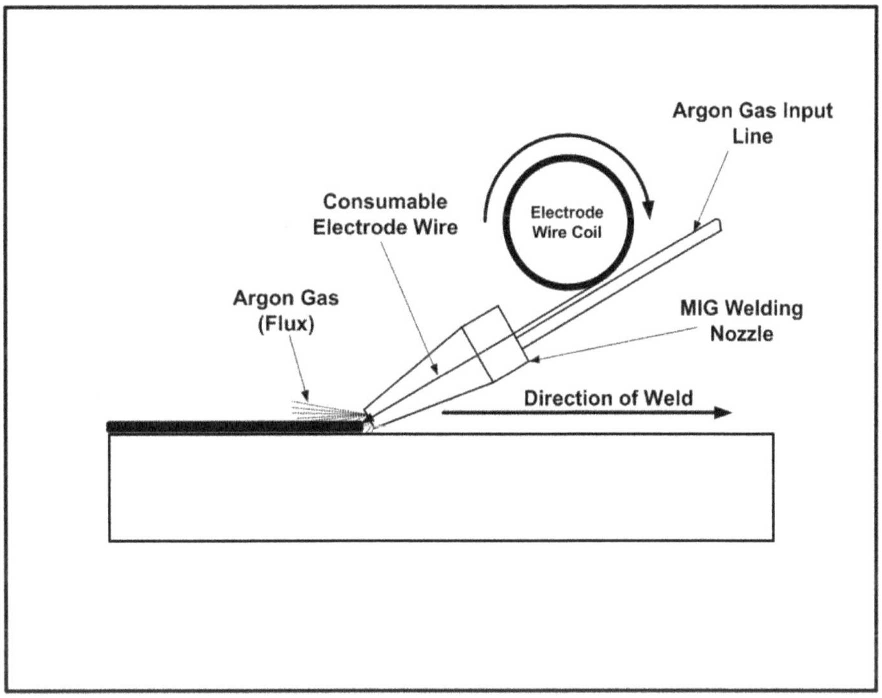

Figure 23.9 GMAW Process and Direction of Weld Flow

23.3.1.3 Gas Tungsten Arc Welding (GTAW)

Gas Tungsten Arc Welding (GTAW) is more commonly referred to as TIG (Tungsten Inert Gas) Welding. This process looks similar to GMAW but utilizes a nonconsumable electrode. The electrode used to create arc is a hard tungsten rod which is inserted into the nozzle. The tip of the electrode can be ground to either a point, or ball shape, or any of most geometric configurations desired, depending on the application. As the nozzle is moved close to the workpieces, the operator presses on a foot peddle similar to that utilized in GMAW, but rather than moving wire through the nozzle, it regulates the power for the weld. The more the peddle is depressed, the more the current applied to the electrode. As with GMAW, an inert gas is used as a flux and is blown through the nozzle during welding.

The welding process can be performed either with or without filler material. A weldment without filler material is performed by merely flowing the base materials together forming a bond. To implement filler material in the GTAW process, it needs to be added separately by hand with the arc to melt the filler rod and combine it with the base materials. These filler rods are of various thicknesses and copper-plated to protect them from corrosion. They are normally two feet in length and a steady hand is required to get the filler rod close to the arc, but not touching the electrode. Figure 23.10 illustrates the GTAW process showing the welding nozzle and optional filler material being added. You will notice that the direction of the weld being performed is opposite to that of both SMAW and GMAW.

The GMAW process is very versatile and can perform weldments that both SMAW and GMAW cannot. A big advantage with the process is that it can weld extremely thin materials. By being able to utilize very light currents, it is possible with a very experienced operator to weld materials as thin as 0.005–0.010 in. Cutting a soda can and welding it can be seen as an example of this welding process. Also, the GTAW process is used for welding materials that cannot be welded with other processes. Aluminum, stainless steel, and chrome-moly steels are all welded with this process. With the ability to control the current extremely low and have the option to not use filler materials makes this process very versatile in the fusion welding arena.

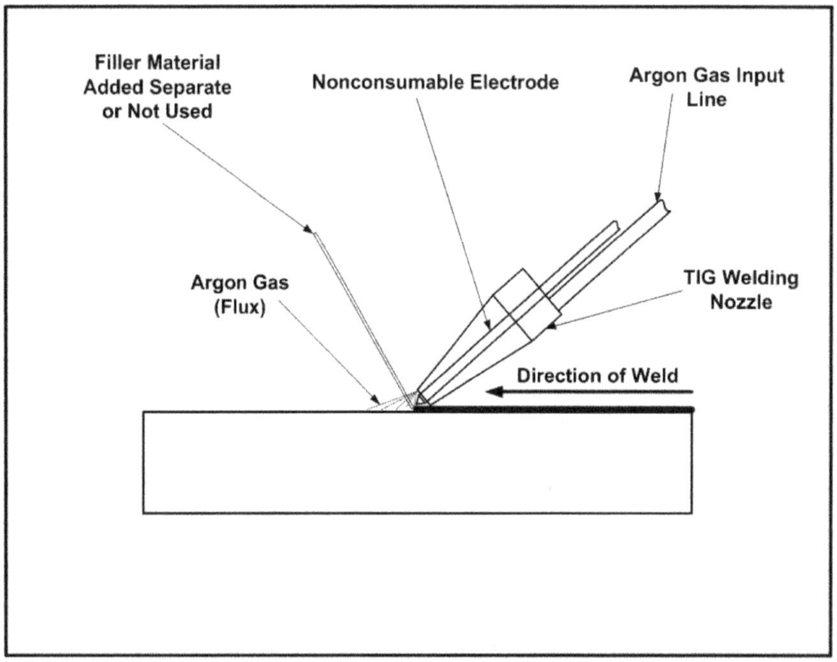

Figure 23.10 GTAW Process and Direction of Weld Flow

23.3.2 Oxyacetylene Welding

Oxyacetylene Welding is more commonly referred to as torch welding. For this process, the heat applied to create the coalescence of the materials is supplied from a torch flame. The flame is produced from a mixture of acetylene, which is the fuel used to produce the flame, and oxygen, which adds the heat. To begin with, the acetylene is turned on and ignited. Extreme caution is required to NEVER run the acetylene above atmospheric pressure of 15 psi. Acetylene is produced in the tank when released, and if the tank is allowed to release above atmospheric pressure, it can and will explode. Once the acetylene has been ignited and the flame adjusted, the oxygen is added to increase the heat. Both the gasses are adjusted to produce a neutral flame. The flame on the torch will have three distinct temperature levels within it, which are the outer envelope, acetylene envelope, and the inner envelope, described as follows.

— *Outer Envelope*: The outer envelope is the coolest portion of the flame. It will have a temperature of around 4,200°F and is used to preheat the material. The flame will have a feather-like appearance at the end of a yellow/orange color.

— *Acetylene Envelope*: The acetylene envelope is next in the heat range with a temperature of around 6,900°F. This is used to maintain the temperature of the material while welding and to assist in the flowing of the base material and the filler. This portion of the flame will be a mixture of yellow and blue and will have a conical structure.

— *Inner Envelope*: The inner envelope is the portion closest to the torch and is the hottest. The temperature of this flame is around 11,400°F and is used for melting of the base material and filler. This envelope is very small and light blue in color with a pointed end.

Figure 23.11 illustrates the neutral flame along with the three distinct envelopes with their temperatures.

Once the flame is set, the process for performing a weld is very similar in nature to GTAW. The flame is put in contact with the base material and is heated up until the material begins to flow. To adjust the heat being applied, the torch is moved either closer or further away from the workpiece as desired. Oxyacetylene welding can be performed either with or without filler material being added. The filler material is added by hand, similar to GTAW. Figure 23.12 illustrates oxyacetylene welding and the direction of the weld being performed. As with GTAW, the direction is the same where the weld is being pushed forward rather than pulled back.

Oxyacetylene welding is by far the most versatile of all the welding processes. With a large torch, heave materials can be welded, and in the same manner, a small torch is capable of welding extremely thin materials.

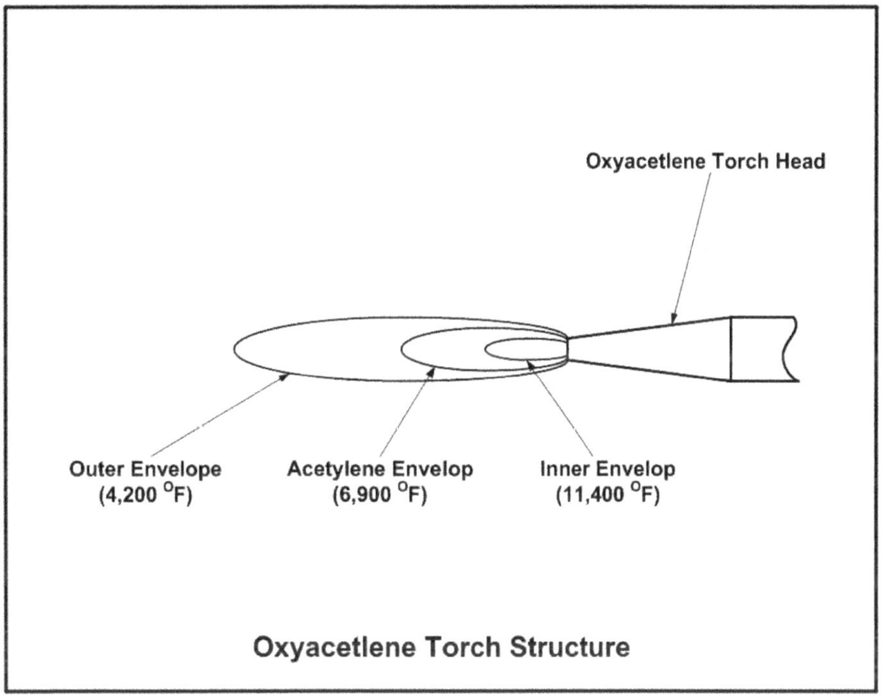

Figure 23.11 Oxyacetylene Torch Structure and Temperature Zones

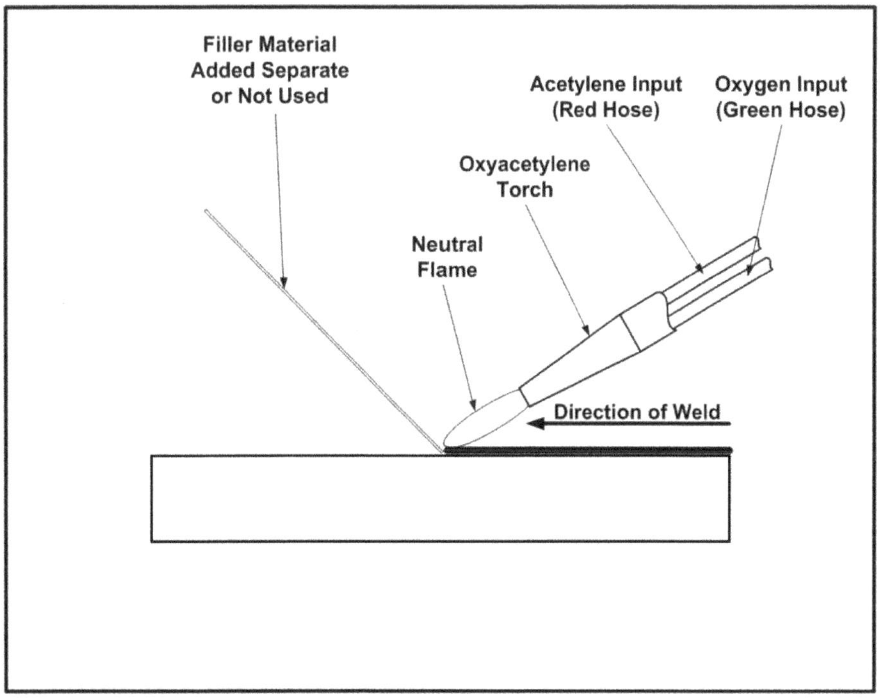

Figure 23.12 Oxyacetylene Torch Welding Process and Direction of Weld Flow

This process with its versatility allows the welding of materials used in fabricating jewelry such as gold and silver, along with other materials not capable of being welded with SMAW and GMAW. In addition to welding, the oxyacetylene process is heavily used for brazing and soldering, which are covered in Chapter 24. Another advantage to this process is cutting. With a special torch, additional oxygen can be injected into the flame producing temperature of above 20,000°F. This can cut through the steel easily to remove unwanted material or to slice material into the desired lengths.

23.3.3 Resistance Welding

The resistance welding process is one that utilizes electrical current to produce the heat. This process is used for welding sheet metal of thicknesses up to about 1/8 in. The material is placed between the electrodes and the current is applied, which flows the material together at the electrode clamping points. In resistance welding, there is *NO FILLER MATERIAL* added in the process. The base material flows together creating the weldment assembly. In resistance welding, there needs to be a sufficient overlap between the materials being joined. This process cannot be utilized for materials that are butted together as with overwelding processes. The overlap creates the area in which the welding will take place and allows for the electrode access and material flow when the current is applied.

There are two categories of resistance welding, which are Spot Welding and Seam Welding. The processes are basically the same, but the applications for each are different. These two resistance welding processes are explained as follows. These processes are extremely fast and are utilized extensively for sheet metal joining. An example of where resistance welding is widely used is in the automotive industry. The majority of the frames and body components on the vehicles being produced are resistance-welded together. Also, this process is easily automated to robotic applications and for manual operations, and once the equipment is set up, operators with minimal skills can perform the procedure.

23.3.3.1 Resistance Spot Welding

With resistance spot welding, the sheet metal is placed between the electrodes. A light clamping force is applied and the current is run through them to produce the material flow. Both the current and time at which it is applied are adjusted in order to produce the required material flow to produce the weldment. Figure 23.13 illustrates the resistance spot welding process showing the material flow between the electrodes.

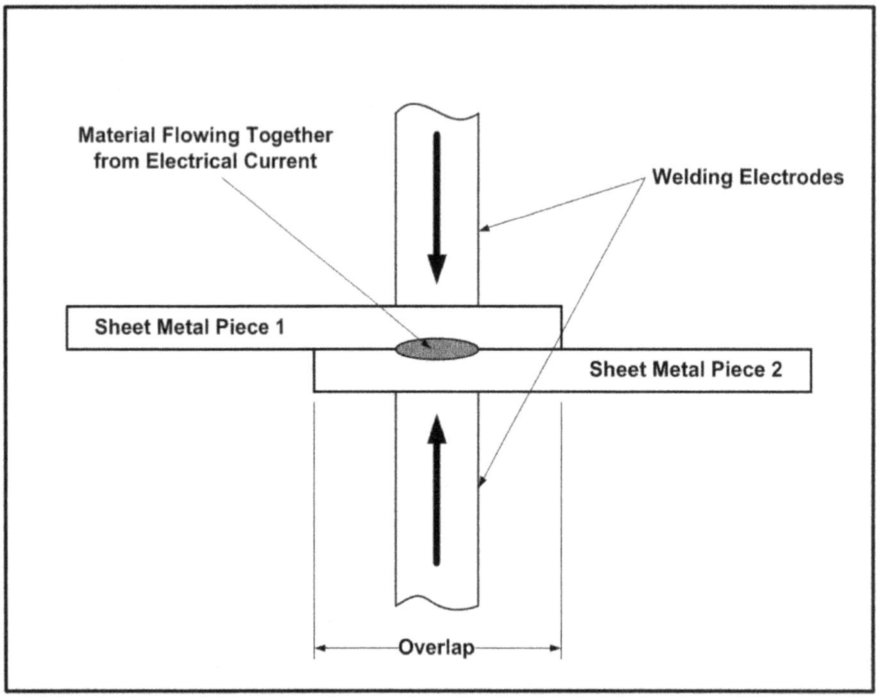

Figure 23.13 Resistance Welding Process and Material Flow

23.3.3.2 Resistance Seam Welding

The resistance seam welding procedure is identical to that of resistance spot welding except that the sheet metal material is run between the electrodes rather than being pinched at a single point. Figure 23.14 shows the resistance seam welding process utilizing round electrodes and moving material under them. This procedure can be performed with a constant current, which will produce a continuous seam weldment, or with a current that is timed and turned on and off as the material runs under the rolling electrodes producing an intermittent seam. Figure 23.15 displays a continuous seam and an intermittent seam produced from resistance seam welding.

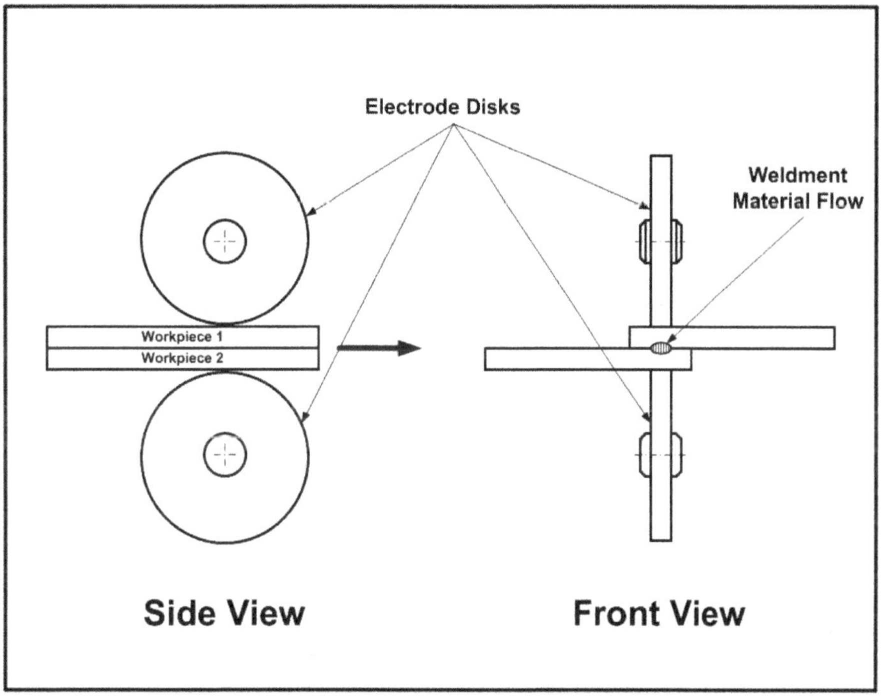

Figure 23.14 Roller Seam Welding Process for Sheet Metal

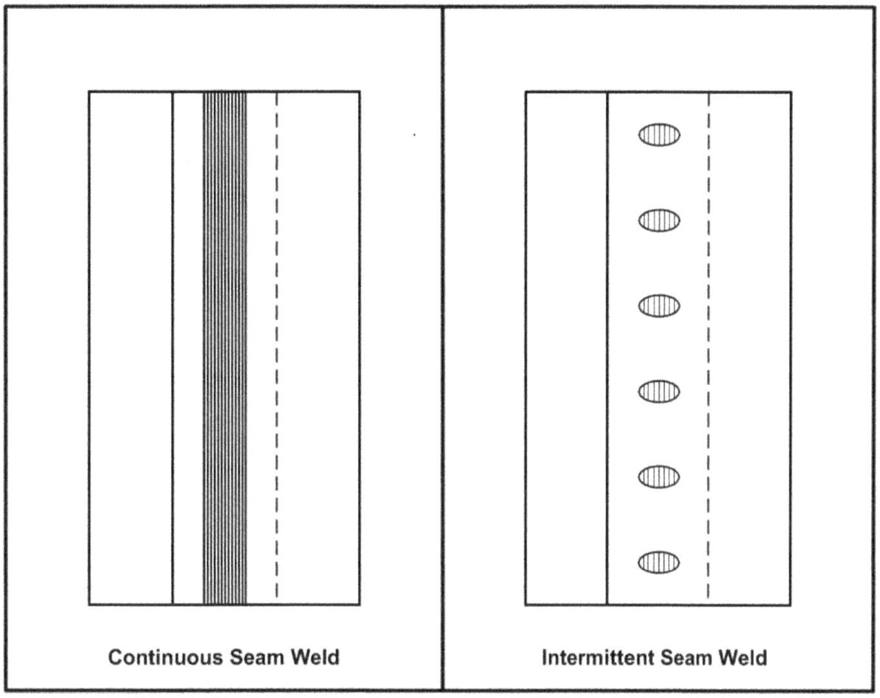

Figure 23.15 Continuous and Intermittent Seam Welding Beads

23.3.4 Laser Beam Welding (LBW)

In laser beam welding (LBW), the heat source is delivered from a highly focused, highly concentrated beam of light. As mentioned in Chapter 17, in nontraditional machining processes, laser beams were utilized to cut through metals. The same theory is used to perform welding by regulating the amount of energy from the laser beam to a concentrated area and allowing the materials to be coalesced at their contact points. As with laser machining, the focal point of the laser beam is critical in order to concentrate the energy properly, so a determined standoff distance needs to be implemented. This welding process utilizes no filler materials during the welding and produces very high quality welds. It is capable of welding extremely thin materials and with powerful beam and can weld to a depth of up to 3/4 in. Figure 23.16 illustrates the laser beam welding process and shows the critical standoff distance required for a concentrated power beam.

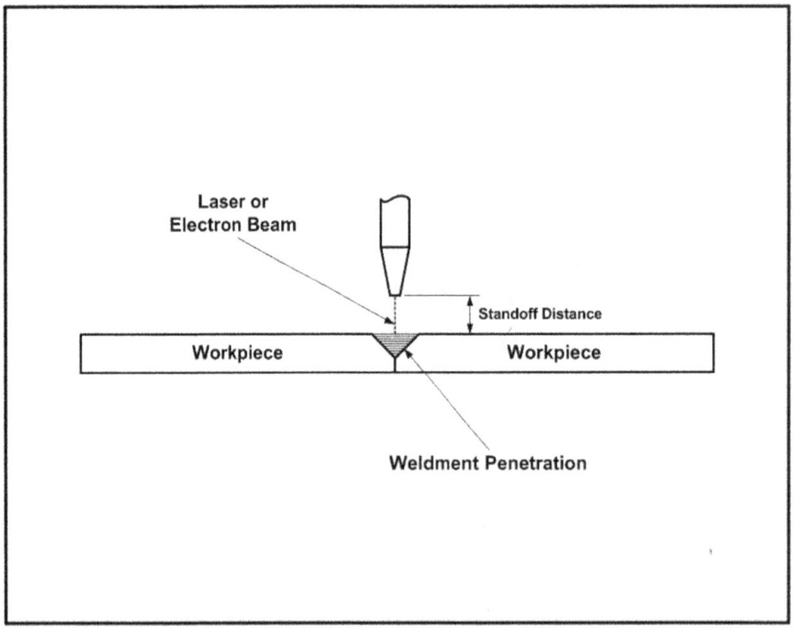

Figure 23.16 Laser or Electron Beam Welding Process

23.3.5 Electron Beam Welding (EBW)

Electron beam welding (EBW) is performed using a beam of extremely high energy electrons, which is focused and directed against the work surface of the components. This process is operated at very high voltages and low currents, which accelerate the electron beam into a high density, creating the heat required. The setup for EBW resembles that of laser beam welding with the standoff distance from the beam nozzle. One drawback of EBW is that this process requires being performed in a vacuum chamber and emits radiation. This is extremely time-consuming in a manufacturing environment due to the time required to establish the vacuum required in the chamber. The chamber also needs to contain the radiation that is a by-product of the electron beam generated. However, the process is capable of welding paper-thin material and can also weld to a penetration depth of up to about 2" with a powerful beam.

23.4 SOLID-STATE WELDING

The solid-state welding category performs the welding of the materials by utilizing either direct pressure or a combination of heat and pressure to flow the materials together. By definition: *Solid-state welding accomplishes the coalescence of the base material by the application of pressure or a combination of heat and pressure to a concentrated area.* Pressure applied to any materials will increase the heat when they are in contact with each other, which is the concept of solid-state welding. By introducing heat with the pressure applied, less pressure is required to accomplish the weldment. A variety of procedures are utilized to perform this type of welding and several are used in manufacturing on a daily basis. In solid-state welding, no filler materials are added to the processes. All the welding bonds are performed by the flowing of the base metals at their contact points. The more common welding processes in this category are explained as follows.

23.4.1 Forge Welding

Forge welding is the oldest of all the welding processes. This process utilizes a combination of both heat and pressure to weld the metals together. It was used by the old-time blacksmiths by heating the two pieces in a forge and then hammering them together on an anvil. By heating the steel to a temperature close to its melting point, and then hammering it, the pressure from the impact raises the temperature at that point and begins to flow the materials together.

The same concept is utilized in production in modern-day forge welding. The hand-swung hammer has been replaced by either a mechanical or a hydraulic press capable of delivering far more pressure to the concentrated area. The materials are heated in a furnace, placed into a fixture to hold them in the proper location, and then loaded into the press to receive the impact required to join them. This may need to be repeated a few times to establish the depth of weld required to support the function of the weldment. Figure 23.17 illustrates the forge welding process with the materials receiving the impact pressure from a hammer press.

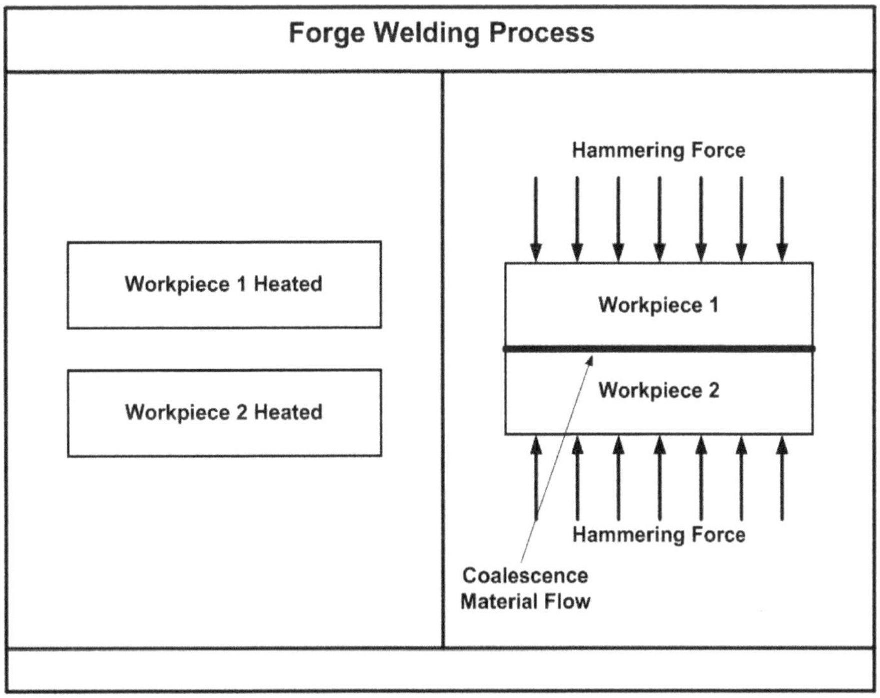

Figure 23.17 Forge Welding Process and Material Flow

23.4.2 Diffusion Welding

Diffusion welding is also a process that utilizes a combination of heat and pressure to flow the materials together but is performed differently than forge welding. In this process, the workpieces are clamped together under extremely high pressure and then the heat is applied. The pressure applied is not in the form of a hammering force but rather a constant force on the clamped components. The heat is then applied to the clamped components causing a flow of the base metals at their contact surfaces. An advantage to this welding process over forge welding is that several weldments can be performed simultaneously in the furnace as opposed to hammering a single assembly at a time. Figure 23.18 illustrates the diffusion welding process showing the clamped components with the heat being added.

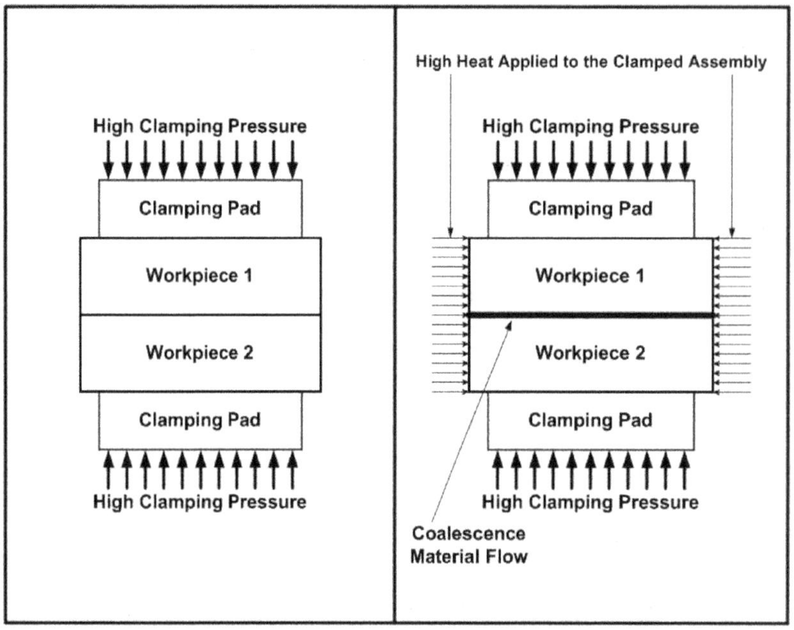

Figure 23.18 Diffusion Welding Process and Material Flow

23.4.3 Friction Welding

The friction welding process is used for welding a pin into the hole of a component workpiece. A pin having a clearance with the hole of between 0.0001 and 0.0003 in. is placed into a holding fixture of the welding equipment and rotated at a given RPM. The pin gets lowered into the hole of the component into which it is to be welded and friction will be generated. The pressure from this friction begins to generate heat between the pin and the hole, causing the material between the two to flow. In friction welding, the fixture holding the pin has a slip clutch that is set to the resistance required for a proper weldment. This is critical to adjust properly because if too much resistance is obtained, the pin may break or deform in the component it is being welded into, creating a defective assembly. In the same manner, if too little resistance is set on the clutch, the weld may not be completed properly and the weldment could fail in operation. Figure 23.19 illustrates how the friction welding process is performed and shows the material flow between the pin and the hole.

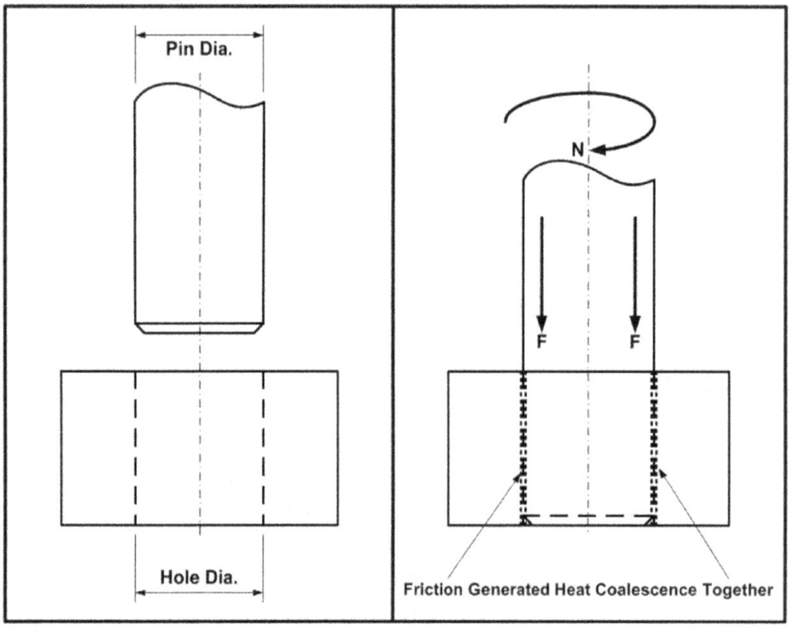

Figure 23.19 Friction Welding Process and Material Flow

23.4.4 Ultrasonic Welding

Ultrasonic welding is primarily utilized to weld plastic components together. The equipment used for this process resembles a setup for spot resistance welding, but no current is applied. The plastic components are placed between two probes called horns and clamped with light pressure. The horns are then activated with a high-frequency vibration of between 20 and 80 KHz, causing friction between the clamped components. This friction allows the plastic to flow together at the contact points, creating the weldment of the plastic. The welding times for this process are extremely fast of about 1 s or less, and the clamping pressure is much lighter than that utilized in resistance spot welding. If the clamping pressure is set too high, or the welding time is set too long, the surfaces at the clamping points will be deformed, creating an unacceptable weldment. Figure 23.20 shows the ultrasonic welding process with the material flow created by the friction generated from the high-frequency vibrations.

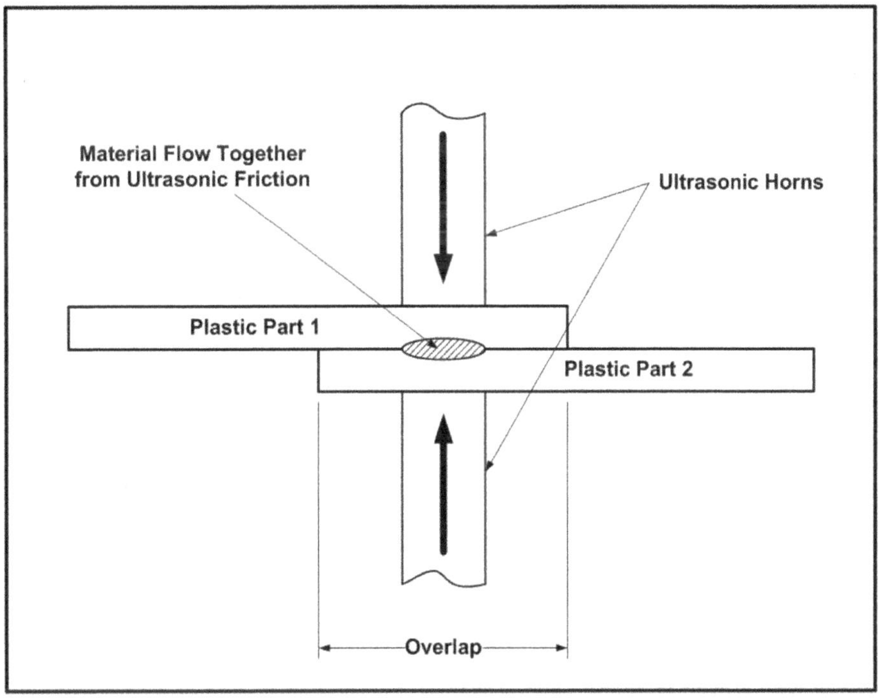

Figure 23.20 Ultrasonic Welding and Material Flow

23.4.5 Explosion Welding

Explosion welding is a highly specialized and unique process. With this process, there is no external heat applied but rather performed by pressure alone. Welding of the components is accomplished from the pressure generated by an explosion sending one of the components into the other at an extremely high velocity. The pressure generated from the impact creates a high enough pressure to weld them to each other. This process has limitations, and even though it falls into the welding arena, it is not widely utilized in a production environment. To perform this procedure, a pressure vessel capable of withstanding an internal explosion created from explosive chemicals is required. The result from the explosion pressure does not melt either of the two materials but rather plasticizes the two surfaces together causing at welded bond. An advantage of this process is that during the explosive force, the entire contact surfaces are bonded, creating a weld area not possible with other welding processes. One major disadvantage is that geometric configurations are limited in size and thickness. The most common application is to weld plates and sheet metal parts.

23.5 WELDING DESIGN CONSIDERATIONS AND DEFECTS

When component parts are being designed for a weldment assembly, an understanding of what will produce a quality weld is required. How the components will connect to each other and the position in which the weld will be performed are critical for both the ease of welding and the overall quality. Following are some design principles for both arc welding and resistance welding that should be understood and implemented.

Arc Welding

— *Good Fit-Up of Components*: The component parts that will be welded together need to have a good fit between them. Gaps between the parts will cause the weldment to have weak points and could cause failure. With a tight fit between the parts, the welding bead will be equal as the welding electrode runs along the seam creating the welding bead.

— *Welding Gun Access*: In order to produce an acceptable and complete weldment, the component parts that are to be welded must allow for access of the welding gun. If the welding gun cannot access the full area where the components are in contact, there will have to be gaps in the weld bead, which may cause a weak area in the weldment depending on its function while operating in the product.

— *Welding Positioning*: The orientation of the components to be welded can allow for either easy or difficult welding. There are three basic welding positions that can be realized, which are *Flat Welding*, *Vertical Welding*, and *Inverted Welding*. Whenever possible, the assembly should be designed to allow for flat welding. This is the easiest of the three and will render fewest defects if performed properly. The next position is vertical welding, which is more complicated and can render an uneven weld as the electrode is drawn down the side of the workpiece. Inverted welding is the most difficult and requires a very skilled operator. Welding from the bottom of the workpiece can also render an uneven weld and has the danger of molten metal dripping down from the weld onto the operator. Figure 23.21 illustrates the three possible welding positions that can be preformed.

Resistance Welding

— *Low-Carbon Steel Is Preferred*: Low-carbon steel will produce the best resistance weldment. Steel with a higher level of carbon will perform localized hardening at the welding points and produce a bond that is brittle and has the potential of failure. Another consideration is the stiffness of the material being welded. Thin metals can warp during resistance welding and produce an assembly that is not flat or straight, which can be unacceptable in the required function.

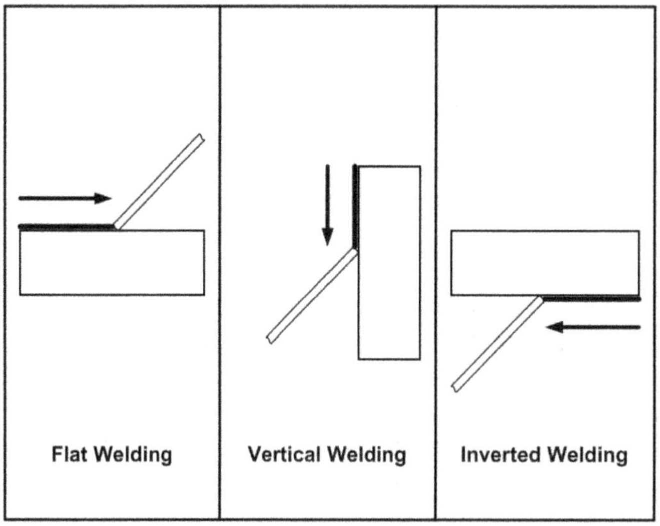

Figure 23.21 Three Possible Welding Positions

— *Electrode Access*: As arc welding requires access for the welding gun, resistance welding requires that the electrodes have proper access where the welds are to be performed. This requires access from the bottom of the two components as well as from the top. Without a direct lineup of the electrodes, the weld will not be properly performed.

— *Component Overlap*: Unlike arc welding, which can be performed by butting two components together, resistance welding requires that the two components have an overlap where the welds are to be performed. The amount of overlap required will be determined by the size of the electrodes being utilized to perform the welding. As a rule of thumb, the overlap of the materials should be roughly ¼" on either side of the electrode at a minimum. This will allow for full access to the welding area and not bleed out from the sides.

As with any fabrication process, defects can be produced and need to be understood. Most of the defects associated with welding can be avoided by having a proper setup and skilled operators. Another issue with welding defects is being able to detect them. Most of the cosmetic defects are visible on the surface, but the majorities are hidden internally in the welded joint. Destructive testing will determine if the welded joint is acceptable, but the weldment is destroyed in the process. Nondestructive testing processes such as X-rays and other techniques can detect defects hidden inside the weld and can predict if there may be a failure in operation. Following are some of the more common welding defects that can be encountered in welding assemblies and joints.

Welding Defects

— *Porosity*: Porosity is air bubbles trapped inside the welded bead. The air pockets inside the weldment will cause weakness due to the interruption of the solid welded area. There are several potential reasons that this can occur, such as improper heat, improper filler materials, and an improper welding setup. Figure 23.22 shows porosity air pockets hidden internally in the welded joint.

— *Inclusions*: Solid inclusions are foreign materials internal in the welded joint. This inclusion can be debris from the surface on the workpiece or flux trapped under the weld. Regardless of the inclusion in the weld, they will cause a weak bond between the two components. The main issue with this defect besides not being able to see it is that the inclusions are solid and not flowed together in the joint.

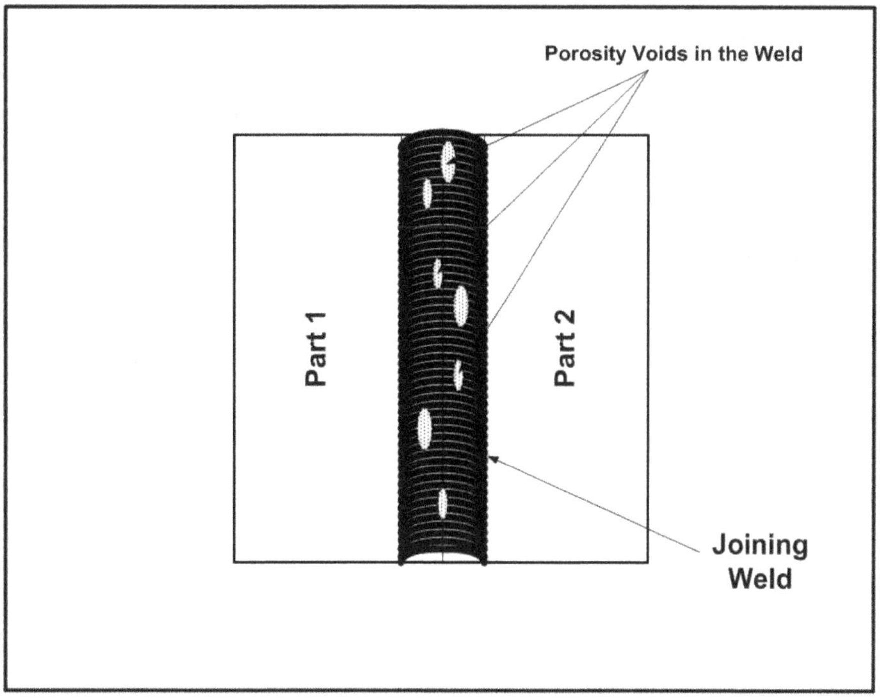

Figure 23.22 Hidden Porosity in a Welded Assembly Joint

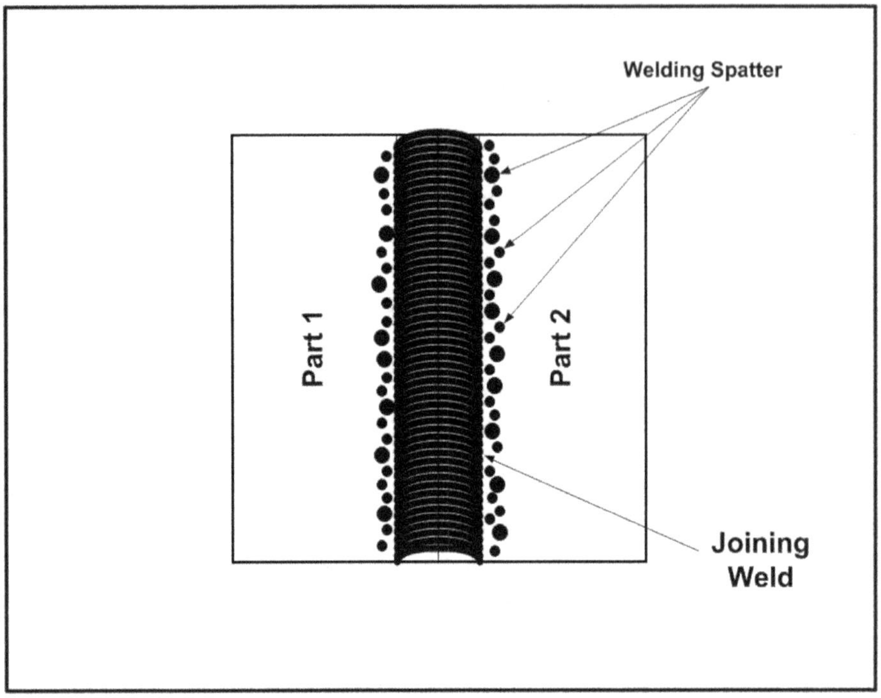

Figure 23.23 Welding Spatter on a Welded Assembly Joint

— *Incomplete Fusion*: This defect is caused from an improper setup of the equipment. Mainly there is not enough heat being applied to the welded bead, causing the penetration to be not deep enough. This will produce a weak welded joint with the potential of failure in operation.

— *Other Defects*: Other defects that are visible are mainly cosmetic but can also produce an unacceptable weld depending on the application of the weldment. One of these defects is a nonstraight contour of the welded bead. Even though there may be sufficient fusion and no other internal defects, it affects the appearance of the weldment, which may be unacceptable to the customer. Another is something called spatter. This is caused in arc welding from the welding gun or electrode being too far away from the area being welded, causing filler material to jump off the bead and onto the side of the weld. Again, the weld can have no other defects, but from a cosmetic standpoint, this could be unacceptable as well. Figure 23.23 shows the spatter that is adhered to the materials on both sides of the welded joint.

23.6 SUMMARY

The welding processes are very widely utilized in the manufacturing arena for joining component parts. From welding, the outcome will be a permanent assembly, which is the strongest of all assembly processes. With a variety of processes to select, whether they be fusion welding or solid-state welding, the proper selection will result in a weld that can be produced under almost any condition. Remembering that a welded assembly cannot be disassembled without destroying it, consideration of the function in the finished product needs to be understood if there is a potential to require it to be disassembled for service.

23.7 QUESTIONS

1. Define the principle of welding.
2. What is the difference between fusion welding and solid-state welding?
3. How can a welded joint be stronger than the base material being welded?
4. Why is it not possible to weld dissimilar materials?
5. What are some of the dangers associated with welding?
6. What is arc welding performed, and what are the three processes of performing it?
7. What are the functions performed by flux in the arc welding process?
8. Why is filler material added to welds, and what does it do?
9. What is the difference between a consumable and nonconsumable electrode, and which processes utilize them?
10. When doing torch welding, what two gasses are utilized and what does each do?
11. In torch welding, what is meant by a neutral flame and what temperatures are reached in the three areas?
12. What precaution needs to be taken when using acetylene?
13. Explain how resistance welding is performed, and what are the two processes and their differences?
14. What are the differences and similarities between laser beam welding and electron beam welding?
15. Both forge welding and diffusion welding processes are performed in the same manner. What is the major difference between the two?
16. What is the concept behind friction welding, and how does it perform a weld between two components?

17. How is the ultrasonic welding process performed, and what materials is it used on?

18. What are the design considerations for both arc and resistance welding, and why are they important?

19. For a welding defect, what is the difference between porosity and inclusions?

20. How is incomplete fusion caused in welding?

CHAPTER

24

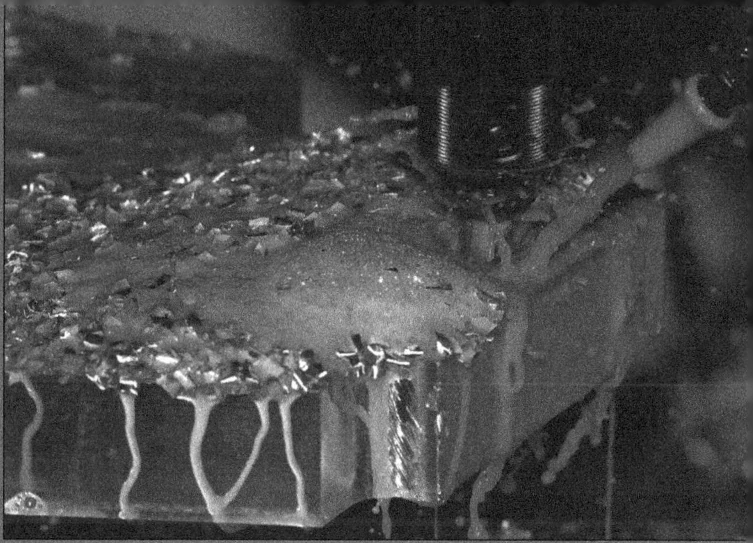

Brazing and Soldering Principles

24.1 INTRODUCTION

Brazing and soldering are very widely utilized in the manufacturing assembly arena. Both of these processes are an alternative to welding but will render a weaker joint than welding does. However, these processes can provide certain characteristics that are either difficult or not possible with the welding processes. A major advantage of these two processes is their ability to join completely different materials together. In welding, the materials being joined require that they be of the same family type and have a melting temperature relatively close. Brazing and soldering are capable of assembling different materials with a strong bond, which will be stronger than other assembly processes with the exception of welding. This chapter covers the principles and processes of both brazing and soldering along with the advantages and disadvantages of both.

24.2 BRAZING CONCEPT

Brazing is a material joining process accomplished by the use of heat and filler material. The principle of this process is to have the filler material transported between the two component workpieces through capillary action. Heat is applied to the workpieces and the filler material is melted, which then flows between the parts. Figure 24.1 illustrates the brazing process where capillary action carries the filler material between the workpieces. The bond is created from the strength of the filler material and the amount that is transmitted between the components. Unlike welding, there is no melting of the base materials during the process. Enough heat is applied to melt and transfer the filler material only. Another area that is critical for brazing is that the surfaces of the material must be clean and free of debris. Any corrosion and oils on the surfaces will affect the

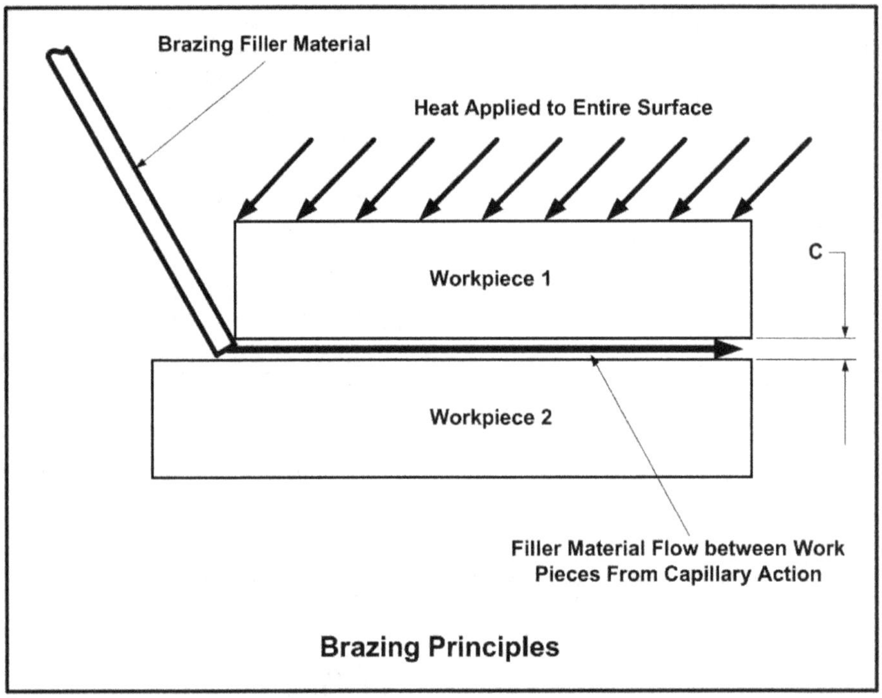

Figure 24.1 Brazing Theory Where Capillary Action Transfers Filler Material between the Workpieces

capillary action and not transfer the filler material between the components properly, resulting in an unacceptable bonded joint.

In order to get a good capillary flow of the filler material between the workpieces, a clearance is required. If there is no clearance and the pieces fit together tightly, no flow can occur and the bond will be extremely weak and sure to fail. Likewise, if the clearance is too large, there will be too much filler between the pieces and also creates a weak joint. For a good brazing bond, a clearance of between 0.001 and 0.010 in. is required with around 0.005 in. rendering the best brazing strength. Figure 24.2 shows a chart with the amount of clearance

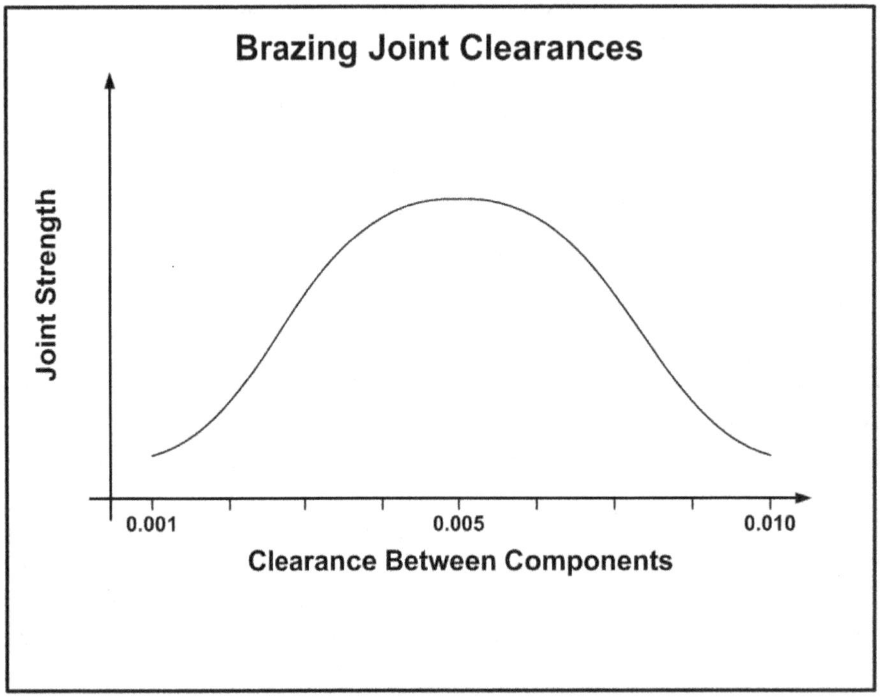

Figure 24.2 Clearances vs Bonding Joint Strength for Brazing

vs the bonding strength of a brazed joint. Some typical brazing joints are either lap joints or butt joints, which are shown in Figure 24.3. Other joints are utilized, but the lap and butt joints allow for the best capillary action and produce the strongest bonds.

The selection of filler material is dependent upon the materials that will be brazed together. There are a variety of fillers available and are all dependent upon the materials being brazed and the type of brazing joint being performed. Figure 24.4 shows a table of some filler materials along with their relative melting temperatures and uses. There are points and considerations that need to be observed when selecting the proper filler material for a brazed joint. These are described as follows.

— **Melting Temperatures:** The melting temperature of the filler material being utilized must be compatible with the melting temperatures of the base material being brazed. The base material will be heated but not to its melting temperature, which requires the filler material to be lower in order to create the brazed joint through capillary action and not through welding.

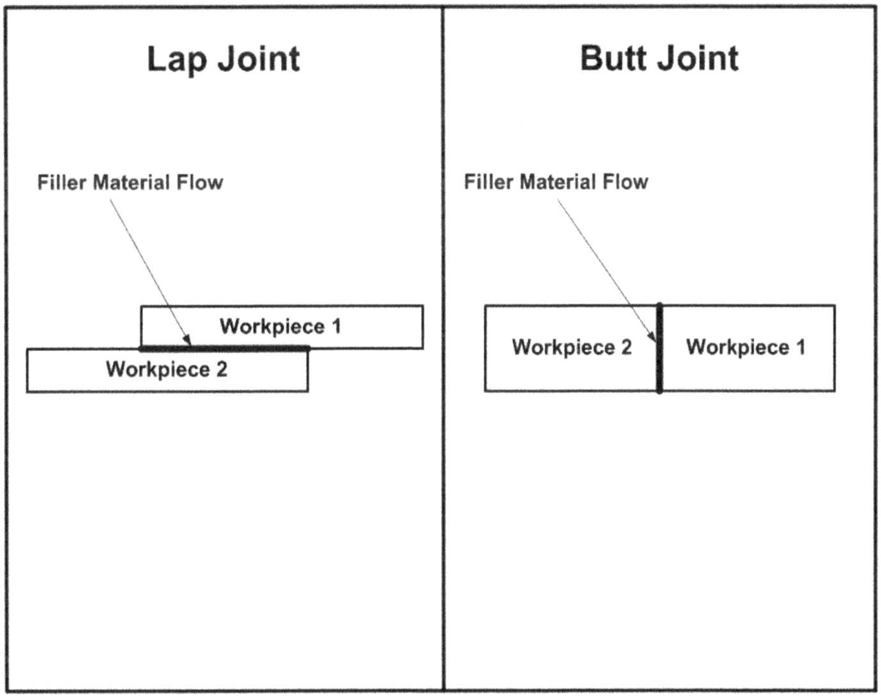

Figure 24.3 Lap and Butt Joint Brazing Joints

Brazing Filler Materials

Filler Material	Temperature °F	Base Materials
Aluminum-Silicon	1,100	Aluminum
Brass	1,700	Steel, Cast Iron, Nickels
Copper	2,050	Nickel, Copper
Copper–Silver	1,900	Copper, Nickel Alloys
Gold/Silver	1,750	Gold, Silver, Stainless Steel
Nickel Alloy	2,100	Stainless Steels, Nickel Alloys
Silver	1,400	Monel, Inconel, Titanium

Source: www.wikapedia.com

Figure 24.4 Various Filler Materials and Their Uses in Brazing

— **Reactions:** With the filler materials being of a different composition than the base materials, caution needs to be taken so that there will not be any chemical reaction between the two materials. These reactions will be violent in nature but will cause the filler material to not bond with the base materials, thus causing a bond that will not hold. Also, this reaction depending on the material mix may degrade over time, causing the bond to fail at a later date.

— **Low Viscosity; High Fluidity:** The fluidity of the filler material when melted is critical to obtaining a good brazed bond. The higher the fluidity of the filler material, the more it will travel during the brazing process and create a stronger bond than a filler with a high viscosity will.

As with welding, flux is utilized in the brazing process as well. Flux materials are available as either a powder or a liquid paste, which can be applied by brushing it onto the joint or squirted on during the process. Some filler rods will be coated with the flux, similar to a welding rod utilized in SMAW. The flux performs very much the same functions in brazing as it does in welding. These are described as follows.

— **Joint Protection:** As in the welding processes, the joint being brazed is very susceptible to the elements. During the brazing process, the flux will shield the joint from the atmospheric elements, protecting from any potential contamination that can affect the quality of the brazed joint.

— **Wetting Ability:** Wetting refers to the ability of the fluxes to create a wet surface for the filler material to flow on. This aids in the overall capillary action as it transfers the filler material between the workpieces being brazed.

— **Melting Temperature:** The flux melting temperature needs to be lower than the melting temperature of the filler material. This allows for the flux to penetrate the brazed joint first and allow for a better flow of the filler between the workpieces being brazed.

— **Low Viscosity:** The flux needs to have very high fluidity so that it will penetrate between the workpieces in order for the filler material to flow. Most of the flux material will be more fluid and have a lower viscosity than the filler material being brazed into the joint.

As with all manufacturing processes, there are both advantages and disadvantages that need to be understood before selection. Brazing is no exception, and both the advantages and disadvantages of the process are described as follows.

Advantages

— *Capillary Action*: Capillary action allows for the filler material to travel between the surfaces to create the brazed joint. This allows for a larger bonding area than welding does. Also, due to the capillary action transporting the filler material throughout the brazed joint, less access for the heat source is required than in welding.

— *Dissimilar Materials*: Because there is no melting of the base materials during the brazing process, dissimilar materials are capable of being joined. In welding, the base materials are melted and flow together, so they need to be of the same material category in order for the welded joint to bond properly. However, when brazing dissimilar materials, they both must have a higher melting temperature than the filler material being utilized to perform the bonding joint.

— *Simultaneous Joints*: Due to the capillary action transferring the filler material between the workpieces, it is possible to braze more than two components at the same time. The proximity of the components needs to be close enough for the filler to travel under each one in order for a proper bond to be established.

— *Thin Materials*: Brazing is capable of applying joints to extremely thin materials. Due to the fact that the base materials are not melted, the capillary action allows brazing to join thin materials easier than welding does.

— *Heat-Affected Zones*: With less heat required for brazing than for welding, the areas that are being joined will experience less heat and, therefore, be less affected. Welding melts the materials together at extreme temperatures, causing stress and areas that can produce defects from the heat. Brazing is performed below the melting temperatures of the base materials, thus greatly reducing the potential of heat affecting the bonding joint areas.

Drawbacks

— *Bonding Strength*: The bonding strength of a brazed joint will only be as strong as the filler material being utilized. Also, the strength of the brazed area will also be dependent on how much capillary flow of the filler material between the workpieces exists. The more filler in the brazed area, the stronger the bond will be.

— *Joint Weakness*: Brazed joints will not be as strong as a welded joint due to the fact that there is no melting of the base materials during the process. Welding produces the strongest joints, and depending on the filler material used or the buildup of filler, the joint can be stronger than the base materials. Brazed joints will not be as strong as the base material, and depending on the application of the assembly, they need to be considered to avoid a failure in operation.

— *Service Temperatures*: The temperatures that a brazed assembly will be operating at needs to be considered in order to avoid a failure. With only the filler material holding the brazed joint together, if the assembly will be subjected to high temperatures that are close to the melting temperature of the filler material, it may begin to flow and weaken the brazed joint. Service temperatures are often overlooked for brazed assemblies and should be a factor when selecting either the brazing process or the filler material to be used.

24.3 BRAZING PROCESSES

There are several processes that can create a brazed joint. They will all accomplish the same result but are performed in different manners. The three most common are *Conventional Brazing*, *Resistance Brazing*, and *Furnace Brazing*, which are explained as follows.

24.3.1 Conventional Brazing

Conventional brazing is performed by applying a heat source to the workpieces and adding the filler material separately. The two common methods for applying the heat are utilizing either an oxyacetylene torch or an electrical arc that uses a nonconsumable electrode as in GTAW welding. Caution needs to be taken as to not melt the base materials but rather to apply enough heat to cause the flux and filler material to melt and transfer between the parts. Even with the heat being applied from the GTAW electrode, which utilizes an inert gas, the brazing flux must be applied in order to achieve a quality braze. Figure 24.5 illustrates the conventional brazing process where the heat is applied and the filler material is added by hand separately.

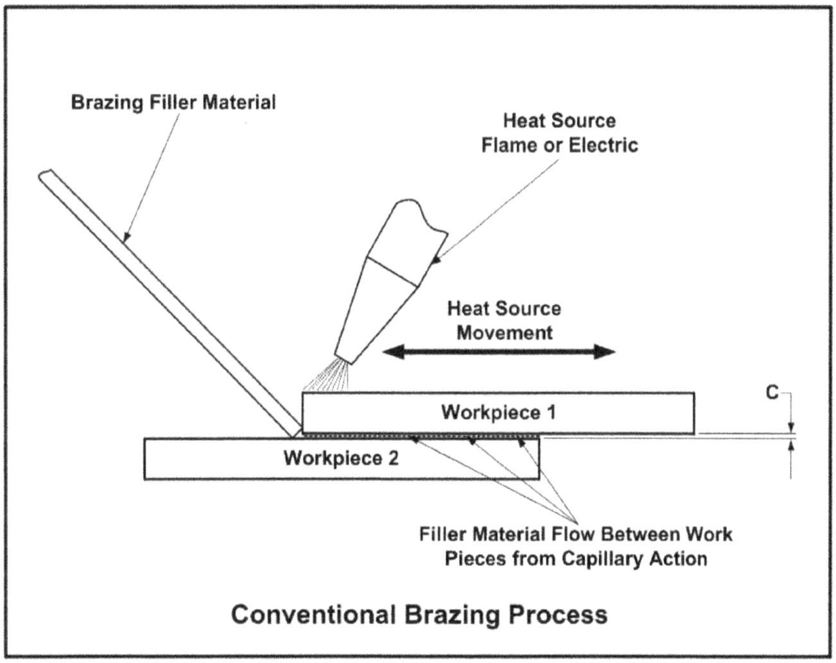

Figure 24.5 Conventional Brazing Process Utilizing Either Torch or Arc Heat Source

24.3.2 Resistance Brazing

Resistance brazing looks very similar to resistance welding, utilizing nonconsumable electrodes, which the workpieces are pinched between and a current will induce the required heat. The difference is that there is a thin sheet of filler material and flux placed between the workpieces and the electrodes are much larger than those used for resistance welding. This is to have the ability of spreading the heat over a much larger area during the process. The current is applied and timed so that the workpieces will reach the required temperature and enable the filler material to flow between the pieces. Resistance brazing has the ability of having more filler material between the surfaces than does conventional brazing and is utilized for sheet metal applications. Figure 24.6 illustrates the resistance brazing process and shows the filler material between the workpieces to be distributed with the applied heat.

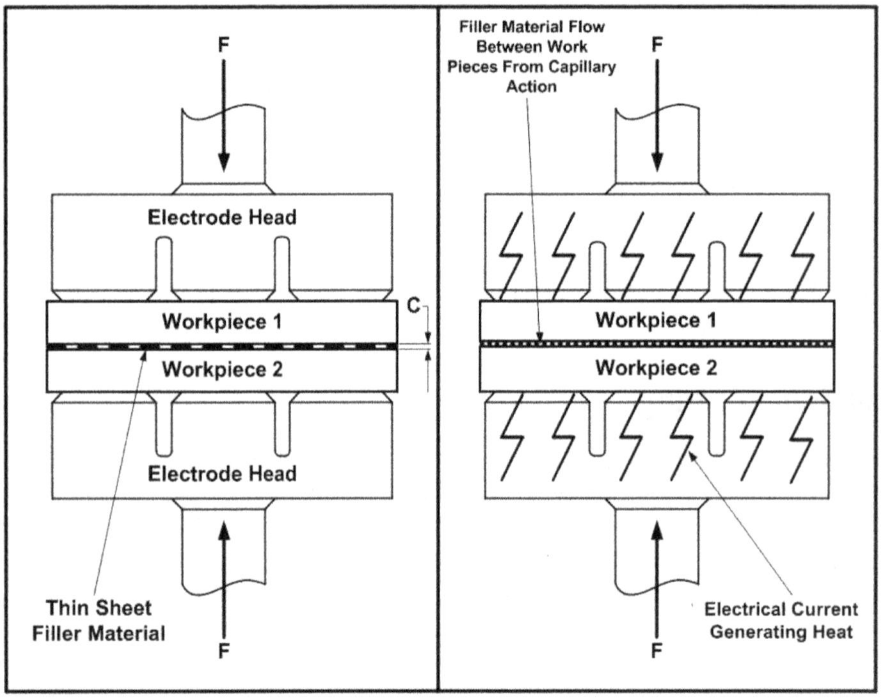

Figure 24.6 Resistance Brazing Showing Filler Material Flowing between the Workpieces

24.3.3 Furnace Brazing

In the furnace brazing process, the two components to be joined are assembled and the filler material with the flux premixed is applied to the sides. This process is very common for brazing pins into other components and will render a strong braze. The filler/flux is preheated to the consistency of a paste and applied to the side where the pin is assembled into the mating part. From there, the assembly is loaded into a furnace where the heat is applied and the filler/flux paste is distributed between the workpieces. Care needs to be taken to apply the proper amount of filler/flux paste. If too little is applied, the joint will not be fully brazed, and likewise, if too much is applied, there will be a overflow of filler that would need to be removed in a secondary operation. Figure 24.7 illustrates the furnace brazing process showing the filler/flux paste applied to the side of the assembly and the heat causing it to flow between them.

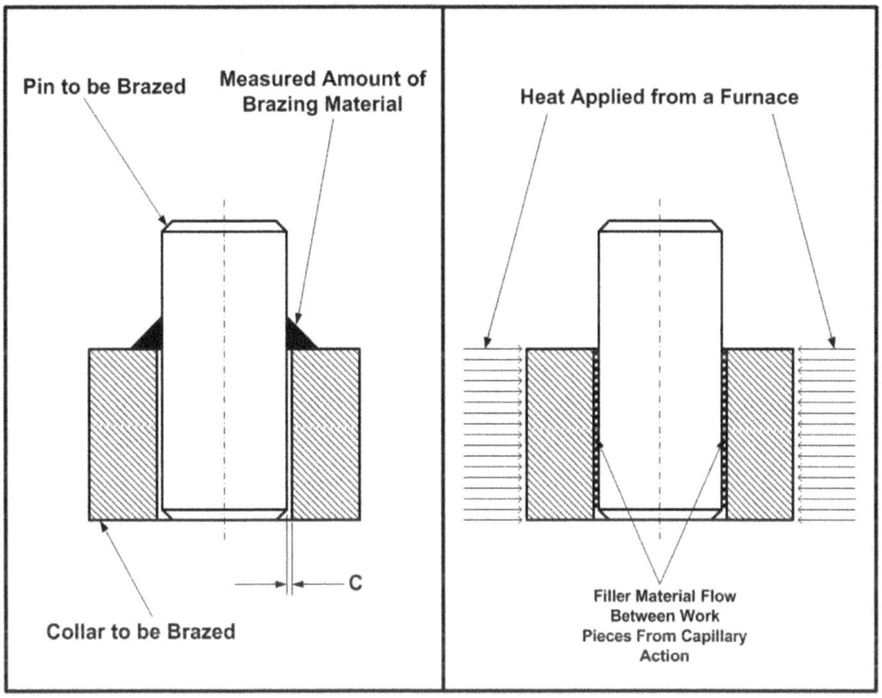

Figure 24.7 Furnace Brazing of a Pin and Collar with Filler Paste Flowing between the Workpieces

This process is fast and can braze several assemblies requiring nothing more than preparing the assemblies and loading them into the furnace. For high-volume furnace brazing operations, a continuous belt furnace is used where the operator applies the filler/flux paste and places them on a timed belt. As the assemblies move through the furnace, the heat causes the filler to flow between the parts creating the brazed joint. The parts then just fall off the end of the belt into a container completed. Figure 24.8 show a continuous brazing furnace with the assemblies loaded into one end and the completed brazed assemblies exiting the other.

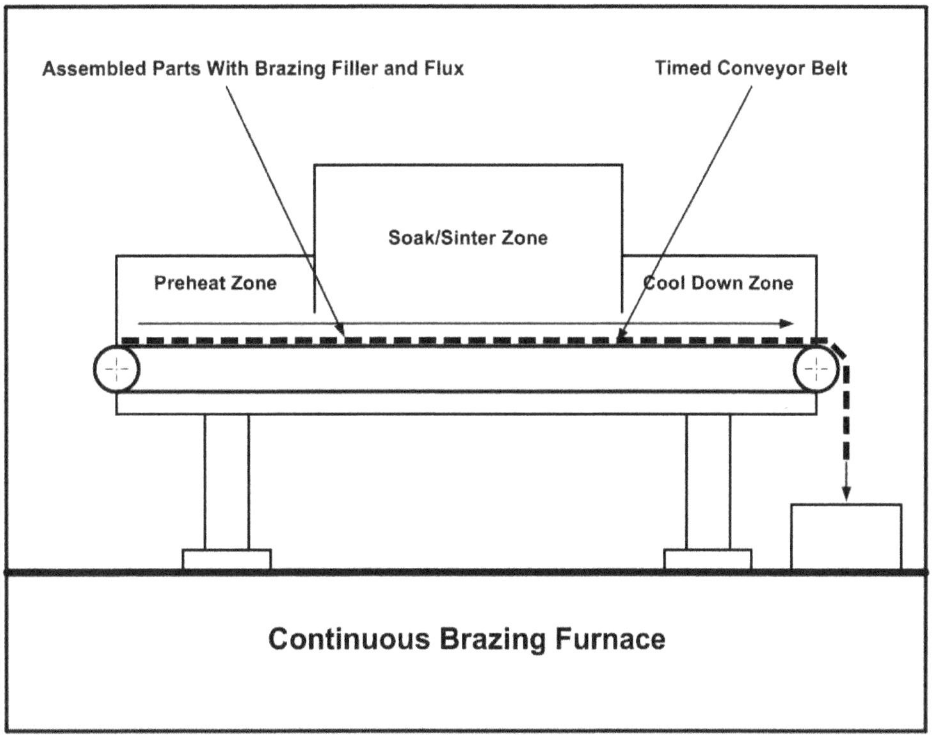

Figure 24.8 Continuous Belt Furnace for High Production Furnace Brazing

24.4 SOLDERING CONCEPT

Soldering is performed identical to brazing but at much lower temperatures and with different filler materials. Heat is applied to the components in a variety of methods, and the filler material is transferred between the surfaces by capillary action creating a bonded joint. As with the brazing process, the surfaces of the component workpieces need to be clean in order for the filler material to flow properly between them. Soldering does require a clearance between the workpieces also, and it is less than the clearance needed for brazing. Figure 24.9 shows a chart of the clearance vs the joint bonding strength, which ranges from 0.001 to 0.005 in. with 0.0025 in. being the optimum for bonding strength. This reduced clearance requirement is for two reasons. The first is that the filler becomes much more fluid when heated and requires less area for the solder to flow between the components and create the bonded joint. Secondly, the filler material does not have the strength that brazing filler does, and with a lesser clearance, it will create a stronger bond. Most of the soldering filler materials are a mixture of tin and another alloy material. Figure 24.10 shows some of the more common fillers utilized in soldering with their relative melting temperatures and applications.

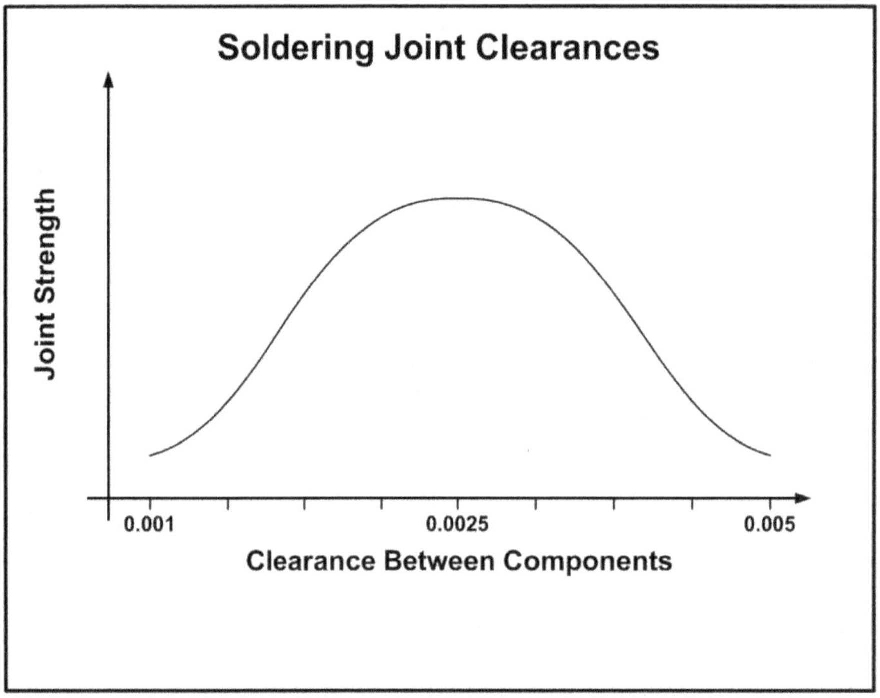

Figure 24.9 Clearances vs Bonding Joint Strength for Soldering

Soldering Filler Materials

Filler Material	Temperature °F	Base Materials
Lead–Silver	580	Elevated Temperature Joints
Tin–Antimony	460	Plumbing & Heating Applications
Tin–Lead	361–405	General Purpose Applications
Tin–Silver	430	Food Container Applications
Tin–Zinc	390	Aluminum Joining Applications
Tin–Silver–Copper	423	Electronics: Surface Mount Technology

Source: Grover

Figure 24.10 Various Filler Materials and Their Uses in Soldering

Soldering does have both advantages and drawbacks. As with any fabrication process, understanding of what the process will yield is critical for a proper outcome. The advantages and drawbacks for soldering are discussed as follows.

Advantages

— *Low-Energy Requirements*: With the melting temperature of the filler materials being far less than that required for brazing, less energy is required to heat the component workpieces to achieve the bonding.

— *Leakproof Joints*: A major advantage of soldering is that it will render a liquid and airtight joint. With the filler material being more fluid than in brazing, it is able to travel between the workpieces further and more uniformly, thus creating a leakproof joint.

— *Easy Repair/Rework*: Because the melting temperature is low, performing any repair or rework is easy with soldering. By reapplying the heat to the joint, the component parts can easily be disassembled, cleaned, and resoldered. Or, if an assembly is simply required to be taken apart for whatever reason, it can be easily accomplished.

Drawbacks

— *Joint Bonding Strength*: Soldering will render a weaker bond than does brazing. This is due to the strength of the filler material being utilized and the lower temperatures required in the process.

— *Service Temperatures*: Due to the low melting temperatures required in soldering, the temperature range that the bonded joint will operate in is important to understand. As with brazing, if the temperature of the atmosphere approaches the melting temperature of the filler, it will begin to flow and weaken the bonded joint.

24.5 SOLDERING PROCESSES

Almost all of the processes utilized for soldering are identical to those used for brazing to apply the required heat to the component workpieces. Heat can be applied either by torch, by electrical arc, in a furnace, or with resistance current. One process that is unique to soldering is the wave soldering process. This is utilized for soldering components onto a printed circuit board. The solder filler material is melted in a vat and allowed to rise up through an opening forming a wave or bubble of filler. The printed circuit boards with the components are then run over the wave, allowing the filler to touch each one and soldering them to the board. This process is extremely efficient for printed circuit boards because it will solder every component onto the board in a matter of seconds, whether there are two or two thousand components. With the components on the printed circuit board having a minimum of two contact points, it would be an extremely time-consuming process to solder each one by hand, and the possibility of missing some contacts is very high. Figure 24.11 illustrates how the wave soldering process is performed with the printed circuit board being run over the wave of soldering filler material to bond them.

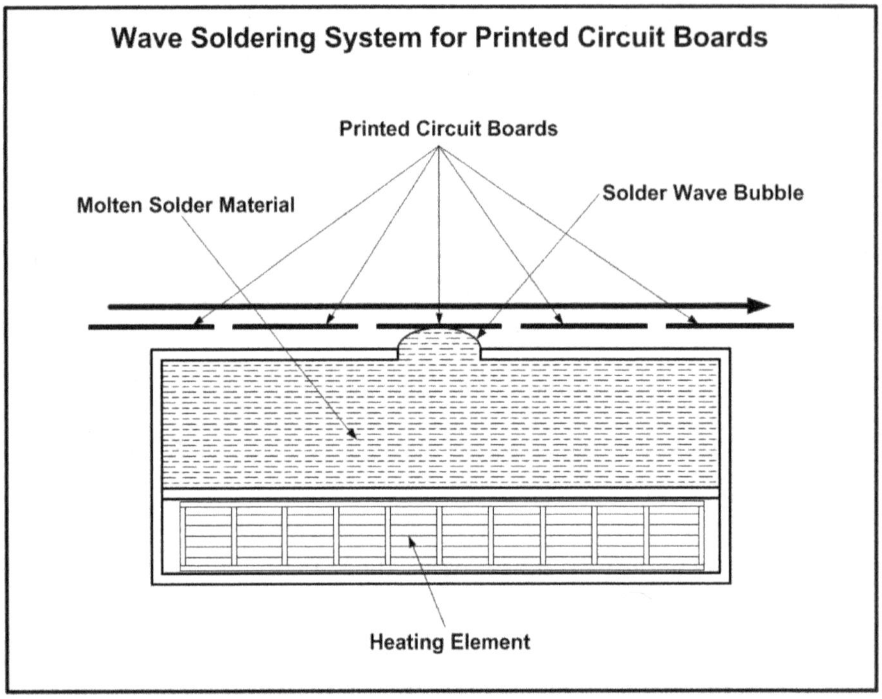

Figure 24.11 Wave Soldering of PC Boards with Solder Wave in Contact with Electronic Components

24.6 DESIGN CONSIDERATIONS

All fabrication processes require an understanding of what is needed in the design portion in order to be successful. Both brazing and soldering are no exception to designing a proper assembly that will function in the assembled product. These considerations during the design phase are listed as follows.

— **Clean Surfaces:** Unlike the welding process, both brazing and soldering require that the surfaces be clean in order for the filler materials to flow between them. With welding, the base materials are melted together and any contamination will be burned away. Here, the filler will be transferred between the surfaces due to capillary action and any contamination will restrict this transfer, thus creating an incomplete bond.

— **Proper Clearances:** Both brazing and soldering require a clearance between the component workpieces to allow for proper flow of the filler materials. Unlike welding where a clearance can cause a defective welded joint, these processes require the clearance for the capillary action to transfer the filler materials between the workpieces. Figures 24.2 and 24.9 show the proper clearance for brazing and soldering, respectively.

— **Proper Filler Selection:** The selection of the filler materials that will be used is important to create a quality bonded joint. This selection is dependent on what materials are going to be joined and also an understanding of whether there will be any chemical reactions between the filler materials and the base materials.

— **Heat Application Access:** With the exception of furnace brazing and soldering, access to apply the heat to the component workpieces is a consideration during the design phase. An area to apply the heat needs to be accessible for either the torch of electrical arc or for the electrodes when performing resistance processes.

— **Service Temperatures:** The temperature range that the assembly will operate within is critical for both brazing and soldering. With the filler materials having a melting temperature lower than those of the base materials, if operating in a range that approaches the melting temperature of the filler, it will cause the bonded joint to weaken and possibly cause a failure.

24.7 SUMMARY

Both brazing and soldering are very viable assembly processes in the manufacturing arena. With the capillary action transferring the filler material between the workpieces, the bond can be accomplished with less effort than with welding. With capillary action transferring the filler materials between the workpieces, the bonded joint can be accomplished with less energy and have the capability of joining more than two components at a time. Although both brazing and soldering will create a weaker joint than welding, if the strength is acceptable, it is a good alternative to consider. The soldering process can render a joint that is both liquid and airtight, which is advantageous for applications where these elements will flow through the assembly. By understanding both the advantages and drawbacks of both processes, and applying the design considerations required for producing an acceptable assembly, brazing and soldering can be a very viable alternative to welding.

24.8 QUESTIONS

1. What is the major difference between the welding processes and the brazing and soldering processes?

2. Why do both brazing and soldering require a clean surface and a clearance between the components that are to be joined?

3. What are the three requirements that need to be considered when selecting a filler material for brazing?

4. Flux is utilized in the brazing process. What is meant by a flux's wetting ability, and what is its function during the process?

5. What are the advantages of the brazing process over the welding process?

6. A drawback to brazing is the service temperature that the brazed joint will operate in. Why is this important to consider for the assembly?

7. What are the heating sources that are utilized when performing conventional brazing?

8. How does resistance brazing differ from resistance welding?

9. Both brazing and soldering can be performed in a furnace. Describe the two processes and how they are performed.

10. Soldering and brazing are performed in the same manner. What are the advantages and drawbacks of soldering over brazing?

11. Explain the wave soldering process, and what is this utilized for?

12. What are the design considerations for developing an assembly that will be either brazed or soldered?

CHAPTER 25

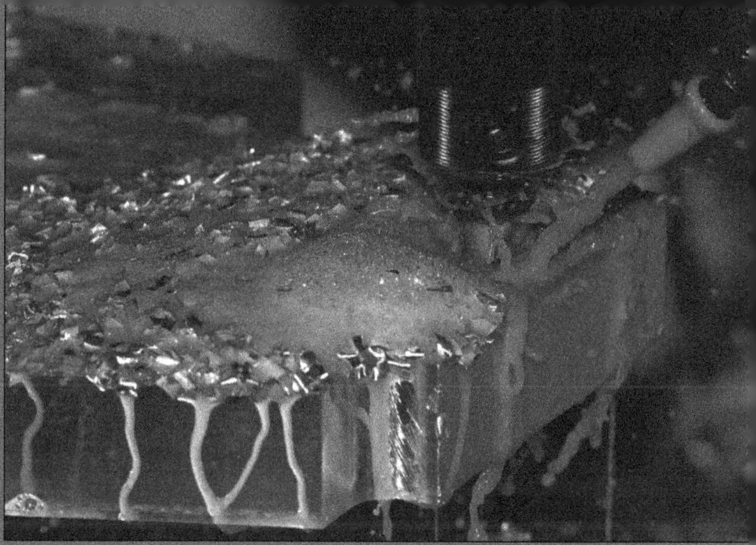

Adhesive Bonding Theory and Principles

25.1 INTRODUCTION

Adhesives play a major role in the manufacturing assembly process. With the advent of newer and stronger adhesives being developed almost on a daily basis, materials can be assembled through bonding through the utilization of various adhesives creating functional assemblies. This can be a very viable alternative to welding, brazing, and soldering, reducing both time and cost during the assembly process. Understanding what adhesive bonding is both capable and not capable of achieving is important to ensure that the assembly being designed will properly function and be of the required quality levels desired. This chapter discusses the theory and principles of adhesive bonding applications in manufacturing along with the considerations that need to be implemented during the design phases.

25.2 ADHESIVE BONDING PRINCIPLES

The adhesive bonding process is one that utilizes some form of adhesive to bond two or more base materials together in assembly. Everyone has performed this at home to bond broken items with some form of glue, but in a manufacturing environment, the process is more complicated if a quality bonded joint is to be attained. Selecting the proper bonding agent for the base material being joined and knowing the function of the assembly being bonded are critical for a quality assembly that will not fail.

Adhesive bonds are more effective in certain applications than in others. When a bonded assembly is operating in an environment where it will be receiving shear or tension forces, the adhesive will hold the components together better. This is because the forces applied are against the entire bonded surfaces, resulting in

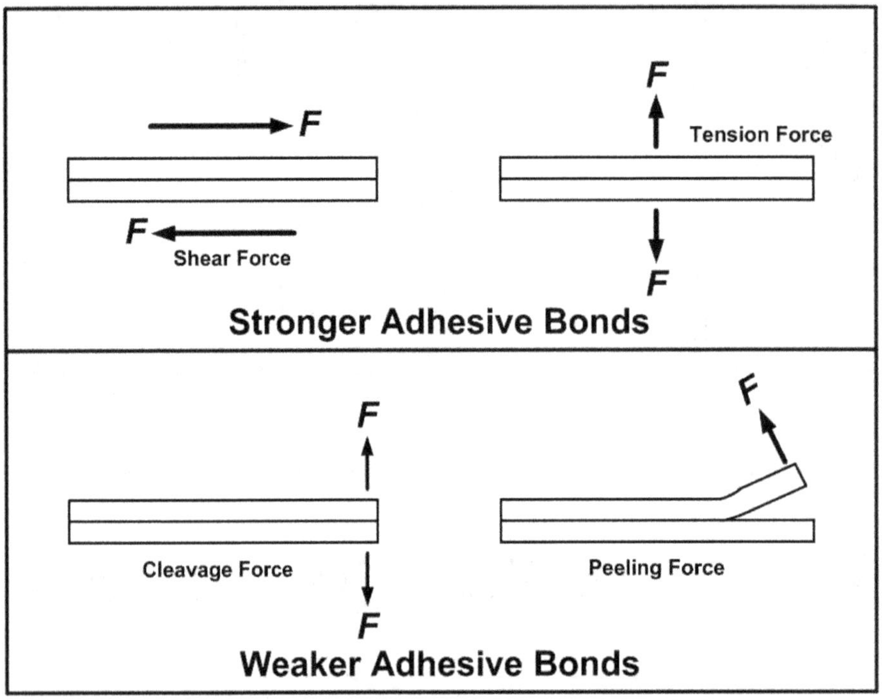

Figure 25.1 Forces Applied to Adhesive Bonded Components and Their Reactions

a stronger holding power between the two components. When an adhesive-bonded assembly is subjected to cleavage and peeling forces, it will have a greater potential to fail. These two forces will be applied to just a single point of the assembled components and the bonded area is more likely to fail. Figure 25.1 shows the forces that are applied to an adhesive bonded assembly and how the assembly will react.

Working with adhesives will have some advantages and drawbacks, which need to be understood during the designing of the assembly. In the right situation, an adhesive bond can be the best alternative for assembly, depending on the applications. Following is a listing of the advantages and drawbacks of adhesive bonding that need to be considered.

Advantages

— *Bonding Surfaces*: Adhesives will offer the largest bonding surface ability of any assembly process. Welding will only join a seam, while brazing and soldering will transfer the filler materials between the components to a point. If a large component surface area needs to be bonded, say one foot square, the adhesive can be applied over the entire surface, creating a large bond.

— *Adhesive Types*: Several types of adhesives are available for given applications. These can vary in holding strength, allowing an application to be either permanent or semipermanent, depending on the assembly requirements. These types of adhesives are discussed later in the chapter.

— *Material Variety*: A big advantage is the variety of adhesives available to use. Today, there is an adhesive that can bond just about any two different materials together. Whether it is plastic to metals, wood to glass, or any possible combination, there is an adhesive that can perform the task.

— *Varying Component Configurations*: The ability of adhesives for bonding components of varying configurations over other bonding processes allows for a wider range of assembly possibilities. It is possible to bond components of various thicknesses easily where welding, brazing, and soldering may not be able to. For example, it one component is 6 in. thick and the other is only 1/16 in. thick, an adhesive bond can be accomplished with no issues where a joining operation requiring heat would be extremely difficult or impossible.

— *Sealing Bonds*: Many adhesive agents will render both a liquid and airtight bond. Several plastic pipes and tubes are bonded together for carrying material where leaks are not permitted. These bonds can also be placed under a limited amount of pressure.

— *Curing Temperatures*: Some adhesives will render a raise in temperature during the curing process. These are extremely minimal, and the assembly can be handled immediately during the curing process with no delay in time for cooling.

— *Simplistic Designs*: The design for adhesives is the simplest possible. All that is required is a large enough area for the adhesive agent to be applied to the components for the required proper bonding strength in the completed assembly.

— *Heat-Affected Areas*: Unlike welding, brazing, and soldering, there is no heat required during an adhesive bonding process. This eliminates any and all potential for a heat-affected area in the assembled components.

Drawbacks

— *Compatible Adhesives*: Selecting the proper adhesive is critical for a quality bond. No one adhesive can perform all bonding required, and the selection of the agent to be used is critical to the operation, depending on the materials being bonded and the function of the assembly in the product.

— *Joint Strength*: The strength of an adhesive-bonded joint will be weaker than in other bonding processes. Understanding what the ultimate bonding strength of the assembly will be is critical for a quality bond, which will not fail during operation. This is dependent on both the adhesive being utilized and the materials being bonded.

— *Clean Surfaces*: Although both brazing and soldering require a clean surface for the process to be performed properly; adhesive bonding requires extremely clean surfaces. In most assembly operations, the surfaces that are to be bonded are cleaned with a cleaning agent such as alcohol to ensure that any and all contaminants are completely removed, so the bonding agent will adhere completely to the surfaces.

— *Service Temperatures*: The temperature range in which the adhesive-bonded joint will operate within is important to know. Adhesives will react differently in elevated and reduced temperatures and could fail during operation.

— *Curing Times*: All adhesives require a curing time in order to complete the bonding cycle. Some will cure within seconds, while others may have a long curing time and require the bonded components to be placed in an oven at set temperatures. These curing times can be as long as 24–48 h to bond completely. This is dependent on the adhesive being utilized and the materials being bonded.

— *Inspection*: Inspecting an adhesive-bonded assembly is extremely difficult if at all possible. For welding, brazing, and soldering, a nondestructive test such as X-rays can be performed to see if the bond is acceptable. Adhesives for the most part are clear and no testing, which can be performed in the other processes, is possible.

25.2.1 Adhesive Types

There is a vast amount of adhesives available to bond just about any materials possible. These are broken down into two major categories, which are straight adhesives and epoxy adhesives. The difference between these two categories is as follows. *Straight Adhesives* are those that are applied to the materials directly. The bond is created between the adhesive and the component parts on which it is applied. These require no mixing to accomplish the bond and will either react with the component materials or perform the bond independently. *Epoxy Adhesives* are a thermoset reaction between the elements mixed together. They are a combination of two materials that need to be properly mixed and then applied to the component parts. When mixed, the two elements have a chemical reaction that allows them to create the bond when applied to the component workpieces. Many of the epoxy adhesives require a long curing time, and some require the curing to be done at a set temperature. Some adhesives are flexible and are utilized in an application where the bonded components will be subjected to a flexing situation. Whether the bonding agent is a straight adhesive or an epoxy adhesive, they will be one of three types. These are Natural Adhesives, Inorganic Adhesives, and Synthetic Adhesives, which are explained as follows.

— *Natural Adhesives*: These types of adhesives are derived from natural sources. Bonding agents like rubber cement is a natural adhesive along with others available. The majority of these are widely utilized within the food processing industry where any nonorganic element can cause issues in the processing and canning of the foods being produced.

— *Inorganic Adhesives*: The inorganic types of adhesives are designed for low-strength bonding applications. These can be either a straight or an epoxy adhesive and are utilized in applications where they need to fail when certain forces are applied. This acts very similar to a shear pin in a mechanical assembly to avoid damage to another part of the product by failing first and avoiding a more costly repair elsewhere.

— *Synthetic Adhesives*: Synthetic adhesives are the strongest bonding agents available. They can be either straight or epoxy adhesives and will render the most powerful bonding strengths. For some of the straight adhesives, the curing time is almost instantaneous and produces a permanent bond. For the epoxy category, they too produce a very strong bond but may require a long curing time to fully set up in the joint.

25.2.2 Bonding Types

Depending on the adhesive being used, the type of bond that will occur will be one of four types. Some adhesives will have a reaction with the component workpieces, creating a bond, while others are just the strength of the adhesives alone. The four types of bonds for adhesives are described as follows.

— **Chemical Reaction Bonding:** A chemical reaction bond is created from the adhesive reacting with the component material. This type of bond is strong as the mixing of the adhesive and component material combine and penetrate deeper than the surface, creating the bond. The adhesive can be either straight or epoxy and both will require a substantial curing time for the reaction to complete the bond. Figure 25.2 illustrates a chemical bond where the adhesive reacts with the base material of the component workpieces.

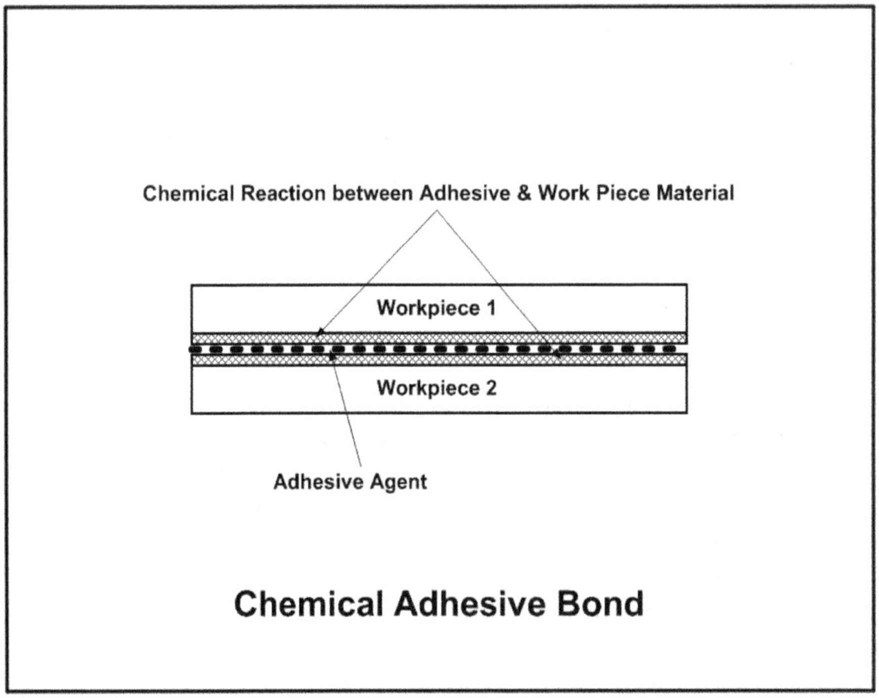

Figure 25.2 Chemical Bonding where a Reaction Occurs between the Adhesive and the Base Materials

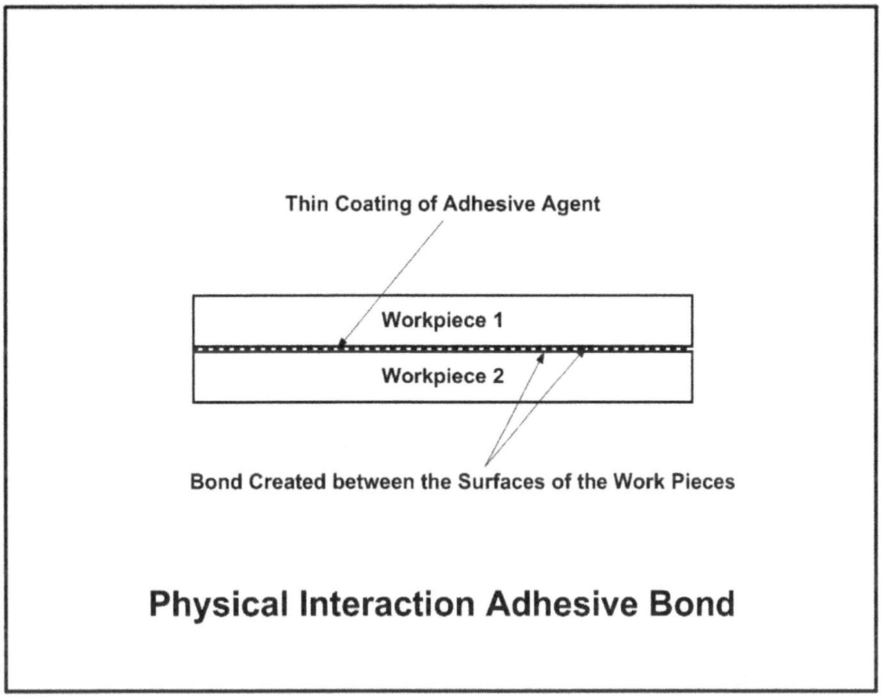

Figure 25.3 Physical Interaction Bonding with the Adhesive Performing the Bonding of the Base Materials

— **Physical Interaction Bonding:** This type of bonding is performed by the adhesive bonding agent alone. The atoms of the bonding agent when cured perform a secondary force between the component workpieces to bond them together. In most cases, a very small clearance is utilized due to the fact that this type of bond is weaker if there is a buildup of adhesive between the components. Many of the synthetic adhesives perform bonding of this type and are normally straight adhesives. Figure 25.3 shows a typical physical reaction bond with a very thin layer of adhesive between the component workpieces.

— **Mechanical Interlock Bonding:** A mechanical interlock bond is one that is a combination of the surface of the component workpieces and the adhesive being applied. The texture of the surfaces on the workpieces can be either extremely rough or fabricated with groves or interlocks, allowing the adhesive bonding agent to be built up between the parts. In this type of bond, the adhesive is a buildup and a heavier amount of the agent is applied. This will connect to both of the surfaces of the component workpieces, creating the bond. The adhesive can be either straight or epoxy for this bonding type, and an extended curing time will be required in order for the adhesive to complete the binding cycle. Figure 25.4 illustrates the mechanical interlock bond showing the surface roughness and the adhesive filling the area as a buildup between the components.

— **Adhesive-Assisted Bonding:** An adhesive-assisted bond is used in mechanical assembly applications. When two components are assembled with either a bolt or a screw, the adhesive is applied to them, offering additional holding power. This is utilized in cases where vibration that the assembly may encounter in operation may cause the bolt or screw to loosen. Figure 25.5 illustrates an assisted bond with the adhesive applied between the screw or bolt.

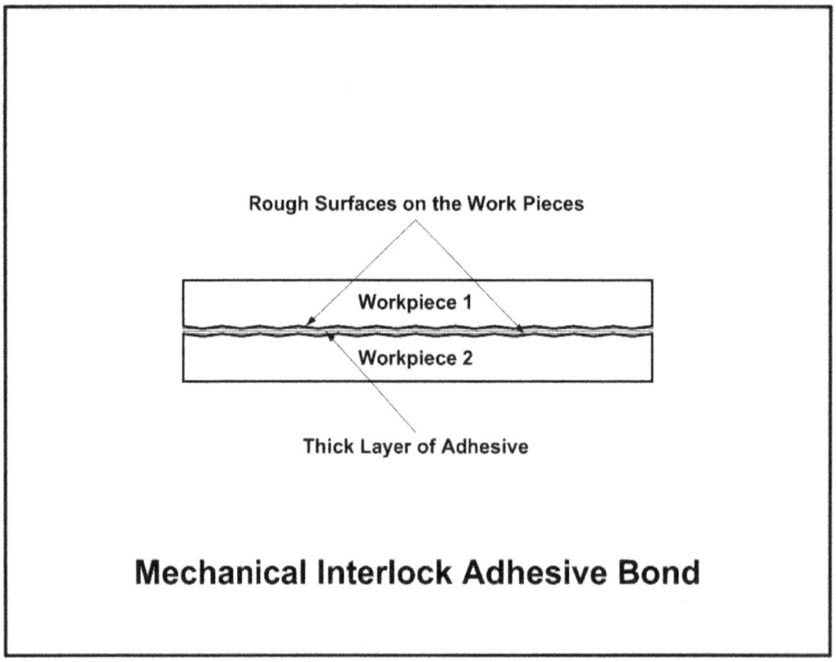

Figure 25.4 Mechanical Interlock Bonding with the Surfaces of the Base Materials Rough to Hold the Adhesive

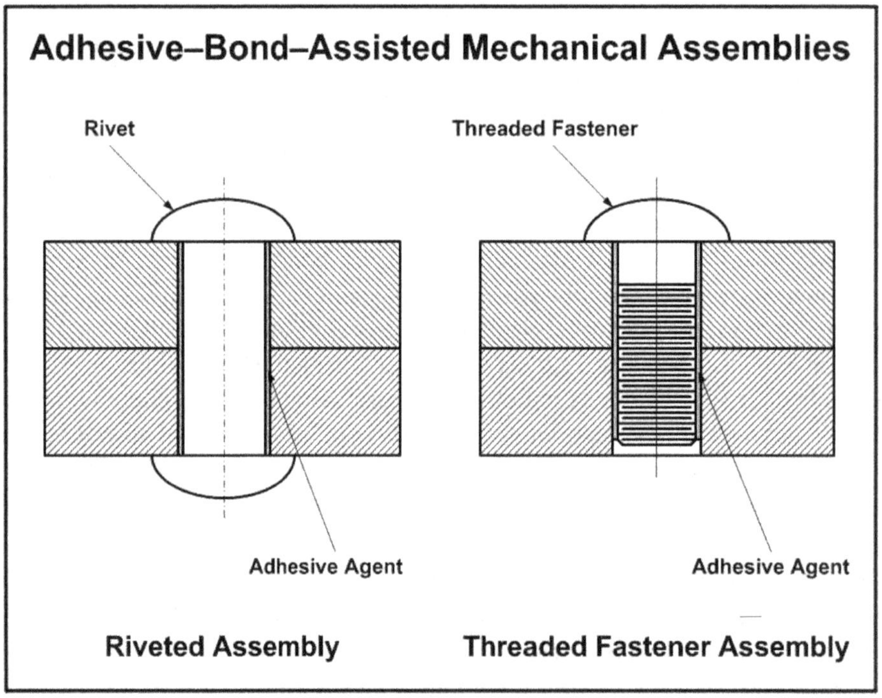

Figure 25.5 Adhesive Agents Utilized to Enhance a Mechanical Assembly

25.3 DESIGN CONSIDERATIONS

Adhesive bonding can be an extremely viable option for joining component workpieces in an assembly. The ease with which an adhesive bond can be performed allows for an extremely simple assembly process and component designing. Understanding the design considerations along with the advantages and drawbacks of adhesive bonding will render a functional design for a product. These are described as follows.

Design Considerations

— *Joining Materials*: Knowing what materials are to be bonded together, and if there may be some type of reaction between them, is critical in order to produce an assembly that will function properly. Some materials when in contact can have a reaction over time, which can cause a failure. Even though adhesives are available to join just about any two different materials together, knowing if there is the potential of some type of negative reaction over time is important in order to avoid a future issue in the product.

— *Joint Design*: Designing for adhesive bonding is possibly the simplest type of assembly design to perform. Allowing for enough area for the adhesive to be applied on is the most critical component of the joint design. By knowing the bonding strength of the adhesive that will be utilized, enough surface area can be implemented into the design for sufficient strength for the assembled components. During the designing of the joint, always allow for the application of the adhesive with ease for the operator performing the bond in assembly.

— *Adhesive Selection*: A major design consideration is the selection of the proper adhesive to accomplish the required bond. No one adhesive will be acceptable for every bond, and an investigation of what adhesives are available is critical for a quality assembly. Understanding the function of the assembly and what conditions it will be exposed to in operation, and if there are any exceptional areas, will allow for the proper selection of the bonding agent for that particular assembly.

— *Design Function*: The function of the assembly is possible the most important factor in the design arena. Areas such as the forces that will be applied to the assembly, the temperature range the bonded joint will be exposed to, and whether or not there will be any flexing in operation need to be considered to avoid any potential failures either instantly or over time.

Advantages

— *Large Bonding Surfaces*: A major advantage is the surface areas that can be bonded. If extremely large components are to have an adhesive bond applied to them, the entire surface can be coated, thus creating a bond across the entire cross section of the assembly.

— *Material Variety*: Adhesive bonding can be performed to join just about any two different materials together. This is an advantage over welding and other applications that require heat to perform the bond.

— *Adhesive Variety*: The variety of adhesives is almost unlimited. There is one for just about any application desired whether it is to be a strong bond or one that is designed to fail at a given force. Adhesives can have extremely strong bonding forces or weaker bonds that can act as a shear pin along with some being flexible, allowing for an almost unlimited amount of possibilities for bonding applications.

— *Sealing Bonds*: Some adhesive bonding agents can create a bond that is both airtight and liquid-tight. For applications that require joints that will have gasses passing through them, a bond can be accomplished to provide a seal that will not leak during the function of the assembly.

— *Simple Designs*: As mentioned in the design considerations, the designing of an adhesive bonded joint is extremely simple. By following the design criteria described earlier, the assembly will be both simple and functional.

— *Curing Temperatures*: As the adhesive bonding agent sets up, the temperature for curing is extremely low. Some, however, do need to be cured at an elevated temperature in an oven, but the temperature requirements are very low.

— *Heat-Affected Areas*: Unlike welding, brazing, or soldering, adhesive bonding will produce no heat-affected areas in the assembly. With no heat being required to bond the components together, there is no potential for the joint to have any effects that can be caused by exposure to heat at all.

— *Component Size Variations*: Along with being able to bond totally different materials together, materials, thickness, and cross sections can vary greatly. Bonding an extremely thin material to an extremely thick material is just as easy as bonding two materials that are of the same thickness.

Drawbacks

— *Joint Strength*: An adhesive bonded joint will not have the strength that can be attained from welding, brazing, or soldering. This type of bond will be the weakest of the mentioned bonds from the other processes. Understanding what forces will be applied to the bonded joint before selecting an adhesive application is critical to produce an acceptable assembly.

— *Difficult Inspections*: Unlike welding, brazing, and soldering, adhesive bonds are extremely difficult to inspect. With the majority of the bonding agents being clear or transparent, nondestructive testing such as X-ray inspection is not possible. Most adhesive joints are bonded and then a destructive test is performed to obtain the results. This proves the design of the assembly but cannot be inspected to ensure that the bond was performed properly before being implemented into the product.

— *Compatible Adhesives*: As mentioned several times, there is an almost unlimited variety of adhesives to select from. No one adhesive will bond everything perfectly, and the selection of the proper bonding agent is critical for a quality assembly. This can lead to having several adhesives in the assembly operation with the potential of selecting the wrong one in the assembly area by error.

— *Service Temperatures*: Adhesives can be affected by the temperature range that the assembly will operate in. Both heat and cold can cause the adhesive to fail and need to be considered before selection an adhesive bond as the assembly option.

— *Surface Conditions*: Surface conditions require more attention that those required for brazing and soldering. Although both of those processes require clean surfaces, adhesive bonding will require surfaces far cleaner. In order for the bonding agent to properly interact with the component workpieces, the surfaces need to be totally void of any contamination matter. This may require an operation to clean the surface with a clean gent first and then allowing the surfaces to dry before the adhesive can be applied.

— *Curing Times*: Some adhesives will require an extended curing time for the bonding agent to completely set up. These can be as long as 24–48 h and need to be considered in the overall assembly times for production.

25.4 SUMMARY

The adhesive bonding process can be a very viable alternative for assembling components. With the variety of adhesive bonding agents available and the ability to bond different materials and components of varying cross sections, this process can offer assembly possibilities that others cannot. Different bonding agents can provide either extremely strong bonds or those that are weaker and flexible, depending on the application of the bonded joint in operation. Care needs to be taken to completely understand the advantages and drawbacks along with the design considerations of adhesive bonding before selection this process for assembly. By following the rules of what adhesive bonding can produce and equally important what it cannot, a quality assembly can be produced very economically.

25.5 QUESTIONS

1. What are the forces that can be applied to an adhesive bond, and which are stronger and why?

2. What are the categories of adhesive bonding agents, and what is the difference between them?

3. What are natural adhesives, and where are they traditionally used?

4. An inorganic adhesive offers a weaker bond than most adhesives. Why is this important, and where would it be used in an assembly?

5. What is a chemical bond, and how is it different from the other bonding types?

6. What is an assisted bond, and why would it be implemented in a mechanical assembly?

7. Why is an adhesive bond difficult to inspect?

CHAPTER

26

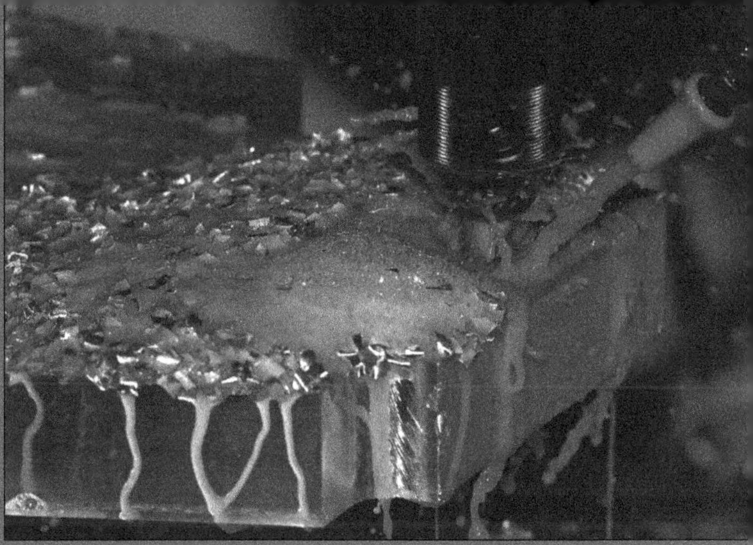

Mechanical Assembly Principles

26.1 INTRODUCTION

A very widely utilized process for assembly involves various mechanical means to join component workpieces together. These assembly processes can be nonpermanent, allowing for disassembly and reassembly over and over easily, allowing the product to be serviced easily. They can be semipermanent where the assembly can be disassembled but not with the ease of a nonpermanent assembly, and some of the components in the assembly will not be reusable after disassembly. Lastly, a permanent assembly can be accomplished, allowing the assembly to remain intact during the life of the product and never intended to be disassembled ever. Depending on the category of assembly being performed, it will utilize either force to join the components such as tightening a bolt or pressing them together or temperature for heating and cooling the components during the assembly. This chapter covers several of the more common mechanical assembly processes utilized in manufacturing along with the design considerations required to produce a quality assembly.

26.2 NONPERMANENT ASSEMBLY PROCESSES

A nonpermanent mechanical assembly is one that is designed where it is intended to be able to be assembled, disassembled, and reassembled. An assembly of this nature is important for any service that will need to be performed on the product where areas of the product can be disassembled in order to replace the worn or defective component. When designing an assembly that will be nonpermanent in nature, the overall function of the product needs to be evaluated for areas that may be exposed to various environmental impacts that could cause it to fail or not function properly. Vibration and alternating forces that would be applied during the operational functioning can loosen the fasteners over time, resulting in a failure.

Two of the more common nonpermanent assembly methods are threaded fasteners and E-clips. Threaded fasteners are the most universal with the ability of attaching both flat and round components, whereas the E-clip fasteners are only utilized on round components. These two methods of mechanical assembly are described as follows.

26.2.1 Threaded Fasteners

Threaded fasteners are probably the most widely utilized means for producing non permanent assemblies in manufacturing. Bolts and screws are available in a large variety of sizes and configurations that can fit just about any design situation possible and are easily replaceable at an extremely low cost. They can be coated for corrosion protection with plating or an oxide finish and can be hardened to provide more strength in the assembled joint. Along with the bolts and screws, support components are available to enhance the assembly and provide more support and strength. The most common thread type utilized on threaded fasteners is the unified thread. This thread type can deliver high holding forces and is readily available for just about any application required. Figure 26.1 shows the construction and terminology for unified threads. These threaded fasteners are also available with both right- and left-hand threads. The standard threads are right-hand threads, but in some special applications where certain forces may loosen a right-hand thread, left-hand-threaded fasteners are utilized to avoid the assembly failing during operation. When selecting the thread type for any screw or bolt, the number of threads per inch (TPI) is important to understand. A fastener with a fine thread will be able to exert a higher force but will have the potential of stripping easier. Likewise, a fastener with a course thread will be able to exert less force but will be more difficult to strip the threads. This is a simple mechanical advantaged

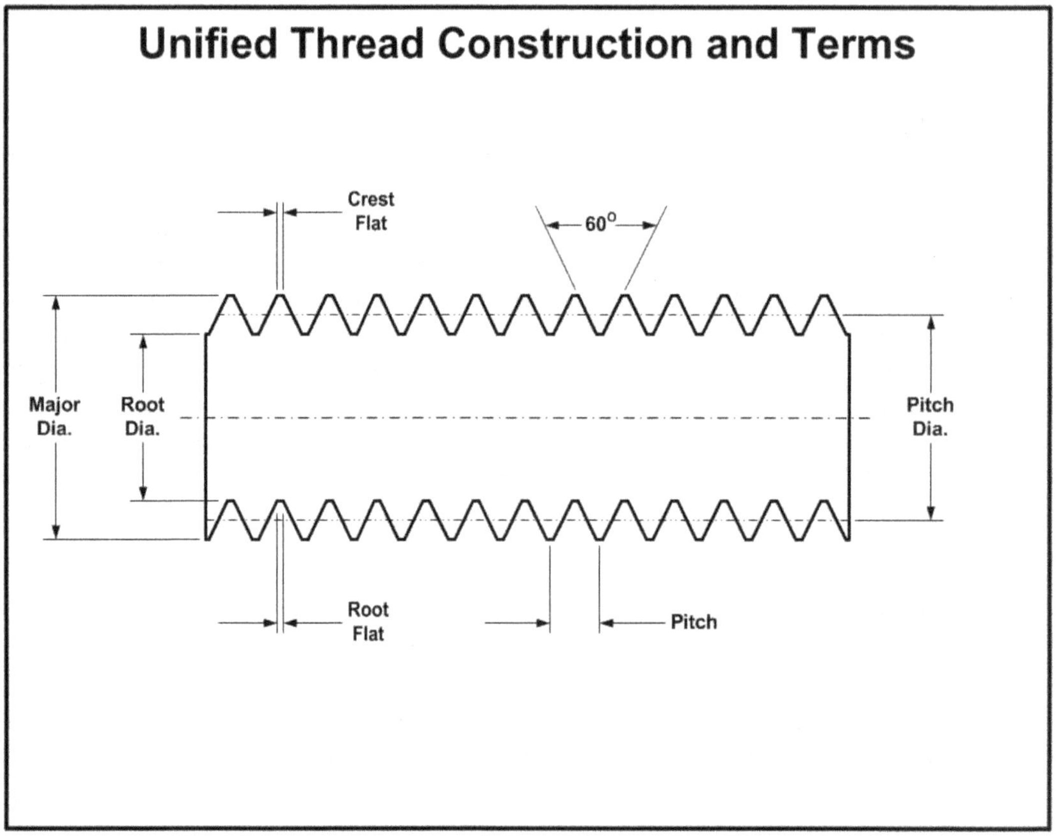

Figure 26.1 Unified Thread Construction and Terminology

calculation to determine the leverage that a given threaded fastener can deliver. For example, a fastener with 32 TIP will deliver a mechanical advantage of 32 – 1, whereas a fastener with 10 TIP will have an advantage of 10 – 1 in relation. Threaded fasteners can fail in the assembly in one of three manners. All the failures of a threaded fastener assembly are the result of applying too much force and torque in the assembly and are listed as follows.

— **External Failure:** The threads on the outside of the screw or bolt are stripped off.

— **Internal Failure:** The threads on the inside of the nut or component workpiece are stripped out.

— **Breakage:** The diameter of the screw or bolt twists apart.

Details of the threaded fastener arena along with the calculations required to ensure the assembly will be functional are given as follows.

26.2.1.1 Screws and Bolts

Both bolts and screws function almost identical in the assembly of the components but are defined differently. Regardless of the size of the threaded fastener, a bolt is a threaded fastener that passes through both component parts having a clearance hole in both and utilizes a nut to tighten and secure them together. A screw will pass through one of the components and thread into the other to secure them together. Figure 26.2 illustrates the difference between a screw assembly and a bolt assembly.

Bolts are available in a variety of grades, which determine the hardness and what forces can be exerted on them. This was developed by the Society of Automotive Engineers (SAE) and is the standard for bolt hardness. The grades range from grade 1, which is the least strong, to grade 8, which is the strongest available. There are gaps in the bolt grades leaving openings for future grades to be developed. Each bolt will have indicator marks on the head to determine the bolt grade for easy identification. A description of the various grades of bolts and their identification markings is outlined as follows.

— **Grade 1 Bolt:** No indicator marks. Grade 1 bolts are made from low-carbon steel and are not heat-treated. These are intended for everyday usage and are the least expensive.

— **Grade 5 Bolt:** Three indicator marks. Grade 5 bolts are made from medium-carbon steel and are heat-treated to increase the hardness for applications requiring higher strength and torque.

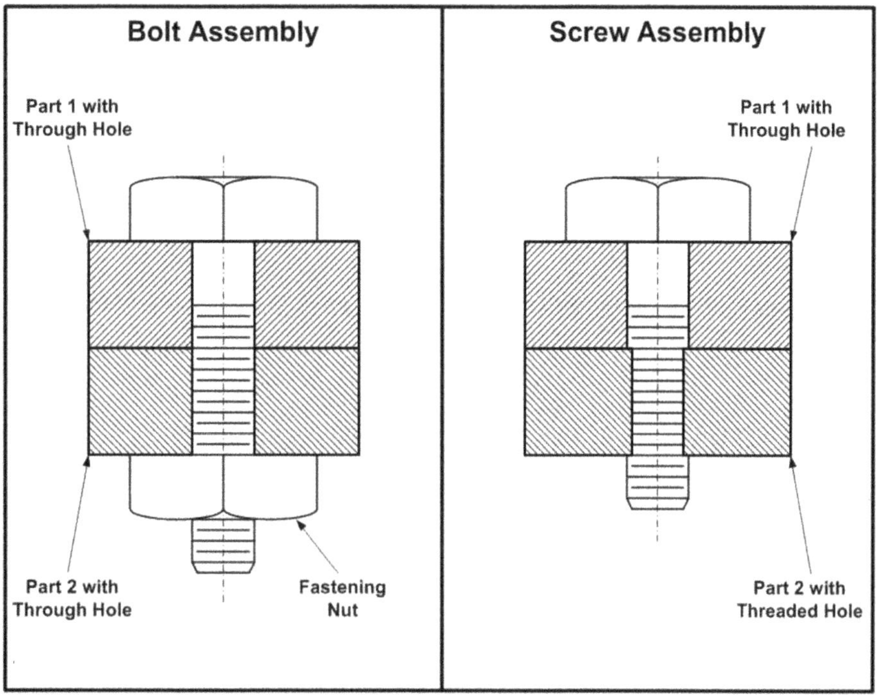

Figure 26.2 Nonpermanent Bolt and Screw Assembly Defined

— **Grade 7 Bolt:** Five indicator marks. Grade 7 bolts are made from medium-carbon alloy steel and are heat-treated to increase the strength. This grade of bolts is harder and stronger than the grade 5 bolts and can deliver higher forces.

— **Grade 8 Bolt:** Six indicator marks. Grade 8 bolts are the highest grade available. They are made of the same medium-carbon steel alloy as the grade 7 bolts and heat-treated to be 15% harder than the grade 7 bolts.

Figure 26.3 shows the indicator marks for the four grades of bolts just described.

In conjunction with the most threaded fastener assemblies, washers are utilized for a variety of reasons. Washers are flat items that are placed between the head of the bolt and the component workpiece and the tightening nut and the component workpiece. Figure 26.4 illustrates how washers are placed between a bolt head and tightening nut in a threaded fastener assembly. Most washers are constructed of steel and can be either coated with a plating or not and are available in a variety of sizes. Some washers are constructed of other materials such as various plastics and will provide a cushion between the bolt and nut against the component workpiece for special applications where any scratching or marring of the workpiece is unacceptable. Other washers such as lock washers are intended to be used to lock the bolt and/or nut in place for applications where vibrations in the assembly can cause loosening. Figure 26.5 shows a flat washer and a lock washer utilized in threaded fastener assemblies.

Washers have several functions in a threaded fastener assembly. A main function is to distribute the force exerted from tightening the bolt/screw against the component workpieces being assembled. They will also receive any scratching and/or marring from the bolt head or nut during tightening to protect the surfaces of the component workpieces, maintaining the surface integrity from damage. Another function is to build up the distance between the head of the bolt/screw or nut in special conditions where space may be an issue in the assembly. For applications where special washers are utilized such as a lock washer to reduce any vibrational loosening of the assembly, they are normally placed between the tightening nut and the flat washer and, when tightened, will exert a force between the two to increase the tension in areas where vibration could cause issues in the assembly.

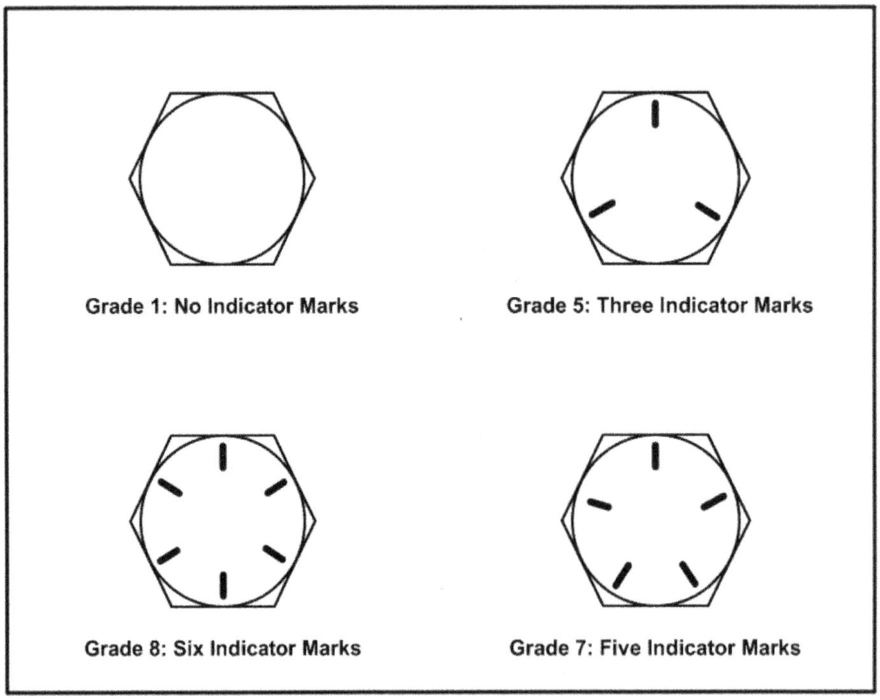

Figure 26.3 SAE Bolt Identification Markings for Grades 1–8

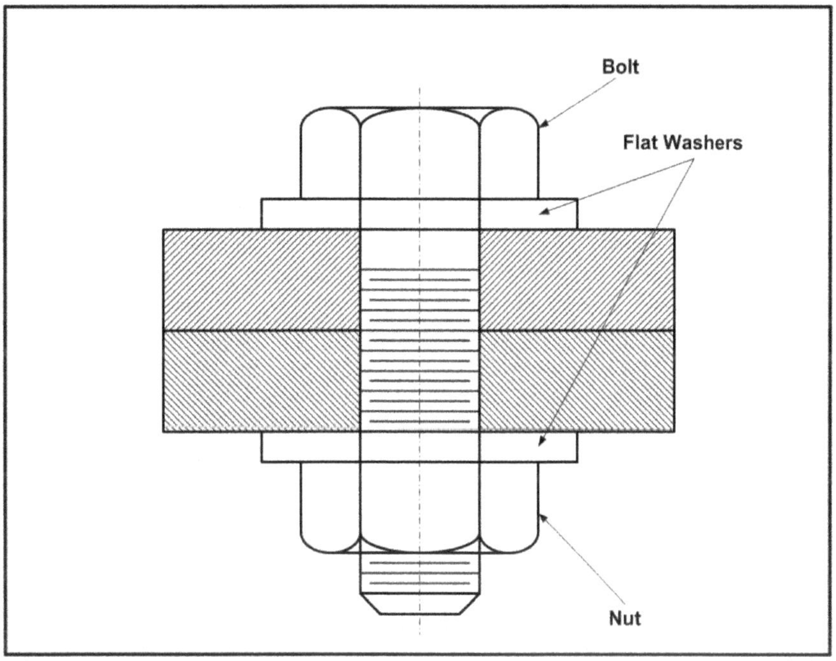

Figure 26.4 Utilization of Washers to Distribute the Holding Forces

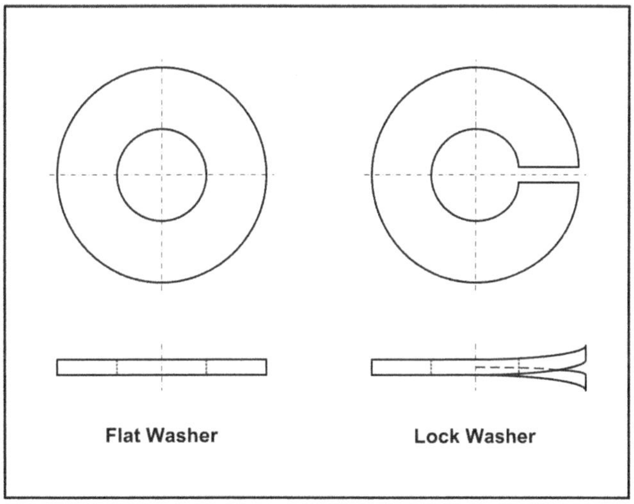

Figure 26.5 Flat and Lock Washers Utilized for Threaded Assemblies

26.2.1.2 Threaded Stud Assemblies

Threaded stud assemblies resemble a combination of a screw assembly and a bolt assembly. The stud is a component that is threaded on both ends and is anchored into one of the component workpieces similar to how a screw threads into the bottom component. Other components are placed over the stud with a through hole and a nut and washer is utilized to tighten the assembly together as in a bolt assembly. Figure 26.6 illustrates how a threaded stud assembly is comprised with the stud anchored into the base component. This will allow for the components to be disassembled similar to a screw or bolt assembly, but the stud is anchored into the base and will not be easily removed. An advantage to this type of threaded fastener assembly is the quick removal of components with the threaded component remaining fixed. Another advantage is if the stack-up of the components being assembled is large, the stud will allow for alignment and if space is a consideration where a long bolt will not be easily implemented, the components can be assembled with only a clearance for the tightening

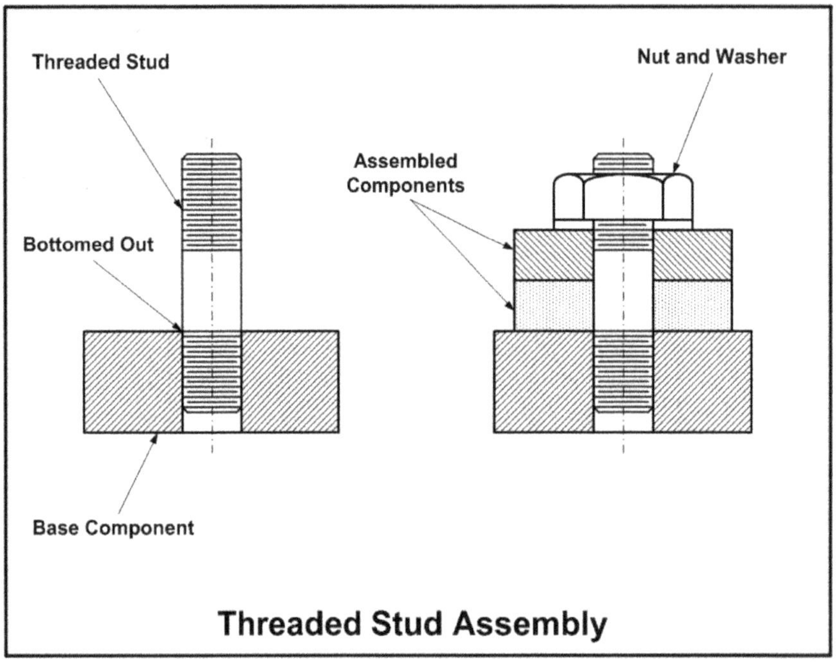

Figure 26.6 Nonpermanent Threaded Stud Assembly

nut required. Studs are available in a multitude of sizes and lengths and follow the same grade scale as do bolts for hardness. They can also have different thread sizes on the ends for special applications requiring a stronger anchor than the tightening forces for the assembly.

Every threaded fastener will be rated for a maximum torque that can be applied to it before a failure will occur. The following equation is utilized for calculating the torque that will be observed in the threaded fastener during assembly to avoid overtightening and failures:

$$T = C_1 \, D \, F$$

T = Torque (in.-lbs)

C_1 = Torque Coefficient (Based on the grade of bolt or material)

F = Preload Force (lbs)

D = Bolt/Screw Diameter

Example Problem

A given threaded fastener assembly will be performed and the torque needs to be calculated in order to not exceed the bolt's maximum requirements. The following criterion for the assembly is: Bolt Diameter (D) = 0.625 in.; Preload Force (F) = 350 pounds; Torque Coefficient (C_1) = 0.83

$$T = C_1 \, D \, F$$
$$T = (0.83)\,(0.625)\,(350)$$
$$T = 181.5625 \text{ in.-lbs}$$

26.2.2 E-Clip Assembly

A common means for attaching a component onto a shaft is by means of an E-clip. This is a device that gets snapped into a groove on a shaft to contain the component being assembled. Figure 26.7 illustrates an E-clip assembly of a component part retained on the shaft. An assembly of this type is utilized to retain the assembled

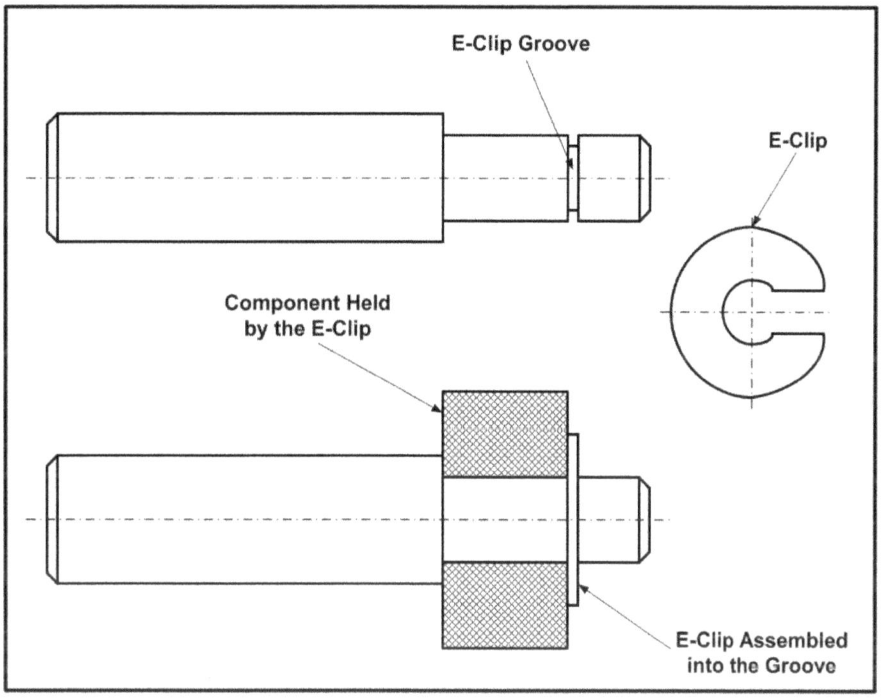

Figure 26.7 Nonpermanent E-Clip Assembly

component onto the shaft in a precise location and can be removed and reassembled easily with a special applicator tool. The advantage or disadvantage with an assembly of this type is that the component retained by the E-clip will not be attached to the shaft in a manner that allows it to rotate and not in a fixed position. This assembly can, however, locate the component in a precise location either against a wall of the shaft or by utilizing an E-clip on both sides.

26.3 SEMIPERMANENT ASSEMBLY PROCESSES

With the semipermanent arena of assembly processes, they are intended to be more permanent in nature but can be disassembled if needed. The issue with disassembly is that either the assembly component or one of the assembled components will be damaged or destroyed in the process. This may involve drilling out the attachment component or the component workpieces in order to remove the assembly. In most cases, if reassembly is required, it will entail either a new attachment component, reworking the component parts, or both. Two very common and widely utilized semipermanent assembly methods are riveting and snap assemblies, which are explained as follows.

26.3.1 Riveted Assemblies

A rivet is an assembly component that has a straight shaft with a head on one end. The shaft passes through the holes in the component workpieces and the straight end is then deformed over to hold the components together. Figure 26.8 illustrates a riveted assembly showing the rivet with a head on one end and deformed on the other, creating the assembly. This type of assembly is mainly utilized to join only two components together, but in some cases, several can be joined in the same manner. In order to disassembly an assembly of this type, the rivet needs to be drilled out carefully in order not to damage the assembled components. Reassembly would require a new rivet and may be difficult to do if there is no access to deform the end required to hold the components together.

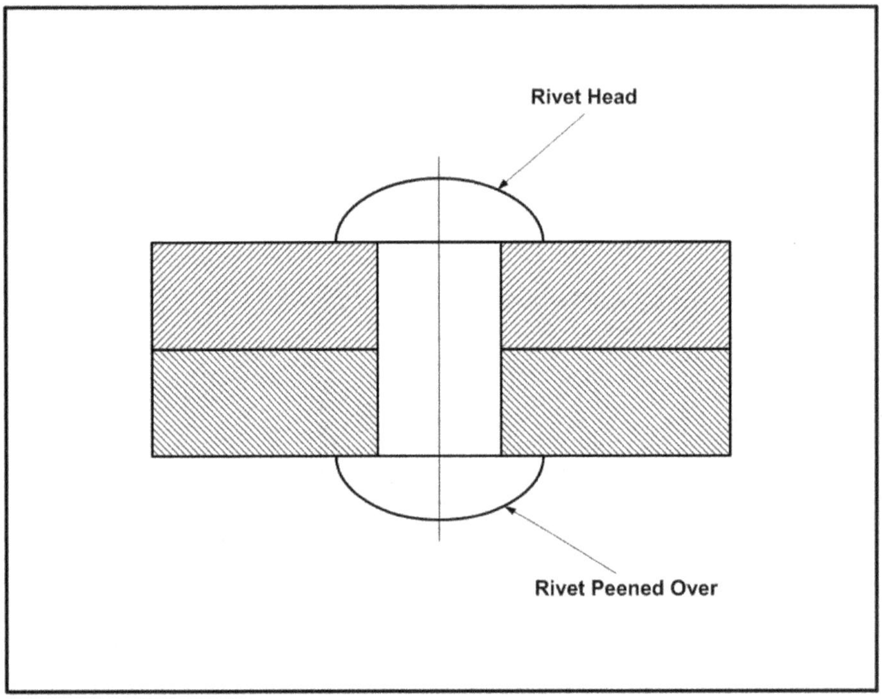

Figure 26.8 Semipermanent Riveted Assembly

26.3.2 Snap-Fit Assemblies

Snap-fit assemblies are most common in the plastics arena. For this type of assembly, the attachment device is part of one of the component workpieces to be assembled. A barbed protrusion is molded into one of the components and the other has a hole with a certain clearance in it. This barbed protrusion needs to have a slot down the center, which will allow it to contract during the assembly process. The barbed protrusion is pressed into the hole until it exits and the barb snaps out, holding the two components together. Figure 26.9 illustrates a snap-fit

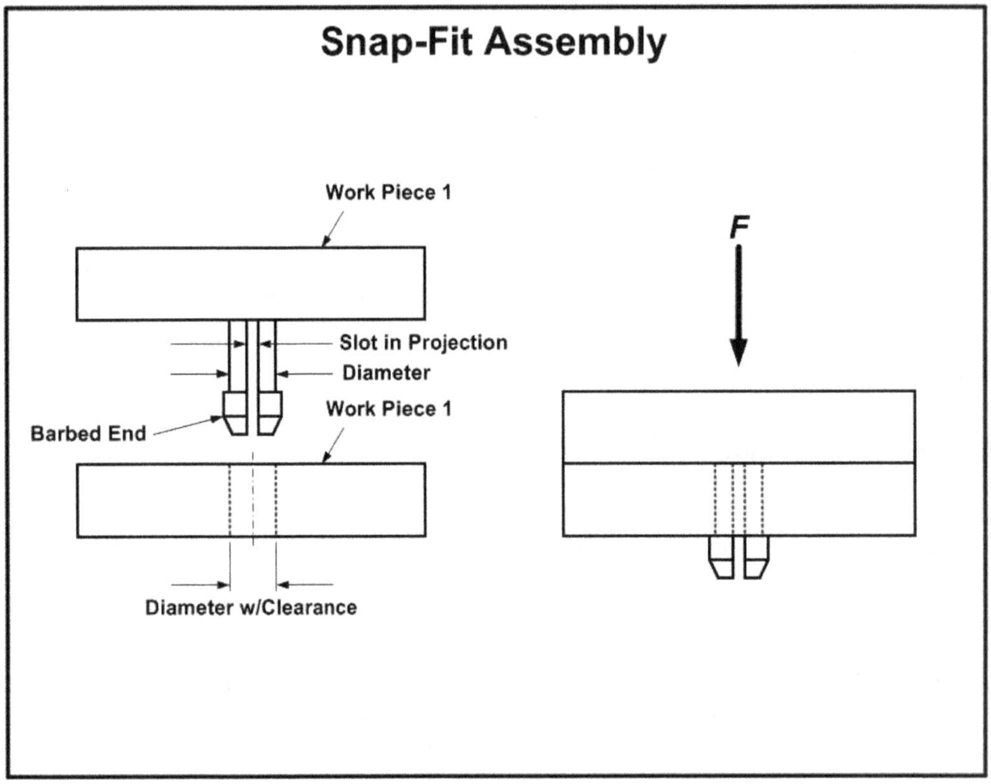

Figure 26.9 Semipermanent Snap-Fit Assembly

assembly showing the barbed protrusion with a slot down the center to allow it to contract when pressed into the hole of the opposing component. If disassembly is required, it can be accomplished in one of two manners. The first is to try and pinch the barbed end to reduce it enough allowing it to repass up through the component. This is extremely difficult especially if the assembly had multiple assembly points as they all need to be done at the same time, and in the process the barbs can get damaged, rendering them useless. The second method is to drill out the barbed protrusions, leaving a hole in the component that once had the attachment devices. Reassembly after this operation would require using some type of threaded fasteners if possible.

26.4 PERMANENT ASSEMBLY PROCESSES

Permanent assemblies are those that are intended to not be taken apart and in doing so, they will be destroyed. These are accomplished by either utilizing force to press them together or applying heat and using the thermal expansion to produce the assembly. An issue with this type of assembly is the stresses that can be building up from two components being joined that have an interference between them. Care needs to be taken to understand how much internal stress the components can accept in order to not have them fail during the assembly process or during the function of the assembly while operating. The three main processes for producing permanent assemblies are interference fits, expansion fits, and shrink fits, which are described as follows.

26.4.1 Interference Fits

This type of assembly is one that requires force to produce the joining to the component workpieces. It is most common on assembling round components together, which are pressed together under force. One component will have a hole and the other component being normally a pin will have a diameter slightly larger than that of the hole. The component to be assembled is then pressed into the hole, and the interference holds the two together in a permanent manner. A pin and collar assembly is the most widely utilized, where a pin is pressed into a component having a thin wall called a collar. Figure 26.10 illustrates how there is an interference between the pin and the hole as they get forced together under pressure. Of the three permanent mechanical assemblies discussed in this section, the interference fit will be the weakest. This is because there can only be a certain amount of interference between the pin and the hole to allow for pressing without causing shearing of the material, which will damage one or both of the component workpieces.

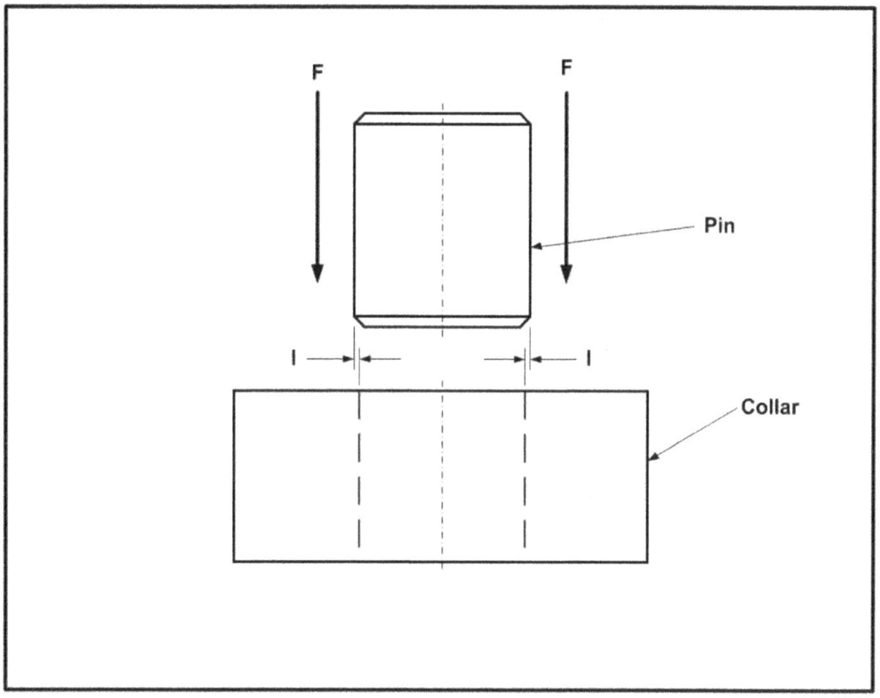

Figure 26.10 Permanent Interference-Fit Assembly

When performing this assembly process, two forces will be exerted on the component workpieces. The first is the pressing force that will be required to push the pin into the hole of the part. This force will be dependent upon the amount of interference existing between the two components and the modulus of elasticity of the material. The second is the radial stress forces that will be built up in the component that the pin is being pressed into. With the pin having a slightly larger diameter than the hole, when it is forces into the hole, it causes stress, which will radiate out from the center of the hole into the component workpiece. Figure 26.11 shows the force exerted while pressing in the pin and the radial stresses, which will be present in the component receiving the pin.

The required pressing force and the radial stress need to be calculated in order to provide enough pressure to push the pin into the hole and to know how much stress will be present in the assemble. If the collar will have to a high stress within it, the walls can crack, resulting in a failure of the assembly. The information required for these calculations are the diameter of the pin to be pressed in, the outside diameter of the collar, the total interference between the hole and the pin, and the modulus of elasticity of the material. Figure 26.12

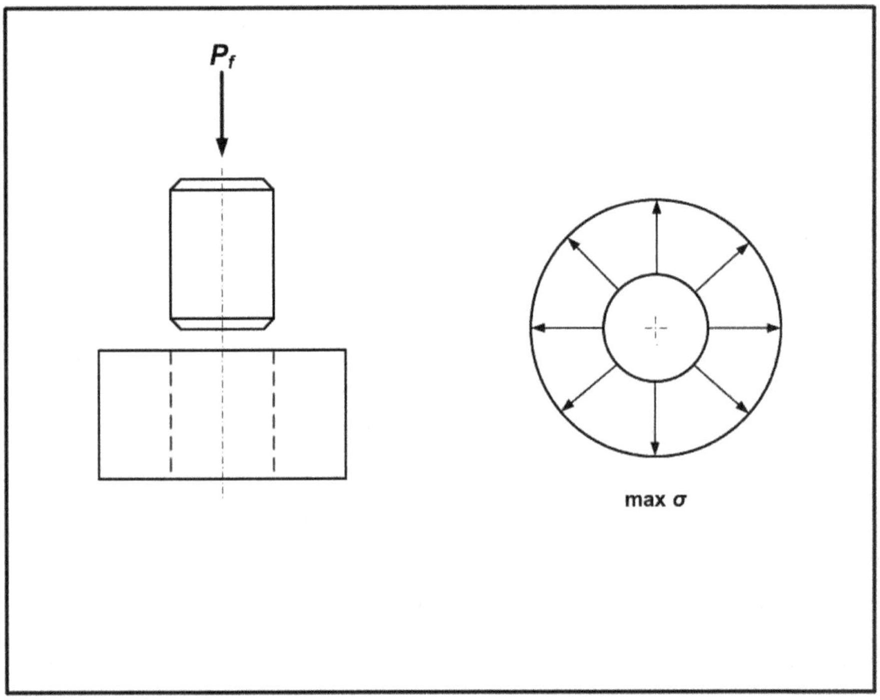

Figure 26.11 Pressing Force and Radial Stress for an Interference-Fit Assembly

Modulus of Elasticity of Selected Materials	
Material	**Modulus of Elasticity Range (PSI)**
Aluminum	9.00×10^6–10.00×10^6
Brass	11.90×10^6–17.00×10^6
Bronze	10.50×10^6–19.80×10^6
Cast Iron	9.14×10^6–34.8×10^6
Chrome–Moly Steel	28.00×10^6–33.90×10^6
Copper	10.40×10^6–21.80×10^6
Low–Carbon Steel	26.50×10^6–30.90×10^6
Magnesium	6.53×10^6
Stainless Steel	11.50×10^6–31.00×10^6
Titanium	28.00×10^6–33.90×10^6

Source: www.matweb.com

Figure 26.12 Modulus of Elasticity Ranges for Some Selected Materials lists the modulus of elasticity of some selected materials. The equations used for these calculations are listed as follows.

Pressing Force

$$P_f = \frac{Ei(D_c^2 - D_p^2)}{D_p(D_c^2)}$$

P_f = Pressing Force (PSI)
E = Modulus of Elasticity (PSI)
i = Interference
D_c = Collar OD
D_p = Pin OD

Radial Stress

$$max\ \sigma = \frac{2P_f D_c^2}{D_c^2 - D_p^2}$$

$max\ \sigma$ = Maximum Radial Stress (PSI)
P_f = Pressing Force (PSI)
D_c = Collar OD
D_p = Pin OD

Example Problem

A pin with an outside diameter of 0.5016 in. is going to be pressed into a collar having an outside diameter of 0.750 in. and a hole diameter of 0.500 in. If the material being used is low-carbon steel and the modulus of elasticity is at the bottom of the range from Figure 26.12, what would the pressing force and radial stress of the assembly be?

Step 1: Calculate the Interference

$$i = 0.5016 - 0.500$$
$$i = 0.0016 \text{ in.}$$

Step 2: Calculate the Pressing Force

$$P_f = \frac{Ei(D_c^2 - D_p^2)}{D_p(D_c^2)}$$

$$P_f = \frac{(26.5 \times 10^6)(0.0016)(0.750^2 - 0.5016^2)}{0.5016(0.750^2)}$$

$$P_f = 46{,}720.0122 \text{ PSI}$$

Step 3: Calculate the Maximum Radial Stress

$$\max \sigma = \frac{2P_f D_c^2}{D_c^2 - D_p^2}$$

$$\max \sigma = \frac{2(46{,}720.0122)(0.750^2)}{0.750^2 - 0.5016^2}$$

$$max\ \sigma = 126{,}794.2584 \text{ PSI}$$

26.4.2 Expansion Fits

Expansion fits are a process where one of the component workpieces is cooled allowing the metal to contract so it will slide into the hole of the other component. When the chilled component returns to the room temperature, it will expand and, with the calculated interference between the two, form a permanent assembly. This will produce an assembly stronger than an interference fit because with the contraction of the material being assembled, a larger interference can be realized. For chilling the component to be assembled into the hole, the most common method is utilizing liquid nitrogen. This will be at a temperature of approximately −350°F and will cause the material to contract based on the coefficient of linear thermal expansion as shown in Figure 26.13 material expansion. The equation that is used to calculate the contraction of the pin when chilled is:

$$D_2 = D_1 + (D_1 \alpha (T_2 - T_1))$$

D_1 = Starting Diameter at Room Temperature
D_2 = Diameter at the Chilled Temperature
T_1 = Start Temperature (°F)
T_2 = Temperature the Pin is Chilled to (°F)
α = Coefficient of Linear Thermal Expansion (in./in./°F)

When constructing the components for an assembly of this nature, the hole will be inserted at the nominal size, which would be D_1 and all of the interference will be implemented into the pin. This is done to be able to utilize a standard drill for the hole while the pin being assembled can be generated at any desired size. Also, a safety factor for the clearance is implemented to avoid the pin only going part way into the hole and not fully. If the pin does not go fully into the hole before it begins to expand, there is no way to remove it and the assembly will be defective and have to be discarded. Figure 26.14 illustrates an expansion fit with the component containing the hole remaining at room temperature and the pin being chilled to contract the diameter.

Coefficients of Thermal Expansion

Material	in./in./°F
Aluminum	0.0000133–0.0000152
Brass	0.0000100–0.0000144
Bronze	0.00000889–0.0000144
Carbon Steel	0.00000611–0.00000922
Cast Iron	0.00000431–0.0000107
Chrom-Moly Steel	0.00000389–0.00000772
Cobalt	0.00000694–0.00000789
Copper	0.0000089–0.0000122
Magnesuim	0.00000767–0.0000156
Stainless Steel	0.00000511–0.00000933
Titanium	0.00000494–0.00000561
Zinc	0.0000108–0.0000222

Source: www.matweb.com

Figure 26.13 Linear Thermal Expansion for Selected Materials

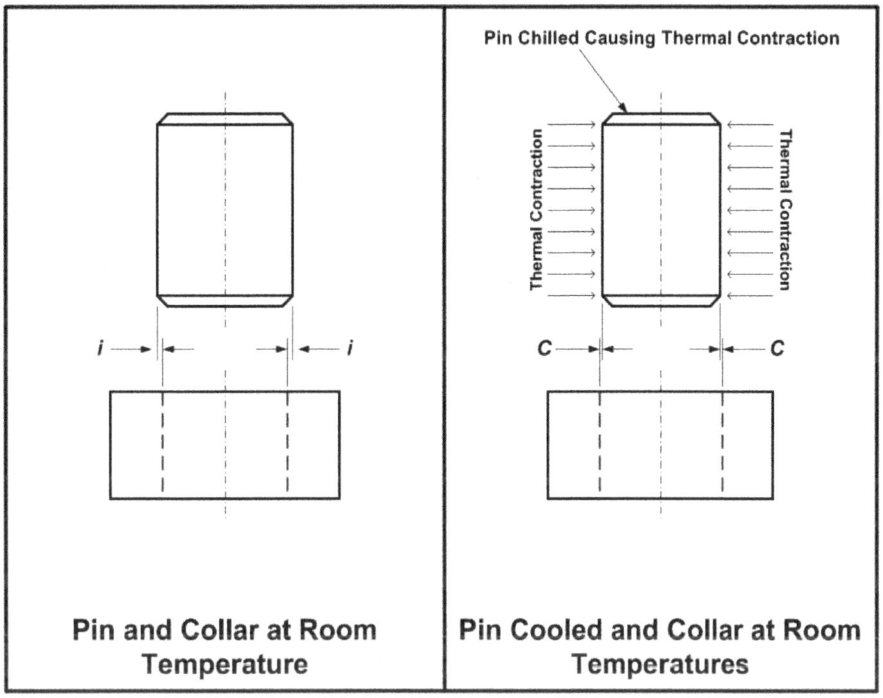

Figure 26.14 Permanent Expansion-Fit Assembly

Example Problem

An expansion-fit assembly is to be performed with both components being constructed of aluminum material. The diameter of the hole is held at 0.625 in., and the room temperature is 68°F. The pin will be chilled to –300°F, and the coefficient of thermal expansion for the assembly will be in the middle of the range for the aluminum material shown in Figure 26.14. If a clearance safety factor of 15% will be utilized, what would the starting size of the pin need to be at room temperature?

Step 1: Calculate the Contraction of the Chilled Pin

$$D_2 = D_1 + (D_1 \alpha (T_2 - T_1))$$
$$D_2 = 0.625 + (0.625)\,(0.00001425)\,(-300-68)$$
$$D_2 = 0.6127 \text{ in.}$$

Step 2: Calculate the Total Interference Realized

$$i = 0.625 - 0.6217$$
$$i = 0.0033 \text{ in.}$$

Step 3: Calculate the Clearance Safety Factor

$$\text{Clearance} = (i)\,(0.15)$$
$$\text{Clearance} = (0.0033)\,(0.15)$$
$$\text{Clearance} = 0.0005 \text{ in.}$$

Step 4: Calculate the Starting Pin Size

$$\text{Pin Dia.} = D_1 + (i - \text{Clearance})$$
$$\text{Pin Dia.} = 0.625 + (0.0033 - 0.0005)$$
$$\text{Pin Dia.} = 0.6278 \text{ in.}$$

26.4.3 Shrink Fits

A shrink fit is very similar to an expansion fit except that the pin will be chilled to allow the material to contract and the component workpiece containing the hole will be heated allowing it to expand. This process will produce the strongest permanent assembly possible because of the extreme interference that can be realized. The equation that is used to calculate the contraction of the pin and the expansion of the hole is:

$$D_2 = D_1 + (D_1 \alpha (T_2 - T_1))$$

D_1 = Starting Diameter at Room Temperature
D_2 = Diameter at the Chilled or Heated Temperature
T_1 = Start Temperature (°F)
T_2 = Temperature the Pin or Hole is Chilled or Heated to (°F)
α = Coefficient of Linear Thermal Expansion (in./in./°F)

As with the expansion-fit process, the hole will be inserted at the nominal size, which would be D_1 and all of the interference will be implemented into the pin. This is done to be able to utilize a standard drill for the hole while the pin being assembled can be generated at any desired size. Also, a safety factor for the clearance is implemented to avoid the pin only going part way into the hole and not fully. If the pin does not go fully into the hole before it begins to expand, there is no way to remove it and the assembly will be defective and have to be discarded. Figure 26.15 illustrates a shrink fit with the pin being chilled to contract the diameter and the component workpiece being heated to expand the diameter.

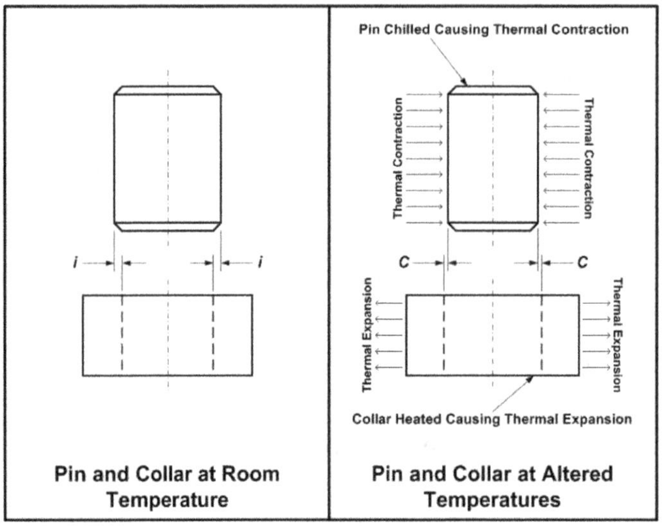

Figure 26.15 Permanent Shrink-Fit Assembly

Example Problem

An expansion-fit assembly is planned for two component workpieces made from carbon steel. The starting diameter will be 0.750 in., and the component containing the hole will be heated to 1450°F while the pin component will be chilled to −275°F. If a coefficient of thermal expansion in the middle of the range for carbon steel is used from Figure 26.14 and a clearance safety factor of 10% is used, what would the starting pin size need to be?

Step 1: Calculate the Contraction of the Chilled Pin

$$D_{2pin} = D_1 + (D_1 \alpha (T_2 - T_1))$$
$$D_{2pin} = 0.750 + (0.750)\ (0.000007665)\ (-275 - 72)$$
$$D_{2pin} = 0.7480 \text{ in.}$$

Step 2: Calculate the Expansion of the Hole

$$D_{2hole} = D_1 + (D_1 \alpha (T_2 - T_1))$$
$$D_{2hole} = 0.750 + (0.750)(0.000007665)(1450 - 72)$$
$$D_{2hole} = 0.7579 \text{ in.}$$

Step 3: Calculate the Total Interference

$$i = 0.7579 - 0.7480$$
$$i = 0.0099 \text{ in.}$$

Step 4: Calculate the Clearance Safety Factor

$$\text{Clearance} = (i)(.10)$$
$$\text{Clcarancc} = (0.0099)(0.10)$$
$$\text{Clearance} = 0.0010 \text{ in.}$$

Step 5: Calculate the Starting Pin Size

$$\text{Pin Dia.} = D_1 + (i - \text{Clearance})$$
$$\text{Pin Dia.} = 0.750 + (0.0099 - 0.0010)$$
$$\text{Pin Dia.} = 0.7589 \text{ in.}$$

26.5 DESIGN CONSIDERATIONS

When designing a mechanical assembly, the criterion is very simple. This is mainly for nonpermanent assemblies, which utilize threaded fasteners alone with the designs for the component parts that will be assembled in the process.

— **Standardize Fasteners:** By doing everything possible to standardize the threaded fasteners and E-clips that will be used in the assembly will eliminate the possibility of the wrong fastener being assembled. Also, make every attempt to reduce the number of threaded fasteners required in the overall assembly. This is for two reasons. The first is, with fewer fasteners, there will be less possibility of missing one in the assembly or using the wrong one. Second, it will require more time to assemble more fasteners, thus costing the operation more.

— **Fewest Parts Possible:** By reducing the assembly to the least number of components in the design, it will require less time to assemble, thus costing less to produce. A complex assembly consisting of several parts will also have a higher probability of assembly errors in the process.

— **Reduce Orientations of Components:** Component workpieces that are no symmetrical in the design will require the operator to orientate them during assembly. This requires more time than parts that are symmetrical, which can be assembled in any configuration and still be correct. There will also be a higher probability of an incorrect assembly with nonsymmetrical components resulting in a faulty assembly. This could be as simple as adding a second hole in the component so that it can be assembled in whichever orientation that the operator picks it from the pile at the assembly station.

— **Avoid Parts that Tangle and Shingle:** Components that tangle or shingle cause issues in assembly. The assembly operator will have to untangle or unshingle the part at the point of assembly, resulting in a longer assembly time and higher costs.

26.6 SUMMARY

Mechanical assembly processes are very widely utilized in product assembly and can be one of the most economical methods. With the options for nonpermanent, semipermanent, and permanent assemblies, there will be an assembly process that will fit just about every requirement in a product. Understanding what each type of mechanical assembly process can provide whether disassembly will ever be required is important in order to not have issues later if service is not able to remove the required components when needed. For assemblies that will be in the permanent arena, the amount of interference between the two components being assembled needs to be precisely calculated. This category of assembly processes will also have residual stresses, which can cause failure issues in the assembly if not properly compensated for based on the materials being used, the size of the components assembled, and the function of the assembly in the product. By a good understanding of that this category of assembly processes can yield and also not produce, the proper process can be selected for the assembly design.

26.7 QUESTIONS AND PROBLEMS

Questions

1. What is the difference between a nonpermanent, semipermanent, and permanent mechanical assembly, and where would each be used?

2. What are the two most common nonpermanent assembly processes utilized?

3. Threaded fasteners utilize one thread type the most. Which thread type is that?

4. Define the difference between a bolt assembly and a screw assembly.

5. Why does a fine thread have more tightening force than a coarse thread?

6. Why can a fine thread strip easier than a coarse thread?

7. What is the difference between a flat washer and a lock washer, and what are the uses for each in a threaded assembly?

8. What are the grades for bolts, and define the difference between the grades.

9. Where are E-clip assemblies used, and why don't they keep the assembled component stationary in the location?

10. What are the two common semipermanent assemblies, and what can happen to each if disassembly is required?

11. Define what an interference-fit assembly is with a pin and a collar.

12. Why is there radial stress in an interference fit, and what can happen if it is too high?

13. What is the difference between an expansion fit and a shrink fit?

14. What precautions need to be calculated when performing an expansion fit or a shrink fit to ensure that the components assemble completely?

Problems

1. A screw is to be tightened into an assembly, which requires a torque of 110.7 in.-lbs. If the screw diameter is 0.250 in. and the torque coefficient is 0.62, what tightening force would be required?

2. A bolted assembly requires that the torque be between 140 and 160 in.-lbs. If the bolt diameter is 0.750 in. with a torque coefficient of 0.94, what is the range that the tightening force needs to be?

3. Two different screws are being considered for a threaded assembly. Both will be 0.500 in. in diameter and will be tightened to 450 PSI. What would the torque range be if one screw had a torque coefficient of 0.82 and the other had a coefficient of 0.76?

4. Two different bolts are being considered in an assembly. One has a diameter of 0.4375 in., and the other has a diameter of 0.375 in. A tightening force of 275 PSI will be used, and the torque requirement for the assembly is 101 in-lbs. What would the torque coefficient for each bolt need to be to meet the torque requirements?

5. An interference fit is planned for an assembly. The collar has an outside diameter of 1.125 in. with a hole diameter of 0.625 in. The pin that will be pressed into it has a diameter of 0.6258 in., and both components are constructed of brass material. If a modulus of elasticity from Figure 26.12 at the top of the range was used, what would the required pressing force be and how much redial stress would remain in the assembly?

6. A pin is to be pressed into a collar for an interference fit, and the diameter of the pin can range from 0.5007 to 0.5011 in. With the collar diameter being 0.875 in., both components are made from low-carbon steel. If a modulus of elasticity in the middle of the range for low-carbon steel is used, what would the difference in pressing force and radial stress be for both pin sizes?

7. A pin with a diameter of 0.250 in. is going to be assembled with an interference fit into a collar with an outside diameter of 0.375 in. Both components are fabricated of stainless steel material and with a modulus of elasticity in the middle of the range. If the maximum radial stress allowed in the assembly is 144,000 PSI, what would the interference need to be between the pin and hole?

8. A expansion-fit process is going to be utilized to assemble a pin into a hole in a block, which has a diameter of 1.125 in. The starting temperature is 70°F in the room, and the pin will be chilled in liquid nitrogen to a temperature of −225°F. The pin material is aluminum, and the coefficient of thermal expansion used will be at the top of the range shown in Figure 26.13. If a safety clearance of 25% is required, what should the size of the pin be manufactured to?

9. An expansion fit will be used to assemble a pin into a block with a hole size of 0.375 in. The starting room temperature is 65°F, and the pin will be constructed of carbon steel. For the requirements of the assembly, the pin requires a starting diameter range of 0.3741–0.3738 in. with no safety clearance factor. What would the temperature range that the pin needs to be chilled to in order to meet these requirements?

10. A shrink-fit process will be used to assemble a pin into a collar with a hole diameter of 0.9375 in. The starting room temperature is 72°F, and the collar will be heated to a temperature of 1,375°F. The pin will be chilled to a temperature of –325°F, and both components are fabricated from cobalt material. If a coefficient of thermal expansion in the middle of the range for cobalt is utilized, what would the diameter of the pin need to be if a 15% clearance safety factor is required?

11. A shrink-fit process is to be used to assemble a pin into a block, which has a 0.375 in. diameter hole in the center. The block is constructed of cast iron, and the pin will be fabricated from titanium. With a starting room temperature of 72°F, the block will be heated to a temperature of 1,725°F, while the pin is going to be chilled to a temperature of –275°F. The coefficient of thermal expansion for the cast iron block used will be in the middle of the range, while the titanium pin will use the top of the range. If a 20% clearance safety factor is required, what should the starting diameter of the pin be?

12. The shrink-fit process will be utilized to assemble a zinc pin into a copper block, which has a hole with a diameter of 0.750 in. With a room temperature of 65°F, the block will be heated to a temperature range of between 1,250 and 1,375°F, while the pin will be chilled to a temperature range of between –250 and –325°F. Utilizing a 15% clearance safety factor, what would the starting pin diameter range be and what is the delta within the range?

SECTION

8

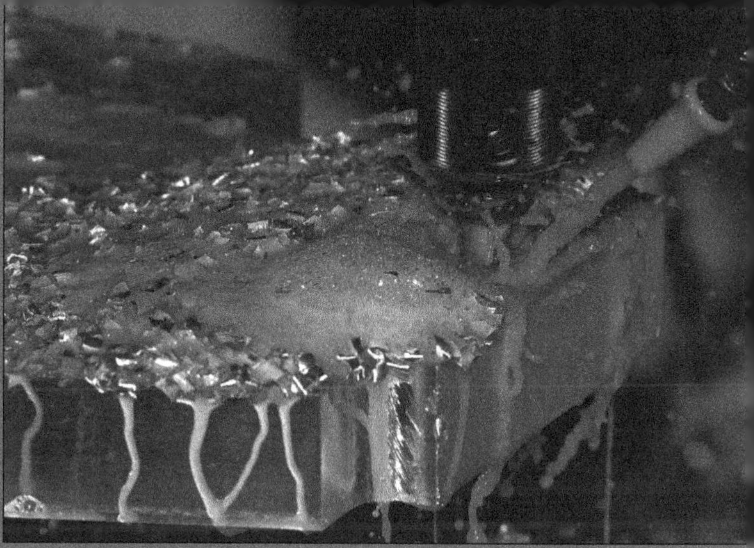

MANUFACTURING CONTROL AND IMPROVEMENT TECHNIQUES

In the manufacturing arena, controlling the processes and being able to predict what may possibly occur is important for producing high-quality products consistently. This involves monitoring the process and implementing improvements and indicators in order to be proactive rather than reactive within the operation. A major area in the process monitoring arena is quality. By simply inspecting the component workpieces after they have been fabricated to confirm if they were produced to the required dimensions is not acceptable in a world-class manufacturing facility. However, by implementing quality indicators for the process and monitoring the output during the fabrication will allow for instantaneous information and avoid the potential of producing a full run of parts that do not meet the requirements. This is accomplished by utilizing what is called Statistical Process Control (SPC) where evaluations are performed during the production run to obtain a "Real-Time" picture of what is being produced. This will also allow the ability to predict what could happen over time, thus allowing the operation to be adjusted before defective components are produced.

Implementing improvements within the operation is a constant, ongoing endeavor. These improvements can be directly related to the component fabrication such as higher quality or faster processing time, which results in lower component cost. They can also be indirectly related to the workpieces in areas like reducing setup times for the equipment. When setup times are reduced, the equipment will be available for production more, resulting in an increase of equipment utilization.

This section looks at areas involving improving and monitoring quality along with the implementation of various automation techniques to improve throughput. Areas of SPC and numerical control (NC) and how they are implemented and the benefits of each are discussed.

CHAPTER

27

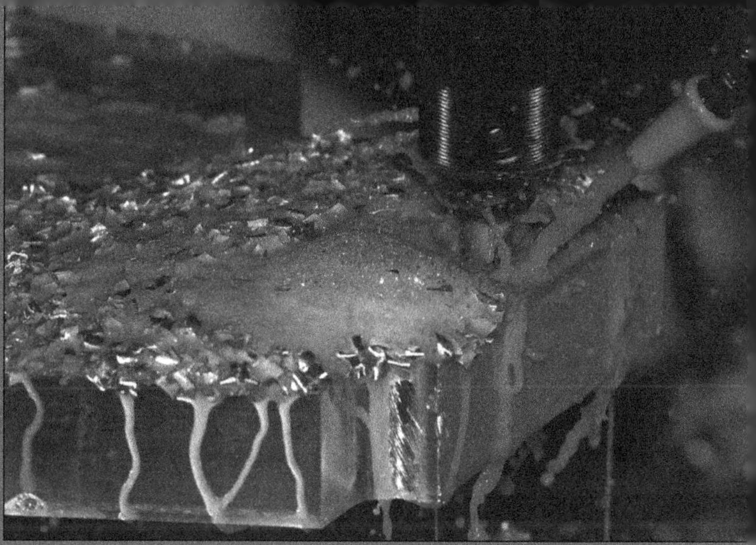

Quality Control Principles and Techniques

27.1 INTRODUCTION

Quality is the main ingredient in manufacturing to produce components and products. This begins with the raw materials being utilized in the fabrication, controlling the processes making them, and the assembly of the completed product. Confirming the quality levels can be performed in several ways depending on the conditions required. From inspection of components to controlling the quality as they are being fabricated, the end results will only produce a product that will be the best possible. This chapter explains quality and the various techniques utilized in the manufacturing arena to assure that everything is being maintained to the highest levels.

27.2 QUALITY DEFINED

Quality is defined as how something measures up to the design and functional requirements for a desired item or product. This can be measured on both a micro level such as an individual component part or product and a macro level like an entire organization. With manufacturing being the topic, the quality focus here will be on products produced and the processes used to produce them. In manufacturing, all the attention needs to be applied to the manufactured product quality. This is explained as follows.

Manufactured Product Quality

— *Product Features*: The features of a product are mainly the result of the design. These features are requested from marketing and built into the product design. Manufacturing has the responsibility of implementing these features into the product in a high-quality manner.

— *Defect-Free Products*: Producing a product with no defects is contingent on the overall quality from start to finish. No customer wants to purchase a product that either does not function properly or will break during operation. This requires a very tight control of the processes from start to finish to product a defect-free product for the customers.

27.2.1 Accuracy vs Precision

Chapter 4 on metrology discussed the difference between accuracy and precision. As a review, accuracy is how the measured values compare with the absolute value, while precision is the amount of repeatability. Figure 27.1 shows the difference between accuracy and precision in a firearm test. This difference is important to understand when measuring process quality within the operation. With a process that is producing precision, any alterations can be adjusted for to bring the components into the center of the range. With accuracy, the spread is so far apart that it may be impossible to adjust the process to produce consistent acceptable components. Repeatability is the key to having a process that is in control, and this is covered later in the chapter.

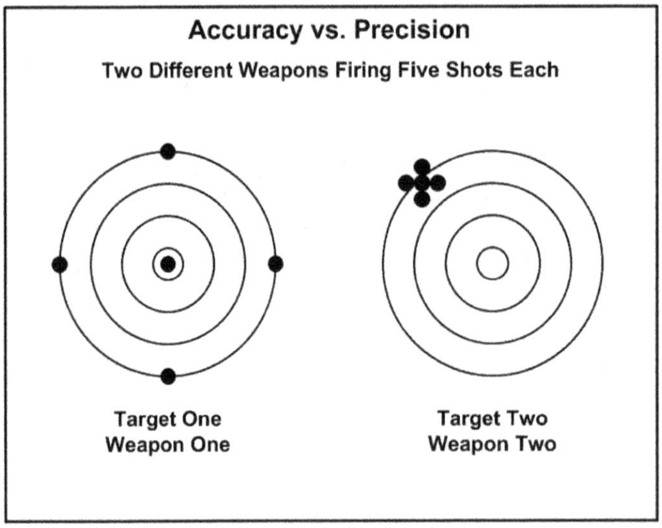

Figure 27.1 Accuracy vs. Precision Example of a Firearm

27.2.2 Total Quality Management (TQM)

The concept of total quality management (TQM) is one that not only deals with the products being produced and the processes fabricating them but also reaches completely through the organization from the highest to the lowest levels. This concept was developed to ensure that a facility will be totally dedicated to every aspect of quality across each function and from all employees within the organization. The structure for TQM is divided into five categories, which are called the five pillars of quality from Bill Creechs' book and are explained as follows.

Five Pillars of Quality

1. *Products*: As mentioned earlier, the product quality needs to include features that the customer desires and be free of defects. This is within the design of the product but needs to be designed in a way that manufacturing can provide the product to the customers and be able to ensure the total quality.
2. *Processes*: Controlling the processes, which are fabricating the components or inspecting components and raw materials being received into the facility, is critical to building a quality product. By implementing proactive measures from start to finish in the manufacturing areas is the key to ensuring that the processes being utilized will produce precise, high-quality components, which will result in a quality product. This chapter deals with the process quality and expounds on the various techniques utilized.
3. *Organization*: The organization of a company needs to be structured in a manner such that it can respond to the needs of the customers and the direction of the marketplace. Not only having the proper number of employees, but also the quality of the employees is equally critical. This includes not only the ability of the individuals but also their attitudes and desire to do the very best job possible.

4. *Leadership*: Quality leadership ensures that the operation will be moving in the proper direction and market requirements change. This requires good vision to the direction the market is taking along with establishing a firm mission of the company in order to be a leader in the marketplace.
5. *Commitment*: Although listed last in the five pillars, commitment is the glue that holds the other four points together. Everyone within the organization needs to be committed to the overall quality within the organization. From the lowest level employee to the COE of the company, quality needs to be their number one focus. Customer satisfaction is the most critical attribute that a company can have and should be the first priority of everyone in the organization.

Figure 27.2 illustrates the TQM structure of an organization where every area is connected in a closed-loop arrangement. In this chapter, we focus on the quality of the manufacturing processes and how to ensure that from starting with the raw materials through the final product, each process will have the proper checks and balances to control every facet, ensuring a high-quality product at the end.

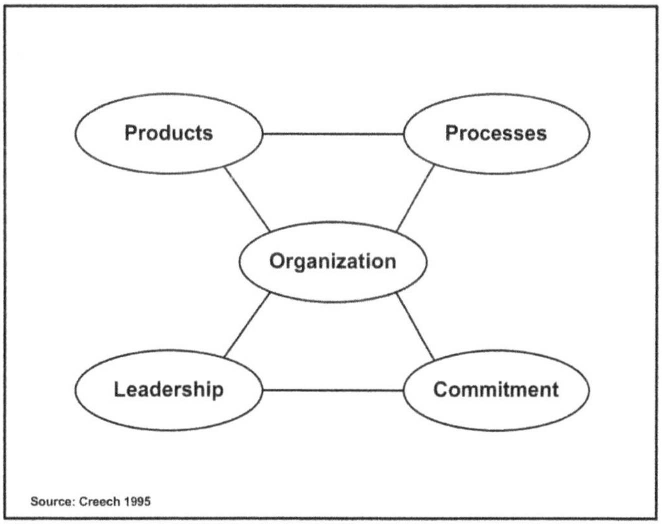

Figure 27.2 Total Quality Management (TQM) Structure

27.2.3 Statistical Distributions

Statistics plays the major role in the process quality arena. This will allow the ability to both control the process while it is fabricating the components and be able to predict when an adjustment to the process will be required. Also, statistical analysis is utilized to inspect components previously manufactured to evaluate what defect rates if any can be expected. This concept is based on a normal statistical distribution or more commonly referred to as a bell curve. It is assumed that if enough samples of anything are taken, the distribution of the items will fall within a normal statistical distribution, allowing this tool to be utilized for a multitude of analysis to predict values and predict future outcomes. Figure 27.3 shows a normal statistical distribution which is divided into ±3 standard deviations (σ) determining the area under the distribution. The determination of the standard deviations is covered later in the chapter, but it is a value based on the number of readings taken, the average value of all the samples, and the spread the values have within the lot. The distribution runs along the bottom axis in both directions but never comes in contact with it. This allows for analysis to the smallest increments for extremely precision values.

The area under the distribution is all dependent on the standard deviation, which will be calculated and, when moved out on both sides from the center, will show what percentage of the items will fall within a given boundary. For most analysis, using $\pm 3\sigma$ will be sufficient, which will render the following areas under the distribution.

$$\pm 1\sigma\text{: } 68.27\%$$
$$\pm 2\sigma\text{: } 95.45\%$$
$$\pm 3\sigma\text{: } 99.73\%$$

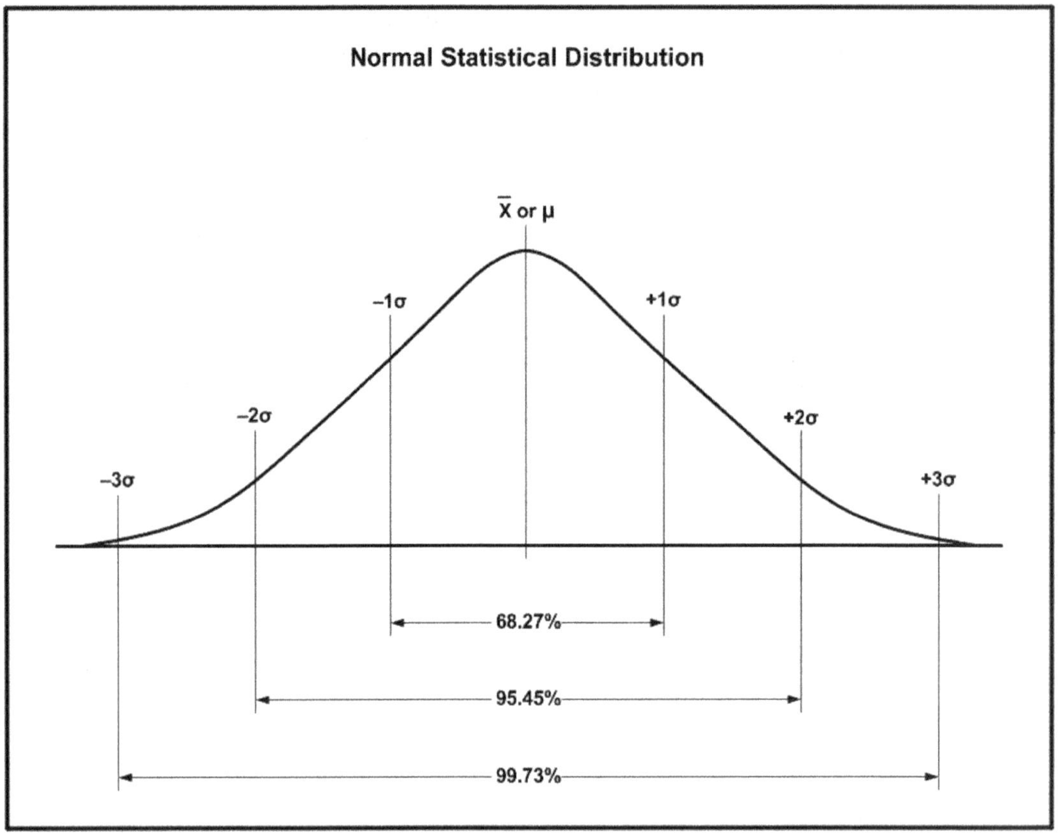

Figure 27.3 Normal Statistical Distribution Showing ±3σ Limits

Figure 27.4 shows a table that, once the value of the standard deviation is determined, can be used to break down the area under the distribution to X.XX standard deviations.

Example

If the average value of a lot of components was 1.250 with a standard deviation of 0.000924, the following percentage of components within the distribution will fall as such:

Step 1: Multiple the standard deviation by 1, 2, or 3, then add and subtract it to the average value

$$+1\sigma = 1.250\ (0.000924 \times 1) = 1.2509$$
$$-1\sigma = 1.250\ (0.000924 \times 1) = 1.2491$$
$$+2\sigma = 1.250\ (0.000024 \times 2) = 1.2518$$
$$-2\sigma = 1.250\ (0.000924 \times 2) = 1.2482$$
$$+3\sigma = 1.250\ (0.000924 \times 3) = 1.2528$$
$$-3\sigma = 1.250\ (0.000924 \times 3) = 1.2472$$

Step 2: Display the range that the expected readings will fall for 1, 2, and 3 standard deviations

$\pm1\sigma$ = 1.2509–1.2491; Range for 68.27%
$\pm2\sigma$ = 1.2518–1.2482; Range for 95.45%
$\pm3\sigma$ = 1.2528–1.2472; Range for 99.73%

Standard Normal Distribution Single Side Areas

Z	.00	.01	.02	.03	.04	.05	.06	.07	.08	.09
0.0	.0000	.0040	.0080	.0120	.0160	.0199	.0239	.0279	.0319	.0359
0.1	.0398	.0438	.0478	.0517	.0557	.0596	.0636	.0675	.0714	.0753
0.2	.0793	.0832	.0871	.0910	.0948	.0987	.1026	.1064	.1103	.1141
0.3	.1179	.1217	.1255	.1293	.1331	.1368	.1406	.1443	.1480	.1517
0.4	.1554	.1591	.1628	.1664	.1700	.1736	.1772	.1808	.1844	.1879
0.5	.1915	.1950	.1985	.2019	.2054	.2088	.2123	.2157	.2190	.2224
0.6	.2257	.2291	.2324	.2357	.2389	.2422	.2454	.2486	.2518	.2549
0.7	.2580	.2612	.2642	.2673	.2704	.2734	.2764	.2794	.2823	.2852
0.8	.2881	.2910	.2939	.2967	.2995	.3023	.3051	.3078	.3106	.3133
0.9	.3159	.3186	.3212	.3238	.3264	.3289	.3315	.3340	.3365	.3389
1.0	.3413	.3438	.3461	.3485	.3508	.3531	.3554	.3577	.3599	.3621
1.1	.3643	.3665	.3686	.3708	.3729	.3749	.3770	.3790	.3810	.3830
1.2	.3849	.3869	.3888	.3907	.3925	.3944	.3962	.3980	.3997	.4015
1.3	.4032	.4049	.4066	.4082	.4099	.4115	.4131	.4147	.4162	.4177
1.4	.4192	.4207	.4222	.4236	.4251	.4265	.4279	.4292	.4306	.4319
1.5	.4332	.4345	.4357	.4370	.4382	.4394	.4406	.4418	.4429	.4441
1.6	.4452	.4463	.4474	.4484	.4495	.4505	.4515	.4525	.4535	.4545
1.7	.4554	.4564	.4573	.4582	.4591	.4599	.4608	.4616	.4625	.4633
1.8	.4641	.4649	.4656	.4664	.4671	.4678	.4686	.4693	.4699	.4706
1.9	.4713	.4719	.4726	.4732	.4738	.4744	.4750	.4756	.4761	.4767
2.0	.4772	.4778	.4783	.4788	.4793	.4798	.4803	.4808	.4812	.4817
2.1	.4821	.4826	.4830	.4834	.4838	.4842	.4846	.4850	.4854	.4857
2.2	.4861	.4864	.4868	.4871	.4875	.4878	.4881	.4884	.4887	.4890
2.3	.4893	.4896	.4898	.4901	.4904	.4906	.4909	.4911	.4913	.4916
2.4	.4918	.4920	.4922	.4925	.4927	.4929	.4931	.4932	.4934	.4936
2.5	.4938	.4940	.4941	.4943	.4945	.4946	.4948	.4949	.4951	.4952
2.6	.4953	.4955	.4956	.4957	.4959	.4960	.4961	.4962	.4963	.4964
2.7	.4965	.4966	.4967	.4968	.4969	.4970	.4971	.4972	.4973	.4974
2.8	.4974	.4975	.4976	.4977	.4977	.4978	.4979	.4979	.4980	.4981
2.9	.4981	.4982	.4982	.4983	.4984	.4984	.4985	.4985	.4986	.4986
3.0	.49865	.4987	.4988	.4988	.4989	.4989	.4989	.4989	.4990	.4990
4.0	.49997									

Figure 27.4 Single Side Table for Area under a Normal Statistical Distribution against Standard Deviations

27.2.3.1 "6" Sigma Distributions

In an attempt to improve quality to the highest degree, the limits on the normal standard distribution have been expended from $\pm 3\sigma$ to $\pm 6\sigma$. Even though $\pm 3\sigma$ will cover limits of 99.73% of the area under the curve, in some instances, this would still produce an unacceptable level of defects. By expanding the statistical limits to $\pm 6\sigma$ will bring the area under the curve for all intensive purpose to 100.00%. This allows for a much higher level of quality measurement and for operations to work toward a 100% quality level for their products. Figure 27.5 shows a breakdown from $\pm 1\sigma$ to $\pm 6\sigma$ with the area contained under the distribution, the expected defect levels, and how many defects would be expected for a population of one million items. In viewing this, $\pm 3\sigma$ at 99.73% would relinquish 2,700 defects for a lot of one million items, while $\pm 6\sigma$ at close to 100% would have a defect rate of 0.0000002% and would produce 0.002 defects for the same lot of one million items.

A good example of where a $\pm 3\sigma$ quality measurement would not be acceptable is the post office. The mail system, which for the most part is taken for granted, is a complex system that runs extremely efficiently. They handle close to fifty million mail pieces per day, which gets delivered in 2 to 3 days. The system begins with the mail being picked up and brought to the local post office. From there, it is sent to a postal hub in the state where it is sorted by the state it is to be delivered to the postal hub for that state. The delivery location then sorts the mail for that state by the town it is to go to and shipped to the local post office of the town. The local

Statistical Defect Rates vs. Standard Deviations

±σ	Acceptable Part Rate (%)	Defective Part Rate (%)	Defective Parts / Million
±1σ	68.27	31.73	317,300
±2σ	95.45	4.55	45,000
±3σ	99.73	0.27	2,700
±4σ	99.9937	0.0063	63
±5σ	99.99994	0.000057	0.57
±6σ	100.000	0.0000002	0.002

Figure 27.5 6σ Area Distribution and Defect Rates for a Normal Statistical Distribution

post office will then sort the mail to the exact route the mail carrier will deliver the mail and gets delivered. This encompasses handling the mail five times:

1. Picking up the mail.
2. Sort by state.
3. Sort by town.
4. Sort by carried route.
5. Deliver to final destination.

Even a mail piece that is being sent to your neighbor living next door will go through these five steps to get it to the final destination. Every time the mail is handled, there is the potential of an error in the system for an undelivered mail piece. Figure 27.6 shows an example of fifty million mail pieces in the system where a ±3σ quality level would result in 135,000 undelivered mail pieces per day for a lot size of fifty million.

Undelivered Mail Pieces for a 50,000,000 Population		
±σ	Defect Percent	Undelivered Mail Pieces
±σ1	31.73	15,865,000
±σ2	4.55	2,275,000
±σ3	0.27	135,000
±σ4	0.0063	3,150
±σ5	0.000057	28.5
±σ6	0.0000002	0.10

Figure 27.6 Example of Undelivered Mail Pieces for a Fifty Million Population Size

27.2.4 Process Capability

The capability of a manufacturing process is critical to know when deciding if it will be acceptable for producing component workpieces for the operation. If a given tolerance on a workpiece is to be held to ±0.001 in., is the process capable of maintaining that on a consistent basis? In order to predict what the process is capable of producing, a process capability study needs to be performed to verify this. This involves running several test pieces and inspecting each part and recording the readings. A capability study normally is performed for an 8-hour period with a given population of components being measured in order with the times they were fabricated. From there, the readings are evaluated to determine the standard distribution over time to determine the capability of the process and the equipment being utilized. This will render a picture of what the process is capable of producing and any drift from one time frame to the other.

When determining the capability of the process, it will relinquish whether it is in statistical control or not. Understanding that issues will occur during the process, they are divided into two categories, which are Random Variations and Assignable Variations. These are described as follows.

Random Variations: These categories of variations are those that can be predicted during the process or are outside of the process capability study for the equipment producing the component workpieces. These may include areas such as material variations, vibration in the setup of the workpieces, and the equipment being operated by a different operator. Random variations can and will occur, and even with these being compensated for, the process is still considered to be in statistical control for producing the workpieces.

Assignable Variations: Assignable variations are exceptions from the normal random variations. These include areas such as mistakes and errors, defective material being utilized, constant tool failures, etc. When assignable variations occur, the process is deemed to not be in statistical control while producing the workpieces.

When a process is operating in statistical control, the normal standard distribution will display a precision closely held in the distribution. Even if the process floats over time, adjustments can be made to bring the average of the distribution back to the center. A process not in statistical control will relinquish a normal standard distribution that does not display consistency from one set of readings to the other. The workpieces may display accuracy from the process but not precision, and adjustments of any type may not render the process back into statistical control. In-control and out-of-control processes are discussed in Section 27.4 on statistical process control.

27.2.5 Statistical Tolerancing

When component workpieces are to be assembled, the dimension and tolerance of each component will add up to the total length of the assembly. This concept of statistical tolerancing is utilized to determine the final tolerance for the assembly when all of the workpieces are assembled. Figure 27.7 illustrates the concept of statistical tolerancing where each component workpiece is being produced to their given dimension and tolerance and the buildup will determine the overall length and tolerance expected in the assembly.

This method for determining the tolerance range for an assembly relies on four assumptions for it to operate correctly. These are described as follows.

Statistical Tolerancing Assumptions

— *Dimensions are Normally Distributed*: For all the component workpieces to be assembled, the dimensions need to fall within a normal statistical distribution. This means that the bell curve displaying the dimensions is not slanted in any manner but is a symmetrical distribution. This ensures that an equal number of components will be represented in the area on each side of the distribution.

— *Distributions Are Independent of Each Other*: Each normal statistical distribution for the components has to not be linked together in any manner. Every part that is going to be assembled needs to have its own independent distribution. This allows for the selection of the parts to move from one distribution into the assembly distribution. Figure 27.8 illustrates that each component being assembled is within its own independent distribution moving into the assembly distribution.

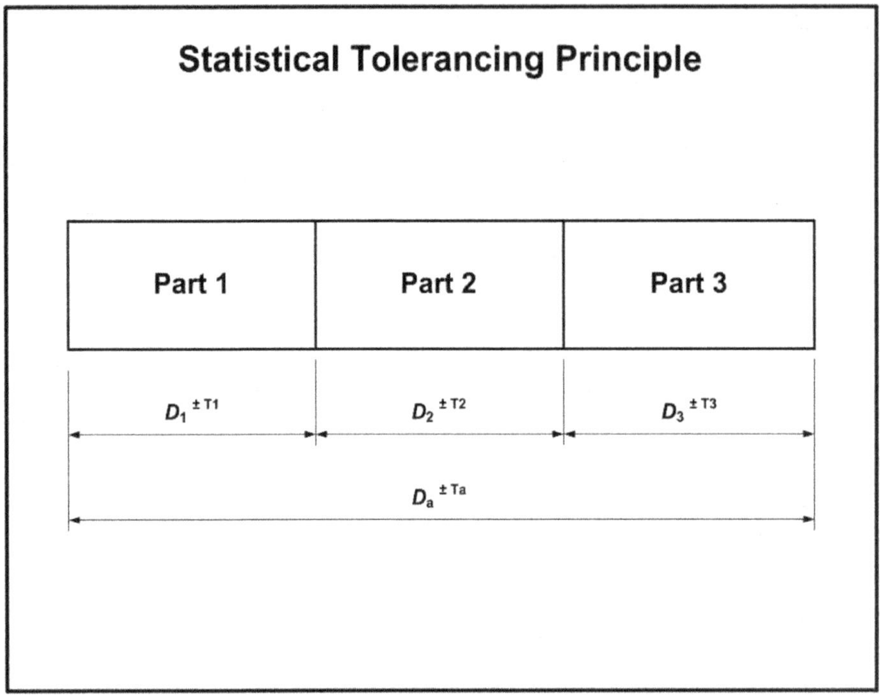

Figure 27.7 Statistical Tolerancing Concept Applied to an Assembly

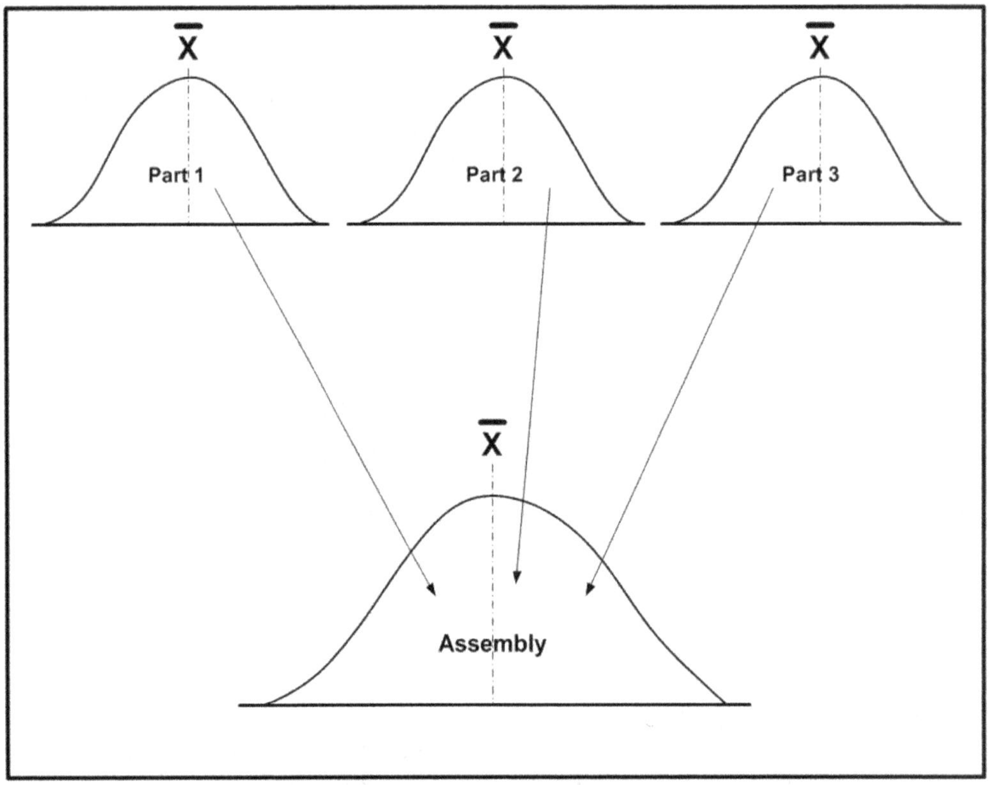

Figure 27.8 Independent Distributions in a Statistical Tolerancing Assembly

- *The Process Producing the Components is in Statistical Control*: In order for the component parts to be manufactured in a way that statistical tolerancing will function, the processes fabricating the parts must be in statistical control. This will render a distribution that is symmetrical with a equal amount of parts under the curve.
- *Parts for the Assembly are* ***RANDOMLY SELECTED***: The main function when performing an assembly with statistical tolerancing is to not screen the components for assembly but to randomly select them

from the pile. With the parts ranging within their independent tolerances, random selection will ensure that they will all fit together and the total tolerance of the assembly will fall in. Even though the part has a tolerance range, statistically a random selection of the components and selecting each one that is on either the high or the low end of the tolerance spectrum is near zero.

The total statistical tolerance for an assembly will produce a reduced range than by simply just adding up all the ± ranges of each component. This allows for the random selection with an extremely low probability of having a selection of each part being on either the high or the low range within their independent distributions. To calculate the statistical tolerance range of an assembly, we will utilize the following equation:

$$T_a = \sqrt{\sum_i^n T_i^2}$$

The standard distribution required for the assembly is determined as the product of the standard distribution for each independent distribution for each component part. The following equation is utilized to determine the standard distribution for a statistically tolerance assembly.

$$\sigma a^2 = \sum_i^n \sigma i^2$$

Example

If three components are to be assembled together and had the following dimensions:

$$1.875^{+0.002\ -\ 0.004}$$
$$0.750^{+0.000\ -\ 0.002}$$
$$2.750^{+0.003\ -\ 0.001}$$

What would the tolerance of the assembly need to be in order to have everything assembled with no issues?

Step 1: Determine the Upper Assembly Tolerance

$$T_a = \sqrt{\sum_i^n T_i^2}$$

$$T_a = \sqrt{\sum_1^3 0.002^2 + 0.000^2 + 0.003^2}$$

$$T_a = 0.0036 \text{ in.}$$

Step 2: Determine the Lower Assembly Tolerance

$$T_a = \sqrt{\sum_i^n T_i^2}$$

$$T_a = \sqrt{\sum_1^3 0.004^2 + 0.002^2 + 0.001^2}$$

$$T_a = 0.0046 \text{ in.}$$

Step 3: Determine the Assembly Length

$$D_a = D_1 + D_2 + D_3$$

$$D_a = 1.875 + 0.750 + 2.750$$
$$D_a = 5.375 \text{ in.}$$

So, for this combination, the assembly dimension and tolerance would be:

$$5.375^{+0.0036\ -\ 0.0046}$$

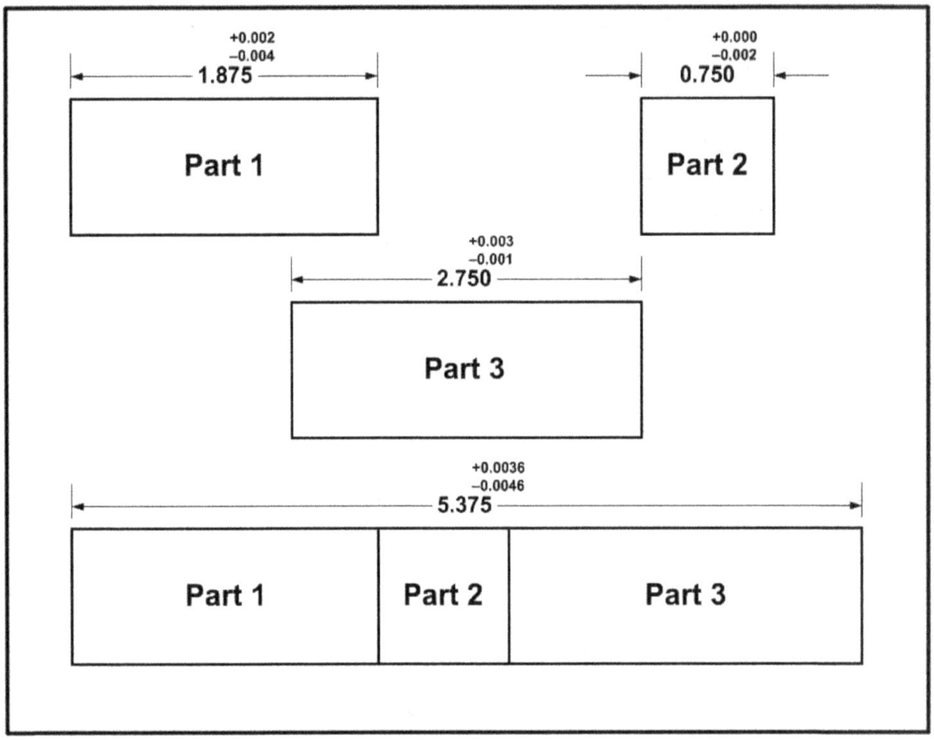

Figure 27.9 Example of Statistical Tolerances Applied within an Assembly

Figure 27.9 shows the three components to be assembled with the assembly dimension and statistically calculated tolerances.

27.3 QUALITY INSPECTION

Inspection techniques are used either to check the quality levels of a given lot of components or to determine the incoming quality from a supplier. Once again, statistical analysis is utilized to determine what the level of quality and the range that can be expected from the components being inspected. These inspection techniques are divided into inspecting a population and inspecting a sample of the part and are explained as follows.

27.3.1 Population vs. Sample

Population inspection and sample inspection are similar in nature with some differences. For a population inspection, every component in the lot will be inspected. This is also referred to as 100% inspection and is used to determine the exact risk of every part in the lot. Sample inspection will only inspect a determined number of components within the lot, which are selected RAMDOMLY from the population. This is used to determine the EXPECTED RISK of the population of the components within the lot. There is a risk with sample inspection over population inspection, but statistically the results should display a very similar normal statistical distribution. Population inspection is normally utilized when a defective component could cause a catastrophic failure of the product or danger to the operator of the product. Figure 27.10 illustrates the difference between a population inspection where all 30 components are inspected and a sample inspection where a 30% random sample is selected to check.

When performing either a population or sample inspection, the standard deviation of the inspected components needs to be calculated. This will display either the exact range of the inspected parts in the case of a population inspection or the expected range for a sample inspection. The equations utilized for determining the standard deviations for both are listed as follows.

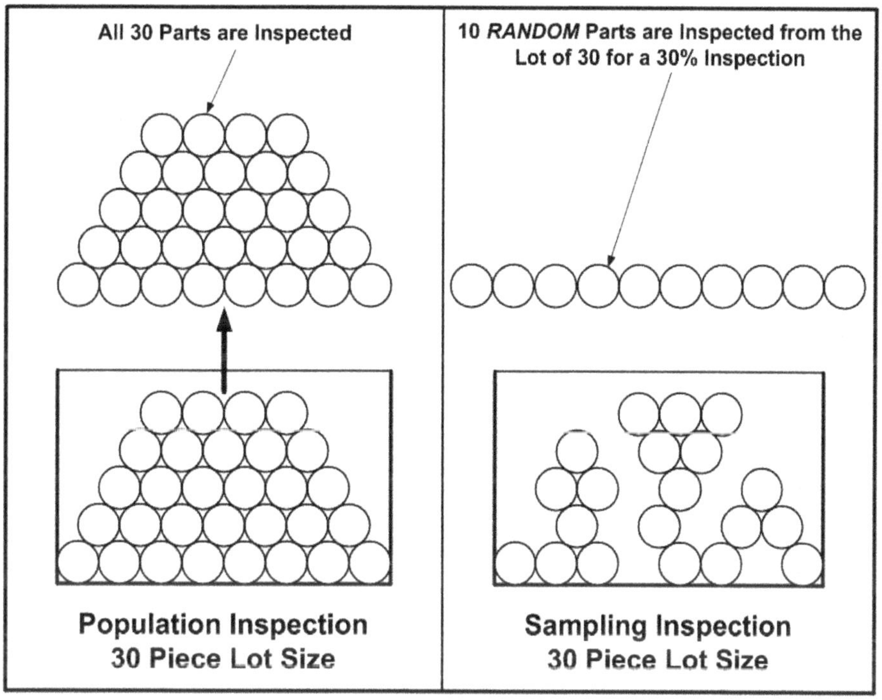

Figure 27.10 Population Inspection vs. Sampling Inspection

Population Inspection Standard Deviation Equation

$$\sigma = \sqrt{\frac{\sum_{i}^{n}(x_i - \mu)^2}{n}}$$

Sample Inspection Standard Deviation Equation

$$\sqrt{\frac{\sum_{i}^{n}(x_i - \bar{x})^2}{n-1}}$$

x_i = Measured reading value
X *or* μ = Average of all readings
n = Number of readings taken

Once it is determined whether a population or sample inspection is required, the components are measured and the readings recorded. From there, the standard deviation is calculated utilizing the correct equation and the dimensional spread is determined. Figure 27.11 shows both population and sample inspections performed on the same lot of 30 components. The values for the sample were randomly selected from the population and the standard deviation for each was determined. Reviewing the data calculated, the following differences are determined.

	Population	Sample	Delta
Average:	0.6251	0.6252	0.0001
$+3\sigma$:	0.6287	0.6284	0.0004
-3σ:	0.6214	0.6221	0.0006

Reviewing both inspections of the average value of 0.625 with an upper range of 0.627 and a lower range of 0.623, both show that the upper and lower ranges will fall out of the acceptable levels. The delta between the population and sample inspections is extremely close, and depending on the critical nature of the components, the sample inspection would render an acceptable determined range.

30 Piece Population Inspection Results

Dimension: 0.625 Tolerance: ± 0.002

No.	Reading	No.	Reading
1	0.6257	16	0.6240
2	0.6231	17	0.6272
3	0.6249	18	0.6248
4	0.6268	19	0.6255
5	0.6251	20	0.6264
6	0.6254	21	0.6247
7	0.6262	22	0.6246
8	0.6258	23	0.6253
9	0.6247	24	0.6266
10	0.6238	25	0.6261
11	0.6229	26	0.6230
12	0.6267	27	0.6232
13	0.6227	28	0.6252
14	0.6255	29	0.6253
15	0.6250	30	0.6262

μ : 0.6251
σ : 0.001222
+3σ: 0.6287
−3σ: 0.6214

10 Piece 30% Sample Inspection Results

Dimension: 0.625 Tolerance: ± 0.002

No.	Reading	No.	Reading
1	0.6249	6	0.6248
2	0.6254	7	0.6247
3	0.6247	8	0.6266
4	0.6267	9	0.6232
5	0.6250	10	0.6262

$\overline{X}$: 0.6252
σ : 0.001056
+3σ : 0.6284
−3σ : 0.6221

Inspection Variation

Population	Sampling	Delta
μ : 0.6251	$\overline{X}$: 0.6252	0.0001
σ : 0.001222	σ : 0.001056	0.000166
+3σ : 0.6287	+3σ : 0.6284	0.0004
−3σ : 0.6214	−3σ : 0.6221	0.0006

Figure 27.11 Population Inspection vs. Sampling Inspection Data Results Comparison

27.4 STATISTICAL PROCESS CONTROL (SPC)

As the equipment is fabricating the components, being able to control the quality in a real time manner is the function of statistical process control (SPC). This technique allows the operation to be proactive in quality control and not reactive where the components would need to be measured and screened for quality. To perform SPC on a process, the fabricated components will be inspected during the process and the measured readings recorded and analyzed. In most cases, SPC utilizes a 95% confidence level, which relates to ±1.96σ from the average. Figure 27.12 shows the 95% confidence level within the normal statistical distribution. This is used to bring the range of the measured components closer together as a safety factor; so if the process begins to slip outside some, the components may still be acceptable for use.

With a statistical analysis of this type, there are two golden rules that need to be adhered to at all times. The first is that the data lot size must remain constant. If it is determined that 10 components need to be checked each time, that cannot be violated. The second is that the time frame must remain constant. If it is determined that the 10 components need to be checked every 30 min, that too must be adhered to. If a time frame is missed for whatever reason, you cannot check 20 pieces in an hour. By complying to these two rules, the results will produce a real-time picture of what the process is doing over a given time and render information that will allow you to forecast future results if the process is operation in control. Section 27.4.1 covers how the data is collected and analyzed in a statistical process control environment.

27.4.1 Date Collection and Analysis

If a component part is being produced in a manufacturing operation and SPC is going to be applied to a dimension of 0.500 in. with a tolerance of ±0.005. Figure 27.13 shows the part drawing for the component for SPC. A data set of 10 pieces needs to be measured with a time frame of every 30 min. The information will be collected are outlined as follows. But note that steps 1 through 6 are performed for each data set as soon as they are measured and recorded.

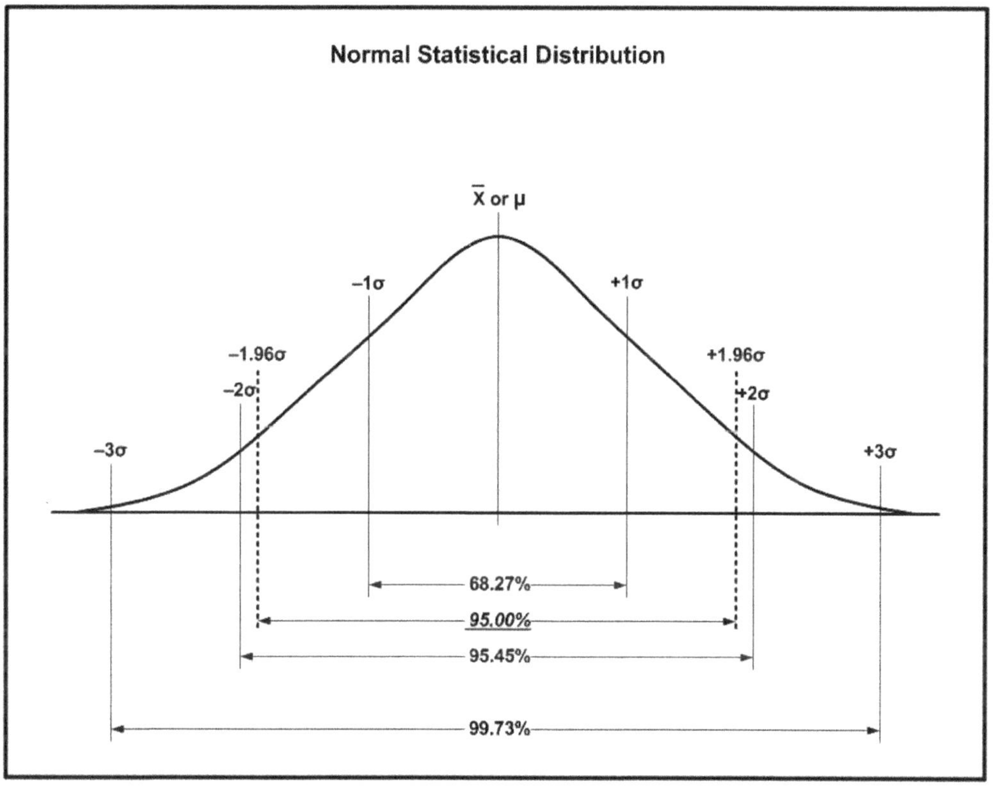

Figure 27.12 Normal Statistical Distribution Showing a 95% Confidence Level

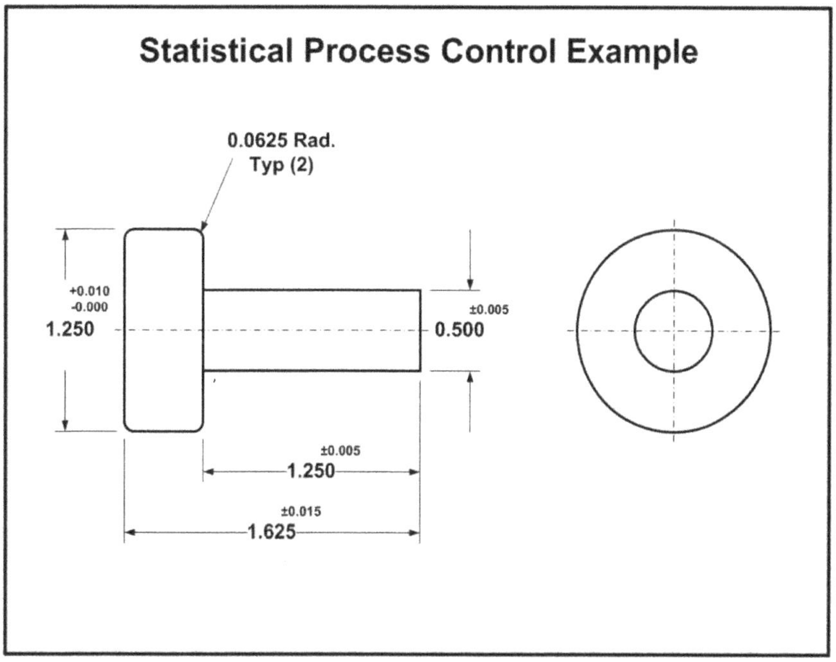

Figure 27.13 SPC Example Print Working with the $0.500^{\pm 0.005}$ Dimension

Step 1: Collect the designated number of components from the process at the proper time interval. For this example, 10 parts are to be selected and they are taken *IN ORDER* off the machine as they are being produced, not randomly selected from the pile already produced. Figure 27.14 shows the SPC data chart and the values collected over the time frames.

SPC Control Data

Dimension: 0.500
Tolerance: ±0.005

Reading	7:00	7:30	8:00	8:30	9:00	9:30
1	0.4967	0.4973	0.4980	0.4984	0.4994	0.5004
2	0.4965	0.4971	0.4983	0.4988	0.4991	0.5008
3	0.4971	0.4974	0.4979	0.4987	0.4993	0.5007
4	0.4966	0.4976	0.4981	0.4983	0.4995	0.5010
5	0.4969	0.4974	0.4980	0.4989	0.4997	0.5008
6	0.4972	0.4977	0.4982	0.4986	0.4994	0.5006
7	0.4970	0.4973	0.4977	0.4991	0.4998	0.5003
8	0.4968	0.4978	0.4983	0.4990	0.4997	0.5011
9	0.4971	0.4975	0.4980	0.4992	0.5000	0.5009
10	0.4968	0.4979	0.4982	0.4987	0.4999	0.5010
Average:						
S.D.:						
Max:						
Min:						
Range:						
+95%:						
-95%:						
Growth:						
Avg. Growth:						
Avg. S.D.:						

	10:00	10:30	11:00	11:30	12:00	12:30
Average:						
+95%:						
-95%:						

Figure 27.14 Step 1. Data Collection at a Given Part Count and Time Interval

Step 2: Calculate the average of the values collected.
Again, this is performed for each data set as the values are recorded. Figure 27.15 shows the average values for each data set calculated.

Step 3: Calculate the standard deviation for the values collected.
Be sure to use the POPULATION standard deviation equation and not the sample equation. This is because the 10 components were taken in order, making each data set a population of 10 and not a sample. Figure 27.16 shows the population standard deviation calculated for each data set.

Step 4: Determine the maximum and minimum values for each data set.
These are simply the highest and lowest values found in the data for each set collected. Figure 27.17 shows the maximum and minimum values for each data set.

Step 5: Calculate the range of the values in each data set.
Subtract the minimum value from the maximum value for each data set. This will provide the range that the data set has and shows the spread that the values are relinquishing. Figure 27.18 shows the ranges for each data set.

Step 6: Calculate the 95% confidence levels for each data set.
Multiply the standard deviation of each set by 1.96 and then either add or subtract it from the average values for each set. This will show the expected spread for the components collected in each data set. Figure 27.19 shows the 95% confidence levels for each data set.

The calculated data for each set begins to paint a picture of how the process is operating and what values can be expected. Keep in mind that each data set is run through Steps 1–6 as soon as the data is collected. Once the data is calculated and the data set is determined to be within control and the values are showing that acceptable components are being produced, the part produced during that time frame are removed from the

SPC Control Data

Dimension: 0.500

Tolerance: ±0.005

Reading	7:00	7:30	8:00	8:30	9:00	9:30
1	0.4967	0.4973	0.4980	0.4984	0.4994	0.5004
2	0.4965	0.4971	0.4983	0.4988	0.4991	0.5008
3	0.4971	0.4974	0.4979	0.4987	0.4993	0.5007
4	0.4966	0.4976	0.4981	0.4983	0.4995	0.5010
5	0.4969	0.4974	0.4980	0.4989	0.4997	0.5008
6	0.4972	0.4977	0.4982	0.4986	0.4994	0.5006
7	0.4970	0.4973	0.4977	0.4991	0.4998	0.5003
8	0.4968	0.4978	0.4983	0.4990	0.4997	0.5011
9	0.4971	0.4975	0.4980	0.4992	0.5000	0.5009
10	0.4968	0.4979	0.4982	0.4987	0.4999	0.5010
Average:	**0.4969**	**0.4975**	**0.4981**	**0.4988**	**0.4996**	**0.5008**
S.D.:						
Max:						
Min:						
Range:						
+95%:						
-95%:						
Growth:						
Avg. Growth:						
Avg. S.D.:						

	10:00	10:30	11:00	11:30	12:00	12:30
Average:						
+95%:						
-95%:						

Figure 27.15 Step 2. Calculated Averages of Each Data Set

SPC Control Data

Dimension: 0.500

Tolerance: ±0.005

Reading	7:00	7:30	8:00	8:30	9:00	9:30
1	0.4967	0.4973	0.4980	0.4984	0.4994	0.5004
2	0.4965	0.4971	0.4983	0.4988	0.4991	0.5008
3	0.4971	0.4974	0.4979	0.4987	0.4993	0.5007
4	0.4966	0.4976	0.4981	0.4983	0.4995	0.5010
5	0.4969	0.4974	0.4980	0.4989	0.4997	0.5008
6	0.4972	0.4977	0.4982	0.4986	0.4994	0.5006
7	0.4970	0.4973	0.4977	0.4991	0.4998	0.5003
8	0.4968	0.4978	0.4983	0.4990	0.4997	0.5011
9	0.4971	0.4975	0.4980	0.4992	0.5000	0.5009
10	0.4968	0.4979	0.4982	0.4987	0.4999	0.5010
Average:	0.4969	0.4975	0.4981	0.4988	0.4996	0.5008
S.D.:	**0.00022**	**0.00024**	**0.00018**	**0.00028**	**0.00027**	**0.00025**
Max:						
Min:						
Range:						
+95%:						
-95%:						
Growth:						
Avg. Growth:						
Avg. S.D.:						

	10:00	10:30	11:00	11:30	12:00	12:30
Average:						
+95%:						
-95%:						

Figure 27.16 Step 3. Calculated Standard Deviations for Each Data Set

SPC Control Data

Dimension: 0.500
Tolerance: ±0.005

Reading	7:00	7:30	8:00	8:30	9:00	9:30
1	0.4967	0.4973	0.4980	0.4984	0.4994	0.5004
2	0.4965	0.4971	0.4983	0.4988	0.4991	0.5008
3	0.4971	0.4974	0.4979	0.4987	0.4993	0.5007
4	0.4966	0.4976	0.4981	0.4983	0.4995	0.5010
5	0.4969	0.4974	0.4980	0.4989	0.4997	0.5008
6	0.4972	0.4977	0.4982	0.4986	0.4994	0.5006
7	0.4970	0.4973	0.4977	0.4991	0.4998	0.5003
8	0.4968	0.4978	0.4983	0.4990	0.4997	0.5011
9	0.4971	0.4975	0.4980	0.4992	0.5000	0.5009
10	0.4968	0.4979	0.4982	0.4987	0.4999	0.5010
Average:	0.4969	0.4975	0.4981	0.4988	0.4996	0.5008
S.D.:	0.00022	0.00024	0.00018	0.00028	0.00027	0.00025
Max:	**0.4972**	**0.4979**	**0.4983**	**0.4992**	**0.5000**	**0.5011**
Min:	**0.4965**	**0.4971**	**0.4977**	**0.4983**	**0.4991**	**0.5003**
Range:						
+95%:						
-95%:						
Growth:						
Avg. Growth:						
Avg. S.D.:						

	10:00	10:30	11:00	11:30	12:00	12:30
Average:						
+95%:						
-95%:						

Figure 27.17 Step 4. Enter Maximum and Minimum Readings of Each Data Set

SPC Control Data

Dimension: 0.500
Tolerance: ±0.005

Reading	7:00	7:30	8:00	8:30	9:00	9:30
1	0.4967	0.4973	0.4980	0.4984	0.4994	0.5004
2	0.4965	0.4971	0.4983	0.4988	0.4991	0.5008
3	0.4971	0.4974	0.4979	0.4987	0.4993	0.5007
4	0.4966	0.4976	0.4981	0.4983	0.4995	0.5010
5	0.4969	0.4974	0.4980	0.4989	0.4997	0.5008
6	0.4972	0.4977	0.4982	0.4986	0.4994	0.5006
7	0.4970	0.4973	0.4977	0.4991	0.4998	0.5003
8	0.4968	0.4978	0.4983	0.4990	0.4997	0.5011
9	0.4971	0.4975	0.4980	0.4992	0.5000	0.5009
10	0.4968	0.4979	0.4982	0.4987	0.4999	0.5010
Average:	0.4969	0.4975	0.4981	0.4988	0.4996	0.5008
S.D.:	0.00022	0.00024	0.00018	0.00028	0.00027	0.00025
Max:	0.4972	0.4979	0.4983	0.4992	0.5000	0.5011
Min:	0.4965	0.4971	0.4977	0.4983	0.4991	0.5003
Range:	**0.0007**	**0.0008**	**0.0006**	**0.0009**	**0.0009**	**0.0008**
+95%:						
-95%:						
Growth:						
Avg. Growth:						
Avg. S.D.:						

	10:00	10:30	11:00	11:30	12:00	12:30
Average:						
+95%:						
-95%:						

Figure 27.18 Step 5. Calculated Range for Each Data Set

SPC Control Data

Dimension: 0.500
Tolerance: ±0.005

Reading	7:00	7:30	8:00	8:30	9:00	9:30
1	0.4967	0.4973	0.4980	0.4984	0.4994	0.5004
2	0.4965	0.4971	0.4983	0.4988	0.4991	0.5008
3	0.4971	0.4974	0.4979	0.4987	0.4993	0.5007
4	0.4966	0.4976	0.4981	0.4983	0.4995	0.5010
5	0.4969	0.4974	0.4980	0.4989	0.4997	0.5008
6	0.4972	0.4977	0.4982	0.4986	0.4994	0.5006
7	0.4970	0.4973	0.4977	0.4991	0.4998	0.5003
8	0.4968	0.4978	0.4983	0.4990	0.4997	0.5011
9	0.4971	0.4975	0.4980	0.4992	0.5000	0.5009
10	0.4968	0.4979	0.4982	0.4987	0.4999	0.5010
Average:	0.4969	0.4975	0.4981	0.4988	0.4996	0.5008
S.D.:	0.00022	0.00024	0.00018	0.00028	0.00027	0.00025
Max:	0.4972	0.4979	0.4983	0.4992	0.5000	0.5011
Min:	0.4965	0.4971	0.4977	0.4983	0.4991	0.5003
Range:	0.0007	0.0008	0.0006	0.0009	0.0009	0.0008
+95%:	**0.4973**	**0.4980**	**0.4984**	**0.4993**	**0.5001**	**0.5012**
-95%:	**0.4964**	**0.4970**	**0.4977**	**0.4982**	**0.4990**	**0.5003**
Growth:						
Avg. Growth:						
Avg. S.D.:						
	10:00	10:30	11:00	11:30	12:00	12:30
Average:						
+95%:						
-95%:						

Figure 27.19 Step 6. Calculated 95% Distribution of Each Data Set

machine and a new container is put in place to collect the parts being produced. This is done so that the process begins to produce defective components after the data set was reviewed, and the good components will not be contaminated by defective parts and need to be screened to remove the bad components.

27.4.2 Range and Control Charting

After Steps 1–6 have been completed for each of the data sets, charting is the next step. The charts display a picture of what the process is doing to give a visual effect of the process. Figure 27.20 illustrates a range chart for the six data sets from 7:00 to 9:30. This displays the spread that the component parts have for each data set. This is utilized to evaluate any deviations from one set to another. Looking at the chart, the data sets for 8:30 and 9:00 show the highest range of 0.0008 in., while the data set for 8:00 shows the lowest range of 0.0006 in. What needs to be evaluated is what was happening at 8:30 and 9:00 to have the highest range and likewise, what was operating better at 8:00, so it can be replicated.

Another chart used for a visual aid is a control chart. This will show the data sets with the average values in the center and the ±95% confidence levels above and below the average line. Figure 27.21 shows a control chart for the six data sets collected with the pennant information important to the process being evaluated. The upper and lower tolerance levels are plotted for the total range from 0.505 to 0.495 in., and the curve is evaluated. From looking at the chart, it is evident that the average readings are growing as time elapses. This indicates that at a given time, the process will begin to produce components that will approach the high limit of the tolerance and an adjustment will need to be performed to bring the average back into an acceptable level.

From evaluating the information calculated and the visual picture displayed in the charts, it is evident that the process is running in statistical control with a well-defined growth of the pieces. With a controlled process, the data can be expanded and a forecast can be produced to predict when the tolerances of the components being produced may run outside of the upper level range. Section 27.4.3 describes how to construct a probability forecast.

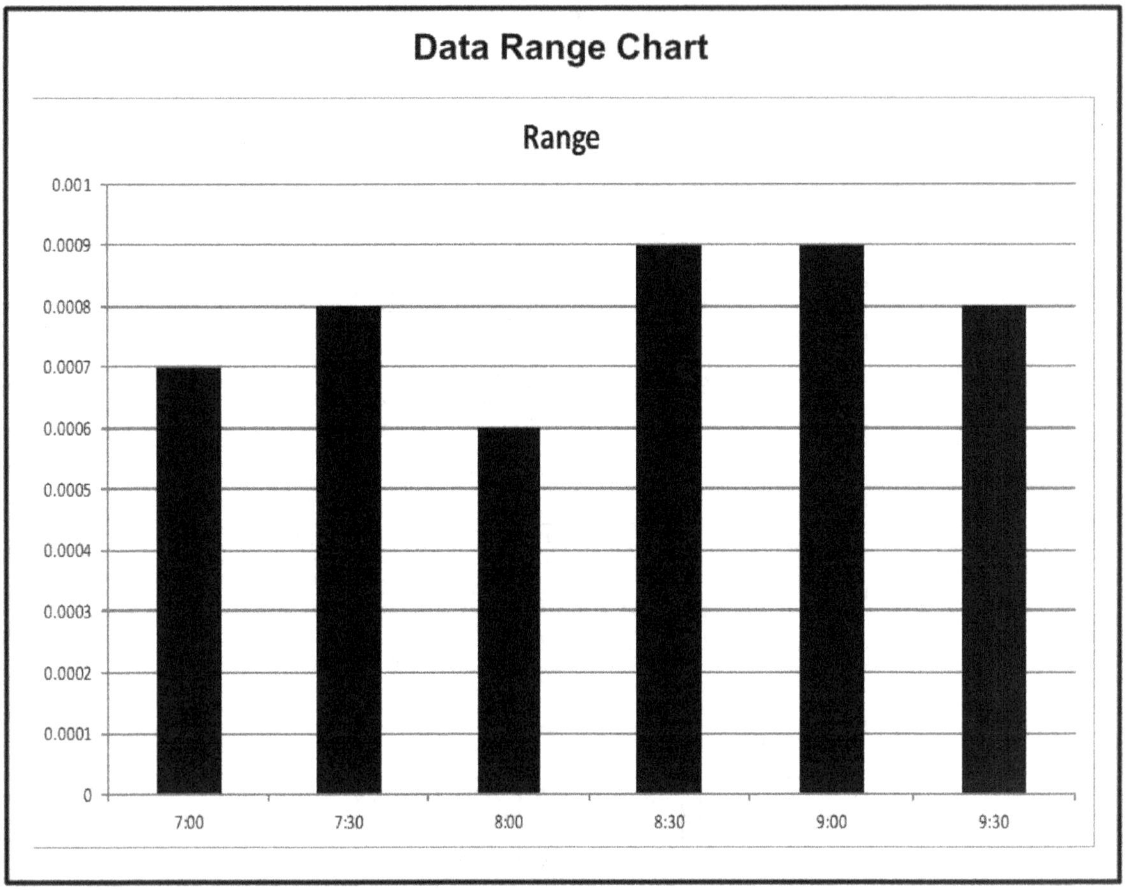

Figure 27.20 Range Chart for the Measured Data Sets

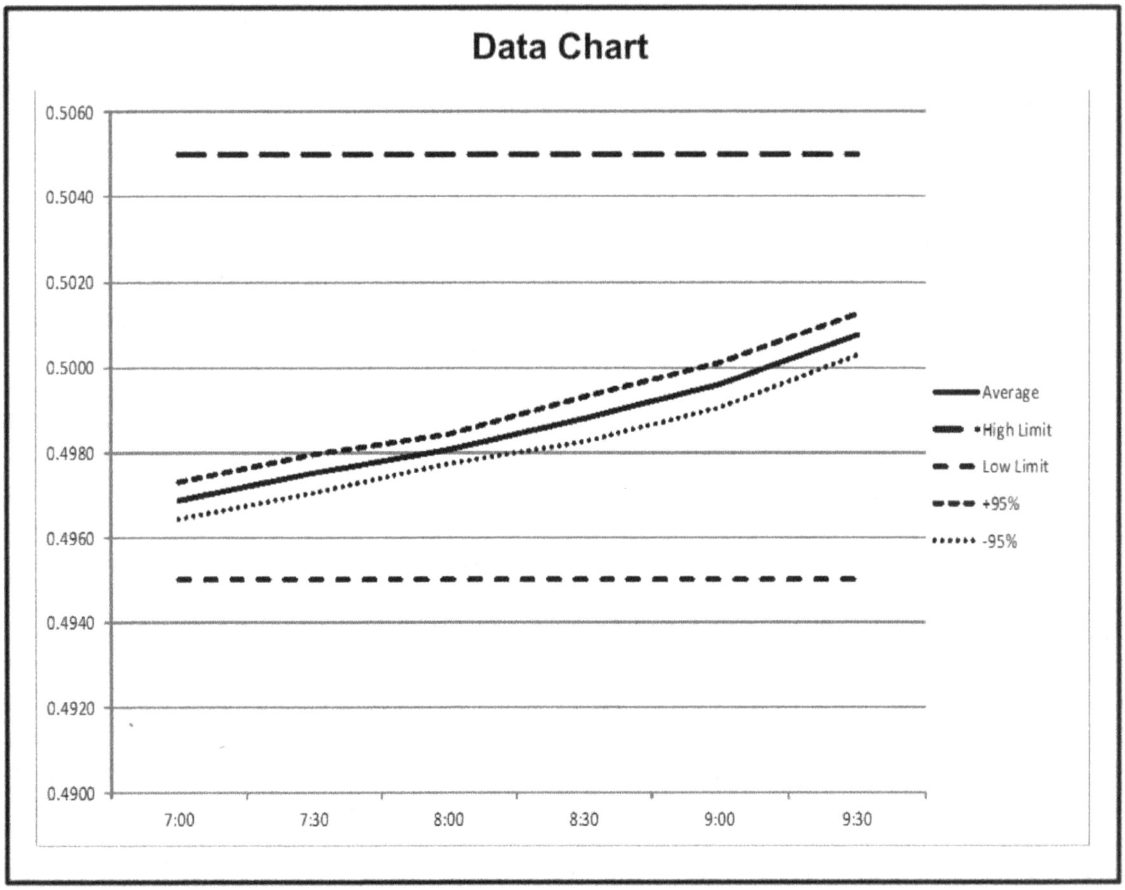

Figure 27.21 Control Chart for the Measured Data Sets

27.4.3 Probability Forecasting

Construction of a probability forecast can only be performed on a process that is in statistical control. Forecasting assumes that the future will represent the past, and if the data from the past is not consistent and in control, it is impossible to predict any future performance. The construction of the probability forecast will be done after Steps 1–6 have been performed on all the data sets taken and is outlined as follows.

Step 7: Determine the growth of the averages for the six data sets.
The growth between the data sets is calculated by subtracting the average from set 1 from the average of set 2. This is carried out for all the data sets, i.e., by subtracting the average from set 2 from the average of set 3, etc. Figure 27.22 shows the growth between each data sets.

Step 8: Determine the average growth and average standard deviation for all six data sets.
Once the growth between each set has been calculated in Step 7, take the average of all the growth calculations. Next, take the average of all the standard deviations of all six data sets. Figure 27.23 shows the average growth and average standard deviations from 7:00 to 9:30.

Step 9: Determine the predicted averages running out past data set 6.
With Steps 7 and 8 completed, the forecasted averages can be determined. This is done by taking the last recorded average from the 9:30 data set and adding the average growth to it. Continue this out by taking the forecasted average just determined and add the average growth to it, etc. Figure 27.24 shows the forecasted averages carried out for the predicted six time frames.

Step 10: Determine the ±95% ranges running out past data set 6.
The final step in the probability forecasting is to calculate what the ±95% spread would be for the future time frames. Utilizing the average standard deviation calculated in step 8, multiply it by 1.96 and either add or subtract it from the calculated averages for the future time frames. Figure 27.25 shows the completed probability forecast with the ±95% values determined.

SPC Control Data

Dimension: 0.500
Tolerance: ±0.005

Reading	7:00	7:30	8:00	8:30	9:00	9:30
1	0.4967	0.4973	0.4980	0.4984	0.4994	0.5004
2	0.4965	0.4971	0.4983	0.4988	0.4991	0.5008
3	0.4971	0.4974	0.4979	0.4987	0.4993	0.5007
4	0.4966	0.4976	0.4981	0.4983	0.4995	0.5010
5	0.4969	0.4974	0.4980	0.4989	0.4997	0.5008
6	0.4972	0.4977	0.4982	0.4986	0.4994	0.5006
7	0.4970	0.4973	0.4977	0.4991	0.4998	0.5003
8	0.4968	0.4978	0.4983	0.4990	0.4997	0.5011
9	0.4971	0.4975	0.4980	0.4992	0.5000	0.5009
10	0.4968	0.4979	0.4982	0.4987	0.4999	0.5010
Average:	0.4969	0.4975	0.4981	0.4988	0.4996	0.5008
S.D.:	0.00022	0.00024	0.00018	0.00028	0.00027	0.00025
Max:	0.4972	0.4979	0.4983	0.4992	0.5000	0.5011
Min:	0.4965	0.4971	0.4977	0.4983	0.4991	0.5003
Range:	0.0007	0.0008	0.0006	0.0009	0.0009	0.0008
+95%:	0.4973	0.4980	0.4984	0.4993	0.5001	0.5012
-95%:	0.4964	0.4970	0.4977	0.4982	0.4990	0.5003
	Growth:	**0.00063**	**0.00057**	**0.00070**	**0.00081**	**0.00118**
Avg. Growth:						
Avg. S.D.:						

	10:00	10:30	11:00	11:30	12:00	12:30
Average:						
+95%:						
-95%:						

Figure 27.22 Step 7. Calculate the Growth Between the Averages of Each Data Set

SPC Control Data

Dimension: 0.500
Tolerance: ±0.005

Reading	7:00	7:30	8:00	8:30	9:00	9:30
1	0.4967	0.4973	0.4980	0.4984	0.4994	0.5004
2	0.4965	0.4971	0.4983	0.4988	0.4991	0.5008
3	0.4971	0.4974	0.4979	0.4987	0.4993	0.5007
4	0.4966	0.4976	0.4981	0.4983	0.4995	0.5010
5	0.4969	0.4974	0.4980	0.4989	0.4997	0.5008
6	0.4972	0.4977	0.4982	0.4986	0.4994	0.5006
7	0.4970	0.4973	0.4977	0.4991	0.4998	0.5003
8	0.4968	0.4978	0.4983	0.4990	0.4997	0.5011
9	0.4971	0.4975	0.4980	0.4992	0.5000	0.5009
10	0.4968	0.4979	0.4982	0.4987	0.4999	0.5010
Average:	0.4969	0.4975	0.4981	0.4988	0.4996	0.5008
S.D.:	0.00022	0.00024	0.00018	0.00028	0.00027	0.00025
Max:	0.4972	0.4979	0.4983	0.4992	0.5000	0.5011
Min:	0.4965	0.4971	0.4977	0.4983	0.4991	0.5003
Range:	0.0007	0.0008	0.0006	0.0009	0.0009	0.0008
+95%:	0.4973	0.4980	0.4984	0.4993	0.5001	0.5012
-95%:	0.4964	0.4970	0.4977	0.4982	0.4990	0.5003
Growth:		0.00063	0.00057	0.00070	0.00081	0.00118
Avg. Growth:		0.00078				
Avg. S.D.:		0.00024				

	10:00	10:30	11:00	11:30	12:00	12:30
Average:						
+95%:						
-95%:						

Figure 27.23 Step 8. Calculate the Average Growth and Average Standard Deviation of the Data Sets

SPC Control Data

Dimension: 0.500
Tolerance: ±0.005

Reading	7:00	7:30	8:00	8:30	9:00	9:30
1	0.4967	0.4973	0.4980	0.4984	0.4994	0.5004
2	0.4965	0.4971	0.4983	0.4988	0.4991	0.5008
3	0.4971	0.4974	0.4979	0.4987	0.4993	0.5007
4	0.4966	0.4976	0.4981	0.4983	0.4995	0.5010
5	0.4969	0.4974	0.4980	0.4989	0.4997	0.5008
6	0.4972	0.4977	0.4982	0.4986	0.4994	0.5006
7	0.4970	0.4973	0.4977	0.4991	0.4998	0.5003
8	0.4968	0.4978	0.4983	0.4990	0.4997	0.5011
9	0.4971	0.4975	0.4980	0.4992	0.5000	0.5009
10	0.4968	0.4979	0.4982	0.4987	0.4999	0.5010
Average:	0.4969	0.4975	0.4981	0.4988	0.4996	0.5008
S.D.:	0.00022	0.00024	0.00018	0.00028	0.00027	0.00025
Max:	0.4972	0.4979	0.4983	0.4992	0.5000	0.5011
Min:	0.4965	0.4971	0.4977	0.4983	0.4991	0.5003
Range:	0.0007	0.0008	0.0006	0.0009	0.0009	0.0008
+95%:	0.4973	0.4980	0.4984	0.4993	0.5001	0.5012
-95%:	0.4964	0.4970	0.4977	0.4982	0.4990	0.5003
Growth:		0.00063	0.00057	0.00070	0.00081	0.00118
Avg. Growth:		0.00078				
Avg. S.D.:		0.00024				

	10:00	10:30	11:00	11:30	12:00	12:30
Average:	0.5015	0.5023	0.5031	0.5039	0.5047	0.5054
+95%:						
-95%:						

Figure 27.24 Step 9. Calculate the Predicted Averages for Extended Time Frames

SPC Control Data

Dimension: 0.500
Tolerance: ±0.005

Reading	7:00	7:30	8:00	8:30	9:00	9:30
1	0.4967	0.4973	0.4980	0.4984	0.4994	0.5004
2	0.4965	0.4971	0.4983	0.4988	0.4991	0.5008
3	0.4971	0.4974	0.4979	0.4987	0.4993	0.5007
4	0.4966	0.4976	0.4981	0.4983	0.4995	0.5010
5	0.4969	0.4974	0.4980	0.4989	0.4997	0.5008
6	0.4972	0.4977	0.4982	0.4986	0.4994	0.5006
7	0.4970	0.4973	0.4977	0.4991	0.4998	0.5003
8	0.4968	0.4978	0.4983	0.4990	0.4997	0.5011
9	0.4971	0.4975	0.4980	0.4992	0.5000	0.5009
10	0.4968	0.4979	0.4982	0.4987	0.4999	0.5010
Average:	0.4969	0.4975	0.4981	0.4988	0.4996	0.5008
S.D.:	0.00022	0.00024	0.00018	0.00028	0.00027	0.00025
Max:	0.4972	0.4979	0.4983	0.4992	0.5000	0.5011
Min:	0.4965	0.4971	0.4977	0.4983	0.4991	0.5003
Range:	0.0007	0.0008	0.0006	0.0009	0.0009	0.0008
+95%:	0.4973	0.4980	0.4984	0.4993	0.5001	0.5012
-95%:	0.4964	0.4970	0.4977	0.4982	0.4990	0.5003
	Growth:	0.00063	0.00057	0.00070	0.00081	0.00118
	Avg. Growth:	0.00078				
	Avg. S.D.:	0.00024				

	10:00	10:30	11:00	11:30	12:00	12:30
Average:	0.5015	0.5023	0.5031	0.5039	0.5047	0.5054
+95%:	**0.5020**	**0.5028**	**0.5036**	**0.5043**	**0.5051**	**0.5059**
-95%:	**0.5011**	**0.5018**	**0.5026**	**0.5034**	**0.5042**	**0.5050**

Figure 27.25 Step 10. Calculate the 95% Distribution for Each Predicted Time Frame

With the forecasted information, you can predict when the tolerances of the fabricated parts may go beyond the boundaries. Looking at the forecasted values, at 12:00, the +95% value reaches 0.5051 in., which is 0.0001 in. above the tolerance limit. Knowing this, an adjustment can be made to the process to bring the average back down around 11:00 to avoid defective components. Adding this forecasted information to the control chart displays a visual aid for what the predicted results could produce. Figure 27.26 shows the original control chart with the estimated values added into it. From looking at the chart, it shows the upper limit starting to pass beyond the tolerance boundary.

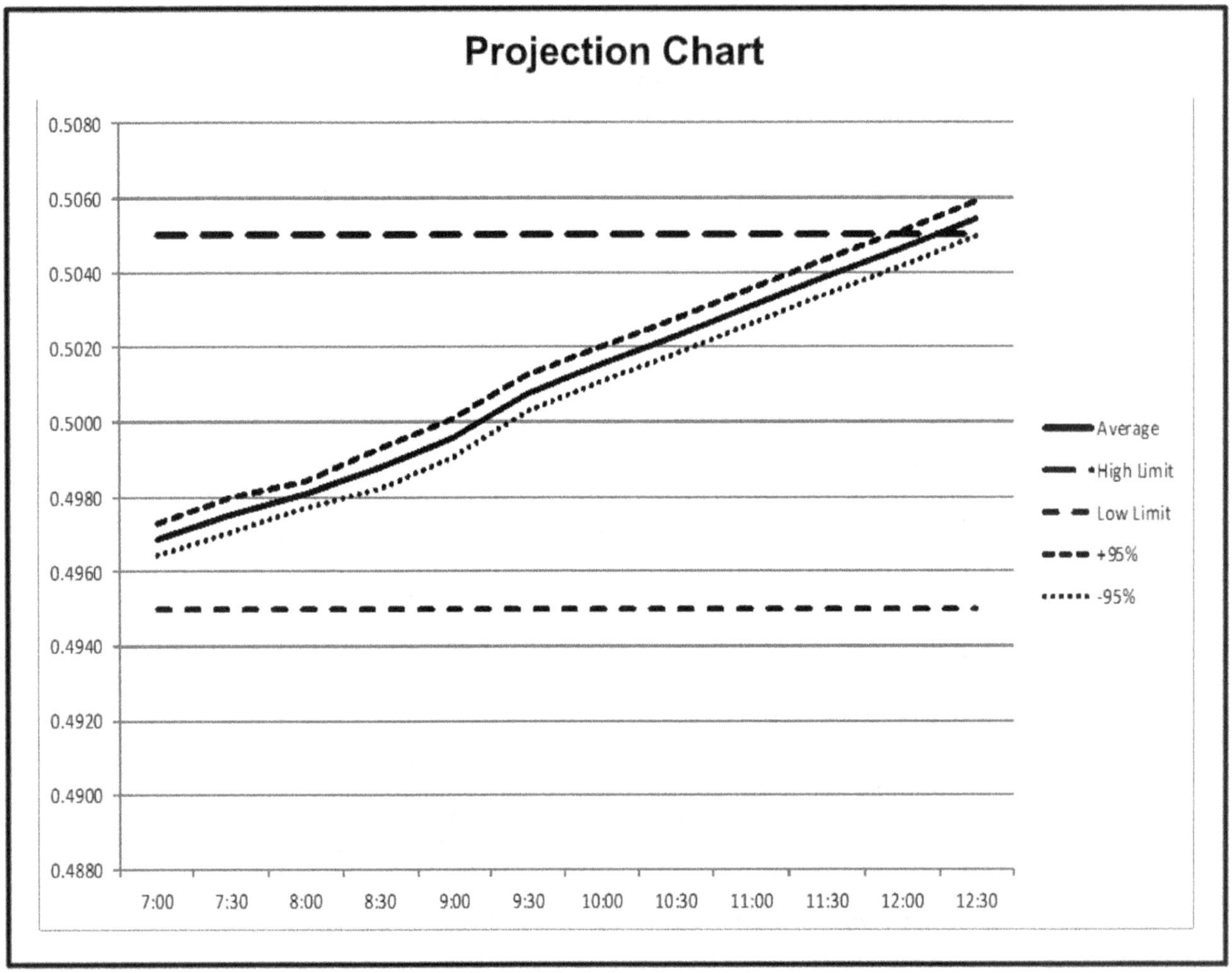

Figure 27.26 Control Chart for the Calculated Predicted Values from the Data Set Values

27.4.4 Out-of-Control Processes

When a process is running out of control, the data being analyzed from an SPC evaluation will not render consistency. Figure 27.27 shows data sets measured from a process that is running out of control. The analysis was carried out through Step 7, and looking at the growth rates from the average values, they vary from positive to negative. With inconsistent results like these, any probability forecasting would never render any useful results. Figure 27.28 shows a range chart for the six data sets analyzed. Looking at this, there is no consistency of pattern to the display. The ranges vary from 0.0006 to 0.0018 in. resulting in very large spread with no logical reasons. Figure 27.29 shows a control chart for the analyzed data with an up and down average with a varying ±95% spread. Any attempt to forecast what the future data sets would produce would be impossible and never render any useful information whatsoever.

SPC Control Data

Dimension: 0.500
Tolerance: ±0.005

Reading	7:00	7:30	8:00	8:30	9:00	9:30
1	0.4967	0.4983	0.4963	0.4988	0.4974	0.5024
2	0.4965	0.4981	0.4963	0.4998	0.4978	0.5028
3	0.4971	0.4984	0.4967	0.4997	0.4973	0.5027
4	0.4966	0.4986	0.4961	0.4993	0.4975	0.5030
5	0.4969	0.4984	0.4964	0.5003	0.4977	0.5028
6	0.4972	0.4987	0.4962	0.4996	0.4974	0.5026
7	0.4970	0.4973	0.4967	0.5001	0.4978	0.5023
8	0.4968	0.4988	0.4963	0.5000	0.4977	0.5019
9	0.4971	0.4985	0.4964	0.5002	0.4980	0.5029
10	0.4968	0.4992	0.4962	0.4997	0.4979	0.5037
Average:	0.4969	0.4984	0.4964	0.4998	0.4977	0.5027
S.D.:	0.00022	0.00047	0.00019	0.00043	0.00022	0.00045
Max:	0.4972	0.4992	0.4967	0.5003	0.4980	0.5037
Min:	0.4965	0.4973	0.4961	0.4988	0.4973	0.5019
Range:	0.0007	0.0019	0.0006	0.0015	0.0007	0.0018
+95%:	0.4973	0.4994	0.4967	0.5006	0.4981	0.5036
-95%:	0.4964	0.4975	0.4960	0.4989	0.4972	0.5018
	Growth:	0.00206	-0.00262	0.00385	-0.00250	0.00551
	Avg. Growth:					
	Avg. S.D.:	0.00033				

	10:00	10:30	11:00	11:30	12:00	12:30
Average:						
+95%:						
-95%:						

Figure 27.27 Data Collected from an Out-of-Control Process

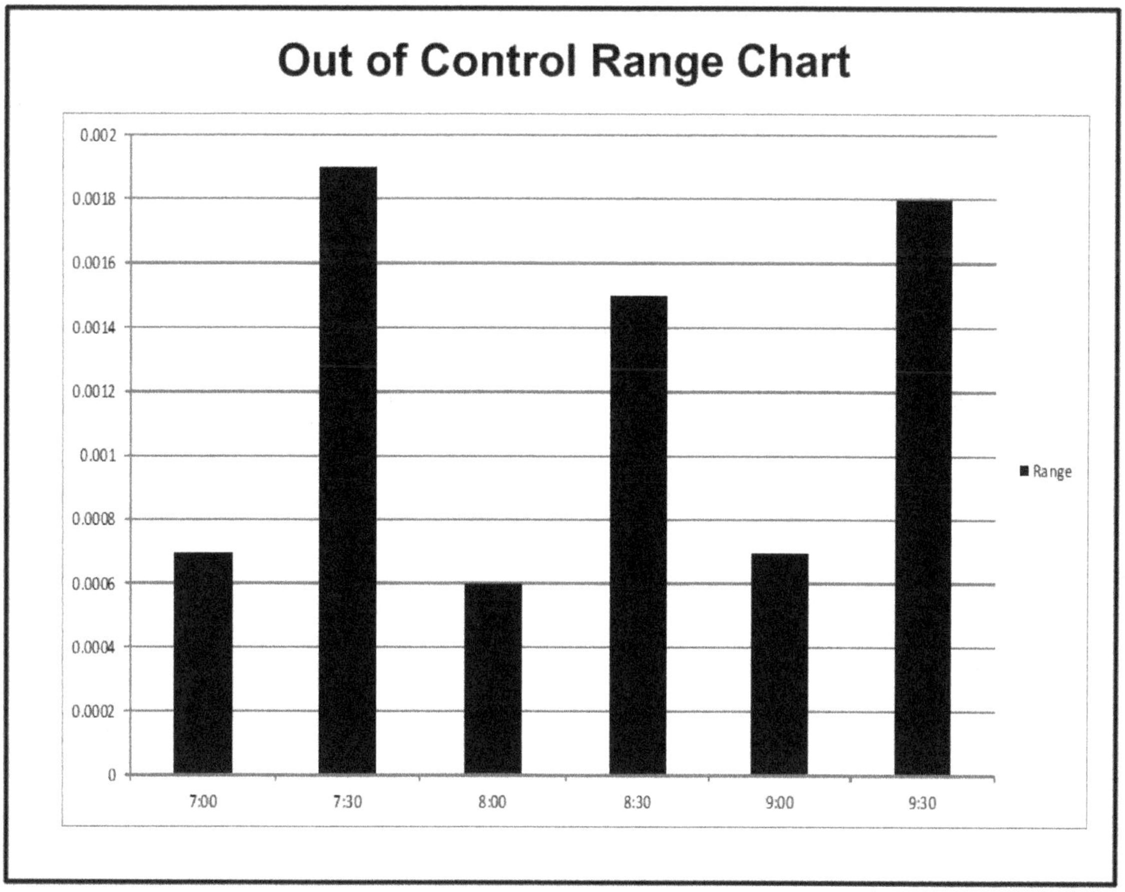

Figure 27.28 Range Chart for the Out-of-Control Process Data

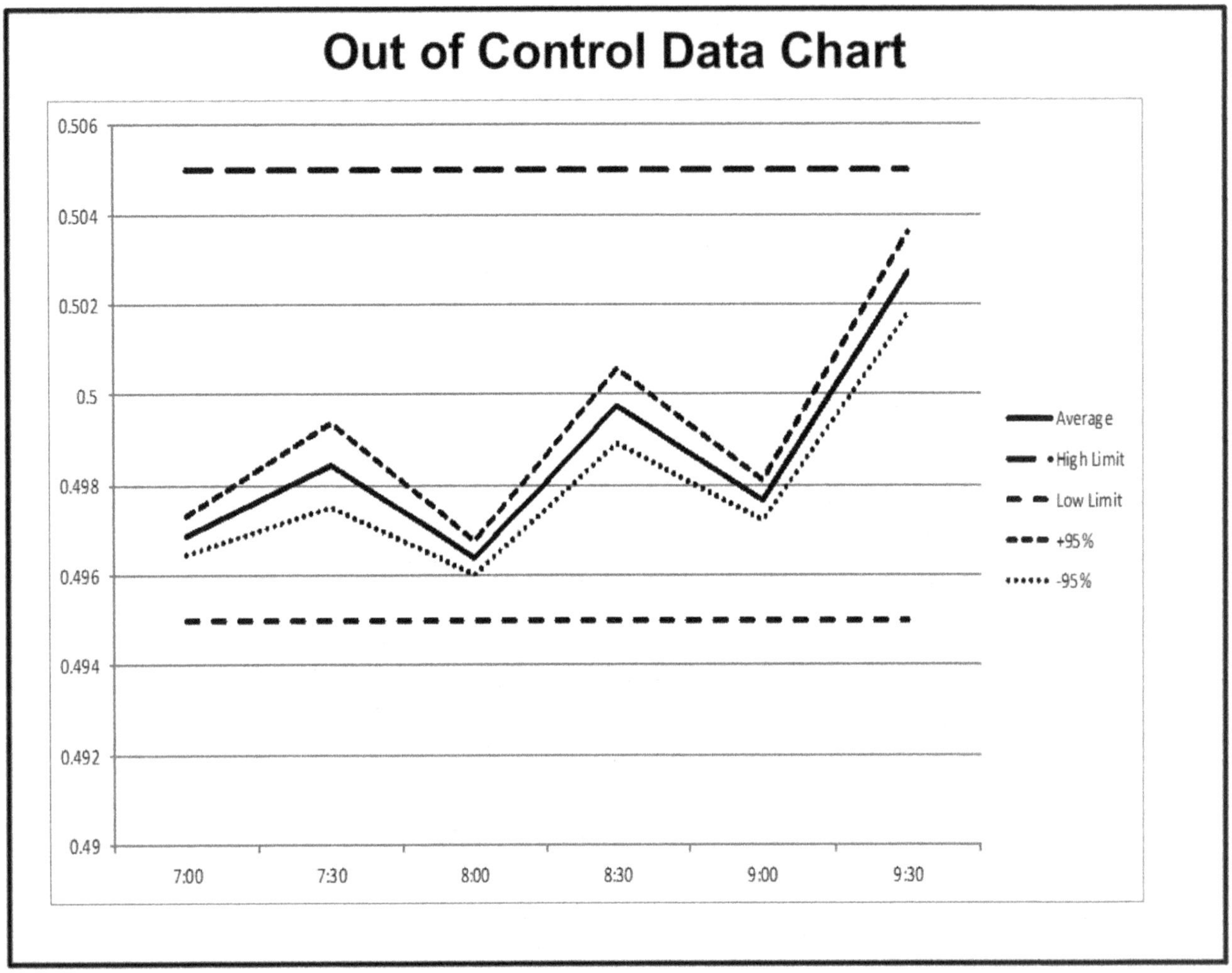

Figure 27.29 Control Chart for the Out-of-Control Process Data

27.5 SUMMARY

Quality is far more than just producing acceptable components; it is a blanket that covers the entire operation. From having a total quality outlook within the operation, the focus from everyone will be on perfection and strive to attain it in every possible manner. Looking at the quality of both the processes and products, understanding what a process is capable of producing consistently is the key to producing quality products. If the requirements for a component part are too close for a given process to maintain, knowing this before utilizing that particular process will avoid producing defective components, which may have to be either remanufactured or sorted to remove the defective parts.

Quality should always be a proactive function and not a reactive one. Monitoring the fabrication of the components during the process will ensure that they are being produced properly. With tools such as statistical process control, evaluations performed at given time frames will relinquish a real-time view of how the fabrication is performed and allow for adjustments before defective components are produced. By applying various simple statistical methods and utilizing a normal statistical distribution, precision assumptions can be made for either real-time SPC or lot sampling of previously produced component workpieces. Understanding the importance of quality from the beginning with the raw material through the final assembly of the process will provide the operation with power to produce the best products possible and be in favor with the customer base that it serves.

27.6 QUESTIONS AND PROBLEMS

Questions

1. Define quality from a product and process view point.

2. What are the two measures for manufactured product quality?

3. From an organizational point, what are the five pillars of quality, and why are they important?

4. What is a normal statistical distribution and a standard deviation, and why are they important from a quality standpoint?

5. What is the percentage under the distribution for $\pm 1\sigma$, $\pm 2\sigma$, and $\pm 3\sigma$?

6. Why would quality ever need to be measured utilizing the 6 sigma spread?

7. Define process capability, and why is this an important area to understand?

8. What is the difference between random variations and assignable variations?

9. What is statistical tolerancing, and what is it utilized for?

10. What are the statistical tolerancing assumptions, and how do they work together?

11. When performing an inspection, what is the difference between a population inspection and a sample inspection, and when would they be utilized?

12. Define statistical process control (SPC), and why is this an important function in component fabrication?

13. What is the confidence level utilized in SPC, and why is this level used?

14. When performing SPC on components, why is the population standard deviation used rather than the sample standard deviation?

15. What do range charts and control chart show in SPC, and why are they important?

16. How is probability forecasting performed in SPC, and what does it do for the process producing the components?

17. What is meant by an out-of-control process?

18. Why can't a process that is out of control have probability forecasting applied to it?

Problems

1. A given lot of components have an average value of 2.250 in. and a standard deviation of 0.008294. What would the values be at ± 1, 2, and 3σ?

2. A part has a dimension of 1.750 in. with a tolerance of ±0.003 in.. If an inspected lot of components rendered a standard deviation of 0.000863, would it produce an acceptance at ±3σ?

3. An inspected lot of parts rendered an average value of 0.758 with a standard deviation of 0.001952. What would the upper and lower values be at +1.65σ and −1.73σ, and what percentage would fall under the distribution at those indicators? Use Figure 27.4 for area under the curve.

4. Statistical tolerancing is going to be applied to an assembly of four components. Each part has the following dimensions and tolerances:

   1. $5.672^{+0.004/-0.002}$ 2. $2.745^{+0.001/-0.002}$ 3. $4.104^{+0.005/-0.005}$

   What would the assembly length and tolerances need to be for it to be acceptable?

5. Four components are to be assembled with the following dimensions and tolerances:

   $0.875^{+0.004/-0.003}$ $0.625^{+0.002/-0.004}$ $1.375^{+0.004/-0.002}$ $1.473^{+0.002/-0.004}$

   If the assembly requires a tolerance of ±0.0055, would statistical tolerancing work for this?

6. An inspection is to be performed on a population of 20 components. The data set collected is as follows:

1) 2.1257	6) 2.1250	11) 2.1238	16) 2.1254
2) 2.1252	7) 2.1241	12) 2.1243	17) 2.1261
3) 2.1245	8) 2.1259	13) 2.1257	18) 2.1243
4) 2.1259	9) 2.1260	14) 2.1258	19) 2.1238
5) 2.1242	10) 2.1256	15) 2.1248	20) 2.1262

   If the dimension on the part drawing is 2.125 with a tolerance of ±0.003, would the statistical analysis say that the lot was acceptable at ±3σ?

7. Instead of performing a population inspection of the 20 components listed in problem #6, what would the statistical results be if a sample inspection was performed where randomly parts 2, 3, 8, 10, 14, 17, and 19 were selected? Would the lot be acceptable at $\pm 3\sigma$?

8. A process is running producing components. SPC is going to be implemented to determine that the process is in control and producing acceptable parts. It was determined to check six parts every hour starting at 7:00 a.m.. $0.875^{\pm 0.008}$ is the dimension and tolerance on the part drawing and the data sets collected from 7:00 to 12:00 are as follows:

7:00	8:00	9:00	10:00	11:00	12:00
0.8741	0.8752	0.8761	0.8774	0.8782	0.8795
0.8745	0.8755	0.8766	0.8777	0.8785	0.8791
0.8743	0.8751	0.8764	0.8775	0.8781	0.8793
0.8746	0.8757	0.8769	0.8771	0.8786	0.8798
0.8742	0.8756	0.8764	0.8773	0.8784	0.8795
0.8747	0.8758	0.8762	0.8772	0.8785	0.8792

With the measured results for each time frame, determine the following:

a) Average value for each set
b) Standard deviation for each set
c) Maximum value for each set
d) Minimum value for each set
e) Range for each set
f) +95% Confidence level for each set
g) −95% Confidence level for each set.

From reviewing the statistical data, would the process be in control or out of control?

9. With the data sets in problem and your calculated statistical values, do a projection forecast for time frames from 1:00 to 6:00. With the information for the forecasted averages and ±95% confidence levels, when would the projected tolerances go outside the tolerance of the part drawing?

CHAPTER

28

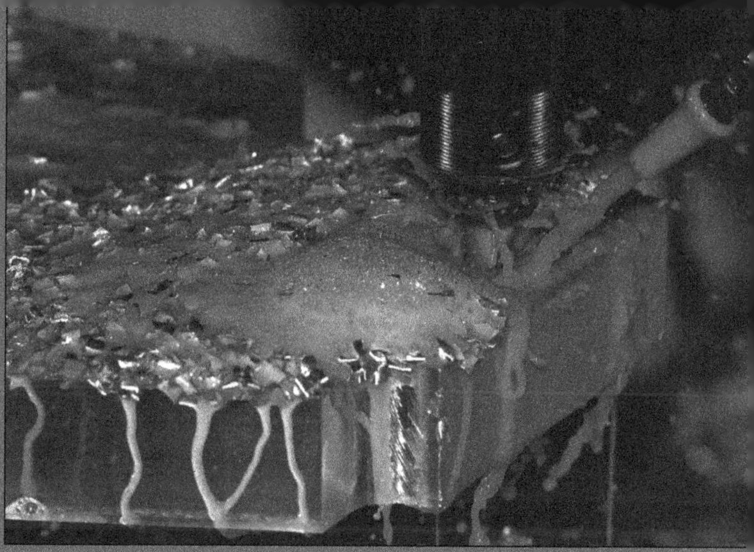

Numerical Control Theory and Concepts

28.1 INTRODUCTION

Numerical control (NC) has grown greatly over the years. This is for all-intensive-purposes manufacturing automation where the fabrication equipment runs unattended to produce the parts. The machine tools that are the most common for NC applications are milling machines, engine lathes, robotics, EDM equipment, laser, plasma, and water jet cutting. Several other pieces of equipment utilize numerical control, but the aforementioned ones dominate the NC arena. With this introduction and refinement of the technology, NC has provided the operations the flexibility of increasing throughput by reducing fabrication times and utilizing less labor. The ability to respond faster to the needs of the market is key to being a leader in the industry and providing the customers with the highest quality products at the lowest possible costs. This chapter provides an overview of the numerical control world and explains how the process works along with the benefits it can render within the operation.

28.2 NUMERICAL CONTROL DEFINED

By definition, numerical control is flexible automation where the mechanical motions are controlled by a program containing alphanumeric data. The program tells the machine tool where to move and at what speed to generate the component workpieces being fabricated in the operation. As the machine reads the program, motion is calculated and movement is produced in the various directions desired allowing the equipment to perform all the desired tasks without any operator intervention.

How the programming information interfaces with the NC equipment is accomplished in two manners. One is Computer Numerical Control (CNC) where the machine has a computer as part of the unit.

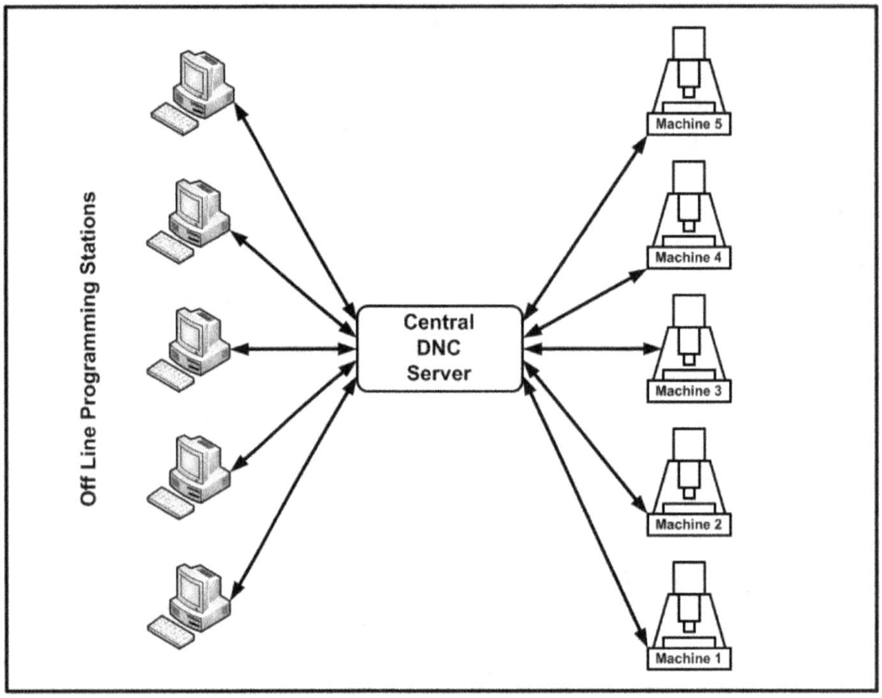

Figure 28.1 Direct Numerical Control System Configuration

All numerical control equipment has a computer as part of the equipment. A program can be either generated on that computer directly or written on a separate computer and loaded into the machine computer by external means. The machine's computer can store several programs depending on the storage memory, which can be called up to fabricate components when required. This is common for operations that have a few machines and is easy to change from one program to another.

Another method is Direct Numerical Control (DNC) where all the programs written are stored on a central server. From the server, a given program can be called up on several machines to run the component parts. This is common in large operation where several NC machines are being utilized in production. A system of this nature allows the program to be called up on different machines, which allows for either multiple machines to run the same operation, or if the machine that it would normally be performed on is occupied with another job, it can be downloaded to alternative machines. Figure 28.1 illustrates the configuration of a direct numerical control system with every NC machine tied to a central server.

28.3 NUMERICAL CONTROL COMPONENTS

For a completely automated manufacturing system like numerical control, the components that operate it are relatively simple and few. The components required to operate an NC system are the part program, the machine control unit, and the fabrication equipment. These items are described as follows.

Numerical Control System Components

- *Part Program*: The part program is what gets generated from the requirements on the part drawing. This tells the machine where to move to, what speed to move at, what cutting tools to use, along with other critical information the operation will require. The program is the first stage in an NC process and is the most critical. Any errors in the program will result in errors in the fabrication. There are several methods to generate an NC program, which are covered in Section 28.6.
- *Machine Control Unit*: A machine control unit is the heartbeat of the NC process. This unit will read the program and transfer it into mechanical motion. The fabrication equipment will receive all of the commands from the machine control unit and respond according to the commands from the program.

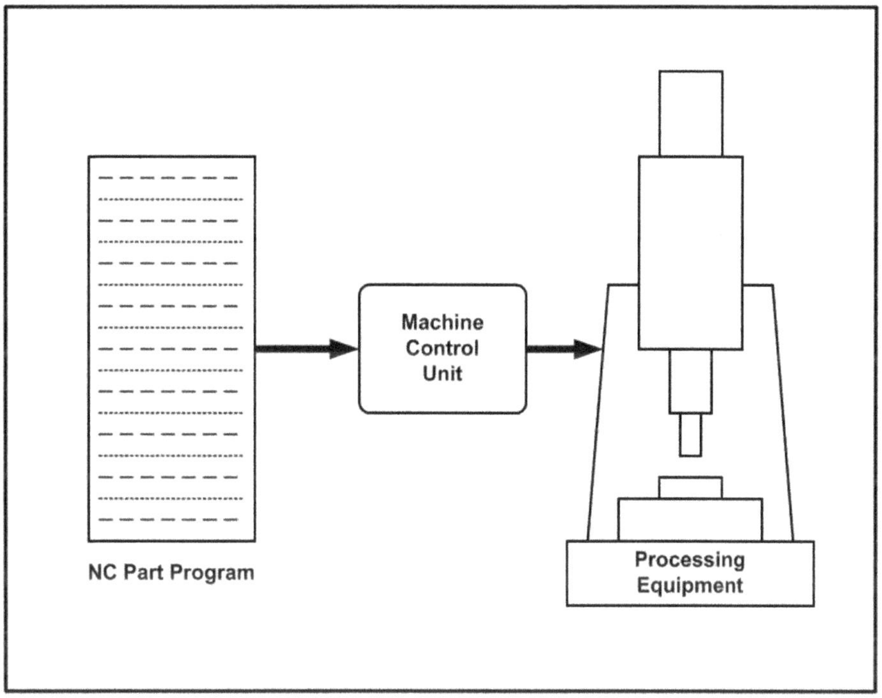

Figure 28.2 Components of a Numerical Control System

— *Fabrication Equipment*: Numerical control equipment can vary greatly depending on what is being controlled. These can be milling machines, engine lathes, robots, burning equipment such as EMD, laser, plasma, and water jet, etc. Basically any equipment that is set up for numerical control input can be operated.

Figure 28.2 illustrates the components of a numerical control system and how they flow from one to another.

28.4 COORDINATE SYSTEMS

A coordinate system defines the range of motions that a given piece of fabrication equipment has. Some may have as few as two ranges of motion, while others can possess up to six ranges of motion. The first category of motion is the Cartesian coordinate system that moves in three directional axes, which are classified as "X," "Y," and "Z". The "X" axis moves the machine left and right, the "Y" axis moves it in and out, and the "Z" axis moves the machine up and down. All NC machines will have these three main ranges of motion, and by combining the movements of two or three axes at the same time, contoured configurations can be produced on the component workpieces. Figure 28.3 illustrates the Cartesian coordinate system with the "X," "Y," and "Z" axes ranges.

The addition of three more ranges of motion allows for more complex and interactive operation to be performed on the NC equipment. These axes provide angular and rotational ranges of motion for the equipment, allowing for producing geometric configurations that would either take multiple operations or not be possible with only "X", "Y," and "Z" axes' capability. The additional axes are Roll, Pitch, and Yawl and are classified as "a", "b," and "c."

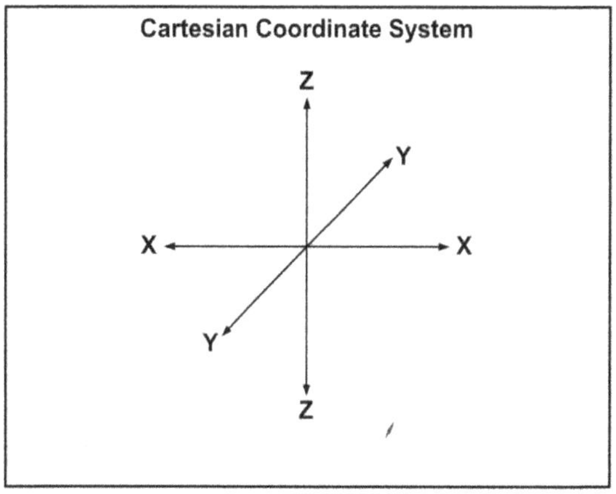

Figure 28.3 Axes 1–3 (X, Y, and Z) Coordinate Motion System

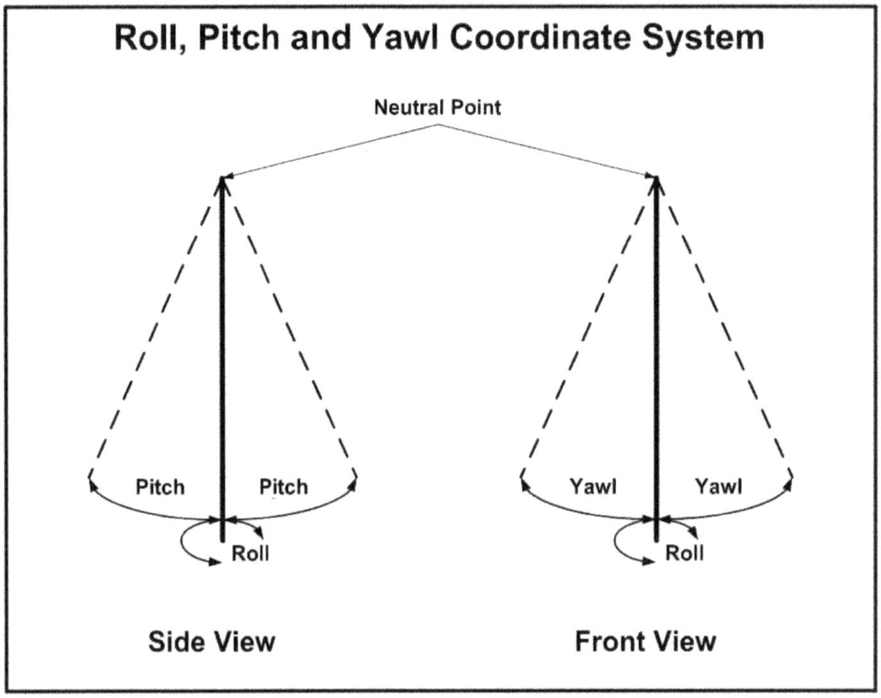

Figure 28.4 Axes 4–6 (Roll, Pitch, Yawl) Coordinate System

A good example of how these axes operate is to look at the lower arm of a person. The roll ("a") would be like rotating the hand. The pitch ("b") and yaw ("c") axes would be moving the hands up and down or side to side. Figure 28.4 illustrates the additional system for a, b, and c axes ranges.

28.5 POSITIONING AND MOTION CONTROL SYSTEMS

Numerical control systems are dependent on the ability to both control the motion of the equipment and provide positioning in order to fabricate the desired geometric configurations required. Positioning is how the machine will interpret the program and move from one location to another, while motion control is how the equipment will move from one location to the next. These are explained as follows.

28.5.1 Positioning Systems

The positioning system determines the movement of the equipment based on the program requirements. This is done in two ways, which are absolute positioning and incremental positioning. With absolute positioning, every move of the equipment is taken from the origin point for X, Y, and Z. Figure 28.5 shows how an absolute positioning system operates with all the movement coordinates reference from the origin location. In an incremental positioning system, the first movement is taken from the origin, and then each movement from the first is taken from the current location. Figure 28.6 illustrates the incremental positioning system where the coordinates are reset from each movement back to zero and the next movement is taken from that point.

It is important to understand how the motion control unit will read the program in either an absolute or an incremental manner in order to generate the program coordinates properly. This is a factor of the equipment, and it will read one or the other but not both. From a precision viewpoint, absolute positioning will produce a tighter range of motion due to the fact that every coordinate is taken from a common point rather than resetting for each future move. However, the precision is also a factor of the motion control systems in the equipment as well.

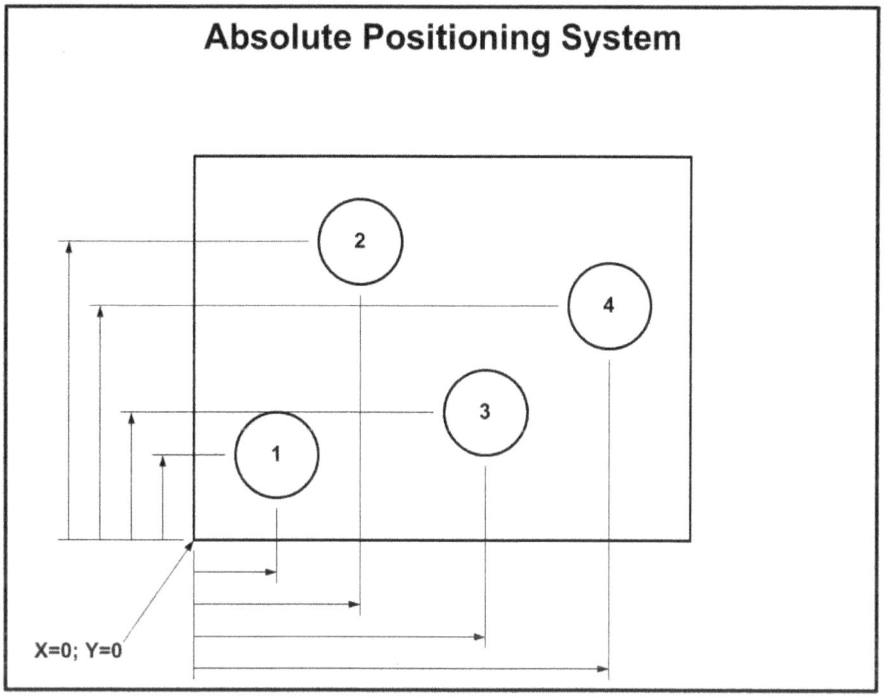

Figure 28.5 Absolute Positioning with all Movements Taken from a Common Zero Point

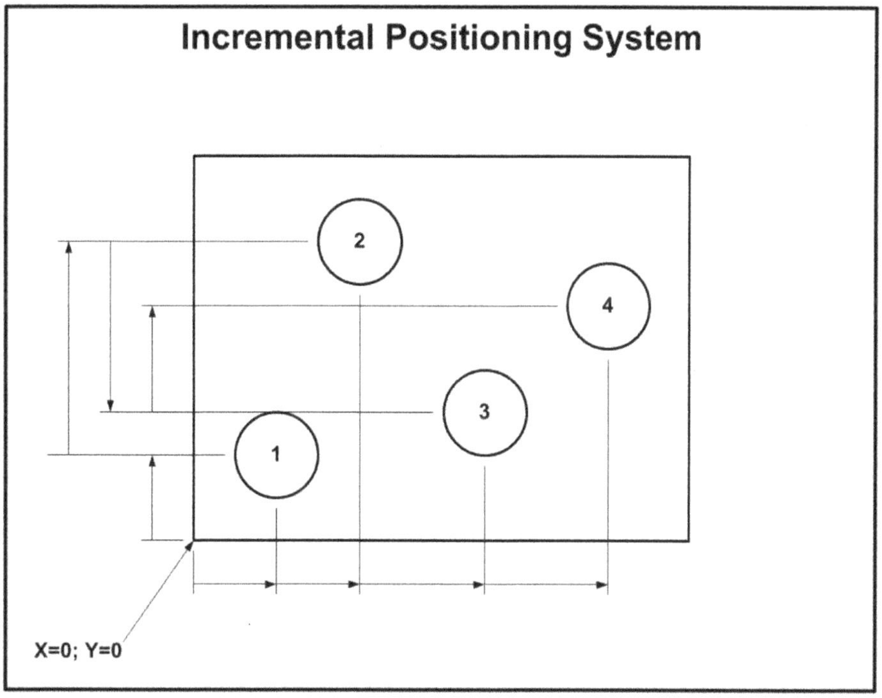

Figure 28.6 Incremental Positioning with Each Movement Taken from the Previous Location

28.5.2 Motion Control Systems

Motion control is divided into two categories. These are how the equipment moves from one point to the other on the workpieces and how the equipment knows it has reached the requested destination from the program. The systems for moving from one location to the other are classified as either point-to-point or continuous path. How the equipment knows it has reached the desired location would be either an open-loop system or a closed-loop system. These motion control systems are defined as follows.

28.5.2.1 Point-to-Point vs. Continuous Path Systems

Both of these motion control systems will move the equipment from one location to the other but will perform in different manners. With a point-to-point system, the equipment will determine how it is going to move to the various locations. As the program is read through the motion control unit, the equipment will make the decision on how to go to the next location. This movement may vary each time it makes the move and the operation will have no control on how it reaches the next destination. Figure 28.7 illustrates a point-to-point system showing a variety of paths the equipment could select for the movement. A system such as this is utilized mainly for operations where the location movement is not critical, such as for drilling holes in various locations on the workpiece.

In a continuous path system, the movement from one location is determined by the operator. This type of system is used to perform contour cutting on the workpiece where the path of the equipment is critical to the geometric configurations of the component being processed. Figure 28.8 illustrates a continuous path system where the movement from one point to the other on the workpiece is maintained and controlled.

Every numerical control machine will offer the option to perform either point-to-point or continuous path movement. Depending on the application of the operation, the program can select which movement system would be acceptable for the operation being performed. Even on the same workpiece, the program can utilize a combination of both point-to-point and continuous path motions, which will be determined by the program and the motion control unit.

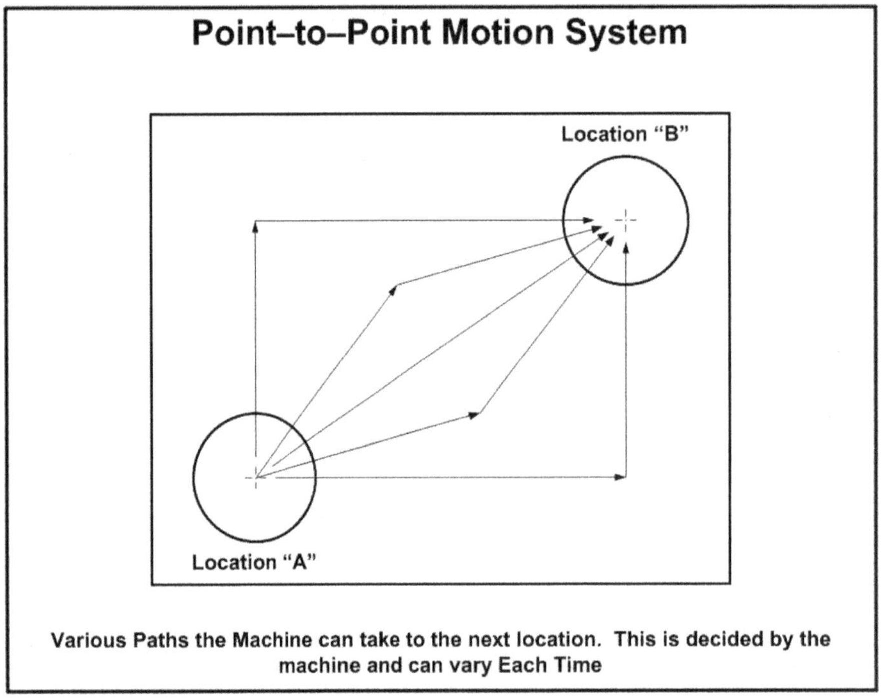

Figure 28.7 Point-to-Point Motion System

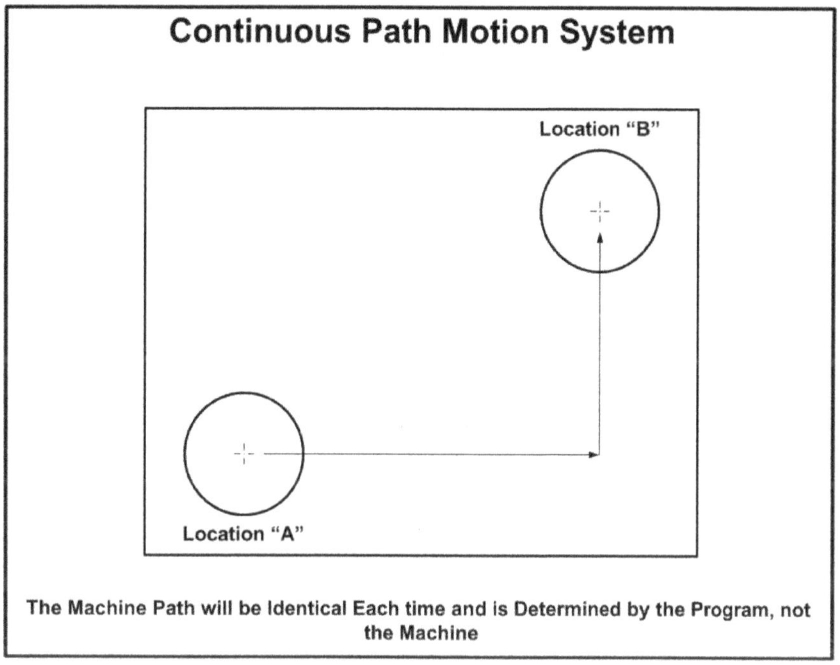

Figure 28.8 Continuous Path Motion System

28.5.2.2 Open-Loop vs. Closed-Loop Systems

The systems determining how the equipment knows it has reached the determined destination can be either an open-loop system or a closed-loop system. Each of these systems will move the equipment, but how it knows the location has been reached is different. These two systems of location determination are defined as follows.

Open-Loop System

Open-loop systems have no feedback mechanism to let the equipment know that it has reached the desired location. This operates using a stepper motor, which rotates in small, incremental steps and with the combination of the number of threads on the lead screw will determine both the location and precision resolution of the movement. Figure 28.9 illustrates an open-loop system where the pulse train input is received from

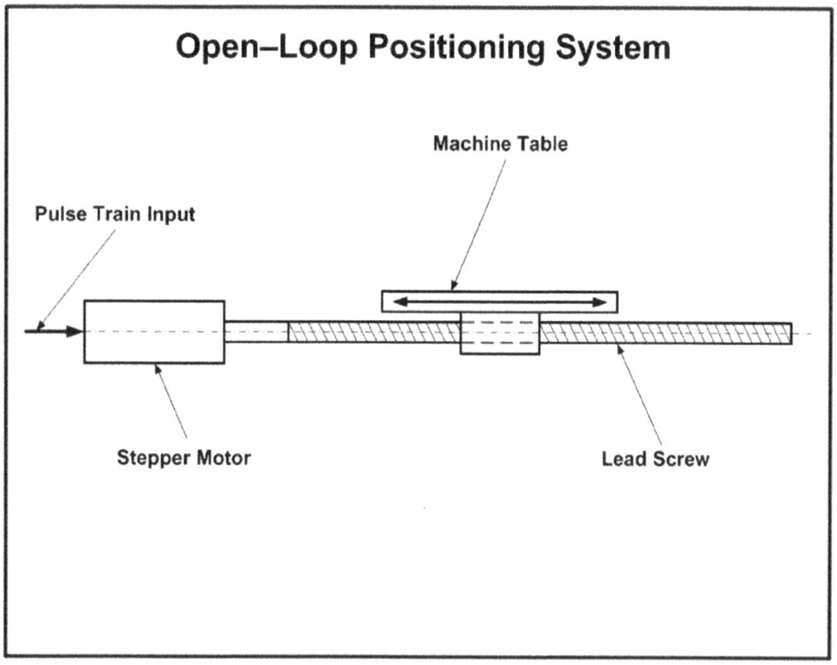

Figure 28.9 Open-Loop Motion Control System

the machine control unit and based on the number of steps required, the equipment will determine when the desired location is reached. The resolution precision of the equipment is explained as follows.

Lead Screw Thread per Inch = 20

Stepper Motor Increments = 75

1/20 = 0.0500 in. Lead Screw Resolution

1/75 = 0.0133 in. Stepper Motor Resolution

0.0500 × 0.0133 = 0.00067 in. Total Equipment Resolution Precision

In a closed-loop environment, the system will receive feedback as to the exact location of the equipment at any point in time and halt when it has reached the required destination. This operates by receiving an input signal into a comparator unit, which is sent to a digital-to-analog convertor (DAC), which directs the servomotor to begin the motion. As the equipment moves, an encoder records the steps and sends a signal back to the comparator unit to verify the exact location of the equipment. When the desired location is reached, the input signal will stop. Figure 28.10 illustrates a closed-loop system where feedback signals are constantly received from the encoder unit to the comparator unit.

An encoder, or digital encoder as it is properly termed, is a disk that rotates and a light beam is sent through the openings in the disk. Each time the light beam is broken, a signal is sent to the comparator unit to verify the location. Figure 28.11 shows how a digital encoder is constructed with the slots for letting the light beam pass through as it rotates. The encoder is attached to the lead screw of the equipment, which is responsible for the motion. In many instances, the encoder will be geared in order for it to rotate faster than the lead screw, allowing for a higher precision resolution. The total equipment precision resolution is a combination of the lead screw threads per inch, the number of openings on the encoder wheel, and the increase of the gearing from the lead screw to the encoder. Figure 28.12 shows the calculations for the resolution precision for a typical closed-loop system.

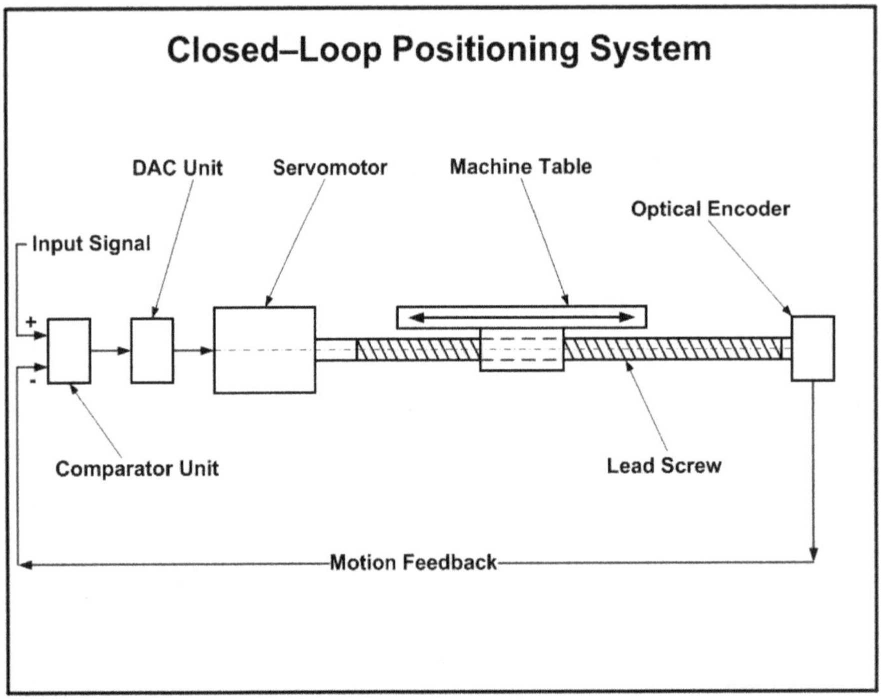

Figure 28.10 Closed-Loop Motion Control System

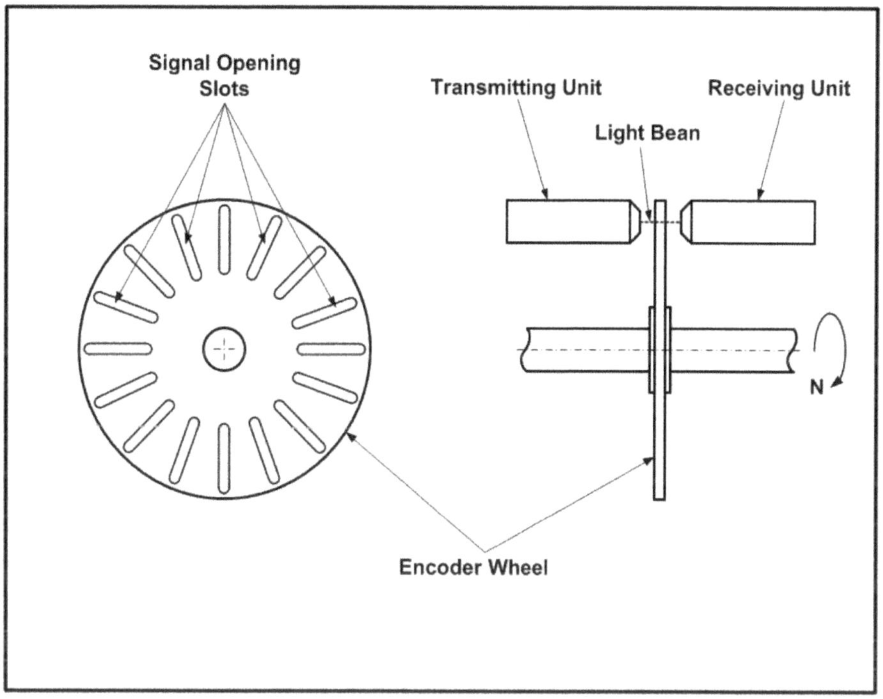

Figure 28.11 Encoder Wheel Construction for Closed-Loop System Feedback

Encoder Resolution and Motion Speed	
Positioning Resolution	
Encoder Signal Slots:	40 0.0250
Lead Screw Threads per Inch:	15 0.0667
Encoder Increased Gearing:	5/1 0.200
Positioning Resolution:	0.00033"
Maximum Motion Speed	
Motor Maximum RPM:	2500
Lead Screw Threads per Inch:	15
Maximum Inches per Min:	166.667
Maximum Feet per Min:	13.889

Figure 28.12 Breakdown of Encoder Feedback Calculations for Positioning and Travel Speed

28.6 NUMERICAL CONTROL PROGRAMMING

The program utilized for all numerical control equipment is the starting point for the operation. A part drawing with the required operations is reviewed for the requirements, and a program is generated for the motion control unit to read and turn it into mechanical motions. These programs will consist of alphanumeric data for each movement that is desired during the processing. There are four methods for generating a numerical control program, which are Manual Data Input, Manual Programming, Computer-Assisted Programming, and CAD/CAM-Assisted Programming, described as follows.

Manual Data Input

The manual data input programming method is used for small programs or for editing areas requiring revision at the point of use. Here, the data required to perform the operations is inputted into the equipment directly. The information is loaded into the control computer located on the equipment line by line to inform the machine control unit what moves are required and how. This is used for small operations along with prototype work where only a few components are required for the operations. Another use of manual data input is to implement small revisions of a program. If a change is required for a large program that was written previously and an error was detected during the initial run of the components, it may be easier to implement the changes at the point of use rather than off line. The programs may be stored in the equipment's computer if future runs will be required.

Manual Programming

Manual programming is performed identical to manual data input programming but is performed off line on a computer not attached to the processing equipment. This is used for simple programs such as drilling holes into a component workpiece and simple material removal operations. Each line of the programming code is written in the sequence required to complete the required tasks of the equipment. Once the program is written, it is loaded into the computer attached to the equipment to run the operation.

Computer-Assisted Programming

With computer-assisted programming, the computer will have several subroutines that can be called up for the program being composed. This form of programming is used for more difficult operations where complex geometric shapes are required and not simple movements. For example, if a circle or an arc is required in the component workpiece, a subroutine can be called up and all this would require is to input the radius of the circle or arc and the computer will calculate the moves required to produce the shapes. In most cases, the subroutines will be moving more than one axis of the machine simultaneously, which would render any form of manual programming extremely difficult to calculate the combinations of moves required.

CAD/CAM-Assisted Programming

With the introduction of Computer-Aided Design/Computer-Aided Manufacturing (CAD/CAM) programming, numerical control equipment has become more efficient and effective. As the component workpieces are designed on a computer in a three-dimensional solid model in the CAD portion of the system, the solid model geometry is called up in the CAM side of the system. From the geometry that was designed, the NC program will be graphically programmed allowing the programmer to see what will be performed and how. The CAM system will poses the same subroutines as computer-assisted programming does, but a big advantage to CAD/CAM programming is simulation. With the three previously described programming methods, the program must be run on the machine to see if it is written correctly. Here, a simulation can be run on the computer to observe things like cutter paths and material removal to provide a visual of what will occur on the equipment during processing. Several errors can be detected in this phase, which is not possible with the other methods of programming. Also, the program will be attached to the solid designed model; so if revisions were to be made to the component part, it will flag the program so that the appropriate revisions can be made.

28.7 INDUSTRIAL ROBOTICS

The introduction of robotics in the manufacturing arena has become more pronounced over the past 10 to 15 years or so. By definition, an industrial robot is a programmable equipment that is capable of performing some human tasks. As technology has improved allowing the industrial robots to be able to perform more complex tasks, they have been implemented throughout the operations. When robots were first developed, they were considered unintelligent and performed very simple tasks such as moving something from one location to another like unloading and loading a machine. For the most part, the early robots were not programmable and constructed with fixed stops for the movements desired.

Today's robots are all numerically controlled and require programming as does the fabrication equipment. This allows them to be utilized for more than one function, and with the technological improvements, the precision resolution is much higher. Most modern-day robots will be equipped six axes of motion allowing for dexterity similar to that of a human arm. The majority of industrial robots will be stationary and work from a fixed location. This is referred to as the working envelope of the robot, and the tasks that the robot will be performing need to be within that area. Some robots are mobile and are used from moving materials from one location to another in the factory. They are either guided on tracks or will follow a magnetic strip on the floor. Depending on the operations to be performed, the robots will fall into one of three categories, which are Processing Operations, Assembly and Inspection, and Material Handling. These are explained below.

Processing Operations

Processing operations are those that will actually perform work on a component part. This can be operations like welding, grinding, and drilling. The manipulator arm of the robot would have the required tooling and, based on the program, perform the required tasks. Some of these tasks can be combined by having several tools that the robot can replace in the manipulator, allowing for multiple operations to be performed at a single workstation.

Assembly and Inspection

Assembly has become a big arena for robotics. As the product being assembled is moved into the work envelope, the robots assemble the various components, which traditionally was performed by human workers. In order for robotic assembly to be effective, the product needs to be designed with automated assembly in mind. For inspection operations, equipment is produced with robotic motions implemented in it and programs are comprised for the dimensions to be checked with the tolerances for each. The components are loaded into the equipment, and the robotic manipulator moves around touching the areas to be inspected with a tactile sensor or a laser. Each reading of the sensor or laser is recorded, and a report is generated with all the statistical analysis provided. Figure 28.13 shows a stationary industrial robot, which is used for loading and unloading production machines.

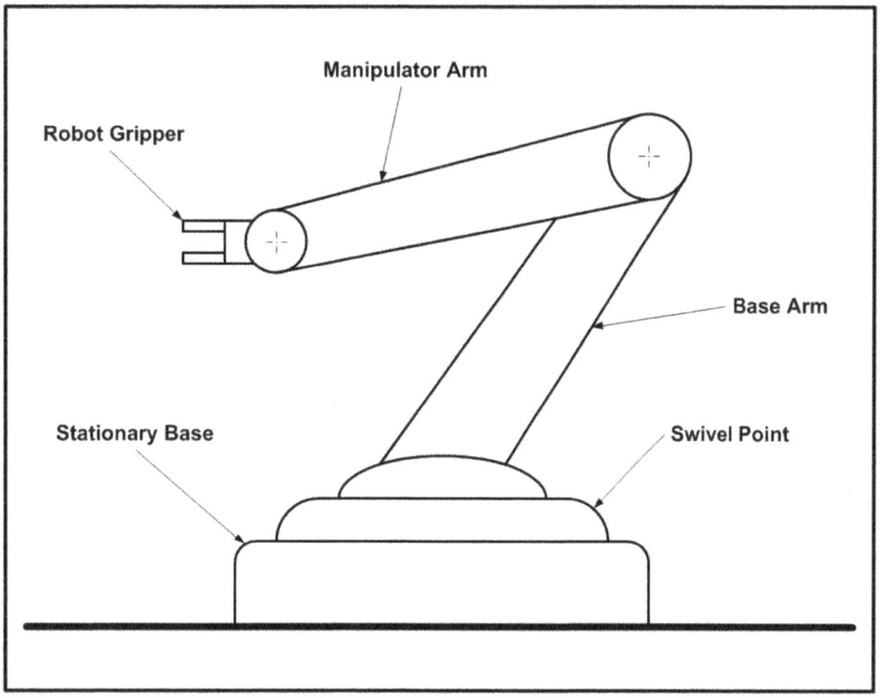

Figure 28.13 Typical Stationary Industrial Robot

Material Handling

Robotic material handling can be divided into two categories. The first is a stationary robot that is loading and unloading a machine. This has become common practice in areas where the unloading can cause harm to humans performing the same tasks. An example of this would be unloading castings that have just been molded. Although they are solidified, the casting can still have a temperature of between 1,000 and 1,800 °F, which can seriously burn a human operator. The other category is mobile robots, which are used to move materials from one location to another in the factory. These are programmed but will either be on tracks or follow at the magnetic tape strip on the floor. This application reduced the requirements of the operators to perform non-value-added time, and their talents can be utilized for more productive activities within the operation.

28.7.1 Robotic Applications

There are countless applications that can have robotics implemented for. However, even though they are programmable to be able to perform different tasks, some important considerations need to be evaluated in order to gain the most efficiency for the operation. These are described as follows.

— *Repetitive Work Cycles*: When tasks are extremely repetitive, those, if possible, are excellent applications for robotics. This is found in all high-production operations where a very high volume of a given product is being produced annually.

— *Long Production Runs*: A long production run refers to how long a given product will be produced. If a product or a portion of a product will be produced for years, it can render a robotic application. However, if the product will only be produced for a short period of time, it may not be feasible to implement an automated robotic application that will require a lot of initial setup and need to be broken down and reset up in a year of two.

— *Multishift Operations*: If an operation is running three shifts per day, robots can be a big advantage. This will eliminate the requirements for more operators to provide non-value-added tasks such as material movement and loading and unloading equipment.

— *Stationary Locations*: With the exception of some material movement robots, the application will be stationary in nature. If an operator is required to be at a given workstation for the entire shift, a robot can be implemented to perform the tasks if feasible.

— *Hazardous Environments*: When an operator is required to work in any type of a hazardous environment, robots are a perfect solution. This can include environments where there are toxic fumes, extreme heat, etc.

— *Difficult Human Tasks*: Difficult tasks are defined as operations where an operator could get injured. Lifting heavy or high-temperature components can injure the person where a robot that is powerful enough can move large, heavy, and cumbersome parts with ease. Also, components at a high temperature such as a casting can be moved without injury to the operators.

— *Established Component Positioning and Orientations*: Unlike a human operator, a robot will require that the components that are to be worked with always be in the exact same location and in the proper orientation. Robots cannot pick up a workpiece from a random pile and orientate it in the way a human could. This will require very expensive support tooling to ensure that every component or tool that the robot will need to utilize performs the required tasks.

28.8 NUMERICAL CONTROL BENEFITS

The introduction of numerical control equipment into an operation can render some very advantageous benefits. With the equipment being program-controlled and flexible, it can replace several manually operated machines and combine operations that previously required to be performed independently. The main advantages to numerical control are explained as follows.

— *Lower Manufacturing Lead Times*: A lead time refers to how long the processing of the component workpieces will take. Numerical control equipment will operate at the optimum speeds for the material being processed, where in most cases, manually operated machines will run below optimum with a safety factor for the operators. Another factor that will reduce lead times with NC equipment is the combination of operations. The NC equipment is capable of combining several operations that were performed independently in a manual mode.

— *Reducing Non-Value-Added Times*: With lowering the manufacturing lead time as stated earlier, the NC equipment will also reduce the non-value-added times within the operation. If three manually performed operations were to be combined into one NC operation, the material would not have to be transported to the different locations required for the manual operations. This reduces the transportation times and waiting times for the operations to be performed.

— *Greater Manufacturing Flexibility*: Numerical controlled equipment is programmed to perform whichever tasks are required for a given component part. When that batch of components is completed, the equipment can be reprogrammed for a new part easily. This is not possible with fixed automation where a piece of equipment is designed to perform one given task for one given component workpiece.

— *Improved Manufacturing Consistency*: Consistency is a major advantage of NC equipment. With each component being produced exactly the same, the quality of the fabricated parts will be more closely controlled than with manually performed operations. With no human intervention in the processing of the components, the quality will be more consistent.

— *Reduction of Human Errors*: Human errors are guaranteed to occur and are unavoidable. With no human intervention in numerically controlled equipment, each component will be produced exactly the same, resulting in a reduction of errors. For most NC operations, the only task performed by the operator is to load and unload the machines, eliminating errors in manual processing operations.

28.9 NUMERICAL CONTROL WORK CELLS

NC equipment is best implemented in what is called a work cell. This has more than one NC machine organized in a circular pattern, allowing a single operator to be responsible for running each machine. Figure 28.14 illustrates a numerical control work cell with a single operator running four machines with an inspection table in the center. Setting up the equipment in the work cell so that there will be minimum down time for the operator is important to gain the most efficiency from the operation. If the four machine work cells shown in Figure 28.14 were all producing the same operation with a fabrication time of 6.75 min each and 6.25 min were required to load, unload, and inspect each part, the starting times for the cycles will be staggered to allow time

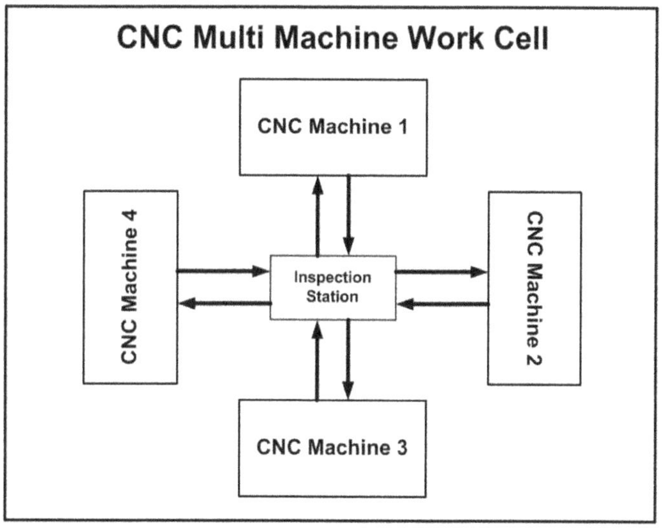

Figure 28.14 Typical Numerical Control Work Cell Layout

CNC Work Cell Cycle Balancing

Machine 1	
Cycle Time:	27.00 min
Load/Unload Time:	2.50 min
Inspection Time:	3.75 min
Machine 2	
Cycle Time:	27.00 min
Load/Unload Time:	2.50 min
Inspection Time:	3.75 min
Machine 3	
Cycle Time:	27.00 min
Load/Unload Time:	2.50 min
Inspection Time:	3.75 min
Machine 4	
Cycle Time:	27.00 min
Load/Unload Time:	2.50 min
Inspection Time:	3.75 min

1. Load, Unload, Inspect:	6.25 min
2. Load, Unload, Inspect:	6.25 min
3. Load, Unload, Inspect:	6.25 min
4. Load, Unload, Inspect:	6.25 min
Machine Cycle Timing	
27.00 / 4 = 6.75	
Each Machine is Started Every 6.75 Minutes Apart	
Machine Timed Cycle:	6.75 min
Handling/Inspection Time:	6.25 min
Time Between Cycles:	0.50 min

Figure 28.15 Numerical Control Work Cell Cycle Balancing Calculations

for the operator to perform their tasks. Figure 28.15 shows the calculations required to gain optimum efficiency with the staggered cycle time. From looking at this, the operator will have 0.50 min or 30 s between machine cycles while allowing them to perform the required tasks.

28.11 SUMMARY

Numerically controlled equipment offers several benefits to a manufacturing operation. With the flexibility and efficiency of the processing machines, they can reduce manufacturing lead times and improve product quality. With the produces being designed with CAD systems, the geometry can be utilized to generate programs with ease in the CAM system and be linked to the design in the event of changes. Simulations can also be run to view the actual fabrication prior to running actual part providing the ability to identify any errors before actually starting manufacturing. These benefits offer the operation several areas to improve the overall efficiency and effectiveness, allowing the operation to produce high-quality products rapidly to respond to the market needs of the customers.

28.12 QUESTIONS

1. By definition, what is numerical control?

2. What is the difference between computer numerical control (CNC) and direct numerical control (DNC)?

3. What are the components of a numerical control system, and what function does each perform?

4. What are Cartesian coordinates, and which axes do they control?

5. When NC equipment has more axes of motion than "X," "Y," and "Z", what are they and how do they operate?

6. What is the difference between positioning and motion control in a numerical control environment?

7. What is the difference between absolute and incremental positioning?

8. Define the difference between point-to-point and continuous path motion control.

9. How are an open-loop and a closed-loop motion control system for NC equipment different?

10. What is a digital encoder, and what function does it perform in an NC motion control system?

11. How are manual input and manual programming different?

12. What advantages do computer-assisted programming and CAD/CAM-assisted programming offer in numerical control?

13. By definition, what are industrial robots?

14. What is meant by a work envelope in industrial robotics?

15. What are the three categories in which industrial robots perform work?

16. Why are component positioning and orientation critical in an industrial robot implementation?

CHAPTER

29

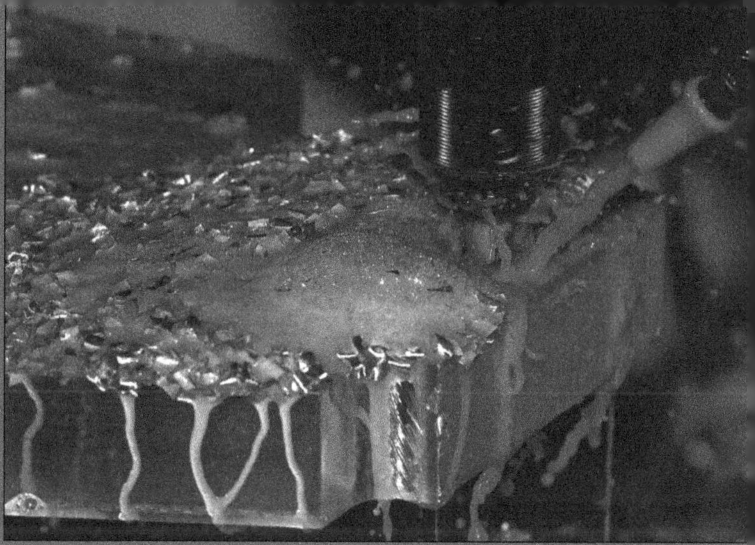

Rapid Prototyping

29.0 INTRODUCTION

Rapid prototyping is not a true manufacturing process, but aids greatly in the product development process. During the design and development stages of a new product, it is critical to build a test unit to prove out various concepts for feasibility and functionality. Having the ability to fabricate prototype components rapidly greatly speeds up the process and provides quick feedback for functionality and feasibility. With the component parts and assembly being designed now in a three-dimensional (3D) solid modeling CAD system, the geometry can be utilized in whichever rapid prototype system that is selected. Before the introduction of various rapid prototyping processes, components needed to be fabricated by hand, which was extremely time consuming, thus delaying the initial testing of the product being designed. The more rapidly functional components can be fabricated and assembled for testing, the feasibility and functionality of the design can be confirmed and alterations can be made. This chapter will cover some of the more popular and readily utilized methods for the rapid fabrication of prototype components that will be used for testing and evaluation during the design process.

29.1 RAPID PROTOTYPE MATERIALS

The rapid prototyping arena has grown dramatically over the years. In its conception, this process would produce a model of a component part that would resemble the desired geometry of the component but was not usable for any stringent testing in an assembly. As these were created from a 3D solid model and were not completely functional, the process was referred to as 3D printing. Today, the processes have become much more technically advanced and can produce component parts that are functional and can be utilized for testing. 3D printing is still used to produce a model to evaluate the configuration, but the advancements in both processes and materials have allowed for the construction of functional parts.

The materials that can be used in rapid prototyping are divided into two main categories: metals and polymers. Metals can be rapid prototyped either from a solid piece of raw material or from metallic powder, depending on the process being used. This offers the construction of a prototype component of the exact material that the design requires. The polymer materials are broadly available. They can be constructed either from a solid piece or from a special process utilizing either a UV light or laser to cure the resin to form the geometry. However, the material being used needs to be a thermoplastic and not a thermoset plastic. (See Chapter 6 for clarification of the materials.) Depending on whether the component being prototyped needs to be functional or not, the material and/or process will be selected for the best results.

29.2 COMPONENT FABRICATION

Constructing prototype components can be performed in three ways. The first for simple parts is by hand utilizing conventional machine shop equipment, such as milling machines and engine lathes, that a typical machine shop would have. These components would be fabricated from solid raw material and require removal from the stock to construct the required geometry. This method can be effective for simple components where the quantity is low, but is not rapid in nature. However, this can be a very effective method for simple parts and will render a completely functional component fabricated from the exact material that the part requires in the design.

More typical processes for fabricating complex prototype components are the multilayer processes and computer numerical control equipment. Both of these processes will utilize the solid model created in the CAD system and generate a program from the model to produce the desired geometry. These two methods will produce extremely complex components that are difficult to be fabricated in a manual manner and can produce multiple parts once the program has been created. These two processes will not require an operator to be at the fabrication point constantly. Both the multilayer and numerical control processes are discussed below.

29.2.1 Multilayer Fabrication

This process is one that takes the geometry from the 3D solid model and generates multiple layers slicing them at a given thickness and creates a program for each layer. Figure 29.1 shows the solid geometry and the layers created for fabrication.

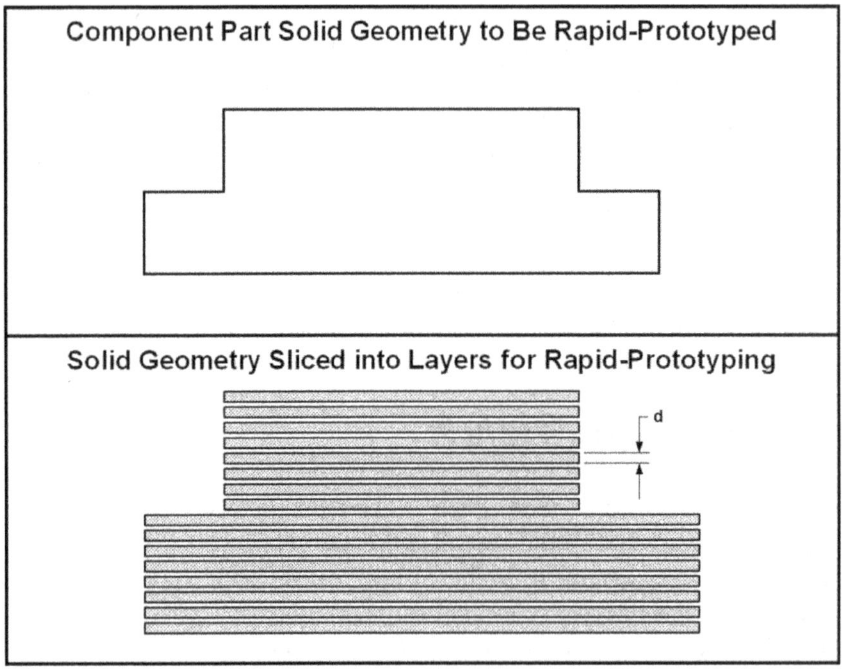

Figure 29.1 Layer Created Geometry Fabrication

Each layer will be fabricated starting from the bottom layer and working up to the top, stacking the layers to construct the desired geometry. Depending on the process and the material being utilized the thickness of each layer being produced will be determined. The layers adhered to one another as they are stacked together and the component is generated starting from the bottom layer working up to the top layer. Figure 29.2 illustrates the layers being produced and the stacking process.

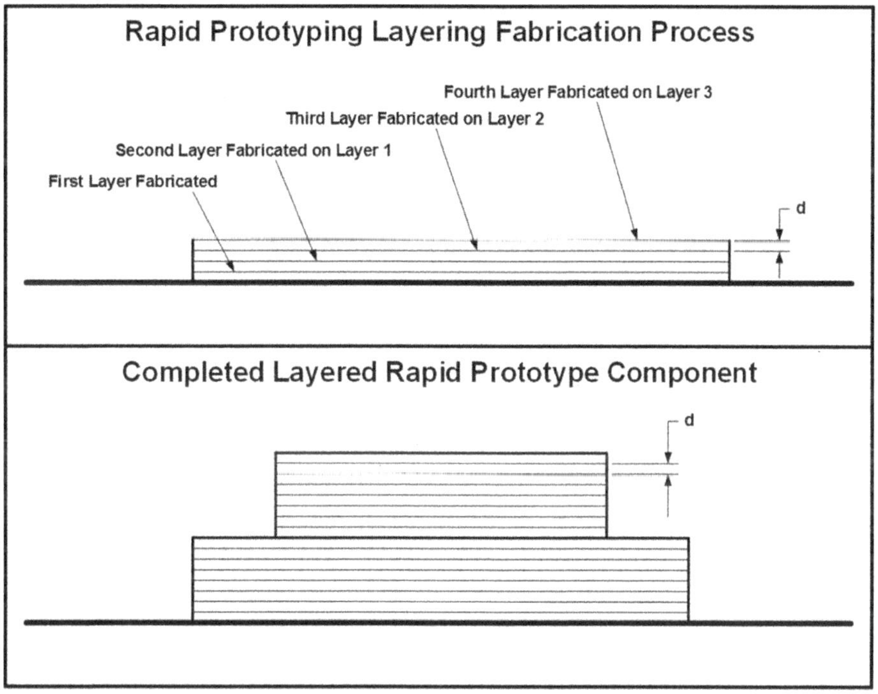

Figure 29.2 Rapid Prototype Layers Fabricated from Bottom to Top

This process works by building the component part from the bottom up. Each layer is an independent program and stacked to create the desired geometry. To prepare a component with layers, the individual slices are programmed and the equipment will run them in order forming the geometry. Figure 29.3 illustrates the programs for selected layers for the rapid prototype process.

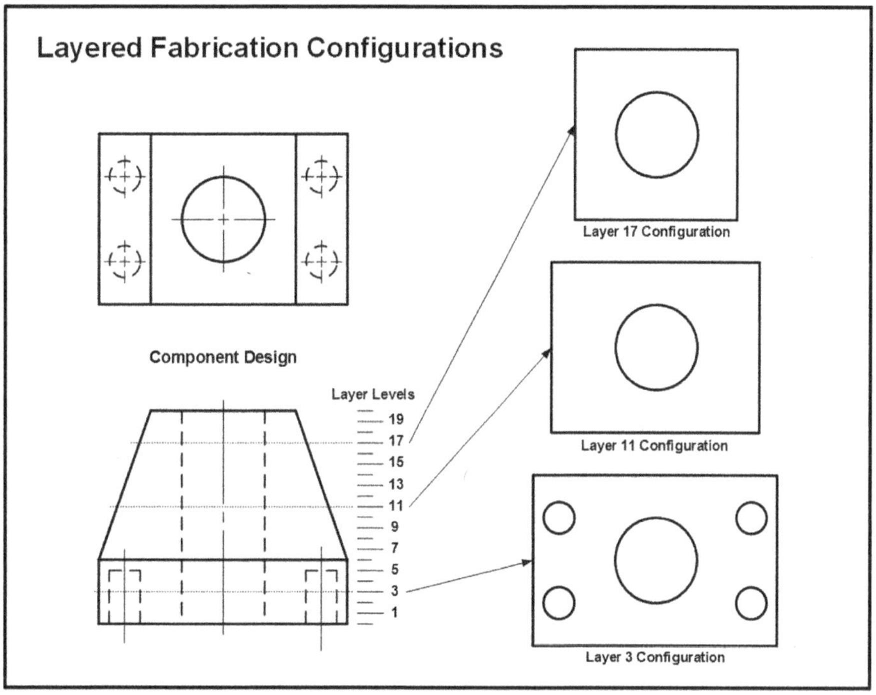

Figure 29.3 Layer Fabrication Individual Programs Stacking

As the layers are produced and stacked, the final configuration will be produced. The working head moves in both "X" and "Y" axes to either cure the plastic material or sinter the metal material, depending on the process being utilized. As each layer is produced, the working table of the machine will index down the amount of each layer until the final layer is produced. Figure 29.4 shows a typical rapid prototype machine and the steps required to produce the final geometry.

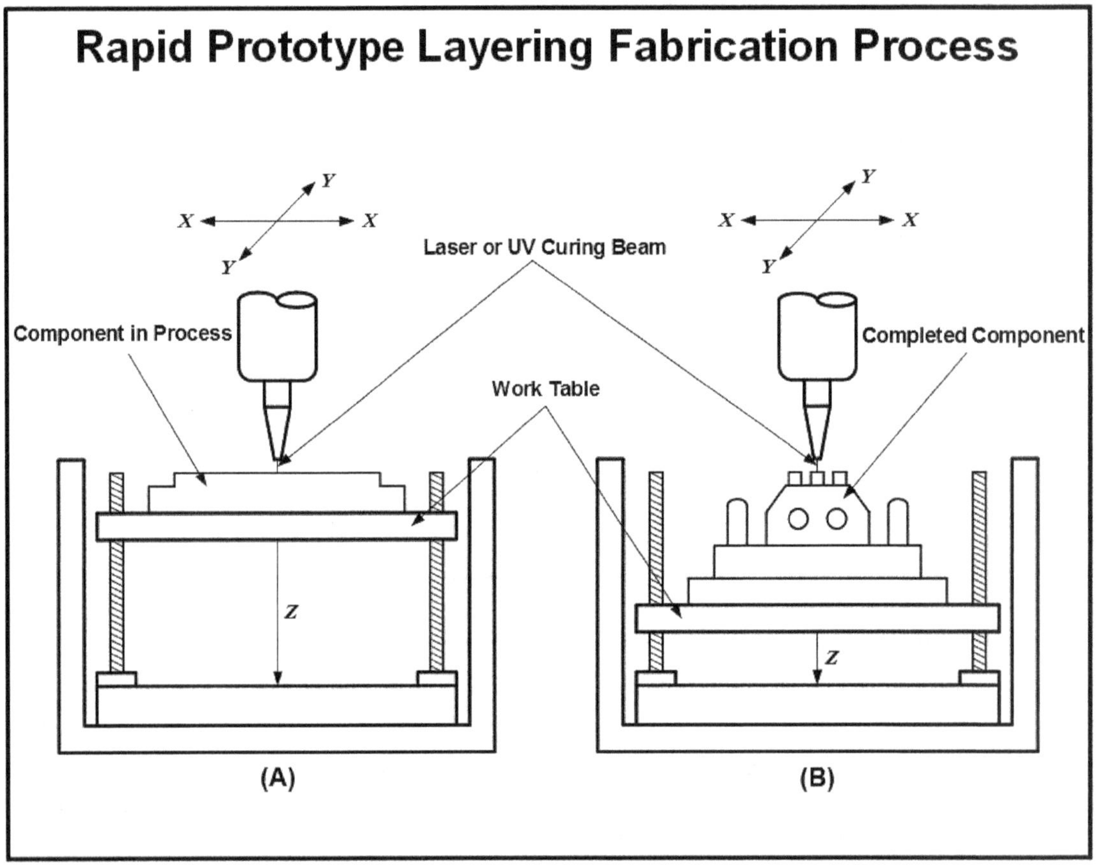

Figure 29.4 Typical Rapid Prototype Machine Fabricating Layers

29.2.1.1 Polymer Material Processes

An extremely popular material utilized in the layering rapid prototype process is polymer plastic. A wide variety of materials are available and some can produce a usable component for testing. The polymer material is in the form of thick syrup and it spreads on the work table, or the previous layer with a blade producing the layer depth desired. After the layer of material is spread evenly, a UV light is traversed along the surface to cure the material. In most cases, the curing time is extremely rapid and another layer can be applied almost instantly. But some materials require a longer curing time so a delay needs to be implemented to allow for proper curing. Figure 29.5 illustrates the steps for applying and curing the various layers to produce the component geometry.

The polymer rapid prototyping process is the most popular and with the large material available has become more than just producing a 3D model, but can produce usable components for assembly and testing in most cases. The materials being utilized in the actual design and be easily replicated. However, if an exact material cannot be fabricated in this process, other processes that are discussed in Section 29.2.2 can be used.

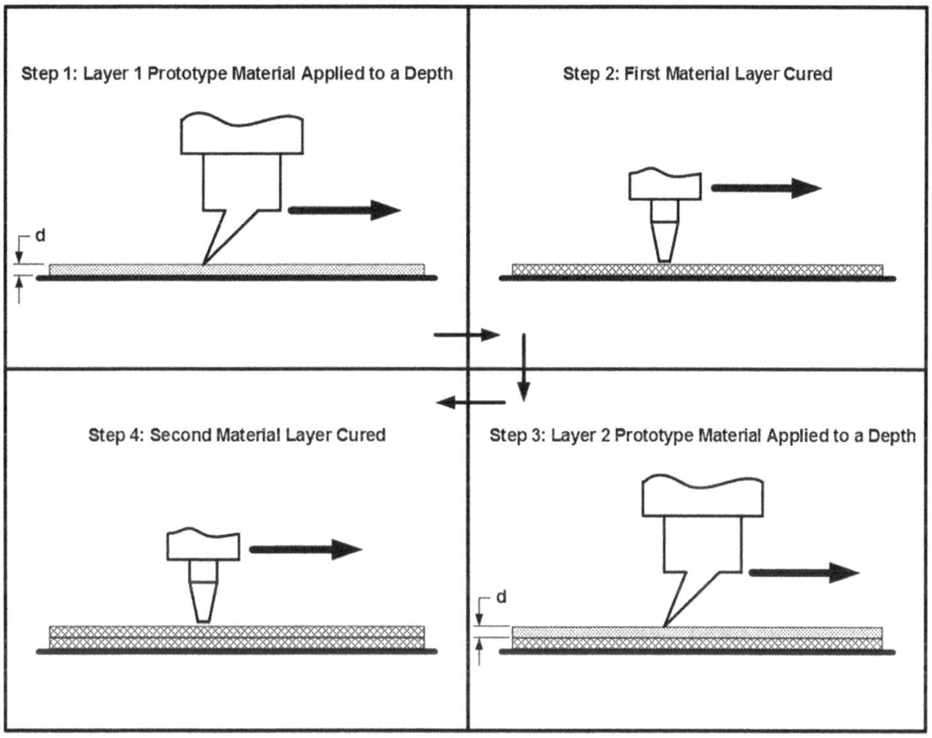

Figure 29.5 Polymer Layering and Curing Process

29.2.1.2 Metal Powder Processes

Producing rapid prototype components from metallic powders works very similar to that of polymers, but utilizes metal powders and a laser to bond the material rather than a UV light for curing. The powered metal gets spread across the work table or previous layer produced to a given depth and then a laser is utilized to move across the desired configuration to sinter the powder particles together. Figure 29.6 shows the process for producing rapid prototype components utilizing the layering procedure.

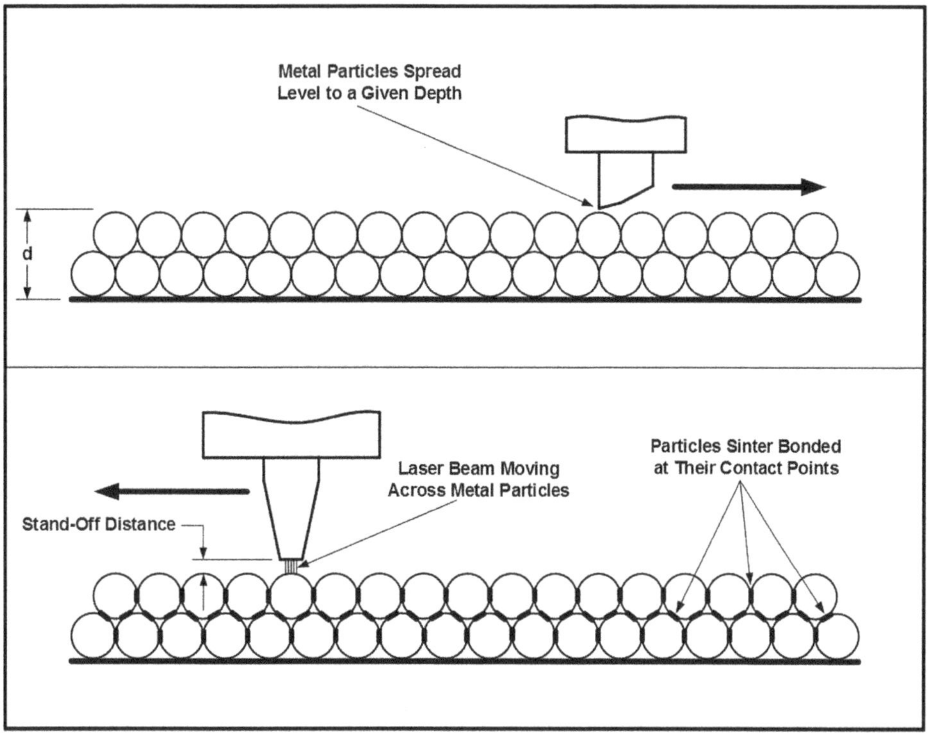

Figure 29.6 Powder Metal Layering and Sintering Process

This process can produce fully functioning components that can be assembled and tested like the desired geometry in the design. Depending on the size of the metallic particles being used, the finished component can have a desired porosity like a conventional powdered metal part as covered in Chapter 9. By using extremely small particles of size 1–2 microns, the process will render a component that can be 98–98% density for a solid part. With a wide variety of powder metals available, components can be constructed easily from aluminum, steels, and titanium simply by adjusting the laser intensity to the proper sintering temperatures.

29.2.1.3 Multilayer Fabrication Drawbacks

As with all fabrication processes, there are some drawbacks that need to be understood. With the component geometry being fabricated in layers, the process will never produce a smooth edge on both tapers and diameters. As the geometry is sliced into layers, the program for each layer will produce a straight edge for that slice. Figure 29.7 illustrates the desired component geometry and the layer slices to produce.

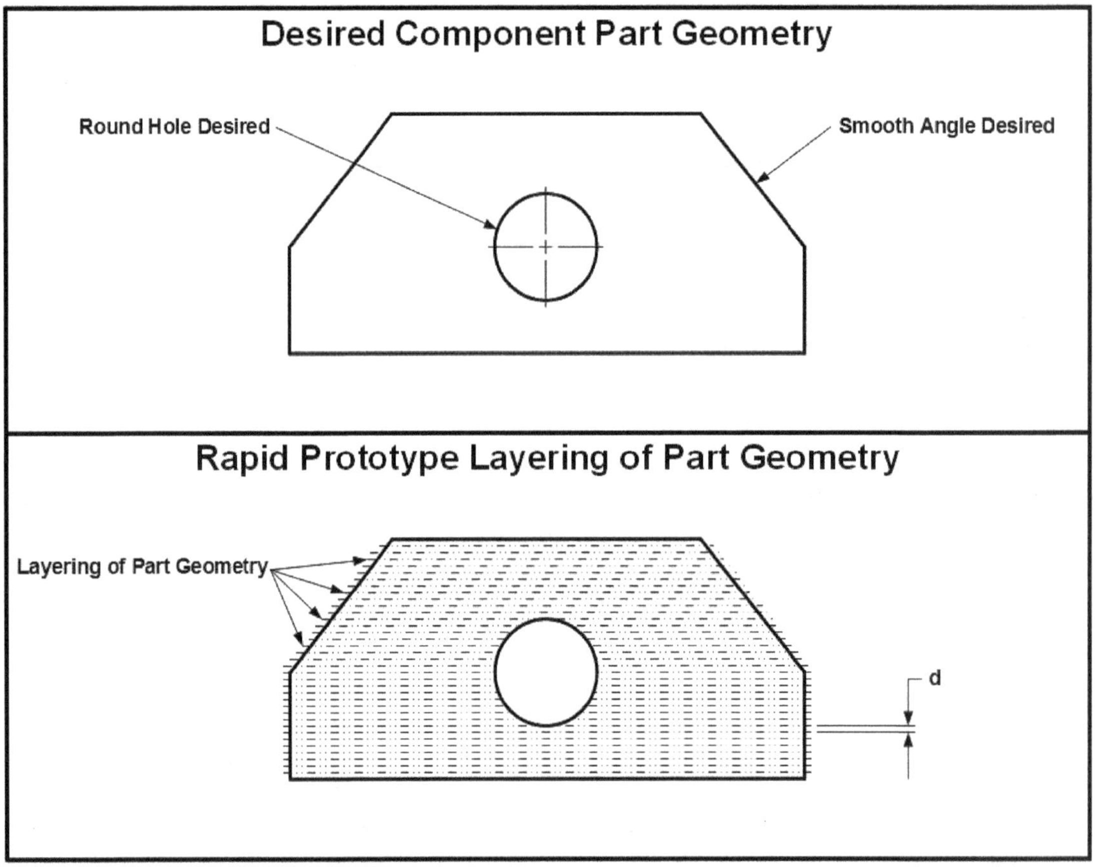

Figure 29.7 Desired Component Configuration for Layering Processes

As these layers are produced and stacked on top of one another, both holes and angles will be produced with steps on them. Depending on the depth of these layers, the steps can be close to unnoticeable, or extremely pronounced. Holes will produce the most variation in the steps as the diameter is produced. Figure 29.8 shows the variation in the steps from the layering process for producing a hole.

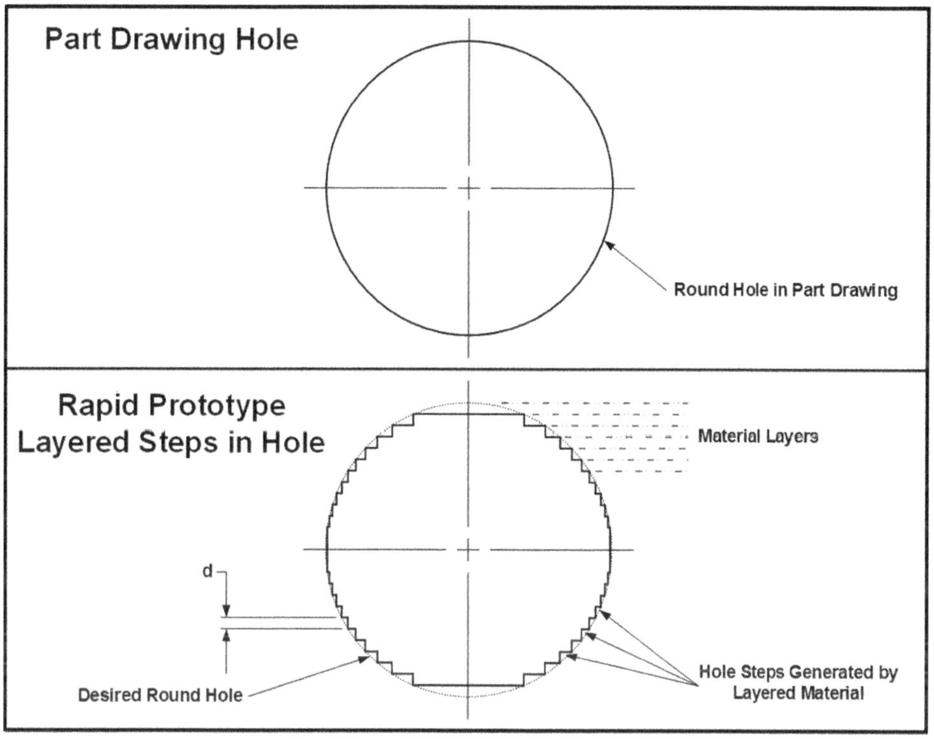

Figure 29.8 Actual Configuration Created in Layering Rapid Prototype Processes

Having these steps on the internal structure of a hole may not be an issue for some components being produced, but if a perfect diameter is desired it can simple be opened with a drill or reamer after fabrication is completed in the rapid prototype process. The side of a component with an angle or taper on it will experience the same step side, and depending on how steep the angle or taper is the severity of the steps will be determined. Figure 29.9 illustrates the step side that will be produced from a taper or angle on a layer-formed prototype component.

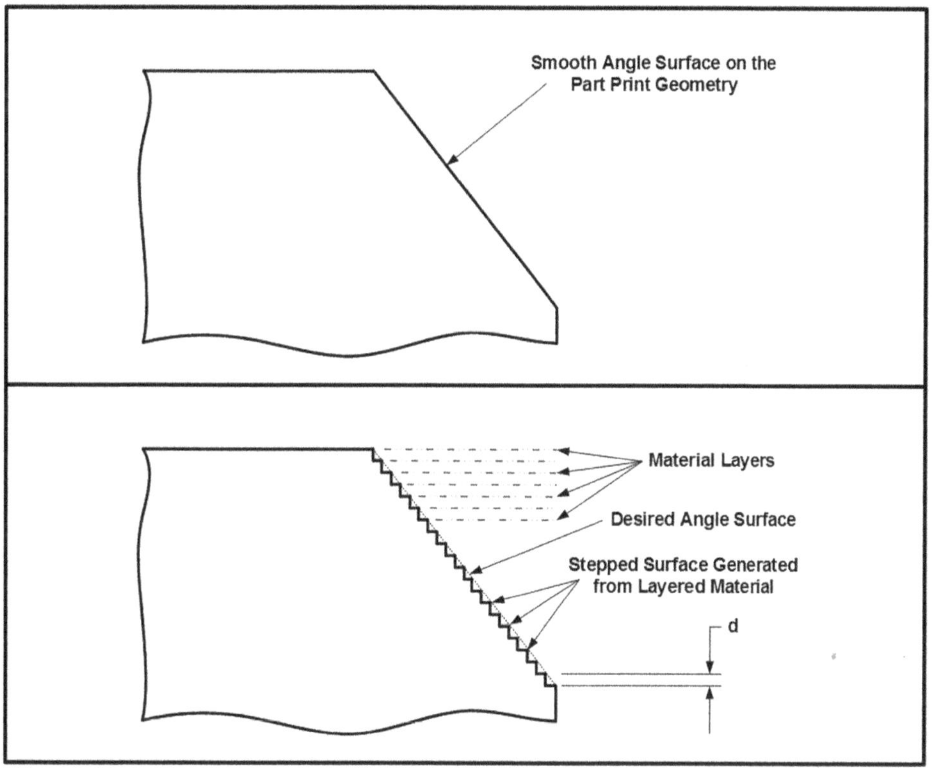

Figure 29.9 Stepped Angles Created with Layering Rapid Prototype Processes

Resolving this issue will not be as easy as for a hole diameter. The component can be filed to produce a smooth surface depending on the amount of the step difference. However if the steps are not critical for the prototype component the steps can remain as produced. Utilizing the thinnest depth for the layers will minimize the step configuration of the component. However, the thinner the layers are, the longer it will take to produce the part.

29.2.2 Numerical Control Processes

Utilizing numerical controlled (NC) equipment is also a viable means for producing prototype parts rapidly. This method is perfect if the components being fabricated have to be constructed of the exact material as the design calls for and will always render a completely usable part for assembly and testing. With the flexibility of NC, complex configurations can be produced and require very little manual intervention and attention once the equipment is set up and operating. As discussed in Chapter 28, the machining times will be maximized resulting in a fast fabrication cycle. With the NC equipment being extremely flexible, close to any desired geometry can be produced rapidly, producing completely functional components. Some of the main NC processes utilized for rapid prototyping are discussed below.

29.2.2.1 Milling Processes

The milling process can produce very complex configurations for prototyping. All NC milling machines have three main axes that can be operated simultaneously, the "X," "Y," and "Z" axes. Figure 29.10 shows the three axes and their motions.

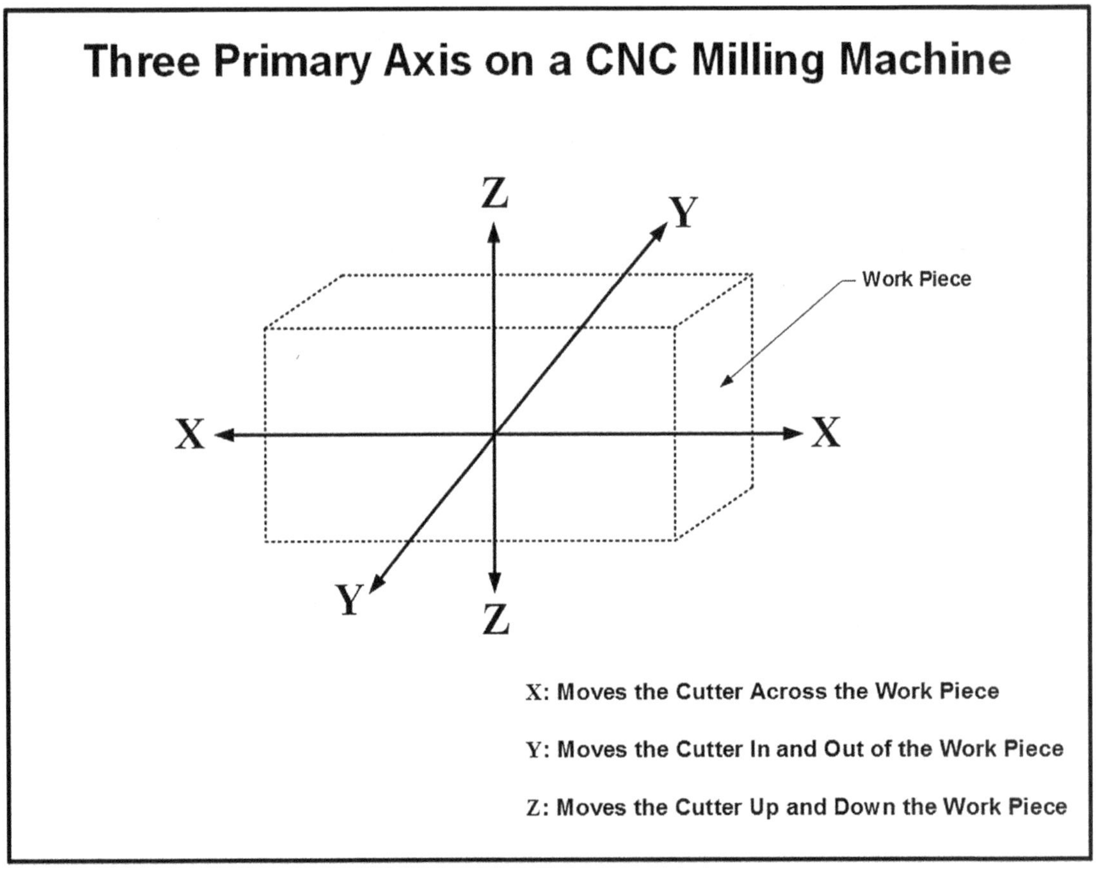

Figure 29.10 Three Primary Axis for NC Milling Equipment

The NC milling machines can also be purchased with two additional axes that will move the normally perpendicular head in two directions allowing for greater fabrication ability. With the inclusion of the two additional axes, the component can be fabricated much more easily and requires least changes in setups over a conventional three-axis NC milling machine. Figure 29.11 illustrates the two additional axes on an NC milling machine.

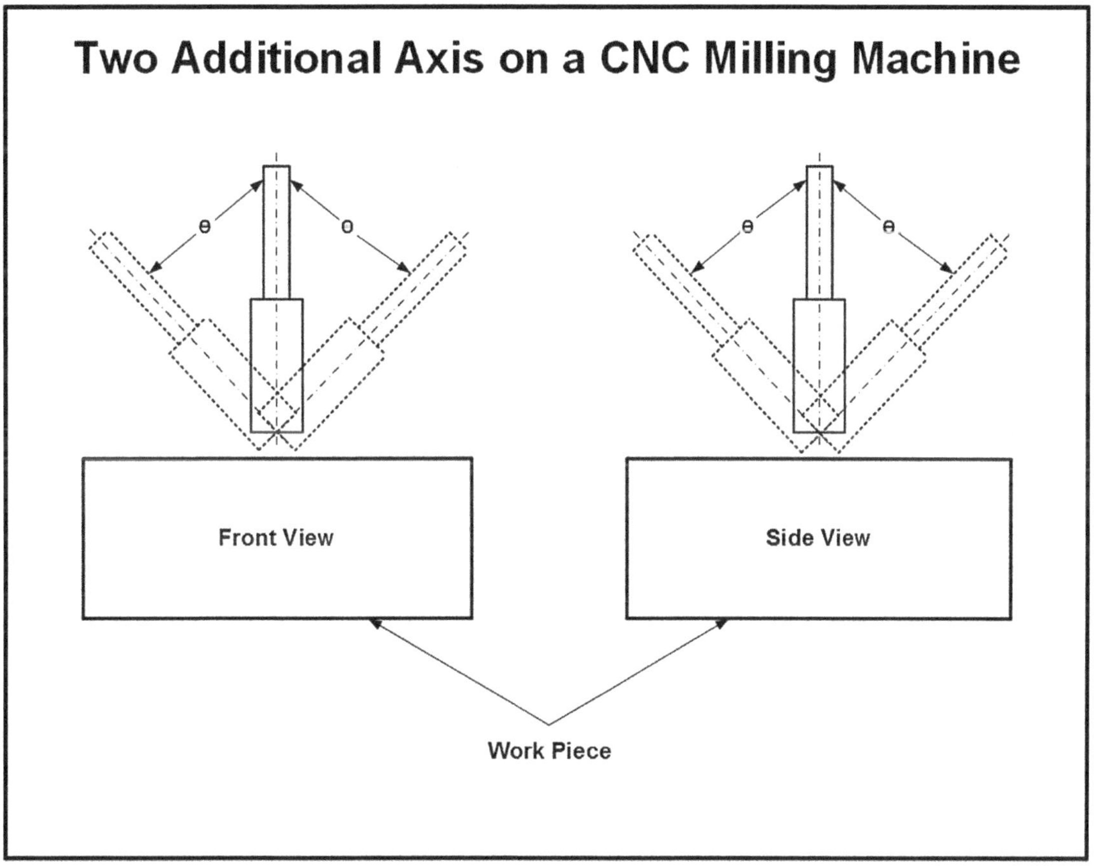

Figure 29.11 Two Additional Optional Axis for NC Milling Equipment

With the NC milling equipment having the ability to load and change several cutting tools, the components can be completely fabricated without interruptions. Also, once a program is written for the part being fabricated, several parts can be produced the same as they can in a production mode. The advantage of utilizing an NC milling machine with five axes of motion is that besides the part being fabricated from the exact material that the design calls for is that the geometry can be perfectly replicated. For example, if a part is slated to be a casting or plastic component, the draft angles can be implemented to represent the exact geometric configurations. Figure 29.12 shows the geometry that can be produced with a five-axis NC milling machine utilizing a single cutter.

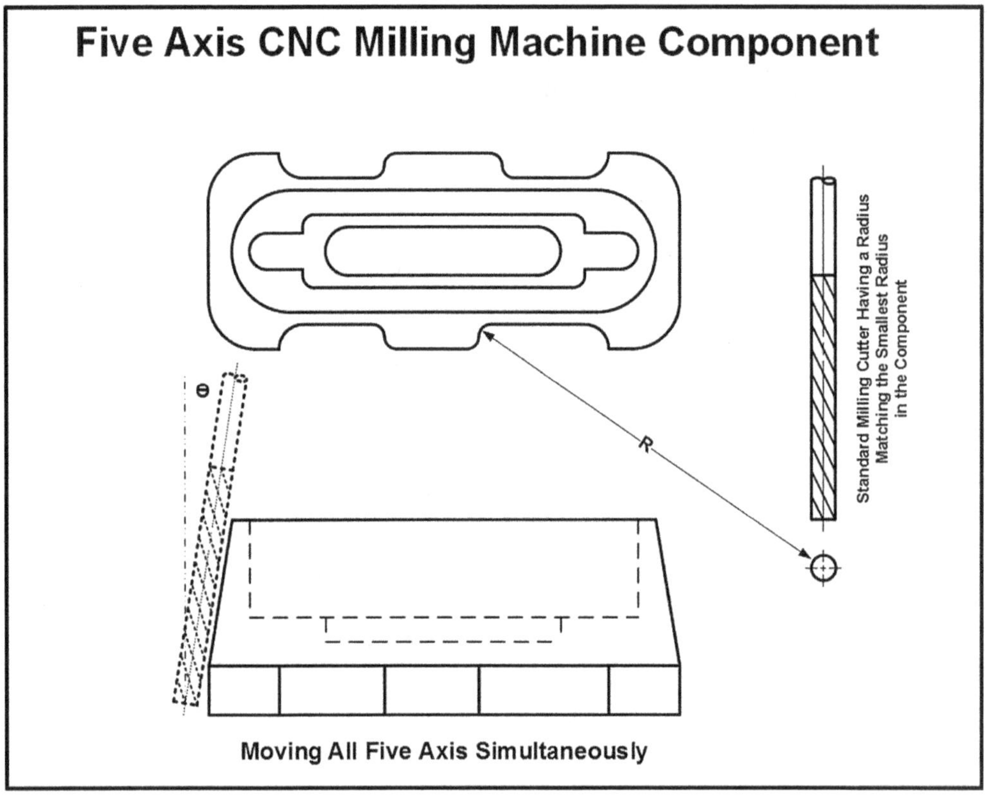

Figure 29.12 Geometry Created Utilizing Five Axis NC Milling

29.2.2.2 Turning Processes

As with the milling processes, the NC turning processes can produce complex round geometric configurations with same ease. All turning equipment, NC or manual, have two primary axes "X" and "Z." The "X" axis will move the cutting tool in and out of the diameter of the work piece while the "Z" axis will traverse the cutting tool along the length of the diameter. Figure 29.13 illustrates the tool movement for both the "X" and "Z" axes.

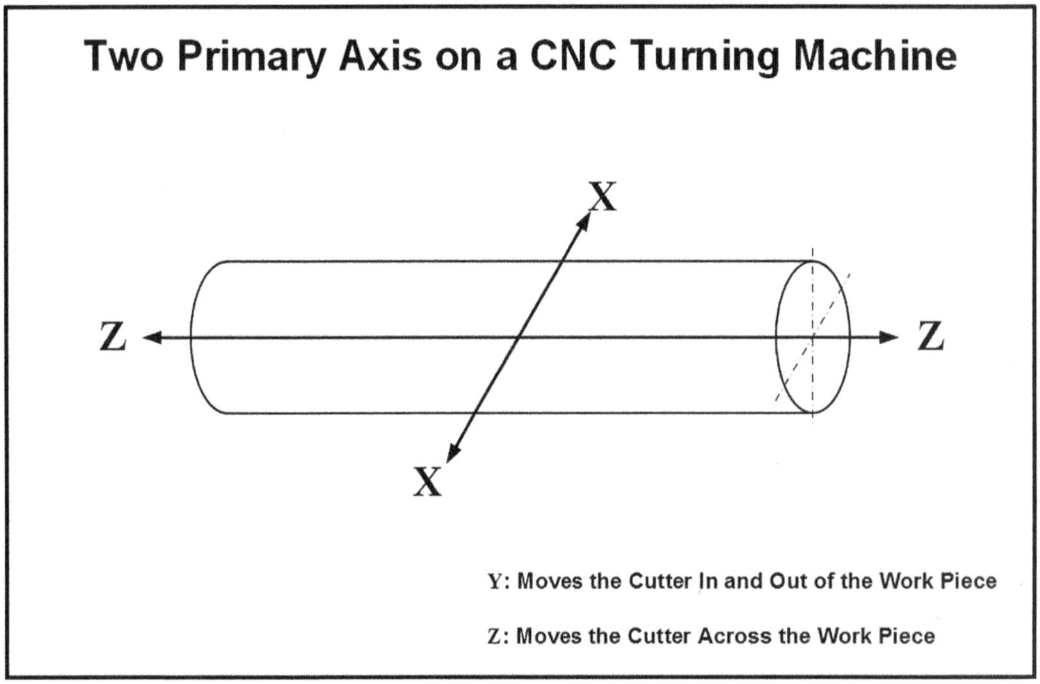

Figure 29.13 Two Primary Axis NC Turning Equipment

With the ability to move both the "X" and "Z" axes simultaneously along the work piece, complex configurations can be produced rapidly. With the NC turning equipment being able to hold and change cutting tools automatically like the milling equipment can, the geometry can be produced and replicated without interruptions, thus producing a completed component with ease. Figure 29.14 shows a typical NC turned component that can be produced in a single setup on an NC turning machine utilizing six different cutting tools.

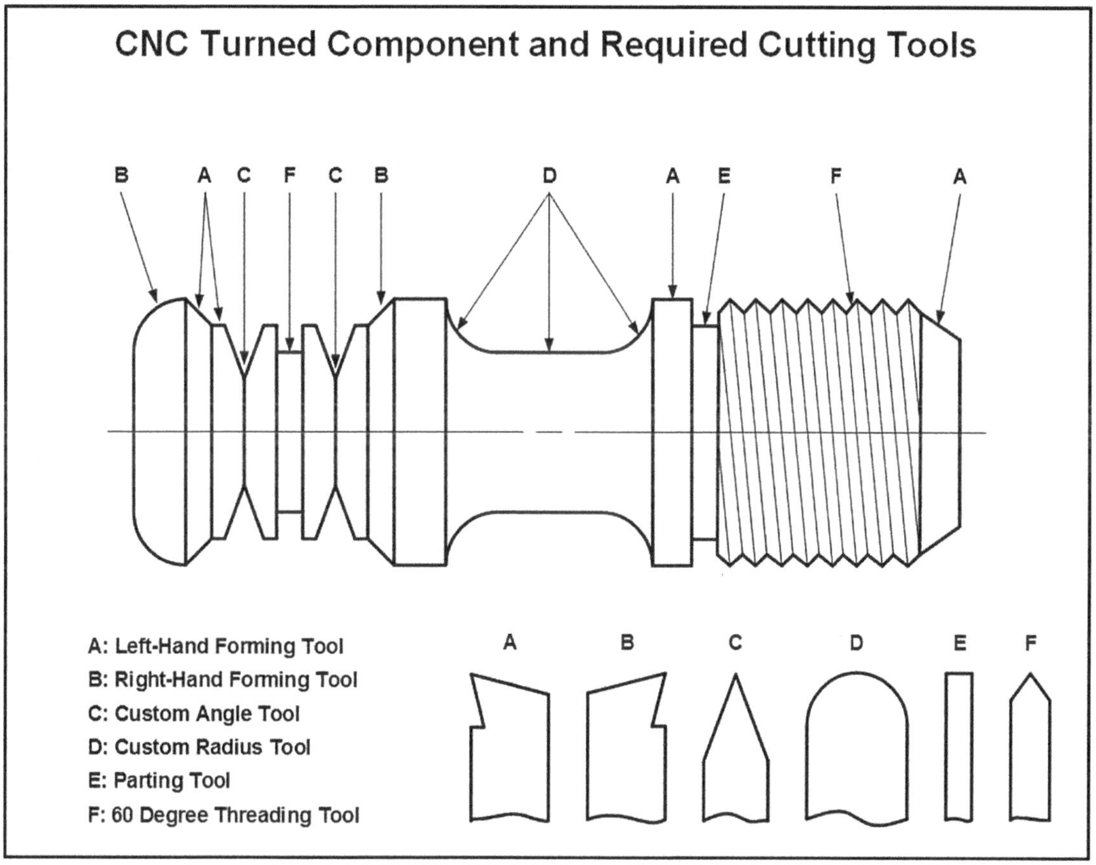

Figure 29.14 Example of Typical NC Turning Geometry Created Utilizing Tool Changes

As with NC milling processes, the exact material called for in the design can be used along with the ability to produce many components once the program for the machine has been compiled. This allows for exact geometric configurations and the ability to test the pieces under the same conditions that they will encounter in the service of the product.

29.2.2.3 Laser Processes

As discussed in Chapter 17, the laser cutting process is a non-traditional manufacturing process, but is extremely viable for producing rapid prototype components from sheet metal that would require a punch and die process in production. This allows for the components to be fabricated from the exact material that the design calls for and will produce a flat blank with ease. If the parts require any bends, those will need to be added in secondary operations at another time. Figure 29.15 shows the "X" and "Y" axes movement of an NC laser cutting machine.

The laser beam intensity needs to be adjusted for the type of material that will be cut along with a critical standoff distance from the material. This will focus the beam to attain its ultimate power to the cutting point. Another area that needs to be considered if close tolerances are to be maintained on the work piece is the overburn from the laser beam and the burn angle. The overburn will be the amount of material that will be

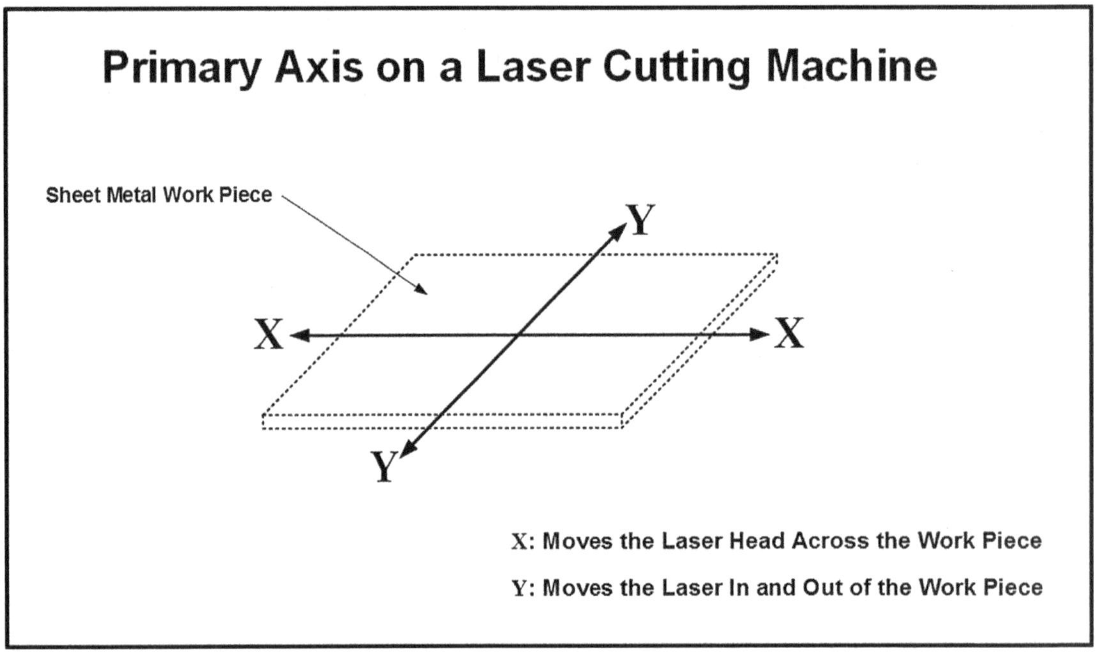

Figure 29.15 Primary Axis for NC Laser Equipment

consumed beyond the beam on each side of the cut while the burn angle will be present from the beam blowing through the material. Figure 29.16 shows the standoff distance of the beam head along with the overburn and burn angle from the process.

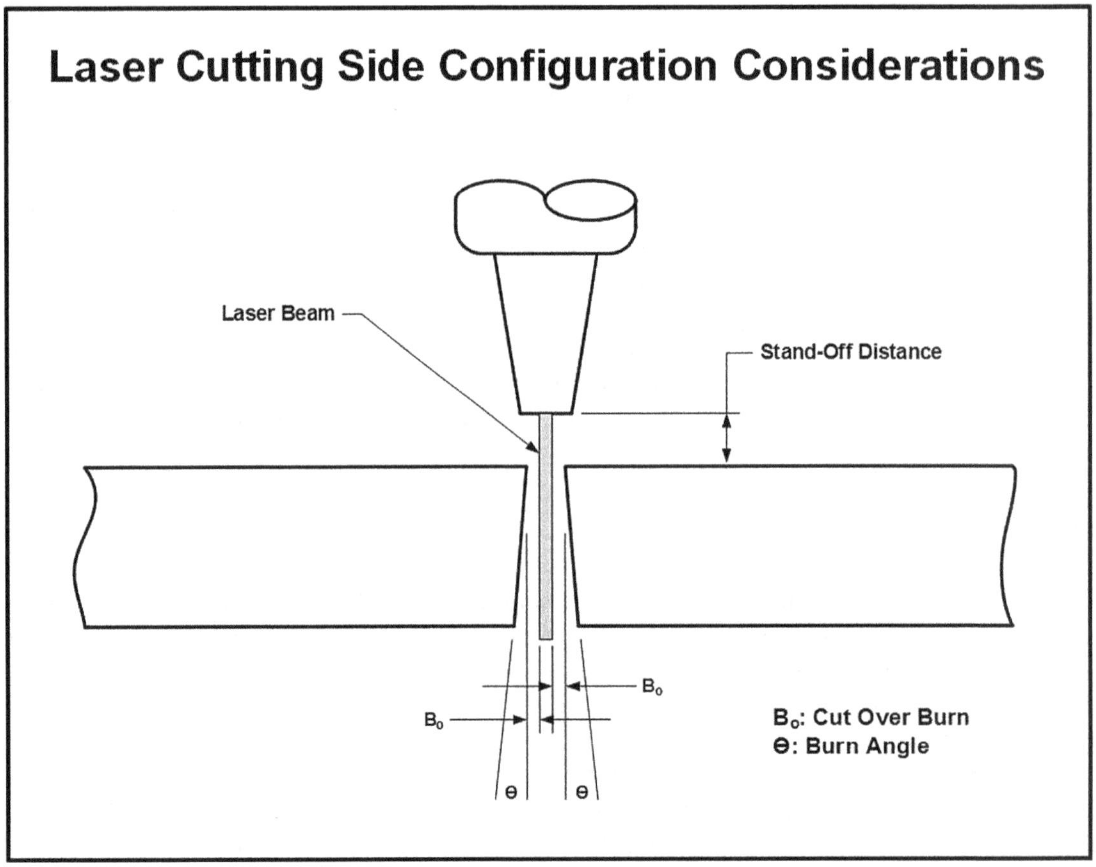

Figure 29.16 Laser Fabrication Blow-Through Cutting Angles

Depending on the type and thickness of the material being cut, the laser burning process is extremely fast for cutting sheet metal components. Both external and internal configurations can be produced rapidly for prototype parts, and with the machine programmed, multiple parts can be produced for both testing and prototype fabrication. As mentioned earlier, if bends are required on the sheet metal components, they would need to be added to the flat blank in an additional operation. Figure 29.17 shows a typical sheet metal component that can be easily produced utilizing the laser burning process.

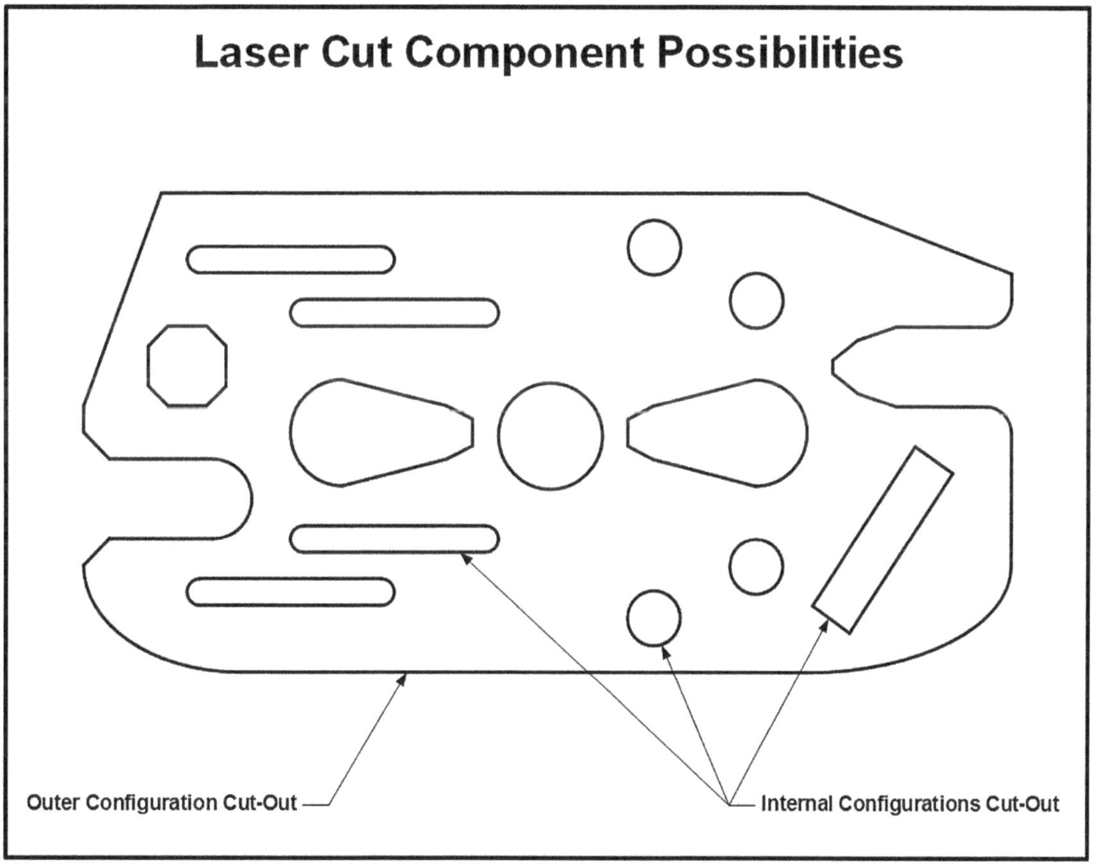

Figure 29.17 Example of configurations Possible Utilizing NC Laser Equipment

Another feature that NC laser equipment is capable of performing is burning a configuration to a given depth. This is accomplished by adjusting the power of the laser beam to partially cut the material being processed rather than cutting completely through. This can be time consuming though because the laser beam will need to pass over the entire surface that is required to be cut to the desired depth. With the laser beam burning an extremely small area, if a large size pocket in the work piece is required, the beam will have to make several passes to accomplish the required fabrication. Another feature that burning to a depth is to engrave. If words or numbers will be required on the finished component, burning to a depth can accomplish this task both easily and rapidly. As with all NC equipment, once the program is written, multiple components can be manufactured easily. Figure 29.18 illustrates how an NC laser machine can cut into the material to a given depth.

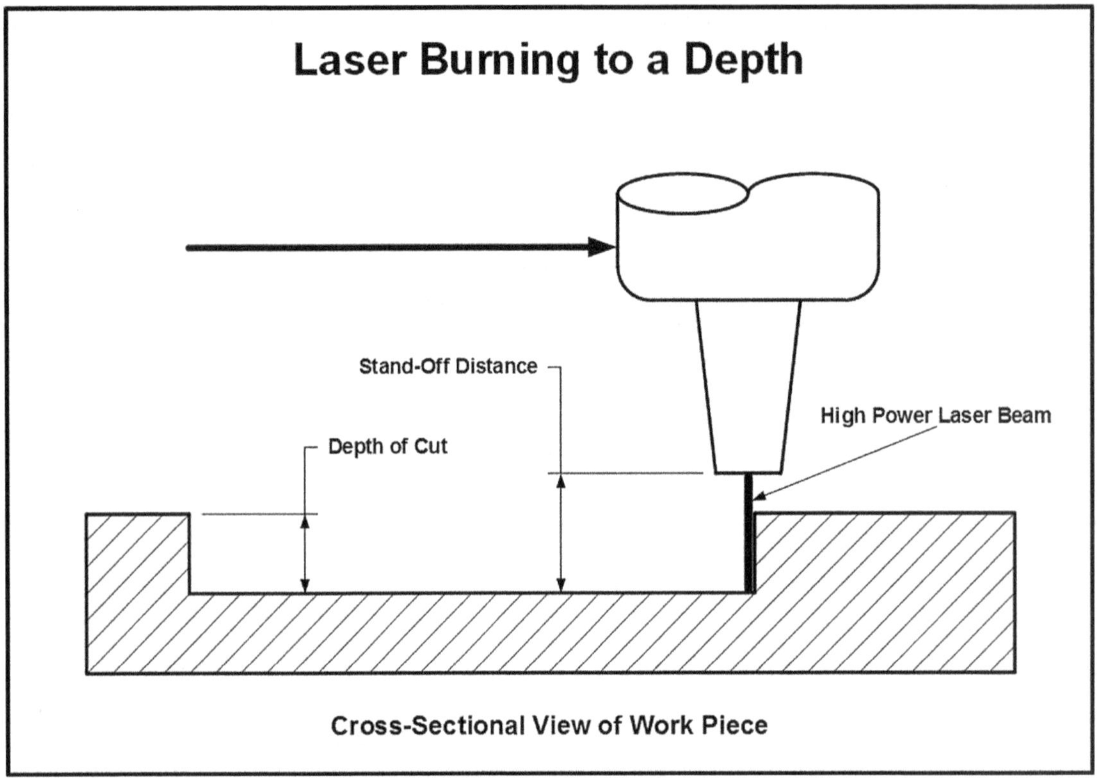

Figure 29.18 Laser Fabrication cutting to a Given Depth

29.3 CONCLUSION

The rapid prototype processes are not typically used to fabricate components for production, unless the volume requirements are extremely low. However, during the design and prototype phases of a new product, the ability to produce components, which can be assembled and tested, is critical to the overall success of the project. With high production component quantities that may require injection molding, casting, or other processes where the tooling expense would be restrictive, the rapid prototype processes can supply the work pieces in the required materials, thus not requiring a high tooling investment to components that the configurations may change before the design of the product is finalized.

CHAPTER

30

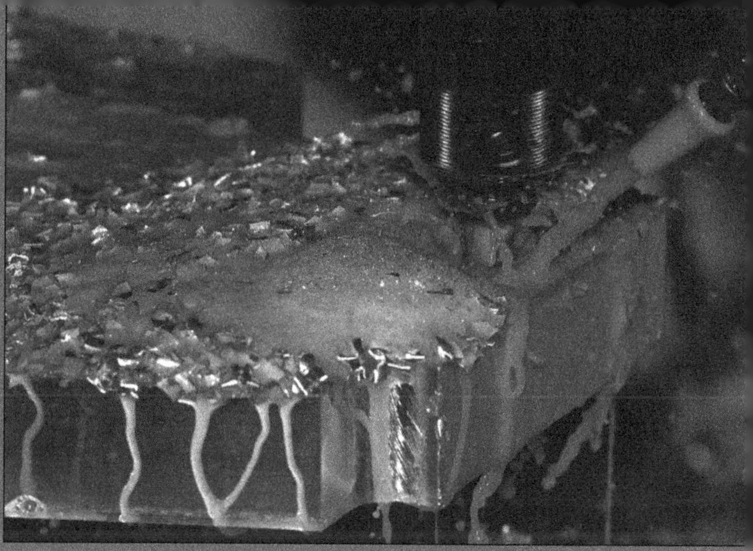

Manufacturing Efficiency

30.0 INTRODUCTION

Ensuring that the operation is running as efficiently as possible is the goal to a successful business. Being able to perform better, faster, and more economical than the competitor will without a doubt increase the overall profits and the bottom line for the operation. The key to this is being able to develop a strategy that will position the business to best serve the current and future markets. Also, setting goals for the operation that can be accomplished or continuously pursued to increase efficiency to the fullest needs to be established. As the operation is very dynamic and not static, a process for continuous improvement on a daily basis will achieve several small steps for improving efficiency and over time results in a large improvement as a whole. Without these steps of constant evaluation and improvement for the operation, things will remain static, and no matter how efficient it was at the point of implementation, in the long run, it will fall off and the competitors will surpass and overtake the high efficiency level that your operation once held. Only through constant evaluation and improvement will the operation remain on the cutting edge of competitiveness.

This chapter will cover various techniques and methods for evaluating areas where efficiency can be improved. Goal setting is discussed along with the law of diminishing returns to guide any improvement arenas to ensure that the efforts being put forth will realize the maximum return on the time invested. A variety of tools for identification and evaluation are discussed along with implementation improvement measures.

30.1 PRODUCTIVITY

The key to every operation is how efficient and effective the productivity is. This measure is used to determine how much product is being produced with the various input resources being utilized. Productivity by definition is "The quantity of good output per the units of each input resources utilized." The overall effectiveness of an operation's productivity can be determined by how close the output is to the inputs. However, the efficiency of an operation's productivity is measured differently. Here, the measure looks at the output of one completed unit per unit of time to produce it. In other words, how many units per minute, hour, or day would be the measure for the efficiency of the productivity of an operation:

$$\text{Productivity} = {}^{\text{Output}}/_{\text{Input}} \quad \text{or} \quad \text{Productivity} = {}^{\text{Productive Time}}/_{\text{Total Time Worked}}$$

Productivity is the main focus for any improvement activities. By improving the productivity, the economics of the operation will improve as a result. If you can improve the output within the same time frame, the operation will be able to reap more profits from the products. Also, if the same products are produced, rendering less scrap and waste, profits will increase. The improvement process is not a onetime event, but rather a dynamic process of continually improving areas within the operation. This process is covered in more detail later in the chapter.

By looking at both measures together, the total productivity can be measured, thus resulting in an improvement for the economics of the operation. If the effective productivity resulted in more output per unit resource, the cost would be reduced due to the fact that the materials are being utilized and transformed more effectively. Likewise, if the productivity efficiency were to go from producing 100 units per hour to 120 units per hour, the cost per unit to produce it would be reduced accordingly resulting in additional savings. By viewing productivity from two dimensions, the efficiency would be maximized to its fullest within the operation.

Example

An operation was producing 500 units per day and working 7 hours of actual production time during a normal 8-hour day. It costs $12.75 to produce one unit with an input cost of $15.25 for raw materials. The productivity of the operation under the current operating conditions would be as follows:

a) Data:

Units/Day = 500
Cost/Unit = $12.75
Input Cost = $15.25

$$\text{Productivity} = {}^{\$12.75}/_{\$15.25} = 72.9\%$$

b) Data:

Units/Day = 500
Work Day = 8 Hours
Productivity Work Time = 7 Hours

$$\text{Productivity} = {}^{7}/_{8} = 87.5\%$$

If the productive time were to increase to 7.5 hours during the 8-hour work day and the input cost dropped to $14.10 per unit from reducing scrap and waste during the process, the changes in productivity for the operating conditions would change as follows:

a) Data:

Units/Day = 500
Cost/Unit = $12.75
Input Cost = $14.10

$$\text{Productivity} = {}^{\$12.75}/_{\$14.10} = 90.4\%$$

b) Data:

Units/Day = 500

Work Day = 8 Hours

Productive Work Time = 7.5 Hours

Productivity = $^{7.5}/_{8}$ = 93.8%

30.2 CONTINUOUS IMPROVEMENT

Every operation needs to be in a state of continuous improvement to maintain efficiency to the fullest. Even if things appear to be operating to their very best, areas for improvement can be identified and implemented regardless of how minimal they may seem. If small improvements are identified and implemented, the sum of several small improvements will total up to a substantial overall gain for the operation.

The first step toward a continuous improvement program is to identify what area(s) within the operation need to be focused on. Depending on the nature of the operation and the conditions surrounding it, the focus may vary greatly. The ultimate goal will always be to decrease operating cost. However, this can be realized through several avenues to achieve the reduction in cost depending on the focus being taken. Some of the main areas that are the target for improvements within the operation are the following.

Time: Time is without a doubt the main ingredient for reducing cost. No matter what is being evaluated, it all boils down to the least amount of time will result in the least amount of cost. The faster something can be processed, the less it will cost the operation. There are some exceptions where all the time components need to be evaluated as a whole to conclude the optimum total time for an operation to ensure that an operation is being timed to the most efficient and effective cycle.

All operations have a fixed rate that they take to operate. This cost is normally broken down to a given cost per hour. The time that is required to process something within the operation is then multiplied by the total processing time, thus resulting in the cost. By controlling the time, the cost will be reduced and allow for a higher profit for the operation.

Product Flow: How the products flow through the operation is important to both minimize time and increase efficiency. If the product that is in processing has to move from one location to another and then back again, there is a lot of wasted movement along with the corresponding time. To become very efficient and effective within the operation, how the product flows from beginning to end can streamline the process and reduce processing times greatly. The product flow will begin at the first operation of the product and end with it being delivered to the customer. Internal flow refers to the processing of the product within the operation from the delivery of the raw material to the first operation through the completed product ready to be sent to the customer. The internal flow process is discussed in Section 30.3, but is a very viable area for a continuous improvement program.

Floor Space: Every operation occupies floor space that needs to be optimized for its usage. The overhead cost of the operation will determine what the cost is to operate the facility. In large operations, this cost will be broken down to a cost per square foot. By knowing what that cost is, being able to reduce the amount of floor space being utilized within the operation will allow for two options. The first is to have excess open space to do additional processing for new/additional products. Second would be to streamline the operation and produce the same volume in less space, opening the option for moving into a smaller facility that would carry less cost. Whatever the reason, evaluating the space requirements will result in improvements in time, product flow, and cost to the operation.

Quality: Quality is one area that by implementing a continuous program may involve some increased cost to the operation, but in the long run will greatly improve customer satisfaction that may result in return business. Every major operation has quality improvement programs stemming from product quality through corporate quality. Chapter 27 covered quality in detail for an operation and discussed various tools for measuring quality throughout the processing sectors along with the total quality management philosophy being adopted widely. The quality is the first thing the customer sees and notices for the products; so, by improving this area customer loyalty will increase. Improvements in the quality arena may not product immediate results, but will reduce cost and increase profits over the long haul.

Continuous improvement programs are the best vehicle for making changes for the betterment of the operation on a day-to-day basis. By setting goals and targets to be achieved, the entire workforce can all work to implement improvements regardless of how small they may seem at the time. But by continually improving, the costs will be reduced and the operation will realize increased efficiency to the fullest.

30.2.1 Value Added vs. Non-value Added

Every operation in the world never operates at 100% efficiency. That doesn't mean that the operation isn't efficient, it is only that there has to be some form of non-productive time throughout the work day. If a normal work day was evaluated, just how much time is spent doing things that add no value to the operation and products within? No one works continually for the entire time he or she at work, and when working, how many tasks is he or she doing that don't directly affect the product? Thinking about this, the work day will have a substantial amount of what's called non-value added time in it. If we were to look at a typical machining operation, the only time spent in that type of processing that is adding value is when the machine is actually cutting the material that will eventually become the product. All the other time involved in the process that needs to be done is the non-value added time. Figure 30.1 shows the tasks involved in a typical operation of which only the machining cycle and assembly times are adding value to the product.

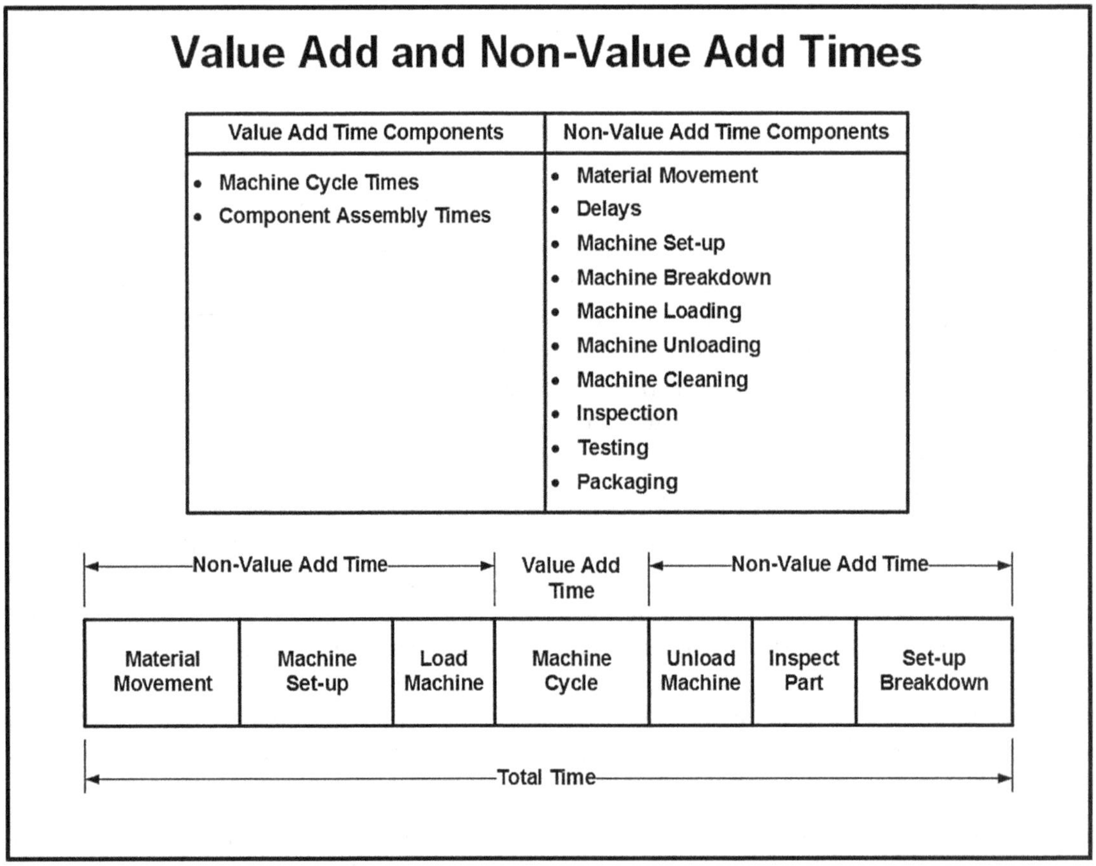

Figure 30.1 Value added and Non-value added times

When the operator is loading and unloading the machine, no value is added. Also, when the machine needs to be changed over from one job to another, the breakdown and set-up time involved add no value. Any material movement within the operation needs to be accomplished, but truly is non-value added time spent.

For an example of value added vs. non-value added time, let's look at a typical football game. The actual "playing time" on the clock is 1 hour. Yet, the game runs for about 3 hours. Keeping in mind that the only time the clock is running is when the ball is in play. When the play stops, so does the clock. All the other time taking up the remaining 2 hours such as the team regrouping for the next play, time outs, changing to the defensive side, and any other activities that don't involve a play can be viewed as non-value added time. With this understood, the only value-added time in the ball game is when the ball is in play and the clock is running. Comparing this to a machining example, the machine cycle time would be the play time of the football game. Loading and unloading the machine, changing cutting tools, setting up and breaking down the machine, and material movement would compare to the team regrouping and time outs. It's not realistic for a football game to play non-stop and having the regrouping activities done instantaneously, but in an operation, what can be done to minimize the non-value added time spent in the process? Working to reduce the non-value added times within the operation as a whole will allow for more time adding value to the operation that will reduce costs and open up unrealized time to be able to take on additional work.

However, evaluation of the time needs to be done looking at the operation from a big picture viewpoint. The ultimate goal is to optimize the process to find the best combinations of the value added and non-value added times. As with a football game, the non-value added time can't be completely eliminated. A balanced mix of the total times needs to be looked at to find the point where the total time and cost will be realized. Figure 30.2 illustrates what a process optimization needs to include for a typical machining process.

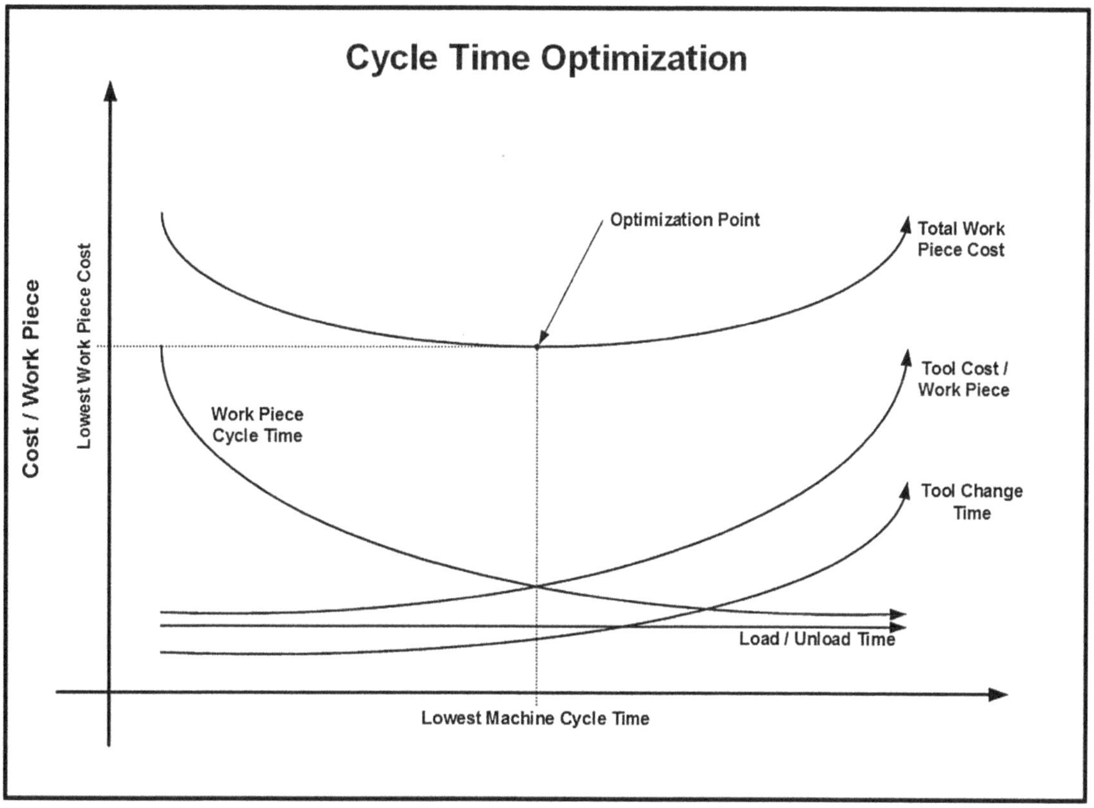

Figure 30.2 Typical optimization curve for machining operations

As Chapter 18 covered, if we look at the components for this operation, there is the machining time/cost, the part handling time/cost, tool changing time/cost, and the tool cost itself. Making the machine run faster and faster will greatly reduce the machining cost, but by doing this, the cutting tool will have to be changed more often resulting in more non-value added time in that arena. Likewise, the part handling time remains constant for loading and unloading the machine regardless how fast it's cutting. Reducing the cycle time will reduce the cost of that component, but when the other components involved in the operation are considered, at a point the cost begins to raise up again.

Machine Total Cycle Time

$$T_c = T_h + T_m + {}^{Tt}/_{np}$$

T_c: Total cycle time

T_h: Material handling time

T_m: Machine cycle time

T_t: Tool change time

n_p: Number of pieces before tool change is required.

Machine Cycle Cost

$$C_t = C_o T_h + C_o T_m + C_o {}^{Tc}/_{np} + {}^{\text{Tool Cost}}/_{np}$$

C_t: Cycle time cost

C_o: Cost rate for performing the work

Tool cost: Cost for the cutting tool utilized.

Example

If a given machine operation had the following times and costs associated with it, what would be the total cycle time and cost for performing the operation?

T_h= 2.18 min

T_m= 12.13 min

T_c= 15.73 min

n_p= 350 pieces

C_o = \$95.00/hour

Tool cost = \$85.00

1. Total Cycle Time

$$T_c = 2.18 + 12.13 + {}^{15.73}/_{200} = 14.355 \text{ min}$$

2. Machining Cycle Cost

a) Cost per minute = $^{\$125.00/\text{hour}}/_{60}$ = \$1.583/minute

$$C_t = (\$1.583)(2.18) + (\$1.583)(12.13) + (\$1.583)({}^{15.73}/_{158}) + ({}^{\$85.00}/_{350})$$

$$C_t = \$30.149/\text{piece}$$

Some changes were made to the machining cycle and the T_m was reduced to 3.57 minutes. However, in doing so, it now requires the cutting tool to be changed every 67 pieces. What would the new cycle time and cost be?

$$T_m = 9.71 \text{ min}$$

$$n_p = 200 \text{ pieces}$$

1. Total Cycle Time

$$T_c = 2.18 + 9.71 + {}^{15.73}/_{200} = 11.969 \text{ Min}$$

2. Machining Cycle Cost

$$C_t = (\$1.583)(2.18) + (\$1.583)(9.71) + (\$1.583)({}^{15.73}/_{200}) + ({}^{\$85.00}/_{200})$$

$$C_t = \$25.360/\text{piece}$$

Even with changing the machining time from 12.13 min to 9.71 min that represents a 21.1% reduction in time, the overall cycle time only changed by 16.6% and the cost by 15.9%.

So, before an improvement program is considered for removing non-value added time, the operation needs to be evaluated as a whole and not as a part. Overlooking the big picture can actually cause more cost to the operation than it is actually saving. Remembering that an operation is a closed loop system, all the components of the system need to be considered when evaluating for opportunities is being considered. Without doing this, the results may seem advantageous at first, but the total cost will reveal that the improvement did more harm than good.

30.2.2 Goal Setting

Before any improvement programs can be implemented, there needs to be a goal that the program can work toward. The goal provides the target to which the improvement program will strive to. Without a goal or target, there is no real measurable improvement that can be substantiated. Setting a goal needs to be a thought through process in itself to derive the expected result. Once an area has been identified for improvement, the first step in the goal setting process is to evaluate what state the area is presently at. This involves a detailed look at exactly how things are being done down to the smallest detail. When the evaluation process is being performed, no improvements are to be done at that time no matter how obvious they are. Once the evaluation process has been completed, the information is reviewed and a target goal can be set. These goals are normally depicted in a percentage for improvement. Examples of these could be to improve quality by 15%, reduce cycle time by 25%, reduce floor space by 10%, and increase throughput by 15%. From establishing a goal, the improvement team can evaluate the process and determine the best method and approach to accomplish the expected results.

Goals don't have to be tied exclusively to exact operational functions, but can be used to larger business targets as well. These goals are far more long term and are normally tied to an operations strategic plan. The targets set here will be broader and require high level decisions that govern the operation as a whole to accomplish. As a goal gets more broad and "shooting for the stars," it normally becomes the vision for the businesses. A vision is a very high level goal for where the operation direction needs to be at some point in the future. From the vision, other lower level goals will be set to work toward the vision of the future. As mentioned, this is part of the operation's strategic plan for the future.

Possibly the most important thing to remember about a goal is to always stay focused on it. Setting a goal and then forgetting about it will be a surefire way to not accomplish it at all. No matter what the goal is, if the focus is lost, the goal will always fail.

> "Keep your mind focused on your goal. Don't let doubts, fears, and 'small thinking' undermine that goal. Don't think too small—if you do, the operation you create will always remain small. Your operation reflects your mind; it reflects your dominate visualization." (Allen, M)

Goals can be either easy or difficult to accomplish. The difficulty of the goal will determine the amount of performance that needs to be put forth to attain the results expected. However, if the goal begins to enter into the impossible range, the performance of those being challenged to accomplish it will degrade. This is due to the fact that they know that there is no way the results can be achieved so they just don't put forth the amount of effort to work toward the final means. Figure 30.3 shows the amount of expected performance against the difficulty of the goal being attempted.

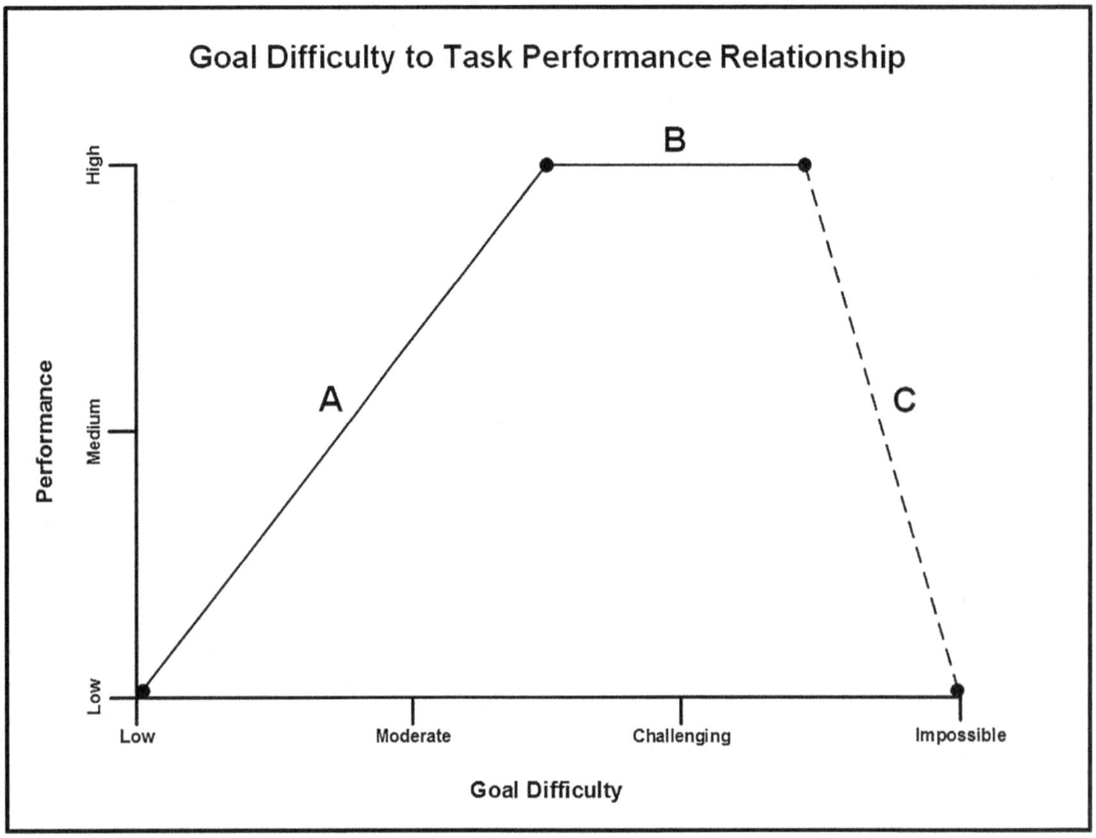

Figure 30.3 Task difficulty vs. performance chart

If the goal is to be realistic, the results need to be achievable by those working on it. However, if the end result is beyond the means of being accomplished, the effort put forth will be minimized because the result is unrealistic and unachievable.

Setting high goals is a positive thing to do. If you think on a large scale, then large results will happen. Likewise, if you think on a small scale, only small results will be realized. Make the sky the limit and then devise a plan to reach it. Also, keep away from people who will undermine the goals you've set. Mark Twain summed this up perfectly by saying:

> "Keep away from people who try to belittle your ambitions. Small people always do that, but the really great make you feel that you, too, can somehow become great"
>
> —Mark Twain

Without goals, the operation may be working well, but possibly not in the right direction for improvements. Staying focused on the end results will ensure that progress is being made in the right directions. However, goals are not a static entity. With changes to the environment surrounding the operation, the goals may need to be altered or even abandoned completely. Only through constant monitoring of the performance indicators that drive the operation can the 3Es be maximized through the goals being set.

30.2.2.1 Goal Brainstorming

Brainstorming is an activity that is utilized to come up with ideas that will be used to develop efficiency goals. The first session is performed by getting a group of people together where everyone throws out ideas that are written down for further evaluation. This activity will always be focused on a specific area of the operation where improvements are required from a variety of reasons such as throughput, quality, and scrap. During this process, the ideas that are brought up are not discussed, but rather just listed. One important thing about a brainstorming session is that there are NO bad ideas, and whatever is brought up does not get any criticism. They are simply listed as they are mentioned. Once a large list of ideas has been compiled, they are documented to be evaluated at a second session where each will be discussed as to the feasibility and implementation required to set the goals into action.

During the second session of the brainstorming activity, the same group will reconvene and each idea that was listed will be discussed in some detail as to the importance and feasibility for further evaluation. Here, the original list will be cut down to smaller ones for consideration, but the ideas are not completely discounted but set aside to be considered at a future time. This reduced list of ideas will be considered for a more in-depth investigation. The group will then develop action items for the ideas now listed to investigate in detail and bring this information back to the third brainstorming session.

At the third session, the group then reviews the investigated ideas and discusses which ones are the most feasible to implement. This decision will be based on the ease of implementation and the return on investment of the ideas being considered. Once the ideas were investigated in depth, one or two ideas from the investigated ideas will be put into motion. A plan is established for the selected ideas for implementation with a time frame and cost for implementation and a team is compiled to carry these ideas into action. Figure 30.4 shows the stages of a brainstorming activity where the original list of several ideas gets reduced down to a few that will be implemented at this time.

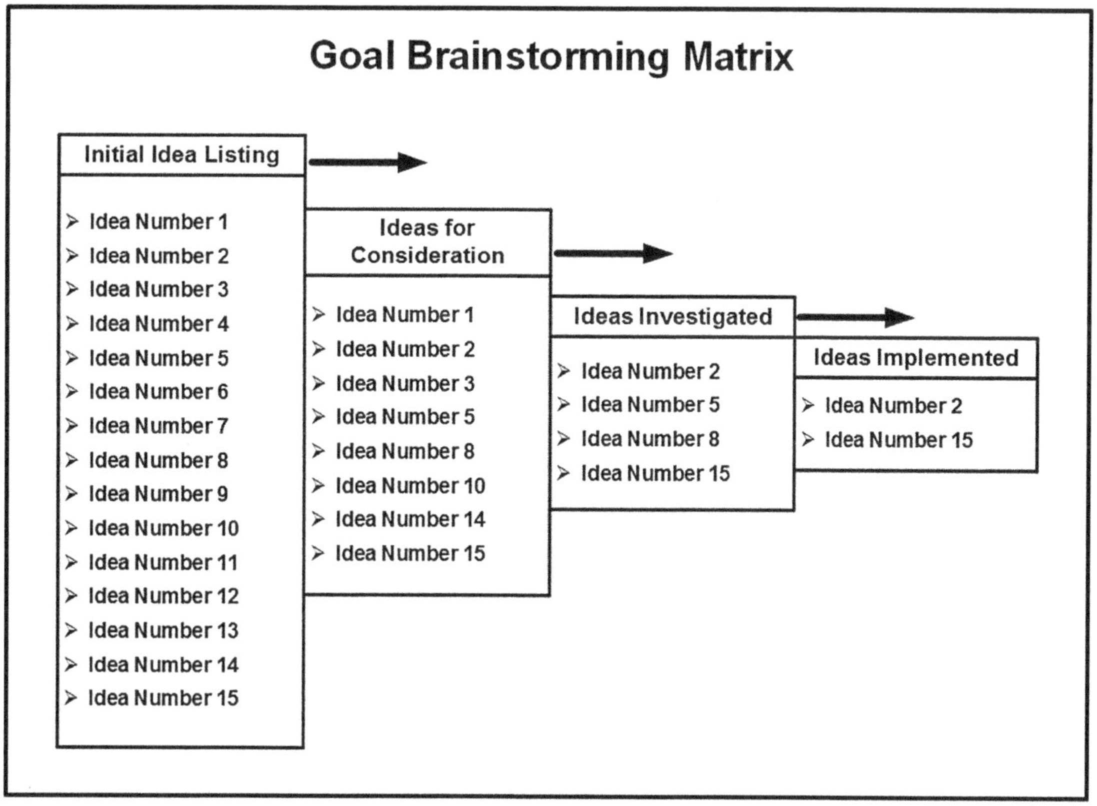

Figure 30.4 Brainstorming matrix for improvement implementation

After the implementation process has been performed, tested and perfected, the list of investigated ideas will be revisited for potential implementation. Also, the ideas for consideration from the second meeting session will be revisited for possible investigation to bring them to the next level. This activity will continue until all feasible ideas have either been implemented or discarded in the area of focus.

30.2.2.2 Realistic Goals

Possibly the most used goals are those that are realistic in nature. Being a realistic goal means that there is a clear, achievable target set with a given end. The goal may be to do a certain action within a given time frame that can be achievable from both the target and time frame viewpoint. An example of a goal of this nature could be to reduce product inventory by 10% over the next 6 months. From that, a plan can be designed after evaluating the existing situation and then executed until the target has been achieved. The goal has a clear starting and ending point that can be measured for its progress along the path toward the final target.

The most common mistake when establishing a realistic goal is not making it challenging enough. If the goal isn't challenging enough, then minimal effort will be put forward toward its accomplishment. Every goal needs to require effort and attention no matter what it is. To accomplish them needs some "stretch" to arrive at the desired target within the given time frame. If the target seems easy, then the time frame should be tighter to maintain focus. Likewise, if the target is difficult in nature, then the time frame should be longer, but not so long that the focus will be lost.

Goals, no matter what they are, need to be constantly reviewed to see what progress is being made. This can be done on an individual basis, by the team working on the goal, or by management. Regardless of who or how the reviewing is being done, a progress evaluation needs to happen to see if there are other elements that need to be considered as the work toward the target is progressing. This may require rethinking the final result or the time frame in which the result is expected. Without reviewing the progress, the goal is hanging out there and the focus of it may be lost.

30.2.2.3 Utopia Goals

What is a utopia goal? This type of goal is one that has no real achievable end, but rather a perfect target to continually work toward. The end target would be something that would place the operation into a perfect state, which will never happen. However, the work being done on a goal of this nature will continually move the operation closer to the perfect state that improves efficiency and effectiveness. An example of a utopia goal could be to reduce the set-up time of a plastic injection molding machine to zero. That is a perfect state, but can never be achieved under any conditions. However, steps toward reducing the set-up time can be taken to reduce the loading time in steps that will result in total reduction time being realized greatly over the long term. Figure 30.5 shows what the possible reduction steps can be from various improvement implementations over time, resulting in a large stride toward the "perfect state" for the operation.

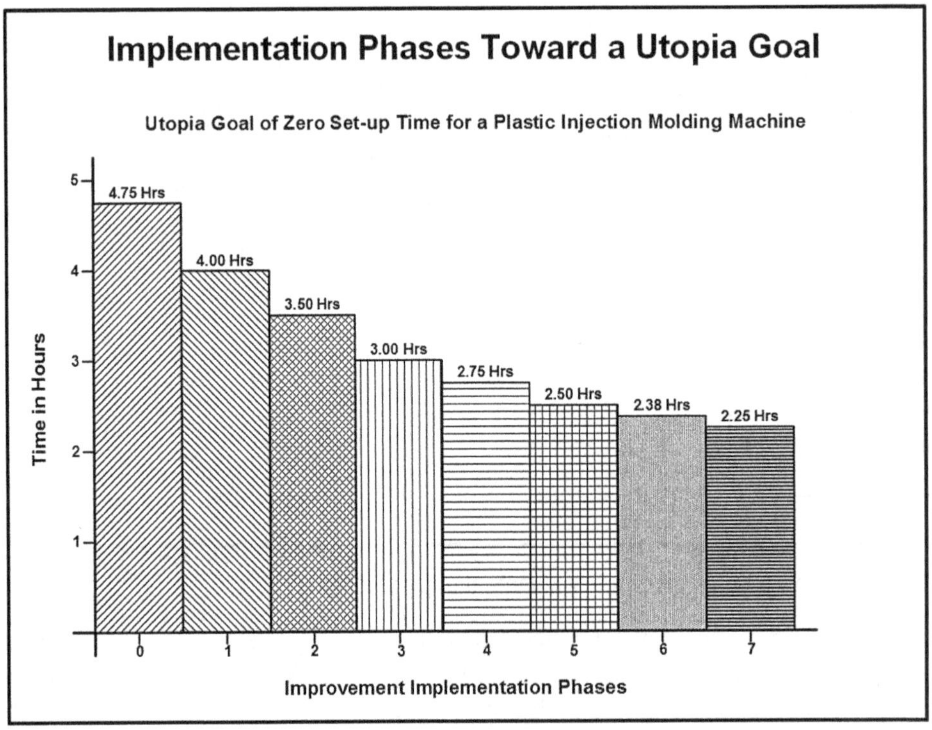

Figure 30.5 Utopia goal implementation phases

Looking at the reduction times in Figure 30.5, each individual step doesn't seem to add up too much, but the sum of them shows a vast reduction in total. Figure 30.6 shows the various improvement phases and the corresponding percentage gains for each phase. Each improvement phase never achieved more than a 15.8% improvement, but over the long term a total of 52.6% was eventually realized.

Set-up Reduction Phases Percentage Statistics

Set-up Reduction Phase	Set-up Time	Phase Reduction	Total Reduction
Current Condition	4.75 Hrs		
First Implementation	4.00 Hrs	15.8%	15.8%
Second Implementation	3.50 Hrs	12.5%	26.3%
Third Implementation	3.00 Hrs	14.3%	36.8%
Fourth Implementation	2.75 Hrs	8.3%	42.1%
Fifth Implementation	2.50 Hrs	9.1%	47.4%
Sixth Implementation	2.38 Hrs	4.8%	49.9%
Seventh Implementation	2.25 Hrs	5.5%	52.6%

Figure 30.6 Improvement statistics toward a utopia goal

When working toward a utopia goal, the improvements will start off greater in the beginning due to the fact that the most obvious will be able to be implemented first and quickly. Then as future phases are implemented, the results will diminish as more complex steps are required to achieve positive gains. As the gained results become less and less, a decision needs to be made as to when enough is enough. This brings into focus the law of diminishing returns, which is covered later in the chapter.

A utopia goal can have some great results in total even though the portions may be less than expected. By continually working to improve the same thing over and over, the compounding of all the improvement phases will allow the efficiency to become much optimized over the long term. However, there has to be a point when the improvement gains need to be halted and the focus taken off the goal. Knowing exactly when this point is will depend on the goal itself, but when the effort and cost being put into achieving results outweighs the gains, the improvement needs to be stopped and you should move on to another area.

30.2.3 Law of Diminishing Returns

Every continuous improvement program will reach a stage where the cost for improvement will outweigh the benefit being gained. When the cost of implementing any improvement overruns the cost to be gained, the improvement has reached the law of diminishing returns. As time goes on with any continuing improvement program, this will eventually happen and the efforts need to be directed into other areas for improvements. This diminishing return law can be seen in Figure 30.7 where the point cost savings being realized begins to flatten out and the implementation cost starts to run high.

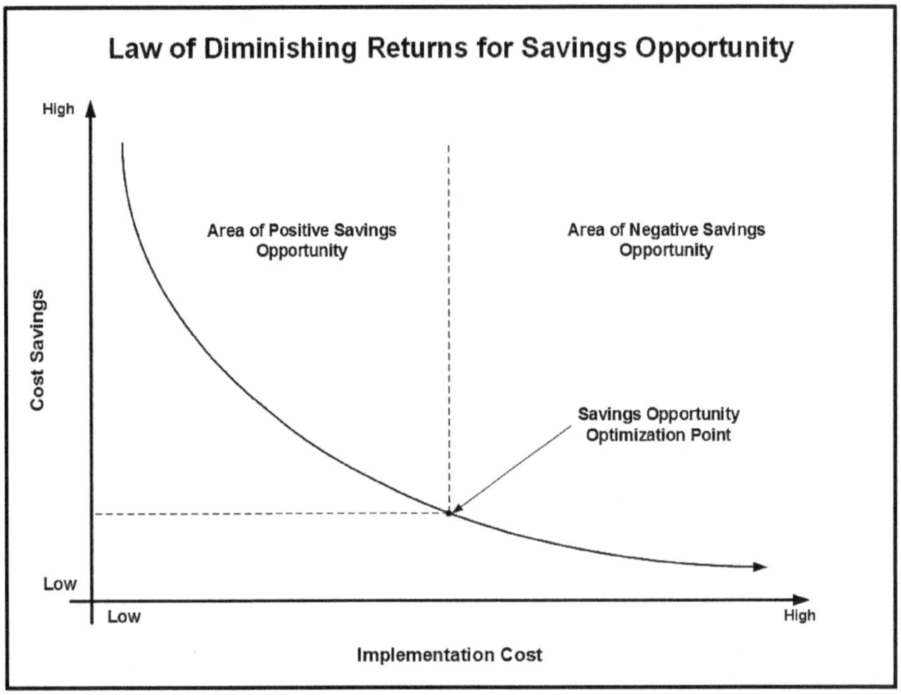

Figure 30.7 Law diminishing opportunity chart

Normally in any improvement program, the 80/20 rule works when the improvements are first started. What this means is that the first 20% of improvements implemented will supply 80% of the benefits to be gained. After that, the returns begin to become less and less as the cost for implementation grows. Knowing when to end a program is critical; so, more cost isn't put into something that is giving minimal returns for the operation. Being able to track the gain vs. cost will be critical to know when the efforts need to be halted in one area of improvement and deployed into other improvement requiring areas. By passing the point of positive savings area and entering the negative savings area, the benefits of any improvement will result in an upward curve for total cost involvement.

If we look back at the utopia goal for reducing the setup for the plastic injection molding machine from a diminishing return view point, the data show that large gains were realized in the early stages as shown in Figure 30.8.

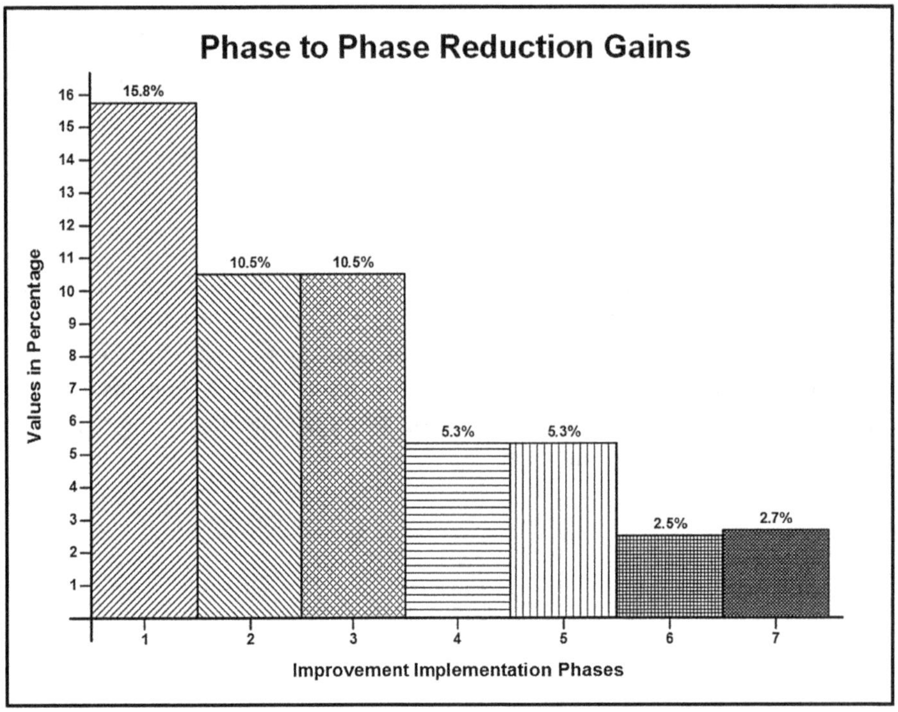

Figure 30.8 Chart of diminishing returns toward a utopia goal

As time progressed, the efforts put forth still produced a reduction in time, but has become less and less for the time spent. Figure 30.9 shows the percentage of reduced time as steps to reduce the setup have been implemented.

Set-up Reduction Phases Percentage Statistics

Set-up Reduction Phase	Set-up Time	Total Reduction	Phase to Phase %
Current Condition	4.75 Hrs	0.0%	0.0%
First Implementation	4.00 Hrs	15.8%	15.8%
Second Implementation	3.50 Hrs	26.3%	10.5%
Third Implementation	3.00 Hrs	36.8%	10.5%
Fourth Implementation	2.75 Hrs	42.1%	5.3%
Fifth Implementation	2.50 Hrs	47.4%	5.3%
Sixth Implementation	2.38 Hrs	49.9%	2.5%
Seventh Implementation	2.25 Hrs	52.6%	2.7%

Figure 30.9 Diminishing return statistics toward a utopia goal

At some point, the amount of return is not worth the time, cost, and efforts being put forth. It is important to know when to not abandon the utopia goal, but rather to focus on another area requiring efficiency improvements. As new techniques are developed or introduced, the goal can be revisited to see if there would be any merit in implementing them to further realize any potential benefits.

30.3 INTERNAL FLOW

Every operation will have an internal flow for various items. These can include materials, products, paper, and information, along with a variety of other areas that flow from one point to another. However, the manner in which the flow is structured can be either an asset or unnecessary burden to the overall efficiency of the business. The fashion in which the operation handles the transfer from one location to another is critical from both a material/product and information view point. Everything needs to move within the operation in an effective and efficient manner to ensure that the end results can be met in the shortest amount of time.

Depending on the products that the operation is handling, and the overall volume of the products, the flow of both information and materials will vary. The volume and processing utilized will determine the best method for developing the overall process flow for the operation. Whether the materials being flowed through the operation are hard materials utilized in manufacturing or information, the process for best moving these is critical to maintain an efficient and effective operation.

30.3.1 Materials

As materials flow through a typical operation, they will fall into one of three categories. These categories are low, medium, and high production flows. Chapter 1 explained the differences between low, medium, and high production operations. They are dependent upon the volume and/or size of the products being processed. Each would be efficient for the appropriate volume required to produce the desired products. However, if a change occurs in the volume, then an alteration in the flow needs to be made to keep the operation running as efficiently as possible. This alteration can be a change required if the volume either increases or decreases.

An increase in the product volume will cause the operation to not meet the requirements based on the existing material flow system. In this case, an investment may need to be made to move the flow system to the next level to better maintain control of the processing and allow for a higher volume of production. Likewise, if the volume decreases, the material flow system should be downscaled to allow for a material flow that will be more efficient for the lower production requirements. This could require the selling off of excess equipment that is no longer required. In either case, the changes made will result in the operation maximizing the efficiency to its fullest and keep the operation at its highest competitive levels.

One type of component flow through a manufacturing operation is batch production. This operates by having departments that specialize in various processes such as milling, drilling, grinding, and turning. The raw materials would begin in the starting department and move from one to the other to have the specialized operations performed. At times, they may move in and out of a given department more than once to accommodate the required geometric configurations. Figure 30.10 shows a typical batch-type factory layout.

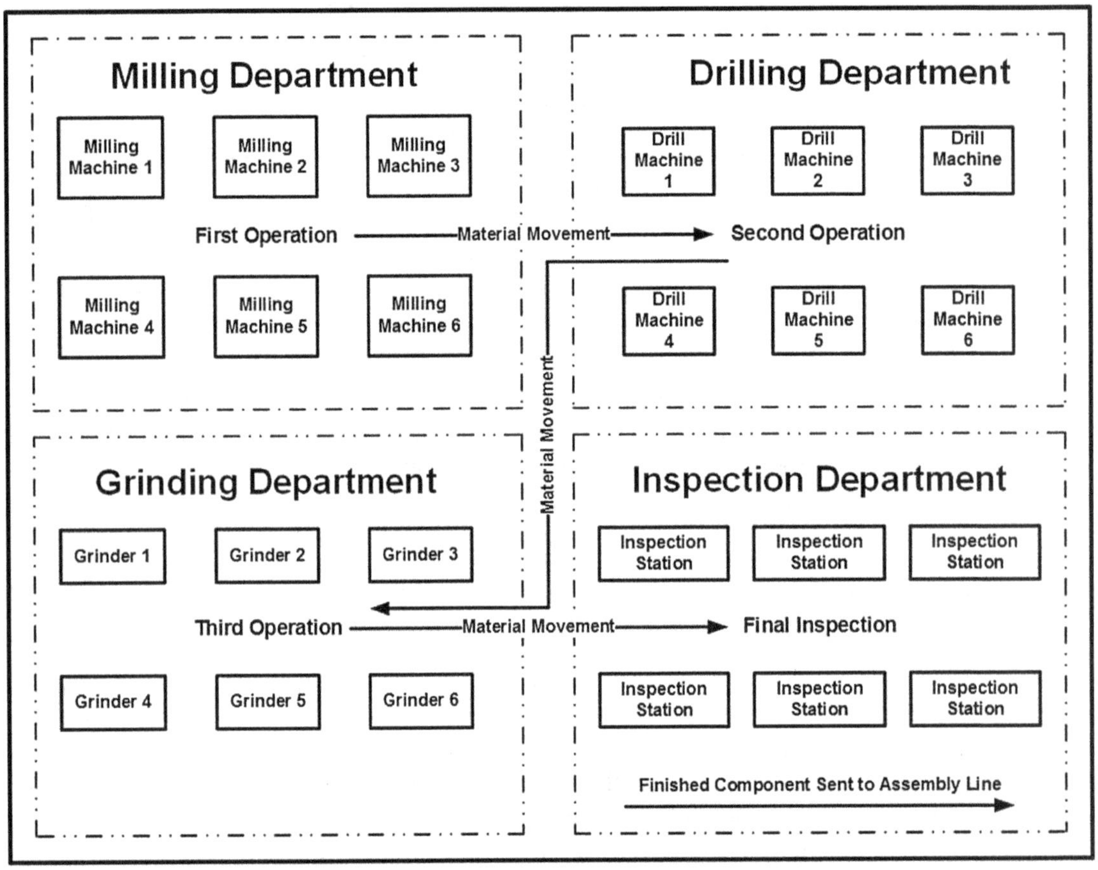

Figure 30.10 Typical batch-type fabrication layout and flow

Although the specialization of batch-type fabrication has its advantages, there are drawbacks as well. For example, if a batch of parts is started in the first department, all of the parts need to have the first operation performed on them before they can move to the next department for the second operation. This greatly extends the overall completion time of the components, even though the utilization of the equipment in the specialized department remains high.

For high volume manufacturing operations, the fabrication flow may vary from the typical batch-type production. Here, dedicated work cells are constructed to fabricate a single component required for the assembly line. A work cell is a group of dedicated machines that are permanently set up for a single operation for the production of a single operation. As the first operation is completed, the work piece is handed to the machine that will perform the second operation and so on until the component is completed. Figure 30.11 shows a typical work cell dedicated to fabricate the same component that was shown in the batch processing in Figure 30.10.

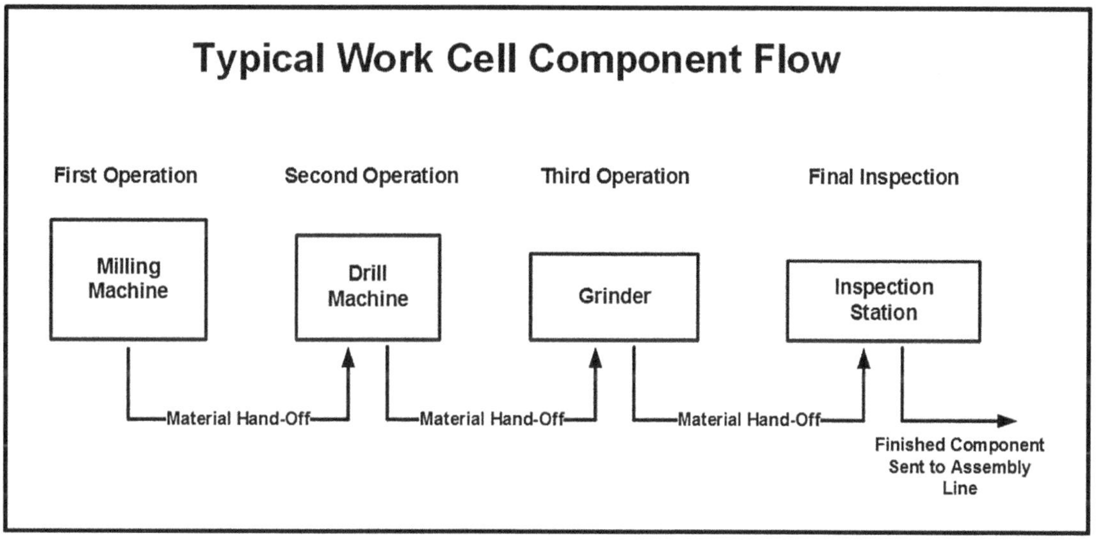

Figure 30.11 Typical work cell fabrication layout and flow

By constructing several work cells for the components required at the assembly line in a high production environment, the flow of work pieces to the assembly line remains constant and no delay is encountered as it could in the batch-type production. Also there will be no large volume of inventory in the operation and the work cells produce only what is required daily to the assembly line. Figure 30.12 shows how the work cells are integrated into the assembly process feeding the line constantly.

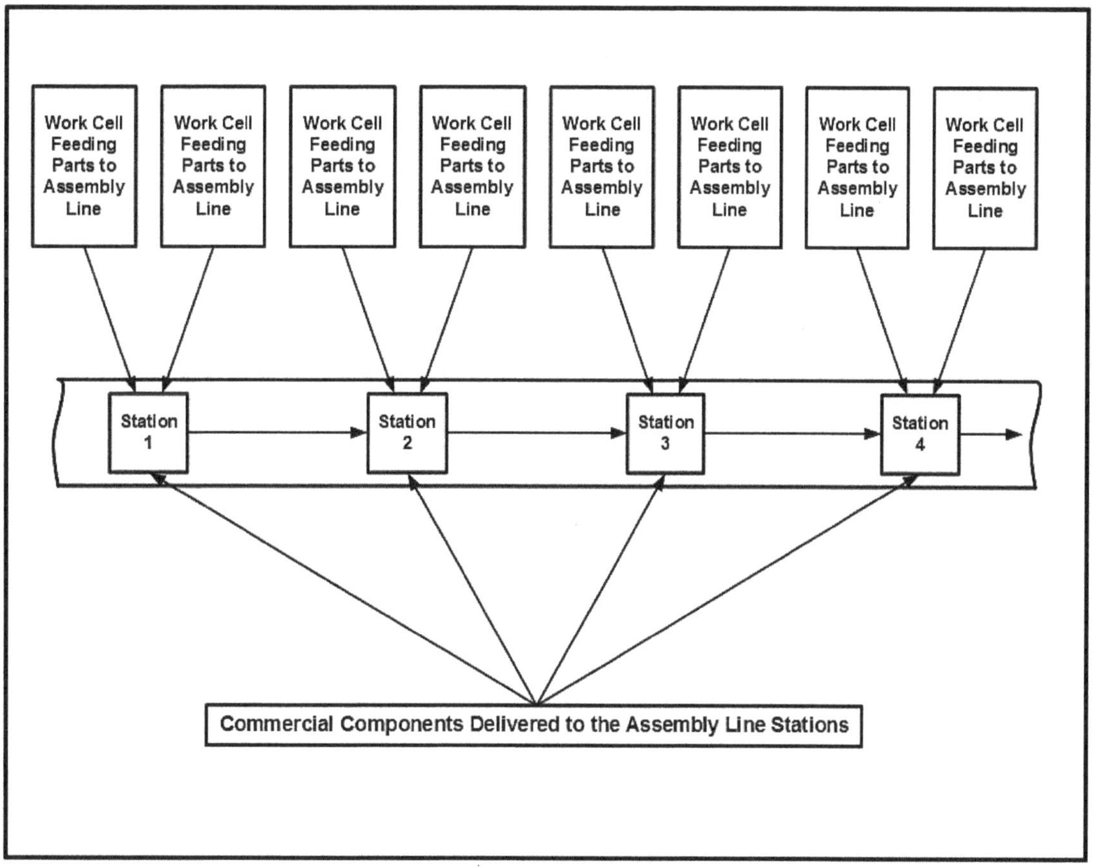

Figure 30.12 Work cells integrated into an assembly line

30.3.2 Information

Information is the heartbeat of any operation if it wants to perform as efficiently as possible. Getting the information required to any point of the operation when it is needed is critical if errors are to be prevented. The fashion in which information is passed along needs to be as streamlined as possible. Without an efficient transfer of information throughout the operation, the overall flow may not be operating to the latest requirements, or be prepared for any upcoming changes that may be present on the horizon.

The transfer of information will predominately be either in written form or utilizing electronic media. Both will be present within the operation and both will be required. Each has its positive and negative aspects and requires a discipline to ensure that it will be effective and efficient. Implementing a system where the important information will flow to the areas that it needs to be will improve the overall process.

Paper Information

The first form of communication throughout operations was paper. This could be formal letters, proposals, part drawings, memos, notes, etc. Even in today's so-called paperless environment there is a vast amount of paper documents that get generated and flow through the system relaying information. Whether it's some form of formal documentation or a simple post-it note, how this is passed through the operation is critical. The paper trail always seems to be huge and endless no matter what the operation is doing. For items such as weekly updates, monthly reports, quarterly reports, and annual summaries, the amount of paper passed around is vast. Being kept up to date throughout the operation seems to generate more and more information that may not be required or not required by everyone who receives it.

Some factors to consider when generating paper information may help to streamline the process flow. Knowing exactly who needs the information and just how much he or she needs will probably eliminate generating piles of documentation that will get distributed to countless people and contain far too much information in them. Below is a quick checklist to review before starting the paper trail in the operation.

Who Needs It: When sending documentation whether it's a formal report or a memo, it's critical to know who needs to receive it, and equally important to know who doesn't. More and more as an operation begins to grow, the distribution list for reports and memos grows with it. Being able to streamline the distribution of the paper being passed around the operation to include only those who need to know the information will greatly reduce both the overall distribution and the processing of the paper for distribution. One of the best ways to see who is actually reading the documentation being sent is to eliminate those who you feel do not need the information. If no one says anything, then most likely he or she wasn't reading it in the first place.

What Do They Need: After you've thinned down the distribution list, now you can begin to streamline the amount of information that the document includes. More than not, reports include far too much detail that the reader isn't interested in. When generating a document, include the critical information that needs to be conveyed. Get right to the point with the important data so the reader won't have to read through pages of documentation to find the results and recommendations that are being sent.

A simple structure is to have an introduction paragraph, body of information describing what was done, no more than three paragraphs, and the findings of two paragraphs. You can always state that more information is available on request and if the reader is interested in knowing more he or she can contact you for in-depth details.

What Path Will It Take: In the case of a proposal, performance review, promotion, or a purchase requisition, levels of signatures are always required. Depending on the size of the operation and the value of the expense, the levels can be many before something is approved. This can take a considerable amount of time as it moves from location to location, waits on desks, or even gets misplaced in the process. With all the red tape and distance a document needs to travel, the chances of having the approvals in time and having everyone know all the information to make the rational clear to everyone are slim. This flow needs to be kept to a minimum and up-front communication with all parties that need to sign can be the solution. One way to make sure everyone who needs to approve would be to have a meeting explaining the proposal with everyone required for signature. This way all involved parties will get the information and be able to have questions answered. This way when it arrives, all they need to do is sign and move it on to the next level.

There is no complete error proof method for this, but by providing up-front information as to what will be headed their way it may eliminate the need for any returns for more details. But by knowing all the channels the document will travel and potential stalling points, actions can be taken to either prevent or reduce any delays.

Electronic Information

In today's electronic means of communication, it is far too easy to send to everyone and include way too much information. Everyone has received e-mails sent to a distribution list including just about everyone possible and containing several attachments to open and review. Just as with the paper documentation, follow the same rules of who needs it and what does he or she need.

However, electronic information is by far the fastest method of transferring information possible. Unlike paper, there is no manual distribution to move the document from one place to another. The document can be sent to the person sitting down the hall or to another country in the same amount of time. If the operation needs to correspond with parties spread out over a large geographical area, this method of information transfer is invaluable. The only downfall is if the document needs to be an original requiring signature. This will still require a single document moving from place to place.

Regardless of whether the information being distributed is paper or electronic, keeping it to a minimum will reduce the massive information overload. Information needs to be documented and distributed, but the amount being sent can easily be minimized. Reducing this information flow will make it more effective and efficient when the proper amount is being received by only those needing to know its contents. Avoiding the massive paper waves in the operation will allow the employees to spend more time on productive work and not generating or reading documentation.

30.4 DELAYS

Possibly the most unproductive things in any operation to reduce effectiveness and efficiency are delays. Regardless what the delay is, the time wasted waiting for anything could be utilized for productive work within the operation. If you think about all the situations a delay is created it really adds up. Some situations are waiting for parts, machine availability, information, truckers, loaders, and service. Each may not be long in duration, but each can hold up production from taking place. The best way to increase efficiency is to remove the amount of delay time during the work day. Some of the delays are unavoidable, but the majority can be either eliminated or greatly reduced.

Just observing the production flow will reveal the delays that occur during the normal processing. The tools for these observations will be covered in detail in Section 30.5.3. Documenting the existing process flow can identify where the delays are taking place and adjusting the process can be made to eliminate or reduce them. Even if the delay seems to be small, the total time in an 8-hour work day will reveal the amount of time wasted. For example, if there was a thirty second delay in a processing operation that only takes two minutes, this would equate to a 75% efficiency rating for that operation. In an 8-hour work day it would be a delay of 2 hours per day. If the operation costs $125.00 per hour to run, then $250.00 is being tied up in delays. Simply by reducing the delay from 30 seconds to 15 seconds, the daily delay time would be only 1 hour with a new efficiency of 87.5% and costs the operation $125.00 per day.

Working to eliminate or reduce delays can increase efficiency for the operation rapidly resulting in greater productivity time and less cost to the business. While no operation can run at 100% efficiency, the more the delay times are reduced regardless of where they occur the overall efficiency will increase as will the profitability. Some delays are easily remedied, while others will take more in-depth evaluation to reduce. But by focusing on the wasted time tied up in delays, the operation can be far more productive and can serve its customers better while reducing the operating costs.

30.4.1 Waiting Line Theory

The waiting line is exactly what it states. How long of a delay is there before service can be rendered to a customer or processing on a production product. If there is no one ahead of you, then the line is non-existent. However, if the area performing the service or processing begins to load up with work, future work needs to wait until that particular station opens up. This staging of future work is known as queue, or staged work waiting to be processed.

No doubt everyone has had to stand in a line for service at one point or another. Some examples of a typical waiting would be at a bank teller, grocery checkout, fast food restaurants, post offices, an amusement park ride, and on the phone on hold. Waiting isn't something unusual, but how it's handled can be the difference between angry customers and a delay in production flow. Being able to minimize the amount of time in the waiting line will do two things. First, it will reduce the amount of angry customers if the line is for being served. No operation wants to have customers aggravated regardless of the reason. The longer someone has to wait, the more aggravated he or she becomes. I witnessed this first hand in a grocery store where the line for checkout was very long and moving slowly and saw some just leave their cart filled with produce and walk out. Even though they most likely headed for another store to buy the same items, and would spend more time overall by having to drive there and fill another cart, the fact that they weren't going to wait on line aggravated them enough to do that. The chance of those customers going back to the original store again is very slim and repeat customers were lost. Second, if the wait is reduced, more processing can take place, thus increasing the overall productivity of the operation. In the case of the grocery store, if additional checkout stations would have been opened, the angry customers could have been accommodated better and the sale wouldn't have been lost. If the wait is for processing products in a factory, more products can be produced simply by increasing the capacity of the problem area.

The concept of a waiting line is simple, too much work requiring processing through a given point. If the work area is performing at 100% of its capacity, the utilization is also at 100% and no more work can flow through that point unless some changes are made. Figure 30.13 shows a typical waiting line where the future work requiring attention is queued up for processing. By addressing these critical points in the processing flow, the time work is waiting for processing can be reduced and a smoother throughput can be realized.

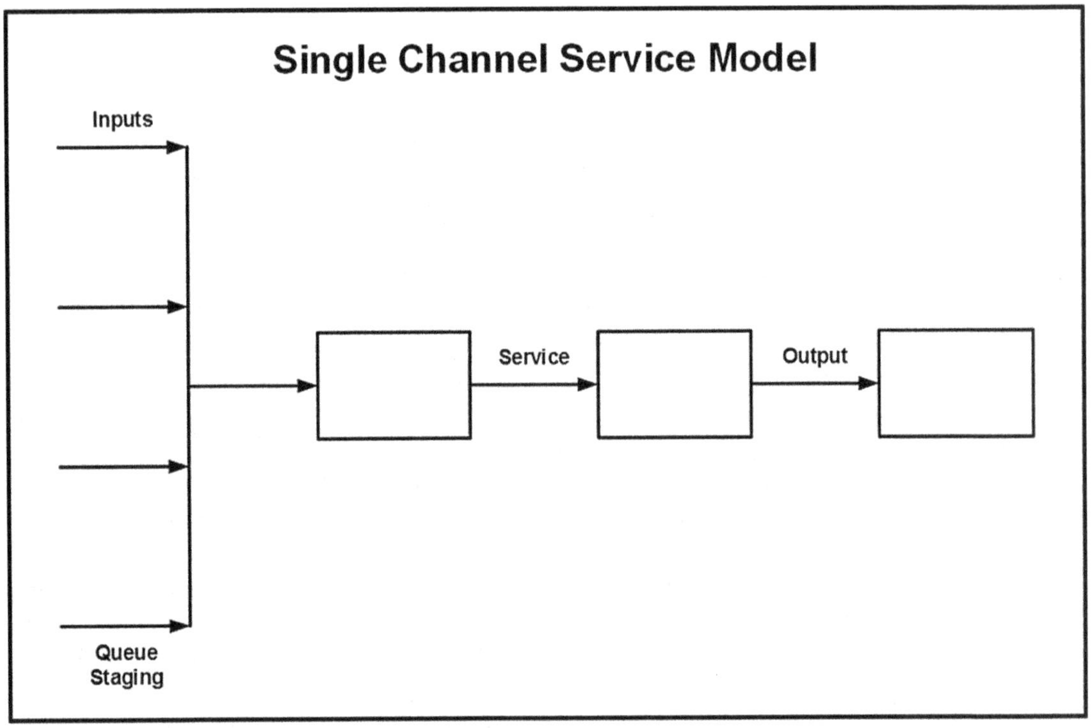

Figure 30.13 Single channel service model for waiting line

Some methods that help to reduce these critical delays can be implemented and for the most part, not add much burden to the operation. If the business operates in some form of cyclical pattern, then proactive steps can be taken to reduce known delays before they occur. In the case of a grocery store, most people will be shopping after work hours rather than in the middle of the day. By opening additional checkout stations before those critical times arrive, customers will be better accommodated. In the case of a production operation working to produce products, if there's a wait for the product to move into a given station then either an additional station would have to be constructed to increase the capacity or some investments for increasing the capacity of the single station would need to be made. However, in some cases if the given workstation cannot be expended other steps can be taken to advise the customer of the waiting time ahead. Probably the best example that comes to mind of this is the waiting line at the amusements in Disney Land. If you're standing in one of those lines, there are signs stating the time you'd have to wait from the point you're at in the line. This way the customer can make the decision whether they are willing to wait that long or not. Also, what I've observed is that in those lincs, the amount of time required to wait is always shorter than the actual time stated on the sign. This does two things for the customer. First, it gives them advance warning on the delay they are facing, and second, by overstating the actual time the customer feels that they were served better by waiting less than the sign stated.

The amount of units staged ahead of a given workstation and the amount of delay can be charted based on the capacity and utilization being performed there. The wait goes up exponentially as the workstation begins to climb in its capacity and utilization increases. Figure 30.14 shows a pictorial view of the amount of units delayed as the area approaches higher capacity utilization.

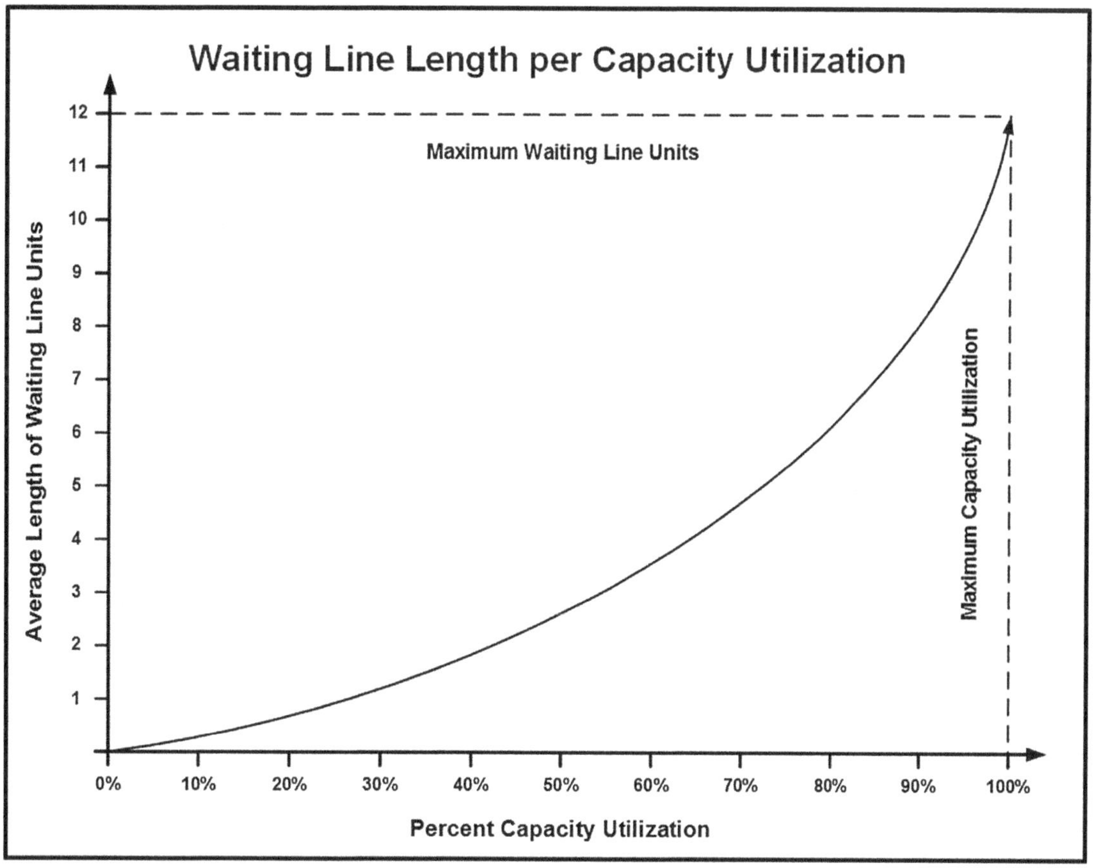

Figure 30.14 Waiting line vs. capacity chart

Although each given workstation would have its own chart based on how long the processing will take, as a rule of thumb the queuing delay will be minimized up to about 70–80% of the overall capacity utilization of the station. After that point, the delay increases dramatically from that point forward.

By being able to identify where the waiting line will establish itself, the operation needs to be proactive in reducing the amount of time to the best of its power. As stated earlier, some resolutions are simple and won't place much burden on the business where others may involve some investments. In the cases where nothing can be done, notification of how long the customer needs to wait is critical. Even if it's being on the phone on hold, some companies have implemented a message that comes up every minute telling the customer how much longer they will be waiting before their call is answered. Keeping the processing as smooth of a flow as possible is critical to the operation, whether it's for customers or product flow.

> The waiting line model demonstrates why a continuous output system needs to be balanced so that work (products, customers, etc.) flows from stage to stage in sequence. If each stage were independent of each other and the flow was random rather than scheduled, full capacity utilization would create the potential for an infinite waiting line in front of every workstation. (Bassett, 1991)

Only through constant evaluation of the reason for delays and waiting can the operation improve its customer service and overall throughput.

30.4.2 Bottlenecks

A bottleneck can best be defined as anything that causes the flow of materials or information to be delayed. Just as the name implies, a bottleneck is where there's too much of anything trying to pass through a given point causing either a slowdown or a complete stop. A good example of this would be a traffic jam on a road. If too many cars are trying to get through a point on the road that can't accommodate the volume, the traffic slows down. If the number of cars is too many, then everything comes to a stop. Everyone has been involved in these at one time or another.

Bottlenecks can be either a fixed feature in the process flow or a temporary one. In the case of a traffic jam, if the road just can't handle the normal volume, then the bottleneck is fixed in nature and only widening the road will resolve the problem. However, if there is an accident in an area that is normally easily passable, then the bottleneck is only a temporary one and will be resolved when the accident is cleared. The temporary situation will resolve itself over time, but the fixed bottleneck needs to be addressed to open the process flow to a smooth, steady state.

By identifying the bottlenecks within the operation's process flows, revisions can be made to resolve the issues causing the delays. Some of these can be very easily fixed while others may be more difficult and require some investments in equipment or labor. Regardless of what actions need to be taken, by removing the bottlenecks the delays will be either eliminated or greatly reduced. From this the efficiency will be maximized in the operation and productivity will increase.

30.5 IMPROVEMENT TECHNIQUES

There are several methods and techniques that are utilized in the improvement process. Each has its own specific place and result depending on the area being evaluated. However, each tool will follow the same basic pattern in the evaluation process that leads to a recommended improvement. The evaluation methods described below are basic tools that will render an evaluation of a process from a high level or a detailed level. These tools can be utilized for either measuring material movement or operator activities for a single operator or multiple operators and/or machines. By using them together, just about any situation can be evaluated and an improved process can be developed.

When developing any improvement for an area in the operation, some basic rules need to be adhered to in order to obtain the best results with the least amount of error. A detailed snapshot of the existing process has to be taken first, showing exactly what is being performed and the related times and distances recorded.

This will provide the basis for evaluation needed to implement various improvements. During the initial evaluation, absolutely no improvements are implemented regardless of how obvious they appear. Doing this will not give a true picture of the existing process and may alter the overall improvement percentage. Basically, you have to know exactly where you are starting from to record the improvement that will be implemented.

Once the existing process has been fully documented, an improved process can be constructed. This is done by looking at the various motions and steps being performed and looking for opportunities to improve or eliminate what is being performed. The section on waste removal covers some evaluation methods and steps that need to be taken to construct and sound proposed improvement process that can be implemented easily and with little chance of error.

One improvement technique that stared in Japan and has been adopted throughout the world is the Kaizen process. Here, everyone in the company is involved in the improvement process and it stretches much further than just the area being evaluated. Kaizen focuses on improvement of life in general from personal to work.

> Kaizen means improvement. Moreover it means continuing improvement in personal life, home life, social life, and working life. When applied to the workplace, Kaizen means continuing improvement involving everyone - managers and workers alike. The Kaizen strategy maintains and improves the working standard through small, gradual improvements and clearly delineates responsibility for maintaining standards to the worker, with management's role being the improvement of standards. (Imai, 1986)

The Kaizen philosophy drives improvement on a macroscale and is a continuous process and not a single event. Employees are rewarded for implementation of improvements within the operation and once the improvement has been implemented, further investigation is performed to see if it can be further improved. Management is totally involved at all levels to support and drive the process from the bottom up. This has proven to be extremely effective and has been implemented in conjunction with other company-wide programs such as Total Quality Management (TQM).

30.5.1 Theory

Improvement of the operation focuses on three major indicators that are efficiency, effectiveness, and quality. Each needs to be considered when an improvement investigation is being performed. The overall theory is to eliminate any wasted time in the process to improve the efficiency. In doing so, the effectiveness and quality need to be considered in conjunction so neither will be diminished in the process. As mentioned earlier in the chapter, the focus is on removing any non-value added time in the process that is taking away from the overall efficiency. In doing so, the percentage of value added time will increase, thus increasing the efficiency. This process can focus on the removal of downtime, delays, transportation times/distances, machine utilization, etc. Depending on the situation at hand, focusing on the removal of non-value added time is the objective of process improvement, whether it is on a large or a small scale. The compounding of several small improvements will result in a major increase for the operation.

30.5.2 Waste Removal

Waste refers to anything that reduces the overall efficiency of the operation. This can include but isn't limited to areas such as time, materials, transportations, floor space, operational delays, and equipment downtime. In a perfect environment the efficiency would be at 100% for the operation 24 hours per day and 7 days per week. Although this is utopia, the removal of waste in the operation will increase the efficiency percentage and thus increase the profitability. Looking at each area mentioned, the importance of removing the waste will present itself.

Time: The time required to produce a product from start to finish is the most critical component for operational efficiency. In general, time equals money. The more value added time being recorded against the overall process will result in producing the product at the least cost to the operation and thus increase the profits that will be realized. The removal of any non-value added time will undoubtedly reduce the production time and increase the efficiency of the operation.

Materials: When looking at the materials being utilized in the processes, two main factors need to be considered. First is the material actually required to produce the product? This can be either the materials being placed in the product or support materials being utilized in the process from an indirect standpoint. Second, can the material that is being used be replaced with a different material having similar properties but costing less? This can be either a less expensive material in general or one that would cost less to fabricate the components in general.

Transportations: The arena of transportation can be looked at from either a low or a high level. This broad area can take in everything from transporting the finished products around the world to the distances the materials are being moved at the assembly bench. Whenever a component or a product is being transported, no value is being added. By evaluating the amount of transportation either at a high or at a low level and either reducing it or, if possible, eliminating it, less time will be wasted in moving items from point to point and more focus can be placed on value added operations.

Floor Space: The square footage of a building has a cost associated with it. If the total cost of running an operation is added up, it can be broken down to a value per square foot of the facility. With that understanding, the more productive activities that can be performed within the given square footage will increase the output per square footage. By streamlining the operational flow, the same production can be performed in less space thus resulting in savings. Also, if floor space can be maximized and additional space opened up, additional production can be brought into that area that will increase the production output without having to acquire additional space, or can be subleased out for additional income.

Looking at an example of two components being produced in an operation and the current and improved floor space, Figure 30.15 shows the reduction in floor space for the first work piece by rearranging the equipment and streamlining the flow, thus reducing the overall footprint.

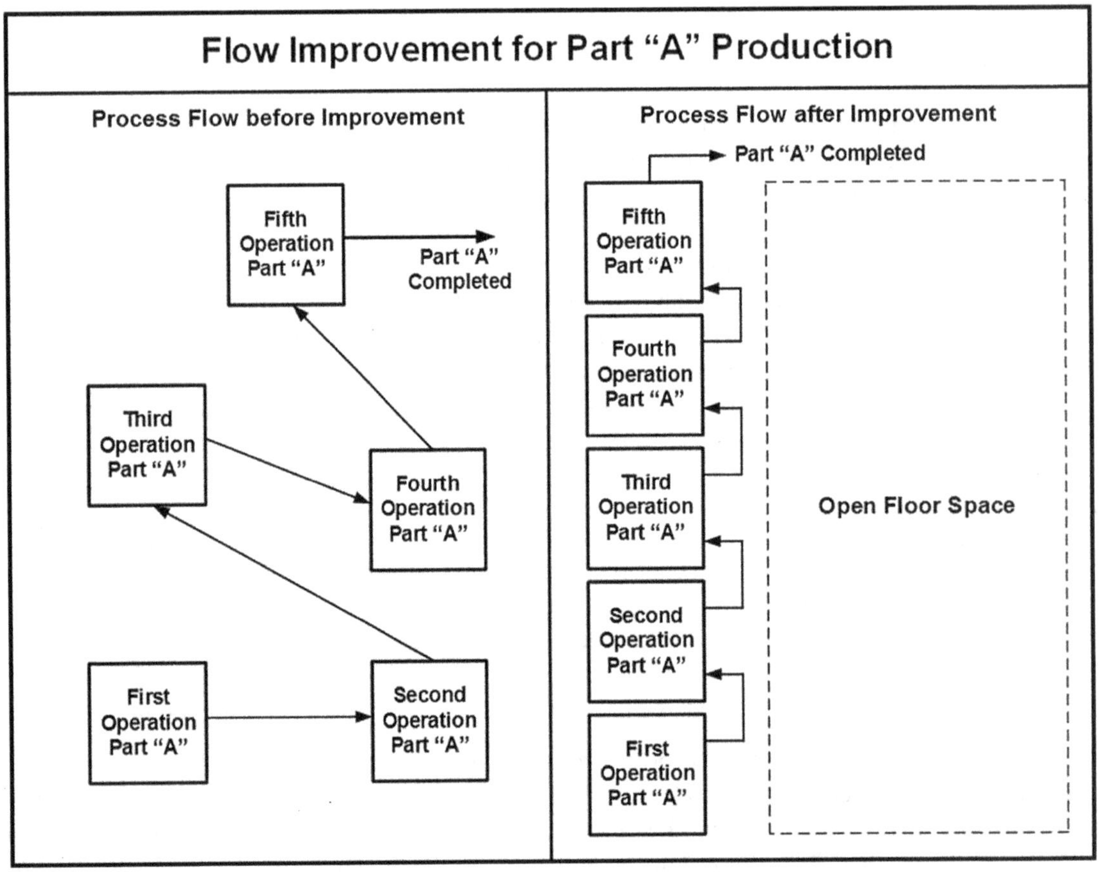

Figure 30.15 Floor space consolidation for Part "A" fabrication

Performing the same rearranging for the second component, the same reduction in floor space can be realized as shown in Figure 30.16.

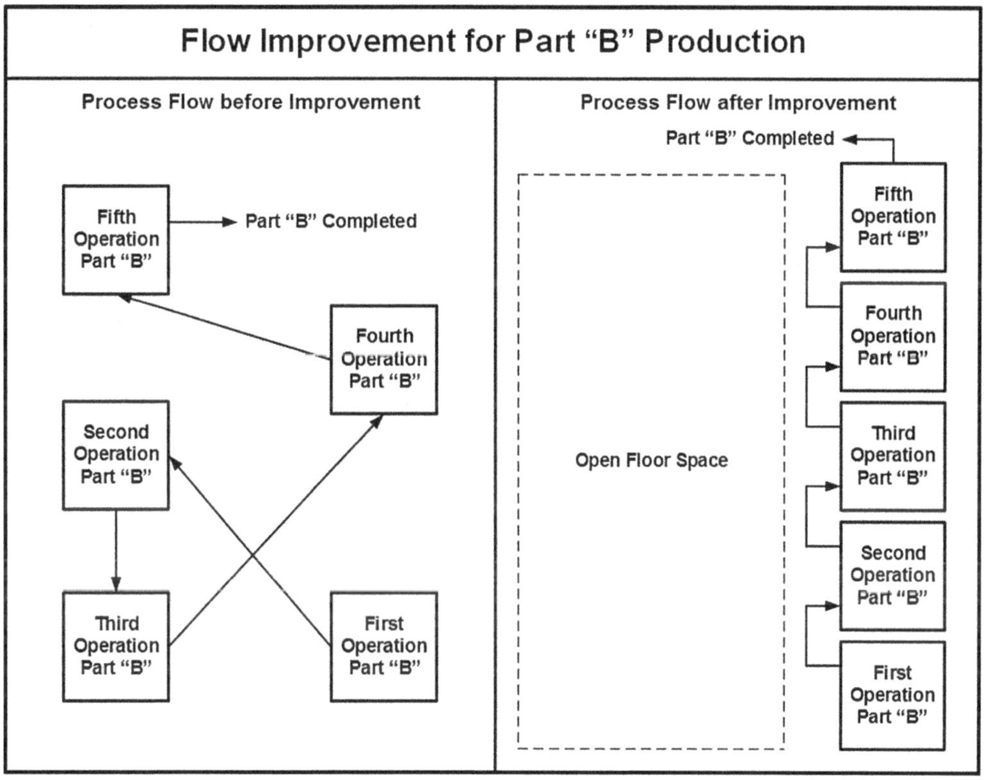

Figure 30.16 Floor space consolidation for Part "B" fabrication

By combining the two component work cells together as shown in Figure 30.17, the result is opening a section of floor space that was originally required to produce one component thus allowing for more capacity in the operation.

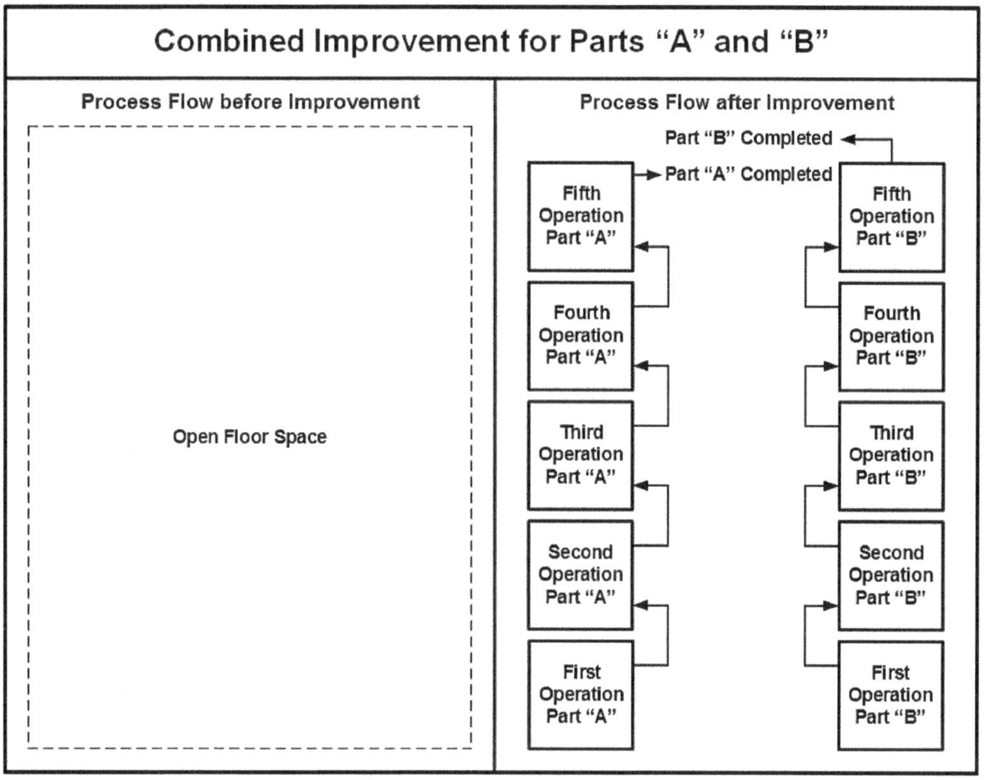

Figure 30.17 Combined floor space reduction for Parts "A and B" fabrication

Operational Delays: By definition, a delay is any time that something has to wait to have processing performed on it. While the item is waiting, no value is being added to it. Looking for areas where delays are being caused and working to resolve them will increase throughput and efficiency. This was covered in Section 30.4.1 for the waiting line theory and bottle necks. Finding techniques to reduce or eliminate any delays will ultimately improve the efficiency of the operation.

Equipment Downtime: When your equipment is down, it's not earning you any money. Finding ways to keep the utilization of the equipment as high as possible will increase the overall efficiency of the operation. With a steady flow of materials entering the equipment with minimal time in between setups, the machinery can do what it was purchased to do. However, unplanned downtime is unavoidable. Equipment breakdowns are impossible to schedule, but with a good preventative maintenance plan, the downtime can be scheduled to perform any maintenance. Even though this will cut into production, the long-term effects of an unplanned breakdown will have more of a negative impact on the operation and production will be halted when it's needed the most.

30.5.3 Evaluation Tools

To perform any type of structured analysis, you need to have tools that will aid you in the process. There are several types of tools and systems available for use. Each one will be utilized for a specific type of analysis to perform on a process being evaluated for improvement opportunities. For our needs, four tools listed below will be covered. All of these are very general tools, but can be utilized across a multitude of process evaluations. These evaluation tools are very widely used throughout operations worldwide and can render very positive results. In some instances a combination of these tools can be used for the same process in evaluation depending on the work being performed.

One critical point to remember when performing any improvement evaluation is that it is always done in three independent phases. The first phase is to record exactly what steps are being performed in the current process. As stated earlier, never try to implement any improvement until the evaluation of the current process has been completed. Regardless of how obvious the improvement appears, always wait until the evaluation is completed before any implementation is done. What seems obvious at first may have a different view with the entire process being viewed in total. The second phase is to evaluate the entire current process to identify all the improvement opportunities. By looking at the current process and all of the steps, no opportunity will be overlooked. If improvements are implemented in phase one, this may hide or alter the total picture for improvement opportunities. Phase three involves the actual implementation of the improvements identified. Sometimes these improvements do exactly what is expected and other times they will render less than expected results. From here, alterations need to be considered to ensure that the improvements are actually making the process more efficient. If alterations are done to a proposed operation, the changes need to be documented in the improvement implementation. An accurate record of before and after is the only way that the improvement can be measured.

30.5.3.1 S.E.A.R.C.H. Analysis

An analysis method that can provide some interesting information is the S.E.A.R.C.H. analysis. What this analysis is used for is to look for areas where any waste can be eliminated. The letters that make up the word SEARCH each stands for an analysis step in the evaluation. Each would be applied to every element in the analysis, and steps for improvement would be applied to see if there are any opportunities for improvement. The S.E.A.R.C.H. analysis is a thought provoker used for identifying improvement opportunities and helps to reduce any waste that is occurring. Each phase in the analysis looks into different opportunity areas of the analysis that was performed. Below are the phases of the S.E.A.R.C.H. method for waste identification.

S.E.A.R.C.H. Analysis Elements

	Description	Examples
S	Simplify	Improve Fixturing, Shorten Motions, Reduce Body Motions, Re-Structure Workstation, etc.
E	Eliminate	Delays, Transportations, Unnecessary Work, Non-Required Tasks, Workstation Motions, etc.
A	Alter Sequence	Re-Arrange Operation Steps, Reduce Idle Times and Delays, etc.
R	Requirements	Are the Requirements too Stringent, Are the Requirements too Loose, Are the Requirements Necessary, etc.
C	Combine	Operations, Value Added Times with Non-Value Added Times, Motions, etc.
H	How Often	Production Lot Sizes, Frequency of the Operation, Frequency of the Maintenance, etc.

Figure 30.18 S.E.A.R.C.H. element symbols and examples

Simplify: Simplify means just what it says. Are there any opportunities to make the operation easier for the operator to perform the required functions? Look at motions being performed, head and body turns, eye focuses, etc. Also look at the structure of the workstation itself to see if anything can be simplified to better organize it for the operator. In general, what can be done to make the process simpler?

Eliminate: Eliminate means to see what can be removed from the process. This could be steps being currently performed, unnecessary tooling, motions, etc. Also look at any transportation that can be removed to make the process more efficient. Another thing that can be evaluated to see if it can be eliminated is the material being used. Can something else be substituted in its place that would make the process easier to perform? In any process improvement, always try to eliminate as much as you possibly can without causing any negative impact to the operation, operator, or products.

Alter Sequence: Altering the sequence in which a process is being performed can sometimes render positive results. Just because something is being done in a certain way doesn't mean it's the most efficient. Look at changing the sequence of events from the current method to see if this will increase the output from the process. Most of the time, this will reduce the workstation downtime, which results in more value added time being performed.

Requirements: Always look at the requirements of the work being performed. Sometimes they may be able to be loosened up some which will reduce the amount of precision required. With less precision, the work can be performed faster without the fear of producing a defective job. Remember, the tighter the requirements of a job, the longer it will take to produce an acceptable product. Also, in some instances it may be necessary to ask for the requirements to be tighter. This happens in several machining operations where a surface that has very loose requirements will be utilized to locate the component for subsequent operations. By increasing the requirements for that given surface, the subsequent operations can be performed more rapidly.

Combine: Evaluating a process should always include seeing what steps can be combined. This may require some sort of tooling or fixturing, but will increase the efficiency. Another step to consider is to see if operations being performed can be more easily combined in the processing station either before or after the one that is currently being evaluated.

How Often: Before any improvements should be implemented, you need to know how often the operation will be run. If the volume is high, it may be something that will run daily. On the other hand, if the area being evaluated is only being run weekly or monthly, you will know how much to invest in the overall improvement. You want to focus on investing the operations time and money in the areas that will show the greatest payback. For some low volume processing operations it may be feasible to simply construct another identical workstation to eliminate the bottleneck that occurs in a low frequency. Table 6.2 shows the S.E.A.R.C.H. analysis elements and some examples of what is considered in each.

Waste removal is a critical step in improving any operation. The more time that is spent performing value added work will increase both the efficiency and utilization. With the focus on this, the efficiency will be maximized within the operation allowing for greater production and higher profits.

30.5.3.2 Process Flowcharts

The process flowchart is possibly the most versatile evaluation tool. This particular tool can be used at either a high or a low level. Also, this tool can be applied to an operator's activities or the material being routed. The concept is a simple one to use and can be mastered in a very short amount of time. When using a process flowchart, the process is evaluated by breaking down the steps into five categories. Figure 30.19 shows the symbols utilized in the flow process chart and examples for each.

Process Flow Chart Symbols

Symbol	Description	Example	Process Value
	Inspection	Measure Material Dimensions; Examine Instructions; Observe Machine Gages; etc.	Non-Value Add Element
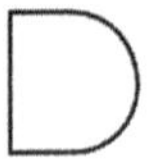	Delay	Wait for an Elevator; Materials Waiting to be Processes; Papers Waiting to be Filed; etc.	Non-Value Add Element
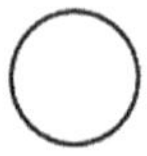	Operation	Machining A Component; Assembling Components; Writing a Letter; Filing Proposals; etc.	Value Add Element
	Storage	Raw Materials in the Warehouse; Finished Product Inventory; Documents in the Filing Cabinet; etc.	Non-Value Add Element
	Transportation	Moving Raw Materials to the First Operation; Moving Work-In-Process from Station to Station; Moving Finished Product to Inventory; etc.	Non-Value Add Element

Figure 30.19 Process flowchart symbols, values, and examples

When a flow process chart is being used, the sequence of events is charted noting the time required for each step along with any distances involved for moves, transports, reaches, etc. The appropriate symbol is indicated as the flow of the process is charted. Figure 30.20 shows a typical flow process chart.

Process Flow Chart

Operation: **Date:**

Machine: **Evaluator:**

Operator: **Proposed:**

Material: **Present:** **Sheet: # 1 of**

	Step Description	Symbol	Dist.	Time	Notes
1		○ ⇨ □ D ▽			
2		○ ⇨ □ D ▽			
3		○ ⇨ □ D ▽			
4		○ ⇨ □ D ▽			
5		○ ⇨ □ D ▽			
6		○ ⇨ □ D ▽			
7		○ ⇨ □ D ▽			
8		○ ⇨ □ D ▽			
9		○ ⇨ □ D ▽			
10		○ ⇨ □ D ▽			
11		○ ⇨ □ D ▽			
12		○ ⇨ □ D ▽			
13		○ ⇨ □ D ▽			
14		○ ⇨ □ D ▽			
15		○ ⇨ □ D ▽			
16		○ ⇨ □ D ▽			
17		○ ⇨ □ D ▽			
18		○ ⇨ □ D ▽			
19		○ ⇨ □ D ▽			
20		○ ⇨ □ D ▽			
21		○ ⇨ □ D ▽			
22		○ ⇨ □ D ▽			
23		○ ⇨ □ D ▽			
24		○ ⇨ □ D ▽			
25		○ ⇨ □ D ▽			
26		○ ⇨ □ D ▽			
27		○ ⇨ □ D ▽			
28		○ ⇨ □ D ▽			
29		○ ⇨ □ D ▽			
30		○ ⇨ □ D ▽			

Figure 30.20 Process flowchart example

When performing an evaluation with a process flowchart, the relevant data need to be included on the form. These being the date, who the operator was, who performed the evaluation, the operation being evaluated, if it is the present or proposed method, and whether the evaluation is looking at the operator activities or the material flow. From there, a careful and detailed observation is performed on the existing process making sure that every element is documented in the exact sequence it is being performed in. As mentioned earlier, absolutely no improvements are to be implemented during this initial evaluation. Upon completion of the initial evaluation of the current process, an improved process can be developed from the present method being performed. This involves developing a second flow process chart that details the proposed improvements in the same detailed structured form as the current method. In conjunction with the flow process chart, a flow diagram needs to be developed for both the current and proposed processes. Figure 30.21 shows a process flow diagram that always accompanies the process flowchart.

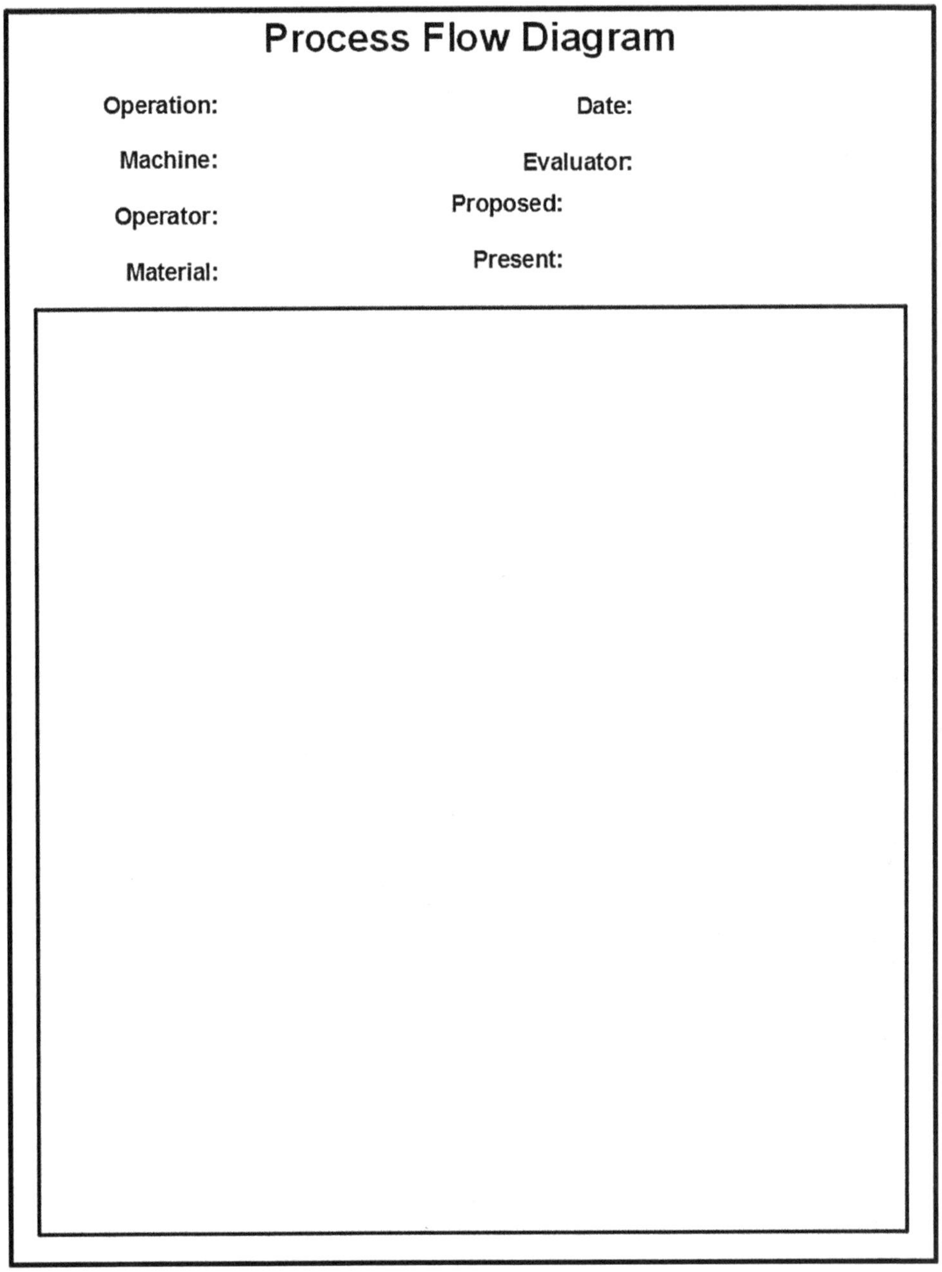

Figure 30.21 Process flow diagram example

If we were to look at an example of a simple machining operation being performed by an operator on a hand drill press, both the present and proposed methods will show improvements that are intended to increase the efficiency of that individual operation. For example purposes, let's say that an operation requires an operator to drill two holes in a block. One hole will be drilled through the part while the other hole will be drilled to a set depth. Both require a different holding fixture that the operator needs to change, and upon completion of the drilling the holes need to be countersunk that will be performed by hand on the same drill press. The machine is currently equipped with a standard drill chuck that requires the operator to loosen and tighten by hand. The fixtures that will be used currently are resting on a table behind the operator.

The first steps are to document the existing process on a flow process chart and develop a flow diagram of the layout. From there, the improvement ideas can be developed and a proposed method can be developed and documented with a new flow process chart and flow diagram. The process flowcharts in Figures 30.22 and 30.23 show the present and proposed processes.

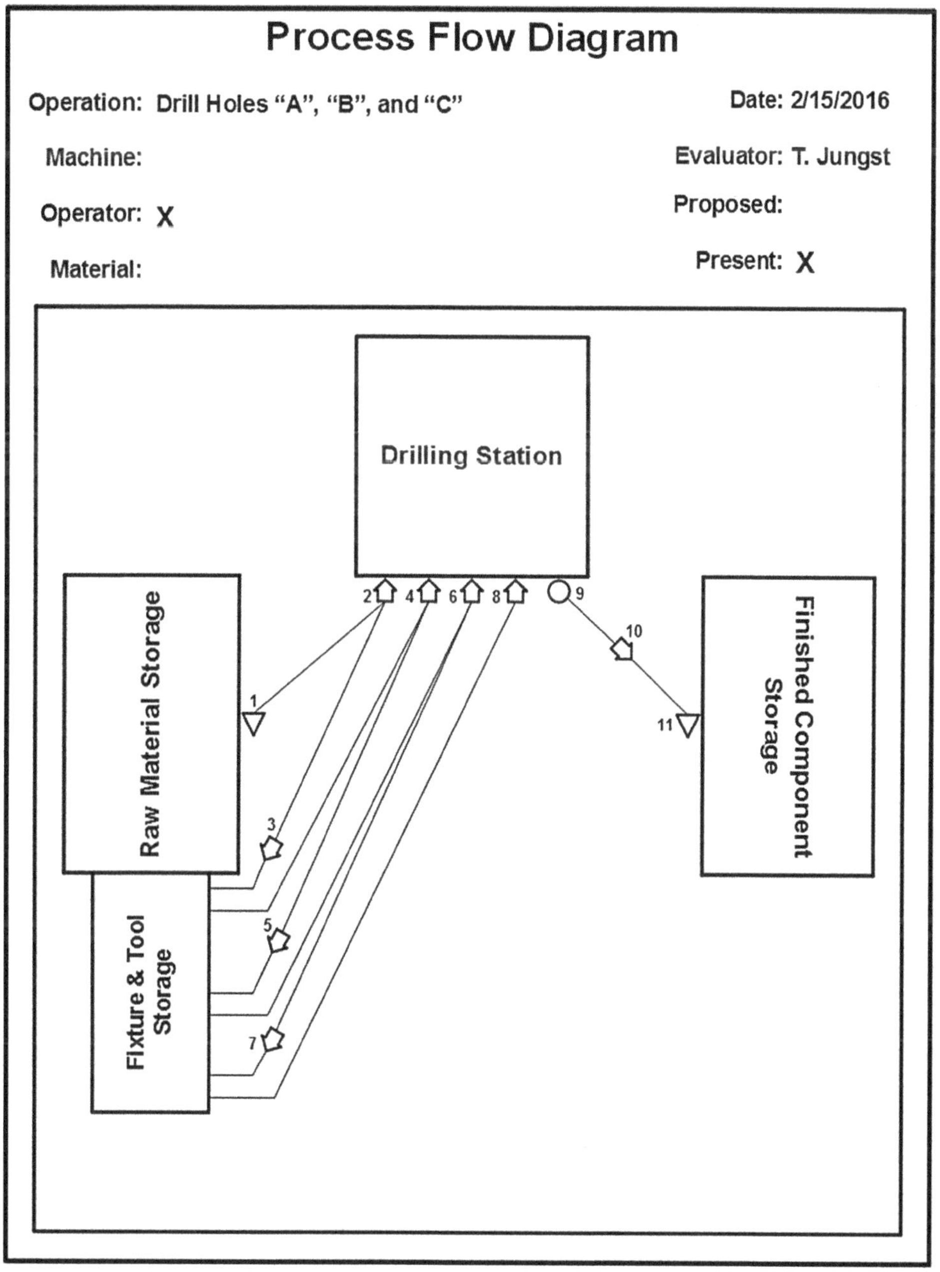

Figure 30.22 Process flow diagram for a present operation condition

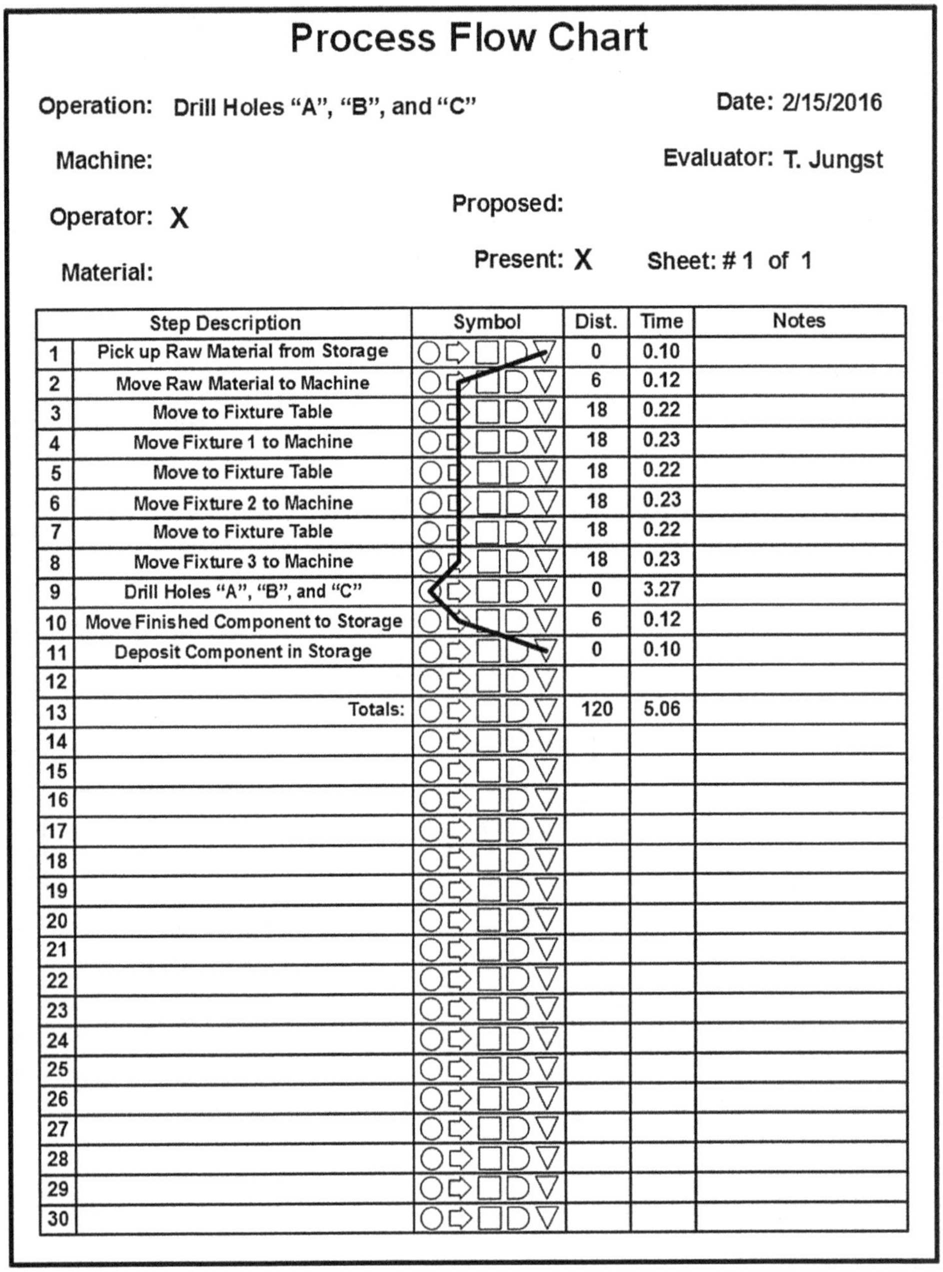

Process Flow Chart

Operation: Drill Holes "A", "B", and "C" **Date:** 2/15/2016

Machine: **Evaluator:** T. Jungst

Operator: X **Proposed:**

Material: **Present:** X **Sheet:** # 1 of 1

	Step Description	Symbol	Dist.	Time	Notes
1	Pick up Raw Material from Storage	○⇨□D▽	0	0.10	
2	Move Raw Material to Machine	○⇨□D▽	6	0.12	
3	Move to Fixture Table	○⇨□D▽	18	0.22	
4	Move Fixture 1 to Machine	○⇨□D▽	18	0.23	
5	Move to Fixture Table	○⇨□D▽	18	0.22	
6	Move Fixture 2 to Machine	○⇨□D▽	18	0.23	
7	Move to Fixture Table	○⇨□D▽	18	0.22	
8	Move Fixture 3 to Machine	○⇨□D▽	18	0.23	
9	Drill Holes "A", "B", and "C"	○⇨□D▽	0	3.27	
10	Move Finished Component to Storage	○⇨□D▽	6	0.12	
11	Deposit Component in Storage	○⇨□D▽	0	0.10	
12		○⇨□D▽			
13	Totals:	○⇨□D▽	120	5.06	
14		○⇨□D▽			
15		○⇨□D▽			
16		○⇨□D▽			
17		○⇨□D▽			
18		○⇨□D▽			
19		○⇨□D▽			
20		○⇨□D▽			
21		○⇨□D▽			
22		○⇨□D▽			
23		○⇨□D▽			
24		○⇨□D▽			
25		○⇨□D▽			
26		○⇨□D▽			
27		○⇨□D▽			
28		○⇨□D▽			
29		○⇨□D▽			
30		○⇨□D▽			

Figure 30.23 Process flowchart for a present operation condition

The improvements for this example are fairly straightforward for this particular operation. First, a speed chuck would be installed on the drill press that will allow for a faster drill change during the operation. Second, a cart was implemented allowing all the tooling to be moved at one time to the workstation, eliminating the turning and carrying of the fixtures to the existing table in the current process. These simple but effective improvements can be calculated to see what the increase in efficiency will be. Below are the results from the analysis performed. Figures 30.24 and 30.25 show the proposed new process flow for the operation and the improvements implemented.

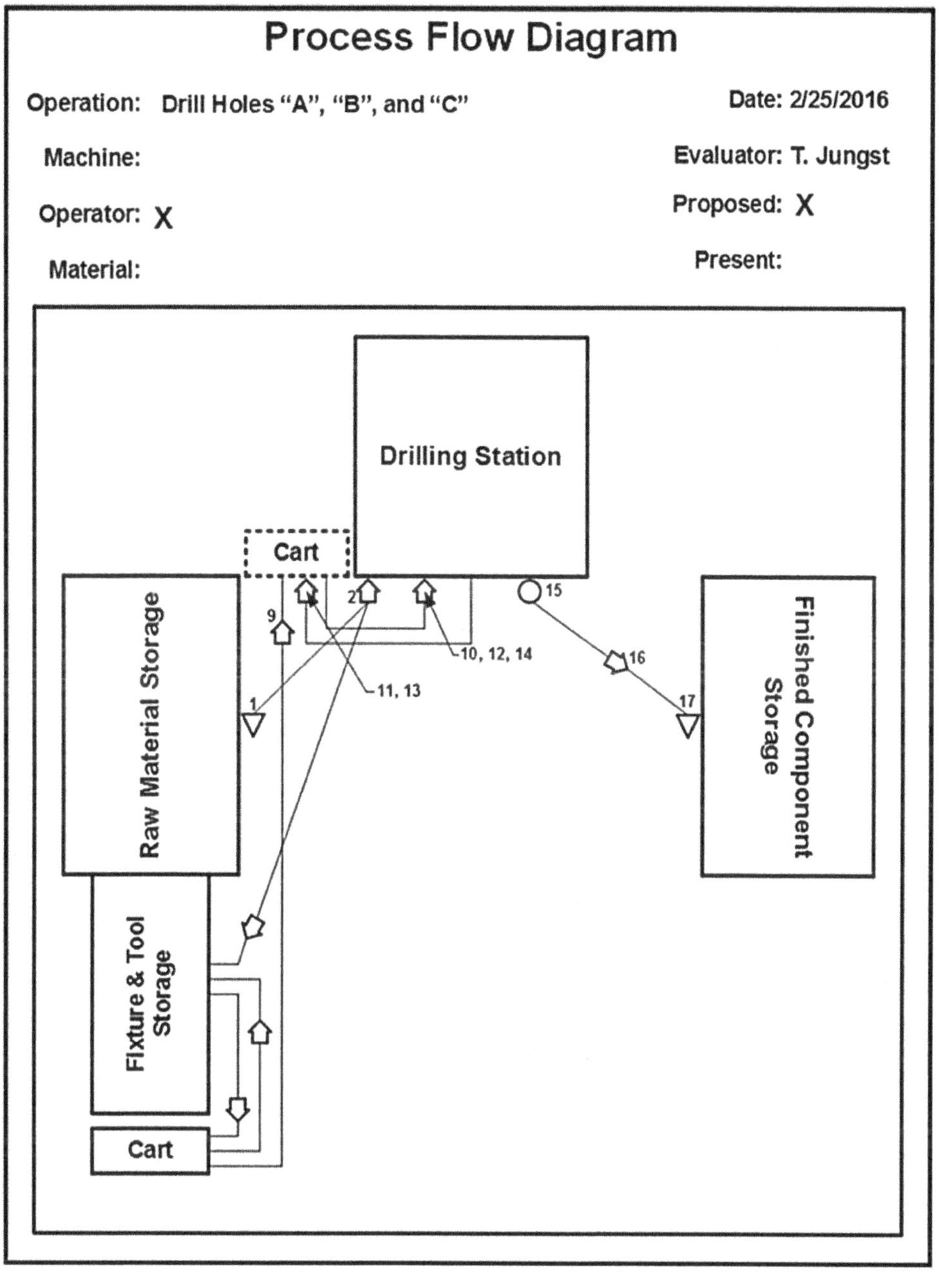

Figure 30.24 Process flow diagram for proposed changes for operation conditions

Process Flow Chart

Operation: Drill Holes "A", "B", and "C" Date: 2/25/2016

Machine: Evaluator: T. Jungst

Operator: X Proposed: X

Material: Present: Sheet: # 1 of 1

	Step Description	Symbol	Dist.	Time	Notes
1	Pick up Raw Material from Storage	○⇨□D▽	0	0.10	
2	Move Raw Material to Machine	○⇨□D▽	6	0.12	
3	Move to Fixture Table	○⇨□D▽	18	0.22	
4	Load Fixture 1 on Cart	○⇨□D▽	2	0.02	
5	Move to Fixture Table	○⇨□D▽	2	0.02	
6	Load Fixture 2 on Cart	○⇨□D▽	2	0.02	
7	Move to Fixture Table	○⇨□D▽	2	0.02	
8	Load Fixture 3 on Cart	○⇨□D▽	2	0.02	
9	Move Fixture Cart to Machine Area	○⇨□D▽	18	0.34	
10	Move Fixture 1 to Machine	○⇨□D▽	4	0.03	
11	Move to Fixture Cart	○⇨□D▽	4	0.03	
12	Move Fixture 2 to Machine	○⇨□D▽	4	0.03	
13	Move to Fixture Cart	○⇨□D▽	4	0.03	
14	Move Fixture 3 to Machine	○⇨□D▽	4	0.03	
15	Drill Holes "A", "B", and "C"	○⇨□D▽	0	3.27	
16	Move Finished Component to Storage	○⇨□D▽	6	0.12	
17	Deposit Component in Storage	○⇨□D▽	0	0.10	
18		○⇨□D▽			
19	Totals:	○⇨□D▽	78	4.52	
20		○⇨□D▽			
21		○⇨□D▽			
22		○⇨□D▽			
23		○⇨□D▽			
24		○⇨□D▽			
25		○⇨□D▽			
26		○⇨□D▽			
27		○⇨□D▽			
28		○⇨□D▽			
29		○⇨□D▽			
30		○⇨□D▽			

Figure 30.25 Process flowchart for proposed changes for operation conditions

Looking at these improvement calculations, it is clear that with the minimal investment of one speed chuck and the implementation of a tooling cart the time was reduced by 10.7% and the distance by 35.0%. From a time saving viewpoint, the optimal output per 8-hour day increased from 94.9 pieces per day to 106.2 pieces per day. With the decrease in distance, time was gained but there are other factors that this improvement will affect. By having the fixtures moved all at once to the workstation, the operator will stay stationary at the workstation. This avoids them having to lift the fixture, turn 180°, walk to the table, lift the next fixture, turn 180°, and walk back to the workstation. Over time, the fatigue of this movement will cause the operator to work slower than the observed current process. By moving all the fixtures at once to the workstation, the operator will expel less energy and not have to perform body turns and walking. Figure 30.26 shows the reduction in time and the increase in efficiency and production.

Process Flow Improvements

	Time	Parts / Min	Parts / Hour	Parts / Day
Old Method	5.06 Min.	0.20	11.9	94.9
New Method	4.52 Min.	0.22	13.3	106.2
Delta	-0.54 Min.	+0.02	+1.4	+11.3
10.7% Increase in Time and Output				

Figure 30.26 Process flow improvement of time decrease and productivity increase

The flow process chart is an evaluation tool that can be used at any level of analysis. In the example case used here, the evaluation was performed on an operator at a low level, meaning it analyzed what was being performed in detail at a single workstation. If a higher level of analysis was desired, an evaluation of the material movement could have been performed. Here, the material is charted from the beginning of its processing to the end. This would entail analyzing the path that it takes from start to finish with less detail. For example, the flow process chart constructed above would only be one line on a high level material analysis. The entry for this would be listed as an operation for drilling and nothing more. This type of analysis normally shows more transports, delays, and inspections than a detailed analysis. In most cases, this type of evaluation would be performed to identify where the delays and the length or major transports are in the overall process. From there, a detailed analysis can be performed in areas showing the highest waste.

30.5.3.3 Activity Charts

An activity chart is another high level observation tool that is utilized to evaluate a work situation where more than one operator is working on the same job. The process flowcharts are excellent tools to evaluate a single operator whether the operator is performing a fully manual operation or one that involves running a machine. However, if there are multiple operators performing a single operation at a workstation, the process flowchart doesn't lend itself to this type of observation very well. For these cases, the activity chart is a very effective tool to observe and document the events and tasks being performed by those operators performing the work. This can be a fully manual operation or one that involves operator(s) and machine(s). The method that this tool uses is to document all the "work" times of either the operators or machines. All times are recorded and any idle or downtimes are colored in to show that no work is being performed. The objective is to observe the present method of work and then evaluate the overall pattern of tasks being performed to see if any shifting of tasks would result in a saving. Figure 30.27 shows a typical activity chart that is used in these evaluations. These charts can be expanded to include as many operators and or machines as required depending on the operation being evaluated.

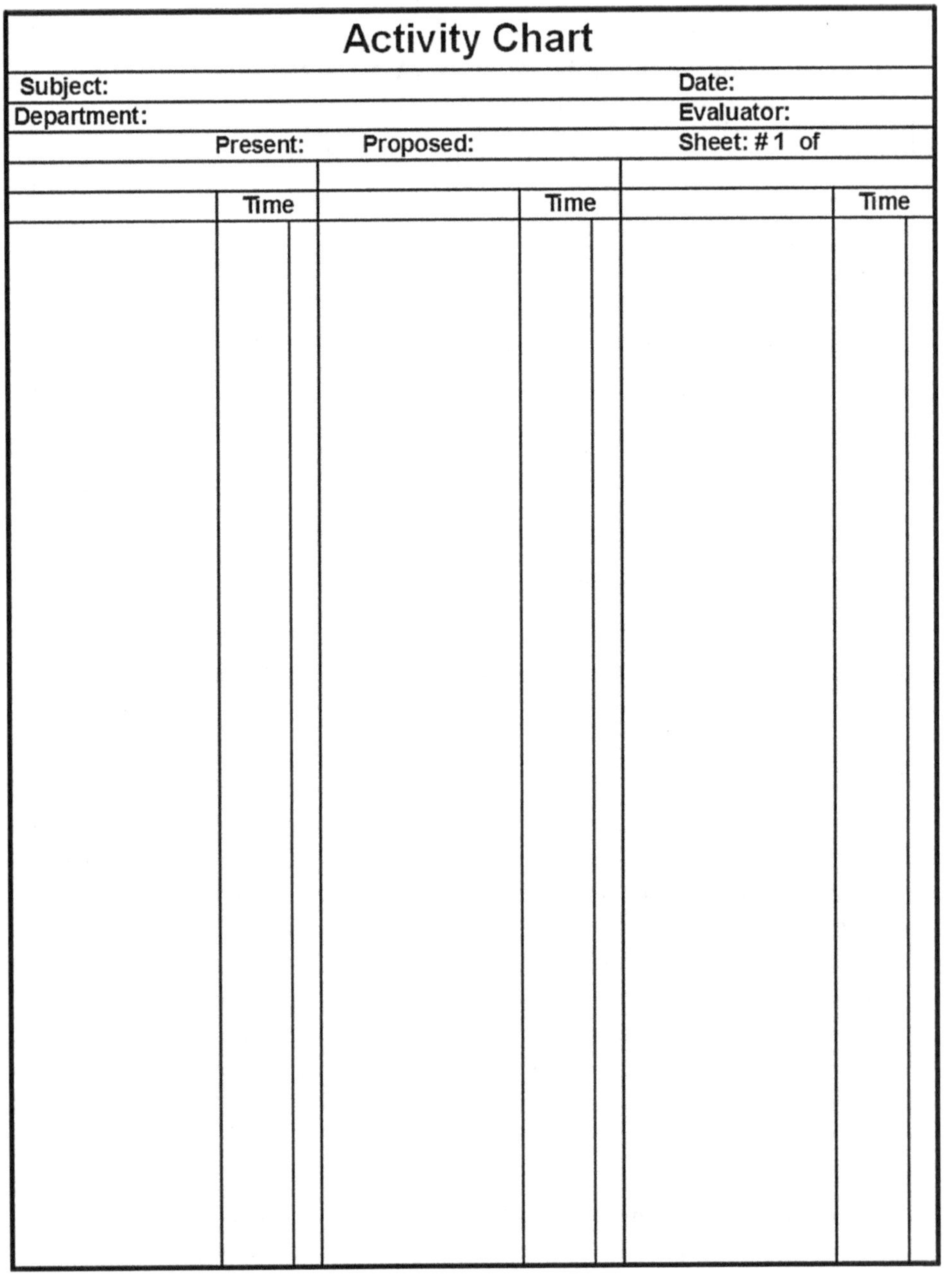

Figure 30.27 Activity chart example

The process for this type of evaluation is relatively simple, but needs to be performed with an observant eye. As with the right and left hand chart, a single operator or machine is observed and documented. All the work tasks and idle times are taken and charted on the activity chart in the appropriate column. By only watching one operator or machine at a time, the information being recorded will have far less chance of containing errors.

Once the events of all involved have been recorded individually, they are merged together on the activity chart showing all the tasks performed and the time frames that correspond with them. Any idle or non-work times are colored in alongside their recorded times. As this unfolds, we get a visual picture of all the idle times during the operation that will allow us to evaluate how, or if there is any way, the work can be redistributed among the operators to eliminate or reduce the downtimes.

Looking at an example of an operator, a helper, and a machine being utilized to perform a given operation, the present activities are charted as shown in Figure 30.28.

Activity Chart					
Subject: Machining Part # 1234567				Date: 5/16/2016	
Department: Milling				Evaluator: D. Miska	
Present: X	Proposed:			Sheet: # 1 of 1	
Operator		Helper		Machine	
Time in Minutes	Time	Time in Minutes	Time	Time in Minutes	Time
Run Machine	1.0	Wait for Part	1.5	Machine Cycle	1.0
Unload Machine	0.5			Machine Idle	5.5
Clean Machine	0.2	De-Burr Part	1.0		
Wait for Helper	4.0	Inspect Part	1.2		
		Pack Finished Part	1.5		
		Unwrap New Part	0.5		
Load Part	0.5	Wait for Part	0.8		
Start Machine	0.3				
Total:	6.5	Total:	6.5	Total:	6.5

Figure 30.28 Activity chart of a current two person machining operation

Both the operator and helper have their times and sequence of events placed on the activity chart along with the machine time. All the idle times have been colored in for all three. By performing some reassignment of the work task that both the operator and helper were performing and resequencing the events, the overall time to perform the operation was greatly reduced. By looking at the present method, it's obvious that the machine had a very high amount of idle time, thus giving it very low utilization. Simply by altering the tasks being performed and shifting some from the helper to the operator, the overall idle time for all three was reduced and the overall performance was increased. Figure 30.29 shows the activities rearranged to redirect the idle time of the operation.

Activity Chart					
Subject: Machining Part # 1234567				Date: 6/4/2016	
Department: Milling				Evaluator: D. Miska	
Present:		Proposed: X		Sheet: # 1 of 1	
Operator		Helper		Machine	
Time in Minutes	Time	Time in Minutes	Time	Time in Minutes	Time
De-Burr Part	1.0	Wait for Operator	1.0	Machine Idle Time	2.3
Unwrap New Part	0.5	Inspect Part	1.2		
Load Part	0.5				
Start Machine	0.3				
Run Machine	1.0	Pack Finished Part	1.5	Machine Cycle	1.0
Unload Part	0.5			Machine Idle Time	0.7
Clean Machine	0.2	Wait for Operator	0.3		
Total:	4.0	Total:	4.0	Total:	4.0

Figure 30.29 Activity chart of proposed changes for a two person machining operation

If we look at both the present and proposed methods from a performance viewpoint, we can evaluate how large of an increase was realized. Figure 30.30 shows the time decrease from the present method to the proposed method.

Activity Chart Improvements

	Time	Parts / Min	Parts / Hour	Parts / Day
Old Method	6.5 Min.	0.15	9.2	73.8
New Method	4.0 Min.	0.25	15.0	120.0
Delta	-2.5 Min.	+0.10	+5.8	+46.2
33.9% Increase in Time and Output				

Figure 30.30 Activity chart improvement of time decrease and productivity increase

This resulted in a 33.9% decrease in time and a 38.5% increase in the daily output of production.

From the use of these simple but powerful tools, it is evident that improvements leading to greater productivity and idle time can be realized. By observing an operation the way it is currently being performed and documenting it, through either altering the workstation, altering the sequence that the work is being performed, or introducing tooling and fixturing, savings in time can be found. This process is known as process improvement and is dynamic and not static. By improving the efficiency of a single operation, and then others in the same manner, other opportunities will become obvious. It's this compounding of improvements, or combination of operations, as the processes become more streamlined that larger overall improvements for the operation can be gained.

30.5.3.4 Right and Left Hand Chart

The right and left hand chart is very similar to the process flowchart, but measures the work being performed at a much finer detail. With this tool, you evaluate what each hand is doing during the operation being observed. In every operation in which work is being performed by an operator, both hands are being utilized simultaneously to accomplish the tasks. The evaluator needs to record the events being performed by both the right and left hands. Although this chart can be utilized for events that require walking to and from areas in the same fashion as the process flowchart, it is mainly used to evaluate the fine details of an operation being performed at a stationary location. The operator would be either sitting or standing at a given workstation and performing his or her work in a fixed location.

When evaluating a work task and using the right and left hand chart, the evaluator needs to focus on each hand to capture the events being performed. A critical point to be remembered by the evaluator is that he or she should focus one hand at a time. The evaluator should never try to capture what both hands are doing at a time, but concentrate on either the right or the left hand at a time. Document the events and work being performed by the right hand, capturing each movement, assembly, delay, grasp, release, etc. and get the sequence of the events documented thoroughly. Observe as many repetitive operations as required to ensure that all the events being performed have been captured and listed. Once the right hand has been fully documented, the evaluator can move to observing the left hand in the same manner. Again, the observer should observe as many repetitive operations as required to ensure that all the events have been captured. After both hands have been observed independently, the evaluator needs to watch the operation being performed as a whole. This is done to be able to capture which events each hand of the operator is performing their independent work tasks independently or simultaneously. This will be able to identify which tasks are, or can be, performed internal to others. Figure 30.31 shows a typical right and left hand breakdown chart that can be utilized for performing this type of evaluation.

Right and Left Hand Activity Chart

Operation: **Date:**

Layout: **Operator:**

Charted By:

Present: **Proposed:**

Sheet: 1 of

Left Hand	Time		Time	Right Hand
		1		
		2		
		3		
		4		
		5		
		6		
		7		
		8		
		9		
		10		
		11		
		12		
		13		
		14		
		15		
		16		
		17		
		18		
		19		
		20		
		21		
		22		
		23		
		24		
		25		
Totals:				

Figure 30.31 Right and left hand chart example

The time for performing each element of the evaluation is recorded for the appropriate hand and totaled at the bottom of the chart. For the operation being performed, the total time with the largest value is the time awarded as the time to perform the operation in total. A common mistake is to add both times together and award that for the total time. The reason this isn't done is because in most all operations both hands are performing work at the same time so the total operation time is given to the hand that requires the longest time to complete the tasks.

Once the evaluation is performed recording exactly what the operator is doing, the current method for performing the work can be evaluated for any improvement opportunities. The chart in Figure 30.32 shows an evaluation of a bench assembly with both the present observation and a proposed improvement.

Right and Left Hand Activity Chart

Operation: Base Assembly

Date: 2/14/2016

Layout:

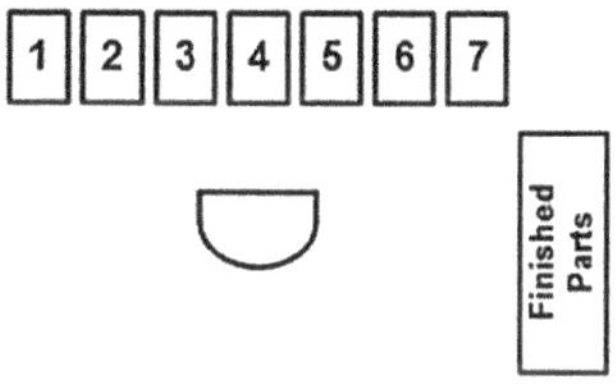

Operator: John Doe

Charted By: K. Cook

Present: X **Proposed:**

Sheet: 1 of 1

Left Hand	Time		Time	Right Hand
Pick up Part 1	2 Sec.	1	4 Sec.	Pick up Part 2
Hold Part 1		2	5 Sec.	Assemble Part 2 to Part 1
↓		3	4 Sec.	Pick up Part 3
↓		4	7 Sec.	Assemble Part 3 to Part 1
↓		5	3 Sec.	Pick up Part 4
↓		6	6 Sec.	Assemble Part 4 to Part 1
↓		7	1 Sec.	Pick up Part 5
↓		8	4 Sec.	Assemble Part 5 to Part 1
↓		9	1 Sec.	Pick up Part 6
↓		10	5 Sec.	Assemble Part 6 to Part 1
↓		11	2 Sec.	Pick up Part 7
↓		12	8 Sec.	Assemble Part 7 to Part 1
Transfer Assembly to Right Hand	1 Sec.	13	1 Sec.	Receive Assembly from Left Hand
		14	8 Sec.	Aside Finished Assembly to Table
		15		
		16		
		17		
		18		
		19		
		20		
		21		
		22		
		23		
		24		
		25		
Totals:	3 Sec.		59 Sec.	

Figure 30.32 Activity chart of a present condition assembly operation

For this simple example, the addition of a holding fixture was implemented to free up the left hand that was being used only to hold the first stage of the assembly while the right hand performed the majority of the work. Figure 30.33 shows the proposed method with the implementation of a holding fixture, allowing both hands to perform work tasks.

Right and Left Hand Activity Chart

Operation: Base Assembly

Date: 2/22/2016

Layout:

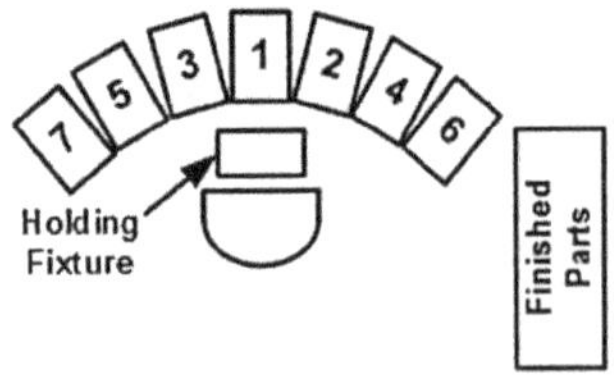

Operator: John Doe

Charted By: K. Cook

Present: **Proposed:** X

Sheet: 1 of 1

Left Hand	Time		Time	Right Hand
Pick up Part 1	2 Sec.	1	2 Sec.	Pick up Part 2
Assemble Part 1 to Fixture	6 Sec.	2		
Pick up Part 3	2 Sec.	3	5 Sec.	Assemble Part 2 to Part 1
Assemble Part 3 to Part 1	7 Sec.	4	2 Sec.	Pick up Part 4
Pick up Part 5	2 Sec.	5	6 Sec.	Assemble Part 4 to Part 1
Assemble Part 5 to Part 1	4 Sec.	6	2 Sec.	Pick up Part 6
Pick up Part 7	2 Sec.	7	5 Sec.	Assemble Part 6 to Part 1
Assemble Part 7 to Part 1	8 Sec.	8	4 Sec.	
		9	5 Sec.	Remove Assembly from Fixture
		10	8 Sec.	Aside Finished Assembly to Table
		11		
		12		
		13		
		14		
		15		
		16		
		17		
		18		
		19		
		20		
		21		
		22		
		23		
		24		
		25		
Totals:	31 Sec.		39 Sec.	

Figure 30.33 Activity chart of proposed changes to the assembly operation

Looking at the time savings, the assembly time went from 59 seconds in the present method to 39 seconds in the proposed method simply by implementing a holding fixture allowing the left hand to perform more tangible work. Breaking this down to a percentage we get a 33.9% decrease in time and a 33.9% increase in productivity. Figure 30.34 shows the decrease in time and increase in productivity from the improvements made.

Right / Left Hand Chart Improvements

	Time	Parts / Min	Parts / Hour	Parts / Day
Old Method	59 Sec.	1.0	61.0	488.1
New Method	39 Sec.	1.5	92.3	738.5
Delta	-20 Sec.	+0.5	+31.3	+250.3
33.9% Increase in Time and Output				

Figure 30.34 Right and left hand chart improvement of time decrease and productivity increase

Sketches of the work area are critical to the evaluation. These should be a close representation of the entire work envelope that the operation is being performed within. Distances need to be measured and recorded also to determine any reach and aside motions. By knowing the layout and distances, improvement ideas can be investigated easily for various new layouts and any additional aids that would make the operator's job both easier and faster. Speed is important because time equals money in the business world, but quality has to be considered too. Being able to produce something faster but to a lower quality level is unacceptable. Quality needs to be the primary goal to every person regardless of what improvement is being considered.

30.5.3.5 Line Balancing

Line balancing refers to making sure that each assembly station will require close to the same amount of time to perform the required tasks. For example, if first station required two minutes to assemble the required components and the second station requires three minutes, there will be a bottleneck where station two cannot keep up with station one. This results in station one having to produce below its maximum output to not flood station two. A typical manual assembly line layout is shown in Figure 30.35 where the product is moved from one station to another as component parts are fed to each station.

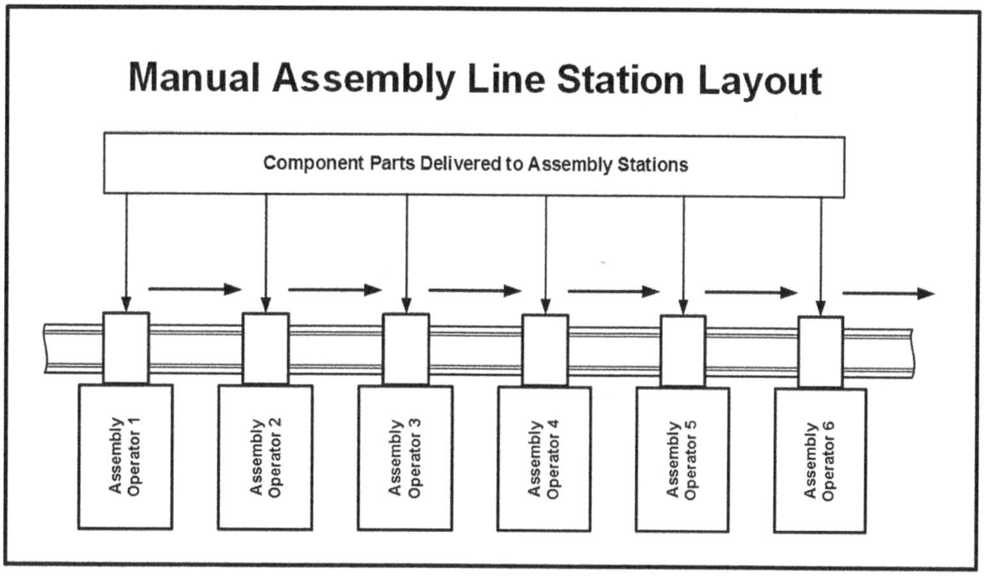

Figure 30.35 Typical six station manual assembly layout

As an example, a new product is going to be assembled on a manual assembly line that has six stations. The first step is to develop a table of elements that need to be assembled with the time required for each along with the time along with which element it needs to be preceded by. Figure 30.36 shows an example of an element table with the assembly times in minutes and what it needs to be preceded by.

Manual Assembly Line Elements

Number	Assembly Description	Time	Preceded By
1	Assemble Base Casting to Pallet and Clamp to Secure	0.24	—
2	Assemble "L" Brackets to Base Casting	0.41	2
3	Assemble Bearing Races to Base Casting	0.36	2
4	Assemble Drive Shafts through Bearing Races	0.73	3
5	Assemble Gears and Pulleys on Drive Shafts	0.58	4
6	Assemble Drive Motor to "L" Brackets	0.87	2
7	Assemble Power Cable & Harness to Base Casting and Drive Motor	0.39	6
8	Assemble Top Deck Unit to Base Casting	0.77	7
9	Assemble Drive Belts and Chains to Drive Shafts and Top Deck Unit	0.51	5, 8
10	Adjust Belt Tension and Tracking	0.44	9
11	Assemble Control Module to Main Harness	0.25	7
12	Assemble Control Module to Top Deck Unit	0.62	8, 10
13	Apply Direction Stickers to Control Module	0.19	—
14	Assemble Unit Cover to Top Deck	0.27	11
15	Clean Wipe Down and Completed Unit	0.16	14
16	Pack Finished Unit in Delivery Box	0.47	15
	Time Values Given in Minutes	7.22	Total

Figure 30.36 Assembly elements with precedence sequence

The next step is to develop a precedence diagram that will graphically display each element showing the order of assembly along with the assembly time for each stage. Figure 30.37 shows a typical precedence diagram for the elements listed in Figure 30.36 and their required order.

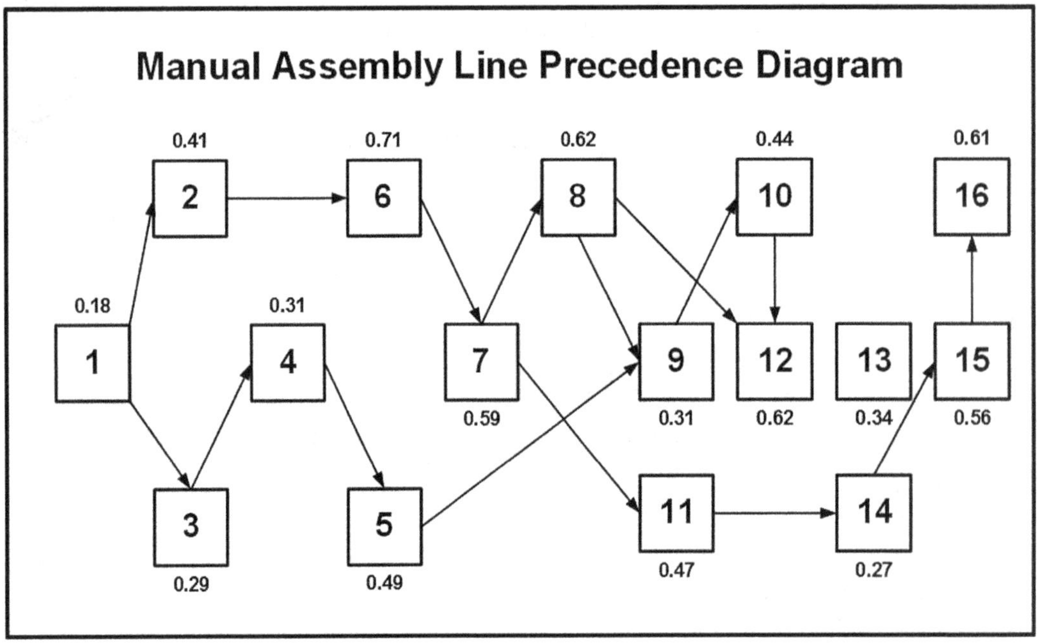

Figure 30.37 Precedence diagram of assembly elements

Once this has been developed, the optimum time per station has to be determined. With the total assembly time being 7.22 minutes and assembly line having six stations, the total assembly time is divided by the number of stations to obtain

$$7.22/6 = 1.203 \text{ Minutes}$$

This provided the time per assembly station in a perfect condition. From this point, the elements are viewed as to their order of assembly and the time required for each. They get added up and are compared against the optimum time to balance each assembly station as closely as possible. This can require having to view the assembly from both an element level and a full assembly at the same time. Figure 30.38 shows each assembly station with the elements that will be assembled at each and the combined element times.

Manual Assembly Line Station Balanced Times

Station	Elements	Time Min.	Total Time	Delta Min.	Delta Sec.
1	1 2 3 4	0.18 0.41 0.29 0.31	1.19	-0.013	-0.80
2	5 6	0.49 0.71	1.20	-0.003	-0.20
3	7 8	0.59 0.62	1.21	+0.007	+0.40
4	9 10 11	0.31 0.44 0.47	1.22	+0.017	+1.00
5	12 13 14	0.62 0.34 0.27	1.23	+0.027	+1.60
6	15 16	0.56 0.61	1.17	-0.033	-2.00

Figure 30.38 Assembly elements time balanced for six stations

From the times per assembly station, each station is close to the optimum time but not exact. In this example the largest difference in times is 3 seconds between station 4 and station 6. This would be a very well-balanced assembly line where minimal delay will be present between stations. Figure 30.39 shows the precedence diagram broken down with the elements being performed at each station.

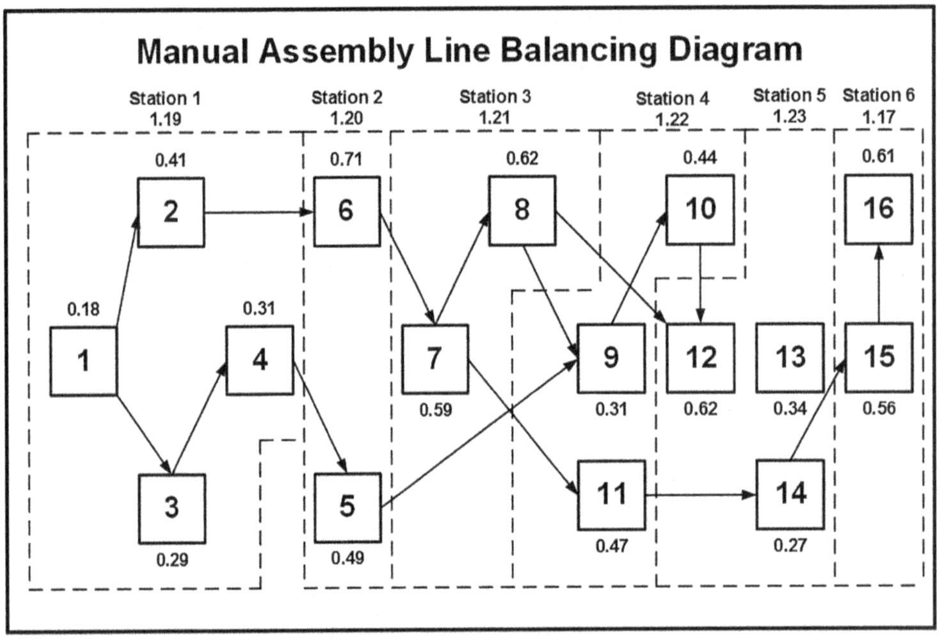

Figure 30.39 Precedence diagram of balanced assembly elements for six stations

From this, the assembly line can be constructed with the balanced elements for each station along with the required tooling and materials supplying each station. Figure 30.40 shows the completed assembly line with the balanced elements, product assembly movement, and component part supply.

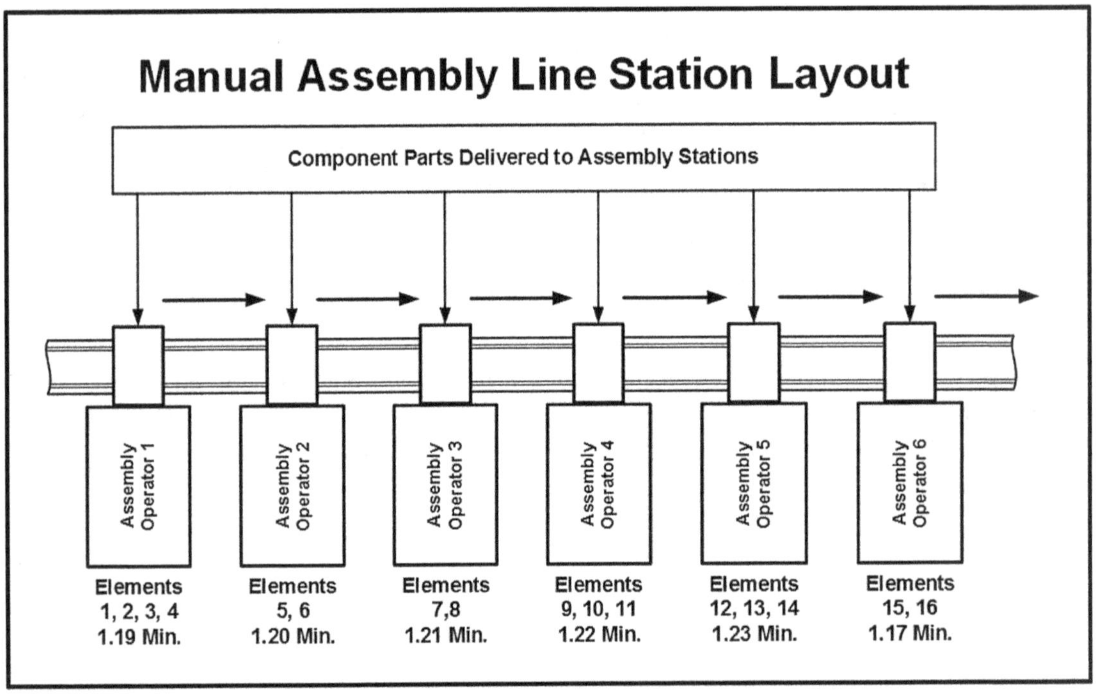

Figure 30.40 Six station manual assembly line with time balanced elements

30.6 CONCLUSION

Operational efficiency is the key to maximizing the profits of the operation. By making the processes, regardless how small they are, as efficient as possible, the non-productive time will be reduced, thus increasing the productivity. Process improvement is not a static event but rather a dynamic one. Once an improvement has been implemented, it gets reevaluated over and over making improvements that build on the previous implementations. The compounding of improvements will allow for several small improvements to add up to a larger, substantial improvement overall.

The more productive the company is, the more products or services it is able to provide to its customer base. With the utilization of the simple tools provided in this chapter, evaluations can be performed and areas of waste can be uncovered and eliminated from the processes. This waste removal is critical to increasing the effectiveness of the business that will allow for better customer service and faster turnaround of products. With the business world being very highly competitive, the more efficient your operation is the better the chance for getting the edge on the competition and a larger share of the markets you are in.

NEW REFERENCES

Allen, M. 1995. Visionary Business Entrepreneur's Guide to Success. New World Library.

Bassett, G. 1991. Management Strategies for Todays Project Shop Economy. Quorum.

Imai, M. Kaizen, 1986. The Key to Japan's Competitive Success. McGraw Hill.

Niebel, B and Freivalds, A. 2003. Methods, Standards, and Work Design. 11th ed. McGraw Hill.

References

Amstead, B., Ostwald, Phillip, and Begeman, Myron. 1987. *Manufacturing Processes*. 8th ed. Wiley Publishing.

Bassett, Glenn. 1991. *Management Strategies for Today's Project Shop Economy*. Quorum.

Black, J. and Kosher, Ronald. 2008. *Materials and Processes in Manufacturing*. 10th ed. Wiley Publishing.

Creech, Bill. 1995. The Five Pillars of TQM, *How to Make Total Quality Management Work for You*. Plume/Penguin.

Doyle, L., Keyser, C., Leach, J., Schrader, G., and Singer, M. 1969. *Manufacturing Processes and Materials for Engineers*. 2nd ed. Prentice-Hall.

Fitzpatrick, Michael. 2011. *Machining and CNC Technology*. McGraw Hill Publishing.

German, Randall. 1990. *Powder Injection Molding*. Metal Powder Industries Federation.

Grant, Eugene and Leavenworth, Richard. 1972. *Statistical Quality Control*. 4th ed. McGraw Hill.

Groover, Mikell. 2008. *Automation, Production Systems, and Computer-Integrated Manufacturing*. 3rd ed. Person/Prentice Hall.

Groover, Mikell. 2007. *Fundamentals of Modern Manufacturing*. 3rd ed. Wiley Publishing.

Groover, Mikell. 2007. *Work Systems and the Methods, Measurement, and Management of Work*. Person/Prentice Hall.

Hayes, Robert, Wheelwright, Steven, and Clark, Kim. 1988. *Dynamic Manufacturing, Creating the Lean Organization*. Free Press.

Kalpakjian, Serope and Schmid, Steven. 2014. *Manufacturing Engineering and Technology*. 7th ed. Prentice Hall Publishers.

Locke, Edwin and Latham, Gary. 1984. *Goal Setting: A motivational Technique that Works*. Prentice Hall.

Niebel, B. and Freivalds, A. 2003. *Methods, Standards, and Work Design.* 11th ed. McGraw Hill.

Pfaffenberger, Roger and Patterson, James. 1987. *Statistical Methods.* 3rd ed. Irwin.

Schonberger, Richard. 1986. *World Class Manufacturing, The Lessons of Simplicity Applied.* Free Press.

Wallace, Thomas. 1994. *World Class Manufacturing.* Omneo.

Vlack, Van. 1975. *Elements of Materials Science and Engineering.* 3rd ed. Addison Wesley.